Fundamentals of Aircraft and Airship Design
Volume I—Aircraft Design

Fundamentals of Aircraft and Airship Design

Volume I—Aircraft Design

Leland M. Nicolai

Lockheed Martin Aeronautics Company
Advanced Development Programs (Skunk Works)
Palmdale, California

Grant E. Carichner

Lockheed Martin Aeronautics Company
Advanced Development Programs (Skunk Works)
Palmdale, California

EDUCATION SERIES

Joseph A. Schetz

Editor-in-Chief
Virginia Polytechnic Institute and State University
Blacksburg, Virginia

Published by
AMERICAN INSTITUTE OF AERONAUTICS AND ASTRONAUTICS, INC.
1801 ALEXANDER BELL DRIVE, RESTON, VA 20191-4344

American Institute of Aeronautics and Astronautics, Inc., Reston, Virginia
6

Library of Congress Cataloging-in-Publication Data
Nicolai, Leland M. (Leland Malcolm)
    Fundamentals of aircraft and airship design / Leland M. Nicolai, Grant Carichner.
        p. cm. — (AIAA educational series)
        Rev. and expanded ed. of: Fundamentals of aircraft design / Leland M. Nicolai. 1975.
    Includes bibliographical references and index.
    ISBN 978-1-60086-751-4 (alk. paper)
    1. Airplanes—Design and construction—History.   2. Airships—Design and construction—History.   3. Aeronautics—Pictorial works.   I. Carichner, Grant.   II. Nicolai, Leland M. (Leland Malcolm). Fundamentals of aircraft design.   III. Title.
    TL671.2.N5 2010
    629.134'1—dc22
                                                                        2010012955

Data and information appearing in this book are for informational purposes only. The authors and the publisher are not responsible for any injury or damage resulting from use or reliance, nor do they warrant that use or reliance will be free from privately owned rights.

To my children—
Jeffrey Stephan, Debra Leigh Nicolai-Moon, Noelle Michelle,
Jeffrey Ray—
and to Carolyn, my best friend and wife of 52 years.

L. M. N.

To my sons Doug and Bryan,
and to a secret editing weapon, my wife Deborah.

Most of all I dedicate this book to Lee, my co-author, work associate, and friend.
Lee's dedication to the teaching of aircraft design has been a lifelong focus;
this book is a tribute to his engineering/teaching career.

G. C.

FOREWORD

What distinguishes this two-volume aircraft design textbook and reference is its unique perspective and the qualification of its authors. Lee Nicolai and Grant Carichner are among a small and shrinking group of active aerospace design engineers who have worked professionally as technical and program leads on many advanced and diverse aerospace projects. Typically, most professionals today work on relatively few projects during their careers often because today's projects are larger and less in number. The authors are among the most widely experienced and productive aerospace engineers in industry today. And, while both have worked on publicly acknowledged and commercial projects, most of their careers have been on military systems that were/are classified within the famed Skunk Works. Their careers started in the era of the majestic SR-71 Blackbird, an aerospace system that is still unparalleled in performance after 50 years. They've led research design teams that introduced NASA to stealth configurations as well as early study vehicles that ultimately led to the F-22 and F-35 fighters. They continue to work on even more advanced programs like unmanned vehicles, military airship programs, and body freedom flutter research.

Moreover, these authors are not without academic credentials either. They have taught Lockheed Martin's internal "knowledge transfer" seminars on stealth, aircraft design, and buoyant vehicle design. Part of Dr. Nicolai's distinguished career was with DARPA and also as an instructor at the U.S. Air Force Academy.

The perspective of these two volumes is unique because it is based on the simultaneous long careers of two practicing engineers who embody what it takes to work at the Skunk Works. Founded in 1943 by aeronautical and program management genius Kelly Johnson, the Skunk Works has been recognized as a crucial national asset by every U.S. president since Eisenhower. The Skunk Works, its engineers, and its projects have won every major aeronautical award and many more than once.

That the Skunk Works developed and maintained a consistent program management approach and culture was, in Kelly's view, the secret of its success. The "Skunks" (a small number of good people, 10–25 percent fewer than normal programs) were unique in that each worked many aspects of many programs, and thus each became something of a generalist. That generalist systems-level thinking is also reflected in the Nicolai–

Carichner approach of these two volumes. This approach embraces the fundamental philosophy of the 80/20 rule: 20 percent of the funding spent results in a solution that contains 80 percent of the requirement. This general philosophy is embraced by the authors, who write about innovation and pursuing excellence.

The Skunk Works written mantra is to be quick; be quiet; be on time. The unwritten rule is that it is better to beg forgiveness than ask permission. Over the years, the fame of the Skunk Works culture has spread and elements of it have been incorporated elsewhere, throughout corporate America and well beyond the original design shop in Southern California.

The authors' great diversity of design and program management experience contributes greatly to the book's practicality, validated correctness, and relevance. These volumes are a great introduction to the fundamentals of aircraft and airship design and belong on the desk top of every engineering college student, practicing aerospace engineer, and project manager.

Ned Allen
Lockheed Martin
Bethesda, Maryland

CONTENTS

LIST OF COLOR PLATES

PREFACE

Volume I is an update and expansion of the 1975 text *Fundamentals of Aircraft Design*. The updated material includes designing for survivability (stealth), solar power aircraft systems, and very high altitude operation with air breathing propulsion. The added material is a discussion of both the science and art of aircraft design and includes a new chapter on materials and structural design. The art of design is captured in the history, the lessons learned, the facts and stories appearing in blue boxes, and the case studies, as well as in the four-color section found at the back of the book. The DC-3 and the F-35B on the title page portray the blend of art and science in aircraft design. The DC-3 represents the *art of design* ingeniously displayed by Donald Douglas in creating an airplane of timeless elegance that set the standard for air transportation in 1935. The F-35B represents the *science of design* in applying the latest in methodology and technology to "create that which never was" (Theodore von Kármán). This work is the result of the collaboration of two practicing engineers with a combined 80 years of design experience. The projects cover aircraft from fast to slow, high to low, big to small. The design of airships is the central theme of Volume II.

The aerospace industry has changed the way it designs aircraft and has expanded the spectrum of vehicle types. Beginning in the late 1970s stealth (RF and IR) became a big part of the aircraft design process. This single technology has enabled the United States of America to have air superiority over every nation for the last 25 years. Volume I includes the unclassified details of incorporating stealth into a viable design. The authors have lived it and now you can read and learn about it.

The text is aimed at upper-level undergraduate and graduate students as well as at practicing engineers. It contains comprehensive treatment of the conceptual design phase, treating civil and military aircraft equally. The book covers all phases of conceptual design, from consideration of user needs to the decision to iterate the design one more time. The book is complete in that the reader should not have to go outside the text for additional information.

The text is structured to lead the reader through one iteration of the conceptual design cycle loop. It can also be used as a convenient reference book for practicing engineers to give them up-to-date methodology in aerodynamics, propulsion, performance, structures and materials, weights, stability and control, and life cycle costing. It can also be used by technical managers to better understand and appreciate the fundamental parameters driving the design of an aircraft and their interplay. It has a rich set of appendixes that puts pertinent data at the designer's fingertips.

The main theme of the text is that the aircraft is only a dust cover. The point being that the designer needs to remember that the aircraft is only a mechanism for transporting the payload (people, cargo, bombs, sensors, and so on). All design decisions must consider the needs of the payload first. The text emphasizes that the aircraft design process is always a compromise and that there is no right answer; however, there is always a best answer based on existing requirements and available technologies. But the best design answer today will probably be different from that of tomorrow.

Chapter 19 about material selection and structural arrangement is completely new. It is written by Walter Franklin, a Lockheed Martin Fellow and practicing structures engineer. It is a wonderful addition to the book as it thoroughly covers the material and structural issues associated with aircraft design. At the end of this chapter is a complete wing design example.

Using design examples throughout the book, the authors guide your journey through the design process as it would happen in the actual design environment. Using color, historical design facts, and case studies, the authors make the journey a real life experience in one of the most important engineering inventions of modern time. Students, practicing engineers, and engineering managers alike will find Volume 1 of *Fundamentals of Aircraft and Airship Design* an indispensible resource and a pleasure to read.

A special thanks to Pat DuMoulin, our AIAA editor, for her consistent efforts to make this book as good as it possibly could be. Thanks also to Becky Rivard and Mike Baden-Campbell for assistance with proof checking.

Leland M. Nicolai
Grant E. Carichner
June 2010

- There Is Never a Right Answer
- Requirements—Requirements Pull & Technology Push
- Always Question the Requirements
- Measure of Merit
- UAVs
- Specs, Standards, & Regulations
- Design Phases
- Scope of Text

First manned, powered controlled flight at Kitty Hawk on 17 December 1903. This photo shows Wilbur running alongside, with Orville at the controls. The flight lasted 12 seconds and covered 120 feet. Three more flights were made that day, with the last one traveling 852 feet!

It's only a dust cover!
Leland Nicolai (1975)

1.1 Aeronautics—The Beginning

A eronautics is a relatively new engineering discipline that is little more than 100 years old. However, serious thoughts about how people could fly have been on the minds of laymen and scholars over the last 500 years. Leonardo da Vinci designed many variations of machines that would allow people to fly. However, serious aeronautical analysis and experimentation did not happen until the early 19th century when Sir George Cayley first started applying the basic laws of flight and the scientific method to the development of manned flight. Cayley's work would influence airplane designers for the next 50 years. He was the first to identify the four fundamental forces of thrust, lift, drag, and weight and their interrelationship in flight mechanics. In particular, he correctly understood and documented that wings should be responsible for lift and engines responsible for thrust. Cayley was also the first to understand and incorporate the concept of cambering to change the lift of a wing. His close observation of bird flight was directly responsible for adding camber to his earliest flying models. He also correctly predicted that continued manned flight could not happen until the development of an engine with a high thrust-to-weight ratio. It would take another 50 years for that engine to be developed. Although history has highlighted the importance of work by Samuel Langley and the Wright brothers, it was Cayley who designed and built the glider that carried the first human aloft. It would take over four decades for someone else to equal this accomplishment.

In the late 1800s many people were attempting to develop efficient gliders to better understand the fundamental principles of flight. Glider designers such as Otto Lilienthal (who would lose his life in a glider accident) and Octave Chanute contributed greatly to the body of knowledge that Langley and the Wrights would use in their pursuit of powered manned flight. Even though December 1903 is heralded as the beginning of powered manned flight, many evolutionary efforts had incrementally led up to this historic event.

As the Wright brothers continued to improve on their design, it created an opportunity for a new engineering discipline—aeronautics. Past efforts had been performed by people who designed, analyzed, and experimentally verified their ideas. Soon, brilliant scholars and teachers such as Ludwig Prandtl and Theodore von Kármán would emerge as pillars of aeronautical thought and principles. At this time, the fluid mechanics principles of Daniel Ber-

"It's only a dust cover." This is not meant to trivialize the external shape of an aircraft design. It is merely a reminder to the designer that protecting and delivering the payload is crucial for mission success. Adding features not related to the payload always results in a more expensive design.

Sir George Cayley's coachman was actually the first person to successfully fly in a glider. This event took place in England in 1853 and the flight covered a distance of approximately 900 feet. It was the only time the coachman flew.

noulli were used by these analytical pioneers as the foundation of the formal engineering discipline *aeronautical engineering*. Prandtl concentrated on subsonic flow and was the first to postulate the existence of the boundary layer and its influence on flow separation. Von Kármán is reknowned for his contributions to the understanding of supersonic and hypersonic flight regimes.

> Let us hope that the advent of a successful flying machine, now only dimly foreseen and nevertheless thought to be possible, will bring nothing but good into the world; that it shall abridge distance, make all parts of the globe accessible, bring men into closer relation with each other, advance civilization, and hasten the promised era in which there shall be nothing but peace and goodwill among all men.
> *Octave Chanute, circa 1895*

1.2 Aircraft Design—A Compromise

The design of an aircraft is a large undertaking requiring the team efforts of many engineers having expertise in the areas of aerodynamics, propulsion, structures, flight control, performance, and weights. As the design takes shape, specialists are called in to design such components as the crew station, landing gear, interior layout, armament location, and equipment installation. The completed aircraft design is a compromise of the best efforts of many talented engineers. The different design groups they represent must work together to produce the most efficient flight vehicle. It should be clear that the design process is a very involved integration effort, requiring the pulling together and blending of many engineering disciplines. The key element in the design process is the design team leader, or Chief Designer, who acts as the integrator and referee. The Chief Designer is usually one who understands and appreciates all of the various disciplines involved in the design process and is often called upon to negotiate compromises between the design groups. For example, the propulsion group might propose an inlet arrangement that aggravates the clean configuration of the aerodynamics group. At the same time, the structures group might recommend a wing thickness ratio of 8% while the aerodynamics group might choose 2%. The flight control group might complicate matters by putting an aft tail on the design and insisting that the wing be moved forward for better balance. The Chief Designer will pull the design groups together to bring about the best compromise toward the design goal. The Chief Designer must prevent any one design group from driving the design, which might otherwise produce one of the designs shown in Fig. 1.1.

> Bernoulli's complex book was titled "Hydrodynamica." As a favor, his close friend and famed mathematician Leonhard Euler rewrote the book to be more understandable. The basis of the Bernoulli principle is established in this work.

1.2.1 Performance vs Cost

Prior to the 1970s, the performance of the aircraft was paramount. All design efforts were

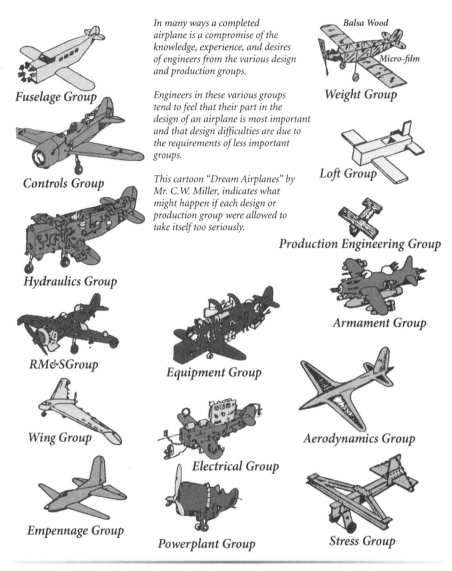

In many ways a completed airplane is a compromise of the knowledge, experience, and desires of engineers from the various design and production groups.

Engineers in these various groups tend to feel that their part in the design of an airplane is most important and that design difficulties are due to the requirements of less important groups.

This cartoon "Dream Airplanes" by Mr. C.W. Miller, indicates what might happen if each design or production group were allowed to take itself too seriously.

Fuselage Group

Controls Group

Hydraulics Group

RM&SGroup

Wing Group

Empennage Group

Weight Group

Loft Group

Production Engineering Group

Armament Group

Equipment Group

Aerodynamics Group

Electrical Group

Powerplant Group

Stress Group

Figure 1.1 Resulting aircraft design if one group is dominant.

focused to produce a vehicle displaying maximum performance for a given aircraft weight or a minimum weight for a specified level of performance. Cost was a consideration after the aircraft design was "locked in." In the late 1960s, the government and the aircraft industry became extremely cost-conscious. The acquisition cost of aircraft systems was skyrocketing and the measure of merit became minimum acquisition cost. The A-10A Thunderbolt II was designed in the late 1960s for an acquisition cost of $3 million. The Lightweight Fighter competition in the early 1970s had a unit cost requirement of $5 million and led to the F-16 and F-18. This cost metric later changed to life cycle cost (LCC; the sum of the development cost, acquisition cost, and operation and support cost) in the late 1970s. This emphasis on design-to-cost [1] brought two new players into the design team: the cost analyst and the manufacturing expert, who were installed in the design team with full voting rights. The cost and performance trade study results have become key considerations in design decisions.

> In many ways a completed airplane is a compromise of the knowledge, experience, and desires of engineers from the various design and production groups.
> Engineers in these various groups tend to feel that their part in the design of an airplane is most important and that design difficulties are due to the requirements of less important groups.
> The cartoon "Dream Airplanes," by Mr. C. W. Miller, (Fig. 1.1) indicates what might happen if each design or production group were allowed to take itself too seriously.

Aircraft program costs spiraled out of control in the 1990s and led to the cancellation of the U.S. Navy A-12 Avenger and TSSAM/AGM-137 Tri-Service Standoff Attack Missile, shown in the photograph (Fig. 1.2). It was at this point in time that the principle of cost as an independent variable (CAIV) was introduced as a legitimate design criterion. The U.S. government instituted the CAIV principle as part of its acquisition regulation for military systems DoD 5000.1 [2], which states that cost must be considered to be equal in importance with performance and that programs must show the cost gradient with respect to performance. Essentially, CAIV is the government equivalent of commercial best business practices.

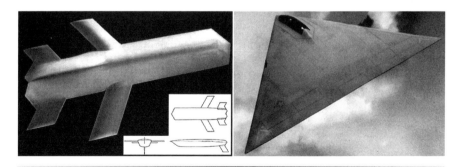

Figure 1.2   A-12 and TSSAM; both programs were cancelled by the Navy due to cost growth.

1.2.2 There Is Never a Right Answer

In the design of an aircraft there is never a right answer—only a best answer at a point in time. The reason is that the design of an aircraft is a balance between the following competing requirements:

- **Technical.** Performance, survivability
- **Signature.** Survivability, appearance
- **Economic.** Cost, LCC
- **Political.** Policy, payback, risk, and so on
- **Schedule.** When needed? The need to be first to market
- **Environmental.** Limited energy source, noise, hydrocarbon emissions

Also, the priorities of these requirements change with time. An aircraft might be designed to certain technical and economic requirements, but if the government administration changes, then the priority requirement becomes political or environmental. The advice to the designer is to remain flexible and develop as robust a design as possible so that it will survive as the requirements change over time. The watchwords are *compromise, balance,* and *flexibility.*

1.3 Overall Design Requirements

Before designing a building, an architect must first establish who and how many will occupy the building, what is its purpose, what are its scale and cost level, and so on. The design of an aircraft is similar in that the aircraft designer must have requirements established before a design can proceed. The requirements define the following: (1) what mission the aircraft will be called upon to perform, (2) how much the aircraft should cost, (3) how the aircraft should be maintained and supported, and (4) the schedule for the aircraft.

1.3.1 Mission Requirements

The mission requirements identify the following:

- **Purpose.** Commercial transport; air-to-air fighter, air-to-ground fighter, bomber; general aviation; intelligence, surveillance, and reconnaissance (ISR); trainer, and so on
- **Crew.** Manned or unmanned
- **Payload.** Passengers, cargo, weapons, sensors, and so on
- **Speed.** Cruise, maximum, loiter, landing, and so on
- **Distance.** Range or radius
- **Duration.** Endurance or loiter (time-on-station)

- **Field length.** Vertical, short, or conventional takeoff and landing (VTOL, STOL, CTOL)
- **Signature level.** Radar cross section (RCS); infrared (IR); visual; and acoustic (noise)

> In 2007 for a typical airline seat, 5% of the ticket went to pay for aircraft maintenance; 25% for taxes and fees; 30% for salaries; 30% for fuel; and 10% miscellaneous.

1.3.2 Cost Requirements

Cost requirements encompass the following:

- Development cost
- Acquisition cost
- Operation and support (O&S) cost
- Life cycle cost (LCC), which is the sum of development, acquisition. and O&S
- Cost as an independent variable (CAIV) for government programs

1.3.3 Maintenance and Support Requirements

The maintenance and support requirements are as follows:

- Maintenance man-hours per flight hour (MMH/FH)
- Ground support equipment (GSE)
- Maintenance levels (i.e., organizational, intermediate, and depot)
- Integrated logistics support (ILS) plan
- Contractor support or user support

1.3.4 Scheduling Requirements

The schedule requirements identify the following:

- Development and test scheduling
- Product availability, that is, when the aircraft should be available for deployment [initial operational capability (IOC)] to the warfighter or the commercial customer

1.3.5 Where Do the Design Requirements Come From?

In the case of a commercial program, the requirements are usually established by the aircraft supplier, such as Boeing, Lockheed Martin, Northrop-Grumman, Gulfstream, Cessna. The aircraft companies perform market analyses to determine what the public's needs or desires will be in the near future. Projections are made for

> For military aircraft ~67% of the LCC is for O&S costs.

future passenger travel, air freight needs, and general aviation aircraft demands. The commercial program is kicked off when a customer steps up and shows serious intent to buy the production airplane. A down payment usually entitles the customer to influence some of the requirements.

The Boeing 777 airliner was a cultural change for Boeing as they invited eight airlines (called the "Gang of Eight") to develop the requirements for the B777 in the early 1990s. It is recommended that the reader review the Boeing 777 Case Design Study in Volume 2.

Careful thought and research must go into establishing the requirements because if they are inappropriate, then the aircraft (if it is even built) may not find a customer or keep its initial customers. Aircraft companies have lost large sums of money because they followed a bad or inappropriate set of requirements. American millionaire Howard Hughes decided in 1942 that the world needed a large plywood flying boat capable of carrying 700 passengers. The U.S. government agreed (initially) but changed its mind and Mr. Hughes was left with one wooden aircraft giant that made one flight of only six seconds. Today, it is generally felt that the technical design of the Hughes Hercules aircraft was sound, but the mission requirements were about 25 years ahead of their time.

Sometimes the requirements are established by a military user such as the U.S. Air Force (USAF), U.S. Navy and Marines (USN), or U.S. Army. These requirements are usually developed to fill a military need (shortfall) or to replace an obsolete system. Such requirements are termed a *requirements pull* because the military need "pulls" the requirements.

Figure 1.3 is an example of an early military requirement for an aircraft. The mission requirements were pretty demanding, in that they required a payload of 350 lb, a top speed of 40 mph, and that the aircraft had to be easily transportable in an Army wagon. This procurement was sole-sourced to two brothers in Dayton, Ohio, named Wilbur and Orville Wright. It is interesting to note that the contract for the heavier-than-air flying machine was for $25,000 and called for delivery in seven months. The photograph (Fig. 1.4) shows a Wright brothers design for Specification 486.

A more modern example of a requirement that met both a military and a commercial requirement was the Request for Proposal (RFP) in the mid-1950s for a utility jet transport. In August of 1956, the USAF addressed an RFP to industry expressing a requirement for a "Twin Jet Aircraft (UTX) to fulfill utility and pilot readiness missions." The letter also stated: "It appears that commercial requirements for such utility and transport aircraft are realistic. ..." In accordance with the USAF's RFP, industry was to develop a prototype for evaluation without government compensation because "... there is a potential commercial market for these aircraft" and the "... estimated costs of development programs of this type are within the capability

SIGNAL CORPS SPECIFICATION, NO. 486

ADVERTISEMENT AND SPECIFICATION FOR A HEAVIER-THAN-AIR FLYING MACHINE

To THE PUBLIC:

Sealed proposals, in duplicate, will be received at this office until 12 o'clock noon on February 1, 1908, on behalf of the Board of Ordnance and Fortification for furnishing the Signal Corps with a heavier-than-air flying machine. All proposals received will be turned over to the Board of Ordnance and Fortification at its first meeting after February 1 for its official action.

Persons wishing to submit proposals under this specification can obtain the necessary forms and envelopes by application to the Chief Signal Officer, United States Army, War Department, Washington, D. C. The United States reserves the right to reject any and all proposals.

Unless the bidders are also the manufacturers of the flying machine they must state the name and place of the maker.

Preliminary.—This specification covers the construction of a flying machine supported entirely by the dynamic reaction of the atmosphere and having no gas bag.

Acceptance.—The flying machine will be accepted only after a successful trial flight, during which it will comply with all requirements of this specification. No payments on account will be made until after the trial flight and acceptance.

Inspection.—The Government reserves the right to inspect any and all processes of manufacture.

GENERAL REQUIREMENTS.

The general dimensions of the flying machine will be determined by the manufacturer, subject to the following conditions:

1. Bidders must submit with their proposals the following:

   (a) Drawings to scale showing the general dimensions and shape of the flying machine which they propose to build under this specification.

   (b) Statement of the speed for which it is designed.

   (c) Statement of the total surface area of the supporting planes.

   (d) Statement of the total weight.

   (e) Description of the engine which will be used for motive power.

   (f) The material of which the frame, planes, and propellers will be constructed. Plans received will not be shown to other bidders.

2. It is desirable that the flying machine should be designed so that it may be quickly and easily assembled and taken apart and packed for transportation in army wagons. It should be capable of being assembled and put in operating condition in about one hour.

3. The flying machine must be designed to carry two persons having a combined weight of about 350 pounds, also sufficient fuel for a flight of 125 miles.

4. The flying machine should be designed to have a speed of at least forty miles per hour in still air, but bidders must submit quotations in their proposals for cost depending upon the speed attained during the trial flight, according to the following scale:

   40 miles per hour, 100 per cent.
   39 miles per hour, 90 per cent.
   38 miles per hour, 80 per cent.
   37 miles per hour, 70 per cent.
   36 miles per hour, 60 per cent.
   Less than 36 miles per hour rejected.
   41 miles per hour, 110 per cent.
   42 miles per hour, 120 per cent.
   43 miles per hour, 130 per cent.
   44 miles per hour, 140 per cent.

5. The speed accomplished during the trial flight will be determined by taking an average of the time over a measured course of more than five miles, against and with the wind. The time will be taken by a flying start, passing the starting point at full speed at both ends of the course. This test subject to such additional details as the Chief Signal Officer of the Army may prescribe at the time.

6. Before acceptance a trial endurance flight will be required of at least one hour during which time the flying machine must remain continuously in the air without landing. It shall return to the starting point and land without any damage that would prevent it immediately starting upon another flight. During this trial flight of one hour it must be steered in all directions without difficulty and at all times under perfect control and equilibrium.

7. Three trials will be allowed for speed as provided for in paragraphs 4 and 5. Three trials for endurance as provided for in paragraph 6, and both tests must be completed within a period of thirty days from the date of delivery. The expense of the tests to be borne by the manufacturer. The place of delivery to the Government and trial flights will be at Fort Myer, Virginia.

8. It should be so designed as to ascend in any country which may be encountered in field service. The starting device must be simple and transportable. It should also land in a field without requiring a specially prepared spot and without damaging its structure.

9. It should be provided with some device to permit of a safe descent in case of an accident to the propelling machinery.

10. It should be sufficiently simple in its construction and operation to permit an intelligent man to become proficient in its use within a reasonable length of time.

11. Bidders must furnish evidence that the Government of the United States has the lawful right to use all patented devices or appurtenances which may be a part of the flying machine, and that the manufacturers of the flying machine are authorized to convey the same to the Government. This refers to the unrestricted right to use the flying machine sold to the Government, but does not contemplate the exclusive purchase of patent rights for duplicating the flying machine.

12. Bidders will be required to furnish with their proposal a certified check amounting to ten per cent of the price stated for the 40-mile speed. Upon making the award for this flying machine these certified checks will be returned to the bidders, and the successful bidder will be required to furnish a bond, according to Army Regulations, of the amount equal to the price stated for the 40-mile speed.

13. The price quoted in proposals must be understood to include the instruction of two men in the handling and operation of this flying machine. No extra charge for this service will be allowed.

14. Bidders must state the time which will be required for delivery after receipt of order.

JAMES ALLEN,
Brigadier General, Chief Signal Officer of the Army.

SIGNAL OFFICE,
WASHINGTON, D. C., *December 23, 1907.*

Figure 1.3 First published mission requirement for a military aircraft, 20 January 1908.

Figure 1.4   Wright Brothers' design for Specification 486.

of industry." The Air Force estimated that the flyaway cost of these aircraft would be about $200,000 each.

The letter was accompanied by a General Design Requirement document that described the requirements of the aircraft in detail. Some of the more pertinent requirements were as follows:

Range	1500 nautical miles
Maximum cruise ceiling	45,000 ft
Service ceiling (one-engine)	15,000 ft
Critical field length	5000 ft
Landing roll (1/2 fuel)	2500 ft
Cruise speed	0.76 Mach
Payload	2 crew, 4 passengers, 240 lb of baggage
Escape	Inflight escape provisions
Fueling	Single point, pressure refueling
Instrumentation	Military-type instruments and avionics
Certification	Suitable for certification by the Federal Aviation Administration (FAA)

North American Rockwell had actually initiated UTX design work in the spring of 1952 so they were more than ready in August of 1956. They firmed up their baseline design and pressed into fabrication of their prototype Sabreliner. The prototype aircraft was completed and the first flight of the Sabreliner was accomplished in September 1958. North American Rockwell won the UTX competition and was awarded a contract in January 1959 for seven flight test aircraft. The aircraft that resulted from the UTX mission requirement is designated the USAF T-39 and is shown in Fig. 1.5. The characteristics of the T-39 are shown in Table 1.1. North American Rockwell has produced several very successful commercial derivatives of the T- 39: the Sabreliner Series 40, Series 50, Series 60, and Series 75. All of the Sabreliner series have similar configuration geometry but incorporate different interior arrangements, special mission features, uprated propulsion units, and different equipment.

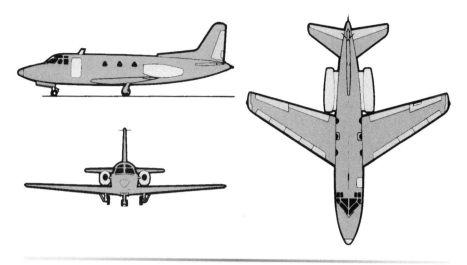

Figure 1.5 North American Rockwell T-39 Sabreliner.

Table 1.1 Characteristics of the U.S. Air Force T-39

Wing span	44 ft 4 in.
Length	43 ft 9 in.
Maximum takeoff weight	19,000 lb
Empty weight	9805 lb
Fuel weight	7122 lb
Crew	2
Payload	9 passengers or 2100 lb
High speed at 41,000 ft	Mach 0.76
Cruise speed at 42,000 ft	Mach 0.7
Service ceiling (one engine)	21,500 ft
Range	2060 miles
Critical field length	4900 ft
Landing distance	2190 ft
Engines	Two JT12A
W/S at takeoff	55 psf
Wing sweep (¼ chord)	28°33′
Wing aspect ratio	5.77
Number built	149 for USAF; 42 for USN

Sometimes a new technology will push the requirements for a new aircraft (termed a *technology push*). The jet engine in the early 1940s, stealth technology in the mid-1970s, and the high-energy airborne laser in the early 1990s are examples of technology push requirements that led to the XP-80, the Have Blue/F-117, and the YAL-1/ABL, respectively.

The requirements usually come with a document called a Concept of Operations, or ConOps for short. The ConOps describes how the aircraft will be deployed, operated, maintained, and supported—essentially all the information the designer needs to complete the design. The ConOps is helpful for a commercial aircraft but is essential for a military aircraft. For example, the military aircraft designer needs to know if the threat defenses will be "up and in-place" or rolled back, if fuel tankers will be available, and what the maintenance concept will be: organization, intermediate, depot, contractor logistics support, and so on.

1.3.6 Need to Question the Requirements

When the requirements arrive, the designer *must* study them, understand them, evaluate them, and question them—and if necessary negotiate them with the customer—because a designer who does not agree with the requirements must walk away. Disagreement with the fundamental requirements will sap the designer's passion and commitment, which are necessary to generate a successful conceptual design that will ultimately be selected to proceed into preliminary design.

Even when the customer tries very hard to generate a credible set of requirements, sometimes they are flawed. History is filled with flawed requirements. Some flawed requirements are discovered and changed, some flawed requirements prevail and designs are produced, and some flawed requirements are ignored (this one is always risky). Three examples of flawed requirements are the following:

- 1932 Transcontinental and Western Air, Inc. (TWA) replacement of Fokker F-10A
- 1985 USAF Advanced Tactical Fighter (ATF)
- 1995 USAF Joint Air-to-Surface Standoff Missile (JASSM)

TWA Specification—1932

In the late 1920s and early 1930s, the flagship of the TWA commercial transport fleet was the three-engine Fokker F-10A Trimotor. In 1931 an F-10A crashed, taking the life of Knute Rockne, the famed Notre Dame football coach [see the photograph (Fig. 1.6)]. Inspectors blamed moisture inside the wooden wing that caused the wing structure of the F-10A to separate. The Aeronautics Branch, Department of Commerce (predecessor to

Figure 1.6 Fokker F-10A.

the FAA), suspended the airworthiness certificate of the F-10A, grounding a major part of the TWA fleet. In August 1932 TWA issued the specification for a modern luxury transport airplane shown in Fig. 1.7. Although the TWA document specified an all-metal three-engine aircraft, the contract winner was the two-engine DC-1. Donald Douglas took a risk and offered up a two-engine design. The TWA specs were an extreme challenge for the time, but were met and exceeded by the DC-1, predecessor to the famous DC-3 and World War II C-47 [see the photograph (Fig. 1.8)]. Even though only one DC-1 was built, and 218 DC-2s, the Douglas Aircraft Company turned out 13,300 DC-3s. The fact that DC-3s are still flying today is a testament to the design genius of Donald Douglas.

USAF ATF Specification—1985

In 1985 the USAF issued an RFP for a new air-to-air advanced tactical fighter to replace the F-15. The RFP requirements called for "supermaneuver" and "supercruise" [the ability to cruise at supersonic speeds on dry power (without afterburners)]—and a modest signature level. The Lockheed Skunk Works challenged the requirements and convinced the USAF that the radar cross section (RCS) levels in their signature requirement

- All Metal Trimotor Monoplane
- Payload : 12 passengers
- Range: 1000 miles
- Crew: 2
- Top Speed, Sea Level: 185 mph
- Cruising Speed, Sea Level: 145 mph
- Landing Speed: 65 mph
- Service Ceiling: 21000 ft
- Rate of climb, Sea Level: 1200 fpm
- Maximum Gross Weight: 14200 lb
- Passenger compartment must have ample room for comfortable seats, miscellaneous fixtures and conveniences
- Airplane must have latest radio equipment, flight instruments and navigational aids for night flying

Figure 1.7 TWA specification for a new transport aircraft (August 1932).

Figure 1.8   DC-3—timeless elegance (courtesy of Gary Shepard).

should be lowered to be a true 21st-century fighter. The USAF agreed and recalled the RFP and reissued it in February 1986 with superstealth as a requirement. Lockheed (YF-22) and Northrop (YF-23) [see the photograph (Fig. 1.9)] won contracts to build two prototypes each and have a fly-off. The rest is history.

USAF JASSM Specification—1995

In the spring of 1995 the Department of Defense (DoD) canceled the stealthy air-launched cruise missile named Tri-Service Standoff Attack Missile (TSSAM/AGM-137) because of excessive unit cost. The mission need for the TSSAM still existed so a draft RFP was issued to industry in the fall of 1995 for JASSM. The JASSM requirements were the same performance as TSSAM but a higher signature (RCS). The unit cost requirement was $400,000—the same as TSSAM. The USAF concluded that the only way to meet the unit cost requirement was to ask for a derivative of an existing cruise missile (forcing the increase in the signature requirement). The 1995 RFP specified a derivative missile. Lockheed Martin questioned the requirements. The Skunk Works convinced the USAF that they could have the same performance, signature, and unit cost as TSSAM with

Figure 1.9   YF-22 and YF-23—ATF competitors.

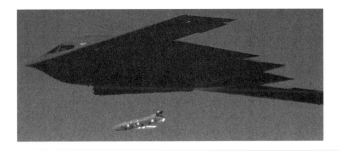

Figure 1.10   B-2 launching a JASSM/AGM-158.

a clean-sheet design. The JASSM RFP was issued in the winter of 1996 asking for a derivative or clean-sheet design with a lower signature. Lockheed Martin and McDonnell Douglas both won contracts, with clean-sheet designs, for further work. Lockheed Martin won the production contract for the AGM-158 [shown in the photograph (Fig. 1.10) being dropped from a B-2] and is now building over 2000 cruise missiles.

1.3.7 Measure of Merit

To be acceptable to the customer, the aircraft design must meet (or exceed) the stated requirements. Meeting the requirements is a necessary condition for being a candidate to proceed to the next phase. If there is a requirement that the designer cannot meet or thinks is unrealistic, then the designer needs to petition the customer for a waiver.

The *Measure of Merit* (MoM; sometimes called figure of merit) is similar to a requirement except that it is initially known only to the customer and is not overtly specified. The MoM is important to the customer and will be used as a "tie breaker" in selecting the winning design. It is often said that meeting the requirements gets you invited to the dance, but meeting the MoM gets you out on the dance floor.

Because the MoM is initially unspecified, the designer (or someone in the design group) must do the homework to understand what the customer is really looking for. Sometimes the MoM is simply that the design must be aesthetically pleasing. This seemed to be a MoM in the selection of the Lockheed Martin X-35 over the Boeing X-32 [see the photograph (Fig. 1.11)] in the Joint Strike Fighter competition. More often, however, the MoM is more substantive and is learned by developing a close rapport with the customer. It goes without saying that developing a design to the wrong MoM will lose the contract.

In the ATF competition mentioned previously, Lockheed determined that what the USAF really wanted was a fighter pilot's airplane with supercruise and superstealth. So Lockheed made "maneuver with reckless

Figure 1.11   X-32 and X-35—JSF competitors.

abandon" their MoM. This MoM actually compromised the supercruise Mach number and RCS, but still met the requirements. The addition of pitch thrust vectoring (not required) added weight and cost to the YF-22 but gave the airplane high angle-of-attack maneuvering that was unprecedented. This feature was demonstrated during the fly-off between the YF-22 and Northrop's YF-23, although it was not required. The YF-23 was a beautiful airplane and actually beat the YF-22 in supercruise Mach and RCS, but the fighter pilots preferred the manueuverability of the YF-22. The F-22A is now operational with the USAF.

1.4 Unmanned Aerial Vehicle

The DoD defines an *unmanned aerial vehicle* (UAV) [3] as a powered aerial vehicle that does not carry a human operator, uses aerodynamic forces to provide vehicle lift, can fly autonomously or be piloted remotely, can be expendable or recoverable, and can carry a lethal or nonlethal payload. With this definition cruise missiles and aerial targets qualify as UAVs. As indicated by Table 1.2, UAVs come in all shapes and sizes [4] [see the photograph (Fig. 1.12)].

The design of unmanned and manned aircraft is the same in that they must obey the same laws of physics, but there the similarity ends. Each has advantages over the other. We should use unmanned aircraft where they have an advantage over their manned counterparts and vice versa.

The main disadvantage of the unmanned aircraft system (and hence the manned aircraft advantage) is that it cannot think for itself and cope with unforeseen or dynamically changing events. No amount of autonomy and artificial intelligence can address all the uncertainties of war. Because of this shortcoming, unmanned aircraft systems will always have off-board human operators in the loop. This means that the unmanned aircraft system must have additional sensors and data-link capability onboard to make the off-board human operator aware of the situation at all times [5].

Table 1.2   Categories of UAVs[a]

UAV Class	Weight Class	Current Vehicles
Micro	Several pounds	DARPA Organic Air Vehicle (Army)
Mini	<100 lb	FPASS (USAF), Pointer (Army), Dragon Eye (USMC)
Aerial target	<2500 lb	BQM-74E (Chukar), BQM-34S (Firebee), MQM-107E, BQM-167 (Skeeter)
Tactical ISR	<2000 lb	RQ-2A (Pioneer), Shadow 200, MQ-5C (E-Hunter), RQ-8A (Fire Scout)
Theater ISR and UAV	>2000 lb	RQ-1/MQ-1 (Predator), RQ-4A (Global Hawk), MQ-9 (Reaper), X-45A (USAF), X-47 (USN)
Cruise missiles	>1000 lb	AGM-84 (Harpoon), AGM-109 (Tomahawk), AGM-129 (Advanced Cruise Missile), AGM-158 (JASSM)

[a]See [3].

The advantages of the unmanned aircraft system are as follows [6]:

1. The design of the unmanned system is not limited by the requirement to carry a human onboard and accommodate human frailties.
2. No human is at risk of capture.
3. No infrastructure is required to recover the crew if the aircraft crashes.
4. The unmanned aircraft does not need to fly for the unmanned system to train or stay proficient.

Figure 1.12   Miscellaneous UAVs.

1.4.1 Design Limitations—Human Operator

Unmanned operation has both plusses and minuses. On the plus side the unmanned vehicle does not have to accommodate a crew station, which gives greater design freedom in locating the inlet, engine, subsystems, and payload. The elimination of the crew station also shortens the aircraft (about 7.5 ft for a single seat and 11 ft for a tandem two seat fighter) and reduces the empty weight by eliminating the crew equipment items, such as seat, cockpit controls, instruments and Environmental Control System (ECS), and the structure for the crew station. Because we buy aircraft by the pound of empty weight (to the first order), the unit cost reduction would be by about the same percentage as the weight reduction. The design freedom that results from not having a crew station in a fighter-size aircraft is very real and should produce a more efficient utilization of internal volume, but it is hard to quantify in terms of weight or cost savings. This advantage, however, diminishes with increasing aircraft size because the crew station weight and volume becomes an ever decreasing portion of larger aircraft, as shown in Fig. 1.13.

Based upon the information in Fig. 1.13 a transport UAV does not make any sense. A transport is a large aircraft and would not have an appreciable empty weight reduction due to eliminating the crew. In addition it does not qualify for any of the UAV advantages because its cargo might be people and it is making revenue flights all the time like an ISR vehicle. On the other

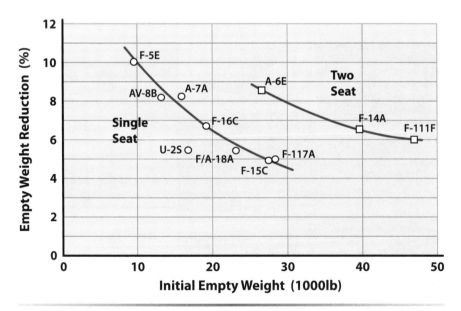

Figure 1.13 Empty weight reduction by eliminating all crew station equipment and shortening fuselage.

hand, an unmanned long range strike (LRS) vehicle may make sense as it qualifies for all the advantages.

Not having to "man-rate" the aircraft will simplify the design and development of the unmanned UAV somewhat. There will be a cost savings due to not having to man-rate the engine (engine testing) and due to the elimination of a crew escape system, canopy, and crew survival equipment (design and testing). Reducing the aircraft factor-of-safety (FS) from 1.5 to 1.25 will let the materials be worked more efficiently in structural design. In addition the safety of flight design criteria can be relaxed and the systems redundancy reduced from quadruple to dual.

Not having to man-rate the aircraft will also open the door for capabilities not available to manned aircraft. For example, the maneuver g and altitude are limited by the mechanical systems, and the endurance is limited by the size, weight, and cost of the unmanned system. This means the maneuver limit load factor can exceed 9 g for a UAV and can have persistence well in excess of 12 hours for an ISR vehicle. In addition, micro-UAVs (weight on the order of a pound) and mini-UAVs (weight on the order of 500 lb) are feasible for limited surveillance missions.

However, the downside to not having a human onboard is the requirement to recover pilot functionality by having an off-board operator who has complete situational awareness. This means increasing the software development for autonomous flight, adding sensors and data links, and of course adding a ground station to the overall system development cost.

The consensus of knowledgeable aerospace professionals is that all of the plusses and minuses together will give only a modest cost reduction in the development and acquisition cost of an unmanned UAV relative to a manned aircraft [7,8]. Many advantages of UAVs tend to be political and will be discussed in the following subsections.

1.4.2 Risk to Human Operator

Because a human pilot is not onboard the UAV, loss of life is not a concern. Thus, the UAV could be assigned missions deemed too risky for its manned counterpart. Examples are the suppression of enemy air defenses (SEAD) mission and the employment of high-powered microwave weapons.

1.4.3 Elimination of Search and Rescue

The elimination of the infrastructure to search for and rescue downed crew members is a real opportunity for cost saving. In addition, the attention given a downed crew results in a significant resource shift away from combat operations. Because the crew has been eliminated from the vehicle, the political sensitivity of the UAV mission is reduced as there is no one to be held hostage and identified with a country (e.g., Gary Powers). In an extreme case, a country could deny ownership of a "laundered" UAV.

1.4.4 Training and Proficiency

The fundamental premise is that the unmanned aircraft does not need to fly for the unmanned system to train or stay proficient. The human operator is off-board and trains by simulation. This means that when the UAV aircraft flies it should be for a revenue flight. A revenue flight for the UAV happens during wartime. During peacetime the vehicle is in some type of "flyable" storage. On the other hand, the ISR UAV flies all the time because its revenue flights are for the purpose of gathering continuous intelligence on target countries. Critics accept this premise but argue that the UAV needs to fly during peacetime as well. As part of a combined arms team, the UAVs have to operate with the manned aircraft as the humans train. This argument fails when the capability of modern air combat simulators is recognized. This notion of no (or at least minimal) peacetime flying presents a tremendous cost-saving opportunity for the UAV.

If we assume four 30-day wars over a 20-year period (typical fighter aircraft system life), the cumulative flying time (and hence design life) for the UAV is less than 1000 hours [9]. This is contrasted with manned fighter aircraft, which are designed for 8000 hours because they have to fly during peacetime. This reduced design life should result in reduced design, development, and production costs.

Design of a UAV with a design life of less than 1000 hours is different than for a manned aircraft with 8000 hours. For starters the UAV does not need to worry about material fatigue. The structure can be designed for strength instead of durability, which yields a lighter structure. In addition, equipment items could be selected with a mean flight time between unscheduled maintenance actions (MFTBUMA) of 1000 hours. The wartime maintenance would be minimal, with the turnaround actions being "refuel, rearm, and go" (true pit stop scenario). There would be a reduced number of access panels with their associated structural cutout penalties. It may be possible for the engine not to need removal during the life of the UAV. The engine removal time on commercial jets is well past 1000 hours. Because equipment items can be "stacked" more than two deep the density of the fuselage would be much greater (better utilization of available volume) than for a manned aircraft. This cost saving for the reduced design life is often difficult to quantify.

The elimination of peacetime flying (or at most minimal flying) would result in a large operation and support (O&S) cost saving relative to a manned fighter squadron. This feature is examined in more detail in

Chapter 24 as part of the discussion on life cycle cost. For the moment suffice it to say that the saving in a UAV squadron O&S costs is on the order of 80% of an equivalent manned aircraft squadron.

The unmanned ISR aircraft, on the other hand, provides continuous (24/365) coverage of a target area during peacetime and war and needs a design life of 90,000 hours over a 20-year life. The U-2S, which has a similar mission scenario, has a design life of 75,000 hours. The annual O&S cost for an unmanned ISR unit would be similar to that for a manned unit.

1.5 Specifications, Standards, and Regulations

The U.S. government regulates the operation of all aircraft in the United States by a system of specifications, standards, and regulations. An aircraft designer must not only meet (or exceed) the requirements discussed earlier, but must also comply with all the appropriate aircraft specifications, standards, and regulations if the aircraft is to be operated in the United States. The regulation of military aircraft is administered by the Department of Defense through the Department of Defense Specifications and Standards System (DODSSS); regulation of civil commercial aircraft is administered by the Department of Transportation through the Federal Aviation Regulations (FAR).

Specifications are procurement documents that describe the essential and technical requirements for aircraft items, materials, or services, including the procedures by which it will be determined that the requirements have been met. Standards establish engineering and technical limitations and applications for items, materials, processes, methods, designs, and engineering practices.

Table 1.3 lists some of the documents in the DODSSS, and [10,11] give a numerical index of the U.S. Government Specifications and Standards. Reading these specifications and standards can be overwhelming because there are over 2500 documents appropriate to a military aircraft and its associated avionics gear. Some of the more important specifications and standards in the DODSSS with which an aircraft designer should be familiar are listed in Table 1.4.

The documentation of the airworthiness standards for civil and commercial aircraft is reported in the Federal Aviation Regulations, Parts 23 and 25 [11]. Part 23 pertains to normal, utility, and acrobatic aircraft; Part 25 considers transport aircraft. The noise standards for these categories are detailed in Part 36. Civilian helicopters are regulated by Parts 27 and 29; moored balloons and kites are considered in Part 101. Parts 23 and 25 are quite thorough in that they set reasonable standards for the performance, stability, control, structure, design, construction, powerplant, equipment, and operating limits for civil and commercial aircraft.

Table 1.3 Documents in the DoD Specification and Standards System

Military specifications	Military handbooks
Military standards	Federal handbooks
Federal specifications	Air Force–Navy aeronautical standards
Federal standards	Air Force–Navy aero design standards
Qualified products list	Air Force–Navy aeronautical specifications
Industry documents	USAF specifications
Aerospace Industries Association (AIA)	Air Force–Navy aeronautical bulletins
Aerospace materials specifications (AMS)	USAF specifications bulletins
American National Standards Institute (ANSI)	USAF regulations
American Society for Testing and Materials (ASTM)	DoD manuals
American Welding Society (AWS)	USAF manuals
Magnetic Materials Products Association (MMPA)	Navy manuals
Tech orders	USAF Systems Command design handbooks

Table 1.4 Partial Listing of Military Specifications and Standards—Aircraft Design

Document Number	Title
MIL-HDBK-1797	Flying Qualities of Piloted Airplanes (replaced MIL-F-8785C)
MIL-F-83300	Flying Qualities of Piloted V/STOL Aircraft
MIL-F-9490	Flight Control Sys-Design, Installation and Test of Piloted Aircraft
MIL-S-8369	Stall/Post-Stall/Spin Flight Test Demonstration Reqs for Airplanes
MIL-C-18244	Control and Stabilization Systems: Automatic, Piloted Aircraft
MIL-D-8708	Demonstration Requirements for Airplanes
MIL-A-8860-64, 70	Airplane Strength and Rigidity
MIL-I-8700	Installation and Test of Electronics Equipment in Aircraft
MIL-P-26366	Propellers, Type Test of
MIL-S-18471	Seat System, Ejectable, Aircraft
MIL-W-25140	Weight and Balance Control Data
MIL-STD-850	Aircrew Station Vision Req. for Military Aircraft
MIL-STD-757	Reliability Evaluation from Demonstration Data
MIL-C-5011	Charts; Standard Aircraft Characteristics and Performance
MIL-STD-881	Work Breakdown Structure (WBS)
MIL-STD-499B	Systems Engineering
MIL-HDBK-516B	Airworthiness Certification—U.S. Tri-Service

It is a fact that the specifications and standards for military aircraft far outnumber the regulations for civil and commercial aircraft. For example, the entire FAR Parts 23 and 25 together form a document about two inches thick, whereas MIL-HDBK-1797, Flying Qualities of Piloted Airplanes, plus its supplement, is over two inches thick alone. It has been asserted that the military specifications and standards are excessive and are part of the reason for the high cost of military aircraft systems. Aircraft companies spend considerable time and money in military "spec" compliance. The authors will remain neutral in this matter but suggest that the reader examine this issue and become involved.

1.6 Aircraft Design Phases

Aircraft design is the name given to the activities that span the creation of a new flight vehicle. It starts as a vision and finishes as metal is being bent or as prepreg cloth for composites is being cut to conform to detail design drawings. It is the most important time in the life cycle of an aircraft as all the features both good and bad are locked in at this point. The design process is usually divided into the following three phases:

- Conceptual design
- Preliminary design
- Detail design

Although the specific activities during these three phases vary from one design group to another, they are generally formed as shown in Fig. 1.14.

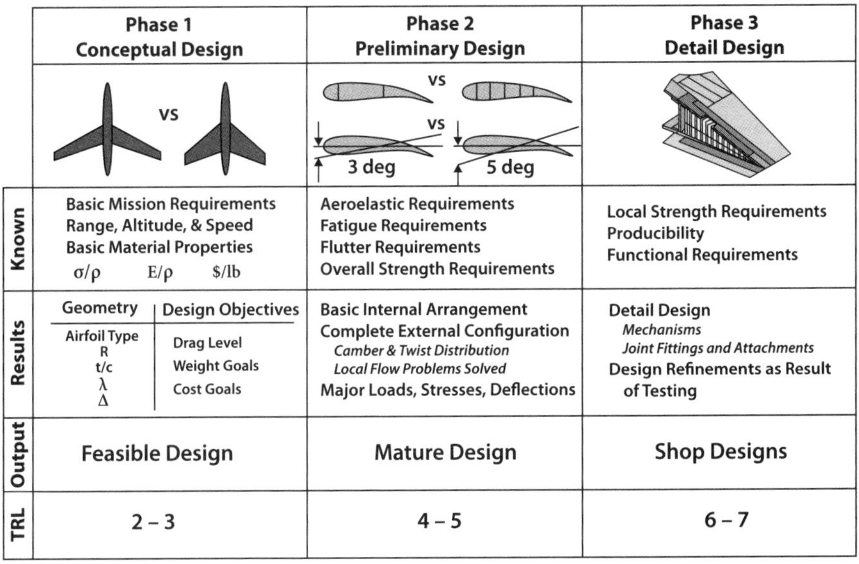

	Phase 1 Conceptual Design		Phase 2 Preliminary Design	Phase 3 Detail Design
Known	Basic Mission Requirements Range, Altitude, & Speed Basic Material Properties σ/ρ E/ρ $\$/lb$		Aeroelastic Requirements Fatigue Requirements Flutter Requirements Overall Strength Requirements	Local Strength Requirements Producibility Functional Requirements
Results	Geometry Airfoil Type R t/c λ Λ	Design Objectives Drag Level Weight Goals Cost Goals	Basic Internal Arrangement Complete External Configuration *Camber & Twist Distribution* *Local Flow Problems Solved* Major Loads, Stresses, Deflections	Detail Design *Mechanisms* *Joint Fittings and Attachments* Design Refinements as Result of Testing
Output	Feasible Design		Mature Design	Shop Designs
TRL	2 – 3		4 – 5	6 – 7

Figure 1.14 The three phases or levels of aircraft design [12].

Conceptual Design Phase

The conceptual design phase determines the feasibility of meeting the requirements with a credible aircraft design. The conceptual design process is shown schematically in Fig. 1.15 and discussed next. The general size and configuration of the aircraft, the inboard profile, and most of the major subsystems are determined during this phase.

The first task is to study, evaluate, understand, question, and if necessary negotiate the requirements (or at least ask for a waiver). The requirements are flowed down to the design group in a document called Design Guidelines. The Design Guidelines lay out the ground rules for the design study along with sensitive information about the MoMs, program strategy, selection criteria, significant design decisions, and assumptions about technologies. The Design Guidelines document is a living document and is changed or updated when appropriate.

It is a good idea at the very beginning to have a brainstorming session to identify all possible solutions to the design problem. This session needs

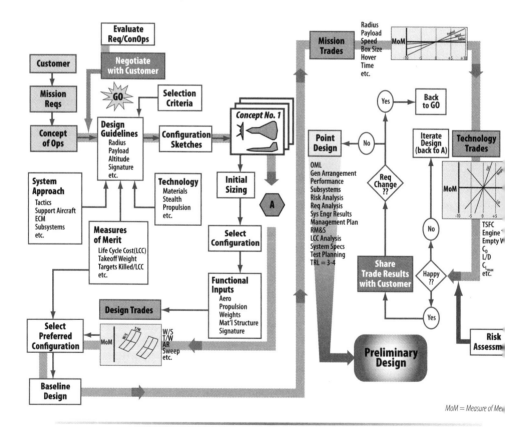

Figure 1.15 The conceptual design process.

to be a wide-open exploration of any and all concepts. Both left- and right-brain thinkers should attend as well as any person who will touch the design, for example, engineers and maintenance, manufacturing, and cost personnel.

Design trade studies are conducted around the more promising concepts using preliminary estimates of aerodynamics and weights to converge on the best wing loading, wing sweep, aspect ratio, thickness ratio, and general wing–body–tail configuration. Different engines are considered and thrust loading approaches explored to find the best airframe–engine match. The control surfaces are sized based upon the static stability and control considerations of a rigid aircraft. The performance requirements are varied (called *mission trades*) to determine the impact of each performance item on the aircraft size, weight, and cost. This information is then shared with the customer to make sure the customer is comfortable with the penalty each requirement imparts to the design. The technologies being considered in the design are examined (called *technology trades*) and estimates made of their "maturity" (probability of success) and the consequence of their not meeting the required maturity level. The results of the technology trades form the design risk analysis (discussed further in Chapter 25). The first look at cost and manufacturing is made at this time. Only gross structural aspects are considered during the conceptual design phase as resources are usually limited and the design is changing weekly. The ability of the design to accomplish the given set of requirements is established during this phase, but the details of the configuration are subject to change. Most of the work done during this phase is on paper and the manpower varies between 15 and 40 people over a year. It should be emphasized that the cost of making a design change is small during conceptual design but is extremely large during detail design.

Preliminary Design Phase

The best configuration in terms of cost and performance from the conceptual design phase is now fine tuned through wind tunnel parametric testing. This fine tuning is accomplished with an expensive wind tunnel model capable of representing the general configuration, with provision for minor variations in wing and tail planform and location. The engine is selected and inlet–engine–airframe problems are considered in detail. If the inlet arrangement is complex, an inlet component wind tunnel test might be warranted. Major loads, stresses, and deflections are determined along with considerable structural design. Aeroelastic, fatigue, and flutter analyses are conducted. Some structural components might be built and tested.

Refined weight estimates are made and a more thorough performance analysis conducted. Dynamic stability and control analysis influences are determined and six-degrees-of-freedom (6-DOF) rigid aircraft simulations are conducted to establish flight control requirements and handling quality levels. If the aircraft is highly flexible (such as a high aspect ratio wing, a high fineness ratio fuselage, low fuselage damping), the simulation might require consideration of more than six degrees of freedom in order to examine the coupling of the rigid aircraft modes and the flexible aircraft modes. The three trade studies (design, mission, and technology) started in the conceptual design phase are continued but with more vigor.

The design is given serious manufacturing consideration with preliminary plans for jigs, tooling, and production breaks. Refined cost estimates are made. Clearly the resources for the preliminary design phase are greater than for the conceptual phase and personnel typically number 100 or more people over several years.

Detail Design Phase

In the detail design phase, the configuration is "frozen" and the decision has been made to build the aircraft. Detailed structural design is completed. All of the detail design and shop drawings of the mechanisms, joints, fittings, and attachments are accomplished. Interior layout is detailed with respect to location and mounting of equipment, hydraulic lines, ducting, control cables, and wiring bundles. Sometimes component mock-ups are built to aid in the interior layout. The drawings for the jigs, tooling, and other production fixtures are done at this time. A detailed cost estimate based upon work breakdown structure (WBS) is made. All equipment and hardware items are specified. Often, system mock-ups (such as fuel system, landing gear, ECS, engine–inlet, and a hardware-in-the-loop flight control system called an iron bird) will be designed, built, and tested during this phase. It is important that from this point on the design changes be kept to a minimum because the cost of making a change is large once the drawing hits the shop floor. The next step is ordering all the equipment items (called Bill of Materials) and the fabrication and assembly of the prototype (usually at least two prototypes are built). Often, the fabrication of some components will be started during this phase as soon as their shop drawings are released.

Figure 1.16 shows the three phases of design in a typical government program acquisition according to DoD 5000.1 [2]. The years shown are extremely optimistic because there are always breaks in the schedule while the government issues a Request for Proposal, industry submits proposals, and the government evaluates the proposals, selects a winner, and gets its funding in place. Commercial programs move much faster because the aircraft builder controls the tempo and funding of the program. Typical times from the decision to build the aircraft [Milestone 1 or B for the government; the start of preliminary design (PD) for commercial] to production is about 10 years for the government and 5 years for commercial.

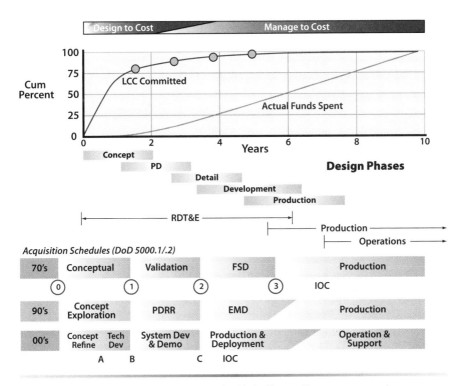

Figure 1.16   Design phases integrated into the entire government program.

Figure 1.16 also shows the importance of the conceptual design phase in that over 70% of the design features that drive life cycle cost (LCC) are selected during that phase. Because it all started with gliders, a number of early examples are shown in Fig. 1.17.

1.7 Scope of the Text

The text is in two volumes:

• Volume 1, Aircraft Design
• Volume 2, Airship Design, Aircraft Design Case Studies, and Photo Gallery

Volume 1 considers the conceptual design phase of the aircraft design process. It is self-contained, and the chapters and appendices lead the reader through one iteration of the conceptual design process. Volume 1 will give the reader an understanding and appreciation of how the different disciplines must blend together to produce an effective aircraft.

Volume 2 is also written as a stand-alone volume and uses rewritten introductory material from Volume 1 that is focused on the unique design

Unpowered gliders of all shapes and sizes were the forerunners of manned powered flight. Otto Lilienthal was the most famous glider designer, builder, and pilot of all. He made over 2000 flights in the 1890s and with his brother Gustav measured lift and drag. Otto would die in a crash of one of his designs in 1896.

Figure 1.17   Various unpowered glider designs.

issues of airships, hybrid airships, and high-altitude balloons. Actual design case studies are also included for both aircraft and airship programs.

In Volume 1 the conceptual design process is shown schematically in Fig. 1.15 and proceeds as follows (this order will be followed throughout the remainder of Volume 1):

1. **Critical requirements.** The mission requirements are studied to identify the requirement that drives the design. For example, will the aircraft be range-dominated, field-length constrained, or required to operate supersonically for extended periods? An early assessment of the driving requirement can help in the proper selection of the wing planform shape and size. The applicable specifications, standards, and regulations should be identified and met throughout the design process.
2. **Initial aircraft sizing.** At this point, the aircraft takeoff gross weight (TOGW) is estimated. If this is the first time through the design loop, many assumptions will have to be made to get started. A first estimate of the fuel weight is made at this point also.
3. **Takeoff wing loading, $(W/S)_{TO}$.** Here the aircraft wing is sized. One or more of the following criteria will size the wing area:
   - Landing and takeoff
   - Operational or intercept altitude and speed (i.e., a low-q or high-q vehicle?)

- Air combat (maneuverability)
- Efficient cruise
- Efficient loiter

4. **Airfoil section and planform shape.** The following items are selected:
   - Airfoil—section, t/c, camber, nose shape
   - Planform—aspect ratio, taper, sweep, fixed versus variable geometry

   This is a major design decision and is not an easy task. The importance of low-speed performance, maximum lift coefficient $C_{L_{max}}$, and other related issues must be determined.

5. **Fuselage sizing and shape.** The fuselage is sized to meet the volume requirements of crew, payload (passengers, bombs, or cargo), engines (external or internal), fuel (in fuselage, wings, or external tanks), avionics, and so on. The fuselage fineness ratio, l/d (ratio of length to maximum diameter), is estimated based on whether the primary requirement of the mission is subsonic, supersonic, or mixed.

6. **Estimation of tail size.** A rough layout of the aircraft is prepared and the wing is positioned on the fuselage. A decision must be made whether the aircraft will be an aft tail, canard, or tailless configuration. The designer will use tail volume coefficients and historical data to prepare an initial estimate of tail sizes.

7. **Configuration aerodynamics.** The designer must estimate the aircraft zero-lift drag C_{D_0}, lift curve slope C_{L_α}, and drag-due-to-lift factor K versus Mach number using the best available methodology. It is important to double-check initial aerodynamic estimates by comparing them with existing aircraft ("sanity check").

8. **Sizing of engines.** The type and number or size (in the case of "rubber" engines) of the engines are determined. The engine thrust-to-weight ratio $(T/W)_{TO}$ is an important design parameter and will be driven by one or more of the following criteria:
   - Efficient cruise or loiter
   - Takeoff
   - Air combat
   - Minimum time to intercept
   - Service ceiling

9. **Design and sizing inlets.** Inlets are located and designed to match engine requirements. Pressure recovery, thrust data corrections, bypass drag, and spillage drag are determined.

10. **Refined fuel estimate.** A refined fuel estimate is made and compared with the original estimate made in step 2.

11. **Component weights and c.g. location.** The fuel is located in the aircraft and a refined estimate of the aircraft component weights is made. The aircraft is balanced and the c.g. travel during the mission is

checked. The updated TOGW is compared with the original estimate made in step 2.

12. **Sizing of tail and control surfaces and determination of trim drags.** A static stability and control analysis (longitudinal, directional, and lateral) is performed to size the control surfaces to meet acceptable requirements. The cruise and maneuver trim drags are determined.

13. **Refined performance analysis.** A refined performance analysis is conducted and compared with the mission requirements.

14. **Cost estimate.** Estimates are made of the prototype and production aircraft costs. If enough information exists, an estimate of operation and maintenance costs should be made to give an indication of the life cycle cost (LCC) for the aircraft.

15. **Design iteration.** At this point, a review of the entire design process is performed, changing and updating design features and results until the design team is satisfied or funds run out, whichever comes first.

Numerous texts dealing with aircraft design are listed at the end of this chapter [12–27]. Corning [14] is a good text dealing with transport design in the early 1950s. Wood [13,15] is a general design text full of design information, including a section on seaplane hull design. Perkins and Hage [16] is an excellent text on performance, stability, and control, but unfortunately it is badly outdated. Stinton [17] gives the aircraft design picture with a British flavor. Torenbeek [18] is an excellent text for transport design, with numerous weight equations. Roskam [22] is an excellent eight-volume series written by a practicing engineer and design professor at the University of Kansas (over 35 years, see [28]). Raymer [23] is a design text used by some U.S. universities. Thomas [26] is an easy-to-read, authoritative text on sailplane design written by a practicing engineer at DLR, Braunschweig, Germany, and a World-class sailplane pilot. Newberry [25] is filled with copies of landmark design papers from the years 1941–1991. The remaining referenced texts and articles are also recommended.

References

[1] "Design to Cost," DoD Directive 5000.28, U.S. Department of Defense, U.S. Government Printing Office, Washington, DC, 2009.

[2] "Acquisition of Major Defense Systems," DoD Directive 5000.1, U.S. Department of Defense, U.S. Government Printing Office, Washington, DC, 2009.

[3] Wilson, J. R., "Unmanned Aerial Vehicles: Background and Issues for Congress," report for Congress, Congressional Research Service Rept. RL31872, 25 April 2003, pp. 30–37.

[4] "2007 UAV Worldwide Roundup," *Aerospace America*, May 2007.

[5] Clough, B., "UAVs—You Want Affordability and Capability? Get Autonomy!", Association of Unmanned Vehicle Systems International paper, 14–17 July 2003, Baltimore, MD.

[6] "Matching Resources with Requirements Is Key to the Unmanned Combat Air Vehicle Program's Success," report to the Subcommittee on Tactical Air and Land Forces, Committee on Armed Services, House of Representatives, General Accounting Office Rept. 03-598, June 2003, p. 3.

[7] Brown, R., "Enhancing UAV Affordability," briefing to the Air Force Scientific Advisory Board, ASC/FB, 19 March 2002.

[8] Brown, R., "DoD Juggles Affordability, Capability as UAV Costs, Budgets Rise" *Aerospace Daily, Aviation Week & Space Technology*, 2 Jan. 2003, p. 7.

[9] "UAV Technologies and Combat Operations," *Panel Reports*, Vol. 2, U.S. Air Force Scientific Advisory Board, SAB-TR-96-01, Dec. 1996, pp. 3–49.

[10] *Department of Defense Index of Specifications and Standards*, U.S. Government Printing Office, Washington, DC, 2009.

[11] "Airworthiness Standards: Part 23—Normal, Utility, Acrobatic, and Commuter Category Airplanes; Part 25—Transport Category Airplanes," *Federal Aviation Regulation*, Vol. 3, U.S. Department of Transportation, U.S. Government Printing Office, Washington, DC, 2009.

[12] Nicolai, L. M., *Fundamentals of Aircraft Design*, METS Inc., San Jose, CA, 2009.

[13] Wood, K. D., *Airplane Design*, Univ. of Colorado, CO, 1947.

[14] Corning, G., *Supersonic and Subsonic Airplane Design*, Edwards Bros., Ann Arbor, MI, 1953 (revised 1964).

[15] Wood, K. D., *Aircraft Design*, Johnson Publ., Boulder, CO, 1966.

[16] Perkins, C. D., and Hage, R. E., *Airplane Performance, Stability and Control*, Wiley, New York, 1949.

[17] Stinton, D., *The Anatomy of the Aeroplane*, American Elsevier, New York, 1966.

[18] Torenbeek, E., *Synthesis of Subsonic Aircraft Design*, Delft Univ. Press, The Netherlands, 1976.

[19] Goodmanson, L. T., and Gratzer, L. B., "Recent Advances in Aerodynamics for Transport Aircraft, Part 1," *Astronautics and Aeronautics*, Vol. 11, No. 12, Dec. 1973, p. 30.

[20] Goodmanson, L. T., and Gratzer, L. B., "Recent Advances in Aerodynamics for Transport Aircraft, Part 2," *Astronautics and Aeronautics*, Vol. 12, No. 1, Jan. 1974, p. 52.

[21] Kuchemann, D., *The Aerodynamic Design of Aircraft*, Pergamon, New York, 1978.

[22] Roskam, J., *Aircraft Design*, Pts. 1–8, Roskam Aviation and Engineering Corp., Lawrence, KS, 1985. [Available via www.darcorp.com (accessed 31 Oct 2009).]

[23] Raymer, D. P., *Aircraft Design: A Conceptual Approach*, AIAA Education Series, AIAA, Washington, DC, 1989.

[24] Whitford, R., *Design for Air Combat*, Jane's Publ., London, 1987.

[25] Newberry, C. F., *Perspectives in Aerospace Design*, AIAA Education Series, AIAA, Washington, DC, 1991.

[26] Thomas, F., *Fundamentals of Sailplane Design*, College Park Press, College Park, MD, 1999.

[27] Corke, C. C., *Design of Aircraft*, Prentice Hall, Pearson Education, Upper Saddle River, NJ, 2003.

[28] Roskam, J., *Roskam's Airplane War Stories*, DARcorp., Lawrence, KS, 2002.

Chapter 2 | Review of Practical Aerodynamics

- Lift & Drag
- Boundary Layers & Skin Friction
- Wings & Bodies
- Subsonic & Supersonic
- Drag Polar
- Transonic Effects
- Finite-Wing Effects
- Swept-Wing Effects

An F-14 flies on a humid day near Mach 0.9 with shock condensation on the top of the wing and canopy. The flow decelerates from supersonic to subsonic through a normal shock with a jump in static pressure that can cause the water vapor in the air to condense into a cloud trailing behind the aircraft.

Eagles don't flock. You have to find them one at a time.
Ross Perot

2.1 Introduction

An aircraft is a unique machine because it lifts itself from the ground. This lift is generated by air flowing over a wing; it is then balanced by smaller amounts of lift (up or down) being generated at the aft end (front end for a canard) and at each wingtip. The cross section of the wing (looking down the span), called an *airfoil section*, is uniquely shaped (see Fig. 2.1). As the air flows over the wing (because the wing is being pushed through the air, as in flight, or because air is being blown over a static wing, as in a wind tunnel; see Fig. 2.2) it accelerates over the upper and lower surfaces so that it meets at the trailing edge (called the *Kutta condition*). This acceleration of the air particles causes the static pressure on the surfaces to drop below the static pressure in the freestream. There is one streamline of air particles (called the *dividing streamline*) that slams into the airfoil leading edge and comes to a stop. This point on the airfoil is called the *stagnation point*, and the pressure at this location is equal to the freestream static pressure plus the dynamic pressure. At this point the reader should review the airfoil appendix, Appendix F.

Lift and drag data for an aircraft throughout the Mach number range of its flight envelope are necessary ingredients for any performance analysis. Moment data about all three axes, as shown in Fig. 2.3, are necessary for stability and control analysis. This chapter will review the fundamental aerodynamic concepts relative to lift, drag, and moment methods. The working equations and methodology for estimating lift and drag are pre-

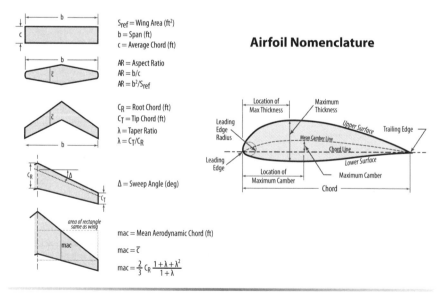

Figure 2.1 Wing geometry and nomenclature.

Figure 2.2 Langley 30 x 60-ft wind tunnel with the X-48B in the test section.

sented in Chapter 13; those for stability and control analysis, in Chapters 21–23.

The pressure distribution over the upper and lower surfaces of an airfoil designed for long endurance missions is shown in Fig. 2.4. The pressure distribution is usually expressed as the pressure coefficient C_p as defined in Fig. 2.4. The shaping of the airfoil is such that there is more suction on the upper surface than on the lower, and a force normal to the

Nondimensionalization Permits Comparison of Forces and Moments Between Dissimilar Vehicles

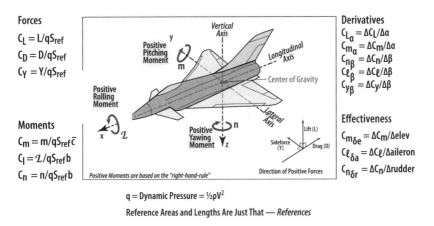

Forces
$$C_L = L/qS_{ref}$$
$$C_D = D/qS_{ref}$$
$$C_Y = Y/qS_{ref}$$

Moments
$$C_m = m/qS_{ref}\bar{c}$$
$$C_l = \mathcal{L}/qS_{ref}b$$
$$C_n = n/qS_{ref}b$$

Derivatives
$$C_{L_\alpha} = \Delta C_L/\Delta\alpha$$
$$C_{m_\alpha} = \Delta C_m/\Delta\alpha$$
$$C_{n_\beta} = \Delta C_n/\Delta\beta$$
$$C_{\ell_\beta} = \Delta C_\ell/\Delta\beta$$
$$C_{y_\beta} = \Delta C_y/\Delta\beta$$

Effectiveness
$$C_{m_{\delta e}} = \Delta C_m/\Delta elev$$
$$C_{\ell_{\delta a}} = \Delta C_\ell/\Delta aileron$$
$$C_{n_{\delta r}} = \Delta C_n/\Delta rudder$$

Positive Moments are based on the "right-hand-rule"

Direction of Positive Forces

$q = $ Dynamic Pressure $= \frac{1}{2}\rho V^2$

Reference Areas and Lengths Are Just That — *References*

Figure 2.3 Major nondimensional aerodynamic parameters and sign convention.

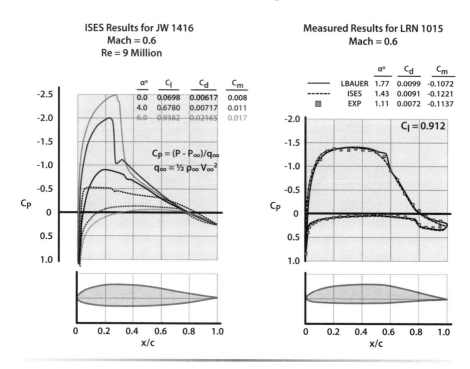

Figure 2.4   Wing surface pressure distributions for airfoil designed for long endurance.

freestream is generated called *lift*. This lift will increase as the angle-of-attack α (the angle between the airfoil chord line and the freestream) is increased until α reaches a value where the flow on the upper surface separates, greatly reducing the lift, and the airfoil stalls. This pressure distribution also produces a pressure drag parallel to the freestream and a pitching moment (usually taken about the quarter chord or aerodynamic center). The *aerodynamic center* (a.c.) is the point on the airfoil at which the value of the pitching moment does not change as the angle-of-attack changes. The aerodynamic coefficients discussed in this chapter are defined as follows:

$$C_L = \frac{L}{qS_{ref}} \tag{2.1}$$

$$C_D = \frac{D}{qS_{ref}} \tag{2.2}$$

$$C_{MA} = \frac{M_A}{qS_{ref}\overline{c}} \tag{2.3}$$

The aerodynamicist usually furnishes the data in coefficient form as a function of Mach number (M), and sometimes Reynolds number (Re). The coefficients of interest are shown in Fig. 2.3, where $q = \frac{1}{2}\rho V_\infty^2$ (dynamic pressure), $\bar{c}$ is the mean aerodynamic chord, and the moments are about a specified point (usually the wing quarter chord, aerodynamic center, or center-of-gravity). Notice that the coefficients are referenced to S_{ref}, which is usually the wing total planform area as shown in Fig. 2.1. The aerodynamic coefficients for three-dimensional bodies (wings, bodies, and combinations) are denoted by capital subscripts; for two-dimensional airfoils, by lowercase subscripts.

2.2 Drag

In aerodynamics, lift is very good, moment is useful, and drag is horrid. Designers spend most of their time trying to maximize the lift, control the moment, and minimize the drag. The *drag* is the aerodynamic force resolved in the direction of the freestream due to (1) viscous shearing stresses, (2) the integrated effect of the static pressures acting normal to the surfaces, and (3) the influence of the wing trailing vortices on the aerodynamic center of the configuration.

- **Skin friction drag.** The drag on a body resulting from viscous shearing stresses over its wetted surface.
- **Pressure drag (or form drag).** The drag on a body resulting from the integrated effect of the static pressure acting normal to its surface resolved in the drag direction.
- **Profile drag.** Usually defined as the sum of the skin friction drag and the pressure drag for a two-dimensional airfoil.
- **Viscous drag-due-to-lift.** The drag that results due to the integrated effect of the static pressure acting normal to its surface (resolved in the drag direction) when the airfoil angle-of-attack is increased to generate lift. (Note: It is present without vortices.)
- **Inviscid drag-due-to-lift (or induced drag).** The drag that results from the influence of a trailing vortex (downstream of a lifting surface of finite aspect ratio) on the wing aerodynamic center. (Note: It is present with or without viscosity).
- **Interference drag.** The increment in drag resulting from bringing two bodies into proximity to each other; for example, the total drag of a wing–fuselage combination will usually be greater than the sum of the wing drag and fuselage drag independent of one another.
- **Trim drag.** The increment in drag resulting from the aerodynamic forces required to trim the aircraft about its center of gravity; usually this takes the form of added drag-due-to-lift on the horizontal tail.

- **Base drag.** The specific contribution to the pressure drag attributed to a blunt afterbody.
- **Wave drag.** Only exists with supersonic flow; this is a pressure drag resulting from noncanceling static pressure components on either side of a shock wave acting on the surface of the body from which the wave is emanating.
- **Excrescence drag (or protuberance drag).** The drag associated with antennas, total pressure probes (part of the air data system), and other protrusions above the exterior of the aircraft; external fuel tanks, missiles, and bombs are also considered excrescence drag items.
- **Cooling drag.** The drag resulting from the momentum lost by the air that passes through the powerplant installation (i.e., the heat exchanger) for cooling the engine, oil, and so on.
- **Ram drag.** The drag resulting from the momentum lost by the air as it slows down to enter an inlet.

2.3 Boundary Layers and Skin Friction Drag

Air molecules flow over a body in layers called *streamlines*. The air molecules in the streamline next to the body surface actually interact with the molecular structure of the surface and come to a stop. This is the "no slip condition" that makes up the foundation of boundary layer theory and is shown in Fig. 2.5. As the streamlines move away from the surface the air molecules speed up, giving an increasing velocity gradient dv/dz. At a distance δ (called the *boundary layer thickness*) from the surface the velocity gradient is zero.

The boundary layer composed of many streamlines can take one of three forms as shown in Fig. 2.5: (1) if the streamlines are smoothly flowing, it is *laminar*; (2) if the streamlines are chaotic and vortical, it is *turbulent*; and (3) if the streamlines are separated from the surface ($dv/dz = 0$ at $z = 0$), it is called a *separated boundary layer*. The character of the separated boundary layer is such that the flow near the surface can actually reverse direction and flow upstream. The shearing action between the streamlines creates a friction force in the streamline direction. At the surface ($z = 0$) this friction force is equal to $\mu \, dv/dz$ times the surface area, where μ is the fluid coefficient of viscosity and acts parallel to the surface. Notice that the velocity gradient dv/dz at the surface is smaller for the laminar boundary layer than the turbulent boundary layer, giving a lower skin friction drag. Notice also that in the separated region $dv/dz = 0$ at the surface there is nearly zero skin friction drag, but at the same time there is a large increase in static pressure, pressure drag, and loss of lift.

The character of the boundary layer is dependent upon a nondimensional parameter called the *Reynolds number, $Re_x = \rho V x / \mu$*, which is a ratio

Friction force ~ μ dv/dz (area) where μ is the fluid coefficient of viscosity and dv/dz is the velocity gradient evaluated at z=0. This force acts parallel to the surface.

Boundary layer starts out laminar and transitions to turbulent at $Re_x \approx 5 \times 10^5$ where Re_x = local Reynold's Number = ρvx/μ

Laminar thickness $\delta_L = 5.2x/Re_x{}^{0.5}$ and turbulent thickness

$\delta_L = 0.37x/Re_x{}^{0.2}$

Flow separates when dv/dz = 0 at the surface.

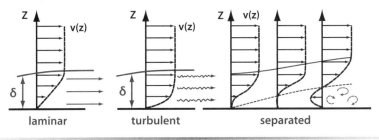

laminar   turbulent   separated

Figure 2.5 Boundary layer profile: three flow conditions.

of the inertia forces in the boundary layer to the friction forces. The boundary layer starts out laminar and transitions to turbulent at a local $Re_x \approx 5 \times 10^5$. The laminar boundary layer is extremely delicate and will transition early if it encounters surface disturbances or increasing pressure gradient. The thickness of the boundary layer is given by the following expressions:

Laminar Thickness:

$$\delta_L = 5.2x / Re_x^{0.5}$$

Turbulent Thickness:

$$\delta_T = 0.37x / Re_x^{0.2}$$

The averaged skin friction coefficient C_F acting on a square unit of the aircraft surface is shown in Fig. 2.6 as a function of Reynolds number, where L is a characteristic dimension of the aircraft such as fuselage length or wing mean aerodynamic chord. The solid lines in Fig. 2.6 assume the boundary layer to be completely laminar or completely turbulent, with the dashed line in the middle denoting a situation of transition. The skin friction drag force of the aircraft component is calculated as follows:

Skin friction drag force = C_F (surface area) (dynamic pressure)

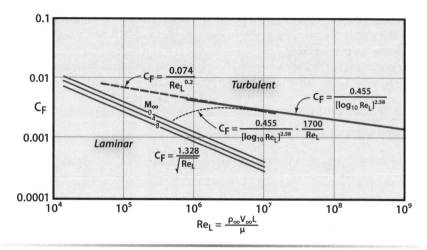

Figure 2.6 Skin friction coefficient over a flat plate.

2.4 Incompressible Airfoil Section Theory

Subsonic thin airfoil theory (incompressible and inviscid) predicts the section lift and moment coefficients quite well for airfoil shapes up to moderate angles-of-attack below stall. The theory predicts (see [1], page 73, and [2], page 34)

$$\frac{dC_\ell}{d\alpha} = m_0 = 2\pi \text{ per radian}$$

Aerodynamic center location = quarter chord = $\frac{1}{4}\bar{c}$

$$\alpha_{ol} = -\frac{1}{\pi}\int_0^\pi \left(\frac{dz}{dx}\right)(\cos\theta - 1)d\theta \tag{2.4}$$

$$C_{m_{a.c.}} = -\frac{1}{2}\int_0^\pi \left(\frac{dz}{dx}\right)(\cos 2\theta - \cos\theta)d\theta \tag{2.5}$$

where dz/dx is the local slope of the camber line, θ is the change of variable $x = c/2(1 - \cos\theta)$, and c is the chord of the airfoil section.

Typical section data are shown in Figs 2.7 and 2.8. References [1], [3], and [4] report experimental section data on many airfoil shapes. Appendix F discusses airfoil nomenclature and presents section data on the more popular airfoil sections.

The $C_{m_{a.c.}}$ is constant with changing C_ℓ or α by definition of the *aerodynamic center* (a.c.). The section lift data up through moderate angles (i.e., the linear lift region) is expressed as

$$C_\ell = m_0 \left(\alpha - \alpha_{0l} \right)$$

Notice that α_{0l} is zero for symmetric or uncambered sections (i.e., dz/dx = 0) such as the NACA 0012 airfoil shown in Fig. 2.7.

The drag coefficient for an airfoil section pretty much has a parabolic behavior with the lift coefficient, except at large α. This parabolic behavior is expressed as

$$C_d = C_{d_{\min}} + k'' \left(C_\ell - C_{\ell_{\min}} \right)^2 \qquad (2.6a)$$

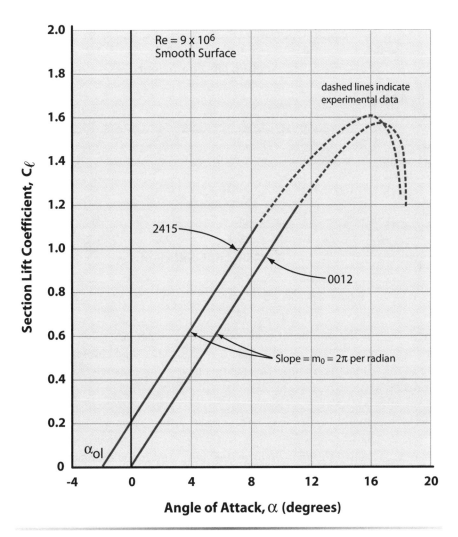

Figure 2.7 Section lift data (from [3]).

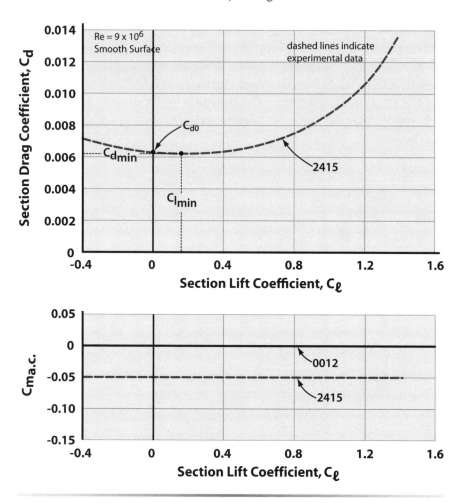

Figure 2.8 Section C_d and $C_{m_{a.c.}}$ data (from [3]).

where $C_{d_{min}}$ is the minimum value of C_d (see Fig. 2.8), $C_{\ell_{min}}$ is the C_ℓ value for $C_{d_{min}}$, and $k'' = \Delta(C_d - C_{d_{min}})/\Delta(C_\ell - C_{\ell_{min}})^2$ is called the viscous drag due to the lift factor. Symmetric sections have $C_{\ell_{min}} = 0$ such that the C_d is expressed as

$$C_d = C_{d0} + k'' C_{\ell}{}^2 \qquad (2.6b)$$

where C_{d0} is called the *zero-lift drag coefficient* and is due to separation and skin friction drag at $C_\ell = 0$. Appendix F shows that the NACA 24XX family of airfoils has a $k'' \sim 0.0047$ at $Re = 3 \times 10^6$.

Thin airfoil theory is an inviscid theory and thus cannot predict section characteristics due to viscous effects (drag and stall characteristics). Results from thin airfoil theory are indicated by solid lines in Figs. 2.7 and 2.8.

The section data shown in Fig. 2.7 must be corrected for Mach number (compressibility), wing sweep, and finite wing effects and are discussed in the next section.

2.5 Subsonic Compressibility Corrections

The subsonic compressibility correction factor is known as the Prandtl–Glauert transformation. Essentially it transforms the compressible flow problem into an equivalent incompressible flow problem. The correction for lift curve slope is as follows:

$$\left(m_0\right)_{M\neq0} = \frac{\left(m_0\right)_{M=0}}{\sqrt{1-M^2}} \tag{2.7}$$

The effect of compressibility is to increase the section lift curve slope. Theoretical and experimental values usually agree well for $0 < M < 0.8$. Beyond $M = 0.8$ the agreement breaks down.

2.6 Finite Wing Corrections

When the airfoil has a finite span (i.e., a wing) the differential pressure on the top and bottom surfaces causes a circular motion of the air about the wingtips. This circular or vortex motion of the air (counterclockwise on the right wing and clockwise on the left wing as you look forward) trails behind the wingtips as trailing vortices. These trailing vortices, called *wake turbulence*, are what forces airports to impose separation distances behind airplanes that are landing and taking off. The trailing vortices induce a downwash at the wing aerodynamic center, which gives a lower effective angle-of-attack and an induced drag. This drag has nothing to do with viscosity and is an inviscid drag-due-to-lift. This decrease in the effective angle-of-attack is shown in Fig. 2.9. Notice that, for a finite wing of aspect ratio = 10, the α must be increased by an amount α_i to give the same lift coefficient as an airfoil of infinite aspect ratio. Finite wing theory (see [1], page 109) gives the following results:

$$\alpha_i = \text{induced angle-of-attack} = \frac{C_L}{\pi \text{AR}}\left(1+\tau\right) \tag{2.8}$$

$$m = \frac{\mathrm{d}C_L}{\mathrm{d}\alpha} = \frac{m_0}{1+\left[m_0\left(1+\tau\right)/\left(\pi \text{AR}\right)\right]} \tag{2.9}$$

$$C_{D_{L_i}} = \text{induced drag} = \frac{C_L^2\left(1+\delta\right)}{\pi \text{AR}} = K'C_L^2 \tag{2.10}$$

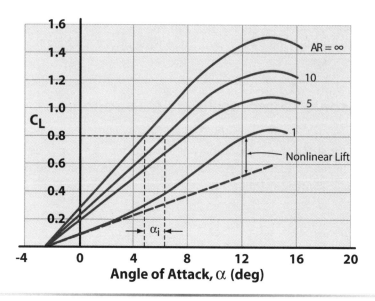

Figure 2.9   Effect of finite span on the lift characteristics of a NACA 65-410 airfoil (Appendix F).

where AR is the *aspect ratio* and τ and δ are correction factors to account for deviations from an elliptical lift distribution. The τ and δ are dependent upon AR and taper ratio and are shown in Fig. 2.10. The K' is called the inviscid drag-due-to-lift factor.

2.7 Sweep Correction

The finite wing correction depends solely on aspect ratio and taper ratio and is based upon the freestream velocity being perpendicular to the quarter chord line. If the wing is swept, the component of velocity perpendicular to the quarter chord establishes the pressure distribution over the wing, and the tangential component flows spanwise along the wing and does not influence the pressure distribution. This is shown in Fig. 2.11.

Empirical data indicates the following corrections for the lift curve slope of swept wings:

$$m = \left(m\right)_{\Delta=0} \cos\Delta, \quad \text{for AR} > 6 \tag{2.11}$$

$$m = \left(m\right)_{\Delta=0} \sqrt{\cos\Delta}, \quad \text{for AR} < 6 \tag{2.12}$$

where Δ is the sweep of the quarter chord or maximum thickness line. Although wing sweep is advantageous for high-speed flight (discussed in the next section) it creates problems for low-speed flight. A swept-wing

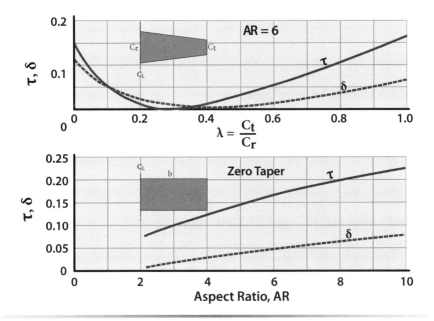

Figure 2.10   Correction factors for nonelliptic lift distribution (data from [4]).

aircraft will be required to land and take off at higher angles-of-attack than a straight-wing aircraft of the same aspect ratio because it has a lower C_{L_α}.

2.8 Combined Effects

The effects of sweep, finite span, and compressibility may be combined into the following useful equation for subsonic lift curve slope:

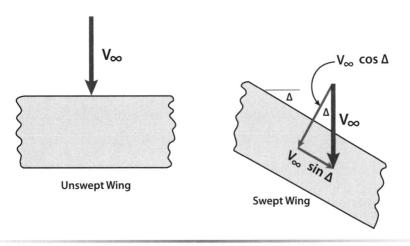

Figure 2.11   Normal component of V_∞ establishes the pressure distribution over wing station.

$$\frac{dC_L}{d\alpha} = C_{L\alpha} = \frac{2\pi\,\text{AR}}{2 + \sqrt{4 + \text{AR}^2\beta^2\left(1 + \left[(\tan^2\Delta)/\beta^2\right]\right)}} \tag{2.13}$$

where AR = aspect ratio = (span)^2/wing area

$$\beta = \sqrt{1 - M^2}$$

Δ = sweep of the maximum thickness line

2.9 Nonlinear Wing Lift and Moment

Usually the wing C_L is thought to be linear in α, that is,

$$C_L = \left(\frac{dC_L}{d\alpha}\right)\alpha$$

However, this is an approximation and is only accurate for high-AR wings. Actually the wing C_L is more correctly written as

$$C_L = \left(\frac{dC_L}{d\alpha}\right)_{\alpha=0}\alpha + C_1\alpha^2 \tag{2.14}$$

where $(dC_L/d\alpha)_{\alpha=0}$ is the wing lift curve slope evaluated at $\alpha = 0$ (or close to it) using Eq. (2.13), and C_1 is the nonlinear lift factor. A pronounced nonlinear relationship between aerodynamic coefficients and α is typical of nearly all planforms when AR is less than approximately 3. The usual lifting surface theories (linear theories) predict linear relationships that underestimate the lift for low-AR wings as shown in Fig. 2.9.

Nonlinear dependence of lift and pitching moment on the angle-of-attack is extremely significant for slender bodies and wings of small aspect ratio. This is because the flow past a slender body or low-AR wing is completely different from the flow past a classic unswept wing of large AR. The characteristic feature of the flow past such a slender body or low-AR wing is the strong cross flow that leads to separation of the flow at the sides of the body or wing edges and to the formation of free vortices on the upper surface, as shown in Fig. 2.12. This formation of free vortices on the upper surface of the low-AR wing or slender body is the reason for the nonlinear relationship between lift and pitching moment and angle-of-attack, α. The nonlinear behavior becomes more pronounced with decreasing AR, and with AR = 1 the nonlinear part is of the same order of magnitude as the linear part, as shown in Fig. 2.9.

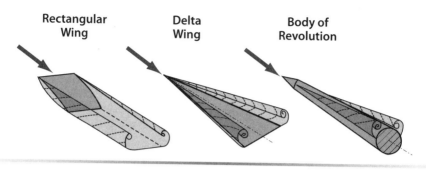

Figure 2.12 Vortex configurations past slender bodies.

The vortex flow pattern is initiated by the flow separation at the edge of the wing and is highly dependent upon the shape of the wing edge. A sharp edge precipitates separation sooner and more cleanly than a rounded edge. Thus, the nonlinear lift and moment contribution from sharp edges are about twice those from round edges. For delta wings the planform tip edges and leading edge are the same; thus, it is logical to assume that the nonlinear behavior is a function of planform (i.e., rectangular, swept, or delta) as well as AR. This nonlinear lift theory is developed in [2,5–8].

The wing C_L is given by

$$C_L = \left(\frac{dC_L}{d\alpha}\right)_{\alpha=0} \alpha + C_1\alpha^2 \qquad (2.14)$$

for α in radians. The moment coefficient about the wing apex is similarly given by

$$C_M = \left(\frac{dC_M}{d\alpha}\right)_{\alpha=0} \alpha + C_2\alpha^2 \qquad (2.15)$$

for α in radians, where the values for C_1 and C_2 are obtained from Fig. 2.13 for sharp-edged wings, and $0.5C_1$ and $0.45C_2$ are used for round leading edges.

This vortex formation and control over low-AR bodies is the reason for the use of fuselage strakes (SR-71 and F-16) and wing leading edge fuselage extensions (F-5, YF-17, and F-18). The strakes and leading edge extensions (LEX) are very low aspect ratio devices and form powerful vortices at moderate angles-of-attack that then reverse course over the top surface of the main wing, sweeping high-energy air into the boundary layer and delaying flow separation. Figure 2.14 visually illustrates the vortex rollup created by the low-AR forebody strakes. Notice also how the "V" tails have been positioned such that these vortices do not impinge on the tail surfaces. The

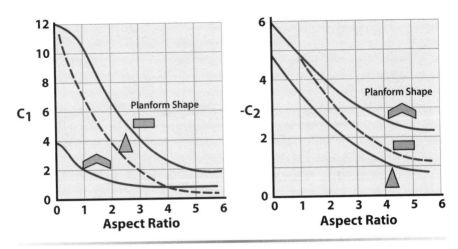

Figure 2.13 Values of C_1 and C_2 for various planform shapes and aspect ratios.

curved shape of the strake tends to tighten the vortex and thus reduce its influence on other parts of the vehicle.

2.10 Total Aircraft Subsonic Aerodynamics

The total wing subsonic drag coefficient is a combination of section and finite wing effects. The wing C_D can be expressed as

$$C_D = C_{D_0} + C_{D_{L_v}} + C_{D_{L_i}} \tag{2.16}$$

where $C_{D_{L_v}}$ is the viscous drag-due-to-lift and $C_{D_{L_i}}$ is the induced (inviscid) drag-due-to-lift. For a cambered wing, Eq. (2.16) is written as

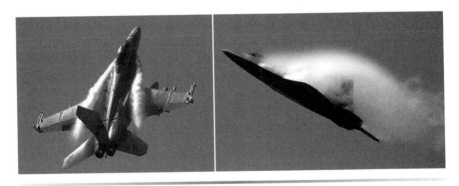

Figure 2.14 a) F-18 in high-*g* maneuver showing vortices rolling up over the LEX, and **b)** F-22 showing vortex shedding from its LEX and the leading edge of its swept back wing.

Tigershark Wing Rock Fix

In 1983 the Northrop F-20A Tigershark (shown in Fig. 2.15) was in flight testing. The aircraft exhibited wing rock at ~18 deg angle-of-attack due to asymmetric flow separation at the wingtips. Because the F-20A was to be an air-to-air fighter with guns and IR missiles this was unacceptable. The normal fix would be to twist the wing, with the LE down, so that the wing root stalls before the tips. This would have been expensive because the wing tools were already fabricated and in place. Because the F-20A always flew with wingtip missile rails, a small LE extension was put on the rails that created a vortex over the wingtip upper surface, which delayed flow separation until about 25 deg.

F-20 Tigershark

Figure 2.15

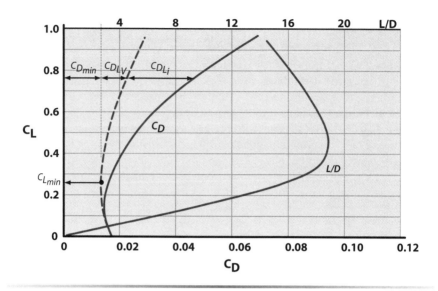

Figure 2.16 Low-speed drag polar ($M \leq 0.4$) for C-141, clean configuration.

$$C_D = C_{D_{\min}} + K''\left(C_L - C_{\ell_{\min}}\right)^2 + K'C_L^2 \tag{2.17}$$

where $C_{\ell_{\min}}$ is the C_ℓ for $C_{d_{\min}}$ from the airfoil drag polar (see Fig. 2.8).

Equation (2.17) is illustrated in Fig. 2.16, which presents the experimental low-speed drag polar for the Lockheed C-141A. The C-141A uses a symmetric airfoil (average section is an NACA 0011) but the wing is at an average +3.2-deg angle of incidence to the fuselage, giving the entire vehicle an effective camber as evidenced by $C_{\ell_{\min}}$ being nonzero.

The wing aspect ratio is 7.9, sweep is 25 deg, and the taper ratio is 0.374, giving a low-speed $C_{L_\alpha} = 0.084$ per degree. For a zero fuselage angle-of-attack, the wing is at a $C_L = 0.27 = C_{\ell_{\min}}$, which results in $C_{D_{\min}} = 0.016$ (see Fig. 2.16). The values for K'' and K' are estimated from the methodology in Chapter 13 to be 0.02 and 0.0407, respectively. Inserting these values into Eq. (2.17) gives a $C_D = 0.0252$ at $C_L = 0.5$, which agrees well with the experimental data. In this example, the $C_{D_{\min}}$ is made up of contributions from the wing, fuselage, tail, engine pods, and pylons; however, the aircraft C_{D_L} is composed primarily of wing drag-due-to-lift. If the wing or entire aircraft is effectively uncambered, then $C_{\ell_{\min}} = 0$ and the expression is simplified to

$$C_D = C_{D_0} + K''C_L^2 + K'C_L^2 = C_{D_0} + KC_L^2 \tag{2.18}$$

Sometimes, early in the design when very little is known about the aircraft configuration, the expression for K will be approximated by

$$K = \frac{1}{\pi ARe}$$

where e is a wing planform efficiency factor (see Fig. G.9).

Equation (2.18) is illustrated in Fig. 2.17. The McDonnell F-4C Phantom II is effectively an uncambered wing aircraft, as shown in Fig. 2.17. The value for K is estimated from Chapter 13 to be 0.169. Figure 2.17 also highlights an interesting drag polar behavior for low-AR aircraft at high angles-of-attack. Aircraft display a parabolic behavior of C_D with C_L up to a lift coefficient C_{L_B}, called the *break C_L*. Above C_{L_B} the C_D departs from classical parabolic behavior. This deviation from parabolic behavior is discussed in more detail in Chapter 13. The symmetric aircraft C_D at large α can be expressed as

$$C_D = C_{D_0} + K C_L^2 + K_B \left(C_L - C_{L_B} \right)^2 \qquad (2.19a)$$

where K_B is called the *break drag-due-to-lift factor* and

$$\begin{aligned} K_B &= 0 \quad \text{for} \quad C_L \le C_{L_B} \\ K_B &> 0 \quad \text{for} \quad C_L > C_{L_B} \end{aligned} \qquad (2.19b)$$

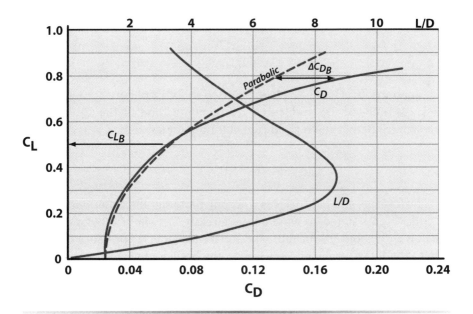

Figure 2.17   F-4C Aerodynamics at Mach 0.8 AR = 2.82, λ = 0.236, t/c = 5%, Series 64A, Δ = 45 deg.

Other fighter aircraft, such as the F-15 Eagle, have carefully tailored wings, resulting in C_{L_B} being very close to the buffet C_L with small values of K_B.

The wing C_{D_0} or $C_{D_{\min}}$ is the same for section and finite wing and is given by

$$C_{D_0} = C_{DP_{\min}} + C_{DF} \qquad (2.20)$$

where $C_{DP_{\min}}$ = pressure drag due to viscous separation (experimentally determined) and small compared to C_{DF}.

$$C_{DF} = \text{skin friction drag coefficient}$$

$$C_{DF} = C_F \frac{S_{\text{wet}}}{S_{\text{ref}}}$$

where S_{wet} is the wetted area of the exposed surface.

If the flow over the wing is laminar, that is, $Re_\ell = \rho V_\infty \bar{c}/\mu < 5{\times}10^5$, use

$$C_F = \frac{1.328}{\sqrt{Re_\ell}} \qquad (2.21)$$

If the flow over the wing is turbulent, that is, $Re_\ell > 5{\times}10^5$, use

$$C_F = \frac{0.455}{\left[\log_{10} Re_\ell\right]^{2.58}} \qquad (2.22)$$

Equations (2.21) and (2.22) are plotted in Fig. 2.6. For thin wings (i.e., thickness ratios of 20% and less) and streamlined bodies, the C_{D_0} is 70% to 80% skin friction. Thus, a good rule of thumb for subsonic C_{D_0} is

$$C_{D_0} \approx 1.2 \, C_{DF}$$

The fuselage and tail surfaces will contribute significant amounts to the aircraft C_{D_0}. As stated above the total aircraft C_{D_0} will be 70% to 80% skin friction with about 5% for mutual interference (adverse pressure gradients due to the interference of the wing on the body and vice versa). The skin friction of each aircraft component is determined and then added together; then a 5% mutual interference factor is added, to give the total aircraft C_{D_0}. The aircraft is broken down as shown in Fig. 2.18.

The C_F is determined for each aircraft component using, for laminar flow,

$$C_F = \frac{1.328}{\sqrt{Re_\ell}} \qquad (2.21)$$

and for turbulent flow,

$$C_F = \frac{0.455}{\left[\log_{10} Re_\ell\right]^{2.58}} \qquad (2.22)$$

and the C_F is always based upon the wetted area of the component. Each component is considered to be a flat plate of equivalent wetted area. The nose of the aircraft is treated as a cone and we use the result

$$C_{F\text{cone}} = \frac{2}{\sqrt{3}} C_{F\text{flat plate}}$$

The total skin friction coefficient of the vehicle is

$$\left(C_{DF}\right)_{a/c} = C_{F\text{fuse}} \frac{S_F}{S_{\text{ref}}} + C_{F\text{nose}} \frac{S_N}{S_{\text{ref}}} + C_{F\text{wing}} \frac{S_W}{S_{\text{ref}}} + C_{F\text{tail}} \frac{S_T}{S_{\text{ref}}} \qquad (2.23)$$

where S_{ref} is the reference area for the C_L and C_D and is usually the total wing planform area.

Finally the total aircraft C_{D0} is

$$\left(C_{D0}\right)_{a/c} = 1.25\left(C_{DF}\right)_{a/c} \qquad (2.24)$$

where we have included a 5% mutual interference effect. This is to be regarded as a rule of thumb for early design estimates. The more accurate methodology of Chapter 13 is recommended for later design work.

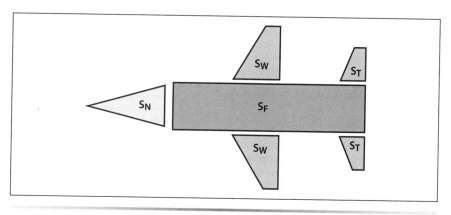

Figure 2.18 Aircraft components for skin friction estimation.

2.11 Transonic Flow and Its Effects

A body is considered to be in the *transonic flow* regime when sonic flow (Mach number greater than 1.0) first occurs somewhere on the body surface. The lower limit of transonic flow is for some M_∞ less than unity and depends upon the thickness of the wing or body. The upper limit is generally considered to be $M_\infty \approx 1.3$ where all surface flows are supersonic.

A conventional subsonic airfoil shape is shown in Fig. 2.19. If this airfoil is at a flight Mach number of 0.50 and a slight positive angle-of-attack, the maximum local velocity on the surface will be greater than the flight speed but most likely less than sonic speed. Assume that an increase in flight Mach to 0.72 would produce the first evidence of local sonic flow. This condition would be the highest flight speed possible without supersonic flow and is termed the "critical Mach number." Thus, critical Mach number is the boundary between subsonic and transonic flow and is an important point of reference for all compressible effects encountered in transonic flight.

As critical Mach number is exceeded, an area of supersonic flow is created on the wing surface. The acceleration of the airflow from subsonic to supersonic is smooth and unaccompanied by any shock waves. However, the transition from supersonic to subsonic occurs through a shock wave and because there is no change in direction of the flow the wave formed is a normal shock wave.

One of the principal effects of the normal shock wave is to produce a large increase in the static pressure of the airstream behind the wave. If the shock wave is strong, the boundary layer may not have sufficient kinetic energy to withstand the large adverse pressure gradient and separation will occur. At speeds only slightly beyond the critical Mach number the shock wave formed is not strong enough to cause separation or any noticeable change in the aerodynamic force coefficients. However, an increase in speed sufficiently above the critical Mach number to cause a strong shock wave will produce separation and yield a sudden change in the force coefficients. Such a flow condition is shown in Fig. 2.19 by the flow pattern for $M = 0.77$. Notice that a further increase in Mach number to 0.82 can enlarge the supersonic area on the upper surface and form an additional area of supersonic flow and a normal shock wave on the lower surface.

As the flight speed approaches the speed of sound, the areas of supersonic flow enlarge and the shock waves become stronger and move nearer the trailing edge (Fig. 2.20). When the flight speed exceeds the speed of sound, the "bow" wave forms at the leading edge as illustrated in Fig. 2.19 for $M = 1.05$. If the speed is increased to some higher supersonic value, all oblique portions of the wave incline more greatly and the detached normal shock portion of the bow shock wave moves closer to the leading edge.

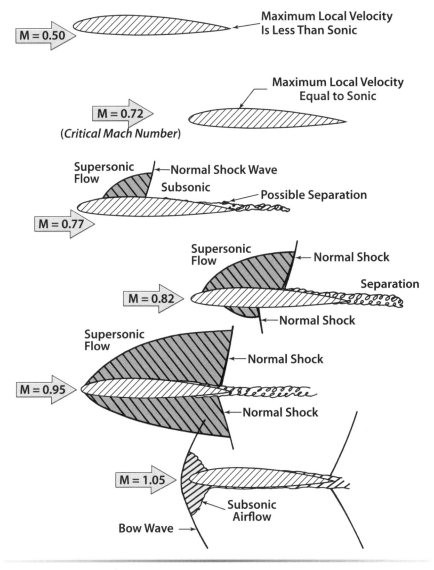

M = 0.50 — Maximum Local Velocity Is Less Than Sonic

M = 0.72 *(Critical Mach Number)* — Maximum Local Velocity Equal to Sonic

M = 0.77 — Supersonic Flow, Normal Shock Wave, Subsonic, Possible Separation

M = 0.82 — Supersonic Flow, Normal Shock, Separation, Normal Shock

M = 0.95 — Supersonic Flow, Normal Shock, Normal Shock

M = 1.05 — Subsonic Airflow, Bow Wave

Figure 2.19 Flow patterns around an airfoil in transonic flow.

The airflow separation induced by the shock wave formation can create significant variations in the aerodynamic force coefficients. Some typical effects are an increase in the section drag coefficient and a decrease in the section lift coefficient for a given angle-of-attack. Accompanying the variations in C_ℓ and C_d is a change in the pitching moment coefficient.

The Mach number that produces a large increase in the drag coefficient is termed the *force divergence Mach number* and for most airfoils exceeds

Figure 2.20 F-14 at M~0.95 with condensation at trailing edge normal shock.

the critical Mach number by 5% to 10%. This condition is also referred to as *drag divergence* or *drag rise*.

Associated with the transonic drag rise are buffet, trim, and stability changes, and a decrease in the effectiveness of control surfaces. Conventional aileron, rudder, and elevator surfaces subjected to this high-frequency buffet may "buzz" and changes in moments may produce undesirable control forces. Also, when airflow separation occurs on the wing due to shock wave formation, there will be a loss of lift and subsequent loss of downwash aft of the affected area. If the wings shock unevenly due to physical shape differences or sideslip, a rolling moment may be created and can contribute to control difficulty. If the shock-induced separation occurs symmetrically near the wing root, the resulting decrease in downwash on the horizontal tail will create a diving moment and the aircraft will "tuck under."

Because most of the difficulties of transonic flight are associated with shock wave induced flow separation, any means of delaying or alleviating this separation will improve the aerodynamic characteristics of an aircraft. Thus, it is important to seek ways to increase the critical Mach number M_{CR} of the aircraft.

The critical Mach number can be increased by the following:

- Decreasing wing thickness ratio
- Increasing leading edge sweep
- Decreasing aspect ratio
- Using a supercritical airfoil

2.12 Wing Thickness Ratio

It is clear from Fig. 2.19 that a smaller thickness ratio will give an increase in M_{CR}. Thus, supersonic aircraft will have small thickness ratios, usually 5% or less, whereas subsonic aircraft will have thicker wings of perhaps up to 18%. Structural considerations prohibit wings of less than 3%. Figure 2.23b (see Section 2.14) and Fig. 7.7 (see Chapter 7) show how M_{CR} increases with decreasing thickness ratio.

2.13 Wing Sweep

One of the most effective means of delaying and reducing the effects of shock wave induced flow separation is the use of sweep. Generally the effect of wing sweep will apply either to sweepback or sweep forward. Although the sweep-forward wing has been used in rare instances, sweep-back has been found to be more practical for ordinary applications.

A method of visualizing the effect of sweepback is shown in Fig. 2.21. The swept wing shown has the streamwise velocity vector resolved into components perpendicular and parallel to the leading edge. The component parallel to the leading edge may be visualized as moving across constant sections and thus does not contribute to the pressure distribution in the wing. The component perpendicular to the leading edge ($V_\infty \cos \Lambda$) is less than freestream velocity and it is this component that determines the magnitude of the pressure distribution and the aerodynamic force coefficients.

Hence, sweep of a surface in high-speed flight produces a beneficial effect, because higher flight speeds may be obtained before components of velocity perpendicular to the leading edge produce critical conditions on the wing. Thus, sweepback will increase the critical Mach number, the force divergence Mach number, and the Mach number at which the drag rise will peak. In other words, sweep will delay the onset of compressibility effects. The critical Mach number M_{CR} is increased by $(M_{CR})_{\Lambda=0}/\cos \Lambda$.

In addition to delaying the onset of compressibility effects, sweepback will reduce the magnitude of the changes in force coefficients due to compressibility. Because the component of velocity perpendicular to the leading edge is less than freestream velocity, the magnitude of all pressure forces on the wing will be reduced (approximately by the square of the sweep angle). Because compressibility force divergence occurs due to change in pressure distribution, the use of sweepback will "soften" the force divergence. This effect is illustrated qualitatively by the graph of Fig. 2.21, which shows the typical variation of drag coefficient with Mach number for various sweepback angles. The straight wing shown begins drag rise at about $M = 0.90$, reaches a peak near $M = 1.1$, and begins a continual drop past $M = 1.1$. Note that use of sweepback then delays the drag rise to some higher Mach number and reduces the magnitude of the rise in drag coefficient. It is evident from the figure that small angles of sweep provide little benefit. If sweep is to be used at all, at least 35–45 deg should be used.

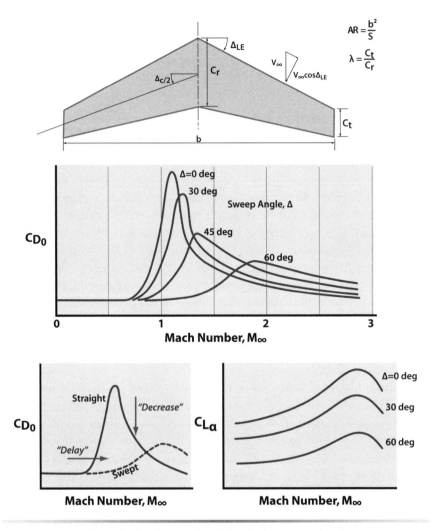

Figure 2.21 General effects of wing sweepback.

A disadvantage of wing sweep is the decrease in wing lift curve slope. This effect can be shown to be

$$\left(\frac{\mathrm{d}C_L}{\mathrm{d}\alpha}\right)_\Delta = \left(\frac{\mathrm{d}C_L}{\mathrm{d}\alpha}\right)_{\Delta=0} \cos\Delta$$

This means that a swept-wing aircraft will have to land and take off at higher angles-of-attack than a straight-wing aircraft.

Other disadvantages to wing sweep are a reduction in $C_{L_{max}}$ and tip stall. Early flow separation at the tip is due to spanwise flow causing a thickening of the boundary layer near the tips and hastening flow separation.

2.14 Supercritical Wing

Another way of delaying the drag rise due to shock wave induced separation is by using an airfoil shape called a supercritical section. The supercritical section is shown in Fig. 2.22 compared with a conventional NACA 64A series section. The supercritical section has a much flatter shape on the upper surface that reduces both the extent and strength of the normal shock, as well as the adverse pressure rise behind the shock, with corresponding reductions in drag. To compensate for the reduced lift on the upper surface of the supercritical airfoil resulting from the reduced curvature, the airfoil has increased camber near the trailing edge.

The advantage of the supercritical airfoil section is shown in Fig. 2.22 and its geometry is shown in Fig. 2.23a. For a given thickness ratio, the critical Mach number stays the same but the divergence Mach number can be delayed as shown in Fig. 2.23b. Most high-subsonic aircraft will cruise near the divergence Mach number. The Boeing Airplane Company engineers were considering the application of a supercritical wing on their 707-320B and they had the choice of cruising at the same Mach number (about Mach 0.82) with an increase in the supercritical wing thickness to 13% or keeping the same thickness and cruising at a higher Mach number. Figure 2.23c shows the savings in wing weight as the wing is made thicker.

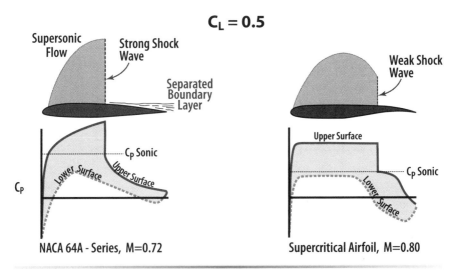

Figure 2.22   Supercritical airfoil flow phenomena.

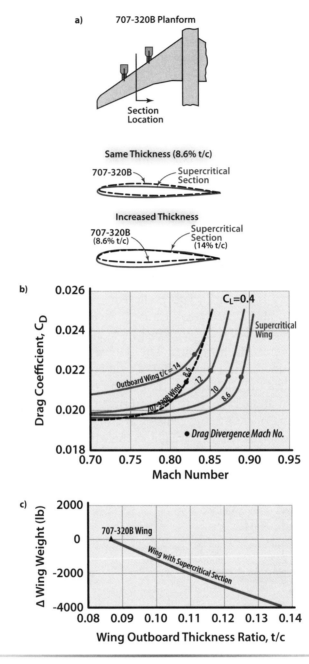

Figure 2.23 Supercritical section, **a)** geometry comparison, **b)** thickness effect on drag, and **c)** wing weight reduction with t/c.

2.15 Wing-Body Combinations for Transonic Flight

The zero-lift drag coefficient, C_{D_0}, of the fuselage will peak about Mach = 1.2. A typical fuselage C_{D_0} curve is shown in Fig. 2.24. The designer should observe the relative magnitude of the body C_{D_0} (referenced to wing area) compared to the wing C_{D_0} [9].

The C_{D_0} of a wing and a body can be added directly (for comparison purposes) to give the wing–body combination drag curve. Notice that the additional drag due to interference is not taken into account. From Fig. 2.24 we can see the advantage of having the wing drag rise occur at a higher

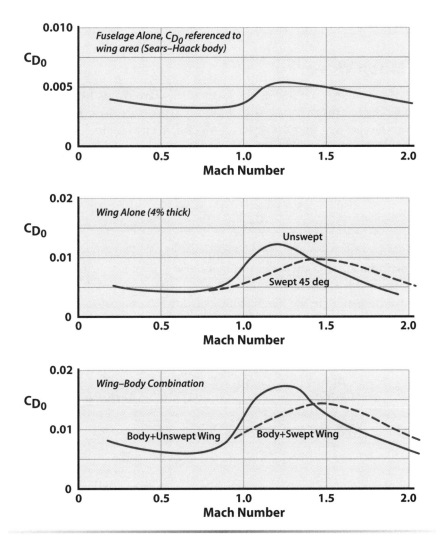

Figure 2.24   Wing, body, and wing–body C_{D_0} vs Mach number.

Mach number. This "adding together" of a wing and a body is shown in the same figure.

The swept-wing–body combination results in a lower drag coefficient and hence lower drag in the critical transonic flight regime. For many propulsion units the thrust around Mach = 1.0 increases more slowly than the drag increases and it is possible for the excess thrust ($T–D$) to be marginal in the transonic region where the aircraft suffers a thrust pinch. There are several design methods to alleviate the thrust pinch. The designer could increase the size of the engines; however, this adds weight and may oversize the engines for other parts of the operating envelope. The wings could be swept as discussed previously and/or made thinner (lower thickness ratio). Another good design practice is to area-rule the aircraft (discussed in detail in Chapter 8 [10]).

2.16 Mach Wave

A *Mach wave* is defined as the locus of all wave fronts from an infinitesimal pressure disturbance. An infinitesimal pressure disturbance propagates at the speed of sound $a = \gamma RT$. A *Mach cone* (three-dimensional Mach wave) is shown in Fig. 2.25.

The angle μ is called the *Mach angle* and is equal to

$$\sin \mu = \frac{1}{M_\infty}, \quad \mu = \arcsin\left(\frac{1}{M_\infty}\right)$$

because the flow normal to the Mach wave is $M = 1$. The infinitesimal pressure disturbance at point A can influence every point inside its Mach cone, called the *zone of activity*. Similarly, the disturbance at point A will not be felt anywhere outside the Mach cone, called the *zone of silence*.

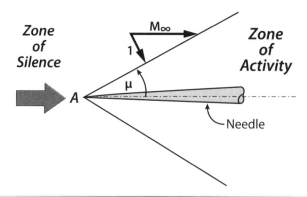

Figure 2.25 Mach cone from an infinitesimal pressure disturbance.

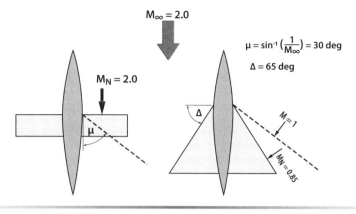

Figure 2.26 Straight and swept wing aircraft in Mach-2 flight.

2.17 Subsonic and Supersonic Leading Edge

Figure 2.26 shows both straight- and swept-wing aircraft where the freestream Mach number (M_∞) is 2.0 and the Mach angle (μ) is 30 deg. The normal Mach number at the leading edge of the straight wing is 2.0 (supersonic) and of the swept wing is 0.85 (subsonic). As was discussed earlier, the pressure distribution over the wing is established by the flow normal to the leading edge. Thus, the wing of the straight-wing aircraft is experiencing supersonic flow. The straight-wing aircraft is said to have a "supersonic leading edge," and the leading edge must be sharp and the section thin for low wave drag, whereas the swept-wing aircraft is said to have a "subsonic leading edge" and its leading edge can be round or blunt and the section fairly thick.

Thus, if $\Delta > (90 - \mu)$, the leading edge is subsonic, and if $\Delta < (90 - \mu)$, the leading edge is supersonic.

The wave drag coefficient for a wing will peak at the M_∞ where the normal Mach ≈ 1.2. This explains the C_D behavior in Figs. 2.24 and 2.29 (see Section 2.20). The F-111 has a round leading edge, characteristic of subsonic leading edges, and the sweep schedule for its variable-sweep wing is such that the normal Mach number is always subsonic, that is, $\Delta > (90 - \mu)$.

Because the lift curve slope decreases with wing sweep, it is not desirable to sweep a wing more than is necessary. Thus, a good rule of thumb for a swept wing with a subsonic leading edge is to sweep it 5 deg behind the Mach line.

2.18 Supersonic Skin Friction

The supersonic flow over a vehicle is very likely turbulent ($Re_\ell > 5 \times 10^5$) so that the skin friction is given by the turbulent skin friction expression

[Eq. (2.22)] corrected for compressibility. The incompressible flat plate turbulent skin friction is given by (for one side of the flat plate)

$$C_{F_i} = \frac{0.455}{\left[\log_{10} Re_\ell\right]^{2.58}} \qquad (2.22)$$

Equation (2.22) is plotted in Fig. 2.6. The compressibility correction is given in [11] as

$$C_F = \frac{C_{F_i}}{\left(1 + 0.144 M_\infty^2\right)^{0.65}} \qquad (2.25)$$

The method for determining the total aircraft skin friction is the same as discussed earlier with the exception of correcting C_{F_i} for compressibility.

2.19 Supersonic Lift and Wave Drag

If the local flow inclination over a body in a supersonic stream is small, then the pressure coefficient at each point is given by the following linear theory result:

$$C_P = \frac{P - P_\infty}{\frac{1}{2}\rho V_\infty^2} = \frac{P - P_\infty}{\frac{1}{2}\gamma P_\infty M_\infty^2} = \frac{2\theta}{\sqrt{M_\infty^2 - 1}}$$

where θ, in radians, is positive for compression regions and negative for expansion regions.

The supersonic section lift and wave drag coefficients are given by supersonic thin airfoil theory (linear theory [12]):

$$C_\ell = \frac{4\alpha}{\sqrt{M_\infty^2 - 1}} \qquad (2.26)$$

$$C_{dw} = \frac{4\alpha}{\sqrt{M_\infty^2 - 1}}\left[\alpha^2 + \overline{\alpha_c(x)^2} + \left(\overline{\frac{dh}{dx}}\right)^2\right] \qquad (2.27)$$

where α is the angle-of-attack (radians), $\alpha_c(x)^2$ is the mean square of the camber line

$$\overline{\alpha_c(x)^2} = \frac{1}{c}\int_0^c \alpha_c(x)^2\, dx \qquad (2.28)$$

and $\left(\overline{dh/dx}\right)^2$ is the mean square of the thickness distribution.

Table 2.1 Section Parameters for Wave Drag

Shape	$\bar{\alpha}_c^2$	$(dh/dx)^2$
Flat plate	0	0
Double wedge	0	t/c^a
Biconvex	0	$\frac{4}{3}\frac{t}{c}$

$^a t/c$ is the max thickness ratio of the section.

$$\left(\overline{\frac{dh}{dx}}\right)^2 = \frac{1}{c}\int_0^c \left(\frac{dh}{dx}\right)^2 dx \tag{2.29}$$

We observe that the supersonic C_{d_w} is made up of drag-due-to-lift, drag due to camber, and drag due to thickness. Table 2.1 gives values of $\bar{\alpha}_c^2$ and $(dh/dx)^2$ for several basic supersonic airfoil sections.

Equation (2.27) is usually rewritten as

$$C_d = C_{d0} + KC_\ell^2 \tag{2.30}$$

where

$$C_{d0} = C_{DF} + \frac{4}{\sqrt{M_\infty^2 - 1}}\left[\bar{\alpha}_c^2 + \left(\overline{\frac{dh}{dx}}\right)^2\right] + C_{dB}\left(\frac{S_B}{S_{\text{ref}}}\right) \tag{2.31}$$

$$K = \frac{\sqrt{M_\infty^2 - 1}}{4} \tag{2.32}$$

C_{dB} = base drag due to flow separation over a blunt base

S_B = base area

At the rear of a wing with a blunt base in supersonic flow, the flow tries to expand 90 deg. Inviscid flow theory would predict that the base pressure P_B would be zero. However, in a viscous fluid the base pressure is not zero but is some value less than ambient pressure p_∞. This is due to the boundary layer bleeding into the separated flow region at the base, giving a turbulent wake and $0 < P_B < P_\infty$. Experimental values of base pressure coefficients C_{pB} for two- and three-dimensional bodies are shown in Fig. 2.27. Notice that the C_{dB} of Eq. (2.31) is equal to the negative of C_{pB} and is referenced to the base area, S_B.

Equation (2.27) assumes that the shock is an attached oblique shock. If the shock is a detached normal shock, an additional drag term due to nose

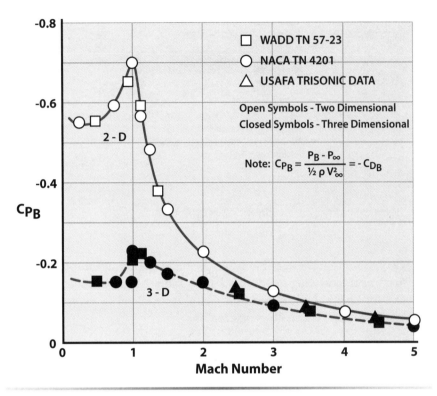

Figure 2.27 Experimental values of base pressure coefficient.

bluntness, $C_{d_{LE}}$, is added to Eq. (2.31). This bluntness term $C_{d_{LE}}$ is discussed more fully in Chapter 13, Fig. 13.13, but it should be pointed out that $C_{d_{LE}} = f[M_\infty, r_{LE}, \cos \Delta_{LE}]$, where r_{LE} is the radius of the leading edge. Thus, for low wave drag at a given M_∞, the best scenario is a wing with small thickness ratio, a small leading edge radius, a low aspect ratio, and lots of sweep. These ideas should be kept in mind when the airfoil and wing planform are selected (see Chapter 7).

2.20 Correction for Three-Dimensional Effects

A fuselage can be approximated by a cone-cylinder. The supersonic wave drag from a cone can be determined using the conical shock charts of Appendix E. The pressure coefficient on the surface of a cone is equal to the wave drag coefficient referenced to the cross-sectional area of the cone. If the cone is blunted, there will be a detached normal shock, and a drag coefficient due to nose bluntness must be determined. This nose bluntness term will be discussed in Chapter 13.

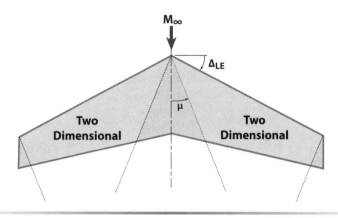

Figure 2.28 Finite span wing in supersonic flow.

The flow field around finite wings is made up of two parts: (1) the region within the Mach cones, and (2) the regions outside the Mach cones where the flow is two-dimensional. This flow pattern is shown in Fig. 2.28.

The analysis of a finite span wing is the determination of the field of three-dimensional flow and its influence on the rest of the wing. The lift and wave drag coefficients for the regions of two-dimensional flow are given by Eqs. (2.26) and (2.27). The three-dimensional wing lift uses supersonic thin airfoil theory and finite wing theory reported in [11–16] and is plotted in Fig. 13.2 for different wing taper ratios. Methods for determining wing–body supersonic C_{D_0} and drag-due-to-lift are discussed in Chapter 13, Sections 13.2.2 and 13.3.6. Figure 2.29 shows a comparison of wind tunnel data and theory for C_{D_0} vs Mach Number for varying 50% chord sweepback angles.

2.21 Sanity Check

It is always a good idea to get a sanity check on any analytical results or estimates (aero, weights, performance, etc.) before using them in the next part of the design process. In industry the sanity check might be performed by the group lead engineer double-checking the numbers. Or it might be done by comparing the estimates with real-world aircraft data. Appendix G contains the measured aerodynamics of many military and commercial aircraft.

The aircraft maximum L/D is a major design parameter as it indicates the aerodynamic efficiency of the configuration. It is important that the $(L/D)_{max}$ be checked with as many sources as possible before moving into the next stage of design. A correlation factor for $(L/D)_{max}$ is developed next.

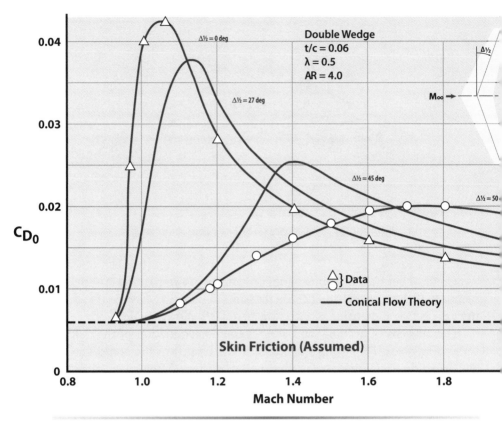

Figure 2.29   Effect of wing sweep on C_{D_0}.

Assume the aircraft wing is uncambered so that the drag coefficient can be expressed as

$$C_D = C_{D_0} + KC_L^2 \quad [\text{see Eq. (2.18)}]$$

where $K = 1/\pi ARe$ and aspect ratio $AR = b^2/S_{\text{ref}}$.

The $(L/D)_{\text{max}}$ is then expressed as

$$\left(L/D\right)_{\text{max}} = 1/\sqrt{4C_{D_0}K}$$

[which is developed in Chapter 3 as Eq. (3.10a)]. As discussed in Section 2.10, the C_{D_0} is *friction + form + interference + excrescence* drag, and the subsonic

$$C_{D_0} \approx 1.25 C_{D_F} = 1.25 C_F \frac{S_{\text{wet}}}{S_{\text{ref}}} \quad [\text{see Eq. (2.24)}]$$

where S_{wet} is defined as the aircraft wetted area and is the total area that would get wet if the aircraft were submerged in water. The correlation of subsonic C_{D_0} with S_{wet}/S_{ref} is shown in the aero data appendix in Fig. G.7. Finally, we can express a correlation factor for subsonic maximum L/D as

$$\left(L/D\right)_{max} \sim b\big/\sqrt{S_{wet}} \qquad (2.33)$$

This correlation of $(L/D)_{max}$ with $b/(S_{wet})^{1/2}$ is shown in the aero data appendix in Fig. G.8. The world's aircraft can be plotted on this figure and will fall between the e/C_f lines shown. The lines of e/C_f represent a ratio of wing efficiency to skin friction. Most sailplanes fall on the upper curve because they have high aero efficiency and low friction drag, and zero interference or excrescence drag is essential for world competition. The lower line represents the "nominal" operational aircraft featuring good wing design but turbulent boundary layers, normal surface roughness, and some excrescence drag.

Life of a Test Pilot (Tongue in Cheek)

Captain Buzz Jetspeed surveyed the sleek, shiny new prototype XRA-5C. This was it—the first flight of the new supersonic fighter. The engine start, taxi, takeoff, and climb to altitude were normal. The excitement grew as Buzz accelerated through Mach 1 to supersonic Mach = 1.6. He knew that he had to be careful here because he was in the region of decreasing minimum C_D (see the drag curve for the RA-5C in Fig. G.5a). If he were to accelerate to Mach = 2, he might not have enough drag to decelerate back through Mach = 1.0 into subsonic flight. But life was good—and so was Buzz. As he started to relax, his knee bumped the throttle and the XRA-5C started accelerating toward Mach = 2.0. He immediately reversed the engines, popped the speed brakes, and deployed the landing gear to generate enough drag to decelerate through the "drag rise." The aircraft was starting to shudder, glow red hot, and melt. The entropy in the engines was decreasing and flowing out of the inlet. The usually cool and collected Buzz Jetspeed was starting to come unglued. Ever so slowly the XRA-5C began inching back toward Mach 1—1.6, 1.5, 1.4, 1.3, 1.2. When the airplane got to Mach = 1.1 the normal shock blew it through Mach 1 to Mach = 0.85 and eventual safety. When Buzz got back to the flight line the XRA-5C was not shiny and sleek anymore; it was bent, with a melted nose and burned leading edges. The floor of the cockpit was covered with a gooey liquid thought to be entropy. Buzz never talked about it.

References

[1] McCormick, B. W., *Aerodynamics, Aeronautics and Flight Mechanics*, Wiley, New York, 1995.

[2] Kuethe, A. M., and Schetzer, J. D., *Foundations of Aerodynamics*, Vol. 2, Wiley, New York, 1959.

[3] Abbott, I. H., Von Doenhoff, A. E., and Stivers, L., Jr., "Summary of Airfoil Data," NACA TR 824, 1945.

[4] Abbott, I. H., and Von Doenhoff, A. E., *Theory of Wing Sections*, Dover, New York, 1959.

[5] Gersten, K., "Calculation of Non-Linear Aerodynamic Stability Derivatives for Airplanes," AGARD Rept. No. 342, 1960.

[6] Gersten, K., "Nonlinear Airfoil Theory for Rectangular Wings in Incompressible Flow," NASA RE-3-2-59W, 1959.

[7] Lawrence, H. R., "Wing-Body Interference at Subsonic and Supersonic Speeds," *Journal of the Aeronautical Sciences*, Vol 21, No. 5, 1954, pp. 289–324.

[8] Gersten, K., "A Nonlinear Lifting Surface Theory for Low Aspect Ratio Wings," *AIAA Journal*, Vol 1, No. 4, April 1963, pp. 924–925.

[9] Mirels, H., "Aerodynamics of Slender Wings and Wing–Body Combinations Having Swept Trailing Edges," NACA TN-3105, 1954.

[10] Whitcomb, R. T., "Supercritical Wing Technology—A Progress Report on Flight Evaluations," NASA Flight Research Center, Edwards, CA, 29 Feb. 1972.

[11] Hayes, W. D., and Probstein, R. F., *Hypersonic Flow Theory*, Academic Press, New York, 1959.

[12] Liepmann, H. W., and Roshko, A., *Elements of Gasdynamics*, Wiley, New York, 1957.

[13] Eshbach, O. W., *Handbook of Engineering Fundamentals*, 3rd ed.,Wiley, New York, 1974.

[14] Jones, R. T., "Properties of Low-Aspect-Ratio Pointed Wings at Speeds Below and Above the Speed of Sound," NACA TR-835, 1946.

[15] Harmon, S. M., and Jeffreys, I., "Theoretical Lift and Damping in Roll of Thin Wings with Arbitrary Sweep and Taper at Supersonic Speeds, Supersonic Leading and Trailing Edges," NACA TN-2114, 1950.

[16] Malvestuto, F. S., Margolis, K., and Ribner, H. S., "Theoretical Lift and Damping in Roll at Supersonic Speeds of Thin Sweptback Tapered Wings with Streamwise Tips, Subsonic Leading Edges, and Supersonic Trailing Edges," NACA TR-970, 1950.

Chapter 3 Aircraft Performance Methods

- Optimum Flight Speed
- Breguet Range & Endurance Equation
- Maximum Range & Endurance
- Turning Performance
- Energy Maneuverability
- Climb & Descent

The SR-71 still holds the world speed record for manned air-breathing operational aircraft set in 1976 at 2194 mph. I was privileged to work on the SR-71, and later on the M-21/D-21 drone-carrying version of the M-21; I remember those times fondly.

Grant Carichner

Good enough is not enough.
It is ever the enemy of the best.

3.1 Introduction

T his chapter will consider steady-state and accelerated performance methods. A large portion of an aircraft's mission profile can be considered as *steady-state* (equilibrium) or a series of near-steady-state conditions. The landing and takeoff phases, the climb–acceleration phase, and the combat phase are not equilibrium conditions and are considered as *accelerated performance* problems. The landing and takeoff analysis is discussed in Chapter 10.

For the discussions in this chapter, the aircraft will be considered as a point mass system with horizontal and vertical translation degrees of freedom and subject to aerodynamic, propulsive, and gravity forces. The force diagram for the aircraft is shown in Fig. 3.1, where the lift and drag forces are normal and parallel to the freestream velocity V_∞, respectively; i_T is the angle (usually small) between the wing chord line (WCL) and thrust vector; and γ is the flight path angle.

3.2 Level Unaccelerated Flight

During level unaccelerated flight, the flight path angle γ is zero and all external forces acting on the aircraft are in balance. Thus, adding forces normal and parallel to V_∞ (the wind axis) yields the following results:

$$L + T \sin\left(\alpha + i_T\right) = W \cos\gamma \qquad (3.1)$$

$$T \cos\left(\alpha + i_T\right) = W \sin\gamma + D \qquad (3.2)$$

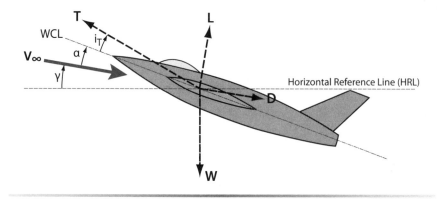

Figure 3.1 Forces acting on aircraft:

Because $\gamma = 0$ and $(\alpha + i_T)$ is usually small during this flight condition, the scalar equations representing level unaccelerated flight for a symmetric (uncambered, $C_{L\text{min}} = 0$) aircraft are as follows:

$$W \approx L = C_L q S \qquad (3.3)$$

$$T \approx D = \left(C_{D_0} + KC_L^2\right)qS \qquad (3.4)$$

where $q = \frac{1}{2}\rho_\infty V_\infty^2$ is the *dynamic pressure* and S is the *reference area* for C_L and C_D (usually the total wing planform area). Because $L = W$ the C_L at which the aircraft must fly at is expressed as

$$C_L = W/qS$$

From Eq. (3.4) the drag determines the *thrust required* T_R:

$$T_R = D = C_{D_0}qS + KW^2/qS \qquad (3.5)$$

The first term on the right-hand side of Eq. (3.5) is the *zero-lift drag*, and the second term is the *drag-due-to-lift* during level unaccelerated flight. For a given aircraft, altitude, weight, and aircraft configuration, the drag or thrust required can be plotted against velocity as shown in Fig. 3.2a.

The *minimum velocity* point for the thrust-required curve in Fig. 3.2a is either stall speed or a minimum control speed. Flight below this speed is not relevant. The intersection of the thrust-available curve (either maximum thrust or military thrust) with the thrust-required curve is the *maximum speed* for the aircraft at that particular power condition (V_{max}). Notice that for speeds less than V_{max} we do not have $T = D$ and the aircraft will accelerate. If the pilot desires to fly at the minimum drag point on the T_R curve, for example, the engine must be throttled back until the available thrust equals T_R for minimum drag. The "thrust pinch point" shown in Fig. 3.2b can be a design challenge as the T–D can become so small that too much fuel is used in accelerating through $M = 1$. The SR-71 (see the SR-71 design case study in Volume 2) had a small T–D near $M = 1$ and had to perform a dive maneuver to accelerate efficiently past $M = 1$.

The point of minimum drag is an interesting flight condition as it represents the velocity for maximum loiter or endurance for a turbine-powered aircraft. At this condition the aircraft's *lift-to-drag ratio L/D* is at maximum, as will be shown in the next section.

The *power required* for a propeller aircraft is given by

$$P_R = DV = T_R V = \left(C_{D_0} + KC_L^2\right)\frac{W}{C_L}\sqrt{\frac{2W}{\rho C_L S}} \qquad (3.6)$$

Power-required curves can be constructed from thrust-required curves and are useful when analyzing reciprocating-engine propeller aircraft. The reciprocating engine fuel flow rate is proportional to power output rather than thrust output as for a jet engine. The power-required curve is constructed similarly to the thrust-required curve. A typical power-required curve is shown in Fig. 3.3. A useful conversion factor that the designer should remember is from horsepower to foot-pounds per second: 1 hp = 550 ft · lb/s.

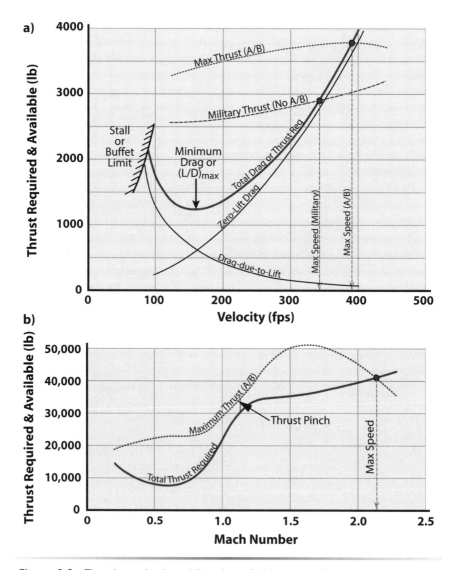

Figure 3.2   Thrust-required and thrust-available curves for typical jet aircraft: **a)** subsonic aircraft at constant altitude, and **b)** supersonic aircraft flying a minimum-time trajectory.

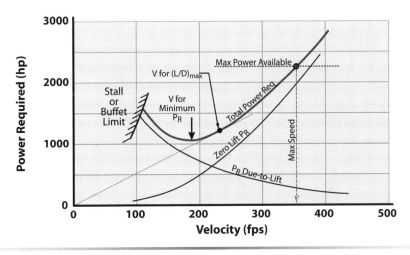

Figure 3.3   Power required for typical reciprocating-engine aircraft at constant altitude.

3.3 Minimum Drag and Maximum *L/D*

The total drag coefficient for an uncambered aircraft may be expressed as

$$C_D = C_{D_0} + KC_L^2$$

and the total drag from Eq. (3.4) as

$$D = \left(C_{D_0} + KC_L^2\right)qS$$

We seek to find the C_L that minimizes the total drag. In other words, the following operation is performed:

$$\frac{\partial D}{\partial C_L} = 0 \qquad (3.7)$$

with $q = (W/S)(1/C_L)$. Performing the operation denoted in Eq. (3.7) (the details are left as an exercise for the reader) gives

$$C_{D_0} = KC_L^2$$

or the zero-lift drag is equal to the drag-due-to-lift. From this relationship, we get the C_L for *minimum drag* (called the optimum C_L) as

$$C_{L_{\text{opt}}} = \sqrt{\frac{C_{D_0}}{K}} \qquad (3.8a)$$

It is also of interest to find the value of the C_L that maximizes L/D or C_L/C_D. In other words, when the following operation is performed

$$\frac{\partial\left(C_L/C_D\right)}{\partial C_L} = 0 \tag{3.9}$$

it is determined that the C_L for *maximum L/D* is

$$C_{L_{\text{opt}}} = \sqrt{\frac{C_{D0}}{K}} \tag{3.8a}$$

the same as for minimum drag. This gives the expression for $(L/D)_{\text{max}}$ as

$$\left(L/D\right)_{\text{max}} = 1\Big/\left(2\sqrt{C_{D0}K}\right) \tag{3.10a}$$

Equation (3.10a) illustrates how the aircraft $(L/D)_{\text{max}}$ is dependent only upon the aircraft aerodynamics. The velocity for maximum L/D or minimum drag is expressed as

$$V_{(L/D)_{\text{max}}} = \sqrt{\frac{2W}{\rho C_{L_{\text{opt}}} S}} = \sqrt{\frac{2W}{\rho S}}\sqrt{\frac{K}{C_{D0}}} \tag{3.11}$$

[shown in Figs. 3.2a and 3.7 (see Section 3.5)].

The total drag coefficient for a cambered airfoil where $C_{\ell_{\text{min}}} \neq 0$ is expressed as

$$C_D = C_{D\text{min}} + K'C_L^2 + K''\left(C_L - C_{\ell_{\text{min}}}\right)^2 \tag{2.17}$$

The C_L for maximum L/D or minimum drag is

$$C_{L_{\text{opt}}} = \sqrt{\frac{C_{D\text{min}} + K''C_{\ell_{\text{min}}}^2}{K' + K''}} \tag{3.8b}$$

and the expression for maximum L/D is

$$\left(L/D\right)_{\text{max}} = \frac{1}{\sqrt{4\left(C_{D\text{min}} + K''C_{\ell_{\text{min}}}^2\right)\left(K' + K''\right)} - 2K''C_{\ell_{\text{min}}}} \tag{3.10b}$$

The velocity for minimum power required is a useful flight condition for propeller aircraft. The P_R is given by Eq. (3.6) and the goal is to find the value of C_L that minimizes P_R, that is,

$$\frac{\partial P_R}{\partial C_L} = 0$$

The result is

$$3C_{D_0} = KC_L^2$$

which means that the aircraft should fly at that flight condition where the zero-lift drag is one-third of the drag-due-to-lift. This gives the required value for C_L as

$$C_{L_{\min}P_R} = \sqrt{\frac{3C_{D_0}}{K}}$$

and the velocity for minimum P_R as

$$V_{\min P_R} = \sqrt{\frac{2W}{\rho S} \sqrt{\frac{K}{3C_{D_0}}}} \tag{3.12}$$

which is 24% less than the speed for maximum L/D as shown in Fig. 3.3.

3.4 Variation of T_R with Weight, Configuration, and Altitude

The effect of changing aircraft weight, configuration (for example, changing the C_{D_0} by lowering the landing gear), and altitude is shown in Figs. 3.4, 3.5, and 3.6, respectively.

The aircraft *load factor n* is defined as

$$n = L/W \tag{3.13}$$

with *level flight* being the condition when $n = 1$. It should be noted that the weight change shown in Fig. 3.4 is equivalent to increasing the load factor from $n = 1$ to $n = 1.5$ for the 15,000-lb aircraft.

The configuration change shown in Fig. 3.5 represents a change in C_{D_0} that comes about when the landing gear or flaps are lowered or external stores are put on a fighter aircraft.

The altitude variation shown in Fig. 3.6 comes about because of the decrease in density ρ with altitude. Notice that the velocity for $(L/D)_{\max}$ is increased with altitude, but T_R does not change. This is an interesting behavior that should be confirmed by close examination of Eq. (3.5).

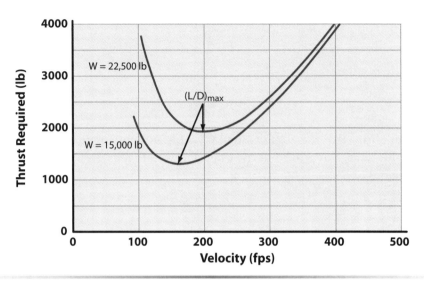

Figure 3.4 Effect on T_R of changing aircraft weight.

3.5 Endurance or Loiter

The aircraft *endurance* or *loiter* can be expressed as

$$E = \int_{t_i}^{t_f} \mathrm{d}t = \int_{W_i}^{W_f} \frac{1}{\mathrm{d}W/\mathrm{d}t}\,\mathrm{d}W \qquad (3.14)$$

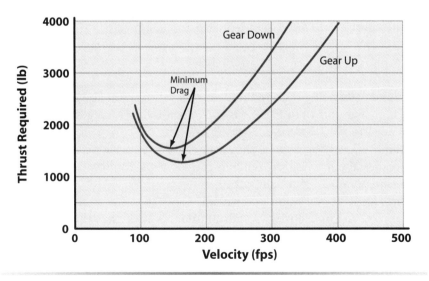

Figure 3.5 Effect on T_R of changing aircraft configuration.

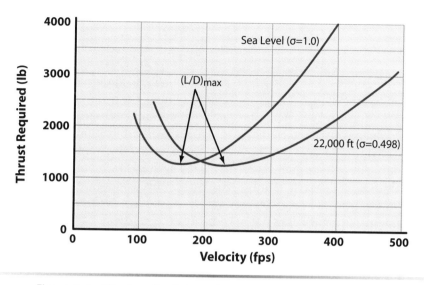

Figure 3.6 Effect on T_R of changing aircraft altitude ($0 = \rho/\rho SL$).

where dW/dt is the rate of change of aircraft weight due to burning fuel and the subscripts are initial and final conditions.

For a turbojet or turbofan aircraft, the dW/dt is *negative* and is expressed as the product of the engine thrust T and the thrust specific fuel consumption C as follows:

$$dW/dt = -T\left(\text{lb}\right)C\left(\frac{\text{lb of fuel}}{\text{lb of thrust}\cdot\text{hour}}\right) \tag{3.15}$$

Putting Eq. (3.15) into Eq. (3.14) gives

$$E = -\int_{W_i}^{W_f} \frac{1}{TC}\,dW = \int_{W_f}^{W_i} \frac{W}{TCW}\,dW$$

Because $L = W$ and $T = D$ for the loiter condition, we can express the relation for E as

$$E = \int_{W_i}^{W_f} \frac{L}{D}\frac{1}{C}\frac{dW}{W} \tag{3.16}$$

For flight at a fixed altitude and Mach number, L/D and C are constant with respect to W so that the expression for the endurance of a jet aircraft (in hours) is given by

$$E = \frac{L}{D}\frac{1}{C}\ln\left[\frac{W_i}{W_f}\right] \tag{3.17}$$

From Eq. (3.17), it is observed that, to obtain maximum endurance for a given weight change (i.e., given amount of fuel), the jet aircraft should fly at that altitude and Mach number such that the endurance parameter (L/D) $(1/C)$, which is often referred to as Range Factor, is a maximum. This usually means flying at or near $(L/D)_{max}$ in the tropopause, where C is a minimum. The designer will normally plot $(L/D)(1/C)$ against Mach number and find the altitude and Mach number that make this endurance parameter a maximum (see Fig. 3.7, for example). It should be emphasized that maximum endurance does not necessarily occur at the velocity for $(L/D)_{max}$ because C is dependent upon maximum Mach number and altitude (see Chapter 14) and a different velocity could give a larger value for $(L/D)(1/C)$. However, the velocity for $(L/D)_{max}$ is close (within 10%) to the velocity for maximum endurance, as illustrated in Fig. 3.7. (The aerodynamic data for Fig. 3.7 are given in Table 3.1.)

The thrust specific fuel consumption (per hour) is sometimes expressed as *specific impulse* I_{sp} (in seconds) as follows:

$$I_{sp} = \frac{3600}{C} \tag{3.18}$$

Table 3.1 Composite Lightweight Fighter Aerodynamic Data for Fig. 3.7 (from Section 5.6, Composite LWF Example)

W_{TO}	15,000 lb	T/W_{TO}	1.2 installed
S_{ref}	349 ft^2	Wing	AR 3.0
W/S_{TO}	43 psf		Δ_{LE} 40 deg
Engine	One F-100-PW-100[a]		λ 0 deg

Cruise and loiter at 36,000 ft and $W/S = 40$			

Mach	C_{D_0}	K	C_D	C (lb of fuel/lb of thrust-h)
0.5	0.0167	0.17	0.056	0.80
0.6	0.0167	0.17	0.035	0.85
0.7	0.0167	0.17	0.027	0.888
0.8	0.0167	0.17	0.023	0.92
0.9	0.0180	0.17	0.022	0.933
1.0	0.0243	0.18	0.027	1.025

[a]see Chapter 14.

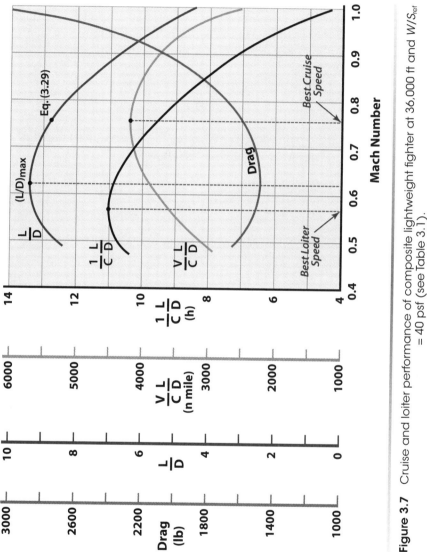

Figure 3.7 Cruise and loiter performance of composite lightweight fighter at 36,000 ft and W/S_{ref} = 40 psf (see Table 3.1).

The jet aircraft endurance equation becomes

$$E = I_{sp} \frac{L}{D} \ln\left[\frac{W_i}{W_f}\right] \tag{3.19}$$

(in seconds). The endurance equation for reciprocating-engine aircraft is determined in the same fashion as the jet aircraft expression using Eq. (3.14). The dW/dt is expressed in the parameters appropriate to propeller aircraft. The *required horsepower* hp_R for the aircraft is given by

$$hp_R = \frac{P_R}{\eta} = \frac{DV}{\eta}$$

where η is the *propulsive efficiency* (see Chapter 14). The required fuel flow rate in pounds of fuel per hour is obtained as follows:

$$-\frac{dW}{dt} = \frac{\text{pounds of fuel}}{\text{hour}} = (C)H_{P_R} = C\frac{P_R}{\eta} = \frac{DVC}{\eta} \tag{3.20}$$

where C is the *brake specific fuel consumption* in pounds per horsepower-hour (lb/hp·h). Equation (3.20) is put into Eq. (3.14) and rearranged as follows:

$$E = -\int_{W_i}^{W_f} \frac{\eta}{DVC} \, dW = \int_{W_f}^{W_i} \frac{\eta}{C} \frac{L}{DV} \frac{dW}{W}$$

$$E = \int_{W_f}^{W_i} \frac{\eta}{C} \frac{C_L}{C_D} \sqrt{\frac{\rho C_L S}{2W}} \frac{dW}{W}$$

$$E = \int_{W_f}^{W_i} \frac{\eta}{C} \frac{C_L^{3/2}}{C_D} \sqrt{\frac{\rho S}{2W_i}} \frac{dW}{W^{3/2}} \tag{3.21}$$

During loiter or cruise, η and C are considered constant for cruising speeds and corresponding engine power. This assumption is valid for a first approximation, as this type of engine usually operates at maximum loiter or range at a constant power with a relatively constant variation in C between 50% and 65% of the normal rated power output of the engine. For corresponding loiter and cruise speeds, the propulsive efficiency is constant within 1% or 2%. If we assume constant altitude and constant C_L and C_D, then Eq. (3.21) becomes

$$E = 26.8 \frac{\eta}{C} \frac{C_L^{3/2}}{C_D} \sqrt{\frac{2\sigma S}{W_i}} \left[\left(\frac{W_i}{W_f}\right)^{0.5} - 1\right] \tag{3.22}$$

(in hours), where $\sigma = \rho/\rho_{SL}$ and C = brake specific fuel consumption (BSFC) in pounds of fuel per brake horsepower-hour.

Because η and C are relatively constant with velocity, it should be clear from Fig. 3.3 that maximum endurance for a reciprocating engine aircraft will occur at minimum P_R. The velocity for this flight condition is given by Eq. (3.12).

3.6 Range

The *specific range* is defined as the distance traveled per pound of fuel consumed:

$$R_S = \frac{dR}{dW_{fuel}} = \frac{dR}{dt}\frac{dt}{dW_{fuel}} = \frac{V}{dW/dt}$$

The *total range* is determined by integrating the specific range over the weight change, that is,

$$R = \int_{W_i}^{W_f} \frac{V}{dW/dt}\,dW \tag{3.23}$$

As discussed before with the endurance equation, the final form of the range equation depends upon the form of dW/dt.

For jet aircraft, Eq. (3.15) is used in Eq. (3.23) and the following is obtained:

$$R = \int_{W_f}^{W_i} \frac{V}{TC}\,dW = \int_{W_f}^{W_i} \frac{V}{C}\frac{W}{T}\frac{dW}{W} \tag{3.24}$$

During cruise flight $T = D$ and $L = W$, so Eq. (3.24) becomes

$$R = \int_{W_f}^{W_i} \frac{V}{C}\frac{L}{D}\frac{dW}{W} = \frac{V}{C}\frac{L}{D}\int_{W_f}^{W_i} \frac{dW}{W} \tag{3.25}$$

If we assume that the aircraft cruises at nearly a constant velocity and that during the weight change the C and L/D are fairly constant, the total weight change can be broken into small weight-change increments in which the assumptions of constant C and L/D are valid; the range increments can then be summed to give the total range. The jet aircraft range equation (called the *Breguet range equation*) is expressed as

The Breguet range equation is named after Louis Charles Breguet, the record-setting aviation designer and builder. Breguet pioneered the development of the helicopter and built the first piloted vertical-ascent aircraft.

$$R = \frac{V}{C}\frac{L}{D}\ln\left[\frac{W_i}{W_f}\right] \qquad (3.26a)$$

$$R = VI_{sp}\frac{L}{D}\ln\left[\frac{W_i}{W_f}\right] \qquad (3.26b)$$

These equations assume flight at constant speed and C_L and change dramatically for propeller powered flight at constant speed and altitude (see Volume 2, "Buoyant Vehicle Design Basics").

Equations (3.26) indicate that the jet aircraft should fly at that altitude and velocity condition such that the range parameter $(V/C)(L/D)$ (often referred to as the *range factor*) is a maximum. This is done by determining the C for the engine when the thrust equals the cruise drag and calculating the value of $(V/C)(L/D)$. This is done for several velocities and altitudes with the weight equal to an average weight during the cruise weight-change increment. The results are plotted as $(V/C)(L/D)$ against Mach number as shown in Fig. 3.8, and the cruise conditions for maximum range are determined. Notice in Fig. 3.8 that, as the cruise altitude is increased, the velocity for maximum range increases. It will be shown later that the altitude for optimum cruise increases as wing loading decreases. Normal range-dominated vehicles will have a wing loading around 120 psf and their optimum cruise altitude is around 36,000 ft. As fuel is burned and the wing loading decreases, the aircraft should climb in altitude to keep optimum

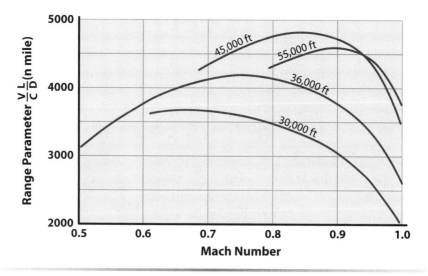

Figure 3.8 Range factor vs Mach number for composite lightweight fighter at $W/S = 40$ psf.

cruise conditions (i.e., constant C_L). This is the familiar "cruise climb" schedule that aircraft follow as fuel is burned.

Further insight into the optimum range conditions can be obtained by examining Eq. (3.24), which can be rewritten as

$$R = \int_{W_f}^{W_i} \frac{a}{C} M \frac{L}{D} \frac{dW}{W} \tag{3.24a}$$

If we substitute

$$a = \left(\gamma R'\theta\right)^{0.5}$$

$$T = D = C_D\left(\gamma/2\right) PM^2 S$$

$$T \approx T_{SL} \frac{P}{P_{SL}} \frac{\theta_{SL}}{\theta}$$

(the approximate expression for thrust; see Chapter 14) into Eq. (3.24a), where θ = static absolute temperature ratio, R' = gas constant, P = static pressure, γ = ratio of specific heats (γ = 1.44 for air), and the subscript SL denotes conditions at sea level, we obtain the approximate expression

$$R \approx \sqrt{\frac{2R'\theta_{SL}}{SP_{SL}}} \int_{W_f}^{W_i} \frac{\sqrt{T_{SL}}}{C} \frac{C_L}{C_D^{3/2}} \frac{dW}{W}$$

If it is assumed that thrust setting and α are constant, the expression for range is

$$R \approx \sqrt{\frac{2R'\theta_{SL}}{SP_{SL}}} \frac{\sqrt{T_{SL}}}{C} \frac{C_L}{C_D^{3/2}} \ln\left[\frac{W_i}{W_f}\right] \tag{3.27}$$

Equation (3.27) indicates that, for efficient cruise, the aircraft should cruise-climb at a constant thrust level and at an angle-of-attack corresponding to maximum $C_L/C_D^{3/2}$. The condition on C_L for maximum $(C_L/C_D^{3/2})$ is

$$C_{D_0} = 2KC_L^2 \tag{3.28a}$$

$$C_L = \sqrt{\frac{C_{D_0}}{2K}} \tag{3.28b}$$

which gives the L/D for maximum $(C_L/C_D^{3/2})$ as

$$L/D = 0.943\left(L/D\right)_{max} \tag{3.29}$$

(see Section 3.11). Equations (3.28) and (3.29) offer some useful "rules of thumb" for determining efficient cruise conditions during the early design phases. One rule is the following:

Efficient cruise will occur near $L/D = 0.943(L/D)_{max}$, which is the velocity and altitude where the zero-lift drag is twice the drag-due-to-lift.

This rule is demonstrated in Fig. 3.9 for the C-141 and F-111A, where the cruise C_L region corresponds to Eq. (3.28a).

Example 3.1

Find the speed for maximum range at 37,000 ft for the composite lightweight fighter shown in Table 3.1.

From Table 3.1, $C_D = 0.0167$, $K = 0.17$, and $W/S = 40.0$. Using Eqs. (3.29) and (3.28b), the cruise $L/D = 0.943(L/D)_{max} = 8.87$ and the required cruise $C_L = 0.222$. From

$$V = \sqrt{\frac{W}{S}\frac{2}{\rho C_L}} = 714 \text{ fps}$$

the cruise Mach number is 0.74. These values are consistent with the more involved analysis depicted in Figs. 3.7 and 3.8.

The term V/TC in the integrand of Eq. (3.24) has the units of distance per pound of fuel. Aircraft companies will often present their range information as shown in Fig. 3.10. This presentation is useful because the total range can be determined from initial and final weight conditions as shown in Fig. 3.10b. Notice that Fig. 3.10 is for one altitude, and a user flying a cruise climb schedule would have to use several charts like Fig. 3.10 for different altitudes in order to develop Fig. 3.10b. Normally, air traffic control requires airlines to fly at constant altitudes so that a single altitude chart is sufficient. The reader should observe that the term V/TC could replace the range parameter in developing Fig. 3.8 to determine cruise conditions.

The range equation for a reciprocating engine aircraft is developed in the same manner as for the jet aircraft. Equation (3.20) is combined with Eq. (3.23) to give

$$R = \int_{W_f}^{W_i} \frac{\eta}{DC}\,\mathrm{d}W = \int_{W_f}^{W_i} \frac{\eta}{C}\frac{L}{D}\frac{\mathrm{d}W}{W}$$

Following the same arguments as for the loiter case we can assume V, C, and L/D constant over the weight-change increment. Thus, the Breguet range equation for reciprocating-engine aircraft is expressed by

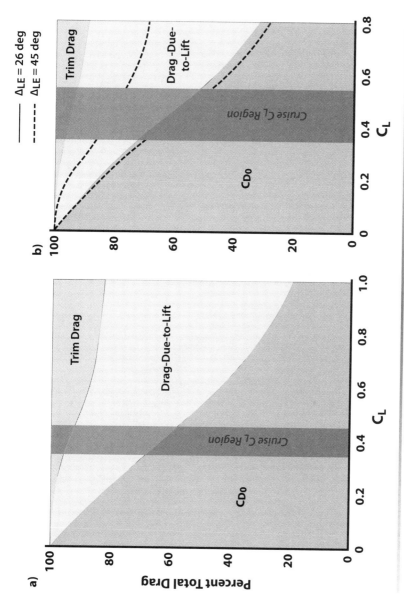

Figure 3.9 Cruise drag for cargo and fighter aircraft: **a)** C-141 at Mach =0.75, and **b)** F-111A.

a)

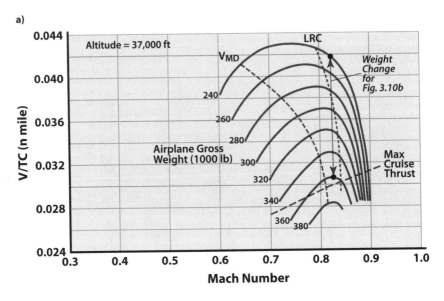

b)

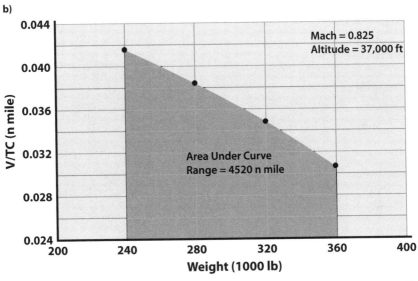

Figure 3.10   Range parameter for Lockheed L-1011 TriStar with Rolls Royce RB 211-22B engines (data from [1]), where V_{MD} is velocity for minimum drag and LRC is the velocity for long-range cruise; for **a)** weight and Mach variation, and **b)** weight change at constant Mach.

$$R = 326 \frac{\eta}{C} \frac{L}{D} \ln \left[\frac{W_i}{W_f} \right] \tag{3.30}$$

(for range in nautical miles). For a reciprocating engine, assuming η/C constant, the maximum range occurs at that velocity for maximum L/D [Eq. (3.11)] as shown in Fig. 3.3. This velocity can be found graphically by constructing a straight line through the origin and tangent to the P_R curve. Proof is left as an exercise for the reader.

3.7 Level Constant Velocity Turn

An aircraft turns by banking and using a component of the wing lift force to turn the aircraft [2,3]. This is shown schematically in Fig. 3.11. The aircraft can also turn with its wings level by deflecting the rudder, but this is inefficient as the vertical tail is not designed to provide rapid heading changes.

If the turn is to be a level turn, it is clear from Fig. 3.11 that the weight W of the aircraft must equal the vertical component of the lift, $nW \cos \phi$. Thus, the angle of bank ϕ for a level turn of $n\,g$ can be approximated by the following (neglecting small thrust component):

$$\phi = \arccos\left(1/n\right) \tag{3.31a}$$

The radius of the turn is then determined as

$$\text{Radius} = \frac{V^2}{n\,g\,\sin\phi} = \frac{V^2}{g\sqrt{n^2-1}} \tag{3.31b}$$

and the time to turn ψ degrees is given by

$$t_\psi = \frac{\text{Radius}\left(\Psi/57.3\right)}{V} \tag{3.31c}$$

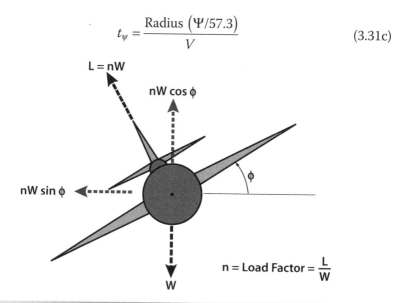

Figure 3.11   Forces acting on an aircraft in banked flight.

The turn rate in degrees/sec is dependent upon the load factor n as follows:

$$\dot{\psi} = \frac{g\sqrt{n^2 - 1}}{V} \tag{3.32}$$

A check should be made to see if the aircraft has enough thrust to overcome the drag while maneuvering at n gs so that the turn is at constant velocity. That is,

$$T_{\text{req}} = qS\left[C_{D0} + K\left(\frac{nW}{qS}\right)^2\right] + D_{\text{trim}} \leq T_{\text{max}} \tag{3.33}$$

where D_{trim} is the trim drag for a load factor of n (see Chapter 22). If we neglect the trim drag, Eq. (3.33) can be rewritten to express the *maximum sustained load factor*

$$n_{\text{MS}} = \frac{q}{W/S}\sqrt{\frac{1}{K}\left(\frac{T_{\text{max}}}{qS} - C_{D0}\right)} \tag{3.34}$$

Putting Eq. (3.34) into Eq. (3.32) will yield the *maximum sustained turn rate, $\dot{\psi}_{\text{MS}}$*.

3.8 Energy-State Approximation (Energy Maneuverability)

The preceding discussions have considered the aircraft to be in a steady-state or near-steady-state condition. The forces on the aircraft were balanced and the system was not accelerating. This point of view was adequate for aircraft cruise, loiter, steady-state turn, and maximum speed.

The remainder of this chapter deals with the aircraft as an accelerating system. Normally for an accelerating system we should consider the governing aircraft equations of motion. These nonlinear, coupled, second-order differential equations are difficult to work with and require the use of high-speed computers. The approach taken in this chapter will be to consider an approximation to the accelerating problem that will give extremely good information without using a computer.

The approach is to cast the non-steady-state accelerating problem into a steady-state problem representing the balance that must exist between the potential and kinetic energy change of the aircraft, the energy dissipated against drag, and the energy derived from the fuel. Thus, we take a different perspective on our accelerating problem and it becomes a steady-state performance problem again [4–6].

The *total energy* of an aircraft in space is its kinetic energy (KE), due to velocity and rotation, and its potential energy (PE), due to its altitude h above mean sea level. The rate of change of the total energy E is the rate at which an aircraft can climb and/or accelerate. *Energy maneuverability* is the name given to an aircraft's ability to change its energy state. It is assumed that the aircraft rotation is zero, $\dot{\gamma} = 0$. The total energy at a point in velocity–altitude space is given by

$$E = Wh + \frac{1}{2}\frac{W}{g}V^2 \tag{3.35}$$

and the *specific energy* h_e (or energy height) is expressed as

$$h_e = \frac{E}{W} = h + \frac{1}{2}\frac{V^2}{g} \tag{3.36}$$

The specific energy has the units of feet and represents the theoretical height (altitude) that an aircraft could reach if all of its KE were converted to PE (i.e., a zoom climb to zero air speed). An aircraft at Mach 2.2 and 70,000 ft would have a specific energy of 140,000 ft. If there were no drag on the aircraft, it could zoom to an altitude of 140,000 ft or dive to sea level (PE = 0) and reach a velocity of 3000 fps. Figures 4.8 and 4.9 in Chapter 4 show the contours of specific energy.

To consider the accelerated performance of an aircraft we need to know the rate at which the aircraft can change its specific energy (i.e., its energy state). This rate of change of h_e is given by

$$\frac{dh_e}{dt} = \frac{dh}{dt} + \frac{V}{g}\frac{dV}{dt} \tag{3.37}$$

The first term on the right-hand side of Eq. (3.37) is the aircraft's rate of climb and the term dV/dt is its acceleration. From Fig. 3.1 the accelerating force is

$$m\frac{dV}{dt} = T\cos(\alpha + i_T) - D - W\sin\gamma \tag{3.38}$$

where m is the mass of the aircraft, W/g. If we multiply Eq. (3.38) by V, divide by W, and rearrange, we have

$$\frac{m}{W}V\frac{dV}{dt} = \frac{T\cos(\alpha + i_T)}{W} - \frac{DV}{W} - V\sin\gamma$$

The term $V \sin \gamma$ is the rate of climb dh/dt. Thus,

$$\frac{V}{g}\frac{dV}{dt} + \frac{dh}{dt} = \frac{V\left[T\cos\left(\alpha+i_T\right)-D\right]}{W} \tag{3.39}$$

Comparing Eqs. (3.37) and (3.39) gives the expression for the rate of change of specific energy (sometimes called *specific excess power* or *specific power*, P_S) with units of feet per second (fps):

$$P_S = \frac{dh_e}{dt} = \frac{V\left[T\cos\left(\alpha+i_T\right)-D\right]}{W} \tag{3.40}$$

Equation (3.40) is the basis for the examination of acceleration performance. The P_S value for an aircraft in space represents its ability to accelerate and/or climb. When $P_S = 0$ the thrust is equal to the drag and we have the situation discussed in Section 3.2. When P_S is negative, the aircraft is slowing down and/or losing altitude. Notice that Eq. (3.40) explains the performance of the aircraft at a point in time along the velocity axis. What the aircraft is doing normal to the velocity axis is expressed through the lift term by the load factor. In other words, using

$$C_L = \frac{nW}{qS}$$

and assuming an uncambered aircraft ($C_{\ell \min} = 0$) the drag is

$$D = qS\left(C_{D_0} + KC_L^2\right) = C_{D_0}qS + \frac{K}{q}n^2\frac{W^2}{S} \tag{3.41}$$

we get [assuming $\cos(\alpha + i_T) \approx 1$]

$$P_S = \frac{dh_e}{dt} = V\left[\frac{T}{W} - \frac{qC_{D_0}}{W/S} - \frac{K}{q}n^2\frac{W}{S}\right] \tag{3.42}$$

Thus, the value of P_S at a point in space is dependent upon the aircraft's load factor n.

An aircraft's capabilities can be determined at a point in space by calculating its value of P_S. For example, the F-104G at $n = 1$, Mach = 0.8 (829 fps) at 20,000 ft has a maximum thrust of 10,000 lb and a drag of 2086 lb. For a weight of 18,000 lb and $\cos(\alpha + i_T) \approx 1$ we have

$$\frac{dh_e}{dt} = P_S = \frac{\left(10,000 - 2086\right)829}{18,000} = 364 \text{ fps}$$

as shown in Fig. 4.8a. The F-104G has an instantaneous rate of climb of 364 fps for this initial point of level flight at $n = 1$. If the F-104F were to perform a level acceleration where $dh/dt = 0$, it could accelerate at

$$\frac{dV}{dt} = \frac{gP_S}{V} = \frac{(32.2)(364)}{829} = 14.1 \, \text{ft}/\text{s}^2$$

A pilot of the F-104G wanting to cruise at Mach = 0.8 and 20,000 ft would have to throttle back or increase drag until $P \approx 0$.

3.9 Energy Maneuverability for Air Combat Assessment

The performance of an aircraft over its entire operating envelope can be displayed by plotting contours of its P_S for constant load factor values. The P_S contours for the F-104G are shown in Fig. 4.8a and a composite lightweight fighter design in Fig. 4.9a.

P_S plots for $n > 1$ are useful in assessing an aircraft's combat maneuverability. For example, the P_S plots at $n = 5$ for two competing aircraft can be laid one over the other and regions of advantage and disadvantage are immediately obvious [7]. Typical P_S charts are shown in Fig. 3.12 for air-

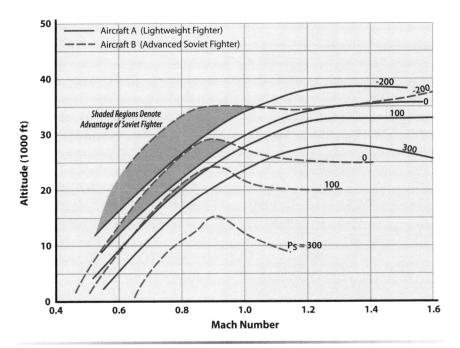

Figure 3.12 P_S contours for aircraft A and B at $n = 5$ g.

craft A and B at $n = 5$. Figure 3.12 shows regions of advantage and disadvantage for the two aircraft. Aircraft B is superior to [i.e., has an energy maneuverability (EM) advantage over] aircraft A at high subsonic speeds above 25,000 ft. The present-day combat arena is around Mach = 0.9 and below 30,000 ft [8]. Thus, aircraft A would appear to be the better air combat fighter based upon the performance comparison at $n = 5$.

Turn rate performance is the primary measure of an aircraft's air-to-air combat effectiveness [8,9] as it indicates the capability of the aircraft to gain a firing position advantage. A value of $P_S = 0$ for $n > 1$ indicates a certain level of maximum sustained turn rate. Thus, $P_S = 0$ contours for $n > 1$ indicate an aircraft's turning performance. The relation between $P_S = 0$ and the maximum sustained load factor is given as follows [solving Eq. (3.42) for n with $P_S = 0$]:

$$n_S = \frac{q}{W/S}\sqrt{\frac{1}{K}\left(\frac{T}{qS} - C_{D0}\right)} = \frac{q}{\sqrt{W/S}}\sqrt{\frac{1}{K}\left(\frac{T}{W}\frac{1}{q} - \frac{C_{D0}}{W/S}\right)} \quad (3.43)$$

and, for the maximum instantaneous load factor,

$$n_{max} = \frac{q C_{Lmax}}{W/S}$$

where

$$\dot{\psi} = \frac{g\sqrt{n^2 - 1}}{V} \quad (3.32)$$

(the turn rate in radians per second). An evaluation of the relative turn performance of two aircraft can be made by comparing their respective P_S vs turn rate curves as shown in Fig. 3.13. Three useful reference points are indicated in Fig. 3.13. The $n = 1$ energy rate (P_S at $\dot{\psi} = 0$) provides a measure of the acceleration or climb performance. The turn rate at $P_S = 0$ is that which can be sustained without energy loss. The maximum instantaneous turn rate corresponds to the maximum usable aerodynamic lift available, or *structural limit*, but is normally accompanied by a high energy loss rate. By having a margin in turn rate at all energy rates, the F-5E is assured a combat advantage. Reference [8] concludes that a desired turn rate margin over a threat aircraft is about two degrees per second. Thus, from Fig. 3.13 we observe that the F-5E should quickly attain a tail aspect relative to the F-5A and gain firing opportunities. The most significant region in the combat hassle occurs at zero to negative energy rates as this is where the majority of the time is spent during a hard turning air-to-air engagement. Even if the F-5A had an energy rate advantage at low turn rates (which it doesn't)

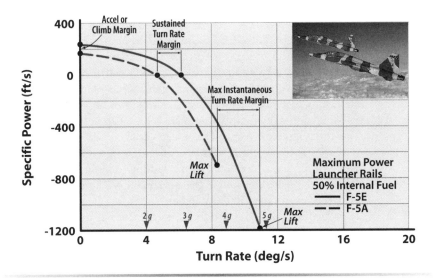

Figure 3.13 Specific power vs turn rate for the F-5A and F-5E at 30,000 ft and Mach = 0.8 (data from [10]).

and chose to disengage the combat, the F-5E could use its higher turn rate capability to gain a missile firing opportunity before sufficient separation distance could be attained.

The effect of speed on turn performance for constant altitude is illustrated in Fig. 3.14 for the F-5A and F-5E. Notice that the F-5A has an instantaneous turn rate advantage at Mach 1.6 due to its higher design load

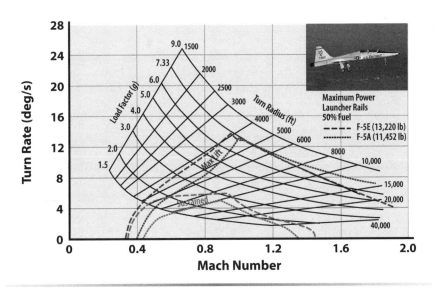

Figure 3.14 Turn performance at 30,000 ft for F-5A and F-5E (data from [10]).

John Boyd: Father of Energy Maneuverability

The energy maneuverability (EM) presented in Sections 3.8 and 3.9 was developed by USAF officer John Boyd. Boyd entered the Air Force in 1951 with a degree in economics and an Air Force ROTC commission. He was an outstanding fighter pilot with an uncanny ability to visualize aerial combat maneuvering in four dimensions (X, Y, Z, and time). While on the faculty at the Air Force Fighter Weapons School (FWS), Nellis AFB, Boyd developed an aerial combat tactics manual that was used in the USAF, USN, US Marine Corps, and most foreign air forces. Boyd moved aerial combat from an art with a "bag of tricks" to a skill. His flying skills were legendary and it is reported that he never lost a "dog fight" in peace-time training or war-time fighting. When Boyd finished his tour at the Air Force FWS he was awarded the Legion of Merit, most unusual for a captain.

In 1960 Boyd entered Georgia Tech to study engineering. He needed the mathematics and physical sciences to put his aerial combat maneuvers on a solid footing and to quantify the positional advantage (expressed as excess specific energy) of one aircraft over another. He graduated three years later and was assigned to Eglin AFB, where he started developing the EM theory presented in this chapter. The development of the full EM theory would consume his free time for the next decade. For the first time the USAF had a tool that could quantify the aerial capability of one aircraft against another. The results of an evaluation of the F-4 against the MIG-25 showed the F-4 at a significant disadvantage and hastened its replacement. John Boyd's EM theory was used to develop the requirements for the A-10 and F-15 in the late 1960s. In 1970 Boyd received a second Legion of Merit for "developing the most powerful evaluative tool for fighter aircraft analysis known to date and providing industry with one of the most effective tools generated in the history of aeronautical engineering" [7].

John was disappointed in the F-15; in his view it was a Cadillac when the Air Force really needed a Corvette. In the late 1960s he formed the "Fighter Mafia" and promoted the development of a small, low-cost, lightweight fighter, once again using his EM theory to develop the requirements. The Lightweight Fighter (LWF) program was initiated in 1971 with funds to build two prototypes. The RFP was issued in January 1972 and called for a 20,000 lb class fighter (half the weight of the F-15) optimized for air combat at speeds of Mach 0.6–1.6 and altitudes of 30,000–40,000 feet. The result was the F-16 for the USAF and the F-18 for the USN. I met John in 1972 when I was a major at Wright-Patterson AFB. For Boyd there were two types of people: fighter pilots and pukes (bomber pukes, desk pukes—it didn't matter to John). I was a design puke and privileged to be a minor member of the "Fighter Mafia."

In June 1975 Colonel Boyd was awarded the Harold Brown Award for the development of the F-16, the highest scientific award granted by the Air Force. In September 1975 John Boyd retired from the Air Force as a colonel. He spent the next twenty years developing the EM theory for land combat. His ideas were embraced by the US Marines and the US Army. The May 6, 1991, issue of *US News & World Report* featured an article about the innovative tactics that won the Gulf War. Credit is given to John Boyd.

Leland Nicolai

factor. However, Mach = 1.6 is well outside the combat arena and this region of advantage would not be very useful. In the combat arena, Mach = 0.7 to 0.9, the F-5E has a decided advantage. The F-5E performance gain is due largely to the incorporation of the J85-GE-21 turbojet engines, which provide 22% more thrust than the J85-GE-13 engines in the F-5A [10].

3.10 Rate of Climb and Descent

The rate of climb for an aircraft is given by $dh/dt = V \sin \gamma$. The expression for P_S is expressed as

$$P_S = \frac{dh}{dt} + \frac{V}{g}\frac{dV}{dt} = \frac{dh}{dt} + \frac{V}{g}\frac{dV}{dh}\frac{dh}{dt}$$

which gives

$$\frac{dh}{dt} = V \sin \gamma = \frac{P_S}{1 + (V/g)(dV/dh)} \qquad (3.44)$$

Values for P_S can be found using Eq. (3.40) along a specified trajectory (i.e., plot of altitude vs velocity; see Chapter 4). Values for dV/dh can be determined because they are the inverse of the slope at any point on the trajectory. Thus, the aircraft rate of climb can be found using the energy state approximation. Notice that if the portion of the trajectory under consideration is for a constant-speed climb, then dV/dh is zero and the rate of climb expression is

$$\frac{dh}{dt} = V \sin \gamma = \frac{V\left[T\cos\left(\alpha + i_T\right) - D\right]}{W} \qquad (3.45)$$

Equation (3.45) with $n = 1$ is quite valid for most subsonic climb situations. However, if the aircraft is a high-performance vehicle capable of accelerating during the climb, then Eq. (3.44) should be used.

When P_S is negative the aircraft is losing airspeed and/or altitude. The pilot usually wants to fly a constant airspeed descent ($dV/dh = 0$) so that Eq. (3.45) is exact. In gliding flight the thrust is zero and the glide is usually a near-equilibrium situation at constant airspeed and $n \approx 1$. Thus, Eq. (3.45) represents the gliding descent quite well.

Figure 3.15 shows the force diagram on an aircraft during gliding flight [11]. The flight path angle during gliding flight is given by

$$\gamma = \arc \sin\left(-D/W\right) \qquad (3.46)$$

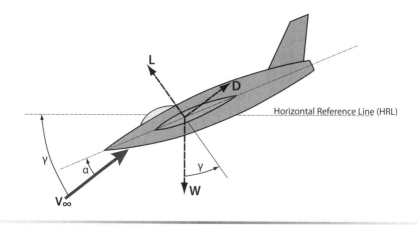

Figure 3.15 Force diagram on aircraft in gliding flight.

and

$$\gamma = \arctan\left(-D/L\right) \tag{3.47}$$

For maximum range during the glide descent (i.e., stretching the glide) the aircraft should be flown at minimum γ, which means flying at $(L/D)_{\max}$. The velocity for maximum gliding range is given by Eq. (3.11)

$$V_{(L/D)_{\max}} = \sqrt{\frac{2W}{\rho S}}\sqrt{\frac{K}{C_{D0}}}$$

The condition for minimum rate of descent (maximum endurance) is different than the condition for maximum range. Assuming that γ is small such that $\sin \gamma \approx \tan \gamma$, the *rate of descent* (ROD) can be expressed as

$$\text{ROD} = V \tan \gamma = -\frac{DV}{L} = -\frac{C_D}{C_L}\sqrt{\frac{2W}{\rho C_L S}} \tag{3.48}$$

The C_L that minimizes Eq. (3.48) is

$$C_L = \sqrt{\frac{3C_{D0}}{K}} \tag{3.49}$$

so that the velocity for minimum ROD is

$$V_{\text{ROD}_{\min}} = \sqrt{\frac{2W}{\rho S}}\sqrt{\frac{K}{3C_{D0}}} \tag{3.50}$$

Table 3.2　Values of C_L for Maximum Range and Endurance

Uncambered Wing			
Mission	**Condition**	**Maximize**	**Value of $C_L^{(a)}$**
Range—jet	Constant altitude	$C_L^{1/2}/C_D$	$\sqrt{C_{D_0}/3K}$
Range—jet	Constant throttle	$C_L/C_D^{3/2}$	$\sqrt{C_{D_0}/2K}$
Range—propeller	Constant altitude	C_L/C_D	$\sqrt{C_{D_0}/K}$
Range—sailplane	Minimum glide angle	C_L/C_D	$\sqrt{C_{D_0}/K}$
Endurance—sailplane	Minimum rate of sink	$C_L^{3/2}/C_D$	$\sqrt{3C_{D_0}/K}$
Endurance—propeller	Minimum power required	$C_L^{3/2}/C_D$	$\sqrt{3C_{D_0}/K}$
Endurance—jet	Minimum thrust required	C_L/C_D	$\sqrt{C_{D_0}/K}$
Use $C_D = C_{D_0}+KC_L^2$ to find L/D or C_L/C_D and $\left(L/D\right)_{max} = 1/(2\sqrt{C_{D_0}K})$			
Maximum jet range, constant throttle	$\dfrac{L}{D} = \dfrac{\sqrt{C_{D_0}/2K}}{C_{D_0} + KC_{D_0}/2K} = \sqrt{\dfrac{2}{9C_{D_0}K}} = \sqrt{\dfrac{8}{9}}\,(L/D)_{max} = 0.943(L/D)_{max}$		
Maximum jet range, constant altitude	$\dfrac{L}{D} = \dfrac{\sqrt{C_{D_0}/3K}}{C_{D_0} + KC_{D_0}/3K} = \sqrt{\dfrac{9}{48C_{D_0}K}} = \sqrt{\dfrac{3}{4}}\,(L/D)_{max} = 0.866(L/D)_{max}$		
Maximum propeller endurance	$\dfrac{L}{D} = \dfrac{\sqrt{3C_{D_0}/K}}{C_{D_0} + K3C_{D_0}/K} = \sqrt{\dfrac{3}{16C_{D_0}K}} = \sqrt{\dfrac{3}{4}}\,(L/D)_{max} = 0.866(L/D)_{max}$		
Cambered Wing			
Use $C_D = C_{D_{min}} + K'C_L^2 + K''\left(C_L - C_{\ell_{min}}\right)^2$　　$(L/D)_{max} = 1/2(AK)^{1/2} - B$			
$A = C_{D_{min}} + K''C_{\ell_{min}}^2$	$B = K''C_{\ell_{min}}$		$K = K' + K''$
Mission	**Condition**		**Value of $C_L^{(a)}$**
Range—Jet	Constant Altitude		$\sqrt{A/3K} - B/3K$
Range—Jet	Constant Throttle		$\sqrt{A/2K} - B/2K$
Range—Prop	Constant Altitude		$\sqrt{A/K}$
Range—Sailplane	Minimum Glide Angle		$\sqrt{A/K}$
Endurance—Sailplane	Minimum Rate of Sink		$\sqrt{3A/K} - B/K$
Endurance—Prop	Minimum Power Required		$\sqrt{3A/K} - B/K$
Endurance—Jet	Minimum Thrust Required		$\sqrt{A/K}$

[a]Fly at prescribed C_L for max range and endurance! Use value of C_L to size wing for the range or endurance phase of the mission.

which is about 23% less than the velocity for maximum gliding range. Notice that the velocity for minimum ROD is the same as for minimum power required, Eq. (3.12).

3.11 Summary for Maximum Range and Endurance

Aircraft C_L is the parameter that is varied to enable flight at a condition of maximum range or maximum endurance. This is because of the direct relationship between lift and drag. The pilot flies at a specific C_L by trimming the aircraft at a specific angle-of-attack.

Table 3.2 shows the values of C_L at which the airplane (jet, propeller, or sailplane) should fly to achieve maximum range or endurance. Notice that the C_L values are different for aircraft with uncambered vs cambered wings. For an uncambered jet aircraft flying a maximum-range mission at constant altitude, the $L/D = 0.866(L/D)_{max}$ and flying at constant throttle (cruise climb) the $L/D = 0.943(L/D)_{max}$.

References

[1]   "L-1011-2 Flight Performance with Rolls-Royce RB. 211-22B Engines," Lockheed Rept. LR 23973, Lockheed-California Co., Burbank, CA, Oct. 5, 1970.
[2]   Perkins, C. D., and Hage, R. E., *Airplane Performance Stability, and Control*, Wiley, New York, 1949.
[3]   Dommasch, D. O., Sherby, S. S., and Connolly, T. F., *Airplane Aerodynamics*, Pitman, New York, 1961.
[4]   Rutowski, E. S., "Energy Approach to the General Aircraft Performance Problem," *Journal of the Aeronautical Sciences*, Vol. 21, No. 3, March 1954, pp. 187-195.
[5]   Bryson, A. E., Desai, M. N., and Hoffman, W. C., "Energy-State Approximation in Performance Optimization of Supersonic Aircraft," *Journal of Aircraft*, Vol. 6, No. 6, Dec. 1969, p. 481.
[6]   Bryson, A. E., and Denham, W. F., "A Steepest Ascent Method for Solving Optimum Programming Problems," *Journal of Applied Mathematics*, Vol. 29, No. 6, June 1962, p. 286.
[7]   Coram, Robert, *Boyd: The Fighter Pilot Who Changed the Art of War*, Back Bay Books, Little, Brown, New York, 2002.
[8]   Fellers, W. E., and Patierno, J., "Fighter Requirements and Design for Superiority over Threat Aircraft at Low Cost," AIAA Paper No. 70-516, AIAA Fighter Aircraft Conf., St. Louis, MO., March 1970.
[9]   Bratt, R. W., and Patierno, J., "Design Criteria for Fighter Aircraft," AGARD Fluid Dynamics Panel on Aerodynamic Design and Criteria, London, England, Sept. 1971.
[10] "F-5E Technical Description," Northrop Rept. NB 71-20, Aircraft Div., Northrop Corp., Hawthorne, CA, July 1973.
[11] McCormick, B. W., *Aerodynamics, Aeronautics and Flight Mechanics*, Wiley, New York, 1995.

Chapter 4 Aircraft Operating Envelope

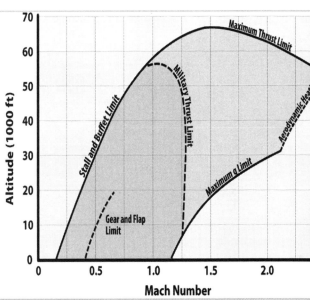

- Stall & Buffet Limits
- Heating & Material Limits
- Thrust & Q Limits
- Propulsion System Limits
- Minimum Time & Fuel Climb
- Optimum Energy Profiles
- FAA Noise Regulations

The F-22 rockets through 30,000 feet, well below its operational ceiling of 65,000 feet.

Time is nature's way of keeping everything from happening at once.
Woody Allen

4.1 Flight Envelope

The aircraft is not free to meander at will in its atmospheric environment. It is constrained to operate within a corridor in space called the *flight envelope* (Fig. 4.1). Mission requirements will usually define where in the airspace the aircraft must operate. The designer must be aware of the potential flight envelope limitations and design accordingly. This operating envelope is determined by aircraft limitations (such as minimum/maximum dynamic pressure and aerodynamic heating) and

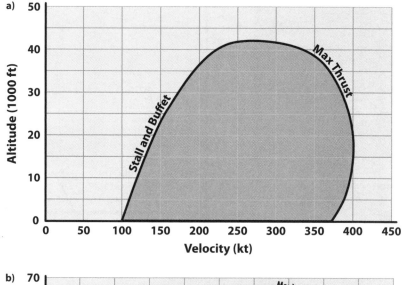

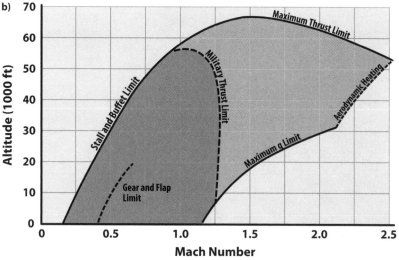

Figure 4.1 Typical aircraft flight envelopes: **a)** subsonic aircraft (Fairchild-Republic A-10), W_{TO} = 30,344 lb with four Mk 82s, and **b)** supersonic aircraft.

operational limitations (such as sonic boom, noise, and air pollution). In addition, engine limitations (such as insufficient thrust, intolerable fuel consumption, and internal pressure and temperature limitations) mold part of the envelope.

4.2 Minimum Dynamic Pressure

The left side of the flight envelopes shown in Fig. 4.1 is determined by the stall and buffet characteristics of the aircraft. *Stall* is the loss of lift from the wing due to the sudden separation of the flow from the wing upper surface, which usually defines the low-speed part of the envelope. This stall boundary is sensitive to aircraft weight and the flap configuration on the aircraft. *Buffet* is caused by the turbulence of the airflow separation shaking some part of the aircraft (usually the wing or horizontal tail). Buffet precedes stall and is more noticeable at higher speeds. An arbitrary assessment of where the buffet becomes objectionable to the pilot is the basis for the buffet limit boundary. The designer can translate this buffet boundary to the left (to lower flight speeds) by selecting a lower wing loading, maneuver flaps, and careful tail-plane location. To develop the stall and buffet boundary, the designer needs information on the maximum usable C_L versus Mach number for the aircraft as shown in Figs. 4.2 and 6.6.

The F-22 Raptor is a single-seat, twin-engine, fifth-generation fighter; its survivability depends on high speed, maneuverability, and a very low signature (stealth). It has an empty weight of 43,430 lb, an air combat mission weight of 64,460 lb, and a max TO weight of 83,500 lb. The propulsion is two PWA F119-PW-100 turbofan engines at 35,000-lb TSLS each (in afterburner and uninstalled), giving an air combat mission thrust/weight of 1.08. The F-22 has a max speed of Mach 2.25, a supercruise (level flight in dry power) speed of Mach 1.82, and a range of 1600 nm.

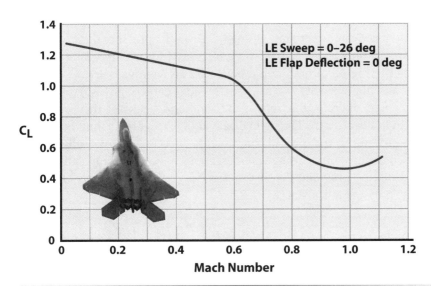

Figure 4.2   Typical variation of the maximum usable C_L with Mach number.

At low speeds this maximum usable C_L is usually close to the $C_{L_{max}}$ of the aircraft. As speed increases and maximum buffet becomes more pronounced, the maximum usable C_L decreases until, in the transonic region, it may drop to one-half or one-third of its low-speed value (review Section 2.11).

4.3 Maximum Thrust Limit

The top and part of the right side of the flight envelope are determined by the engines' maximum thrust. This boundary is where the thrust available equals the thrust required, as discussed in Section 3.2. The maximum altitude that can be reached is called the *absolute ceiling* of the aircraft. Remember that the absolute ceiling and the remainder of the thrust limit line are dependent upon the weight and external store configuration of the aircraft. The operational ceiling is where the rate of climb is 100 ft/min.

4.4 Maximum Dynamic Pressure

The maximum dynamic pressure limit for an aircraft is a structural limitation. Flight at high dynamic pressures introduces aeroelastic problems of flutter and engine inlet static pressure (for more information about aeroelasticity and flutter, see [1,2]). The designer must be well aware of the structural limitations involved in the aircraft's operating envelope and give them due consideration in the preliminary design phase. Current aircraft are generally designed for maximum q limits of about 1800 psf. Figure 4.3 shows the variation of q with altitude and velocity. There are both advantages and disadvantages to increasing the dynamic pressure limit of an aircraft. One advantage of a high q limit is increased survivability for a military aircraft. The high-q aircraft could penetrate an enemy's defenses at a low altitude and high velocity and avoid early radar detection. Also, the high-q aircraft could operate in a region (altitude and velocity) that would be denied to low-q aircraft. The main disadvantage to a high q limit is that the aircraft structural and propulsion weights increase significantly, which decreases performance through an increase in W_{TO}. An aircraft can operate at any desired q provided that it is designed for that condition. The flight dynamic pressure q is given by

$$q = \tfrac{1}{2}\rho_\infty V_\infty^2 = \left(\gamma/2\right)P_\infty M_\infty^2 \tag{4.1}$$

where gamma is the ratio of specific heats (1.4 for air). The freestream total pressure is given by the isentropic relation (see Appendix C):

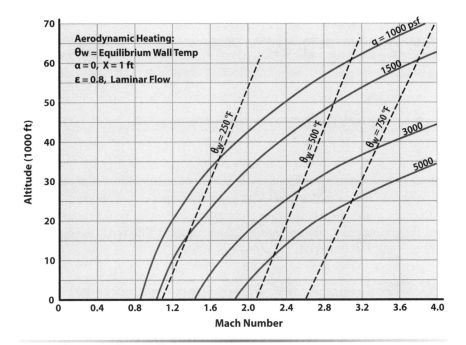

Figure 4.3 Trajectory limits of dynamic pressure and aerodynamic heating.

$$P_{0\infty} = P_\infty \left[1 + \frac{\gamma - 1}{2} M_\infty^2 \right]^{\gamma/(\gamma-1)} \qquad (4.2)$$

As the airflow is decelerated in the engine inlet to approximately Mach = 0.4 at the compressor face, the static pressure of the air increases. The static pressure at the compressor face can be many times greater than the ambient static pressure P_∞ if q_∞ is large. For example, consider a level flight at 25,000 ft at dynamic pressures of 1700 psf and 5000 psf. Assume the total pressure recovery for the inlet (see Chapter 16) is 90% for q = 1700 psf (M_∞ = 1.75) and 75% for q = 5000 psf (M_∞ = 3.0). The resulting static pressure at the compressor face is shown in Table 4.1. An inlet designed for the p_c associated with a q of 1700 psf would be blown apart by the static pressure associated with a q of 5000 psf.

4.5 Aerodynamic Heating

An aircraft flying at high Mach numbers (i.e., 2.0 or greater) heats up due to the conversion of the kinetic energy of the air into thermal energy. This thermal energy represents an elevated temperature, and heat is transferred to the aircraft by convection. The critical regions on an aircraft are

Table 4.1 Inlet Static Pressures for Different Dynamic Pressure Conditions

q (psf)	Altitude (ft)	Mach	$P_\infty{}^a$(psf)	$P_{0\infty}$ (psf)	$P_c{}^b$(psf)
1700	25,000	1.75	786.3	4,180	3,360
5000	25,000	3.0	786.3	28,900	19,300

[a]From Appendix A.
[b]P_c is the static pressure at the compressor face where the Mach number is 0.4.

the stagnation points and lower surfaces. The limiting skin temperatures of the aircraft vary with the material used. At temperatures above 250°F the aluminum alloys display a rapid degradation in mechanical properties [3], and the designer should consider other materials. The temperature limits for other aircraft materials are shown in Table 4.2.

4.5.1 Stagnation Point Heating on Nose and Swept Wing Leading Edge

The expression for heating rate [in British thermal units per square foot per second (Btu/ft²·s)] is

$$\dot{q}_{conv} = 15 \left(\frac{\rho_\infty}{R_0} \right)^{0.5} \left(\frac{V_\infty}{1000} \right)^3 \left(\cos \Delta \right)^{1.5} \tag{4.3}$$

where ρ_∞ is the density [in slugs per cubic foot (slug/ft³)], V_∞ is freestream velocity (in feet per second), R_0 is the radius of the nose or leading edge (in feet), and Δ is the sweep of the leading edge ($\Delta = 0$ for body nose).

We assume a heat balance such that

$$\dot{q}_{conv} = \dot{q}_{radiated}$$

or

$$\theta_w = \left[\frac{\dot{q}_{conv}}{\varepsilon v_{SB}} \right]^{1/4} \tag{4.4a}$$

where ε is the emissivity of the surface (approximately 0.8), v_{SB} is the Stephan–Boltzmann constant (0.481×10^{-12} Btu/ft²·s·°R), and θ_w is the equilibrium wall temperature (in °R).

4.5.2 Lower Surface Heating

This analysis is based upon the Reynolds analogy between skin friction and heat transfer. The local surface heat transfer can be approximated by

		Room Temperature Properties				Temperature Limitation (°F)	
	Condition	Ultimate Tensile Strength (ksi)	Yield Tensile Strength (ksi)	Compression Modulus (10⁶ psi)	Density (lb/in.³)	Primary Structure	Secondary Structure
Beryllium	SR	78	57	42.0	0.066	1000	1350
Ti-6Al-5Zr-4Mo-1Cu-0.2Si	STA	200	177	16.5	0.164	800	800
Ti-6Al-6V-2Sn	STA	170	160	16.5	0.164	800	800
Ti-8Mo-8V-2Fe-3Al	STA	180	165	16.6	0.175	600	600
Ti-6Al-2Sn-4Zr-6Mo	TA	170	160	16.5	0.169	1000	1000
Ti-6Al-4V	STA	157	143	16.4	0.160	800	900
Ti-6Al-6V-2Sn	A	155	145	15.0	0.164	800	800
PH 14-8Mo	STA	240	225	28.0	0.278	1000	1000
Ti-8Al-1Mo	STA	133	121	18.0	0.156	1000	1100
Ti-6Al-4V	A	134	126	16.4	0.160	1000	1000
Inco's "1000°F Alloy"	STA	228	19.1	29.0	0.267	1000	1000
Ti-5Al-2.5Sn	A	120	113	15.5	0.161	900	1100
Inconel 718	STA	210	185	29.0	0.297	1300	1800
Rene' 41	STA	184	145	31.9	0.298	1550	1800
2219-T81 (Aluminum)	STA	60	45	10.8	0.102	400	500
L-605 (Cobalt)	CR	185	145	32.6	0.330	1800	2000
TD NiC	SR	138	94	21.9	0.306	2200	2400
Haynes Alloy No. 188	A	130	67	34.5	0.333	2000	2000
Hastelloy X	A	114	55	28.6	0.297	2000	2100
TZM (Molybdenum)	SR	140	117	40.0	0.369	3200	3400
B66 (Columbium)	A	106	81	14.6	0.305	2600	2800
TDNi	SR	85	68	22.0	0.322	2000	2200
Cb-752 (Columbium)	A	81	70	17.0	0.326	2400	2800
T-222	A	120	110	29.0	0.605	3000	3400

SR, stress relieved; A, annealed; DA, duplex annealed; TA, triplex annealed; STA, solution treated and aged; CR, cold-rolled.

$$\dot{q}_{\text{surf}} = 3.21 \times 10^{-4} C_f \rho V_\infty^3 \qquad (4.5)$$

where C_f is the local laminar skin friction coefficient at a distance x feet from the leading edge. Normally x is taken as 1.0 ft. The other quantities in Eq. (4.5) are the same as for Eq. (4.3).

As before, we assume a balance between the reradiated heat and the local surface heat transfer so that the equilibrium wall temperature is given by

$$\theta_w = \left[\frac{\dot{q}_{\text{surf}}}{\varepsilon V_{\text{SB}}} \right]^{1/4} \qquad (4.4b)$$

Figure 4.3 shows some lines of constant lower surface equilibrium temperatures for $\varepsilon = 0.8$ and $x = 1$ ft.

4.6 Sonic Boom

An aircraft flying at supersonic speed will create pressure waves on the ground associated with its shock system. If the aircraft altitude is low and/ or the Mach number is large (greater than 2.5), this pressure on the ground (called *overpressure*) can cause discomfort and damage [4,5]. To avoid this ground overpressure from becoming excessive, a minimum altitude at which the aircraft may fly at supersonic speeds must be fixed. The supersonic transports (SSTs), Concorde and TU-144, were never able to negotiate these operating limits with those countries they overflew.

4.7 Noise and Pollution Limits

In 1969, the FAA found it necessary to establish the FAR-36 regulations [6] concerning noise. The rules limit the engine and aircraft noise, expressed in terms of effective perceived noise level in decibels (EPNdB), allowed at three reference locations shown in Fig. 4.4. On approach the measuring point is 1 nautical mile (n mile) before touchdown. When approaching on a 3-deg glide slope, the aircraft at this point has an altitude of 370 ft. Steeper glide slopes can reduce the noise measured at this point. On takeoff, the measuring point is 3.5 n mile from the point of brake release. Altitude at this point depends upon the particular aircraft and flight procedures used. The third measuring location is the sideline after liftoff at a distance of 0.35 n mile for four-engine aircraft and 0.25 n mile for three-engine aircraft. The FAR-36 specifies the permissible noise levels at each of these points as a function of aircraft gross weight. These permissible levels are shown in Fig. 4.5.

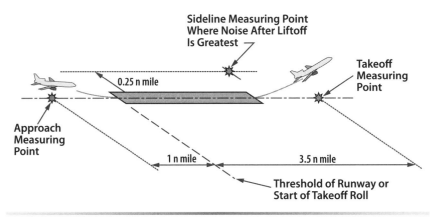

Figure 4.4 Noise measuring locations for FAR Part 36.

4.7.1 Regulations

Noise regulations in FAR Part 36 Stage 3 include restrictions on noise under three conditions. The *takeoff noise* (Fig. 4.5c) is defined as the noise measured at a distance of 21,325 ft (6500 m) from the start of the takeoff roll, directly under the airplane. The *sideline noise* (Fig. 4.5b) is measured 1476 ft (450 m) from the runway centerline at a point where the noise level after liftoff is greatest. The *approach noise* (Fig. 4.5a) is measured under the airplane when it is at a distance of 6562 ft (2000 m) from the runway thresh-

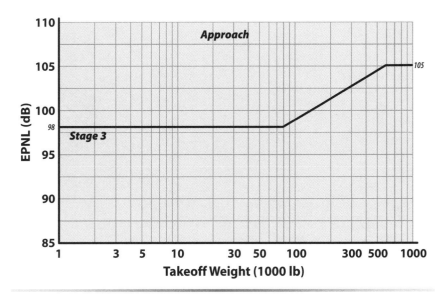

Figure 4.5a Maximum noise limits—approach.

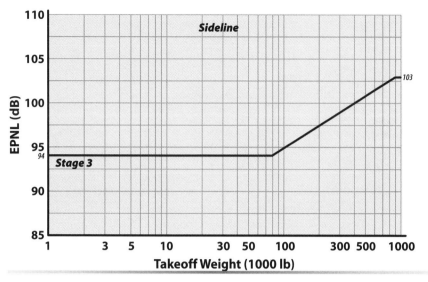

Figure 4.5b   Maximum noise limits—sideline.

old. For each of these conditions the maximum noise level is a function of maximum takeoff gross weight; for the takeoff case the limits depend also on the number of engines.

Stage 4 noise regulations are applicable to new-type designs introduced after 1 January 2006. Existing aircraft will be able to operate under Stage 3 regulations. This new standard will be "Chapter 4" in the International

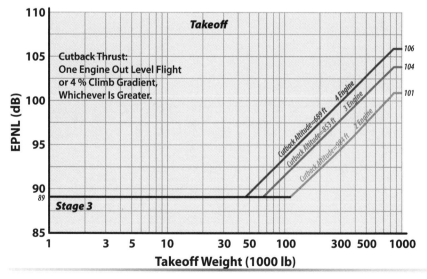

Figure 4.5c   Maximum noise limits—takeoff.

Civil Air Organization (ICAO) Annex 16 and is related to the Stage 3/ Chapter 3 regulations as follows:

* There is a cumulative margin of 10 dB relative to Chapter 3.
* There is a minimum sum of 2 dB at any two conditions.
* No trades are allowed

4.7.2 Estimating Aircraft Noise for Advanced Design

We start with a measurement of the noise due to a known engine at a known distance away. For example, a 25,000-lb thrust [sea level static (SLS) takeoff thrust] turbofan engine with a bypass ratio of 6 produces a noise of about 101 EPNdB at a distance of 1000 ft. This assumes some level of noise suppression (about 5 EPNdB).

Most of the internal noise comes from the high-speed rotating blades and is commonly called *turbomachinery* noise [7]. External noise arises from shear and eddy phenomena during mixing of the high-velocity jets with the ambient air. The fan noise can be effectively suppressed by acoustical treatment added to the engine nacelles, but the jet noise is hard to suppress. One alternative is to transfer a larger portion of the engine energy into the fan stream. This means larger bypass ratio turbofan engines. When the tradeoffs are made between noise suppression, engine efficiency over the mission profile, and inlet or nacelle drag, the optimum bypass ratio comes out around 5.5 [7].

The takeoff and sideline noise levels are established by the engine. However, the approach level is defined by the aircraft. During landing the engines are at low power and actually generate less noise than the turbulence from the aircraft flaps, landing gear, and wheel wells.

Although pollution regulations do not presently exist for aircraft, it is reasonable to assume that they will come into force in the future. These regulations will specify limits on engine emissions with emphasis on operation in the airport area and in the stratosphere. The airport area is of principal environmental concern because of its proximity to large population centers. Engine pollutants during idle result from inefficient combustion during off-design operation of a relatively simple combustor that cannot adapt to large changes in overall fuel–air ratios. The high-power condition during takeoff is another problem for the engine designer. Although hydrocarbon and carbon monoxide emissions can be currently brought to acceptable levels, the nitrogen oxide emission will plague engine designers for some time [7–10].

The operation of fleets of SSTs in the stratosphere could substantially reduce the ozone layer around the earth [9]. It has been concluded that a reduction in the ozone layer would increase the incidence of skin cancer. The reduction of the ozone is due mainly to the introduction of nitric oxide from engine exhaust into the stratosphere. So, once again, the engine designer, with help from the aircraft designer, has the responsibility to produce better, more efficient engine and aircraft designs [9].

4.8 Propulsion Limits

The aircraft designer has nine propulsion devices from which to choose. Seven of these devices are shown in Fig. 4.6 [scramjet and pulse detonation engine (PDE) are not shown]. The decision of which propulsion device to use is not difficult as each device has its own preferred operating regime. The measure of merit for a propulsion device is its *thrust per engine weight* (T/W) and its *thrust specific fuel consumption* (TSFC). As shown in Fig. 4.7, it is hard to beat the reciprocating-engine–propeller combination at low speeds (Mach = 0.5 and less). At higher subsonic speeds the turboprop engine is optimum but is limited to propeller tip speeds less than Mach = 1.0. At high subsonic speeds the turbofan is the best but loses out to the turbojet at low supersonic speeds because of the increased drag of the fan. At speeds around Mach 2 the turbojet engine has to have its thrust augmented by afterburning. Above Mach 3 the ramjet is very efficient [11]. If an effective coupling of a ramjet and a turbojet engine could be achieved (called a *turbo-ramjet*), the best of two worlds could be obtained. At speeds below Mach 2 the turbojet would operate with the inlet to the ramjet closed. Between Mach 2 and 3 both ramjet and turbojet would operate, and above Mach 3 the operation would be on ramjet with the inlet to the turbojet closed.

Because of the self-contained nature of the rocket, it does not care whether it is at Mach 0 or Mach 10. Thus, its characteristics are constant, as shown in Fig. 4.7. Figure 4.7 shows a TSFC of 16 for the rocket, which is typical of current solid-fuel rocket motors. Liquid-fuel rocket motors, such as the boost stages for the Space Shuttle and other spacecraft, use liquid hydrogen and oxygen and have TSFCs on the order of 9 (for more about rockets, see Section 18.12).

4.9 Optimal Trajectories

The *optimal trajectory*, that is, the one that minimizes time or fuel burned, can be determined using the methods of calculus of variations [12] or gradient [13]. These methods yield optimal path solutions but require large complex computer programs and high operator familiarity. The energy-state approximation, discussed in Chapter 3, Section 3.8, yields rapid useful results that agree well with the exact results of the calculus of variations.

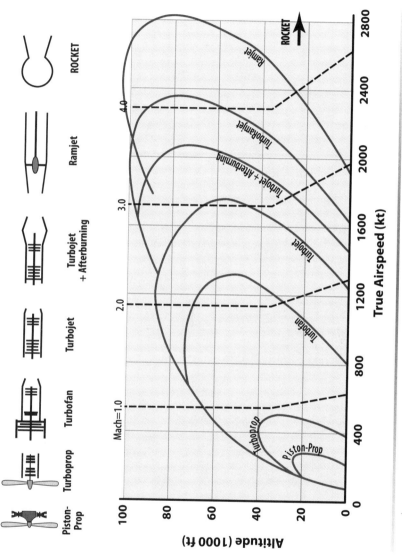

Figure 4.6   Level-flight propulsion options.

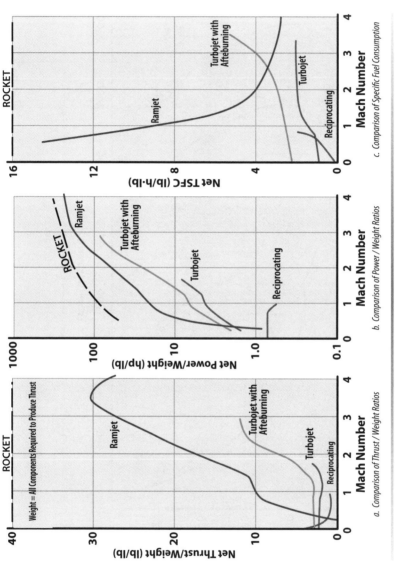

Figure 4.7   Comparison of design characteristics of different propulsion engines as a function of flight Mach number. Each point on the graphs represents an engine design point with a structure that is critical at that point for pressure loads.

From Chapter 3, Section 3.8, the excess specific power P_S (in feet per second) is given by Eq. (3.40),

$$P_S = \frac{dh_e}{dt} = \frac{V(T\cos\alpha - D)}{W} \tag{3.40}$$

and expresses the rate of change of the aircraft's specific energy

$$h_e = h + \frac{V^2}{2g} \tag{3.36}$$

where h is the altitude above mean sea level. The quantity

$$f_S = \frac{dh_e}{dW_f} = \frac{dh_e/dt}{TC} \tag{4.6}$$

where T is the thrust and C is the specific fuel consumption, expresses (in feet per pound) the change of specific energy with respect to a change in fuel weight. Figure 4.8a shows a plot of constant h_e that is independent of the aircraft–propulsion system. Figure 4.8 a and b shows plots of constant P_S and f_S, respectively, for the supersonic F-104G at a load factor of 1. Figure 4.9 shows the P_S and f_S curves for an advanced fighter (composite lightweight fighter). Notice that the F-104G is typical of supersonic aircraft in the 1960–1970 time period and displays a thrust pinch (marginal excess power) around Mach 1.0. This characteristic results in the curves of constant P_S being closed loops in some regions and discontinuous in others, as shown in Fig. 4.8a. Modern propulsion systems and high $(T/W)_{\text{TO}}$ eliminate this thrust pinch, and the curves of constant P_S are as shown in Fig. 4.9a. We shall see later that this difference results in two very different optimal trajectories for minimum time. For reference, Table 4.3 shows the data for the constant energy lines (h_e).

The time for an aircraft to move from one point of altitude and velocity to another point (i.e., to change its energy state h_e) along a trajectory A is given by the line integral

$$\Delta t = \int_{h_{e1}}^{h_{e2}} \frac{1}{P_S} dh_e \tag{4.7}$$

Similarly, the fuel burned along trajectory A from h_{e_1} to h_{e_2} is

$$\Delta W_f = \int_{h_{e1}}^{h_{e2}} \frac{1}{f_S} dh_e \tag{4.8}$$

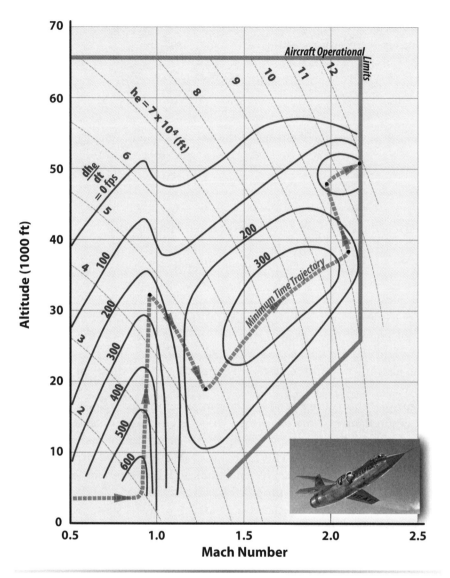

Figure 4.8a   Excess specific power, $P_S = dh_e/dt$, for the F-104G at $n = 1$, 18,000 lb, and maximum power [$W_{TO} = 26,000$ lb, $(W/S)_{TO} = 133$, one J79-GE-11A at $T_{SLS} = 15,800$ lb, $(T/W)_{TO} = 0.61$, maximum Mach = 2.2].

During a climb–acceleration, the aircraft load factor n varies. However, [14] concludes that the average value of the load factor during a climb–acceleration maneuver is such that the P_S and f_S in Eqs. (4.7) and (4.8), respectively, should be for 1 g.

From Eq. (4.7) it follows that the minimum time path for maximum energy change will be the one in which, at each value of specific energy, the aircraft flies at that combination of velocity and altitude that gives the maximum rate of change of specific energy in the direction of changing h_e [15]. Thus, the optimal path for maximum energy change is the locus of the points at which the lines of constant h_e are tangent to lines of constant 1-g P_S. This minimum time path is shown in Fig. 4.8a for the F-104 and in Fig. 4.9a for an advanced fighter aircraft of high $(T/W)_{TO}$. It must be pointed out that this method for finding optimal trajectories by establishing the path

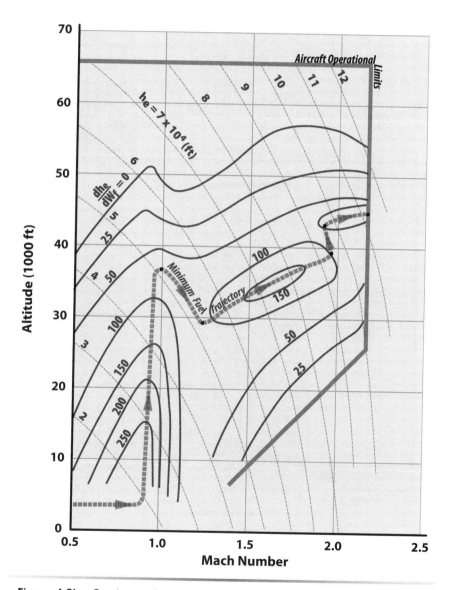

Figure 4.8b Contours of constant $f_s = dh_e/dW_f$ for the F-104G at n = 1 and maximum power.

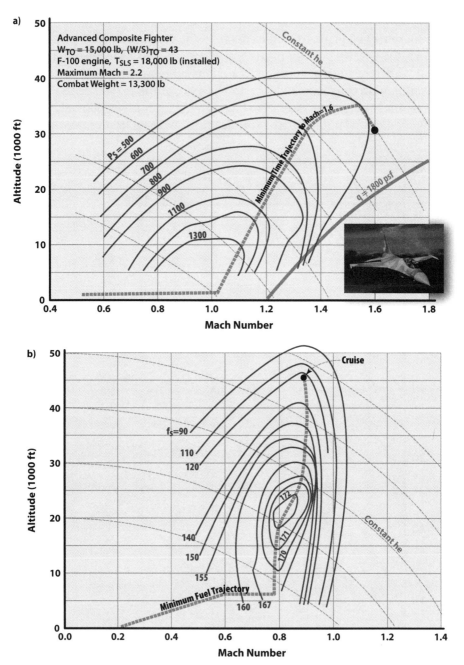

Figure 4.9   a) Excess specific power, $P_S = dh_e/dt$, for the LWF at n = 1 and maximum power (afterburner) with W/S = 38.1 psf, and **b)** contours of constant $f_s = dh_e/dW_f$ for the LWF at n = 1 and military power with W/S = 48.1 psf.

h_e (ft)

Mach	10,000	20,000	30,000	40,000	50,000	60,000	70,000	80,000	90,000	100,000	110,000	120,000
0	10,000	20,000	30,000	40,000	50,000	60,000	70,000	80,000	90,000	100,000	110,000	120,000
0.1	9,819	19,833	29,846	39,854	49,854	59,854	69,853	79,851	89,849	99,847	109,844	119,838
0.2	9,275	19,328	29,382	39,417	49,417	59,417	69,414	79,406	89,398	99,390	109,375	119,352
0.3	8,357	18,478	28,599	38,689	48,689	58,689	68,684	78,665	88,647	98,628	108,598	118,547
0.4	7,051	17,269	27,486	37,670	47,670	57,670	67,663	77,630	87,598	97,565	107,517	117,426
0.5	5,335	15,679	26,024	36,359	46,359	56,359	66,355	76,304	86,253	96,202	106,141	115,999
0.6	3,179	13,683	24,186	34,690	44,757	54,757	64,757	74,690	84,617	94,544	104,470	114,275
0.7	544	11,242	21,940	32,638	42,864	52,864	62,864	72,792	82,692	92,593	102,493	112,264
0.8		8,311	19,243	30,175	40,679	50,679	60,679	70,613	80,484	90,355	100,225	109,979
0.9		4,831	16,040	27,249	38,203	48,203	58,203	68,161	77,998	87,834	97,671	107,436
1.0		726	12,263	23,799	35,336	45,436	55,436	65,436	75,239	85,038	94,837	104,636
1.1			7,822	19,743	31,664	42,377	52,377	62,377	72,214	81,972	91,730	101,488
1.2			2,606	14,979	27,352	39,028	49,028	59,028	68,930	78,643	88,357	98,070
1.3				9,373	22,278	35,183	45,387	55,387	65,387	75,060	84,725	94,390
1.4				2,752	16,285	29,817	41,454	51,454	61,454	71,229	80,843	90,457
1.5					9,162	23,440	37,231	47,231	57,231	67,160	76,719	86,278
1.6					624	15,797	30,971	42,716	52,716	62,716	72,362	81,864
1.7						6,534	22,791	37,910	47,910	57,910	67,782	77,223
1.8							12,734	30,325	42,812	52,812	62,812	72,364
1.9							139	19,399	37,424	47,424	57,424	67,299
2.0								5,388	26,789	41,744	51,744	61,744
2.1									11,088	35,321	45,772	55,772
2.2										17,576	39,510	49,510
2.3											25,485	42,956
2.4												36,111

with the most favorable gradient does not take into account the possibility of increasing performance on one portion of the trajectory by decreasing performance along another segment. Allowing for this possibility might result in a net improvement in performance.

Fortunately, the exact solution and the energy-state approximation for optimal trajectories are quite close [15]. A very important point is that the exact solution is often not available to the designer. The difficulty, time, and expense involved in obtaining the exact solution may not warrant solving the problem. When this is the case, solution by the energy-state approximation is satisfactory and very valuable.

The minimum fuel trajectory is determined in the same manner as the minimum time trajectory, that is, the locus of the points where the constant h_e lines are tangent to lines of constant f_s. The engine power setting for minimum climb–acceleration fuel performance is usually something less than the maximum power condition, whereas, for the minimum time, the aircraft should be at maximum power. The advanced fighter depicted in Fig. 4.9 uses military (continuous) power for minimum climb–acceleration fuel performance but maximum power (afterburner operation) for minimum-time climb–acceleration.

The time to climb and fuel burned can be calculated on a hand calculator to surprising accuracy by graphically integrating Eqs. (4.7) and (4.8). A typical plot of $1/(dh_e/dW_f)$ versus h_e is shown in Fig. 4.10; the area under the

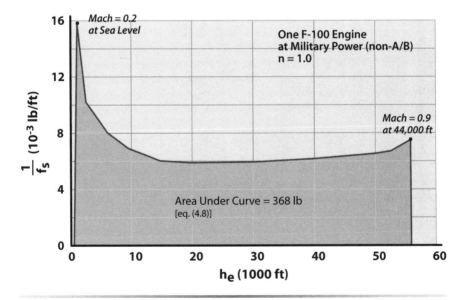

Figure 4.10   Plot of $1/f_s$ vs h_e for composite LWF along minimum fuel trajectory from Mach = 0.2 to Mach = 0.9 (see Fig. 4.9b).

Figure 4.11 F-16 with fuel tanks, sensors, and missiles.

curve between h_{e_1} and h_{e_2} is the fuel burned during a climb–acceleration from h_{e_1} to h_{e_2}. This operation is expressed as

$$\Delta W_f \approx \sum_{j=1}^{h} \left[\frac{1}{\mathrm{d}h_e / \mathrm{d}W_f} \left(h_{e_i} - h_{e_f} \right) \right]_j \tag{4.9}$$

where h_{e_i} and h_{e_f} are the initial and final values of specific energy for the jth interval. If small intervals are considered ($\Delta h_e \leq 1000$), the results using Eq. (4.9) compare quite well with computer results. The LWF ultimately became the F-16; Fig. 4.11 shows a modern F-16 with missiles, fuel tanks, and sensors.

The energy approximation method is not valid for trajectories along constant h_e contours as Eqs. (4.7) and (4.8) would yield zero time and zero fuel to move from one point of velocity and altitude to another.

References

[1] Fung, Y. C., *An Introduction to the Theory of Aeroelasticity*, Wiley, New York, 1955.
[2] Bisplinghoff, R. L., Ashley, H., and Hoffman, R. L., *Aeroelasticity*, Addison Wesley, Reading, MA, 1957.

[3] "Military Standardization Handbook, Metallic Materials and Elements for Aerospace Vehicle Structures," MIL-HDBK-5B, Superintendent of Documents, U.S. Government Printing Office, Washington, DC, 1 Sept. 1971.

[4] Henshaw, J. T., *Supersonic Engineering*, Wiley, New York, 1962.

[5] "Aircraft Engine Noise and Sonic Boom," *AGARD Conference Proceedings No. 42*, Clearinghouse for Federal Scientific and Technical Information (CFSTI), Springfield, VA, May 1969.

[6] "Noise Standards: Aircraft Type Certification," *Federal Aviation Regulations*, Vol. 1, Pt. 36, U.S. Government Printing Office, Washington, DC, Dec. 2009.

[7] Beheim, M. A., Anti, R. J., and Povolny, J. H., "Advanced Propulsion, Cleaner and Quieter," *Astronautics and Aeronautics*, Vol. 10, No. 8, Aug. 1972, p. 37.

[8] Jackson, R. P., "Putting All Our Noise Technology to Work," *Astronautics and Aeronautics*, Vol. 12, No. 1, Jan. 1974, p. 48.

[9] Johnson, F. S., "SST's, Ozone and Skin Cancer," *Astronautics and Aeronautics*, Vol. 11, No. 7, July 1973, p. 16.

[10] "Conference on Aircraft and the Environment," Society of Automotive Engineers, New York, 1971.

[11] Harned, M., "The Ramjet Power Plant," *Aero Digest*, July 1954, p. 38.

[12] Elsgolc, L. E., *Calculus of Variations*, Addison Wesley, Reading, MA, 1962.

[13] Bryson, A. E., and Denham, W. F., "A Steepest Ascent Method for Solving Optimum Programming Problems," *Journal of Applied Mathematics*, Vol. 29, No. 6, June 1962, pp. 4–21.

[14] Nash, R. C., "Analytical and Experimental Investigation of Optimum Fighter Aircraft Maneuvers," U.S. Air Force Flight Dynamics Laboratory Technical Rept. 73–145, Wright–Patterson AFB, OH, July 1974.

[15] Bryson, A. E., Desai, M. N., and Hoffman, W. C., "Energy-State Approximation in Performance Optimization of Supersonic Aircraft," *Journal of Aircraft*, Vol. 6, No. 6, Dec. 1969, p. 481.

Chapter 5

Preliminary Estimate of Takeoff Weight

- Predicting Empty-Weight Fraction
- Estimating Mission Segment Fuel Fraction
- Predicting Takeoff Weight
- Examples

LWF competition finalists General Dynamics YF-16 and Northrop YF-17 fly side-by-side in 1975. The YF-16 was a conventional metal design with an empty weight of 12,500 lb and a mission weight of 18,500 lb. If the YF-16 had been designed using advanced composites, the mission weight could have been reduced (see Example 5.1).

If you can find a path with no obstacles,
it probably doesn't lead anywhere.

Frank A. Clark

5.1 Introduction

The design process begins with the estimate of the *takeoff weight* W_{TO}. The W_{TO} is a very important design parameter as it sizes the vehicle. If this is the first iteration, that is, the first time through the design loop, the designer knows very little about the aircraft except for the requirements. The designer must decide what propulsion system will be used: this does not necessarily mean the size of the engine(s), but specific fuel consumption data are needed. In addition, the designer must select the material type (i.e., conventional metal structure, advanced composites, etc.).

Many people have difficulty getting started at this point because they must make some assumptions based upon very little information. Perhaps the best rule here is to "assume something even if it's wrong" because the process cannot really begin until an estimate of the W_{TO} is made.

It is important to remember that the conceptual phase is actually a looping or iterative process in which the assumptions are refined on subsequent passes through the design loop and the design converges to a feasible baseline point. Aircraft companies have the material in the text computerized, so that they can loop through the conceptual phase in a matter of seconds and quickly "home in" on a baseline point design.

We consider the *takeoff gross weight* (TOGW), or W_{TO}, to be

$$W_{TO} = W_{fuel} + W_{fixed} + W_{empty} \tag{5.1}$$

5.2 Fixed Weight

The *fixed weight* W_{fixed} (usually called *payload*) consists of the following:

1. Nonexpendable
   • Crew plus equipment
   • Sensors
2. Expendable
   • Bombs
   • Missiles
   • Cannon plus ammunition
   • Passengers
   • Baggage and/or cargo
   • Food and drink

5.3 Empty Weight

The *empty weight* (W_{empty}) of the aircraft includes structure, propulsion, subsystems, avionics, instruments, and so on.

Table 5.1 Summary of Empty-Weight Trend
Line Equations

Aircraft Type	Constant	*XX*
Fighter		
Air-to-air or developmental	1.2	0.947
Multipurpose	0.911	0.947
Air-to-ground	0.774	0.947
Bomber and transport	0.911	0.947
Light general aviation	0.911	0.947
Composite sailplane	0.911	0.947
Military jet trainer	0.747	0.993
High-altitude ISR	0.75	0.947
Unmanned air vehicles		
Propeller, endurance > 12 h	1.66	0.815
Propeller, endurance < 12 h	2.18	0.815
Turbine ISR	2.78	0.815
Turbine maneuver UCAV	3.53	0.815
Air-launch cruise missiles and targets	1.78	0.815

The designer will soon discover that this preliminary estimation of the aircraft empty weight is the weakest part of the conceptual design analysis and it has tremendous leverage on the aircraft takeoff weight. It is almost impossible to estimate the empty weight of something that has not been built (usually with new subsystems and structural materials) with any degree of accuracy. However, it is important to press on or the aircraft will never be designed.

The empty weight is determined at this point by using historical data and trends. Appendix I contains the historical weight trends that we will use in our preliminary estimate of takeoff weight. The empty weight data in Appendix I are presented by class of aircraft: fighters (broken out by type): general aviation; bombers and transports; jet trainers; intelligence, surveillance, and reconnaissance (ISR) aircraft; unmanned (combat) aerial vehicles (UAVs and UCAVs); air-launch cruise missiles; aerial targets; and sailplanes. Each aircraft has a trend line given by an equation of the type $W_{empty} = (\text{constant})(W_{TO})^{xx}$, with the data points making up the plotted trend line. A summary of the Appendix I weight-trend charts is given in Table 5.1. At this point Appendix I should be studied to appreciate the importance of having good historical weights data. Appendix I has weight summaries of many existing aircraft so that it can be seen how the component weights roll up into the empty weight. Later in the design process, as more information is available on the aircraft, the empty weight will be determined by estimating the component weights using weight-estimating relationships (see Chapter 20).

5.4 Fuel Weight

The mission is divided into the eight phases shown in Fig. 5.1. The approach will be to determine the *fuel fraction* (ratio of the final to initial weight, $W_{n+1}/W_n < 1$, where n is the number of the phase) for each phase and then multiply them together to get the fuel fraction for the entire mission. If a specific mission does not have one or more of the phases shown in Fig. 5.1, the fuel fraction for the missing phase is set equal to 1. Also, the phases can be put in any order (e.g., phase 7 ahead of phase 6) and repeated as needed to fit the mission profile.

Phase 1. Engine Start and Takeoff

In this phase, the fuel fraction for the start of the engines, taxi, takeoff, and climb out is determined. A reasonable first estimate is to assume that this phase burns about 2.5% to 3% of the takeoff weight in fuel. In other words,

$$\frac{W_2}{W_1} = 0.97 - 0.975$$

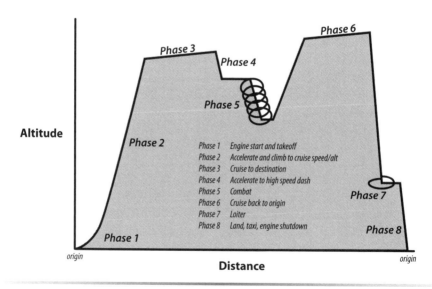

Figure 5.1 Typical military mission profile.

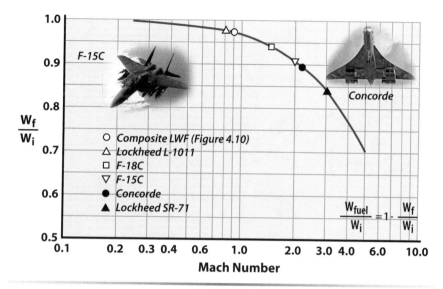

Figure 5.2 Weight fractions for climb–acceleration phase.

Phase 2. Climb and Accelerate to Cruise Conditions

Fuel is used during this phase to increase both the kinetic and the potential energies of the aircraft. If the aircraft is a low-performance aircraft, cruising at less than Mach = 0.4, the fuel burned during this phase is not significant for the first iteration. If the aircraft cruises at speeds greater than Mach = 0.4 and/or altitudes greater than 10,000 ft, then the fuel burned in climbing and accelerating to cruise altitude and Mach number is significant. Figure 5.2 shows the ratio of W_3/W_2 for typical high-performance aircraft. Because propulsion and climb–acceleration trajectories are similar for a wide variety of aircraft, it is possible to collapse the data onto a single curve as shown in Fig. 5.2. The use of Fig. 5.2 is recommended during the first iteration. Subsequent iterations can use a more refined method such as the energy-state approximation discussed in Chapter 4.

Phase 3. Cruise Out

For this phase the designer needs to estimate the cruise speed and altitude, the configuration aspect ratio and wing sweep, and the cruise fuel consumption (for turbojet or turbofan aircraft or for propeller aircraft).

The fuel fraction for phase 3 is determined using the Breguet range expressions presented in Chapter 3. The expression for a turbojet or turbofan aircraft is

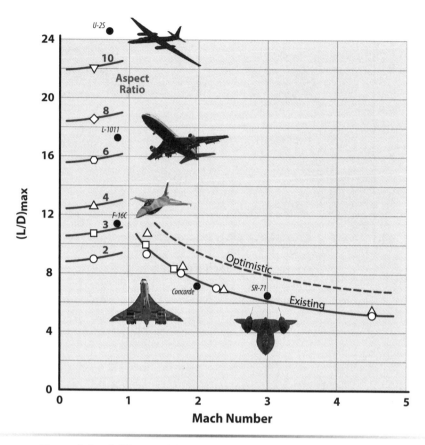

Figure 5.3 $(L/D)_{max}$ vs Mach number for typical cruise aircraft.

$$R = \frac{V}{C}\frac{L}{D}\ln\left[\frac{W_3}{W_4}\right] \qquad (5.2)$$

and for a propeller aircraft (piston or turboshaft) is

$$R = \frac{\eta}{C}\frac{L}{D}\ln\left[\frac{W_3}{W_4}\right] \qquad (5.3)$$

Equation (5.2) or (5.3) is solved for the weight ratio W_3/W_4. For maximum range for a turbojet or turbofan aircraft, the cruise L/D will be close to (but less than) the value for $(L/D)_{max}$. For a constant altitude cruise the L/D would be 86.6% of $(L/D)_{max}$ and for a cruise climb the L/D would be 94% of $(L/D)_{max}$ from Chapter 3, Table 3.2. Values for the *thrust specific fuel consumption* (TSFC) should be determined from appropriate engine data (see Chapter 14 or Appendix J). The subsonic turbojet or turbofan will cruise most efficiently above 30,000 ft and at a partial power setting of 75%

to 100% of normal rated thrust. If there is enough information on the aircraft, the designer should develop a chart similar to Fig. 3.8 to determine the optimum cruise condition and maximum value for the range factor VL/D.

For maximum range the propeller aircraft cruise L/D will be $(L/D)_{max}$. The propulsive efficiency η can be obtained from Chapter 18 or assumed to be 0.85 for preliminary sizing. Values of the *brake specific fuel consumption* (BSFC) should be determined from appropriate engine data. The value for η/C is fairly constant such that maximum range occurs at the velocity for maximum L/D. Thus, the propeller aircraft would prefer to cruise at the velocity given by the expression (from Chapter 3)

$$V_{maxL/D} = \left(\frac{W}{S}\frac{2}{\rho}\right)^{1/2}\left(\frac{K}{C_{D0}}\right)^{1/4} \tag{3.11}$$

The $(L/D)_{max}$ can be estimated several different ways. The first way is to select a wing aspect ratio and obtain an historical value for $(L/D)_{max}$ from Fig. 5.3. The subsonic cruise L/D can also be estimated by using the expression for $(L/D)_{max}$ for a symmetrical aircraft:

$$\left(L/D\right)_{max} = \frac{1}{2\sqrt{C_{D0}K}} \tag{5.4}$$

A value for the subsonic cruise C_{D0} can be estimated from Table 5.2 or by looking at data for similar aircraft in Appendix G. An estimate for the subsonic *drag-due-to-lift factor K* can be obtained from the expression

Table 5.2 Representative Values for Subsonic C_{D0}

Aircraft Type	Subsonic C_{D0}
High-subsonic jet transport	0.014–0.02
Supersonic fighter aircraft	0.014–0.022
Blended wing–body (tailless) jet aircraft	0.008–0.014
Large turboprop aircraft	0.018–0.024
Low-altitude subsonic cruise missile (high W/S)	0.03–0.04
Small single-engine propeller aircraft	
Retractable gear	0.022–0.030
Fixed gear	0.026–0.04
Agricultural aircraft	
With spray system	0.07–0.08
Without spray system	0.06
High-performance sailplane	0.006–0.01

$$K = \frac{1}{\pi \mathrm{AR} e} \qquad (5.5)$$

where e is the wing efficiency factor obtained from Fig. G.9. During subsequent iterations, the designer can reduce the cruise range by the distance covered during the climb–acceleration to cruise (phase 2) if this distance is known. Also, the range can be broken into smaller segments (as discussed in Chapter 3) for more refined range segment fuel fractions.

Phase 4. Acceleration to High Speed

The weight fraction for acceleration from a cruise condition to a high-speed dash can be estimated from Fig. 5.2. For example, assume we want the weight fraction for an acceleration from a cruise at Mach = 0.9 to a dash at Mach = 2.5. From Fig. 5.2 we have

$$\frac{W_f}{W_i}(M = 0.1 - 0.9) = 0.975, \quad \frac{W_f}{W_i}(M = 0.1 - 2.5) = 0.86$$

Thus,

$$\frac{W_5}{W_4} = \frac{W_f}{W_i}(M = 0.9 - 2.5) = \frac{0.86}{0.975} = 0.882$$

Phase 5. Combat

The mission requirements for the combat phase usually specify one (or both) of the following:

1. Specified number of minutes at maximum power and a particular Mach number and altitude
2. Specified number of turns at maximum power and a particular load factor, Mach number, and altitude

Both of these requirements translate into a loiter time condition at maximum power. [From Chapter 3, equation 3.32,

$$\text{Turn rate} = \frac{g\sqrt{n^2 - 1}}{V}$$

(in radians per second), where n = load factor and g = 32.174 ft/s^2; time = (number of turns) (360 deg)/(turn rate).]

Then, the combat fuel is determined using

$$\text{Combat fuel} = (\text{TSFC})(\text{maximum thrust})(\text{time})$$

where TSFC is the maximum-power thrust specific fuel consumption. The mission requirements only size the fuel that must be available for combat. The performance level of an aircraft with this given amount of fuel is determined much later in the design cycle.

> Generally, equations will use C to represent either "thrust specific fuel consumption" (TSFC) for turbine engines or "brake specific fuel consumption" (BSFC) for propeller engines. Context is usually sufficient to determine which term C represents.

Phase 6. Cruise Back

The fuel fraction for this phase may be treated in the same way as the cruise out, phase 3. The return cruise altitude will probably be higher than cruise-out altitude because the aircraft is lighter. Also, if the aircraft dropped external weapons or tanks, the return cruise L/D will be slightly higher than the cruise-out case.

Phase 7. Loiter

Once a loiter is established, the fuel-weight fraction may be found using the endurance equation of Chapter 3 for jet or propeller aircraft. The *endurance equation* for jet aircraft is given by Eq. (3.17):

$$E = \frac{1}{C}\frac{L}{D}\ln\left[\frac{W_7}{W_8}\right] \tag{5.6}$$

where E is the endurance (or loiter time) in hours and C is the thrust specific fuel consumption in pounds of fuel per hour per pound of thrust [(lb of fuel)/(lb of thrust·h)]. The maximum endurance will occur at the flight condition where $(L/D)/C$ is a maximum. This is usually less than but near the velocity for $(L/D)_{\text{max}}$ so the designer can use loiter $L/D \sim (L/D)_{\text{max}}$. If loiter flight conditions are specified it is very probable that the aircraft will not be operating at $(L/D)_{\text{max}}$ and/or minimum C. Thus, conservative estimates for L/D and C should be made. The endurance equation for propeller aircraft is given by Eq. (3.22) in Chapter 3. This equation calls for a fair amount of detail to be available on the aircraft. If this is the first iteration, this detail may not be known and a more simple form of Eq. (3.22) is preferred. Thus, the following expression for propeller aircraft endurance is recommended:

$$E = \frac{\eta}{C}\frac{L}{D}\frac{1}{V}\ln\left[\frac{W_3}{W_4}\right] \tag{5.7}$$

Again, the designer estimates the parameters with the highest accuracy possible.

Phase 8. Reserve and Trapped Fuel

Mission requirements will often specify a reserve fuel. This fuel, normally 5% of the fuel required for the mission, is strictly for reserve and cannot be used for any part of the mission.

There is usually a small fraction of the oil and fuel that is trapped in lines and pumps and is really not available as fuel or lubricant. This trapped oil and fuel usually amounts to about 1% of the mission fuel.

5.5 Determining W_{TO}

The W_{fuel} is determined as a fraction of the takeoff weight, that is,

$$\left[\frac{W_{final}}{W_{TO}}\right] = \left[\frac{W_8}{W_1}\right] = \left[\frac{W_2}{W_1}\right]\left[\frac{W_3}{W_2}\right]\cdots\left[\frac{W_8}{W_7}\right] \quad (5.8)$$

$$W_{fuel} = \left(1 + reserve + trapped\right)\left(1 - \frac{W_8}{W_1}\right)W_{TO} \quad (5.9)$$

If there are any weight discontinuities (such as bombs dropped or missiles fired) and/or fuel phases that cannot be expressed as a weight fraction (such as combat fuel) during the mission, then Eq. (5.9) must be solved in segments. Example 5.1 will demonstrate this situation. The W_{fixed} is determined by the requirements of the mission and will be a fixed number independent of W_{TO}. A value of W_{TO} is assumed and the available empty weight determined using the expression

$$\left(W_{empty}\right)_A = W_{TO} - W_{fuel} - W_{fixed}$$

The required empty weight, $(W_{empty})_R$, is determined from the empty-weight figures in Appendix I with appropriate adjustments for advanced materials and/or concepts. An iteration on W_{TO} ensues until $(W_{empty})_A$ and $(W_{empty})_R$ are close.

Usually the iteration continues until

$$\left|\left(W_{empty}\right)_A - \left(W_{empty}\right)_R\right| \leq 0.01\left(W_{empty}\right)_R$$

This iteration for W_{TO} is illustrated in Fig. 5.4, which shows the empty-weight fraction and fixed-weight fraction as monotonic functions of W_{TO}. The fuel-weight fraction is independent of W_{TO} as depicted in Fig. 5.4

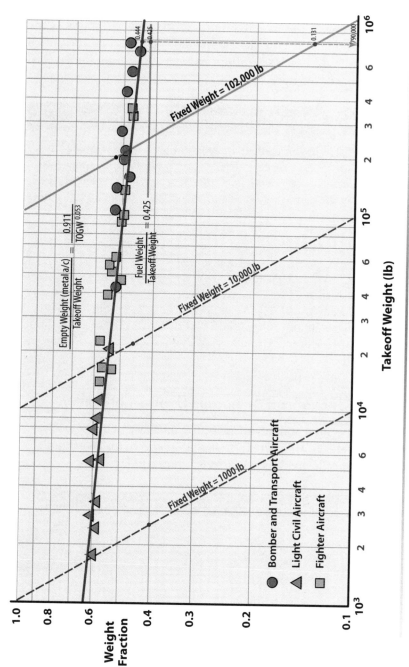

Figure 5.4 Empty-weight fraction for current aircraft (fighters, bombers, and light aircraft).

(0.425, which is the fuel-weight fraction from Example 5.2). The solution is found where the three fractions W_{fuel}, W_{empty}, and W_{fixed} all sum to 1.0.

The selection of an initial estimate for W_{TO} must include the gross assumptions inherent in the first iteration. Subsequent iterations refine these assumptions and converge on a refined W_{TO}. However, this initial estimate of W_{TO} should be close to the final value, because a great deal of work will be done based upon this initial W_{TO}.

USAF Invites Competition for LWF

As a result of the pressure brought about by John Boyd's "Fighter Mafia," the USAF launched an LWF program in 1972. The LWF was to be a low-cost F-15 20,000-lb class day fighter with good turn rate, acceleration and range, and optimized for air combat at speeds of Mach 0.6–1.6 and altitudes of 30,000–40,000 ft.

Competition finalists were General Dynamics' YF-16 (single F100-PW-100 afterburning turbofan at 23,000 lb TSLS) and Northrop's YF-17 (two YJ101-GE-100 turbofans at 14,400 lb TSLS each). The YF-16 was selected: the USAF liked its maneuverability, and the LCC for a single-engine fighter was less than that of a twin-engine craft. General Dynamics/Lockheed Martin has produced over 4500 F-16 Fighting Falcons. Meanwhile, the Navy liked the two-engine performance of the YF-17; it became the basis for the carrier capable F-18 Hornet. McDonnell/Boeing has produced over 1500 F-18s.

Example 5.1   Advanced Composite Lightweight Fighter

Determine the initial weight estimate for an advanced composite lightweight fighter (LWF). The LWF is a good choice for an example as its mission profile involves every phase shown in Fig. 5.1.

USAF LWF Mission Requirements	
Purpose	Lightweight, low-cost, air-superiority, day fighter
Radius (internal fuel only)	250 n mile
Maximum speed	Above Mach 1.6
Cruise condition	Not specified
Combat fuel sizing	One acceleration from Mach = 0.9 to Mach = 1.6 at 30,000 ft
	Four minutes of maximum afterburner (A/B) operation at Mach = 0.9 30,000 ft
Loiter	Sea level at Mach = 0.35 for 20 min
Reserve	5% of fuel
Crew	One

Weapons	Two AIM 9 missiles
	One 20-mm M-61 cannon
Engine	One F100 (see Chapter 14) or two YJ-101 afterburning turbofan engines
Structures	Limit load factor of 7.33 at 80% internal fuel with missiles and full ammunition load
	Placard structural and flutter limit Mach = 1.1 at sea level
	Advanced composites used throughout the vehicle where applicable
LWF Fixed Weights	
Pilot plus gear	200 lb
2 AIM 9 missiles plus racks	472 lb
M-61 cannon plus accessories	485 lb
560 rounds of 20-mm shells	320 lb
Total	1477 lb
W_{fixed}	1500 lb

The LWF fuel fraction followed the mission profile shown in Fig. 5.1.

Phase 1. Takeoff and Climb Out

Assume fuel burned during this phase to be 2.5% of W_{TO},

$$\frac{W_2}{W_1} = \frac{W_2}{W_{TO}} = 0.975$$

Phase 2. Climb–Acceleration to Cruise

Assume cruise is near the tropopause at Mach = 0.9. From Fig. 5.2, the following fuel fraction is obtained:

$$\frac{W_3}{W_2} = 0.975$$

Phase 3. Cruise Out 250 n Mile

Assume cruise is between 36,000 ft and 45,000 ft at Mach = 0.9. (This can be checked later when aerodynamic data are available and this weight fraction refined during subsequent iterations.) From the F100 engine data in Chapter 14 (Fig. 14.7 d and e), it may be observed that a reasonable estimate for the TSFC is 0.93. Because our aircraft is a fighter with the requirement to go supersonic, our fixed wing (another assumption!) aspect ratio will be low. Assuming an AR = 3 gives an $(L/D)_{max} \sim 10$ from Fig. 5.3. Thus, a realistic cruise L/D is about 9. This estimate for cruise L/D can be checked by using Eqs. (5.4) and (5.5). From Table 5.2 or Fig. G.6 we obtain an estimate of $C_{D_0} = 0.016$. Using AR = 3 and $e = 0.8$ (from Fig. G.9) in Eq. (5.5) gives $K = 0.1326$. Finally Eq. (5.4) gives $(L/D)_{max} = 10.8$ (checks with the F-16 in Fig. G.3). Thus, a cruise L/D of 9 is conservative but realistic.

Using the Breguet range equation gives our weight fraction as

$$\frac{W_i}{W_f} = \frac{W_3}{W_4} = \exp\left[\frac{\text{Radius} \times C}{V \times L/D}\right]$$

$$= \exp\left[\frac{(250)(0.93)}{(516.24)(9)}\right] = 1.05$$

Phase 4. Combat Acceleration at 30,000 ft

The fuel fraction for the acceleration from Mach = 0.9 to Mach = 1.6 can be determined from Fig. 5.2 or using an energy-state approximation discussed in Section 4.8. Using Fig. 5.2 gives

$$\left(\frac{W_f}{W_i}\right)_{\text{to } M=0.9} = 0.975, \quad \left(\frac{W_f}{W_i}\right)_{\text{to } M=1.6} = 0.952$$

Thus,

$$\frac{W_5}{W_4} = \frac{0.952}{0.975} = 0.976$$

Phase 5. Combat Turns at Maximum Power

The fuel burned during the four minutes of combat at 30,000 ft cannot be expressed as a fuel fraction. From the F100 engine data in Chapter 14 (Fig. 14.7b), the thrust and TSFC for Mach 0.9 and 30,000 ft are

$$\text{Thrust} = 12,000 \text{ lb} \quad \text{TSFC} = 2.17$$

Thus, the weight of fuel burned during 4 minutes at maximum power is given by

$$\text{Combat } W_{\text{fuel}} = (\text{Thrust})(\text{TSFC})(\text{Time}) = 1740 \text{ lb}$$

Phase 6. Cruise Back 250 n mile

Assume that the cruise back is similar to the cruise out (this is not really true because the aircraft will be lighter and will cruise at a higher altitude) and use the same fuel-weight fraction:

$$\frac{W_6}{W_7} = 1.05$$

Notice that the fuel expended for climb to cruise is not taken into account nor is the distance traveled during this climb. This is an approximation but is still within the noise level of this initial estimate.

Phase 7. Loiter at Sea Level, Mach = 0.35

The 20-minute loiter at sea level uses the Breguet endurance equation. The F100 partial power data yield an estimated TSFC = 0.84. Assuming a loiter $L/D = 8.7$ gives a fuel fraction of

$$\frac{W_f}{W_i} = \exp\left[\frac{\text{Endurance} \times C}{L/D}\right] = \exp\left[\frac{(20)(60)(0.84)}{(8.7)(3600)}\right] = 1.033$$

Iterate for W_{TO}

The determination of W_{TO} is an iterative process whereby the available empty weight obtained from the fuel-weight fractions is balanced by the required empty weight obtained from Fig. I.1 using the upper trend line.

Assume

$$W_{TO} = W_1 = 13,000 \text{ lb}$$

The weight at beginning of combat, W_5, is

$$W_5 = \left(\frac{W_2}{W_1} \frac{W_3}{W_2} \frac{W_4}{W_3} \frac{W_5}{W_4}\right) W_1 = (0.8839)(13,000) = 11,491 \text{ lb}$$

Subtract the following:

- Missiles, 348 lb
- 20-mm ammo, 320 lb
- 4 minutes of combat fuel, 1740 lb
- Total = 2408 lb

from W_5 to get the weight at beginning of cruise back, W_6.

$$W_6 = 9083 \text{ lb}$$

The weight at landing, W_8, is

$$W_8 = \left(\frac{W_7}{W_6} \frac{W_8}{W_7} \right) W_6 = 8379 \text{ lb}$$

The fuel weight required for the mission is

$$\left(W_{\text{fuel}} \right)_{\text{mission}} = W_{\text{TO}} - W_8 - \text{missiles} - \text{ammo} = 3953 \text{ lb}$$

and the total fuel is the fuel required plus the reserve fuel (5%) and trapped fuel (1%). Thus,

$$W_{\text{fuel}} = 1.06 \left(W_{\text{fuel}} \right)_{\text{mission}} = 4190 \text{ lb}$$

Now the available empty weight is

$$\left(W_{\text{empty}} \right)_A = W_{\text{TO}} - W_{\text{fuel}} - W_{\text{fixed}} = 13{,}000 - 4190 - 1500 = 7309 \text{ lb}$$

Because advanced composites are being used, it is assumed that the conventional metal empty weights from Appendix I may be reduced by 16% (this assumption is very important and, right or wrong, it needs to be documented with the rationale used). The required empty weight is obtained from Fig. I.1 using the trend line for high-$(T/W)_{\text{TO}}$ and low-$(W/S)_{\text{TO}}$ fighter aircraft [i.e., $W_{\text{empty}} = 1.2(W_{\text{TO}})^{0.947}$] reduced by 16%. Thus,

$$\left(W_{\text{empty}} \right)_R = 7932 \text{ lb}$$

The conclusion is that a $W_{\text{TO}} = 13{,}000$ lb solution cannot accomplish the required mission because the available empty weight is 623 lb less than required. The next iterations would increase the W_{TO} until the difference between the available and required empty weights is within a specified limit such as $0.01(W_{\text{empty}})_R$. This iteration process is shown in Fig. 5.5. From Fig. 5.5, we observe that the initial estimate of W_{TO} to perform the LWF mission is 15,400 lb. (Note: further refinements decreased the W_{TO} to 15,000 lb, which is the value used in Table 3.1.)

Mission Trades—Influence of Payload and Radius on W_{TO}

It is useful at this point to ask some "What if" questions about the mission requirements. What would the impact be on W_{TO} if the

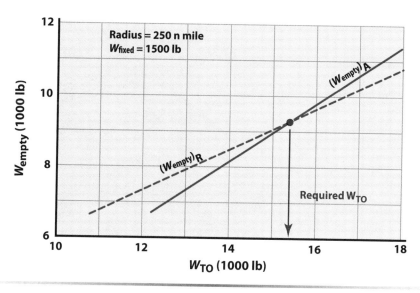

Figure 5.5 Determination of required W_{TO} for composite LWF.

payload weight or mission radius were changed? The answers to these types of questions are called *mission trades*, one of three very important trades studies that should be done during conceptual design (the other two are design trades and technology trades; see Fig. 1.15).

The iteration scheme just discussed can now be used to examine the influence of payload and radius variations on the required W_{TO}. First, consider increasing the payload (i.e., fixed weight) by 500 and 1000 lb. This could be expendable ordnance, increased avionics gear, or merely a fixed-weight increase in components or structure. The influence on W_{TO} due to changing the fixed weight (or payload) is shown in Fig. 5.6. Notice that an increase in fixed weight of 500 lb results in a W_{TO} increase of 1900 lb, or a weight sensitivity ratio of 3.8/1.0. This is because the extra 500 lb requires additional fuel to carry it out and back (or carry it out and drop it), and this extra fuel requires structure to contain it, which requires more fuel, and so on. Thus, the designer, especially the structural designer, must be aware that an extra pound of component or structure is magnified to obtain the new takeoff weight.

The weight sensitivity ratio is often called the *aircraft growth factor* because it reflects how the aircraft takeoff weight increases as the payload increases. The growth factor is defined as $\Delta(\text{TOGW})/\Delta(\text{payload weight})$ and depends primarily on the character of the mission and the payload fraction (the larger the payload fraction, the larger the growth factor). For example, a growth factor of 3.5 to 4.0 is typical for fighter aircraft with payload fractions of 10% to 15%. For

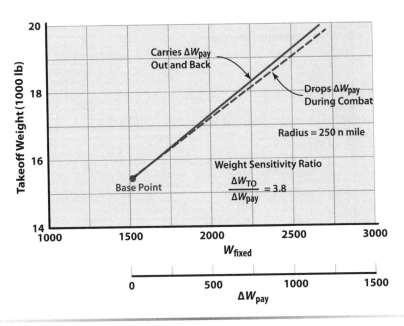

Figure 5.6   Influence of W_{fixed} on W_{TO} for composite LWF.

an air-launched cruise missile the payload fraction is typically 15% or more and the growth factor is approximately 4. For long endurance ISR aircraft (such as the RQ-4A Global Hawk) the payload fraction is less than 10% and the growth factor is approximately 2.5. The influence on W_{TO} of changing radius can be examined by determining new values for W_3/W_4 and W_6/W_7 and performing the iteration process. The result of changing mission radius is shown in Fig. 5.7. An additional 50 n mile in radius will cost an extra 1000 lb in takeoff weight.

5.6 Range- or Payload-Dominated Vehicles

Aircraft that have long range (or endurance) requirements and/or large payload requirements will be large in size. This is because a long range requirement (over 3000 n mile, for example) results in a large fuel fraction and a large payload (over 50,000 lb, for example) directly results in a large fixed-weight fraction. These two large fractions result in a large W_{TO}.

As W_{TO} increases, the empty-weight fraction decreases, as shown in Fig. 5.4. This behavior is welcome because without it long-range, high-payload missions would not have a solution. Empty-weight fraction improves for larger aircraft for two reasons. First, structural efficiency of larger vehicles is better (lower structural-weight fraction); second, many systems and components (such as avionics, hydraulic, actuators, and

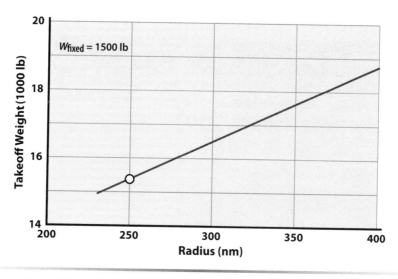

Figure 5.7 Influence of mission radius on W_{TO} for composite LWF.

control units) do not scale with weight and thus become a smaller fraction of the total aircraft weight. Figure 5.4 shows the empty-weight fraction decreasing from an average value 0.63 at $W = 1000$ lb to about 0.44 at 10^6 lb takeoff weight for conventional metal structures. Advanced materials and structural concepts provide an even greater decrease.

Example 5.2 Extended Range or Payload Aircraft

Consider the following:

Mission Requirements	
Purpose	Long-range transport aircraft
Range (unrefueled)	6000 n mile
Payload	100,000 lb
Crew	Nine
Cruise	Subsonic
Maximum speed	Subsonic
Field length	Not specified
Service ceiling	Not specified
Reserve	5% of mission fuel
Assume 2000 lb for crew and equipment	gives a $W_{fixed} = 102,000$ 1b
Assume high-bypass turbofan	gives cruise specific fuel consumption of 0.6 at Mach = 0.8 and 36,000 ft (see Tables 5.3, 14.8, and J.1)
Assume AR = 7	gives approximate $(L/D)_{max}$ from Fig. 5.3 of about 18 and a cruise L/D of 17

Table 5.3 Information on 747 and C-5 Aircraft

Aircraft	W_{TO} (lb)	Range (n mile)	Payload (lb)	Engine Type	Cruise TSFC (Table J.1)
747–200	775,000	3744	200,000	JT9D-7	0.62
C-5A	728,000	3050	220,000	TF39-GE-1	0.582
C-5A	728,000	5500	112,600	TF39-GE-1	0.582

Phase 1

Assume $W_2/W_1 = 0.97$

Phase 2. Weight Fraction

From Fig. 5.2 the weight fraction for climb–acceleration to Mach = 0.8 is

$$W_3/W_2 = 0.978$$

Phase 3. Cruise Fuel Fraction

From the preceding assumptions, the cruise fuel fraction is

$$\frac{W_3}{W_4} = \exp\left[\frac{(6000)(0.6)}{(459)(17)}\right] = 1.583$$

Phases 4 through 7

These phases are not appropriate for this aircraft; thus

$$\frac{W_5}{W_4} = \frac{W_6}{W_5} = \frac{W_7}{W_6} = \frac{W_8}{W_7} = 1.0$$

Fuel Fraction

The ratio of landing weight to takeoff weight is

$$\frac{W_8}{W_1} = (0.97)(0.978)\frac{1}{1.583} = 0.599$$

The resulting fuel fraction, considering reserve and trapped fuel, is

$$\frac{W_{fuel}}{W_{TO}} = 1.06\left[1 - \frac{W_8}{W_1}\right] = 0.425$$

Iteration for W_{TO}

The iteration for W_{TO} proceeds as described in Section 5.5: Assume

$$W_{\text{TO}} = 790,000 \text{ lb}$$

From Fig. 5.4 the empty-weight fraction is approximately 0.444. Thus,

$$\left(W_{\text{empty}}\right)_A = 790,000 \left(1 - 0.425\right) - 102,000 = 352,250 \text{ lb}$$

$$\left(W_{\text{empty}}\right)_R = \left(0.444\right)\left(790,000\right) = 350,760 \text{ lb}$$

The difference is within 1% so that 790,000 lb is a good estimate for the aircraft takeoff weight. Thus, a takeoff weight of 790,000 lb is a realistic design solution for the requirement of 6000 n mile range with 102,000 lb of payload. This range can be traded for additional payload as shown in Fig. 5.8. There is usually some upper payload limit resulting from fuselage volume constraints and/or floor load bearing limits. A payload limit of 250,000 lb is assumed for this example. The results of this example agree well with published weight and performance for the Boeing 747-200 Freighter and the Lockheed C-5A Galaxy. Information on these two jumbo jet aircraft is given in Table 5.3. Figure 5.8 shows good agreement between Example 5.2 and C-5A Ops Planning data.

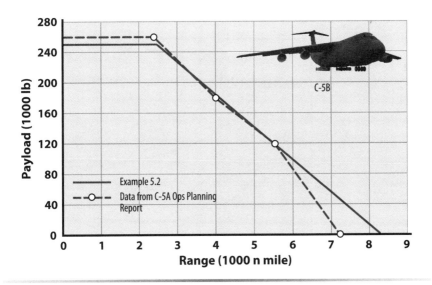

Figure 5.8   Variation of payload with range for large transport aircraft.

Example 5.3 Low-Altitude, Subsonic Cruise Missile

The driving requirements for cruise missiles are long range and high penetration survivability. To minimize the probability of being detected by enemy defenses, the cruise missile will be designed for stealth (low observables) and to cruise at low altitude so that it can take advantage of the decreased horizon range, terrain masking, and background clutter necessary to defeat enemy radar and other threat sensors. Flight at high altitude does not have these advantages and the detection problem is relatively simple for enemy defenses. Low altitude is a poor flight condition for efficient long-range cruise as turbofan engines operate best (low TSFC) at altitudes above 30,000 ft.

Cruise Missile Mission Requirements	
Range	1900 n mile
Payload	W-80 nuclear warhead, 300 lb
Speed	Mach 0.7 at 200 ft above ground level (AGL)
Engine	Williams F 107 turbofan (Fig. 5.9)
Profile	Terrain follow at Mach 0.7 and 200 ft AGL with limit load factor $n = +1.8, -0.5$ g
TSFC	1.15 (average installed value bet. cruise and max. thrust from Fig. 5.9)
C_{D_0}	0.035 (from Table 5.2)
Wing AR	6 with zero sweep
K	0.059 for an $e = 0.9$ (from Fig. G.9)
$(L/D)_{max}$	$1/(4C_{D_0}K)^{1/2} = 11$

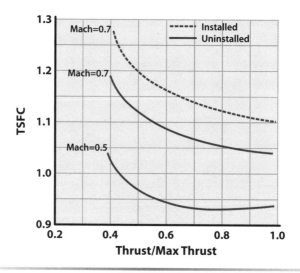

Figure 5.9 F-107 turbofan engine at sea level.

From Table 3.2 for a constant-altitude cruise the C_L for maximum range is

$$C_L = \left(C_{D_0} / 3K\right)^{1/2} = 0.44$$

$$\text{Cruise } L/D = 0.866 \left(L/D\right)_{max} = 9.5$$

In determining the TOGW, the only factor to be considered is the fuel fraction for the cruise out (phase 3), because all other fuel-weight fractions are equal to 1. The phase 3 fuel-weight fraction is determined using the Breguet range equation (5.2),

$$W_3/W_4 = \exp[(1900)(1.15)/(463)(9.5)] = \exp(0.497) = 1.644 \qquad (5.10)$$

Assume a launch weight $W_L = W_3 = 1800$ lb:

$$W_{fuel} = W_3 - (W_4/W_3)W_3 = 1800 - 1095 = 705 \text{ lb}$$

$$(W_{empty})_A = W_3 - W_{payload} - W_{fuel} = 795 \text{ lb}$$

Using Fig. I.7,

$$(W_{empty})_R = 801 \text{ lb}$$

Thus, an 1800-lb cruise missile can accomplish the mission!

At this point our design should be given a sanity check by comparing it with other existing nuclear, low-altitude, subsonic cruise missiles. The Convair/Raytheon AGM-109A Tomahawk has the following characteristics:

Range	1900 n mile
Engine	Williams F107 turbofan
Payload	W-80 warhead at 300 lb
Launch weight	2860 lb
Fuel weight	1200 lb
Empty weight	1350 lb
Wing AR	6 with zero sweep
Wing area	12 ft²
C_{D_0}	0.035
K	0.059

When the 1800 lb estimated launch weight is compared to the actual Tomahawk missile weight of 2860 lb from the preceding list, the question is: Why? Actually, the design analysis has a fundamental flaw in it that will be discussed in more detail in Chapter 6.

5.7 High Altitude Atmospheric Research Platform

In the late 1980s, the world was concerned about the depletion of the ozone layer over the South Pole. The details of the depletion were not understood and there was no evidence that the action could be reversed. In situ measurements needed to be made at altitudes up to 120,000 feet. The U-2S could take measurements up to 70,000 feet and the Condor up to 66,000 feet. In 1990, the Lockheed Skunk Works was commissioned by NASA to develop a manned aircraft that could make atmospheric measurements at 100,000 feet worldwide. That program was referred to as the High Altitude Atmospheric Research Platform (HAARP) and will be used here as an unmanned-aircraft example [1].

Example 5.4   HAARP Requirements

The HAARP mission objective was to conduct atmospheric research over the South Pole using an unmanned aircraft. The mission profile is shown in Fig. 5.10. The other requirements were as follows:

Range	6000 n mile with 5000 n mile at 100,000 ft and Mach = 0.6
Payload	2500 lb at 500 watts
Propulsion	Turbocharged piston engine(s)
Aerodynamics	Efficient wing design at flight $Re < 1.0 \times 10^6$
Structure	Metal [for sizing use $W_{empty} = 0.911 \, (TOGW)^{0.947}$]
BSFC	0.42 lb of fuel per hour/hp
Propeller efficiency @100,000 ft/Mach = 0.6	0.90
Engine start, takeoff, climb, and acceleration weight fraction	$W_3/W_1 = 0.93$
Aspect ratio	25
Span	< 150 ft for operability at airports worldwide

The BSFC = 0.42 is a realistic value for available piston engines operating at sea level (14.7 psia pressure). At 100,000 ft, the static pressure is 0.158 psia. Thus, a turbocharger is needed that will boost the pressure by a (surprisingly large) factor of 93. Whatever the turbocharger design, there will be considerable heating of the air, which will need to be cooled before it goes into the engine. Thus, we will assume wing-mounted, ram air heat exchangers, which will decrease the aircraft L/D by 22%.

The HAARP configuration is shown in Fig. 5.11. The aerodynamics for the aircraft are as follows:

Airfoil (generated using ISES airfoil design code from MIT)	
Maximum t/c	12.2% and located at 50% chord
$C_{l_{min}}$	0.4
Camber	4%
K''	0.006
K'	$1/AR\, e\pi = 0.013$ for $e = 0.97$ (from Fig. 13.5)
$C_{D_{min}}$	0.014
Maximum L/D	27 [from Eq. (3.10b) with 22% penalty]
Cruise C_L	0.89 [$C_{L_{opt}}$ from Eq. (3.8b)]
Cruise-out Weight Fraction $W_3/W_4 = \exp[(5000)(0.42)/(326)(0.9)(27)] = 1.3$	
W_8/W_1	$(0.93)(1/1.3) = 0.72$
W_{fuel}/W_{TO}	$1.06(1.0 - 0.72) = 0.30$
Assume TOGW	16,000 lb
$(W_{empty})_A$	$16,000 - 4800 - 2500 = 8700$ lb
$(W_{empty})_R$	$0.911\,(TOGW)^{0.947} = 8726$ lb
Difference	< 1%

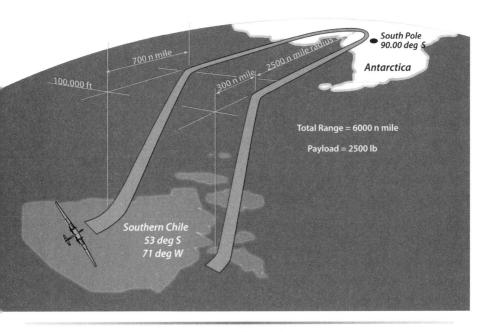

Figure 5.10 HAARP mission profile.

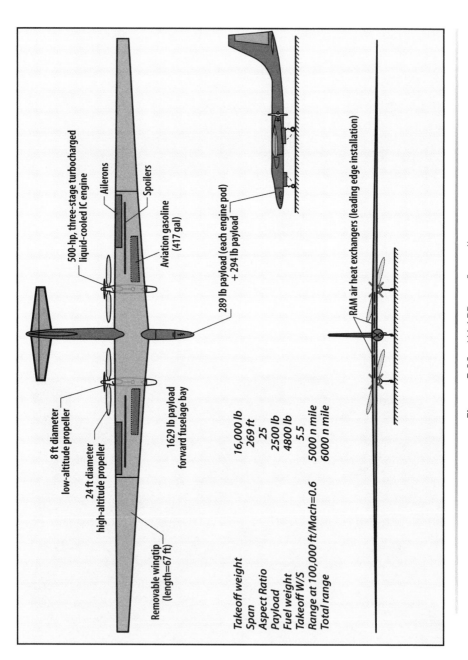

Figure 5.11 HAARP configuration.

500-hp, three-stage turbocharged liquid-cooled IC engine

Ailerons

Spoilers

Aviation gasoline (417 gal)

289 lb payload (each engine pod) + 294 lb payload

RAM air heat exchangers (leading edge installation)

8 ft diameter low-altitude propeller

24 ft diameter high-altitude propeller

1629 lb payload forward fuselage bay

Removable wingtip (length=67 ft)

Takeoff weight	16,000 lb
Span	269 ft
Aspect Ratio	25
Payload	2500 lb
Fuel weight	4800 lb
Takeoff W/S	5.5
Range at 100,000 ft/Mach=0.6	5000 n mile
Total range	6000 n mile

The HAARP aircraft will be carried as an example through the following chapters:

• Chapter 6 (Section 6.6.1) for estimating the wing loading
• Chapter 14 (Section 14.2.1) for a discussion of turbocharger design
• Chapter 18 (Section 18.10) for piston engine sizing
• Chapter 19 (Section 19.14) for wing structural design

Reference

[1] "Global Stratospheric Change—Requirements for a Very High Altitude Aircraft for Atmospheric Research," workshop report Truckee, CA, 15–16 July 1989, NASA Conf. Publ. No. CP-10041, 1989.

Chapter 6 Estimating the Takeoff Wing Loading

- Wing Sizing
- Takeoff & Landing *W/S*
- Air–Air Combat *W/S*
- Long-Endurance *W/S*
- High-Altitude *W/S*
- Examples

Built in the late 1940s, the USAF Boeing B-47 and the RAF Avro Vulcan B-1 were both designed for the same mission—long-range interception of Soviet manned bombers, but with noticeably different configurations and wing loadings. Both aircraft achieved a high cruise L/D as detailed in Example 6.3.

Nothing in this world can take the place of persistence.
Calvin Coolidge

6.1 Introduction

The *takeoff wing loading* $(W/S)_{TO}$ is a very important design parameter as it sizes the wing and locks in the dominant performance features of the aircraft. It is determined by considering the following mission requirements:

1. Range (cruise efficiency)
2. Endurance (loiter efficiency)
3. Landing and takeoff
4. Air-to-air combat (maneuverability)
5. Air intercept (minimum acceleration time)
6. High altitude
7. High altitude, long endurance
8. Low-altitude ride quality

Here the designer faces a real dilemma as these mission requirements are in conflict with one another. For example, good cruise efficiency usually drives the $(W/S)_{TO}$ to high values (i.e., in excess of 100 psf), whereas good combat maneuverability requires low $(W/S)_{TO}$. Thus, the designer must consider the $(W/S)_{TO}$ needed to meet the different mission requirements and then decide upon an appropriate compromise. Remember that there is no right answer and what is appropriate for today might not be appropriate for tomorrow. It is important for the designer to give due priority to the dominant mission phase but also to look at the entire mission to assure that the performance features are acceptable in all phases. An example of a poor design compromise would be a highly swept Mach 3 transport that had fantastic cruise efficiency but required the entire length of the San Diego freeway for takeoff. Table 6.1 shows the wing loading trends for different mission requirements.

6.2 Range-Dominated Vehicle (Cruise Efficiency)

It is important to select the $(W/S)_{TO}$ so that the aircraft can fly at conditions for maximum range for a given amount of fuel, or for a given range with a fuel load such that the aircraft weight (and cost) will be minimized.

Turbine aircraft flying at maximum cruise efficiency should fly at conditions where $(V/C)(L/D)$ is a maximum [see Eq. (3.26a)]. At this maximum cruise condition, the cruise L/D is close to but less than $(L/D)_{max}$ (87% for constant altitude cruise and 94% for a cruise climb; see Fig. 3.2).

The condition for maximum range for a propeller aircraft is to fly at conditions such that $(1/C)(L/D)$ is a maximum [see Eq. (3.30)]. Thus, the propeller aircraft would fly at conditions for $(L/D)_{max}$ as indicated by Table 3.2.

Table 6.1 Takeoff Wing Loading Trends

Dominant Mission Requirement	$(W/S)_{TO}$	Example
High-altitude, long-endurance solar-powered ISR[a]	0.5–3.0	Helios
Competition sailplanes	7–12	ASW 17
Light civil aircraft with short range and field length	10–30	C-172
High-altitude, long-endurance hydrocarbon-powered ISR	25–50	RQ-4A
STOL[b] and utility transports	40–90	C-130
Short or intermediate range with moderate field length	50–90	Learjet 35
Long-range transports and bombers (>3000 n mile)	110–150	B 747
Fighter, high-altitude	30–60	F-106
Fighter, air-to-air	50–80	F-15A
Fighter, close air support	65–90	A-10A
Fighter, strike interdiction	90–130	F-4E
Fighter, interceptor	120–150	F-104G
Low-altitude subsonic cruise missiles	200–240	AGM-109

[a]Intelligence, surveillance, and reconnaissance.
[b]Short takeoff and landing.

For a turbine airplane the cruise altitude should be around 35,000 ft, where the TSFC is near minimum (see Chapter 14). If the airplane is dominated by a long range requirement, its fuel fraction will be about 0.4. This means that its wing loading will change (decrease) approximately 40%. If the flight profile is at constant altitude, the aircraft will have to slow down to keep $(W/S)/q$ a constant so that the $C_L = (C_{D_0}/3K)^{1/2}$ according to Fig. 3.2. It is instructive to return to the air-launched cruise missile example of Chapter 5.

Example 6.1   Air-Launched Cruise Missile

Why does a cruise missile design that closed with a launch weight of 1800 lb do the same mission as the Tomahawk missile, which weighs 2860 lb?

The cruise missile and Tomahawk designs are very similar: same wing aspect ratio (AR), missile C_{D_0}, K, payload, engine, and empty-weight trends. Flying at maximum range at constant altitude gives a constant $C_L = (C_{D_0}/3K)^{1/2} = 0.44$, which results in a wing loading at launch of $W/S = qC_L = (725)(0.44) = 319$ psf, and a 5.64-ft^2 wing.

The $(L/D)_{max} = 11$ at $C_L = 0.77$ yields a cruise $L/D = (0.866)(11) = 9.5$. The L/D vs C_L relationship is shown in Fig. 6.1. Notice that C_L at launch for $L = W$ is 0.44 and the L/D is indeed 9.5. However, as fuel is burned, the C_L must decrease for $L = W$ so that at the end of the mission $C_L = 0.27$ and $L/D = 6.9$ at Mach = 0.7 and 200 ft.

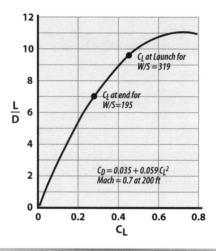

Figure 6.1 Cruise missile L/D throughout the entire C_L range.

Because slowing down or climbing is not permitted (for surviv-ability), C_L and L/D decrease as shown in Fig. 6.1. Thus, in this exercise, the mistake was assuming in Eq. (5.10) that $L/D = 9.5$ for the entire 1900 n mile. If the average L/D of 8.2 had been used, the vehicle would have sized out near the Tomahawk launch weight of 2860 lb.

Example 6.2   Long-Range Subsonic Transport

Determine the best $(W/S)_{TO}$ for a turbine-powered, long-range subsonic transport with a fuel fraction of 0.4. Assume a cruise climb (constant throttle setting) flight profile and

Start of cruise	Mach = 0.8 and 30,000 ft
C_{D_0}	0.018 (from Fig. G.1)
AR	7.25 wing with 25-deg sweep (from Fig. G.9, $e = 0.8$)
K	0.0549
$(L/D)_{max}$	16
cruise L/D	15
cruise C_L	$(C_{D_0}/2K)^{1/2} = 0.41$ (from Table 3.2)
q	282 psf
start-of-cruise W/S	$qC_L = 116$ psf
$(W/S)_{TO}$	≈ 120 psf

The question is, what is the altitude at the end of cruise? The wing loading at end of cruise would be approximately $(120)(1 - 0.4) = 72$ psf; C_L should still be 0.41 (i.e., C_{D_0} and K have not changed). Thus,

$q = (W/S)(1/C_L) = 175$ psf. The altitude at which $q = 175$ psf and Mach $= 0.8$ is 40,000 ft, which is a reasonable altitude for the end of cruise.

Note, the aircraft would like to climb at a constant Mach cruise as the fuel burns off and the wing loading decreases. However, the air traffic control regulations restrict airliners to constant even or odd 1000-ft altitude corridors. Thus, the actual long-range cruise climb profile would resemble stair steps of constant-altitude segments and 2000-ft climb segments.

From these two examples it would appear that range-dominated aircraft always have high wing loadings, as indicated in Tables 6.1 and 6.2. There are two rules that must be learned in the design of aircraft:

1. There are no right answers, only a best answer.
2. There are no rules.

Example 6.3   Boeing B-47 and Avro Vulcan B-1 Comparison

In the early 1950s both the United States and the United Kingdom built bomber aircraft for strategic nuclear missions into the USSR. The result of their different design approaches is shown in Fig. 6.2. Both the USAF B-47 and the Royal Air Force (RAF) Vulcan B-1 were designed for high cruise efficiency, which meant high cruise L/D. The B-47 was designed as a high wing loading, high aspect ratio, wing/body/tail configuration to fly high subsonic at 30,000–40,000 feet. The B-1 also had the requirement to penetrate at high altitude to survive against the Soviet fighters. The Vulcan B-1 was designed as a low wing loading, low aspect ratio, blended wing/body tailless configuration to fly high subsonic at 50,000 feet. A low W/S leads to a

Table 6.2   Takeoff Wing Loading and Fuel Fraction for Various Cruise-Dominant Aircraft

Aircraft	$(W/S)_{TO}$	W_{fuel} / W_{TO}
C-5A	117	0.417
KC-135	124	—
B 747	141	0.428
L-1011	124	0.352
DC-10	153	0.42
B-52G	122	0.62
C-17	152	0.34
B-1B	244	0.47
Tomahawk	213	0.47

	Boeing B-47	Avro Vulcan
Wing Area (ft²)	1430	3446
Total Wetted Area (ft²)	7070	9,500
Span (ft)	116	99
Wing Loading (lb/ft²)	140	43
Span Loading (lb/ft)	1750	1520
Aspect Ratio	9.43	2.84
$C_{D_{min}}$	0.0198	0.0069
$K = 1/(\pi AR\, e)$	0.0425	0.125
Value of e	0.8	0.9
Max L/D	17.25	17.0
$C_{L_{opt}}$	0.682	0.235
Max Cruise C_L	0.48	0.167
$C_{D_{min}}S_{ref}$	28.3	23.8
Wetted Area / S_{ref}	4.9	2.8

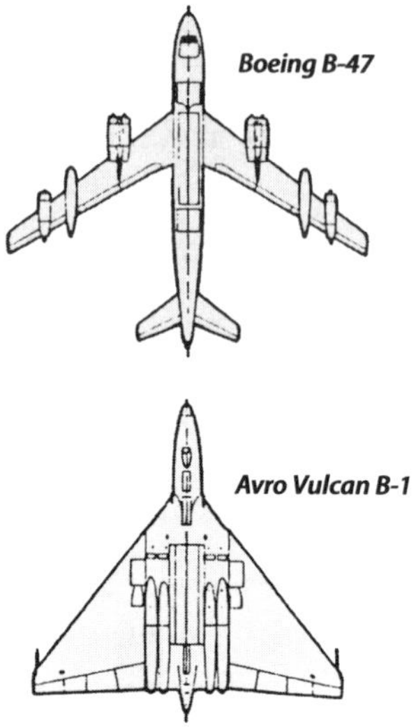

Figure 6.2 Design characteristics comparison—Boeing B-47 vs Avro Vulcan B-1.

large wing, which can mean a large C_{D_0} due to skin friction unless the wing is blended with the fuselage to reduce the configuration wetted area. Notice that the Vulcan B-1 is a blended wing–fuselage configuration with a low-AR wing (high value for K) but a low wetted area relative to S_{Ref} (low C_{D_0}; see Fig. G.7). Thus, the Vulcan B-1 achieved a high maximum L/D by having a low C_{D_0} in spite of the high K. The B-47 was the more traditional approach with a high-AR wing attached to a fuselage, giving a higher C_{D_0} but low K. The $(L/D)_{max}$ for both aircraft was almost the same.

6.3 Endurance or Loiter

For maximum loiter efficiency the turbine aircraft should fly at conditions for maximum L/D and the propeller aircraft for maximum $(L/D)/V$. The conditions on C_L are given in Table 3.2.

Example 6.4   High-Altitude, Long-Endurance ISR

Find the $(W/S)_{TO}$ for the RQ-4A Global Hawk starting loiter at Mach 0.6 and 55,000 ft. The Global Hawk has a very highly cambered airfoil identified as the LRN 1015 (shown in Fig. 2.2). The aero data for the Global Hawk is as follows:

Wing AR	25 with 8-deg sweep
$C_{D_{min}}$	0.019 (from Fig. G.2)
$C_{L_{min}}$	0.3
$K = K' + K''$	0.0165 (the $e = 0.77$ from Fig. G.9)
K''	0.01 (from Fig. 13.6)

Because the Global Hawk wing is highly cambered we need to use the $C_{L_{opt}}$ and $(L/D)_{max}$ for a cambered aircraft [see Eqs. (3.8b) and (3.10b)]. Using these equations we get $C_{L_{opt}} = 0.91$ and $(L/D)_{max} \approx 36$, which agrees with the data in Fig. G.4.

At Mach = 0.6 and 55,000 ft, q = 48 psf. This gives a start-of-loiter W/S = 43.7 psf. Global Hawk weighs 25,600 lb at takeoff and has a 540-ft^2 wing, giving a $(W/S)_{TO} = 47$ psf. It continuously climbs during loiter, reaching altitudes above 60,000 ft.

6.4 Landing and Takeoff

If landing and/or takeoff are important mission requirements, the reader would do well to become familiar with the material in Chapters 9 and 10 before going on.

The wing loading of the aircraft influences the landing and takeoff distances through the stall speed

$$V_{stall} = \sqrt{\frac{W}{S} \frac{2}{\rho C_{L_{max}}}} \qquad (6.1)$$

Takeoff is the distance required for an aircraft to accelerate from $V = 0$ to $V = 1.2 V_{stall}$ and climb over a 50-ft obstacle. Landing is the horizontal distance required for an aircraft to clear a 50-ft obstacle at an approach speed of $1.3 V_{stall}$, touch down at $V_{TD} = 1.15 V_{stall}$, and brake to a complete stop.

The takeoff distance is dependent upon the *takeoff parameter* (TOP):

$$\text{TOP} = \frac{W}{S} \frac{1}{C_{L_{max}}} \frac{1}{T/W} \frac{1}{\sigma} \qquad (6.2)$$

where $\sigma = \rho/\rho_{SL}$. This takeoff parameter is shown in Fig. 6.3. The takeoff distance can be estimated at this point in the design by using the approximate expression (in feet)

$$S_{TO} = 20.9 \frac{W/S}{\sigma C_{L_{max}} (T/W)} + 69.6 \sqrt{\frac{W/S}{\sigma C_{L_{max}}}} \qquad (6.3)$$

or by using Fig. 6.3. It should be clear that a short takeoff distance can be achieved with a high wing loading if $C_{L_{max}}$ and T/W are large.

The landing distance is dependent upon the *landing parameter* (LP):

$$LP = \frac{W/S}{\sigma C_{L_{max}}} \qquad (6.4)$$

The landing distance for conventional takeoff and landing (CTOL) aircraft can be estimated from the approximate expression

$$S_L = 79.4 \frac{W/S}{\sigma C_{L_{max}}} + 50/\tan\theta_{app} \qquad (6.5)$$

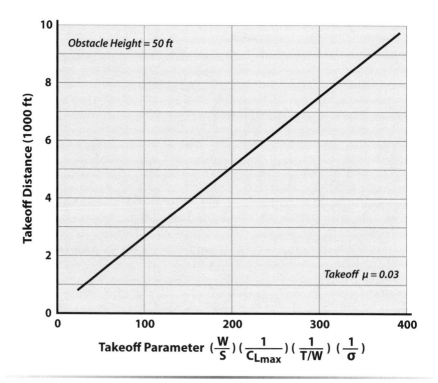

Figure 6.3 Takeoff distance vs the takeoff parameter (TOP).

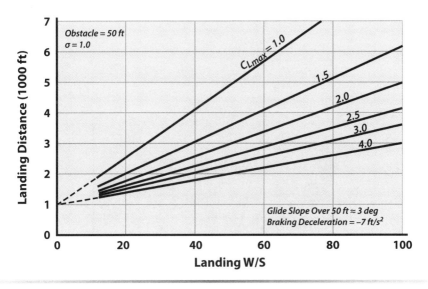

Figure 6.4 Landing distance for CTOL aircraft.

where S_L is in feet. Equation (6.5), shown in Fig. 6.4, assumes a glide slope over 50 ft of approximately 3 deg (~950 ft) and a braking deceleration of -7 ft/s^2.

Short takeoff and landing (STOL) aircraft approach at steeper angles (on the order of 7 deg) and employ thrust reversers and other ground braking devices to shorten the landing field distance. Figure 6.5 shows the impact of landing W/S and approach C_L ($0.8C_{L_{max}}$) on landing field distance.

Mechanical high-lift devices have an upper $C_{L_{max}}$ limit of about 4.0 (see Fig. 9.7), with powered lift devices extending up to about 12.0 (Chapter 9, Section 9.6). If STOL is a dominant feature of the mission requirements, the designer should give considerable thought to the expected performance from high-lift devices because W/S and $C_{L_{max}}$ are partners in the landing and takeoff problem. Selecting a takeoff W/S without due consideration of $C_{L_{max}}$, and even T/W, may lead to an impossible design later.

6.5 Air-to-Air Combat and Acceleration

The ability of a fighter aircraft to accelerate at a point in space is given by its value of excess specific power P_S, for $n = 1$, and to maneuver by its P_S for $n > 1$. Thus, fighter aircraft should have their design parameters selected to maximize the value of P_S for critical mission phases. From Chapter 3, the expression for P_S [assuming $\cos(\alpha + i_T) \sim 1$] is given by

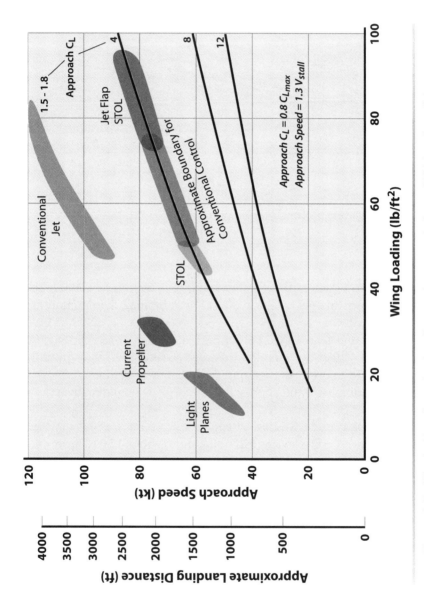

Figure 6.5 Effect of landing wing loading and approach C_L on STOL landing distances.

$$P_S = V\left[\frac{T}{W} - \frac{D}{W}\right] \qquad (6.6)$$

For a certain T/W, the value for P_S can be maximized by finding the conditions for minimum D/W. It is of interest to find the *wing loading* W/S that will minimize D/W. Using*

$$D = qS_{\text{ref}}\left[C_{D0} + KC_L^2\right] \qquad (6.7)$$

and

$$C_L = \frac{W}{S}\frac{n}{q} \qquad (6.8)$$

we find the condition for W/S that sets $\partial(D/W)/\partial(W/S) = 0$. This condition is expressed as

$$\frac{W}{S} = \frac{q}{n}\sqrt{\frac{C_{D0}}{K}}$$

where the reader should recognize $(C_{D0}/K)^{1/2}$ as the C_L for minimum drag or $(L/D)_{\max}$ from Chapter 3. Thus, Eq. (6.8) expresses the condition on W/S that gives maximum P_S at a point in velocity–altitude space for a given T/W and load factor.

Chapter 4 pointed out that during a minimum-time trajectory (an acceleration) the load factor n is close to 1. Thus, Eq. (6.8) with $n = 1$ expresses the best wing loading for acceleration-dominated aircraft.

Example 6.5   Acceleration-Dominated Aircraft

Find the best wing loading for a fighter interceptor at Mach = 0.8 and 25,000 ft using the aerodynamics of the F-4C shown in Fig. 2.17.

At Mach = 0.8 and 25,000 ft, q = 352 psf. From Fig. 2.17 the result is C_{D0} = 0.022, K = 0.169, and K_B = 0 because C_L < 0.5 (check later). Using Eq. (6.8) gives

$$\frac{W}{S} = \frac{352}{1}\sqrt{\frac{0.022}{0.169}} = 127 \text{ psf}$$

---

* It is valid to assume K_B from Eq. (2.19), Chapter 2, is equal to zero, as will be shown later.

Thus, a high wing loading is very desirable for acceleration. The required $C_L = 0.378$ is less than C_{L_B} from Fig. 2.15, so that indeed $K_B = 0$.

Example 6.5 points out that the requirement for minimum acceleration leads to a high wing loading. This should be clear because the goal is to minimize the drag at $n = 1$, and the skin friction is a large part of this drag at all Mach numbers. Thus, decreasing the aircraft wetted area by decreasing the wing area gives a decrease in the zero-lift drag.

Example 6.6   Air-to-Air-Combat Aircraft

Here the issue is to find the best wing loading for an air-to-air fighter at Mach = 0.8 and 25,000 ft and for $n = 5$ using the aerodynamics for the F-4C from Fig. 2.15.

Once again, $q = 352$ psf, $C_{D_0} = 0.022$, $K = 0.169$, and assume $K_B = 0$. Using Eq. (6.8) with $n = 5$ gives $W/S = 27$ psf. Thus, the mission requirement is low wing loading for air-to-air combat. The required $C_L = 0.378$ is less than the C_L for buffer onset, C_{L_B}, C_{L_B} so that the assumption of $K_B = 0$ was valid.

Example 6.6 points out that low wing loading is desired for an air combat fighter, just the opposite of the result for the air interceptor. For air combat, $n > 1$, the drag-due-to-lift is the major part of the drag and it goes as the square of the required C_L. The required C_L is decreased for a given load factor by decreasing the wing loading. Previous research concludes that it is more efficient in terms of W_{TO} to improve the P_S at $n > 1$ (i.e., improve the turn rate, $\dot{\psi}$) of an aircraft by decreasing the wing loading rather than increasing the thrust-to-weight ratio. Because air-to-air combat is a series of hard turning maneuvers and accelerations (to gain back the energy lost during the hard turns), the air combat fighter will have low wing loading for turning flight and high T/W for the accelerations.

The result from Example 6.6 is unconstrained and does not take into consideration the large wing weight and the relatively poor cruise performance associated with a low W/S. For these (and other) reasons the designer would probably select a somewhat higher value of wing loading than is indicated by Eq. (6.8). It is important for the designer to select the wing loading only after giving due consideration to all phases of the mission profile.

Another consideration is that air-to-air combat usually takes place in the transonic regime and altitudes from 10,000 to 35,000 ft. The maximum usable transonic lift coefficient is limited by the onset of buffet and, thus, $C_{L_{\max}} < 1$ is typical for transonic maneuvering. Figure 6.6 shows maximum C_Ls due to buffet obtained during flight tests.

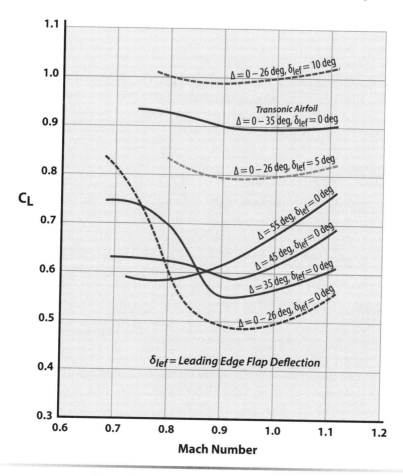

Figure 6.6　Maximum C_L due to buffet.

Therefore, an estimate of combat wing loading can be obtained from

$$\frac{W}{S} = \frac{qC_{L_{max}}}{n} \tag{6.9}$$

where the n is that load factor required for a particular turn rate. The desired maximum sustained turn rate $\dot{\psi}_{MS}$ should be about 2 deg/s better than the enemy aircraft.

6.6 High Altitude

The wing loading for a high-altitude reconnaissance aircraft can be determined from

$$\frac{W}{S} = C_L q$$

Usually the altitude and velocity are specified (hence q is fixed), and realistic estimates of the maximum usable lift coefficient can be made so that the required W/S can be determined. The requirement for high altitude drives the $(W/S)_{TO}$ to low values.

6.6.1 HAARP Wing Loading

The HAARP vehicle discussed in Section 5.8 and shown in Fig. 5.11 will have a wing loading driven by the requirement to fly 5000 n mile at 100,000 ft and Mach = 0.6. The aerodynamics for the HAARP are shown in Fig. 6.7. The aircraft has a maximum L/D = 27 over a C_L range of 0.75–0.9. The wing area will be sized for a start-of-cruise C_L = 0.9. This gives an $S_w = S_{Ref}$ = 2884 ft^2 and a takeoff W/S = 5.5 lb/ft^2. With the 30% fuel fraction the aircraft wing loading at end of the mission is 3.88 lb/ft^2. During the 5000 n mile cruise at 100,000 ft the C_L will be reduced to 0.75 and the altitude increased to 102,000 ft in order to keep the cruise L/D = 27.

The aspect ratio is 25, giving a span of 268 ft. Notice that a wing span less than 150 ft is desired, to facilitate operation from airports worldwide. HAARP will have 67-ft detachable outer wing panels, which will reduce the wing span to 134 ft. The detachable outer wing panels will result in a

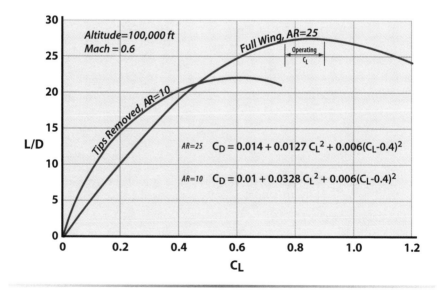

Figure 6.7 HAARP aircraft aerodynamics.

heavier wing than if the wing were continuous from tip to tip. The outer wing panels do not have any control surfaces or carry any fuel.

Removing the outer wing panels gives HAARP an S_w = 1789 ft², an aspect ratio of 10 and a maximum L/D = 22.3.The aerodynamics for the short-wing HAARP are shown in Fig. 6.7. With the 4800 lb of fuel in HAARP, the short-wing version can fly 5500 n mile at 45,000 ft and Mach = 0.6. The concept of operations would be to transport the outer wing panels to the deployment area in a transport airplane. The short-wing HAARP would fly to the deployment area, attach the outer wing panels, and then conduct the atmospheric collection over the South Pole at 100,000 ft.

The design of the HAARP three-stage turbocharger will be discussed in Chapter 14.2.1.

6.7 High Altitude, Long Endurance

The requirement for high altitude and long endurance is especially challenging because not only is low wing loading (large wing) a requirement, but the fuel fraction must be large as well. The Global Hawk has an endurance of one-and-a-half days above 55,000 ft and requires a fuel fraction of 57%.

An endurance of several weeks or more will require frequent resupply of fuel or a propulsion system that can regenerate itself. The sun is a fine source of energy for aircraft: sunlight can be converted into electrical energy through photovoltaic cells (28% efficiency in 2010; discussed more in Chapter 14). However, the sun can only provide energy during the daytime. Thus, solar energy must be collected during daytime and stored to be used to support the aircraft during the night. The next example will examine a solar-powered ISR aircraft.

Example 6.7 considers solar power with storage. The solar energy is collected by thin-film photovoltaic cells covering the external surface of the airplane. Excess solar energy is collected during the daytime and stored in hydrogen–oxygen fuel cells or rechargeable batteries (discussed in Chapter 14) to be used to operate the aircraft during the night.

Example 6.7 High-Altitude, Long-Endurance Solar-Powered ISR

Determine the wing loading and required power for the Solar Snooper. The configuration is the wing–body–tail design shown in Fig. 6.8.

The requirement for this aircraft is 4 weeks at 65,000 ft, in mid-latitudes during the summer conducting ISR. The payload is 500 lb at

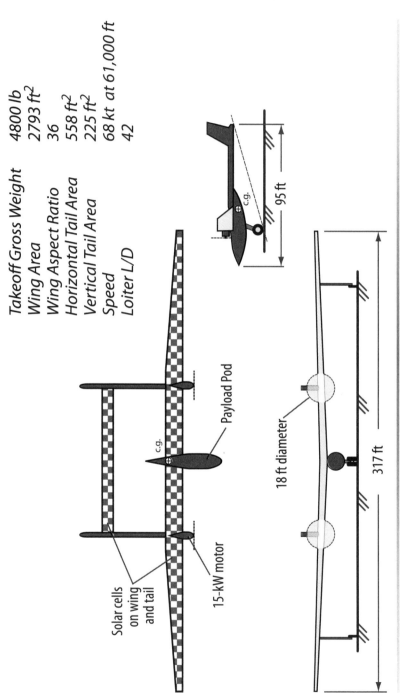

Takeoff Gross Weight	4800 lb
Wing Area	2793 ft²
Wing Aspect Ratio	36
Horizontal Tail Area	558 ft²
Vertical Tail Area	225 ft²
Speed	68 kt at 61,000 ft
Loiter L/D	42

95 ft

c.g.

Payload Pod

c.g.

Solar cells
on wing
and tail

15-kW motor

18 ft diameter

317 ft

Figure 6.8 Solar Snooper configuration.

1 kW. The 4-week endurance immediately eliminates hydrocarbon (gasoline) propulsion systems. The propulsion candidates are solar power with storage and nuclear power.

The political issues connected with flying a nuclear-powered airplane in the atmosphere over another country are overwhelming and politically unacceptable.

It is assumed that the cruise speed at 64,000 feet is 68 kt (115 ft/s, Mach 0.12) and the TOGW equals 4800 lb. The assumed 4800 lb TOGW will be validated in Chapter 18. The-solar powered aircraft has the following characteristics:

Wing aspect ratio	36 with zero sweep (similar to the Boeing Condor, which is discussed in Appendix G.2), shown in Fig. 6.9
$C_{D_{min}}$	0.0085 (assumed)
$C_{L_{min}}$	1.0 (assumed)
Wing efficiency e	0.60 (from Fig. G.9)
K	0.0147 ($K = 1/\pi ARe$)
K''	0.002 (assumed)
$K' = K - K''$	0.0147 − 0.002 = 0.0127
Maximum L/D	46 [using Eq. (3.10b) @ C_L = 0.845]
Best loiter L/D	41 at C_L = 1.33 (from Table 3.2)
Propeller efficiency, η_p	0.85 (assumed, based on Helios report)
Electric motor efficiency, η_{EM}	0.97 (vendor data)

The power required is given by

$$\text{Power required} = [\text{drag}(\text{lb})][\text{speed}(\text{ft/s})](0.745/550)/(\eta_p \eta_{EM}) \quad (6.10)$$

where the 550 converts foot-pounds per second into horsepower and the 0.745 converts horsepower into kilowatts. Using drag = TOGW/ (L/D) = 4800/42 = 114 lb in Eq. (6.10) gives a propulsion power required of 21.6 kW during loiter.

Figure 6.9   Condor (courtesy of The Boeing Company).

The total power required is then

- Propulsion (electric motors), 22 kW
- Payload, 1 kW
- Aircraft operation, 1 kW
- Total = 24 kW

The problem is to find the W/S that will let us fly at 115 ft/s. The dynamic pressure $q = 1.29$ psf (pretty low), such that $W/S = qC_L = (1.29)(1.33) = 1.72$ and the wing area $S_w = 2793$ ft^2. The electric motors will be derated 25% for reliability, resulting in two 15-kW motors.

The design challenge from here is to install the solar cells onto a very lightweight 2793-ft^2 wing, integrate the energy storage and payload into a very lightweight fuselage, and install the engines—all within 4800 lb. Also, we need to check that there are enough solar cells to collect enough solar energy during daylight to operate at night. We will return to the Solar Snooper in Chapters 14 and 18 when we discuss propulsion concepts and engine sizing.

As a sanity check, we will look at the AeroVironment Helios solar-powered UAV. Helios is a flying wing (span loader) with a wing span of 247 ft, aspect ratio of 31, wing area of 1976 ft^2,and weight 2048 lb, giving it a wing loading of 1.04 psf. At 90,000 ft, its speed is 148 kt, or Mach = 0.25 ($q = 1.58$ psf). It flies at $C_L = 0.5$ because it does not have a tail to trim out the pitching moment from a higher C_L. The solar cells (19% efficiency) are on the upper surface of the wing. On a bright sunny day Helios collects approximately 37 kW. It needs less than 20 kW to fly and stores the rest. In June 2003 Helios was on a path to demonstrate "24/7" flight using fuel cell storage when, on its second flight, it encountered turbulence at 3000 ft and broke up in midair over Kauai, Hawaii (see Fig. 6.10). The mishap investigation team concluded that when Helios encountered the turbulence it morphed into an unexpected, persistent, high-dihedral configuration. The aircraft became unstable in a very divergent pitch mode with airspeed excursions from nominal flight speed doubling every oscillation. The aircraft's design speed was exceeded and the resulting high dynamic pressures caused the wing leading edge secondary structure on the outer wing panels to fail and the solar cells and skin on the upper surface to rip off.

6.8 Low-Altitude Ride Quality

Flight at low altitude is bumpy due to the gusts encountered close to the ground. If the flight speed is greater than 250 kt, the ride can be very uncomfortable and make the passengers sick. This is a problem that special operations forces (SOF) troops would have flying low and fast to evade

Normal 1-g flight

Maximum dihedral occurs

Leading edge breaks apart

Crash site

Figure 6.10 Helios crash sequence.

detection. It is often said that the thing astronauts and SOF troops have in common is that they both get sick going to work.

Government specification MIL-F-9490D defines a discomfort index D_V and specifies the time duration for different D_V levels. The ride quality is strongly related to wing loading and lift curve slope C_{L_α}. The design features to meet this requirement are high wing loading (above 100 psf) and low AR and/or wing sweep to reduce the C_{L_α}.

Chapter 7

Selecting the Planform and Airfoil Section

- Measure of Merit
- Maximum Thickness Ratio
- Location of Maximum Thickness
- Camber
- Wing Aspect Ratio
- Wing Sweep
- Wing Taper Ratio
- Compromise

The forward swept wing on the DARPA/Grumman/USAF X-29 demonstrates high subsonic/transonic, high angle of attack maneuvering, and positive control for a fighter aircraft. This type of wing design is compared with other planform candidates in this chapter (see Section 7.7).

It is better to beg forgiveness than ask permission.
Skunk Works slogan

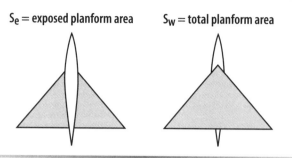

S_e = exposed planform area S_W = total planform area

Figure 7.1 Definition of wing reference area.

7.1 Introduction

The *planform* of a vehicle means collectively the sweep of the leading edge (delta), the aspect ratio (AR), the taper ratio (λ), and the general shape of the top view of the wing. These *planform parameters* are defined in Fig. 2.1. The airfoil is selected when we decide upon the series designation (e.g., NACA 2415) or determine the maximum thickness ratio (t/c), the location of the maximum thickness ratio, the leading-edge radius (r_{LE}), and the camber (usually expressed in percent chord). Appendix F discusses airfoil nomenclature and presents section data on many popular airfoils. References [1–3] are also very useful in selecting an airfoil.

Planform selection is especially important as it influences vehicle aerodynamics significantly and gives the aircraft its characteristic shape. Both the airfoil and planform selection are strictly dependent upon the aircraft mission requirements. Figure 7.1 shows the two most popular methods of computing reference wing area, S_{ref}. (At the end of this chapter Fig. 7.11a–c presents a large selection of planforms for comparison.)

At this point in the design we select the general shape of the wing planform and select the airfoil section, keeping in mind that subsequent design iterations will refine the selection. Generally, an airfoil section and planform are selected to give the following design measures of merit:

- High
  - C_{L_α}
  - $C_{L_{\max}}$
  - Wing fuel volume
- Low
  - C_{D_0}
  - K'
  - Wing weight

The designer quickly learns that this design is impossible because of conflicting conditions. For example, a low K' means a high AR, but low wing weight requires a low AR. Thus, the selection of the airfoil and planform is a compromise of the priorities established by the mission requirements.

7.2 Effect of Airfoil: Maximum Thickness Ratio

At low speed the t/c influences the maximum lift coefficient with the $C_{\ell_{max}}$ increasing as t/c increases. This behavior is shown in Fig. 7.2. The subsonic zero-lift drag coefficient increases slightly as t/c increases (see Fig. H.6).

The critical Mach number M_{CR} usually represents the maximum speed attainable for high-subsonic aircraft due to the increase in thrust required for flight past M_{CR}. Thus, a subsonic vehicle strives to push this upper speed limit as far as possible. A supersonic aircraft also has a large wing M_{CR}, or more correctly, larger than the fuselage M_{CR}. All of the vehicle components have a M_{CR}, and flight past this limit is accompanied by a large drag rise. The individual component drag rises are additive for the most part. Their sum at any Mach number represents the minimum thrust required for a vehicle to accelerate past that Mach number. By having the wing and fuselage drag rises peak at different Mach numbers, the thrust requirement to accelerate past $M = 1$ is lessened. An example of this is shown in Fig. 2.20.

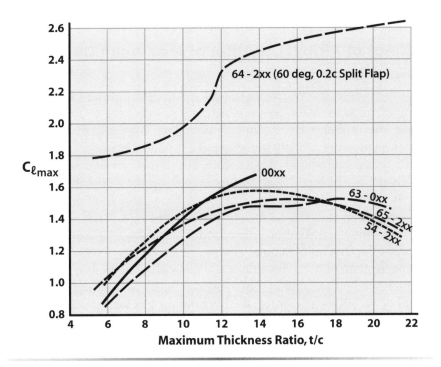

Figure 7.2   Maximum lift coefficient vs airfoil thickness ratio (data from [2]).

The t/c influences M_{CR} as shown in Figs. 2.23b and 7.9 (see Section 7.7). As the t/c increases, the M_{CR} decreases as supersonic flow occurs earlier on the upper surface, leading to the presence of a normal shock and flow separation (see Fig. 2.19).

In supersonic flight the wave drag increases approximately as the square of the t/c [see Eq. (2.31)]. This behavior is illustrated in Fig. H.6. If the aircraft will spend much of its mission at supersonic speeds, the t/c should be small (from 4% to 6%). A value of 3% for an airfoil is attractive from the wave-drag viewpoint; however, 3% represents a lower bound for t/c as it results in a heavy wing with little wing volume available for fuel. The B-58 Hustler supersonic bomber had a NACA 0003 airfoil and had to carry a large external fuel pod to get to the target. The fuel pod also housed the nuclear weapon and was dropped over the target. Figure 7.10 (see Section 7.7) shows a comparison of wing volume available for fuel relative to t/c.

Low wing weight is just as important as low wing drag. A relative comparison of wing weight versus t/c is shown in Figs. 2.23c and 7.10. Chapter 20 presents empirical equations for estimating the wing weight for fighter, bomber–transport, and general aviation aircraft. A quick examination of Eqs. (20.1), (20.2) and (20.69) reveals that t/c is in the denominator of the wing weight equations and has a significant effect on wing weight.

As t/c increases the wing volume available for fuel, landing gear, surface control actuators, and so on, also increases (as shown in Fig. 7.10).

7.3 Effect of Airfoil: Location of Maximum Thickness

The location of the maximum t/c determines the end of the favorable pressure gradient (decreasing pressure) and the start of the adverse pressure gradient as the static pressure increases to match the freestream pressure at the trailing edge. A laminar boundary layer cannot tolerate an increasing pressure and will transition to a turbulent boundary layer at the maximum thickness point (if not before, due to surface roughness or other disturbances). The further back the maximum t/c, the longer the boundary layer is apt to be laminar, thus producing lower skin friction drag. This behavior is shown in Fig. F.3 for the front-loaded NACA 63_2-015 and the aft-loaded laminar NACA 66_2-015 airfoil. The NACA 64, 65, and 66 series represent the families of laminar airfoils.

The location of the maximum t/c (along with camber) also determines whether the section pitching moment coefficient will be negative (nose down) or positive (nose up). If the maximum t/c is forward of the airfoil aerodynamic center (a.c.; about 25% chord), it is termed a *front-loaded section* and will produce a nose-up pitching moment. If the maximum t/c is aft of the a.c., it is called an *aft-loaded section* and produces a nose-down pitching moment. The selection of the maximum t/c location is important

Figure 7.3   Global Hawk.

to the designer when the decision is made as to how the aircraft will be trimmed. For example, an aft-loaded airfoil (nose-down C_m) on a tailless airplane will need to be trimmed with a down load at the trailing edge, reducing overall lift.

Figure 2.3 shows two airfoils designed for intelligence, surveillance, and reconnaissance (ISR) aircraft. They are about the same thickness (16% for the JW 1416 and 15% for the LRN 1015) but very different locations for maximum t/c. The JW 1416 was designed for low positive C_m at $C_l \approx 0.9$ and was used on a high-AR, swept-wing tailless airplane called Polecat and built by Lockheed Martin in 2004. The LRN 1015 was designed for a laminar boundary layer back to about 55% chord at $C_l \approx 0.9$ and used on the RQ-4A Global Hawk (Fig. 7.3), which has an aft horizontal tail (so that the negative C_m could be trimmed out without affecting overall lift). The two airfoils worked well on their respective aircraft, with the JW 1416 having low positive C_m and the LRN 1015 having low drag at the required high section lift coefficients. Each designed for its specific mission.

7.4 Effect of Airfoil: Leading Edge Shape

The leading edge (LE) shape can vary from sharp ($r_{LE} = 0$) to round. Some popular sharp leading edge airfoils, designed primarily for supersonic flight, are shown in Fig. 7.4.

These airfoils have poor low-speed characteristics. However, the sharp leading edge airfoils can have their low-speed performance improved by

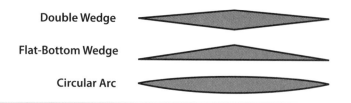

Figure 7.4 Sharp leading edge airfoils.

the use of high-lift devices such as trailing edge (TE) or leading edge flaps. Experimental data in Fig. 7.5 shows the double wedge airfoil with deflected leading and trailing edge flaps. Notice that the basic airfoil section $C_{l_{\max}}$ is about 0.83.

The round-nosed airfoils exhibit much better low-speed characteristics than the sharp-nosed airfoils. The general trend is an increase in $C_{l_{\max}}$ for the larger r_{LE}. All of the airfoils in Appendix F are round-nosed airfoils. A quick examination shows that practically all the airfoils in Appendix F have a higher $C_{l_{\max}}$ than the double wedge. The round-nosed airfoils can also be fitted with slots, slats, and leading and trailing edge flaps to improve their low-speed characteristics (discussed in Chapter 9).

Subsonic C_{D_0} is primarily skin friction and is not influenced by the nose shape of the airfoil section. The subsonic viscous drag-due-to-lift is influenced by r_{LE} in that the smaller leading edge radius promotes earlier flow separation (i.e., more LE separation drag at angle-of-attack). Thus, the aircraft subsonic viscous drag-due-to-lift factor K'' is slightly higher for the smaller r_{LE} (discussed in Section 13.2.1).

In supersonic flight the *wave drag coefficient* is expressed as:

$$C_{DW} = C_{D_{\mathrm{LE}}} + \frac{B}{\beta}\left(\frac{t}{c}\right)^2 \qquad (7.1)$$

where

$$\beta = \sqrt{M^2 - 1}$$

B = constant, dependent upon thickness distribution (discussed in Chapter 13)

The $C_{D_{\mathrm{LE}}}$ is the *leading edge bluntness* term introduced in Chapter 2. The expression for $C_{D_{\mathrm{LE}}}$ is (from [4])

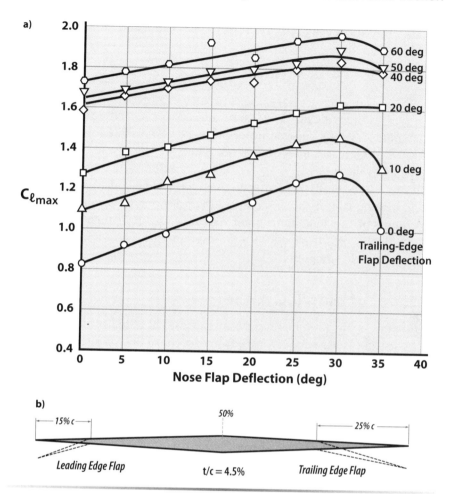

Figure 7.5 4.5% double-wedge airfoil: **a)** variation of $C_{\ell_{max}}$ with nose and TE flap deflections (data from [5]), and **b)** cross-section showing LE and TE flap deflection.

$$C_{D_{LE}} = \frac{2.56}{b}\left[\frac{r_{LE}\,AR\cos^2\Delta}{1+\left[1/(M_\infty^3\cos^3\Delta)\right]}\right] \qquad (7.2)$$

where b is the span of the wing and $C_{D_{LE}}$ is referenced to the exposed planform area of the wing. The effect of r_{LE} on $C_{D_{LE}}$ is readily apparent from Eq. (7.2). If the LE is supersonic (swept ahead of the Mach line such that $M_\infty\cos\Delta > 1$) the LE radius should be small. Typical LE radii for a supersonic leading edge vary from 0 to about 0.25% of the chord. An example is the F-104 with $\Delta_{LE} = 0$ and $r_{LE} = 0$.

The supersonic drag-due-to-lift factor K is expressed as [6]

$$K = \frac{1}{C_{L\alpha}} - \Delta N \qquad (7.3)$$

where $C_{L\alpha}$ is the wing lift curve slope and ΔN is the leading edge suction parameter; $\Delta N = 0$ for supersonic leading edges. For subsonic leading edges $\Delta N > 0$ and increases as r_{LE} increases. Thus, the supersonic K decreases slightly as r_{LE} increases.

If $M_\infty > 2.5$, aerodynamic heating will have to be considered. Aerodynamic heating will dictate a larger r_{LE} than desired, to accommodate the heat inputs at the stagnation point on the wing LE. If flight speeds will exceed Mach = 2.5, the reader should examine the discussion on aerodynamic heating in Chapter 4.

7.5 Effect of Airfoil: Camber

Camber is the amount (in percent chord) that a line equidistant from the upper and lower surface varies from the chord line (positive camber is shown in Fig. 2.1). Deflecting a trailing edge flap is the same as putting aft camber in the airfoil. Aft camber has a powerful effect on changing the lift of an airfoil at a specific angle-of-attack. Positive camber will shift the $C_{l\alpha}$ curve to the left. Camber at the leading edge has almost no effect on changing the lift. The primary purpose of LE camber (i.e., deflecting a LE flap) is to delay flow separation over the forward part of the airfoil, producing higher maximum lift coefficients.

All subsonic low speed airfoil sections have about the same lift curve slope of 2π per radian. The amount of camber determines the value of the angle for zero lift, α_{0L}. The zero-camber airfoil, or symmetric airfoil, has $\alpha_{0L} = 0$. Airfoils with positive camber have negative values of $C_{m_{a.c.}}$, whereas a symmetric section has $C_{m_{a.c.}} = 0$. If the wing must have positive $C_{m_{a.c.}}$, such as a tailless design for static longitudinal stability, then the camber must be negative. The B-58 Hustler has a swooped-up trailing edge, called *inverse camber* or *reflexed trailing edge*, to give it negative camber and a positive $C_{m_{a.c.}}$.

Positive camber gives an increase in section $C_{\ell\text{max}}$. For example, a camber of 6 % of the chord at 30% chord gives an increase in $C_{\ell\text{max}}$ of about 0.4 over an equivalent symmetric section. This behavior should be obvious because a small flap deflection is equivalent to a positive increase in camber.

The effect of increasing camber on C_D is to translate the drag polar to higher values of $C_{L\text{min}}$ (i.e., that C_L for $C_{D\text{min}}$). This behavior is shown in Fig. 7.6 for the NACA 65_3-X18 laminar flow airfoil sections and Fig. 7.7 for a complete aircraft. The designer should determine the lift coefficient that

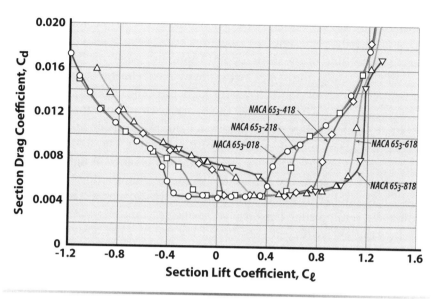

Figure 7.6   Drag characteristics of some NACA 65-series airfoil sections of 18% thickness with various amounts of camber, $Re = 6 \times 10^6$ (data from [2]).

the aircraft requires during a critical mission phase, such as cruise. Then the camber is selected to give a $C_{L_{min}}$ (called the *design* C_L) close to the required C_L. Consider the USSR sailplane Antonov A-15, which has an aspect ratio of 26.4, a wing loading of 6 psf, and a best glide ratio of 40 : 1 at 54 kt. At 5000 feet the required $C_L = 0.7$. The Antonov A-15 uses the NACA

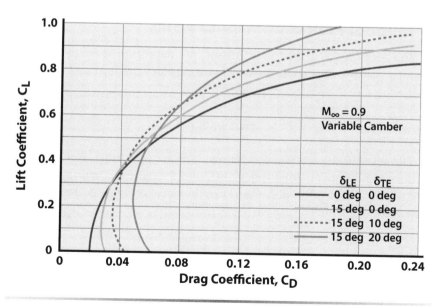

Figure 7.7   Aircraft drag polar with variable-camber wing.

65_3-618 airfoil at the root and NACA 65_3-616 at the tip. From Fig. 7.6 the NACA 65_3-618 has a design $C_\ell = 0.6$ with a broad drag bucket for C_ℓ excursions to either side of the design point (note: the subscript 3 denotes the range of C_ℓ in tenths above and below the design C_ℓ for the drag bucket).

Camber can be used effectively for high-subsonic cruise and transonic maneuvering. Figure 7.7 indicates the behavior of camber for a complete aircraft at Mach = 0.9. A range-dominated aircraft cruising at Mach = 0.8 would normally require a cruise C_L of 0.3–0.4 (see Fig. 3.9). Thus, the designer selects an airfoil to give a design C_L in this range. A fighter aircraft maneuvering transonically at high C_L (in the neighborhood of 0.8) would have a low-camber airfoil (the F-16 uses a 64-204 airfoil) but employs leading and trailing edge flaps (called *maneuver flaps* or *variable camber*) to reduce the level of drag-due-to-lift during combat. The designer must be careful in the selection of positive camber as it is accompanied by a negative $C_{m_{a.c.}}$ and must be trimmed by a down aft tail load for statically stable aircraft. Camber is normally not used for supersonic flight because of the wave drag penalty associated with camber [see Eqs. (2.28) and (2.31), Chapter 2].

The airfoils used on high-altitude ISR aircraft are usually highly cambered because their optimum C_Ls (the C_L for maximum endurance) are high, typically 0.7–1.0 (see Fig. G.4). The Lockheed Tier 3-minus "Darkstar" cruised at $C_L = 0.53$ with its straight-wing tailless design, low levels of pitch control power, and flight altitude of less than 50,000 ft. By comparison the Boeing Condor started its loiter at a $C_L = 1.33$.

7.6 Effect of Planform: Aspect Ratio

This section discusses the effects of the planform aspect ratio on the design measures of merit. If the selected wing planform is a delta shape, the AR is related to the LE sweep by the expression

$$AR = 4 \cot \Lambda \qquad (7.4)$$

The aspect ratio is a major design parameter as it has a large influence on the wing lift curve slope [Eq. (2.13)] and the subsonic cruise efficiency through its relationship with the subsonic inviscid drag-due-to-lift factor K',

$$K' = \frac{1}{\pi \, AR \, e} \qquad (2.18)$$

This relationship is shown in Fig. 7.8 for the wing–body combinations of Appendix H. Figure 7.8 also shows the reduction in K' through the tran-

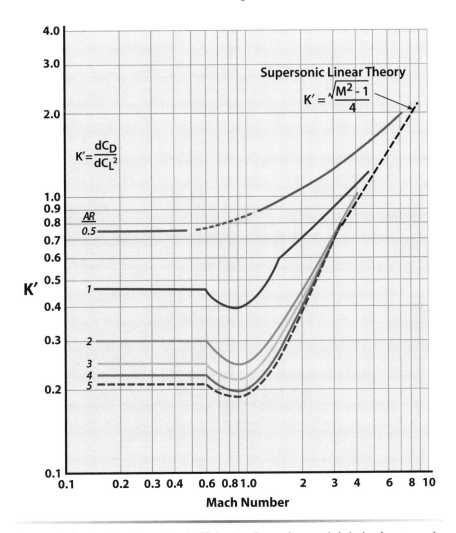

Figure 7.8 Inviscid drag-due-to-lift factor (based upon total planform area) for wing–body combinations with delta planforms and LE radius = 0.45%.

sonic region due to an increase in C_{L_α} and the reduced influence of AR on K' in the supersonic region.

The effect of AR on wing weight is shown quantitatively in Eqs. (20.1), (20.2), and (20.69) [and qualitatively in Fig. 7.10 (see Section 7.7); the effect on wing volume is shown qualitatively in Fig. 7.10].

For low-speed flight there is little effect of AR on C_{D_0} as the drag is primarily skin friction and independent of planform shape. However, supersonically the C_{D_0} increases with increasing AR as shown in Fig. H.6 and limits supersonic design to AR < 5.

7.7 Effect of Planform: Sweep

For the most part the effects of wing sweep are independent of whether the sweep is forward or aft. However, forward sweep does introduce the problem of *aeroelastic divergence*, which results from an increase in section angle-of-attack along the span (from root to tip) as the wing is deflected upward. As the forward-swept wing is loaded along the span the upward deflection increases the angle-of-attack of the tip section, which further loads the tip resulting in a divergent situation if the structural elastic restoring forces cannot halt the wing twist. One solution is to tailor the stiffness of the LE downward twist.

Designing to Counter the Aeroelastic Effect

With its forward swept wing, the X-29 demonstrated excellent roll control up to 60 deg; however, this type of wing exhibits aeroelastic divergence, which can quickly lead to structural failure. Part of the technology goal while designing the X-29 was to demonstrate tailoring of the carbon fiber composite material to counter this aeroelastic effect without a large wing structure penalty. The tailoring of the advanced composite material was successfully demonstrated on the X-29 and later used on the forward swept wing Advanced Cruise Missile AGM-129 (see Fig. 12.22). The X-29 first flew December 1984; it demonstrated subsonic and supersonic high alpha maneuvering from 1985–1991. See an X-29 at the USAF Museum (Dayton, Ohio) or at the NASA Dryden Flight Research Center (Edwards AFB, California).

Although a forward-swept wing would weigh more than an aft-swept wing, the forward sweep offers several advantages. A forward-swept wing should offer improved area rule distribution, longer lever arm between the wing and tail mean aerodynamic chord (mac), and reduced tip stall. This last advantage is quite important as it decreases stall-spin departure tendencies and gives lower landing speeds.

Essentially, the pressure distribution over a chordwise section of the wing is a function of the Mach number normal to the LE (Fig. 2.19). If the wing sweep results in a normal Mach number less than 1.0, the wing is said to have a subsonic LE at that freestream Mach number. If the normal Mach number is greater than 1.0, the wing has a supersonic LE (Fig. 2.24).

Wing sweep can delay and soften the transonic drag rise, as shown for a wing alone in Fig. 2.27 and for wing–body combinations in Fig. H.4. In Fig. H.4 the wing–body combinations represent wing sweeps of 0, 45, and 60 deg. Wing sweep permits subsonic aircraft to cruise at higher subsonic Mach numbers before encountering compressibility effects. Essentially the wing and fuselage drags are additive (plus some interference effects), with the fuselage C_{D_0} peaking at about Mach 1.2 (see Fig. 2.22).

At high subsonic speeds the effects of compressibility must be considered. The M_{CR} represents the onset of compressibility and the upper speed boundary for subsonic aircraft. As discussed earlier, it is desirable to keep M_{CR} as high as possible. The M_{CR} can be increased by increasing wing sweep, as shown in Fig. 7.9.

The peak wing C_{D_0} occurs during the transonic regime. An unswept wing has its peak C_{D_0} occur at a Mach number of about 1.1. The swept wing has its peak C_{D_0} occur at approximately

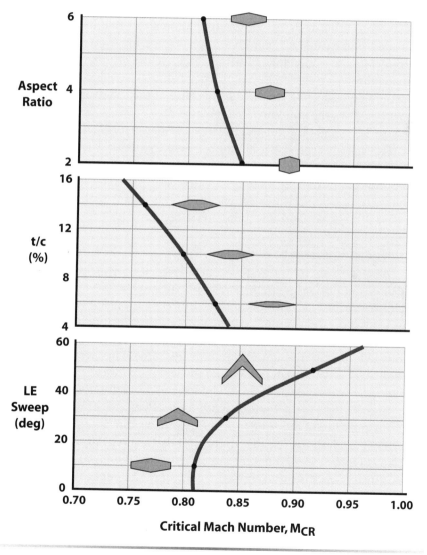

Figure 7.9 Effect of LE sweep, t/c, and AR on the critical Mach number.

$$M_{C_{D\text{peak}}} = \frac{1.2}{\sqrt{\cos \Delta_{t/c}}} \qquad (7.5)$$

where $\Delta_{t/c}$ is the sweep of the maximum thickness line. The value of the peak C_{D_0} is decreased by sweep. This C_{D_0} softening is estimated by

$$C_{D_{\text{peak}\,\Delta_{t/c}}} = \left(\cos \Delta_{t/c}\right)^{2.5} C_{D_{\text{peak}\,\Delta_{t/c}=0}} \qquad (7.6)$$

For supersonic flight the designer must decide whether the leading edge will be subsonic or supersonic. The selection of the leading edge radius rests largely on this decision. Equation (7.2) indicates that increasing wing sweep and decreasing AR will give lower values for $C_{D_{\text{LE}}}$. However, the resulting wing has poor low-speed qualities. If the leading edge is to be subsonic, the sweep should be about 5 deg behind the Mach line. The rule is, "just enough sweep to do the job," because of the disadvantages associated with high sweep and low AR at low speed.

Figure 2.29 shows the general influence of sweep on C_{D_0} for supersonic flight. The airfoil is the sharp LE double wedge in Table 2.1 so there is no drag due to LE bluntness.

A disadvantage of wing sweep is the decrease in wing lift curve slope as given by Eq. (2.13) and shown in Fig. 2.21. This means that a swept-wing aircraft will have to land and take off at higher angles-of-attack than a straight-wing aircraft.

Other disadvantages to wing aft sweep are a reduction in $C_{L_{\max}}$ and tip stall. The early flow separation at the tip is due to the spanwise flow causing a thickening of the boundary layer near the tips and hastening flow separation. This can be troublesome during low-speed flight because the roll control surfaces (the ailerons) are located near the tips. A forward-swept wing will have the opposite situation in that the root will stall early but the ailerons will operate in high-energy attached flow. This premature tip or root stall can be controlled by twisting the wing tip or root.

The wing sweep and aspect ratio interact to influence the pitch-up tendency of the wing. As the tip region of a high-AR, aft-swept wing stalls, the center-of-pressure of the wing moves forward, producing a pitch-up. This behavior is undesirable as the aircraft tends to pitch up violently, with disastrous results. Several fighter aircraft (such as the F-101 Voodoo) had horns, buzzers, or stick shakers that would warn the pilot of entry into the wing stall region. Figure 21.14a shows a pitch-up boundary developed by NASA from extensive wind tunnel and flight test data. Planforms in Region I are pitch-up prone and should be avoided for fighter aircraft. If a Region I planform is used, then the aircraft must have an aft horizontal tail that is located outside the wake of the stalled wing in order to arrest the divergent motion. Figure 21.14b illustrates four regions of horizontal tail location with general recommendations regarding pitch-up. Figure 21.14 can be used by the designer for general guidance in fighter aircraft planform selection and horizontal tail location.

The effect of wing aft sweep on wing weight is shown quantitatively in Eqs. (20.1), (20.2), and (20.69) and qualitatively in Fig. 7.10. The effect of forward sweep would be even greater because extra structure is added to stiffen the wing to arrest the aeroelastic divergence. The effect of wing sweep on wing volume is shown qualitatively in Fig. 7.10 to be negligible.

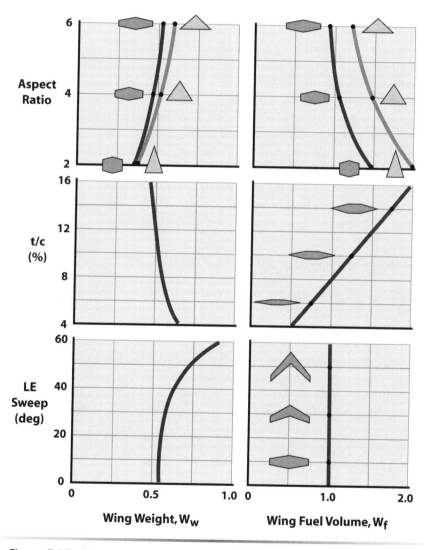

Figure 7.10 Effect of LE sweep, t/c, and AR on wing weight and wing fuel volume.

Wing weight is dependent to a large degree on the weight that is required to take out the bending moment. For a given wing area and moment about the centerline, this weight is a function of the structural span divided by the root thickness. The structural span can be considered as twice the length of the line bisecting the leading and trailing edge angle. If our structural span is high, our weight is high. If wing root thickness is high, wing weight is low. In the first chart of Fig. 7.10, LE sweep is doing one thing. It is increasing the structural span for the wing in question, thus increasing wing weight. Notice in the chart how the projected span remains the same, but the structural weight increases. The weight shown in Fig. 7.10 is nondimensionalized but is relative to the fuel weight, which we will take up shortly. On the aspect ratio plot is shown the wing weight of a delta wing at the same t/c as the straight wing. Reducing AR involves increasing the LE sweep, but here structural span is also reduced, and the root maximum thickness is increased. There is very little difference between a delta-wing weight and a straight-wing weight at the same AR and t/c.

On any long-range vehicle, the wing is extremely important not only as a device for holding the vehicle in the air, but also for carrying fuel because the wing is physically large and has a considerable amount of usable volume. If AR and wing area remain constant, the volume and, thus, fuel weight will not be affected. Therefore, sweep will not affect volume (bottom-right chart of Fig. 7.10). The volume will vary linearly with our thickness ratio, and fuel volume will decrease as we increase aspect ratio.

7.8 Effect of Planform: Taper Ratio

Taper ratio (ratio of tip chord to root chord, $\lambda = C_T/C_R$) has a fine-tuning effect on the wing performance. As the taper ratio increases from zero (a delta planform) toward 1.0 (a rectangular planform) it passes through a nearly elliptical lift distribution at $\lambda = 0.35$, which gives minimum finite-span downwash effects and minimum induced drag. For a given wing area and thickness ratio a delta wing planform will have a larger root chord than a rectangular planform, resulting in approximately 40% more volume available in the delta for fuel (see Fig. 7.10). From the wing weight equations of Chapter 20 it is observed that decreasing taper ratio from 1 to 0 gives a decrease in the wing weight due to the increased root depth and decreased tip loading.

7.9 Variable Geometry

The preceding discussions have pointed out that for good low-speed performance the aircraft should have low sweep and high AR, whereas, for good supersonic cruise the aircraft should have high sweep and low AR.

This conflict in design conditions is a real dilemma for the designer. For a fixed-wing aircraft there is no real solution. The design answer is one of compromise, trying to select the fixed-wing planform that will fit both ends of the performance spectrum with minimum degradation of the mission.

The key term here is "fixed wing" (see Fig. 7.11a). Why does the wing geometry have to be fixed? In fact, the wing geometry can be variable, capable of being adjusted to fit several different performance requirements. This is the idea behind the variable-geometry, or variable-sweep, wing.

The wing panels would be pivoted at the root so that each wing could swing from a low-sweep to a high-sweep condition. This idea is not new. The Bell X-5 and Grumman XF10F were early (1951–1952) research aircraft that demonstrated the versatility of high-speed and low-speed performance with variable sweep. Variable geometry is being used on current aircraft such as the Lockheed Martin (General Dynamics) F-111, Dassault Mirage IIIG, Sukhoi SU-7B, and the Mikoyan Flogger (see Fig. 7.11c).

The idea of variable geometry is a good one, but there are some disadvantages. The main disadvantage is weight. The variable-sweep wing, because of the wing pivot structure and associated machinery, is much heavier than a fixed-geometry wing. Estimates are that the variable-geometry wing weighs about 20% more than a comparable fixed wing. This 20% is significant. For example, the McDonnell F4C wing weighs 4600 lb for an aircraft gross weight of about 50,000 lb. A 20% increase in the wing weight would seriously cut into the aircraft performance.

Another disadvantage is the large shift in the aircraft aerodynamic center as the wings are swept back. This causes large stability and control problems. The wing glove (fixed portion of wing at the root) helps to eliminate part of this aerodynamic-center shift but it is still a problem. Also, if external stores are carried on wing pylons, the wing pylons must be able to swivel the stores for zero yaw angle at all wing sweep angles. This swivel capability also means additional weight.

Variable geometry is a good design feature but not for all aircraft. If an advanced design has conflicting mission requirements that cannot be compromised, then variable geometry should be considered. If the performance gains from using variable geometry outweigh the disadvantages, then the variable geometry "buys its way" onto the airplane. A good example is the U.S. Navy F-14 Tomcat shown in Fig. 7.12. This aircraft required a 120-kt approach speed for carrier landings and a supersonic acceleration to Mach 2.5. These and other performance requirements drove the F-14 design to variable geometry. The F-14 is shown in three-view in Fig. 7.13. The Tomcat protected the Navy battle groups for four decades; it was retired in 2007.

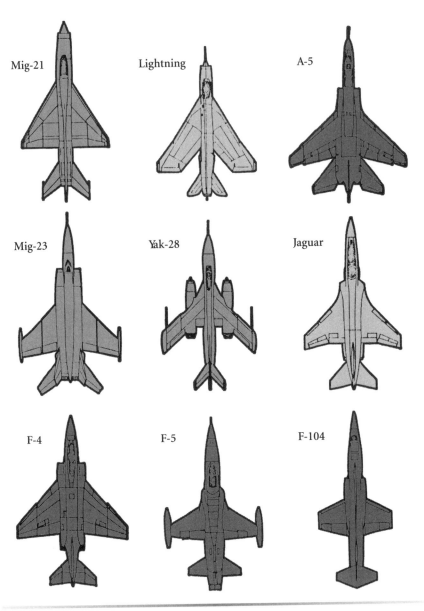

Figure 7.11a   Typical wing planform shapes for fixed-wing, conventional-tail aircraft.

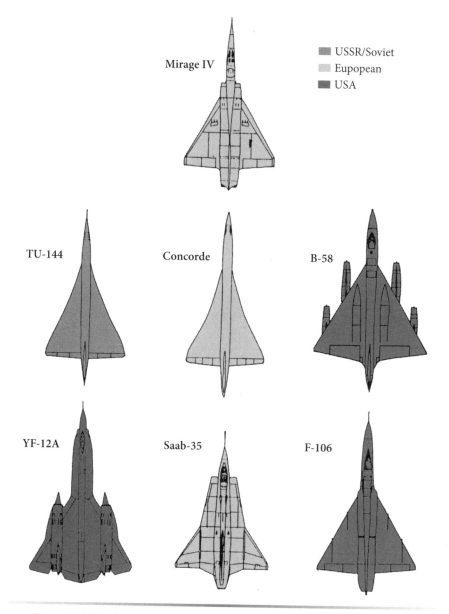

Figure 7.11b Typical wing planform shapes for tailless delta aircraft.

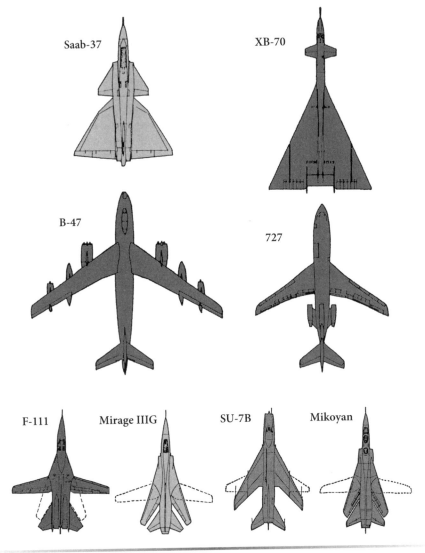

Figure 7.11c Typical wing planform shapes for canard, subsonic-cruise, and variable-sweep aircraft.

Another example of a current aircraft design that was forced into a variable-sweep wing is the Rockwell (North American) B-1 Lancer bomber shown in Fig. 7.14. The mission requirements called for an extended cruise range at Mach = 0.8, which dictated moderate sweep and high aspect ratio

Figure 7.12 USN Grumman F-14 Tomcat.

to give a good $(L/D)_{max}$. In addition, the B-1 must "dash on the deck" at Mach = 0.9 and be able to accelerate to Mach 2.2 at altitude. These two requirements call for low aspect ratio and high sweep. The high sweep alleviates the aircraft gust response during the low-altitude dash and results in a much smoother ride. The B-1 is shown in 3-View on Fig. 7.15.

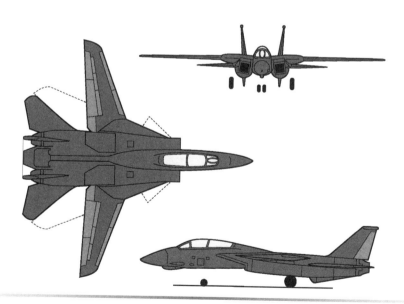

Figure 7.13 Three-view of U.S. Navy Grumman F-14 Tomcat, a two-seat carrier-based swing-wing fighter (W_{TO} = 54,000 lb, length = 62 ft, extended wing span = 64 ft, and sweep = 20–68 deg).

Figure 7.14 USAF/Rockwell B-1 Lancer.

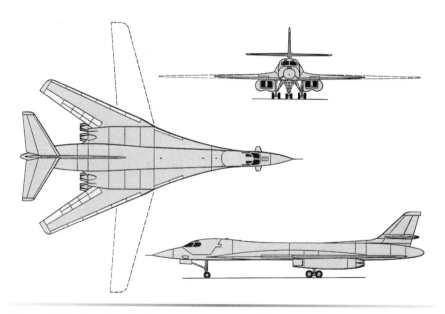

Figure 7.15 Rockwell B-1 strategic bomber (W_{TO} = 400,000 lb, length = 143 ft, extended wing span = 137 ft, and sweep = 15–65 deg).

Table 7.1   Summary of Airfoil and Planform Effects

Increase In / Changes	C_{D0} Subsonic	C_{D0} Supersonic	K	$C_{L\alpha}$	C_{Lmax}	Wing Wt	Wing Vol
Aspect Ratio	No effect	↑	↓	↑	↑	↑	↑
Wing Sweep	No effect	↓	↑	↓	Aft ↓ / Fwd No effect	↑	No effect
Taper Ratio	No effect	No effect	↑↓	↑	No effect	↑	↑
Airfoil Thickness Ratio	No effect	↑	No effect	No effect	↑	↓	↑
Leading Edge Radius	No effect	↑	↑	No effect	↑	No effect	No effect
Camber	↑	↑	↑	No effect	↑	No effect	No effect

7.10 Summary

Selecting the right airfoil and planform is a difficult task. Even experienced designers can be confused, but the point to remember is that there is no right answer—only a best answer at a point in time. And the best answer will involve making compromises in the measures of merit listed at the beginning of this chapter.

Table 7.1 is a summary of the airfoil and planform effects discussed in this chapter. The measures of merit are listed along the top and the airfoil–planform features are along the left-hand side. The table indicates the effect on the measure of merit if the airfoil or planform feature is *increased*.

References

[1] Abbott, I. H., and Von Doenhoff, A. E., *Theory of Wing Sections*, Dover, New York, 1959.

[2] Abbott, I. H., Von Doenhoff, A. E., and Stivers, L., Jr., "Summary of Airfoil Data," NACA TR-824, 1945.

[3] Riegels, F. W., *Aerofoil Sections*, Butterworth, London, 1961.

[4] *Evolution of Aircraft Wing Design Symposium*, AIAA, Reston, VA, 1980.

[5] McCullough, G. B., and Gault, D. E., "Examples of Three Representative Types of Airfoil-Section Stall at Low Speed," NACA TN-2502, 1951.

[6] Simon, W. E., Ely, W. L., Niedling, L. G., and Voda, J. J., "Prediction of Aircraft Drag Due to Lift," U.S. Air Force Flight Dynamics Laboratory AFFDL-TR-71-84, July 1971.

Chapter 8

Preliminary Fuselage Sizing and Design

- Passenger Seating
- Supersonic Area-Ruling
- Body Fineness Ratio
- Integrating Propulsion
- Integrating Crew
- Integrating Landing Gear
- Initial Fuselage Length
- Initial C.G. Location

The Convair F-102 Delta Dagger was built in the 1950s as part of the USAF's air defense. The prototype (left) could not go supersonic, but using area rule theory (see Section 8.4) to revise the aircraft design solves the problem and gives the fuselage a "Coke bottle" appearance at the wing/fuselage intersection (right).

"I must do something" will always solve more problems than "Something must be done."

T he aircraft is now beginning to take shape. In Chapter 5 a preliminary estimate of the takeoff weight and fuel weight was established.
The takeoff wing loading was also determined, in Chapter 6, so
that the wing is sized. The airfoil section and planform shape derive from
the work in Chapter 7. Designing and sizing the fuselage is the next step.

8.1 Fuselage Volume

Preliminary estimates of fuselage volume requirements are relatively
easy to assess, provided the designer has decided upon its contents. The
cutaway drawings in this chapter will help remind the designer of what is
usually put into the fuselage. Reference [1, Vols. 3 and 4] is an excellent
reference for fuselage design. This chapter focuses on two important design items: an initial fuselage length and the c.g. location. The following
points should be considered when locating all equipment, payload, subsystems, fuel, and structure in the aircraft.

Going Supersonic with Area Rule Theory

The Convair F-102 Delta Dagger was an interceptor aircraft built in the late
1950s as part of the USAF's air defense. Entering service in 1956, its main
purpose was to intercept invading Soviet bomber fleets. The 1951 RFP called
for a high-altitude, supersonic interceptor armed with guided missiles to
replace the F-86 Saber Jet and the F-89 Scorpion.

Convair won the competition and developed the prototype YF-102 (see
chapter opener art), which flew in 1954 and had dismal performance. Despite
Convair's prediction, the airplane could not go supersonic. It could not even
reach Mach = 1 because of excessive transonic drag [3]. To solve this problem
of higher-than-expected transonic drag, Convair engineers employed an
emerging NASA technology developed by Richard Whitcomb called *area
ruling* (discussed in Section 8.4).

Area rule theory dictates that cross-sectional area distribution from nose
to tail should be smooth and continuous to give low wave drag, meaning that
the fuselage cross-sectional area in the vicinity of the wing should be reduced
to accommodate the cross-sectional area of the wing.

These criteria gave the fuselage a "Coke bottle" appearance at the wing/
fuselage intersection. The pinched-in waist of the redesigned Delta Dagger is
typical of an area-ruled design. Once this design adjustment was made, the
YF-102 accelerated out to Mach 1.22 at 53,000 feet. Convair went on to build
1000 Delta Daggers.

Table 8.1   Passenger Compartment Requirements

	Long Range	Short Range
Seat width (in.)	17–22	16–18
Seat pitch (in.)	34–36	30–32
Headroom (in.)	>65	>65
Aisle width (in.)	18–20	>12
Aisle height (in.)	>76	>60
Passengers per attendant	31–36	<50

8.1.1 Passengers

The passenger seating in a transport aircraft varies depending upon whether the section is first class, business class, or coach. The first-class and business-class sections have no typical seating arrangement. The arrangement in these more expensive sections depends upon the airline's desire to attract a more affluent clientele. In long-range transports these sections often feature sleeper seats, lounges, and high attendant-to-passenger ratios. Typically these sections accommodate less than 10% of the total number of passengers because the revenue per volume (space required) used is much less than for the coach section.

The seating arrangement in the coach section is driven by a desire to have the most passengers per cubic foot of volume, while still maintaining passenger comfort. Each coach passenger should have the volume shown in Table 8.1 and Fig. 8.1a. Each passenger is assumed to weigh 180 lb (this includes an allowance for carry-on baggage). Each passenger is allowed ~40 lb of baggage (a volume of approximately 15 ft^3) on domestic flights and 65 lb (approximately 25 ft^3) on international flights. In the coach section the number of seats across depends on the size of the aircraft and the selection of a single or double aisle. Table 8.2 shows the numbers of passenger seats across the aircraft for current aircraft.

The passenger section of the aircraft usually has a round cross section because that is the most structurally efficient shape for a conventional metal structure. However, the most volumetrically efficient cross section is an oval, which is the shape of the Boeing 787 composite fuselage. The passengers are located in the upper half of the cross section; baggage and cargo, in the lower half.

The passenger and cargo sections are pressurized to 6500-ft pressure altitude. The pressure differential between 6500 ft and the typical operating altitude of 35,000–40,000 ft is a major factor in the structural and fatigue design criteria for the aircraft. For each row of seats it is desirable to

a)
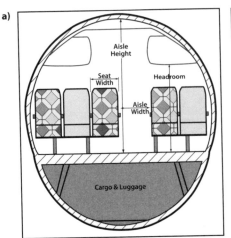
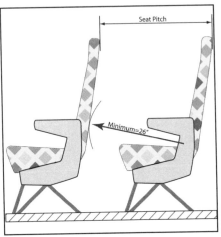

b)

Figure 8.1 Commercial transport seating arrangement: **a)** schematic; **b)** compare the empty cabin (left, courtesy of Sharam Sharifi) to the full interior (right, courtesy of Wikimedia Commons).

Table 8.2 Aisle and Passenger Distributions for Various Commercial Transports

Aircraft	Aisle	No. Passengers on Sides	No. Passengers in Middle
Boeing 727	Single	3 and 3	0
Boeing 737	Single	3 and 3	0
Airbus 300	Double	2 and 2	4
Boeing 747	Double	3 and 3	4
Boeing 757	Single	3 and 3	0
Boeing 767	Double	2 and 2	3
Boeing 777	Double	2 and 2	5
Lockheed L-1011	Double	2 and 2	4
DC-10	Double	2 and 2	5
Boeing 787	Double	2 and 2	5

have a window that is as large as possible. Typical window spacing is 40 inches for 14×10 inch windows.

Reference [1, Vol. 3], has considerable information on passenger section layout for commercial transports.

8.1.2 Lavatories, Galleys, and Emergency Exits

The size of these items will vary depending on the aircraft and number of passengers. There should be one lavatory for approximately every 20 passengers. The number and type of emergency exits required for passenger transport aircraft is defined in the Federal Aviation Regulations (FAR Part 25.807) [2].

8.1.3 Passenger Cargo

The passenger luggage and revenue cargo for the aircraft listed in Table 8.2 is preloaded into standard-size cargo containers and carried beneath the passengers in the cargo compartment. Table 8.3 lists the dimensions for the most widely used cargo containers. Smaller, short-range aircraft do not use cargo containers, but rather have space only for bulk cargo with a volume that is based upon $6-8$ ft^3 per passenger.

8.1.4 Military Cargo and Equipment

The military cargo is preloaded onto flat pallets, tied down and covered with a tarp. The most common is the 463L pallet which measures 108×88 inches. MIL-STD-1791 requires 6-in. clearance in all directions between the cargo and the aircraft interior. Military transports must have their cargo compartment floor approximately $4-5$ feet off the ground to allow for direct loading and unloading of the cargo pallets from a truck bed at air bases without cargo-handling facilities.

Table 8.3   Cargo Container Specifications

Type	Height (in.)	Width (in.)	Depth (in.)	Weight (lb)
LD-2[a]	64	61.5	60.4	2700
LD-3[a]	64	79	60.4	3500
LD-4	64	96	60.4	5400
LD-5	64	125	60.4	5400
LD-8[b]	64	125	60.4	5400

[a]Lower corner chamfered 30 deg.
[b]Both lower corners chamfered 30 deg.

The military transports have the dimensions of their cargo compartments sized by the equipment they need to carry. Typical equipment includes jeeps, humvees, armored personnel carriers, special forces boats, and numbers of 463L pallets. The C-130 has a cargo bay that measures 10 ft 3 in. wide × 9 ft 2 in. high × 41 ft 5 in. long. The C-5 and C-17 were developed to carry "outsize" cargo such as M-60 tanks, helicopters, and large trucks. The C-5 has a cargo bay with dimensions of 19 ft wide × 13 ft 6 in. high × 121 ft long. These military transports need a ramp at the rear of the aircraft to load and unload the equipment.

8.1.5 Crew Compartment

The size of the crew compartment varies depending on the aircraft. For long-range military or commercial transports it is recommended that the crew compartment have a length approximately 150 inches for a crew of four, 130 inches for a crew of three, and 100 inches for a crew of two. This gives the crew room to get out of their seats and stretch their legs as well as room to store their map cases and flight bags.

The size of the cockpit for a fighter depends on the number of crew and whether the seating arrangement is tandem or side-by-side. The cockpit arrangement uses a typical 13-deg seatback angle, although angles up to 30 deg have been used (i.e., F-16) to provide better "*g*" tolerance for the pilot during air combat. Typical fighter single-seat crew station dimensions are 30 in. wide × 50 in. to the top of the canopy × 60 in. from the foot pedals to the back of the seat. An ejection seat is required for safe escape when flying at a dynamic pressure greater than 230 psf (equivalent to 260 knots at sea level). At speeds approaching Mach = 1 at sea level (dynamic pressure = 1480 psf), even an ejection seat is unsafe and an encapsulated seat or crew capsule must be used. The FB-111 and B-1A flew fast and low and used a separable crew capsule. The crew capsules were heavy and complex but gave the crew a chance of surviving a high-q ejection. Reference [3] has an excellent chapter on fighter crew station design.

MIL-STD-850B defines the vision requirements for various classes of military aircraft in terms of over-nose and over-side vision angles. These angles are important during low-altitude maneuvering and allow the pilots to see the runway threshold during landing approach. Because all landing approaches are different and all aircraft have different approach angles-of-attack, the minimum over-nose angles shown in Table 8.4 are typical recommended values. An over-nose analysis would put the aircraft at the α for $0.8C_L$max and an approach angle of 3 deg for commercial and U.S. Air Force aircraft, 4 deg for a carrier approach in a Navy or Marine aircraft, and 7 deg for a short takeoff and landing (STOL) aircraft. The supersonic transports Concorde and Russian TU-144 drooped the entire nose, as shown in Fig. 8.2, during landing to provide pilot vison.

Table 8.4 Typical Minimum Over-Nose and Over-Side Pilot Viewing Angles

Aircraft	Over-Nose	Over-Side
Military transports and bombers	17 deg	35 deg
Commercial transport	11–20 deg	35 deg
Fighter	11–15 deg	40 deg
General aviation	5–10 deg	35 deg

8.1.6 Armament

The number and size of bombs must be determined and they must be located in or on the aircraft. Arrangements for storage, positioning, and release of missiles must be considered.

Guns or cannon can be carried in gun pods external to the aircraft or mounted internally. Armament carried external to the aircraft will provide lower fuselage volume requirements but will cause greater drag on the aircraft. Reference [1, Vol. 4] has an excellent discussion on armament integration and size–weight information on air-to-air and air-to-ground weapons.

8.1.7 Landing Gear

The size and location of the landing gear will vary depending on the aircraft. A good first estimate can be made by examining existing aircraft in the same weight class as you are designing. References [1,4] have good discussions on landing gear design.

Figure 8.2 Concorde with nose drooped for landing (courtesy of Wikimedia Commons).

The main landing gear is located relative to the c.g. of the aircraft based upon the following two considerations:

1. The airplane should not fall on its tail or nose at any possible loading condition.
2. The moment to rotate the aircraft about the main gear to $0.8C_L$max at V_{TO} and worst c.g. location should not size the horizontal tail (or canard or wing flaps).

Based upon these considerations the main gear wheel (with strut depressed) is located behind the most forward c.g., for a tricycle gear, and in front of the most aft c.g., for a tail dragger, by the angles shown in Table 8.5. For carrier-suitable Navy aircraft 15 deg behind c.g. is necessary based on the real possibility of a 5-deg pitching deck. This main gear–c.g. geometry is shown in Figs. 8.3 and 11.1.

A good rule of thumb for the nose gear is to have 20% of the takeoff gross weight (TOGW) on the nose wheel for good nose-wheel steering. The nose gear is located so that the aircraft does not tip over during high-speed taxi turns. This tip-over geometry and typical tip-over angles are shown in Fig. 8.3.

As the aircraft rotates about the main gear for takeoff or touches down for a landing the aft end of the fuselage must not strike the ground. The angle between the main gear in an extended strut position and the aft end of the fuselage is called the *tip-back angle*, shown in Fig. 8.3. The tip-back angle is determined by rotating the aircraft to the α for $0.9C_L$max about the main gear in the extended strut position and observing if any part of the aft fuselage touches the static ground line (see Fig. 8.3). During normal takeoff and landing the aircraft is rotated to $0.8C_L$max. The value $0.9C_L$max is used in the tip-back analysis to account for pilot overshoot.

Aircraft typically have a 3-deg glide slope during approach and then flare over the threshold, giving a 10-ft/s sink rate at touchdown. The flare is an imprecise maneuver and results in considerable dispersion about the touchdown point and great variance in landing distance from one pilot to another. Carrier-suitable Navy and Marine aircraft approach the carrier at about 4 deg and not flare at touchdown (reducing the dispersion distance), giving a sink rate of 24 ft/s and a much heavier landing gear weight require-

Table 8.5 Angles for Location of Main Landing Gear

Gear Type	Aircraft	Angle for Main Gear
Tail dragger	All	15 deg forward of c.g.
Tricycle	All	10 deg behind c.g.
Tricycle	Carrier suitable	15 deg behind c.g.

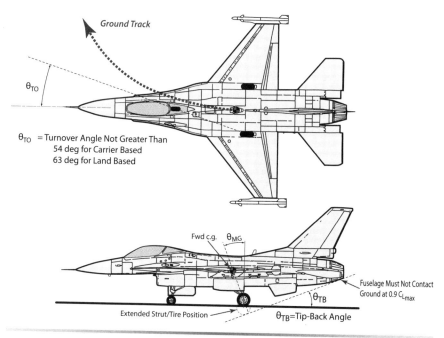

Figure 8.3   Aircraft turnover and tip-back angle definitions.

ment. Supporters of unmanned aerial vehicles (UAVs) argue that autonomous landing systems result in a highly consistent and low touchdown sink rate of 5 ft/s and should therefore have lighter landing gears.

8.1.8 Wing Carry-Through

Although not visible externally, volume must be made available for the wing carry-through. Because the wing is thickest at the root chord this will account for a considerable portion of the fuselage volume requirement. The carry-through may be either a straight carry-through of the wing center section or a ring-type construction following the fuselage cross section outer contour (see Figs. 8.4–8.10 and Figs. 19.1 and 19.2).

8.1.9 Propulsion Integration

Engines may be mounted internally as in the F-15 (Fig. 8.4) and F-18 (Fig. 8.5), partially embedded in the fuselage as with the F-4 (Fig. 8.6), or completely external as with the DC-9 (Fig. 8.7). The internal and partially internal arrangements are difficult to assess because engine size and number are not yet known. If the propulsion units will likely be internal, the designer should reserve some fuselage volume for the engines. First estimates can be

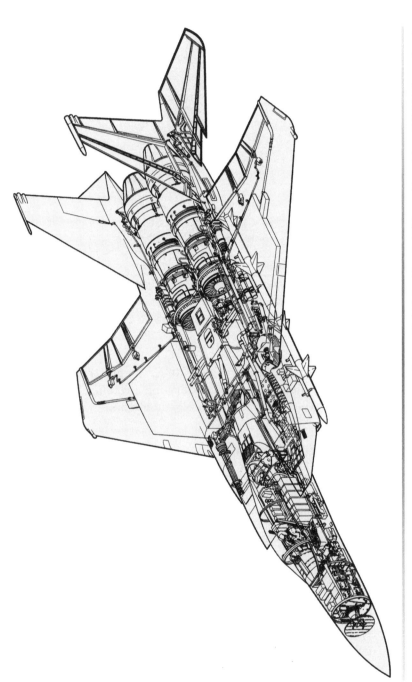

Figure 8.4   Boeing F-15A Eagle internal arrangement ($W_{TO} = 40{,}000$ lb, AR = 3, $W/S_{TO} = 66$ psf, two PW F-100 engines). (Courtesy of The Boeing Company.)

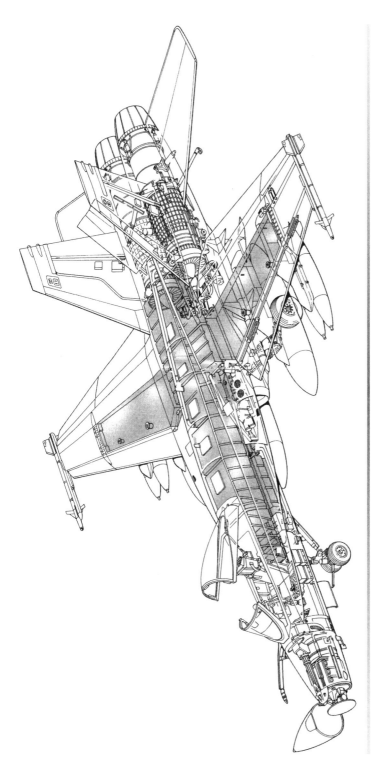

Figure 8.5 Boeing F-18C Hornet internal arrangement ($W_{TO} = 56,000$ lb, AR = 3.5, $W/S_{TO} = 140$ psf, two GE F404-400 turbofans). (Courtesy of The Boeing Company.)

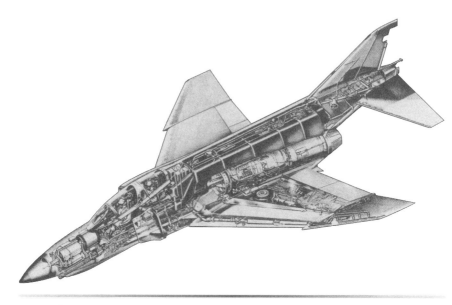

Figure 8.6   F-4 Phantom II internal arrangement, a high-performance fighter–bomber (W_{TO} = 54,000 lb, length = 58 ft, wing span = 38.5 ft, maximum Mach = 2.1). (Courtesy of The Boeing Company.)

made from Figs. 8.4 and 8.5. Internal arrangements for the B-1, F-16, and Piper Comanche are given as further examples in Figs. 8.8–8.10.

If jet engines are mounted in or partially within the fuselage, the volume required for the inlet must be reserved. A first estimate of the inlet volume

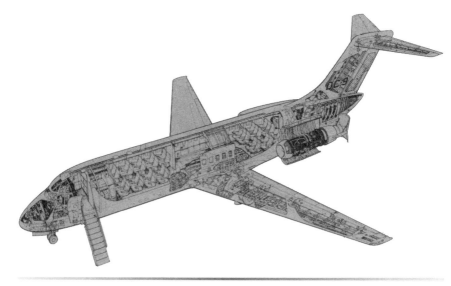

Figure 8.7   DC-9 internal arrangement, a subsonic, short- to medium-range jet transport (W_{TO} = 114,000 lb, 70–125 passengers, two JT8D turbofans). (Courtesy of The Boeing Company.)

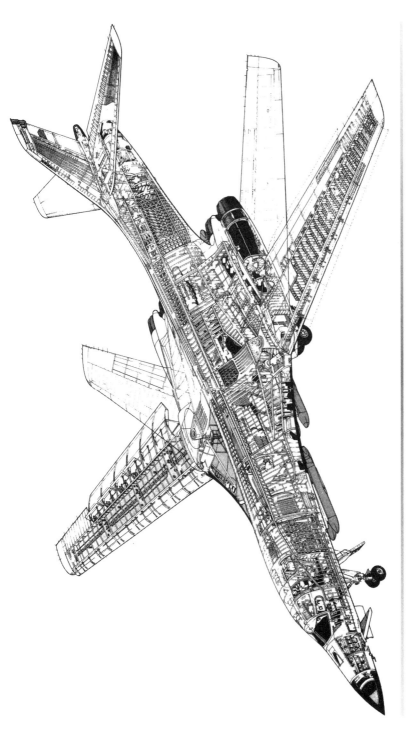

Figure 8.8 Boeing B-1B Lancer internal arrangement [W_{TO} = 477,000 lb, AR(ext) = 9.6, AR(swept) = 3.1, four GE F-101-102 turbofans]. (Courtesy of The Boeing Company.)

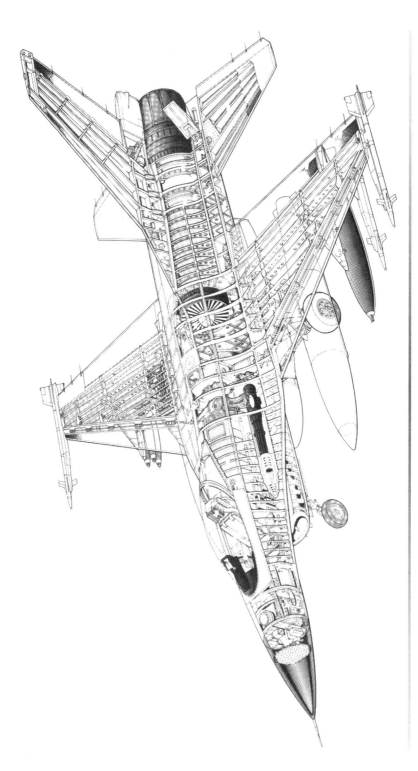

Figure 8.9 Lockheed Martin F-16A Fighting Falcon internal arrangement (W_{TO} = 33,000 lb, wing span = 32.8 ft, length = 49.3 ft, one PW F-100-100 turbofan).

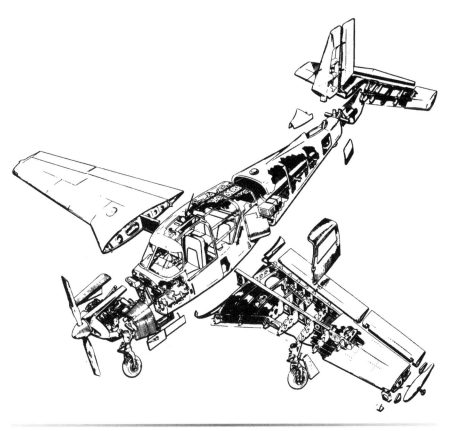

Figure 8.10   Piper Comanche internal arrangement, a four-place general aviation aircraft (W_{TO} = 2800 lb, W_{empty} = 1600 lb, length = 24.9 ft, wing span = 36 ft) (courtesy of Piper Aircraft Corp.).

required can be determined by using the diameter of the engine compressor face and a length equal to six-tenths of the engine length.

8.1.10 Fuel

The large quantity of fuel required for the mission will be carried within the fuselage, the wing structure, or both. The decision must be made about where to store fuel and how "wet" the wing structure will be. Final fuel placement will depend a great deal upon weight and balance requirements plus vulnerability to enemy fire. Use the fuel densities in Table 8.6 to determine the fuel volume requirements.

Locate the fuel around the c.g. with provision for pumping fuel to keep the c.g. envelope small. The wing is usually a good place to put the fuel as the wing is always located close to the c.g., resulting in a small c.g. movement as the fuel is burned. The fuel tank volume required to house the fuel

Table 8.6 Fuel Densities

Fuel	Gallon Weighs (lb)	Cubic Foot Weighs (lb)
JP-4	6.5	48.6
JP-5	6.8	51.1
JP-8	6.7	50
Aviation gas	6.0	44.9

is determined using *packaging factors*, which account for the increased tank volume to accommodate structure, pumps, baffles, fuel lines, and general fuel tank inefficiencies:

Tank volume = (fuel volume)/(packaging factor)

When locating the fuel tanks in the aircraft, use the following packaging factors:

Tank Type and Location	Packaging Factor
Integral tank	
Shallow fuselage	0.8
Deep fuselage	0.85
Wing	0.75
Bladder tank	
Fuselage	0.75
Wing	0.65

8.1.11 Avionics

The *avionics equipment* consists of the communications and navigation gear, radar, fire control system, penetration aids, autopilot, and instrumentation. This equipment may be included in the mission specifications or left up to the designer. Table 8.7 lists current avionics weights and volumes for common avionics systems. If the weight of the avionics gear is known, the volume required can be estimated by assuming an avionics equipment density of 45 lb/ft^3. If the power or volume requirements are known, the avionics equipment weights can be determined from Table 8.8.

Avionics equipment must be maintained frequently, so the equipment must be located for easy access by the ground crew. The equipment must not be stacked, which would require that a good piece of equipment be removed to get to the faulty piece. McDonnell did not adhere to this rule in the F-4 and located some avionics equipment under the rear ejection seat. U.S. Air Force maintenance records reported the rear ejection seat as a "high maintenance" item when in fact it was getting a bad rap: it was being removed frequently only to get to the low-reliability avionics equipment [5].

In many cases the avionics equipment will need to be cooled, so make provision for good cooling by locating the items on cooling plates and separating the items for good air circulation.

Table 8.7   Weights and Volumes for Common Avionics Equipment

Item[a]	Model Designation	Volume (ft³)	Weight (lb)
Intercom system	AIC-25	—	19.2
UHF communications	ARC-109	—	51.0
	ARC-150	0.21	11.0
UHF DF horning	705CA	—	5.0
Air-to-ground IFF	APX-64	—	53.0
	APX-92	0.11	13.0
TACAN	ARN-52	—	61.0
	ARN-100	1.1	46.0
ILS-VOR	ARN-584	—	27.0
	RCS-AVN-220	0.05	3.5
Gyrocompass	ASN-89	0.21	8.4
Inertial navigation system	AJQ-20	—	207.0
	LN-30	1.08	44.0
High-frequency radio	ARC-123	—	78.4
Autopilot system	—	—	168.5
Air data computer	AXC-710	0.5	14.0
Radar warning and homing	APS-109	—	182.0
	APR-41	0.17	22.0
ECM equipment	ALQ-103	—	637.0
Countermeasures dispensing set I	ALE-28	—	117.0
Countermeasures receiving set	ALR-23	—	94.0
Radar altimeter	APN-167	—	38.2
Attack radar	APQ-113	—	387.2
Range-only radar	SSR-1 (GE)	0.55	25.0
Terrain-following radar	APQ-110	—	249.0
Head-up display	TSP-2199	1.6	37.0
Gun camera	16-mm Telford	0.03	2.0
Lead computing optical sight	ASG-23	—	5.0
Flight data recorder	—	0.3	15.6

[a]Abbreviations: UHF, ultrahigh frequency; DF, direction finder; IFF, identification, friend or foe; TACAN, tactical air navigation; ILS-VOR, instrument landing system, very-high-frequency omnidirectional radio; ECM, electronic countermeasures.

Table 8.8   Statistical Methods for Estimating Avionics Weight Given Volume or Power

Radar Systems:
$Wt = 0.431(Power)^{0.777}$ $Wt = 38.21(Volume)^{0.873}$
for radar weight (less antenna) in pounds, power in watts, and volume (less antenna) in cubic feet
Doppler Navigation Systems:
$Wt = 0.408(Power)^{0.868}$ $Wt = 29.67(Volume)^{0.662}$
for weight in pounds, power in watts, and volume in cubic feet
Inertial Navigation Systems:
$Wt = 0.465(Power)^{0.848}$ $Wt = 51.85(Volume)^{0.738}$
for weight in pounds, power in watts, and volume in cubic feet
TACAN Systems:
$Wt = 13.61 + 0.104(Power)$ $Wt = 0.311(Volume)^{0.704}$
for weight in pounds, power in watts, and volume in cubic inches
Receiver Systems:
$Wt = 6.3 + 0.17(Power)$ $Wt = 44.5(Volume)^{0.737}$
for weight in pounds, power in watts, and volume in cubic feet
Transmitter Systems:
$Wt = 0.73(Power)^{0.610}$ $Wt = 6.4 + 40.2(Volume)$
for weight in pounds, power in watts, and volume in cubic feet
Identification Systems:
$Wt = 0.607(Power)^{0.724}$ $Wt = 0.069(Volume)^{0.868}$
for weight in pounds, power in watts, and volume in cubic inches
Computers:
$Wt = 2.246(Power)^{0.630}$ $Wt = 0.123(Volume)^{0.817}$
for weight in pounds, power in watts, and volume in cubic inches
Electronic Countermeasures (ECM):
$Wt = 0.429(Power)^{0.771}$ $Wt = 0.055(Volume)^{0.912}$
for weight in pounds, power in watts, and volume in cubic inches

8.1.12 Carrier Suitability

Carrier suitability is the term for the design attributes that let an aircraft operate from a Navy carrier (CVN class). It covers the high sink rate (24 fps) landing gear, the wing fold to accommodate the elevator/below deck dimensions and increase spotting factor; it also includes forebody and nose gear structure to react catapult loads, the aft body structure to absorb the arrested landing loads, resistance to the salt water environment, and many other unique maritime features. These and other carrier suitability requirements are discussed in References 11 and 12.

8.1.13 Wrap It Up

Once the required volume has been determined for each of the fuselage sections, the fuselage can be "packaged" (locating all the internal items in the fuselage) and the initial length determined. All equipment and subsystems should be designed for easy access. The rule for good "design for maintainability" is as follows:

- Place equipment one deep—do not stack or hide.
- Place equipment chest high—to minimize the need for stands and ladders on the flight line.
- Make all replaceable equipment less than 40 lb—to minimize the need for more than one person or special equipment to remove or replace equipment (engines are exempt).

Consider the fuselage to be a cone-cylinder shape and assume a diameter. Then determine the length required for each of the fuselage sections. This will give the initial fuselage sizing requirement. The fuselage length and diameter can then be juggled to give the desired fuselage fineness ratio as discussed in Section 8.2.

Locate the tail at the aft end of the fuselage and estimate the empennage (horizontal plus vertical tails) weight from Table 8.9 (data from Appendix I).

Determine an initial c.g. location as follows: Assign a weight to every item of the aircraft except the fuselage, the wing, and any item on the wing (fuel, engines, weapons, etc.) and determine the c.g. of the ensemble. Chapter 20 contains weights of many of the minor items such as crew seats, passenger seats, galleys, and lavatories. The reason the fuselage is excluded is that the c.g. of the fuselage is typically about the c.g. of the aircraft, and the wing will be located at the c.g.

Table 8.9 Initial Estimation of Empennage Weight

Aircraft Type	Empennage Area per Wing Area	Empennage Weight per Area
Jet transports	0.44	5.0
Business jets	0.43	4.3
General aviation		
Single engine	0.3	1.1
Twin engine	0.45	1.44
Intelligence, surveillance, and reconnaissance	0.2	3.0
Supersonic fighters		
Land based	0.39	7.0
Carrier based	0.48	6.0

The wing is now located on the fuselage such that the c.g. is at approximately 30% of the wing mean aerodynamic chord. This location will be refined later as more becomes known about the airplane. The designer can now draw the complete airplane and locate the landing gear based on the guidelines in Section 8.1.7. The tip-back and tip-over angles discussed in Section 8.1.7 should be checked.

8.2 Fuselage Fineness Ratio

The *fuselage fineness ratio* is defined as the fuselage length divided by its diameter, l/d. The optimum l/d for the fuselage is different for subsonic and supersonic flow:

- **Subsonic flight.** The subsonic C_{D_0} for a fuselage is a compromise between skin friction drag coefficient C_F and the pressure drag coefficient due to viscous separation $C_{D_{P_{min}}}$. The variation of subsonic C_{D_0} (based upon maximum cross-sectional area) with the inverse of fineness ratio, d/l, is shown in Fig. 8.11 [6]. A $d/l = 1$ is a sphere and Fig. 8.11 shows that C_{D_0} is predominately viscous separation. The C_{D_0} has a minimum value at a d/l of approximately 0.33. Thus, a fineness ratio of 3 gives near-minimum C_{D_0} for subsonic flight.
- **Supersonic flight.** For supersonic flow the C_{D_0} on a streamlined body (i.e., no blunt base) is a compromise between skin friction C_F and wave drag C_{D_w}. The variation of supersonic C_{D_0} with d/l is shown in Fig. 8.11 [6]. From Fig. 8.11 the minimum C_{D_0} occurs for a fineness ratio of approximately 14.
- **Mixed subsonic and supersonic flight.** If the aircraft spends the majority of its flight at either subsonic or supersonic speeds, then the fineness ratios just discussed should be used. An SST is an aircraft that should have a fineness ratio of 14, for example. However, if the aircraft spends about half of its flight at subsonic and the other half at supersonic speeds, then the fuselage fineness ratio should be a compromise between the two conflicting criteria. The F-15 and other fighter aircraft should have a fineness ratio of 8–10 for minimum fuselage C_{D_0}.

8.3 Fuselage Shapes

The fuselage should be a streamlined shape with a tapered aft end. A blunt aft end would cause the flow to separate, with a large increase in C_{D_0} due to the afterbody separation drag (called *base drag* in supersonic flow). Some possible shapes for the fuselage are discussed in the following subsections.

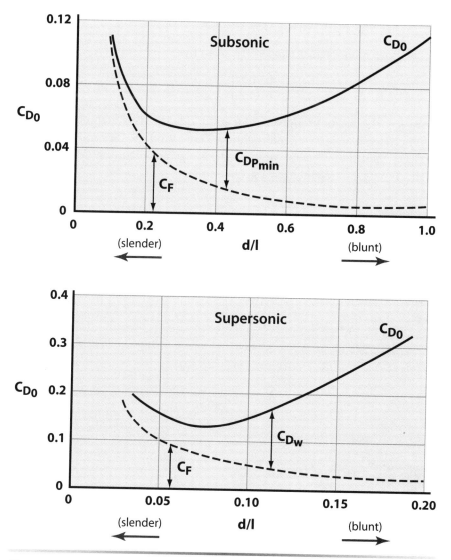

Figure 8.11 Subsonic and supersonic zero-lift drag for various fineness ratios.

8.3.1 Cone-cylinder

The C_{D_0} is easy to determine because subsonically it is primarily skin friction and supersonically $C_{D_w} = Cp$. The pressure coefficient Cp can be determined from the conical-shock charts in Appendix D.

8.3.2 Ogive-cylinder

The ogive is similar to a cone except its shape is formed by segments of arcs rather than by straight lines. It is better than a cone in that it has a greater volume for a given base diameter and length.

8.3.3 Power Series–cylinder

The power series nose shapes are given by

$$\left(\frac{R}{d/2}\right) = \left(\frac{x}{\ell}\right)^n$$

where R is the radius at a given x location, d is the base diameter, and ℓ is the length from nose to end of forebody. There is much data available in the literature for different values of n. An $n = 3/4$ gives the minimum wave drag for this family. Notice that $n = 1$ gives a cone.

8.3.4 Von Kármán Ogive

This is for a half-body of given length and diameter. The following results are from [7]:

$$\left(\frac{R}{R_{max}}\right)^2 = \frac{1}{\pi}\left[\frac{2x}{\ell}\sqrt{1-\left(\frac{2x}{\ell}\right)^2} + \cos^{-1}\left(\frac{-2x}{\ell}\right)\right]$$

for $-\ell/2 \leq x \leq \ell/2$. The volume is $(1/2)\ell S_{max}$ and

$$C_{D_W} = \frac{4 S_{max}}{\pi \ell^2}$$

8.3.5 Sears–Haack Body

This is for a complete body of given length and volume. The following results are from [7,8].

The area distribution is shown in Fig. 8.12 and is described by the equation

$$\left(\frac{R}{R_{max}}\right)^2 = \left[1-\left(\frac{2x}{\ell}\right)^2\right]^{3/2} \quad \text{for } -\frac{\ell}{2} \leq x \leq \frac{\ell}{2}$$

The volume is $(3/16)\ell S_{max}$ and

$$C_{D_W} = \frac{9}{2}\frac{\pi}{\ell^2} S_{max}$$

where the wave drag coefficient C_{D_W} is referenced to the maximum cross-sectional area S_{max}. The wetted area $= 1.8667(\text{length} \times \text{volume})^{1/2}$.

The designer should not worry too much about the entire fuselage shape at this point. The supersonic wave drag is dependent upon the cross-sectional

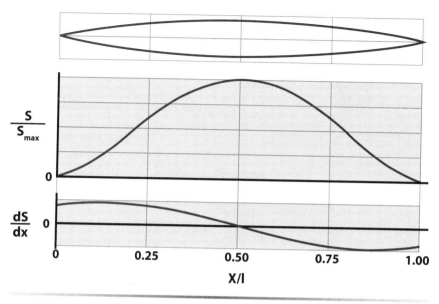

Figure 8.12 Sears–Haack body geometric characteristics.

area distribution of the fuselage plus wing together. Thus, the fuselage is often indented or bulged out to give a smooth wing–body cross-sectional area distribution. This practice is called *area-ruling* and is discussed in the next section.

8.4 Transonic and Supersonic Area-Ruling

Wave drag interference effects in the transonic and supersonic range are greater than those in the subsonic region because of the higher local velocities of the individual components and the greater propagation of these perturbations from this source. The most successful and by far the most systematic method for predicting the transonic and supersonic wave drag is the area-rule concept.

The *area-rule method* is based upon the supersonic slender body theory discussed in [7,9]. It can be assumed that at large distances from the body the disturbances are independent of the arrangement of the components and are only a function of the cross-sectional area distribution. This means that the drag of a wing–body combination can be calculated as though the combination were a body of revolution with equivalent-area cross sections. This is shown in Fig. 8.13 for Mach = 1.

For Mach $\geq$ 1.0, the cross-sectional areas $S(x)$ are along planes inclined at the angle $\mu = \text{arc sin}(1/M_\infty)$ to the x axis. There is a different $S(x)$ for each roll angle ϕ. This is shown in Fig. 8.14.

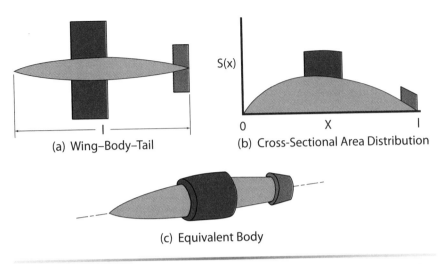

(a) Wing–Body–Tail

$S(x)$

(b) Cross-Sectional Area Distribution

(c) Equivalent Body

Figure 8.13 Equivalent body for a wing–body–tail combination at $M_\infty = 1$.

Once the area distribution $S(x)$ is determined [one $S(x)$ for each ϕ angle for $M_\infty > 1$] the wave drag is calculated using the following expression [9] developed from supersonic slender body (linear) theory:

$$C_{D_W} = \frac{-1}{2\pi S_{\text{ref}}} \int_0^\ell \int_0^\ell \frac{\mathrm{d}^2 S}{\mathrm{d}x^2} \frac{\mathrm{d}^2 S}{\mathrm{d}\xi^2} \ln(x - \xi)\,\mathrm{d}x\,\mathrm{d}\xi \tag{8.1}$$

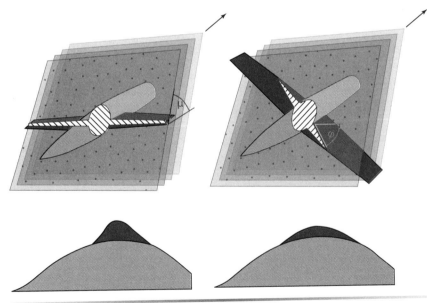

Figure 8.14 Area distribution given by intersection of Mach planes for $M_\infty > 1$.

For $M_\infty > 1$ the C_{D_W} is determined for each roll angle ϕ and then averaged. Application of the area-rule method usually requires automatic computing equipment.

The area-rule method suggests the most desirable way to arrange the vehicle components for minimum wave drag at a particular M_∞. A study of Eq. (8.1) indicates that at the very least it is desireable to have a smooth cross-sectional area distribution dS/dx as any discontinuities would give large values of d^2S/dx^2 or $d^2S/d\xi^2$. The most common example of this is to indent or "coke bottle" the fuselage enough to permit the wing to be added without a sharp discontinuity appearing on the $S(x)$ distribution. If the cross-sectional area distribution of a wing–body combination at a particular M_∞ is the same as a Sears–Haack distribution (see Fig. 8.12), the configuration produces minimum wave drag at that M_∞. Thus, a wing–body can be configured to give minimum wave drag at one Mach number but will usually aggravate the wave drag at other Mach numbers. Figure 8.15 demonstrates this (data from [10]).

An aircraft before area-ruling might have cross-sectional area distribution for $M_\infty = 1$ as shown in Fig. 8.16 Aircraft are usually area-ruled for $M_\infty = 1$ because of the tendency for a "thrust pinch" at $M_\infty = 1$. However, if the aircraft will spend a major portion of its mission at $M > 1$ (such as the Concorde) it should be area-ruled along planes inclined at the Mach angle as shown in Fig. 8.14. It should be pointed out that the cross-sectional area does not include the area of the air flow through the engine. The designer would take the area distribution of Fig. 8.16 and massage it until it is free of discontinuities and looks more like that of a Sears–Haack body than it originally did.

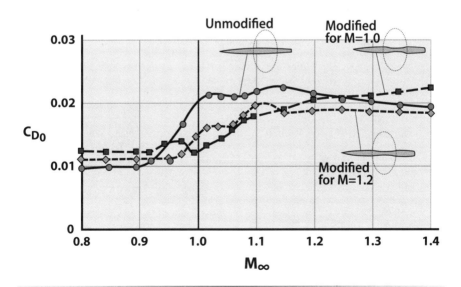

Figure 8.15 Wave drag of bodies with elliptic wings and with or without area-ruling.

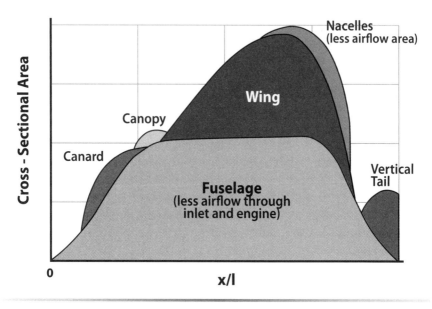

Figure 8.16   Typical cross-sectional area distribution for an aircraft before area-ruling.

References

[1] Roskam, J., "Airplane Design," Roskam Aviation and Engineering Corp., Ottawa, KS, 1989. [Available via www.darcorp.com (accessed 31 Oct. 2009).]

[2] "Airworthiness Standards: Part 25—Transport Category Airplanes," *Federal Aviation Regulation*, Vol. 3, U.S. Department of Transportation, U.S. Government Printing Office, Washington, DC, 2009.

[3] Whitford, R., *Design for Air Combat*, Jane's Publ., New York, 1987.

[4] Wood, K. D., *Aircraft Design*, Vol. 1, Johnson, Boulder, CO, 1966.

[5] *Aerospace Daily*, Vol. 135, No. 15, Sep. 1985, p. 113.

[6] Miele, A., *Flight Mechanics*, Vol. 1, Addison Wesley, Reading, MA, 1962.

[7] Ashley, H., and Landahl, M., *Aerodynamics of Wings and Bodies*, Addison Wesley, Reading, MA, 1965.

[8] Sears, W. R., "On Projectiles of Minimum Drag," *Quarterly Mathematics Series*, Vol. 4, No. 4, 1947, p. 361.

[9] Liepmann, H. W., and Roshko, A., *Elements of Gasdynamics*, Wiley, New York, 1957.

[10] Nelson, R. L., and Walsh, C. J., "Some Examples of the Application of the Transonic and Supersonic Area Rules to the Prediction of Wave Drag," NASA TN-D-446, Sept. 1960.

[11] "General Specification for Performance, Design Characteristics and Construction of Naval Aircraft Weapon systems," SD-24M, Dept of the Navy, Naval Air systems command, Feb. 1994, Arlington VA.

[12] "Aircraft Carrier Reference Data Manual," NAEC-MISC-06900, Naval Air Warfare Center, Aircraft Division, July 1957, Lakehurst NJ.

Chapter 9

High-Lift Devices

- Mechanical High-Lift Devices
- Leading- & Trailing-Edge Flaps
- Slots & Slats
- Conversion of Airfoil to Wing Data
- Powered High-Lift Devices for STOL
- Powered High-Lift Devices for VTOL
- Vectored Thrust
- Lift Fan

The McDonnell Douglas YC-15 features externally blown flaps. This aircraft met Advanced Medium STOL Transport requirements but never saw production (see Section 9.6). One of the YC-15 prototypes is on display outside the main gate of Edwards AFB in California.

Quality is never an accident; it is always the result of intelligent effort.

John Ruskin

9.1 Introduction

To increase the lift of a wing, the suction on the upper surface must be increased relative to that on the lower surface (see Fig. 2.3) and separation delayed or prevented. The suction may be increased by increasing the wing angle-of-attack and by making the airfoil camber more positive in the region of the trailing edge (TE). A TE flap [and to a small extent a leading edge (LE) flap] effectively increases the airfoil camber and increases the air flow acceleration on the upper surface (and overall wing circulation) resulting in an increase in C_L. This increase in C_L is observed as an increase in the magnitude of the angle for zero lift, α_{0L}. Separation is prevented by reducing the adverse pressure gradient over the top of the airfoil or by stabilizing the boundary layer using suction or blowing.

High-lift devices fall into two distinct categories: unpowered or mechanical high-lift devices and powered-lift devices. Mechanical high-lift devices are of two types: (1) TE flaps, which operate by increasing the camber of the airfoil, and (2) separation delay devices. The separation delay devices most commonly used are LE flaps, slats, or slots plus boundary layer control. This chapter considers mechanical high-lift devices in detail. The more practical powered-lift concepts are internal and external blown flaps, deflected slipstream and upper surface blowing, jet flap, lift-fan, tilt wing, direct jet lift, and augmentor wing. These powered-lift concepts are designed for application to *vertical or short takeoff and landing* (V/STOL) and are discussed briefly at the end of the chapter.

9.2 Mechanical High-Lift Devices: Trailing Edge Flaps

Trailing edge flaps operate by changing the camber of the airfoil section as shown in Fig. 9.1. The camber is made more positive in the region of the trailing edge, which has a powerful influence on making α_{0L} more negative [1–3]. A camber change in the region of the leading edge has only a small influence on α_{0L}. The section lift coefficient is expressed as

$$C_\ell = \frac{dC\ell}{d\alpha}\left(\alpha - \alpha_{0L}\right) = C_{\ell\alpha}\left(\alpha - \alpha_{0L}\right) \qquad (9.1)$$

where α is the angle between the unflapped section chord line and the freestream velocity. The performance of trailing edge flaps is shown qualitatively on Fig. 9.2.

Figure 9.1 shows some typical TE high-lift devices. Notice that an aileron is nothing more than a plain flap that operates with both positive

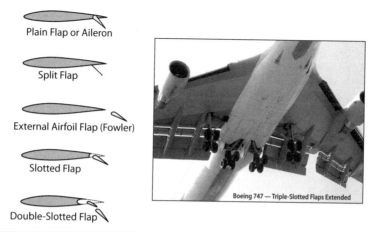

Figure 9.1   Typical TE high-lift devices.

and negative deflections. Also note that the effective area of the wing is increased slightly for Fowler or double-slotted flaps, but the $C_{L_{max}}$ is referenced to the baseline unflapped wing reference area.

Trailing edge flaps do not prevent flow separation; in fact, they aggravate flow separation slightly (decrease α_{stall} slightly as shown in Fig. 9.2) due to the increase in upwash at the leading edge due to increased circulation. Wing sweep promotes stall as discussed in Chapter 2. Thus, trailing edge flaps become less effective as the wing sweep is increased. (This effect is shown as a correction in Fig. 9.23.) Trailing edge flaps are very effective on wings swept up to about 35 deg.

9.3 Mechanical High-Lift Devices: Separation Delay Devices

Flow separation from the top of the airfoil, that is, stall, results from the loss of the kinetic energy in the boundary layer due to viscous shear and an

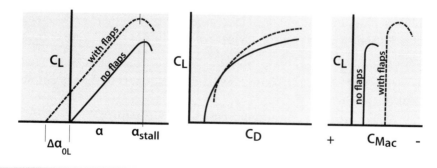

Figure 9.2   Characteristics of TE flaps.

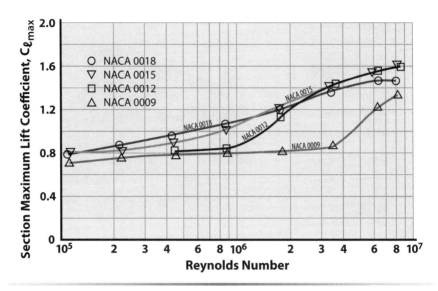

Figure 9.3 Variation of maximum section lift coefficient with Reynolds number.

adverse pressure gradient [4]. A turbulent boundary layer is better able to delay flow separation than a laminar boundary layer because of the higher energy associated with the turbulence. For this reason it is better to have a turbulent boundary layer over the airfoil for lift at high alpha. Vortex generators are put on the top surface of a wing for the purpose of forcing the early transition of the boundary layer from laminar to turbulent. The effect of Reynolds number on $C_{\ell max}$ is shown in Fig. 9.3. Remember that the boundary layer usually transitions from laminar to turbulent at *Re* of one million.

9.3.1 Boundary Layer Control

Boundary layer control (BLC) consists of energizing the boundary layer by either suction or blowing so that the boundary layer can "fight" the adverse pressure gradient with increased energy and thereby delay separation. Boundary layer control devices will not be discussed here because:

1. Numerous papers on this subject exist and are easily available.
2. Significant operational issues become a major part of the system, such as the following:
   - There are large power requirements for the pumps.
   - Much maintenance is needed to keep suction holes and slots free and open.
   - The suction holes and slots cause a rough surface and give large drag at high speeds (if the system is not operative).

9.3.2 Slots and Slats

A *slot* or *slat* (movable slot) operates as shown in Fig. 9.4. The LE shape in the slot or slat is more blunt such that the air flowing through the slot or slat is accelerated and moves farther toward the rear of the airfoil section before slowing down and separating from the surface. Operation of the slat is either manual or automatic. The automatic slat operates at high C_L by the suction in the vicinity of the the LE. The Douglas A-4 Skyhawk had automatic slats that worked quite well during high-C_L maneuvering [5]. Occasionally during a tight turn the slat on one wing would pop out and the slat on the other wing would not. The result was a rapid roll and a surprised pilot. The slats on the A-4 could be manually locked in or out for landing and takeoff.

The principal disadvantage of slots and slats is that a high α is required for $C_{\ell_{max}}$. Also, it is hard to put slots or slats on very thin wings. They are best if used full span, but the main advantage is they protect the outboard wing by reducing tip stall. Slots and slats continue to give beneficial results for sweep greater than 45 deg because they reduce separation near the tip and thus reduce tip stall.

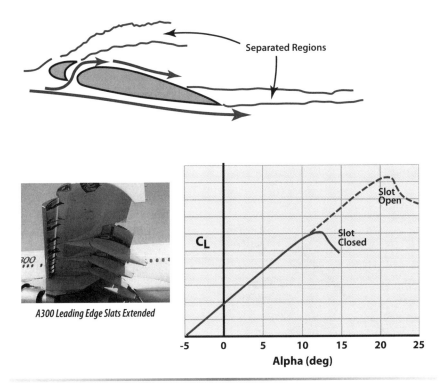

Figure 9.4   Characteristics of slots and slats.

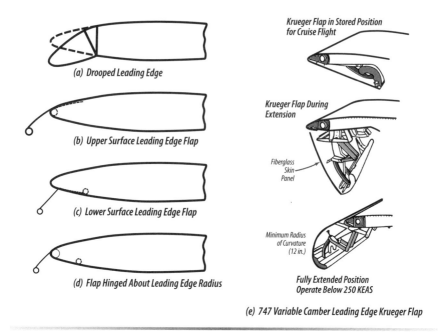

(a) Drooped Leading Edge

(b) Upper Surface Leading Edge Flap

(c) Lower Surface Leading Edge Flap

(d) Flap Hinged About Leading Edge Radius

Krueger Flap in Stored Position
for Cruise Flight

Krueger Flap During
Extension

Fiberglass
Skin
Panel

Minimum Radius
of Curvature
(12 in.)

Fully Extended Position
Operate Below 250 KEAS

(e) 747 Variable Camber Leading Edge Krueger Flap

Figure 9.5   Various LE flap devices.

9.3.3 Leading Edge Flaps

Various types of LE flaps are shown in Fig. 9.5. They operate by making the leading edge more rounded and work well on sharp-nosed airfoil sections as shown in Fig. 7.6 of Chapter 7. Because LE flaps change the camber of the section there is a slight change in α_{0L} as shown in Fig. 9.6. Leading edge flaps are more effective than slots on highly swept wings [6]. They are usually employed over the outer half-span to reduce tip stall. Typically, optimum flap deflections are between 30 and 40 deg.

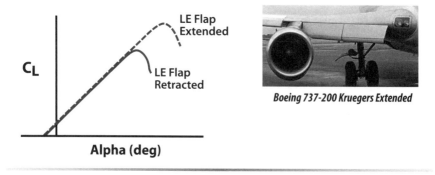

Boeing 737-200 Kruegers Extended

Figure 9.6   Characteristics of LE flaps.

9.3.4 Practical Mechanical High-Lift Systems

The mechanical high-lift devices just discussed are integrated into practical high-lift systems for aircraft to meet the requirements for takeoff, landing, and maneuvering. The low-speed $C_{L_{max}}$ is usually driven by takeoff and landing requirements to meet the runway length available at airports of interest. Remember that when designing the high-lift system for takeoff and landing the $C_{L_{max}}$ has to be usable. That means that the $C_{L_{max}}$ must be available within the limits of the aircraft over-nose vision angle and the tip-back angle (review Chapter 8, Table 8.4 and Fig. 8.3). A $C_{L_{max}}$ of 3.0 at an α of 30 deg is of little value if the aircraft's aft fuselage strikes the ground at an α of 16 deg. Typical limits for takeoff and landing angles-of-attack are 12–16 deg for fighter aircraft, 10–14 deg for transports, and 8–12 deg for general aviation (GA) aircraft due to tip-back angles and over-nose vision.

Fighter aircraft usually have a high T/W (typically > 0.5) so that takeoff is not a problem and the landing distance establishes $C_{L_{max}}$. Transport aircraft have lower T/W (typically <0.35) and either takeoff or landing will determine $C_{L_{max}}$. General aviation aircraft with their low wing loadings do not need much $C_{L_{max}}$ to operate in and out of local airports with 3000-ft fields.

Typical measured data on slots, slats, LE flaps, and TE flaps is presented in Tables 9.1, 9.2, and 9.3. Figure 9.7 shows the practical low-speed $C_{L_{max}}$ limits for mechanical high-lift systems (note that the reference area remains the original wing area and does not change to reflect an increased area for an extended LE or TE flap). Increasing wing sweep and decreasing wing aspect ratio (AR) both decrease the efficiency of the high-lift system. The Airbus A321-200 with its double-slotted TE Fowler flaps and full span LE slats (see the photograph in Fig. 9.4) sets the standard for the transport community with a $C_{L_{max}}$ of 3.2. The A321 flap system was designed to operate from the shorter runways at regional hub airports. The transport aircraft typically have thick airfoil sections (~10% or greater) and can accommodate the internal machinery required for the more sophisticated high-lift systems. The fighter aircraft shown in Fig. 9.7 have thinner airfoil sections and incorporate the more simple high-lift devices such as split or plain TE flaps and drooped LE flaps. General aviation aircraft (such as the Piper PA-30 and Cessna 177 Cardinal) have takeoff wing loadings less than 20 psf and operate from short-field airports using simple plain flap systems. The U-2S with its takeoff $T/W > 0.35$ and $W/S < 40$ psf does not need a high $C_{L_{max}}$ to meet landing and takeoff requirements. The U-2S has a partial-span, simple hinged plain flap, giving it a $C_{L_{max}}$ of 1.2 at a 15-deg flap deflection. The U-2S has a unique bicycle landing gear and does not rotate for takeoff (same for the B-52 and B-47).

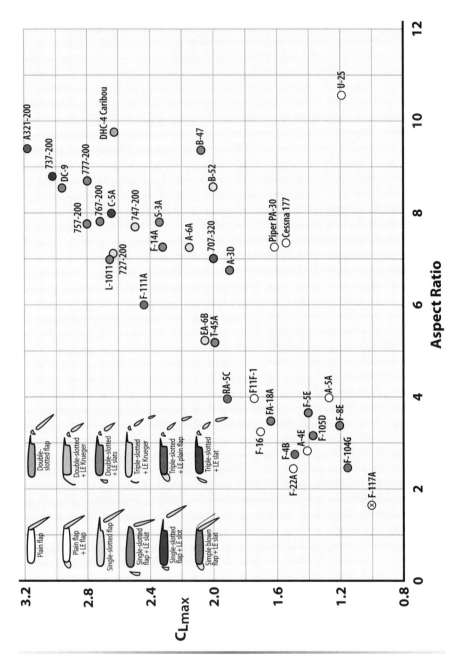

Figure 9.7   Practical low-speed $C_{L_{max}}$ limits for mechanical high-lift systems (data from [7]).

Table 9.1 Mechanical High-Lift Systems and Maximum Lift Summary of Current Aircraft

Current Aircraft Mechanical High-Lift Systems				
Aircraft	**AR**	**$C_{L_{max}}$**	**Leading Edge**	**Trailing Edge**
707-320	7.0	2.0	Full-span plain flap	Triple-slotted Fowler
E-6A	7.0	2.16	Improved 707-320 system	
727-200	7.1	2.62	1/3 Krueger, 2/3 span slats	Triple-slotted Fowler
737-200	8.83	3.05	Krueger IB, slats OB[a]	Triple-slotted Fowler
747-400	7.7	2.5	Krueger IB, slats OB	Triple-slotted Fowler
757-200	7.77	2.8	Full-span slats	Double-slotted Fowler
767-200	7.9	2.75	Full-span slats	Double slot IB, single slot OB
777-200	8.7	2.8	Full-span slats	Double slot IB, single slot OB
787	Var.	NA	Krueger IB, slats OB	Triple-slotted Fowler+variable camber
A321-200	9.5	3.2	Full span slats	Double slotted Fowler+drooped ailerons
L-1011	6.95	2.65	Full-span slats	Double-slotted Fowler
S-3A	7.8	2.36	Slats OB of engine	Single-slotted Fowler
DC-9	8.5	2.96	Ful-span slats	Full-span double-slotted flap
DHC-4	9.9	2.63	None	Full-span double-slotted flap
C-5A	8.0	2.64	Slots IB+slotted slats OB	Partial-span single-slotted Fowler
U-2S	10.6	1.21	None	Partial-span simple hinge flap
PA-30	7.3	1.6	None	Half-span plain flap
Cessna177	7.4	1.55	None	Half-span plain flap
B-47	9.42	2.05	Full-span slat	Partial-span Fowler
B-52G	8.56	2.0	None	Partial-span Fowler
F-16C	3.2	1.7	Full-span maneuver flap	Half-span plain flap
F-22A	2.36	1.48	Full-span maneuver flap	Full flaperon+drooped aileron
A-3D	6.75	1.9	Full-span slats	Partial-span single-slotted flap
F-4B	2.78	1.4	Full plain flap (blown)	Partial-span blown plain flap
A-4E	2.9	1.42	Automatic LE slats	1/2 split flap+drooped ailerons
RA-5C	4.0	1.9	Full-span plain flap	Partial-span plain flap (blown OB)
F-5E	3.7	1.4	Full-span plain flap	Partial-span single-slotted flap
A-6A	5.3	2.05	Full-span plain flaps	Partial-span Fowler flap
F-14A	7.25	2.35	Full-span LE slats	Full-span slotted flaps
F-111A	6.0	2.45	Full-span LE slats	Partial-span blown plain flap
F-117	1.65	0.95	None	None
F-18A	3.5	1.62	Full-span plain flap	Half-span single-slotted TE flap
F-105D	3.18	1.38	Full-span plain flap	Partial-span single-slotted flap
F-104G	2.45	1.12	Full-span plain flap	Blown flap+drooped aileron
T-45A	5.0	2.0	Full-span plain flap	2/3 span double-slotted flaps
F-8E	3.5	1.2	Full-span plain flap	2/3 plain flap+variable-incidence wing
F-11F	3.95	1.75	Full-span slats	Full-span plain flaps

Abbreviations: IB, inboard; OB, outboard.

Table 9.2 Summary of Maximum Lift Coefficient Obtained with Various Types of High-lift Devices (data from [5,8,9])

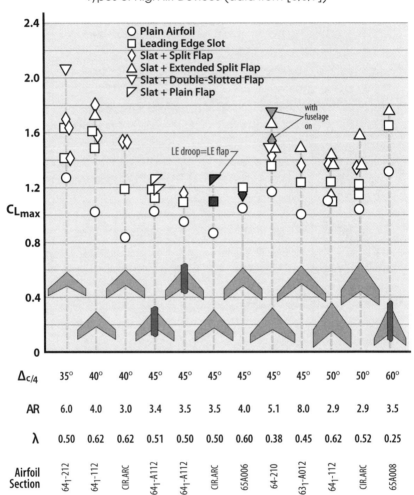

Figure 9.7 shows a trend of increasing $C_{L_{\max}}$ for increasing AR and flap system sophistication. The transport aircraft shown in Fig. 9.7 (i.e., the A321-200) represent the current practical limit in mechanical high-lift system sophistication.

9.4 Methods for Determining Maximum Subsonic C_L of Mechanical Lift Devices

The method presented here is empirical and will give satisfactory results for the first iteration of the design loop [10]. The method involves determining the C_L vs α curve for the basic wing and then correcting it for the effects of the mechanical high-lift devices.

Table 9.3   Typical High-Lift Device Data

$\Delta = 35$ deg, AR = 5.76, $\lambda = 0.54$ Airfoil section: 10% symmetrical		
Arrangement	$C_{L_{max}}$	α_{stall}
Plain wing	0.90	16
20% full-span split flap, $\delta f = 60$	1.45	10.6
20% full-span slat	1.38	23.6
20% full-span LE flap	1.49	26.5
20% full-span split flap + 20% full-span LE flap	2.01	19.7
$\Delta = 0$ deg, AR = 4.0, $\lambda = 1.0$ $Re = 10^5$ Airfoil section: NACA 0010		
Arrangement	$C_{L_{max}}$	α_{stall}
Plain wing	0.80	13
30% full-span split flap, $\delta f = 40$ deg	1.52	10
20% full-span slat	1.36	24

The airfoil section behavior with TE flaps is determined first. The construction of the C_ℓ vs α curve is shown in Fig. 9.8.

Values of $\Delta\alpha_{0L}$, $\Delta C_{\ell_{max}}$, and $\sim\Delta\alpha_{stall}$ are needed to complete the construction of Fig. 9.8. The first step is to obtain the section α_{0L}, C_{ℓ_α}, and α_{stall} from experimental data (i.e., Appendix F or [2,11,12]). If experimental data on

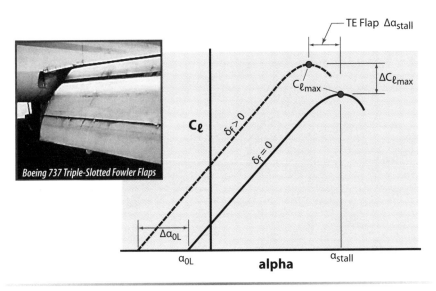

Figure 9.8   Construction of section lift curves for TE flaps.

the selected airfoil section cannot be found, use $C_{\ell_\alpha} = 2\pi$ per radian and compute α_{0L} using Eq. (2.4) of Chapter 2. Estimate $C_{\ell_{max}}$ from Figs. 7.2 or 9.3, then use Eq. (9.1) to determine α_{stall}.

Decide upon the type of TE flap, the flap-to-chord c_f/c ratio (see Fig. 9.9), and the flap deflection δ_f (positive for downward deflection). The $\Delta\alpha_{0L}$ is determined using the method outlined next (from [10]):

1. **Plain TE flaps.** Calculate the change in α_{0L} for flap deflection δ_f:

$$\Delta\alpha_{0L} = -\frac{dC_\ell}{d\delta_f}\frac{1}{C_{\ell_\alpha}}\delta_f K_f' \qquad (9.2)$$

where

C_{ℓ_α} = section lift curve slope (per radian) from Appendix F
K_f' = correction for nonlinear effects, Fig. 9.9
$dC_\ell/d\delta_f$ = change in C_ℓ for a change in δ_f, Fig. 9.10

2. **Single-slotted flaps.**

$$\Delta\alpha_{0L} = -\frac{d\alpha}{d\delta_f}\delta_f \qquad (9.3)$$

where $d\alpha/d\delta_f$ is obtained from Fig. 9.11.
3. **Fowler flaps.** Use single-slotted flap method.
4. **Split flap.**

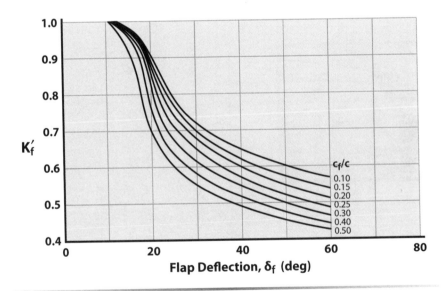

Figure 9.9 Nonlinear correction for plain TE flaps (adapted [10]).

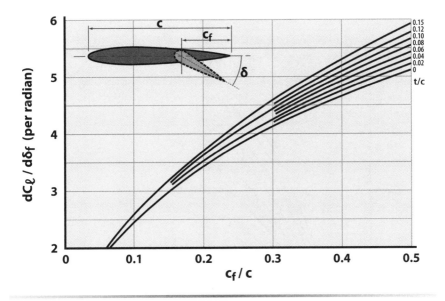

Figure 9.10 Variation of $dC_\ell/d\delta_f$ with flap chord ratio (adapted [10]).

$$\Delta\alpha_{0L} = \frac{k}{C_{\ell_\alpha}}(\Delta C\ell)_{c_f/c=0.2}$$

where k and $(\Delta C_\ell)_{c_f/c=0.2}$ are obtained from Fig. 9.12.

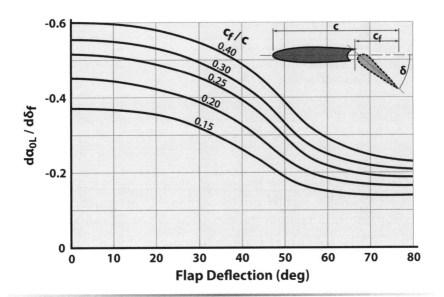

Figure 9.11 Section lift effectiveness parameter for single-slotted flaps (adapted [10]).

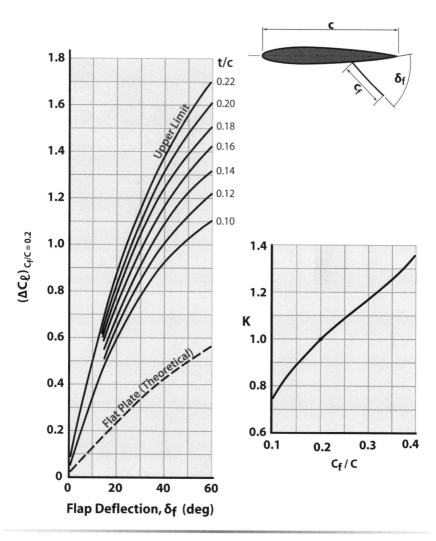

Figure 9.12 Empirical constants for split flap analysis (adapted [10]).

Now construct the C_ℓ vs α curve as shown in Fig. 9.8. The TE flaps aggravate separation slightly and the section α_{stall} decreases. This α_{stall} is obtained from Fig. 9.13. Estimate $\Delta C_{\ell_{max}}$ from the completed curve.

At subsonic speeds a distinction is made between low- and high-AR wings. This is because two different sets of parameters are required to describe the wing characteristics in the two AR regimes. The $C_{L_{max}}$ of a high-AR wing is determined by the properties of the airfoil section, whereas, the $C_{L_{max}}$ of a low-AR wing is primarily dependent upon its planform shape.

The high-AR wing is defined by

$$AR > \frac{4}{(C_1 + 1)\cos \Delta_{LE}} \qquad (9.4a)$$

and the low-AR wing by

$$AR < \frac{4}{(C_1 + 1)\cos \Delta_{LE}} \qquad (9.4b)$$

where C_1 is a function of taper ratio and is obtained from Fig. 9.14.

For high-AR wings the $C_{L\text{max}}$ and α_{stall} for the basic wing are determined from

$$C_{L\text{max}} = \frac{C_{L\text{max}}}{C_{\ell\text{max}}} C_{\ell\text{max}} \qquad (9.5)$$

$$\alpha_{\text{stall}} = \frac{C_{L\text{max}}}{C_{\ell\alpha}} + \alpha_{0L} + \Delta\alpha_{C_{L\text{max}}} \qquad (9.6)$$

where

$(C_{L\text{max}}/C_{\ell\text{max}})$ is obtained from Fig. 9.15
$C_{L\alpha}$ = the wing lift curve slope from Eq. (2.13) of Chapter 2
α_{0L} = the section angle for zero lift
$\Delta\alpha_{C_{L\text{max}}}$ is obtained from Fig. 9.16
$C_{\ell\text{max}}$ = the unflapped section maximum lift coefficient from construction of Fig. 9.8

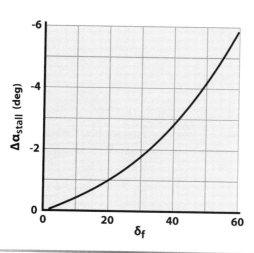

Figure 9.13 Decrease in stall angle with flap deflection (data from [2]).

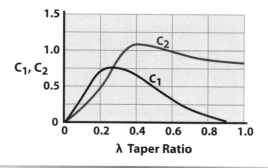

Figure 9.14 Taper ratio correction factors (adapted [10]).

Figures 9.15 and 9.16 make use of a Δy. This Δy is a leading edge sharpness parameter presented in Fig. 9.17.

For low-AR wings the $C_{L_{max}}$ and α_{stall} for the basic wing are determined from

$$C_{L_{max}} = \left(C_{L_{max}}\right)_{base} + \Delta C_{L_{max}} \tag{9.7}$$

$$\alpha_{stall} = \left(\alpha_{C_{L_{max}}}\right)_{base} + \Delta\alpha_{C_{L_{max}}} \tag{9.8}$$

where

$(C_{L_{max}})_{base}$ is obtained from Fig. 9.18
$\Delta C_{L_{max}}$ is obtained from Fig. 9.19
$(\alpha_{C_{L_{max}}})_{base}$ is from Fig. 9.20
$\Delta\alpha_{C_{L_{max}}}$ is from Fig. 9.21

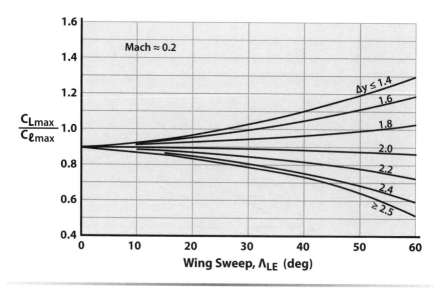

Figure 9.15 Subsonic maximum lift of high-AR wings (adapted [10]).

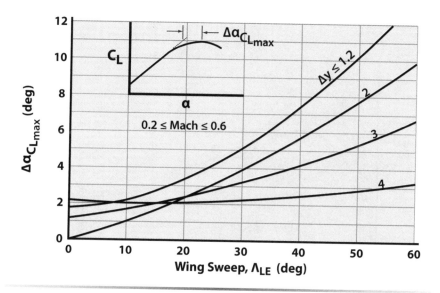

Figure 9.16 Angle-of-attack increment for subsonic maximum lift of high-AR wings (adapted [10]).

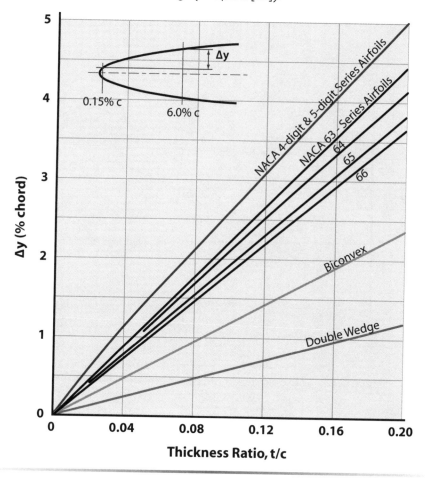

Figure 9.17 Variation of LE sharpness parameter with airfoil thickness ratio (adapted [10]).

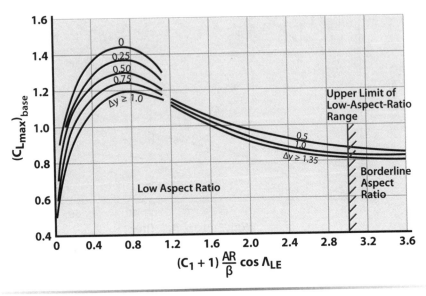

Figure 9.18 Subsonic maximum lift of low-AR wings (adapted [10]).

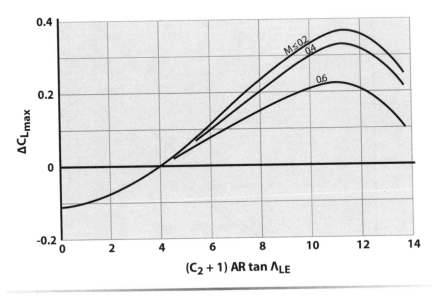

Figure 9.19 Subsonic maximum-lift increment for low-AR wings (adapted [10]).

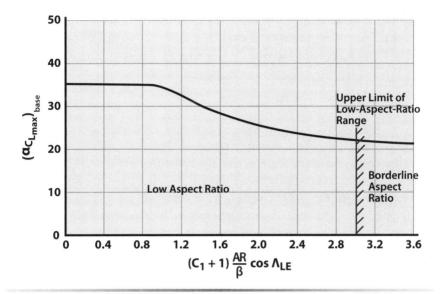

Figure 9.20   Angle-of-attack for subsonic maximum lift of low-AR wings.

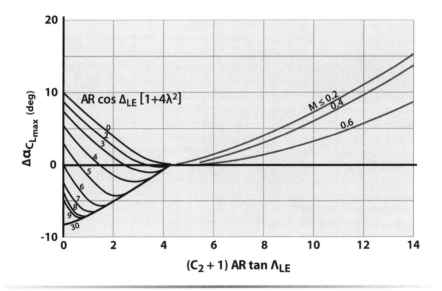

Figure 9.21   Angle-of-attack increment for subsonic maximum lift of low-AR wings.

Now the basic wing C_L vs α chart can be constructed as illustrated in Fig. 9.22. The α_{0L} for the wing is the same as for the airfoil section.

The finite wing increase in $C_{L\max}$ due to a TE flap is obtained from (9.9),

$$\Delta C_{L\max} = \Delta C_{\ell\max} \frac{S_{WF}}{S_W} K_\Delta \qquad (9.9)$$

where K_Δ is an empirical sweep correction (Fig. 9.23), $\Delta C_{L\max}$ is obtained from the construction of Fig. 9.8, and S_{WF} is defined in Fig. 9.24. The $\Delta C_{L\max}$ is added to the basic (unflapped) wing $C_{L\max}$ and the final flapped wing curve is drawn in Fig. 9.22. The $\Delta\alpha_{0L}$ for the flapped wing is the same as for the flapped airfoil section determined earlier. Notice that TE flaps are not particularly effective on highly swept wings.

There is no method to predict the $\Delta C_{L\max}$ for a wing with LE devices. Here the designer should use experimental data such as that presented in Tables 9.1 and 9.2 or Fig. 9.7. For example, for the wing in Table 9.1 a 20% full-span slat gives a $\Delta C_{L\max} = 0.48$ and a 20% full-span LE flap gives a $\Delta C_{L\max} = 0.59$. These values can now be added to the $C_{L\max}$ of similar wing shapes to give the $C_{L\max}$ for a wing with LE flaps or slats. The $\Delta\alpha_{\text{stall}}$ can be determined similarly.

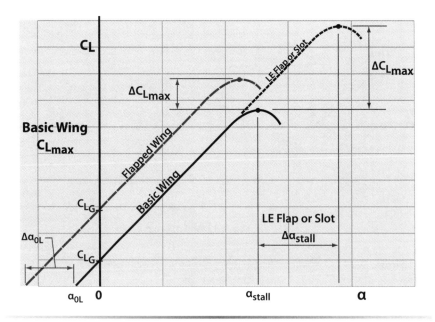

Figure 9.22 Construction of wing lift curves for mechanical high-lift devices.

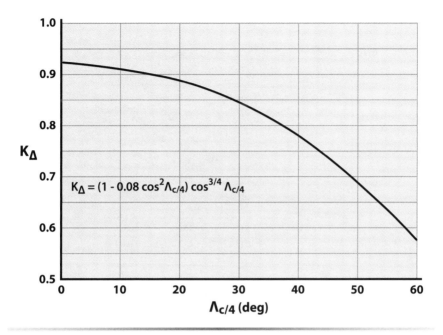

Figure 9.23 Planform correction factors for TE flaps (adapted [10]).

$$K_\Delta = (1 - 0.08 \cos^2 \Lambda_{c/4}) \cos^{3/4} \Lambda_{c/4}$$

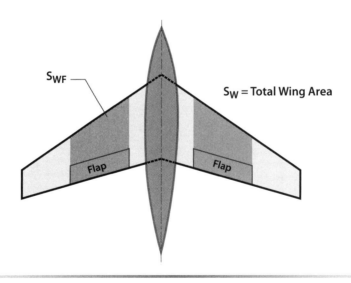

Figure 9.24 Schematic showing flapped wing area.

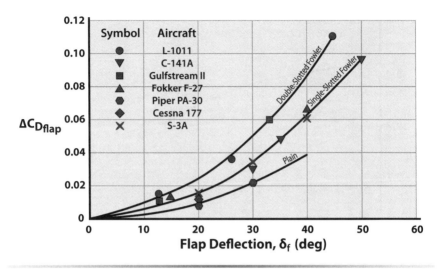

Figure 9.25 Trailing edge flap drag coefficient increment (referenced to wing area).

9.5 Subsonic Drag Due to Flap Deflection

The drag of the deflected flaps must be considered in the landing and takeoff analysis. The designer can get a first-order estimate for the drag of a slotted or plain flap from Fig. 9.25.

A more refined estimate of the drag coefficient for split, plain, and slotted flaps is given in [13]. The influence of the flap chord ratio and flap area ratio is determined as follows:

$$\Delta C_{D_{\text{flap}}} = k_1 k_2 \frac{S_{\text{WF}}}{S_W} \tag{9.10}$$

where k_1 is a function of the ratio c_f/c and is obtained from Fig. 9.26; k_2 is dependent upon δ_f and is presented in Fig. 9.27; and S_{WF}/S_W is the ratio of the flapped wing area to the total wing area (see Fig. 9.24).

9.6 Powered High-Lift Devices for STOL

The characterization of a *short takeoff and landing* (STOL) aircraft is not well defined at present. However, it is generally agreed that the lower limit for a STOL aircraft would be a landing and takeoff distance, over a 50-ft obstacle, of 1000 ft (air distance + ground roll).

This 1000-ft restriction for landing means a steep descent (7 deg) over the obstacle to shorten the air distance and a low touchdown speed with high braking coefficients to keep the ground run short. The air distance

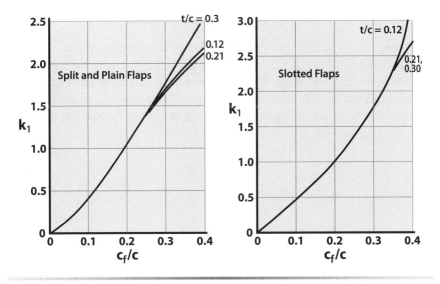

Figure 9.26 Factor k_1 to calculate drag increment due to flaps (data from [13]).

over 50 ft for a 7-deg glide slope is ~400 ft, leaving only 500 ft for the ground roll. The touchdown speed is defined as $1.15V_{stall}$ and the approach speed over 50 ft as $1.3V_{stall}$. For takeoff the aircraft must accelerate to takeoff speed, which is $1.2V_{stall}$. Thus, the stall speed is the primary takeoff performance parameter for STOL aircraft.

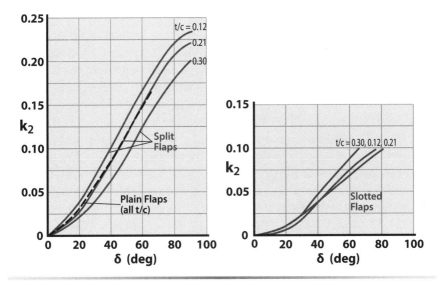

Figure 9.27 Factor k_2 to calculate drag increment due to flaps.

Example 9.1   Wing Loadings for STOL Aircraft

Figure 6.5 (from [14]) indicates that routine landing in a 1000-ft field requires an approach speed of 50 kt, or 84.5 ft/s. Using the Federal Aviation Administration requirement for approach speed equal to 1.3 times the stall speed (Chapter 10) the stall speed is approximately 65 ft/s, which is the result of using the one-g lift expression at sea level ($V_{\text{stall}} = 65$ ft/s):

$$\frac{W}{S}\frac{1}{C_{L_{\max}}} = 5.0 \qquad (9.11)$$

Using the practical upper limit of $C_{L_{\max}}$ for mechanical lift devices of 4.0 (see Fig. 9.7), Eq. (9.11) indicates that the wing loading would have to be 20 psf! This is an appropriate wing loading for light utility aircraft but not for commercial short-haul STOL transports. Transport aircraft cannot operate economically at such a low wing loading (poor cruise efficiency), and passengers would not like the bumpy ride. The Breguet 941, a STOL commercial short-haul transport ($W_{\text{TO}} = 48{,}000$ lb), lands in 1000 ft with a wing loading of about 45 psf. The Breguet 941 employs the deflected slipstream concept, thrust reversers, and oversize brakes. The point here is that commercial STOL operation with field distances of 1000 ft must use powered-lift devices as well as aerodynamic high-lift devices.

9.6.1 Deflected Slipstream

In a *deflected slipstream* system, lift is produced at low speed by deflecting the propeller slipstream or jet exhaust downward by a wing-flap arrangement (as used in the Breguet 941). The slipstream of the four propellers blows over the entire span of the wing and is deflected by slotted flaps at the trailing edge. A derivative of this system is the upper surface blowing (USB) employed on the Boeing Advanced Medium STOL Transport (AMST) YC-14 demonstrator in the 1970s. The effectiveness of a USB flap, shown schematically in Fig. 9.28, in turning a jet exhaust flow depends on a principle known as the *Coanda effect*, which describes how a jet airflow adheres to the outside of a convex curved surface. This phenomenon was first systematically investigated by Henri Coanda prior to WWII. He found that a high-velocity jet will adhere to an adjacent convex surface, provided that the jet depth is not large compared with the radius of the turn. In fact, blowing boundary layer control at the knee of a flap is a practical application of this principle. The YC-14 (Fig. 9.29a) locates its CF6-50 turbofan engines on top of the wing and bathes the inboard upper surface of the

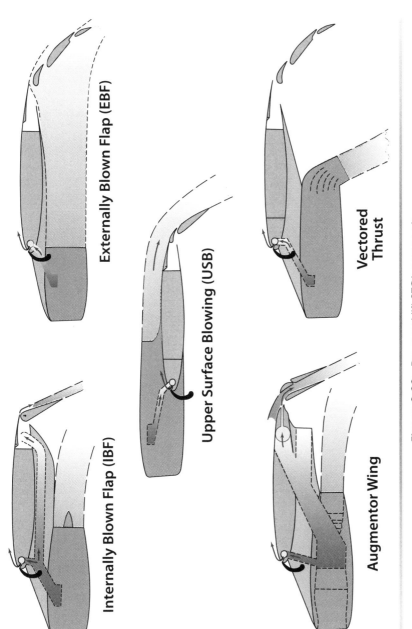

Externally Blown Flap (EBF)

Internally Blown Flap (IBF)

Upper Surface Blowing (USB)

Augmentor Wing

Vectored Thrust

Figure 9.28 Powered-lift STOL concepts.

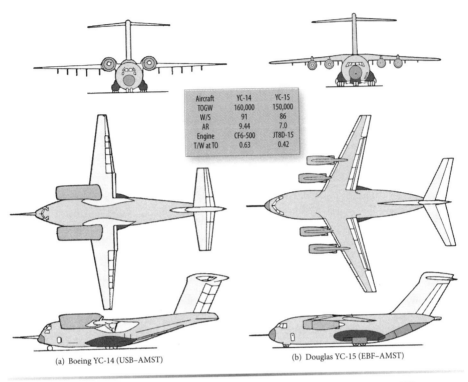

Aircraft	YC-14	YC-15
TOGW	160,000	150,000
W/S	91	86
AR	9.44	7.0
Engine	CF6-500	JT8D-15
T/W at TO	0.63	0.42

(a) Boeing YC-14 (USB–AMST) (b) Douglas YC-15 (EBF–AMST)

Figure 9.29 Prototype advanced medium STOL transports, AMST (data from [13,14]).

wing and TE flap with the jet exhaust. Performance of the YC-14 USB arrangement (Fig. 9.30) depends upon the jet coefficient

$$C_j = \frac{\text{Thrust}}{q\, S_{\text{ref}}} \qquad (9.12)$$

Notice that the YC-14 also employs blowing of the LE flap. The amount of blowing is expressed by the blowing coefficient

$$C_\mu = \frac{\dot{m}_B V_e}{q\, S_{\text{ref}}} \qquad (9.13)$$

where $\dot{m}_B$ is the mass flow rate of the blowing device and V_e is the exhaust velocity of the blowing jet.

9.6.2 Externally or Internally Blown Flap

In the *internally blown flap* (IBF) and the *externally blown flap* (EBF), high-energy jet exhaust is blown over a slotted TE flap arrangement, providing both thrust vectoring and boundary layer control. In the IBF concept, jet exhaust (all or part) is ducted from the engine, through the wing, and

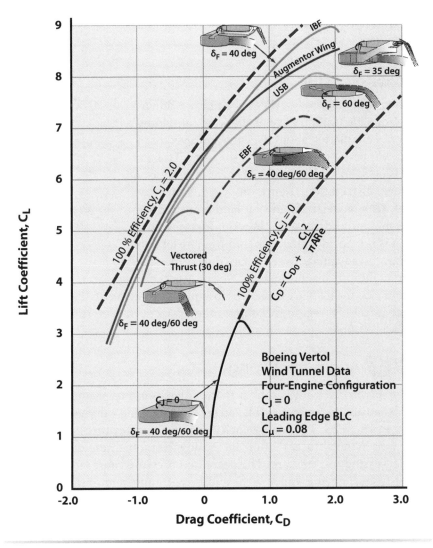

Figure 9.30   Low-speed drag polars for various powered-lift concepts.

exhausted over the TE flap as shown in Fig. 9.28. There is usually a cross-over ducting arrangement so that one engine can feed the flaps on both sides of the aircraft. This ducting is heavy and complicated but the arrangement solves the one-engine-out problem. The performance of the IBF is shown in Fig. 9.30.

The EBF concept is shown in Fig. 9.28 and has the principle advantage of being light and simple. No internal ducting or thrust deflection mechanisms are required other than the flap system itself. However, the flap system is subjected to severe temperature and load environments and one-engine-out is a major design problem. Despite these drawbacks, the EBF concept is popular and was employed on the YC-15 (see Fig. 9. 29b) and the more recent USAF

C-17 transport. The performance of a typical EBF arrangement is shown in Fig. 9.30.

9.6.3 Jet Flap

The *jet flap* is simply a sheet of air blown downward from the trailing edge of the wing, providing increased circulation about the wing and a vectored thrust component. In contrast to the usual TE flap, there is no solid surface in the jet sheet to support a pressure distribution, hence no drag from the device. Drag is desirable during landing to reduce the air and ground distance; thus, this feature of the jet flap is not an advantage for landing. The jet flap does have an advantage over TE flaps during maneuvering flight at transonic speeds. Not only is there the absence of the flap drag but also the jet flap extends the low-pressure region on the wing upper surface in the TE region that moves the upper surface normal shock aft, which delays flow separation.

In 1972, the USAF started an AMST program that called for operating a 27,000-lb payload into a 2000-ft semi-prepared field. Boeing's YC-14 featuring USB competed with McDonnell Douglas's YC-15 featuring EBF. Both companies' prototypes met AMST requirements, but neither aircraft saw production. McDonnell Douglas, however, incorporated the YC-15 EBF into their winning C-17 design.

9.6.4 Augmentor Wing

The *augmentor wing* is similar to the IBF except that the ducted engine air is exhausted between two TE flap sections forming a diffuser section as shown in Fig. 9.28. The high-velocity engine air mixes with stagnant secondary air in the diffuser section, increasing the momentum of the mixture and decreasing the pressure. The decreased pressure causes more secondary air from the wing upper surface to be entrained and to move into the diffuser section. The result is an augmentation of the thrust from the primary engine exhaust by as much as a factor of 2.

9.7 Powered High-Lift Devices for V/STOL

The first thing to understand about V/STOL is that the high lift during vertical ascent or descent does not come from air flowing over a wing, because the "V" means zero air speed. The high lift comes from directing the force from a propulsion unit downward. A good rule of thumb is that you need a force of ~1.2 times the weight of the aircraft directed downward to achieve VTOL. The extra 20% is needed for three-axis control and to overcome the "suck down" (suction is usually present beneath a VTOL aircraft). In addition to the three-axis control, there must be provision for fore and aft translation during hover.

The candidate V/STOL concepts are shown in Figs. 9.31 and 9.32. Figure 9.31 was generated by the McDonnell Aircraft Company in 1968

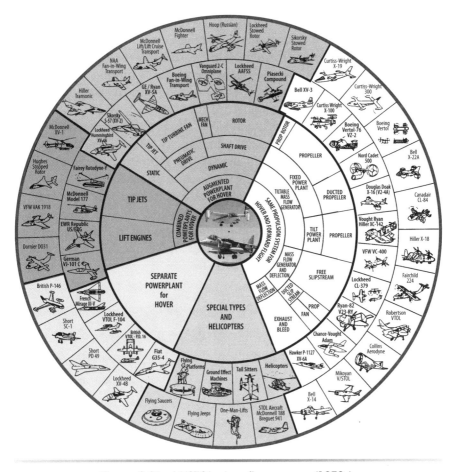

Figure 9.31 V/STOL aircraft summary (1970s).

and shows the state of the art at that time. Figure 9.31 is often called the Wheel of Misfortune because not every aircraft shown in the figure was actually built and flown, many of them crashed during testing, and only the P-1127/XV-6A saw production—as the Harrier. Figure 9.32 shows the more successful V/STOL aircraft up to the present. The only aircraft from Fig. 9.32 that have gone into production are the Yak-38 (lift + lift/cruise), V-22 (tilt rotor), AV-8 Harrier (vectored thrust), and F-35B (fuselage fan + vectored thrust). There is a good chance that a derivative of the XV-15 will be produced for the commercial sector as the Bell/Boeing 609 and for the Coast Guard as the Bell Eagle Eye UAV.

The augmentor (ejector) wing can be used as a VTOL device, for example, the Rockwell International XFV-12A (see [7]) for the U.S. Navy, using all the engine air or as a STOL device using only part of the engine air. The ejector concept was first used in the Lockheed Hummingbird XV-4A in the early 1960s, with limited success. The XV-4A had two 3300-lb thrust

PW JT 12A-3 turbojets that could be exhausted rearward for forward flight or diverted to feed the ejector system located in the fuselage. The fuselage ejector was replaced with lift engines in the XV-4B to demonstrate the lift plus lift–cruise concept. The XV-4B had four J85-19 turbojets in the fuselage to provide vertical lift only and two J85s in external engine pods that could be vectored for lift or forward thrust.

Tilt-wing concepts such as the XC-142 have propellers mounted on the wing and the entire wing is rotated up to 90 deg while keeping the fuselage horizontal. The major portion of the wing is immersed in the propeller slipstream and does not stall during the wing rotation. Several tilt-wing V/STOL prototypes can be seen in Fig. 9.32. In the tilt-rotor/propeller concept the wing is fixed and only the propeller or rotor tilts. The XV-15 was one of the more successful tilt-rotor prototypes due mainly to Bell's 40 years of persistence and commitment to the concept. It will see military service as the V-22 and very likely commercial service.

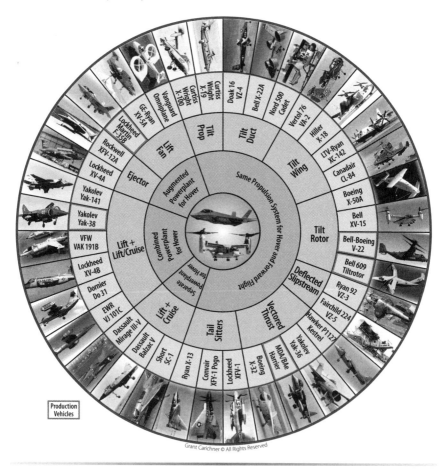

Figure 9.32   V/STOL aircraft summary (2008).

Some V/STOL designs use direct engine thrust to provide additional lift force during takeoff and landing. This direct engine thrust can come from dedicated lift engines, such as on the Lockheed XV-4B, or from cruise engines capable of vectoring their thrust such as on the AV-8B Harrier, or from a combination (called lift plus lift–cruise) such as the German VAK-191B and Russian Yak-38. The lift engine or vectored thrust concepts have the advantage of being mature, straightforward, low-risk concepts but suffer the disadvantage of producing a field of very hot gas beneath the aircraft that can be ingested by the engine thereby reducing its thrust.

Energy can be extracted from the cruise engine to power a lift fan. This energy extraction can be shaft power driving a lift fan or hot exhaust gas driving a tip-driven lift fan. In either case the lift fan exhausts downward with a lift force greater than the vectored thrust of the cruise engine. In other words, the lift fan is able to augment the thrust from the cruise engine. The lift-fan concept was demonstrated in the Ryan XV-5A in the 1960s. The XV-5A had two gas-driven lift fans in the wings and one in the nose of the aircraft. This arrangement gave the XV-5A three-axis control by modulating the nose fan thrust for pitch, modulating the wing fan thrust for roll, and deflecting the nose fan lift sideways for yaw control. Fore and aft translation was available by deflecting the wing fan thrust fore and aft. With the hot exhaust gas from the two J85-GE-5 turbojets (rated at 2650 lb static thrust each) driving the three lift fans, the XV-5A was able to generate 13,886 lb of vertical thrust, or an augmentation ratio of 2.62. The disadvantage of the gas-driven lift fan concept is the volume required for the hot gas ducting and the vulnerability of this ducting to small arms ground fire.

In 1996 the U.S. government awarded a contract to Boeing and Lockheed Martin to build two Joint Strike Fighter (JSF) prototypes each. The prototypes were to demonstrate three JSF variants, a conventional takeoff and landing variant for the U.S. Air Force, a carrier-suitable variant for the U.S. Navy, and a *short takeoff vertical landing* (STOVL) variant for the U.S. Marine Corps. The STOVL variant was the discriminating factor in the program. The JSF engine was the PWF119 afterburning turbofan engine with ~32,000 lb of vectored thrust available for VTOL. Boeing selected the low-risk vectored thrust concept for their X-32B demonstrator (see Fig. 1.11). Lockheed Martin selected the higher risk, but more capable *shaft-driven lift fan* (SDLF) plus vectored thrust concept, shown in Fig. 9.33 for their X-35B demonstrator. Most of the risk was in developing a gear box to drive the SDLF, a clutch and a functional, lightweight fan. The SDLF was located behind the cockpit. The 3 axes control arrangement modulates the lift fan thrust for pitch, deflects the lift fan thrust sideways for yaw, and uses wing tip mounted reaction control jets for roll. At sea level on a 75°F day the X-35B was able to generate 39,100 lb of vertical lift distributed as

Hover Testing

Figure 9.33   X-35B showing shaft-driven lift-fan (SDLF) and auxiliary inlet.

follows: 16,411 lb from the three bearing swivel nozzle, 3,607 lb from the wingtip roll control jets, and 19,082 lb from the lift fan. The lift fan achieved an augmentation ratio of about 1.6. The SDLF plus vectored thrust concept of the X-35B had considerable margin over the vectored thrust concept of the X-32B and won the JSF competition. The aircraft is now in production

F-35 and the Joint Strike Fighter Program

The F-35 has three variants: the F-35A conventional TO and landing for the USAF, the F-35B STOVL for the U.S. Marine Corps, and the F-35C carrier suitable for the U.S. Navy. This aircraft has a single PWA F135 turbofan with 43,000 lb TSLS in afterburner. The specifications for the three variants are as follows:

	F-35A	F-35B	F-35C
Wing area (ft^2)	460	460	668
Empty weight (lb)	29,300	32,000	34,800
Max TOGW (lb)	70,000	60,000	70,000
Range (nm)	1200	900	1400
Combat radius (nm)	610	500	640
Max speed (Mach)	1.67	1.67	1.67

The program is joined by eight international partners: the United Kingdom, Italy, the Netherlands, Canada, Turkey, Australia, Norway, and Denmark.

and will see service in the Air Force, Navy, and Marine Corps as the F-35. The F-35 Case Study in Volume 2 is recommended reading.

There are several excellent references on V/STOL aerodynamics and aircraft technology that are recommended to the reader. Reference [15] is a good text on the theoretical aspect of high-lift devices with supporting experimental data. Reference [16] is an excellent summary report on STOL aerodynamic technology, with [17] being a summary of the USAF Advanced Medium STOL Transport Program. Reference [18] is an excellent design text for V/STOL aircraft. Reference [7] is a superb historical account of VTOL military research aircraft.

References

[1] Kuethe, A. M., and Schetzer, J. D., *Foundations of Aerodynamics*, Wiley, New York, 1959.

[2] Abbott, I. H., and von Doenhoff, A. E., *Theory of Wing Sections*, Dover, New York, 1959.

[3] Furlong, G. C., and McHugh, J. G., "A Summary and Analysis of the Low-Speed Longitudinal Characteristics of Swept Wings at High Reynolds Number," NACA TR-1339, 1957.

[4] Thomson, L. P., "A Review of Leading Edge High Lift Devices," Dept. of Supply, Army Research Laboratory Rept. A-77, 1951.

[5] Gambucci, B. J., "Section Characteristics of the NACA 0006 Airfoil with Leading-Edge and Trailing-Edge Flaps," NACA TN-3797, 1956.

[6] Menees, G. P., "Lift, Drag and Pitching Moment of an Aspect-Ratio-2 Triangular Wing with Leading-Edge Flaps Designed to Simulate Conical Camber," NASA Memo 10-5-58A, Dec. 1958.

[7] Rogers, M. J., *VTOL Military Research Aircraft*, Orion Books, New York, 1989.

[8] Nonweiler, T., "Flaps, Slots and Other High Lift Aids," *Aircraft Engineering*, Vol. 6, Sept. 1955, pp. 19–23.

[9] Cahill, J. F., et al., "Aerodynamic Forces on a Symmetrical Circular Arc Airfoil with Plain LE and TE Flaps," NACA TR-1146, 1953.

[10] Ellison, D. E., "USAF Stability and Control Handbook (DATCOM)," U.S. Air Force Flight Dynamics Laboratory, AFFDL/FDCC, Wright–Patterson AFB, OH, June 1969.

[11] Abbott, I. H., Von Doenhoff, A. E., and Stivers, L., Jr., "Summary of Airfoil Data," NACA TR-824, 1945.

[12] Riegels, F. W., *Aerofoil Section*, Butterworth, London, 1961.

[13] Young, A. D., "The Aerodynamic Characteristics of Flaps," NACA Ames Research Center ARC R&M 2622, 1953.

[14] Kuhn, R. E., "Takeoff and Landing Distance and Power Requirements of Propeller Driven STOL Aircraft," IAS Preprint 690, presented at 25th Annual Meeting, New York, 28–31 Jan. 1957.

[15] McCormick, B. W., *Aerodynamics of V/STOL Flight*, Academic Press, New York, 1967.

[16] May, F., and Widdison, C. A., "STOL High-Lift Design Study, Vol. 1, State-of-the-Art Review of STOL Aerodynamic Technology," U.S. Air Force Flight Dynamics Laboratory Rept. AFFDL-TR-71-26, Wright–Patterson AFB, OH, April 1971.

[17] Oates, G. S., Brown, S., and Nicolai, L., "STOL Tactical Aircraft Investigation, Executive Summary," U.S. Air Force Flight Dynamics Laboratory, AFFDL TR-74-125, PTC, Wright–Patterson AFB, OH, July 1975.

[18] Kohlman, D. L., "Introduction to V/STOL Airplanes," Iowa State Univ. Press, Ames, IA, 1981.

Chapter 10 — Takeoff and Landing Analysis

Landing the Lockheed L-1011 Tristar. The L-1011 with Rolls Royce RB 211 turbofans compared favorably with its competition. The high lift system of the L-1011 is discussed in Section 10.7.

Creative people who know exactly what they are doing aren't creative people.

10.1 Introduction

This chapter considers aircraft takeoff and landing performance in detail. The discussions of Chapter 6, Section 6.3, and of Chapter 9 addressed this performance aspect in general terms and provided initial estimates. This chapter assumes that the aircraft design is defined in fair detail and the designer is ready for a thorough takeoff and landing analysis.

The ground rules governing takeoff and landing are reported in MIL-C-5011A [1] for military aircraft and in the Federal Aviation Regulations (FAR) Parts 23 and 25 [2] for civil and commercial aircraft. The U.S. Navy uses AS-5263 [3] in lieu of MIL-C-5011. Before discussing these regulations it is useful to consider the following definitions for *conventional takeoff and landing* (CTOL):

$V_{stall} = V_S = 1\ g$ stall speed out of ground effect (sometimes called *minimum flight speed* or *control speed, V_{min}*) [Eq. (6.1)]
V_{TO} = takeoff or liftoff speed
V_{CL} = climb-out speed during takeoff
V_{EF} = one-engine-failure speed
V_1 = decision speed (continue or brake)
V_R = rotation speed (speed at which an aircraft is rotated, during the ground run)
V_{OBS} = speed over the obstacle [obstacle height = 50 ft for military and 35 ft for Federal Aviation Administration (FAA) commercial] for takeoff
V_{50} = speed over 50-ft obstacle during landing
V_{TD} = speed at touchdown during landing
V_{app} = approach speed for carrier aircraft (120 kt from [3])

A summary of CTOL ground rules for military and civil aircraft is shown in Tables 10.1 and 10.2. The reader should note that in all cases the speeds specified are minimum speeds.

The takeoff and landing analyses, in Sections 10.3 and 10.4 respectively, will assume realistic speeds compatible with these minimum speeds. There are no satisfactory military or civilian ground rules for STOL aircraft at present. STOL flying qualities are addressed in military specification MIL-F-83300 but this document is currently under revision and only alludes to STOL takeoff and landing ground rules. References [4,5] make recommendations for realistic STOL takeoff and landing ground rules. The analyses in Sections 10.3 and 10.4 are appropriate for STOL performance with consideration given to realistic speed ground rules and the features peculiar to STOL operation (for example, dirt-strip operation giving rolling $\mu = 0.04$ and braking $\mu = 0.30$).

Table 10.1 Summary of CTOL Takeoff Rules

Item	MIL-C-5011C (Military)	FAR Part 23 (Civil)	FAR Part 25 (Commercial)
Speeds	$V_{TO} \geq 1.1 V_S$ $V_{CL} \geq 1.2 V_S$	$V_{TO} \geq 1.1 V_S$ $V_{CL} \geq 1.1 V_S$	$V_{TO} \geq 1.1 V_S$ $V_{CL} \geq 1.2 V_S$
Climb gradient[a]	Gear up: 500 fpm at SL (AEO) 100 fpm at SL (OEI)	Gear up: 300 fpm at SL (AEO)	Gear up: 3% at V_{CL} (OEI) Gear down: 0.5% at V_{TO}
Rolling coefficient	$\mu = 0.025$	Not specified	Not specified
Field-length definition	Takeoff distance over 50 ft	Takeoff distance over 50 ft	115% of takeoff distance over 35 ft *or* critical field length (see Section 10.6)

[a]SL = sea level; AEO = all engines operating; OEI = one engine inoperative.

10.2 Ground Effects

As the aircraft flies close to the ground, the ground interferes with the horseshoe vortex system trailing behind the wing (see Section 2.3 and Fig. 10.1). Ground effects are often analyzed by putting an *image horseshoe vortex* system of equal but opposite strength at the same distance *h* below the ground that the wing is above the ground. This image vortex system effectively cancels the wing's induced velocities normal to the ground surface, the necessary boundary condition at the ground surface. This image vortex system induces velocities at the wing aerodynamic center (a.c.), which decreases the strength of the downwash at the wing a.c., thereby decreasing the induced angle-of-attack, α_i. Thus, the wing C_L is increased (or more correctly, the lift curve slope increases, giving an increase in C_L for the same geometric angle-of-attack, α) and the induced drag is decreased. This influence of the ground effect is a function of how close the aircraft is to the ground and of the size of the wing. The effect of

Table 10.2 Summary of CTOL Landing Rules

Item	MIL-C-5011C	FAR Part 23	FAR Part 25
Speeds	$V_{OBS} \geq 1.2 V_S$ $V_{TD} \geq 1.1 V_S$	$V_{OBS} \geq 1.3 V_S$ $V_{TD} \geq 1.15 V_S$	$V_{OBS} \geq 1.3 V_S$ $V_{TD} \geq 1.15 V_S$
Braking coefficient	0.30	Not specified	Not specified
Field-length definition	Landing distance over 50 ft	Landing distance over 50 ft	Landing distance over 50 ft divided by 0.6

Figure 10.1 B-767 generates a strong trailing vortex system during a landing approach (courtesy of Ray Nicolai).

the proximity of the ground on the wing can be thought of as an increase in the wing geometric aspect ratio (AR) to some effective aspect ratio AR_{eff}. Figure 10.2a shows the variation of AR/AR_{eff} with the nondimensional wing height parameter h/b, where h is the height of the wing above the ground and b is the wing span. Figure 10.2a can be used to obtain the effective aspect ratio AR_{eff}, which is then put into Eq. (2.13) of Chapter 2 to obtain the wing lift curve slope in ground effect.

Figure 10.2b shows the influence of ground effect on an AR = 4 wing. Here, the nondimensional wing height parameter is $h/\bar{c}$, where $\bar{c}$ is the wing mean aerodynamic chord. An $(h/\bar{c})$ value of ∞ means that the wing is out of ground effect or in free air. Notice that the wing lift curve slope increases as the ground is approached and the angle for zero lift, α_{0L}, becomes less negative. Approaching the ground essentially decreases the effect of camber. Camber causes the circulation about the airfoil to increase, which introduces a curving upward of the flow streamlines as they approach the airfoil (an upwash ahead of the airfoil). The proximity of the ground straightens the flow streamlines (or the image vortex system decreases the upwash), thereby decreasing the magnitude of α_{0L}. The C_L for $\alpha = 0$ is approximately the same value for all values of $h/\bar{c}$. This fact should be used when generating a curve similar to Fig. 10.2b to determine the ground C_L during the takeoff and landing analysis.

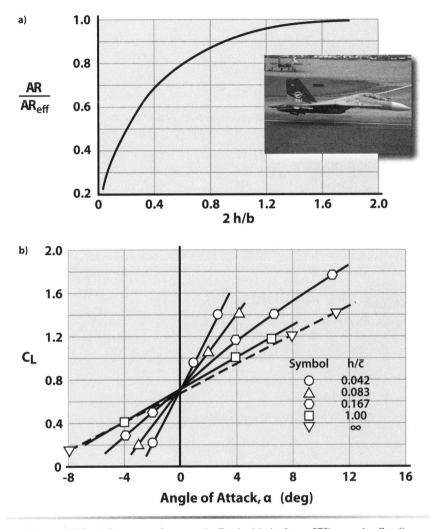

Figure 10.2 Influence of ground effects (data from [7]) on **a)** effective aspect ratio, AR_{eff}, and **b)** an $AR = 4$ wing.

As the wing approaches the ground, the induced drag is decreased but the zero-lift drag is essentially unchanged. The reduction in induced drag of the aircraft in ground effect can be expressed as (from [6])

$$\Delta C_{D_i} = -\sigma C_L^2 / \left(\pi AR \right)$$

where σ is defined as the ground influence coefficient. At values of h/b between 0.033 and 0.25, σ can be estimated from

$$\sigma = \frac{1 - 1.32\ h/b}{1.05 + 7.4\ h/b}$$

10.3 Takeoff Analysis

Takeoff is the distance required for an aircraft to accelerate from $V = 0$ to takeoff speed and climb over a 35- or 50-ft obstacle. Figure 10.3 shows a schematic of the takeoff problem.

The takeoff distance is the sum of the ground distance (S_G), rotation distance (S_R), transition distance (S_{TR}), and climb distance (S_{CL}).

It is assumed that the aircraft accelerates to a takeoff velocity $V_{TO} = 1.2V_{stall}$ and at that speed the aircraft is rotated to an angle-of-attack such that the $C_L = 0.8 C_{L_{max}}$. The aircraft then leaves the ground and transitions from horizontal to climbing flight over the distance S_{TR}.

10.3.1 Ground Distance S_G

The *ground distance* S_G is defined as

$$S_G = \int_0^{V_{TO}} \frac{V dV}{a} = \frac{1}{2} \int_0^{V_{TO}} \frac{dV^2}{a} \qquad (10.1)$$

where a is the acceleration during S_G and

$$V_{TO} = 1.2\ V_{stall} = 1.2 \sqrt{\frac{W_{TO}}{S_{ref}} \frac{2}{\rho C_{L_{max}}}} \qquad (10.2)$$

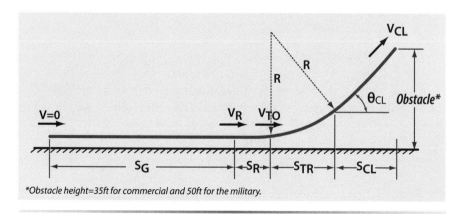

Obstacle height=35ft for commercial and 50ft for the military.

Figure 10.3 Schematic of an aircraft takeoff analysis.

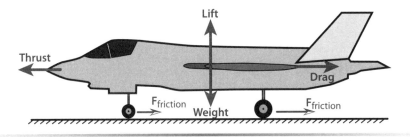

Figure 10.4 Force diagram during ground run.

and $C_{L_{max}}$ is for a particular flap setting. The $C_{L_{max}}$ is determined using the methods of Chapter 9.

From the force diagram shown in Fig. 10.4,

$$a = \frac{g}{W}\left[T - D - F_f\right] = \frac{g}{W}\left[T - D - \mu\left(W_{TO} - L\right)\right] \qquad (10.3)$$

where μ is the coefficient of friction for brakes off (see Table 10.3). The lift and drag during the ground run are given by

$$D = \left(0.5\right)\rho V^2 S_{ref}\left[C_{D0} + \Delta C_{D_{flap}} + \Delta C_{D_{gear}} + KC_{LG}^2\right]$$

$$L = \left(0.5\right)\rho V^2 S_{ref} C_{LG}$$

where C_{LG} is the lift coefficient during the ground run for a particular flap setting (see Fig. 9.22, Chapter 9). There is a difference in the C_L vs α curve

Table 10.3 Coefficients of Friction for Various Takeoff and Landing Surfaces

Type of Surface	Brakes Off, Average Ground Resistance Coefficient	Brakes Fully Applied, Average Wheel-Braking Coefficient
Concrete or macadam	0.015–0.04	0.3–0.6
Hard turf	0.05	0.4
Firm and dry dirt	0.04	0.30
Soft turf	0.07	0.5
Wet concrete	0.05	0.2
Wet grass	0.10	0.2
Snow- or ice-covered field	0.01	0.07–0.10

for an aircraft in ground effect (IGE) and out of ground effect (OGE). This difference is slight and will be ignored for the moment but is discussed in Section 10.7 for the L-1011. The $\Delta C_{D_{flap}}$ is determined from data in Fig. 9.25 and $\Delta C_{D_{gear}}$ from Fig. 10.5.

The *landing gear drag coefficient* $\Delta C_{D_{gear}}$ is a difficult term to determine. The landing gear drag is dependant on the flap setting, the aircraft C_L, and sometimes the spoilers. It is a nightmare in computational fluid dynamics and is seldom attempted. Wind tunnel testing usually underestimates the $\Delta C_{D_{gear}}$ because the full-scale landing gear is often not known at this point in the design and it is expensive to model all of the gear's detail. Thus, determining the landing gear drag is usually delayed until flight testing, and then it is a low priority because the landing gear is already built and the aircraft is flying. Figure 10.5 and Table 10.4 present the landing gear drag coefficients of current aircraft. Based upon this discussion the data needs to be used carefully.

The landing gear is designed to be functional, reliable, and lightweight—but not low drag. The reason is that most landing gears are retracted inside the aircraft during up-and-away flight. If the gear is not to be retracted, it is streamlined by shaping the struts and fitting a fairing over the wheels (sometimes called *wheel pants*). During landing and takeoff the landing gear is extended and the drag must be accounted for. During takeoff the landing gear drag is a nuisance because it reduces the acceleration to V_{TO}.

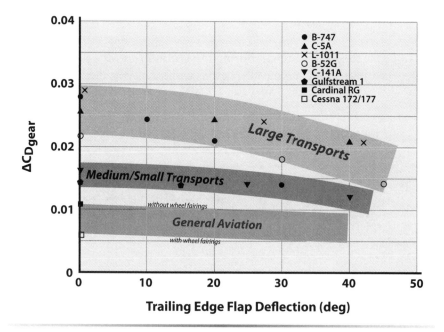

Figure 10.5 Drag of landing gear.

Table 10.4 Landing Gear Drag Coefficients

Aircraft	Reference Area (ft²)	ΔC_{Dgear}	Landing Gear Configuration[a]
Fighters			
A-7	375	0.028	Two-wheel NLG, two one-wheel MLG
F-104	196	0.035	One-wheel NLG, two one-wheel MLG
F-16A1B	300	0.0325	One-wheel NLG, two one-wheel MLG
F-22	840	0.014	One-wheel NLG, two one-wheel MLG
U-2S	1000	0.0045	One dual-wheel MLG, large tail wheel, and two wingtip pogos
Large transports			
L-1011	3456	0.028–0.0205	Two-wheel NLG, two four-wheel trucks MLG
C-5A	6200	0.0257–0.021	Four-wheel NLG, four four-wheel trucks MLG
B-747	5500	0.028–0.014	Two-wheel NLG, four four-wheel trucks MLG
B-52G	4000	0.024–0.0155	Quadricycle with wingtip gear, four dual-wheel MLG
Medium transports			
P-3	1300	0.020	Two-wheel NLG, two two-wheel MLG
L-1049 Connie	1650	0.024	Two-wheel NLG, two two-wheel MLG
B 727	1650	0.017	Two-wheel NLG, two two-wheel MLG
DC-8	2771	0.012	Two-wheel NLG, two four-wheel trucks MLG
C-141A	3228	0.0165–0.012	Two-wheel NLG, two four-wheel trucks MLG
Small transports			
S-3A	598	0.023	Two-wheel NLG, two one-wheel MLG
Gulfstream I	615	0.015	Two-wheel NLG, two one-wheel MLG
Fokker F-27	754	0.024	One-wheel NLG, two dual-wheel MLG
General aviation			
Cessna 172	226	0.006[b]	One-wheel NLG, two one-wheel MLG
Cessna 177	174	0.006[b]	One-wheel NLG, two one-wheel MLG
Cardinal RG	174	0.011	One-wheel NLG, two one-wheel MLG

[a]Abbreviations: NLG, nose landing gear; MLG, main landing gear.
[b]Fixed landing gear with wheel fairings.

However, during landing the landing gear drag is useful as it shortens the air distance and helps slow the aircraft down during braking.

The data in Fig. 10.5 and Table 10.4 reveal several interesting things. First, the drag coefficient due to the landing gear is huge: equal to or greater than the $C_{D_{min}}$ of the entire aircraft in most cases. Second, the $\Delta C_{D_{gear}}$ decreases as the flaps are deflected. This is because the flap deflection increases the circulation about the wing, and the landing gear beneath the wing is in a lower dynamic pressure field than the rest of the aircraft. Figure 10.5 shows that this is more prevalent on transport aircraft, which typically have more sophisticated flap systems to achieve higher maximum C_L (see Fig. 9.7). Fighter and general aviation aircraft have simpler flap systems, have lower maximum C_L, and exhibit a fairly constant $\Delta C_{D_{gear}}$. The fixed gear on general aviation aircraft is typically a simple strut-and-wheel (with wheel pants) arrangement referenced to a large wing area (low W/S) giving a relatively low $\Delta C_{D_{gear}}$. Fighter aircraft on the other hand have beefy retractable gear and small wing area (high W/S) giving a relatively large value for $\Delta C_{D_{gear}}$.

The designer should pick a value for $\Delta C_{D_{gear}}$ from Fig. 10.5 or Table 10.4 for their class of aircraft to use in the ground C_D build-up. Remember that the flaps are usually retracted during the braking ground run (landing) and set to the takeoff position for the acceleration ground run (takeoff).

There are several ways to calculate S_G:

1. Approximate the integral, Eq. (10.1).
2. Assume the acceleration a is constant, equal to the value at $V = 0.707 V_{TO}$.
   Then solve Eq. 10.3 for a at $V = 0.707 V_{TO}$ and then

$$S_G = \frac{1}{2} \frac{V_{TO}^2}{a_{(at\ 0.707\ V_{TO})}} \qquad (10.4a)$$

3. Stepwise integration of Eq. (10.1), that is, calculate a for values of $V = 0, 20, 40, ... , V_{TO}$ and then plot $1/2a$ versus V^2 as shown in Fig. 10.6. The value for S_G is the area under the curve. Putting Eq. (10.2) into Eq. (10.4a) gives

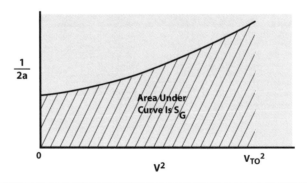

Figure 10.6 Stepwise integration for ground roll distance S_G.

$$S_G = \frac{1.44 \left(W / S_{ref} \right)_{TO}}{g \rho C_{L_{max}} \left[\left(T / W \right) - \left(D / W \right) - \mu \left(1 - L / W \right) \right]} \qquad (10.4b)$$

where the lift, drag, and thrust terms are evaluated at a speed of $0.707 V_{TO}$. From Eq. (10.4b) it can be seen that there are several factors the designer can influence to vary the takeoff ground run: that is, takeoff wing loading $(W/S)_{TO}$, takeoff thrust-to-weight ratio (T/W), retardation force-to-weight ratio $(D + F_f)/W$, and maximum wing lift coefficient.

Low values of $(W/S)_{TO}$ can shorten S_G but will result in poor riding qualities (due to sensitive gust response), less efficient cruise, and poor acceleration. Achieving a short takeoff through low wing loading is considered acceptable only when other performance requirements are lower priority.

A high thrust-to-weight ratio will shorten the takeoff roll but the engines should not be sized by the takeoff requirement alone. If the thrust is installed solely to expedite takeoff, cruise efficiency is likely to suffer. The $(T/W)_{TO}$ must be selected by considering all the mission requirements (see Chapter 15).

An increase in $C_{L_{max}}$ is the one method by which the designer can influence just the takeoff (or landing) performance. The idea is to obtain a large increase in $C_{L_{max}}$ for a small increase in drag because the quantity $\max(D + F_f)/W$ decreases the ground run acceleration. The $C_{L_{max}}$ and aircraft C_D both increase with flap deflection. Thus, the δ_f for minimum ground run is a compromise between lowering V_{TO} (by increasing $C_{L_{max}}$) and lowering the ground run acceleration (by increasing C_D). High $C_{L_{max}}$ is the answer for STOL operation and is receiving considerable attention from government agencies and industry.

10.3.2 Rotation Distance, S_R

Rotation distance is the distance in which the aircraft (still on the ground) is rotated to that α such that $C_L = 0.8 C_{L_{max}}$. For the military this rotation takes two seconds, whereas for FAA–commercial aircraft this time is established during flight testing:

$$S_R = 2V_{TO} \qquad (10.5)$$

(where V_{TO} is in feet per second). Note that the geometry of the aircraft should be checked to insure that the tail of the aircraft does not strike the ground during this rotation. For airplanes that are "geometry limited" during ground rotation the FAA applies different criteria for establishing the various takeoff speeds.

10.3.3 Transition Distance, S_{TR}

In the *transition distance* the aircraft flies a constant-velocity arc of radius R (see Fig. 10.3). The load factor on the aircraft is

$$n = 1 + \frac{V_{TO}^2}{Rg} = \frac{L}{W} = (0.8)(1.2)2 = 1.15$$

and solving for

$$R = \frac{V_{TO}^2}{0.15\,g} \qquad (10.6)$$

(in feet). The aircraft is assumed to be in an unaccelerated climb such that

$$\text{Rate of climb} = V_{TO}\,\sin\theta_{CL} = \frac{V_{TO}\,(T - D)}{W}$$

and then becomes

$$\text{Rate of climb} = V_{TO}\,\sin\theta_{CL} = \frac{V_{TO}\,(T - D)}{W} \qquad (10.7)$$

Now the transition distance is (see Fig. 10.7)

$$S_{TR} = R\sin\theta_{CL} \qquad (10.8)$$

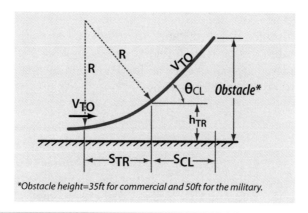

Obstacle height=35ft for commercial and 50ft for the military.

Figure 10.7 Schematic for transition and climb distance.

10.3.4 Climb Distance, S_{CL}

The *climb distance* is

$$S_{CL} = \frac{50 - h_{TR}}{\tan\theta_{CL}} \tag{10.9}$$

where h_{TR} is shown in Fig. 10.7. If $h_{TR} > 50$ ft, then $S_{CL} = 0$.

10.3.5 Time During Takeoff

Assume the ground run acceleration a is constant, equal to $0.707V_{TO}$. Then

$$\text{time (ground)} = t_g = \frac{1}{a}\int_0^{V_{TO}} dV = \frac{V_{TO}}{a}$$

The rotation time is assumed to be two seconds. The time for transition and climb can be approximated by

$$\text{time (transition)} = t_{TR} = \frac{S_{TR} + (S_{CL}/\cos\theta_{CL})}{V_{TO}}$$

10.4 Landing Analysis

The *landing distance* is the horizontal distance required to clear a 50-ft obstacle, free roll, and then brake to a complete stop. Figure 10.8 shows the schematic for the landing analysis.

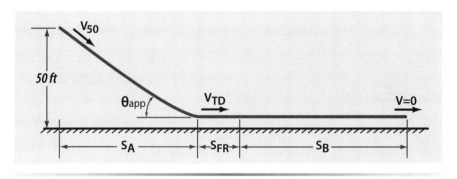

Figure 10.8   Schematic for landing analysis.

It is assumed that the velocity over the 50-ft obstacle $V_{50} = 1.3V_S$ and the touchdown velocity $V_{TD} = 1.15V_S$. The V_S is for the aircraft in its landing configuration, that is,

$$W_L = \text{aircraft weight with 1/2 fuel remaining}$$
$$C_{Lland} = C_{Lmax} \text{ for flaps in landing configuration}$$

The landing distance is the sum of the *air distance* S_A, the *free-roll distance* S_{FR}, and the *braking distance* S_B.

10.4.1 Air Distance S_A

The change in kinetic energy (KE) plus potential energy (PE) is

$$KE + PE = (\text{retarding force})S_A$$

Assume θ_{app} to be small such that $\cos\theta_{app} \approx 1$. Then

$$\frac{W}{g}\left[\frac{V_{50}^2}{2} + 50\,g - \frac{V_{TD}^2}{2}\right] = FS_A$$

$$S_A = \frac{W}{F}\left[\frac{V_{50}^2 - V_{TD}^2}{2g} + 50\right]$$

because $W \approx L$ and $F \approx D$

$$S_A = \frac{L}{D}\left[\frac{V_{50}^2 - V_{TD}^2}{2g} + 50\right] \qquad (10.10)$$

or $S_A = 50/\tan\theta_{app}$, where θ_{app} = approach glide slope (3 deg for typical CTOL, 7 deg for STOL) and where $L = W_L$ and $D = C_D q S_{ref}$,

$$C_D = C_{D0} + KC_{Lmax}^2 + \Delta C_{Dflaps} + \Delta C_{Dgear} \qquad (10.11)$$

10.4.2 Free Roll Distance S_{FR}

The *free-roll distance* S_{FR} is the distance covered while the pilot reduces the power to idle, retracts the flaps, deploys the spoilers, and applies the brakes. Free roll is *assumed* to be three seconds so that

$$S_{FR} = 3V_{TD} \qquad (10.12)$$

10.4.3 Braking Distance S_B

Let a = deceleration. Then

$$S_B = \int ds = \int_{V_{TD}}^{0} \frac{V dV}{a} = \frac{1}{2}\int_{V_{TD}}^{0} \frac{d(V^2)}{a}$$

Using the force diagram in Diagram 10.1 yields

$$a = (1/m)\left[T - D - F_f - R\right]$$

where R is a reverse thrust during braking and $T = 0$. Also, F_f is the force of friction, $F_f = \mu(W_L - L)$, and μ is given in Table 10.3.
 Also,

$$C_D = C_{D0} + KC_{LG}^2 + \Delta C_{D\text{flaps}} + \Delta C_{D\text{gear}} + \Delta C_{D\text{misc}} + \Delta C_{D\text{spoilers}}$$

where

C_{LG}	$= C_L$ at the α during braking (see Fig. 9.22).
$\Delta C_{D\text{flaps}}$	$= \Delta C_D$ due to flaps (from Fig. 9.25). If flaps are retracted during braking , $\Delta C_{D\text{flaps}} = 0$ and $C_{LG} \approx 0$
$\Delta C_{D\text{spoilers}}$	$= \Delta C_D$ due to spoilers
Use $\Delta C_{D\text{spoilers}}$	$= 0.006 - 0.007$
$\Delta C_{D\text{misc}}$	$= \Delta C_D$ due to miscellaneous items such as a drag chute and high energy absorption brakes (Fig. 10.9)
Use $\Delta C_{D\text{misc}}$	$= 1.4$ for drag chutes based upon their inflated frontal area [8].

By neglecting reverse thrust R and setting $T = 0$,

$$-a = \frac{g}{W_L}(F_f + D) = \frac{g}{W_L}\left[\mu(W_L - L) + C_D q S_{\text{ref}}\right]$$

Figure 10.9 Brakes in action: **left)** SR-71 during braking ground roll with drag chute deployed, and **right)** typical commercial transport main gear during heavy braking.

$$S_B = \frac{W}{2g} \int_{V\text{TD}}^{0} \frac{\mathrm{d}(V^2)}{\mu W_L + (\rho/2)S_{\text{ref}}V^2\left(C_D - \mu C_{LG}\right)}$$

$$S_B = \frac{W_L}{g\mu\rho S_{\text{ref}}\left[(C_D/\mu) - C_{LG}\right]} \ln\left[1 + \frac{\rho}{2}\frac{S_{\text{ref}}}{W_L}\left(\frac{C_D}{\mu} - C_{LG}\right)V_{\text{TD}}^2\right] \quad (10.13)$$

An examination of the landing analysis reveals that short landing distances require a low V_{stall} and a large retardation force (unlike the takeoff). A low V_{stall} is the result of a low wing loading at landing and a high C_L. Fortunately, the high C_L will also give a large drag. Thus, high $C_{L_{\max}}$ is extremely important for short loading distances.

10.5 Aircraft Retardation Devices

In designing an aircraft, the method selected to stop the vehicle on landing is dependent upon weight and maintenance penalties, and hence cost, that one is prepared to pay. To dissipate the kinetic energy the aircraft possesses at touchdown, the engineer may select to use mechanical means (wheel brakes or arresting gear), propulsive means (reverse thrust), aerodynamic means (drag chutes or speed brakes/spoilers), or some combination of these methods, but the ultimate decision must be based on a consideration of the overall weapon or transportation system and not just the airframe itself.

Wheel brakes have been the traditional means for halting landing rolls; no matter what other braking devices may be selected by the designer, all aircraft must have some wheel braking capability for use while taxiing and during ground maneuvering. For a given brake life, the weight of the brake assembly is a function of the energy that must be absorbed. The level of this energy can be calculated using the predicted landing weight and a touchdown velocity equal to $1.15V_{\text{stall}}$ for that weight; the brake assembly weight can then be estimated using Fig. 10.10, from [9]. Notice the effect that brake lining life has on brake weight. Some tradeoff must be made between airframe cost–performance and maintenance (hence, total system) costs. A large aircraft requiring frequent full-stop landings and low maintenance, such as an airliner, could not depend on wheel brakes for its entire stopping capability without paying a penalty in payload capacity.

With the trends to higher landing weights and speeds of modern aircraft, energy dissipation requirements have become so large that the weight of a brake assembly that would provide the entire retardation effort would be prohibitive even for low-life brake linings. Some additional means of energy dissipation must be provided to supplement the wheel brakes, and aerodynamic and propulsive retardation devices provide this capability. A

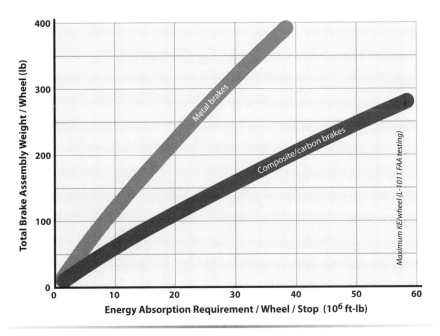

Figure 10.10   Weight of main gear wheel brakes based on maximum energy "rejected takeoff" (RTO).

secondary requirement for these devices has been generated by the desire to shorten landing distances by supplying some stopping force soon after touchdown, when velocities are still high and wheel brakes are least effective. Again, drag chutes and thrust reversers meet this need. However, the B-2 shown in Fig. 10.11 dissipates energy using its split ailerons. It has no drag chute.

Thrust reversers on jet aircraft provide a significant ground deceleration force with high reliability and little additional maintenance with repeated application. On some transport aircraft (DC-8, C-5) the thrust reversers on inboard engines may be activated in flight to function as speed brakes. A jet or turbofan engine with thrust reversers is capable of providing up to approximately 40% of the rated takeoff thrust for braking.

Figure 10.11   B-2 taxiing.

However, the relatively high weight of these devices has all but prohibited their use on combat aircraft. The major exception to this is the Swedish Viggen, a STOL fighter whose thrust reverser is activated automatically upon touchdown for rapid stopping. A thrust reverser may be designed by the engine manufacturer or may be provided by the airframe designer. The weight of a thrust reverser installation will vary with the rated thrust of the engine and may be estimated from Fig. 10.12, which has been compiled from both airframe and engine manufacturers' data.

The provision for reversible pitch on propeller aircraft is an inexpensive, lightweight method of retardation for this class of aircraft. Turboprop engine–propeller combinations have a capability of providing a reverse force of up to 60% of their rated static thrust, whereas reciprocating engines can provide a lower retardation force (approximately 40% of static thrust) and require a longer delay between touchdown and the actuation of full reverse thrust. Reversible propellers also give an aircraft the ability to taxi backward, an extremely useful asset in combat airlift operations.

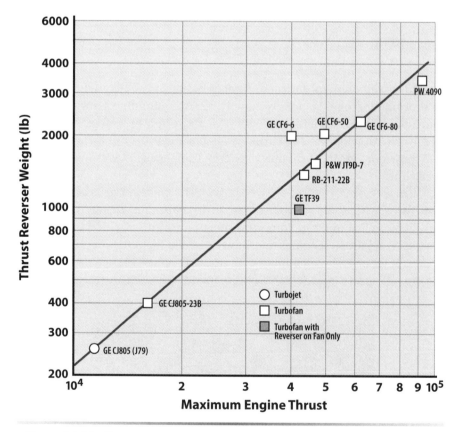

Figure 10.12 Turbojet and turbofan thrust reverser weight as a function of maximum engine thrust.

Drag chutes have received almost universal acceptance as landing retardation devices on high-performance combat aircraft (the F-117A used one). They are lightweight, reliable, and if one is willing to ignore the problems of repacking, relatively simple. Drag parachutes are most effective at higher velocities and, thus, nicely complement the capabilities of wheel brakes. Reference [8] gives an average drag coefficient based on projected (inflated) cross-sectional area of 1.4 for nylon or ribbon-type parachutes. The parachute weight (in pounds) can be predicted using the formula

$$W_P = 6.5 \times 10^{-2} \left(d_{\text{chute}} \right)^2$$

where d_{chute} is the maximum chute diameter in feet. The weight of the entire chute and rigging is approximately 3–4 times this figure.

10.6 Critical Field Length (Balanced Field Length)

The *critical field length* (sometimes called *balanced field length*) is a balance between the distance required to accelerate to V_1 and then either continue the takeoff over 35 ft with one engine inoperative or brake to a stop. The definition of takeoff field length for commercial transport aircraft is 115% of the distance over 35 ft with all engines operating or the critical field length, whichever is greater.

The analysis is one of selecting a failure recognition speed V_{EF} and then calculating two distances. The first distance, called LAB, is the distance required to accelerate to V_{EF}, free roll for 3 seconds at V_{EF}, and then brake to a full stop. The second distance, called LAC, is the distance required to accelerate to V_{EF} and then continue the takeoff over 35 ft with one engine out. During the one-engine-out phase, the thrust is reduced accordingly and the drag is increased to account for a windmilling engine. There is a unique value for V_{EF} such that LAB = LAC. This value for V_{EF} is also called the *refusal speed* and the distance LAB = LAC is called the *critical field length* or *balanced field length*. This analysis is shown in Fig. 10.13. If an engine should fail below V_{EF}, the pilot should brake to a stop; if one engine fails above the refusal speed, the pilot should continue the takeoff with one engine out. Figure 10.14 illustrates several high performance takeoffs.

10.7 Comparison of Analytical Estimates With L-1011 Flight Test

This section will determine the takeoff and landing distances for the L-1011 using the methods in this chapter and Chapter 6, and compare the results with flight test data.

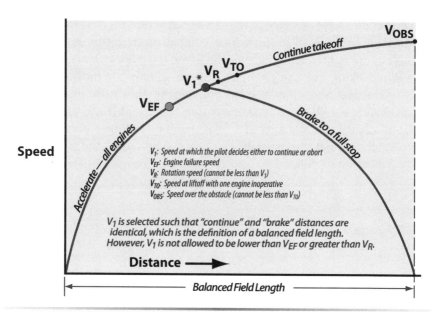

Figure 10.13 Schematic of balanced field length.

The Lockheed L-1011 TriStar is a subsonic wide-body commercial transport aircraft designed to operate over transcontinental, short- and medium-length airline routes with passenger payloads of up to 260 passengers in a 15/85 mixed seating and with passenger payloads of up to 345 in an all-economy-class seating. The L-1011 received FAA certification in April 1972. Lockheed produced 250 aircraft between 1968 and 1984. The airplane is powered by three Rolls Royce RB.211–22 high-bypass-ratio turbofan engines. Two engines are mounted in underwing pylons and the third engine is mounted in the fuselage aft body. Figure 10.15 shows the L-1011 with emphasis on the high-lift system, and Table 10.5 gives dimensions and geometry for the aircraft.

The high-lift system consists of double-slotted Fowler TE flaps and full-span LE slats. Slat and flap extension is manually controlled by the pilot using the flap handle on the center console. Fully extended, the three inboard slat panels deflect 28 deg and the four outboard slat panels deflect 30 deg.

The trailing edge flaps consist of four double-slotted Fowler-type flap surfaces on each wing. The following discrete flap positions may be selected for the corresponding takeoff, landing, or cruise segment:

- Flaps up, cruise
- Flaps down 4 deg, transition
- Flaps down 10 deg, takeoff and alternate approach

- Flaps down 18 deg, takeoff
- Flaps down 27.5 deg, takeoff
- Flaps down 33 deg, alternate landing
- Flaps down 42 deg, landing

Figure 10.14   Mig 31, F-15, B-1B, and L1011 takeoffs.

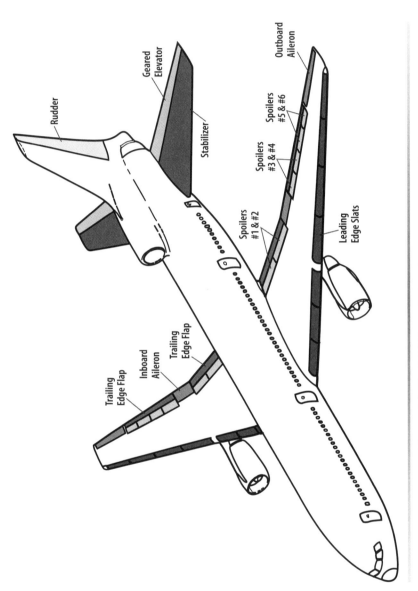

Figure 10.15 L-1011 control surfaces arrangement.

Table 10.5   Summary of L-1011 Dimensions and Geometry

Geometry	Dimensions
Maximum takeoff weight (lb)	430,000
Maximum landing weight (lb)	358,000
Maximum fuel weight (lb)	159,560
Wing span (ft)	155.3
Wing area (ft^2)	3456
Wing sweep (quarter-chord)	35 deg
Aspect ratio	6.95
Mean aerodynamic chord (mac; ft)	24.5
Location of mac leading edge	FS 1143
Spoiler area/span (each side ft^2/ft)	
1	27/10
2	34/12
3	12/5
4	16/7
5	7/3
6	11/5
TE flap area/span/chord (each side ft^2/ft/ft)	
Inboard	145/22/6.5
Outboard	123/22/30% chord
LE slat area/span (each side ft^2/ft)	
Inboard	61.74/22
Outboard	118/38.67
Horizontal tail	
Area (ft^2)	1282
Aspect ratio	4.0
Taper ratio	0.33
mac (ft)	19.5
Location of mac leading edge	FS 1885
Vertical tail	
Area (ft^2)	550
Aspect ratio	1.6
Taper ratio	0.3
mac (ft)	20.3
Location of mac leading edge	FS 1924

The TriStar also has six pairs of spoilers on each wing to dump the load during landing to reduce the lift on the aircraft, thereby putting more weight on the wheels for greater braking. The spoilers are manually deflected up to 60 deg.

Example 10.1 Takeoff Distance Comparison

We calculate the takeoff distance using the method discussed in this chapter, with the following initial conditions:

Weight	358,000 lb
TE flap deflection	27.5 deg
Maximum C_L	2.0 (IGE at $\alpha = 16$ deg, limited by tail strike)
Rolling friction coefficient, μ	0.015 (measured during flight test)
Ground C_L, C_{L_G}	0.32 (value at $\alpha = 0$ deg and $\delta_f = 27.5$ deg from flight test)
Thrust, all three engines	100,022 lb
$\Delta C_{D_{flap}}$	0.042 (from Fig. 9.25)
$C_{D_{min}}$	0.0175 (from Fig. G.1)
$\Delta C_{D_{gear}}$	0.024 (from Fig. 10.5)
Air density, ρ	0.002378 slug/ft^3
Ground run drag-due-to-lift factor, K	0.0468 (from flight test)

The value $K = 0.0468$ can be estimated considering the drag-due-to-lift in the presence of ground effects. Using Fig. 10.2a and assuming the $2h/b = 0.12$ during the ground run, the effective AR for the L-1011 is approximately 17. From Fig. G.9 the wing efficiency factor for this class of aircraft can be estimated to be $e = \sim 0.4$, giving $K = 0.0468$.

Assuming $V_{TO} = 1.2 V_S$ gives a takeoff speed of 250 ft/s, or 148 kt, which is under the flap and slat placard limit of 200 kt for the L-1011. The drag coefficient during the ground run is determined as follows:

$$C_{D_{ground}} = C_{D_{min}} + KC_{L_G}^2 + \Delta C_{D_{flap}} + \Delta C_{D_{gear}} = 0.0883$$

The dynamic pressure at $0.7 V_{TO}$ is 36.4 psf, which gives $L = 40,270$ lb and $D = 11,107$ lb. Using Eq. (10.3) gives an acceleration at $0.7 V_{TO}$ of 7.57 ft/s^2 ($\sim 1/4$ g) and, from Eq. (10.4), a ground run distance $S_G = 4172$ ft.

At V_{TO} the aircraft rotates to an α such that $C_L = 0.8 C_{L_{max}}$. It is assumed that this rotation takes two seconds. The aircraft geometry should be checked to make sure the aircraft can rotate to this α without striking the tail on the ground. Thus, $S_R = 500$ ft.

Table 10.6   Comparison of Takeoff Analysis

Phase	Eq. (10.4) Eq. (10.5) Eq. (10.8)	Eq. (6.3)	Flight Test (1972)
Ground run	4172	3867	3632
Rotate and transition	875	500	1225
Takeoff distance	5047	4367	4857

The transition distance analysis indicates that the L-1011 clears the 35-ft obstacle about midway through the transition phase to a steady-state climb rate of 61.6 ft/s. Thus, it is assumed that the L-1011 spends about 1.5 s in the transition phase and covers another 375 ft in distance. *The total takeoff distance for the L-1011 is 5047 ft.*

The comparison with the simplified method of Chapter 6 [Eq. (6.3)] and flight test results are shown in Table 10.6. It should be recognized that the actual takeoff distance depends on several factors, such as winds, runway characteristics, aircraft aerodynamic performance, and pilot technique. Equation (6.3) should be used for early analysis and sizing the wing loading. The analysis in this chapter should then be used to refine the takeoff analysis as more information is available on the aircraft design.

Example 10.2   Landing Distance Comparison

We calculate the landing distance using the method discussed in this chapter, with the following initial conditions:

Weight	345,100 lb
TE flap deflection	42 deg
Maximum C_L	2.63 (IGE from flight test)
Braking friction coefficient, μ	0.32 (measured during flight test)
Ground C_L, C_{L_G}	−0.18 (value at $\alpha = 0$ deg, $\delta_f = 0$ deg, and spoilers deployed, from flight test)
$\Delta C_{D_{flap}}$	0.095 (from Fig. 9.25 during approach); 0 (flaps retracted during ground braking)
$C_{D_{min}}$	0.0175 (from Fig. G.1)
$\Delta C_{D_{gear}}$	0.021 (from Fig. 10.5 for $\delta_f = 42$ deg); 0.029 (from Fig. 10.5 for $\delta_f = 0$ deg)
$\Delta C_{D_{spoiler}}$	0.0065 (from [10])
Air density, ρ	0.002378
Average approach L/D	5.0 (IGE from flight test)
Drag-due-to-lift factor, K	0.057 (IGE from flight test)

Assuming $V_{TD} = 1.15V_S$ gives a touchdown speed of 199 ft/s (124 kt), and $V_{50} = 1.3V_S$ gives a speed over 50 ft of 225 fps (133 kt), which are both under the flap and slat placard limit of 164 kt for the L-1011.

The average approach $K = 0.057$ can be estimated assuming an average value for $2h/b = 0.32$. From Fig. 10.2a the effective aspect ratio during approach is approximately 10.7. Estimating the wing efficiency factor using Fig. G.9 for this class of aircraft and AR = 10.7 gives $e \sim 0.52$ and $K = 0.057$.

The air-distance phase is one of the few times that drag is good. The larger the value of D/L, the steeper the glide slope and the shorter the air distance over the 50-ft obstacle. The drag coefficient on approach is

$$C_{D_{Approach}} = C_{D_{min}} + KC_{L_{max}}^2 + \Delta C_{D_{flap}} + \Delta C_{D_{gear}} = 0.527$$

The value of L/D for the L-1011 during landing approach is 5.0. Using Eq. (10.10) gives an air distance of 1106 ft.

During the free-roll phase the L-1011 pilot is busy applying the brakes, putting the engines at idle, retracting the TE flaps, and deploying the spoilers. It is assumed that this takes three seconds, so that $S_{FR} = 597$ ft. During the flight test of the L-1011 the test pilots routinely took under two seconds for this phase.

The drag coefficient during the ground braking phase is determined as follows:

$$C_{D_{ground}} = C_{D_{min}} + KC_{LG}^2 + \Delta C_{D_{flap}} + \Delta C_{D_{gear}} + \Delta C_{D_{spoiler}} = 0.0548$$

During ground braking the flaps are retracted and the spoilers deployed to dump the lift and put as much weight on the wheels as possible, to increase the friction force. The spoilers actually produce a small downward force coefficient of −0.18. Equation (10.13) is used to calculate a braking distance of 1778 ft. *The total landing distance is 3481 ft.*

The comparison with the simplified method of chapter 6 [Eq. (6.5)] and flight test results are shown in Table 10.7. The discussion for Table 10.7 is the same as that for Table 10.6 presented earlier. The dependence of the analysis on pilot technique is even more pronounced for landing because pilots (carrier pilots excluded) will flare the aircraft just prior to touchdown to give a soft landing and minimize tail strike. The distance covered during the flare varies significantly between pilots. The analysis in this chapter does not assume a flare.

Table 10.7 Comparison of Landing Analysis

Phase	Eq. (10.10) Eq. (10.12) Eq. (10.13)	Eq. (6.5)	Flight Test (1972)
Air distance	1106	954	953
Free roll	597	Included below	315
Braking distance	1778	3014	1898
Landing distance	3481	3968	3166

The L-1011 compared favorably with its competition (the DC-10 and the Boeing 747), but a Rolls Royce bankruptcy in 1971 caused the L-1011 to arrive late in the marketplace and only 250 were built. Even so, the aircraft is a significant design achievement: The L-1011 is the first airplane to certify Category 3C landing capabilities using its spoilers to provide excellent control of lift during landing approach.

0.8 Airport Operations

Airport personnel have to be vigilant in maintaining safety of operations around airports. The presence of birds and wake turbulence can pose a real hazard to airport operations. Aircraft are especially vulnerable during the ATC (air traffic control) part of their mission as they are low and slow, high angle-of-attack, and the crew is in a high workload environment.

The vortex system generated by an aircraft during landing and takeoff (Fig. 10.1) is both beautiful (on a humid day) and dangerous. This vortex system is generated by a wing as it develops lift (see Section 2.6) and should not be confused with prop wash (the rotating air mass behind a propeller) or jet wash (the swirling hot gas from a jet engine exhaust), which are both of shorter duration and extent than the trailing vortex system. The trailing vortex generated by an aircraft wing (called wake turbulence) can cause a trailing aircraft in close proximity to the vortices to be flipped upside down. For this reason airports have landing and departure time intervals that gives the vortex system time to dissipate. The time intervals are dependant upon the temperature and wind conditions, and the relative size of the trailing aircraft to the lead aircraft that is generating the trailing vortices. A typical departure time interval is 3 minutes for a smaller aircraft taking off behind a larger aircraft. The landing time interval is in terms of separation distance. The separation for two aircraft of the same size (for example, two B 767s) is 4 nm and increases to 8 nm for a small aircraft landing behind a large aircraft (such as a Cessna Citation behind a B 747).

Birds around an airport are a problem as they can be ingested into a jet inlet or impact a propeller and shut down an engine. On January 15, 2009 an Airbus A320 struck a flock of Canadian Geese during departure from LaGuardia Airport, New York, which resulted in an immediate loss of thrust from both turbine engines. Due to the masterful piloting by the pilot the aircraft was ditched in the Hudson River with no loss of the 155 passengers and crew. Airports go to great lengths (scare crows, sound systems emanating shrill and annoying noise, and professional hunters) to discourage the bird population around airports.

References

[1] "Charts; Standard Aircraft Characteristics and Performance," Military Specification MIL-C-5011A, 14 Feb. 2003.

[2] "Airworthiness Standards: Part 23—Normal, Utility and Acrobatic Category Airplanes; Part 25—Transport Category Airplanes," *Federal Aviation Regulation*, Vol. 3, U.S. Department of Transportation, U.S. Government Printing Office, Washington, DC, Dec. 1996.

[3] "Guidelines for Preparation of Standard Aircraft Characteristics Charts and Performance Data of Piloted Aircraft (Fixed Wing)," AS-5263, Naval Air Systems Command, Patuxent River, MD, 23 Oct. 1986 (for Navy use in lieu of MIL-C-25011).

[4] Davenport, F. J., Rengstorff, A. E., and Van Heyningen, V. F., "Takeoff and Landing Performance Ground Rules for Powered Lift STOL Transport Aircraft," U.S. Air Force Flight Dynamics Laboratory AFFDL-TR-73-19, Vol. 3, Wright–Patterson AFB, OH, May 1973.

[5] Hebert, J., Moorehouse, D., and Richie, S., "STOL Takeoff and Landing Ground Rules," U.S. Air Force Flight Dynamics Laboratory AFFDL-TR-73-21, Vol. 3, Wright–Patterson AFB, OH, May 1973.

[6] Carter, W. E., "Effect of Ground Proximity on the Aerodynamic Characteristics of Aspect Ratio 1 Wings with and without End Plates, II," NASA TN D-970, Langley Research Center, Oct. 1961.

[7] Fink, M. P., and Lastinger, J. L., "Aerodynamic Characteristics of Low Aspect Ratio Wings in Close Proximity to the Ground," NASA TN D-926, Langley Research Center, July 1961.

[8] Brown, W. D., *Parachutes*, Pitman, London, 1951.

[9] Schaezler, G. B., "Synthesis of Ground Flotation Requirements and Estimation of Aircraft Running Gear Weight," Proc. 3rd Weight Prediction Workshop, Aeronautical Systems Division, Wright–Patterson AFB, OH, Oct. 23–27, 1967.

[10] "FAA Type Certification Report, Model L-1011-385–1 with Rolls-Royce RB.211-22 Engines," Lockheed Rept. LR 25089, 14 July 1972.

Chapter 11

Preliminary Sizing of the Vertical and Horizontal Tails

- Tail Volume Coefficient Method
- Vertical Tail Sizing Criteria
- Horizontal Tail Sizing Criteria
- Location of Horizontal Tail Criteria
- Location of Vertical Tail Criteria

During flight this B-52 encountered severe CAT and lost most of its vertical tail. Yaw stability was lost, making the aircraft directionally unstable. Directional control was maintained through differential thrust from its four engines. The pilot was able to steer and land the airplane successfully!

Good instincts usually tell you what to do long before your head has figured it out.

Michael Burke

11.1 Tail Volume Coefficient Approach

At this point the aircraft has been sized [a takeoff gross weight (TOGW) has been estimated] and it has a wing–body configuration. Tails are now added to the configuration and their associated aerodynamics are determined. Before tail aerodynamics can be incorporated it is necessary to know the tail's location (front, back, or no tail at all), shape, and size.

What the tails look like and where they are located is a design decision: The designer needs to decide what the tails will look like, determine their mean aerodynamic chord (mac), and decide where they will be located. Then the designer estimates the distance from the initial c.g. location (estimated in Chapter 8) to the macs of the vertical and horizontal tails (mac or $\bar{c}_h$ and $\bar{c}_v$). These distances are the moment arm length of the pitch and yaw stability and control devices and are shown in Fig. 11.1.

The sizing of the tail surfaces is a lot of work. It requires a precise knowledge of the c.g. location and is the subject of Chapters 21–23. Unfortunately, the c.g.'s location depends upon knowing the weight (size) of the tail surfaces. Thus, at this point the tail surfaces are sized using a shortcut technique called the *tail volume coefficient* approach. It is based on the observation that values of these volume coefficients are similar for like classes of existing aircraft [1–3].

11.2 Sizing the Vertical Tail

The vertical tail provides directional control and stability (motion about the Z axis). It may be sized by one or more of the following criteria (discussed in Chapter 23) [1,2]:

1. **Landing and takeoff.** Low-speed one-engine-out or severe crosswind conditions.
2. **Maneuverability.** The required maneuverability for a fighter aircraft may size the vertical tail.
3. **Subsonic cruise directional stability.** Both MIL-HDBK-1797 and FAR Parts 23 and 25 require that the directional static stability derivative $C_{n_\beta} > 0$ for normal cruise. Typical C_{n_β} values for business jets and commercial transports is 0.08–0.17 per radian at Mach = 0.8.
4. **High-speed directional stability.** For high speed ($M > 2.0$) the directional static stability derivative C_{n_β} decreases so that the vertical tail might be sized to give a minimum value of $C_{n_\beta} = 0.08$ at high speed.

At this point in the design there is insufficient information to size the vertical tail by any of these four criteria. Tail size and c.g. location have to

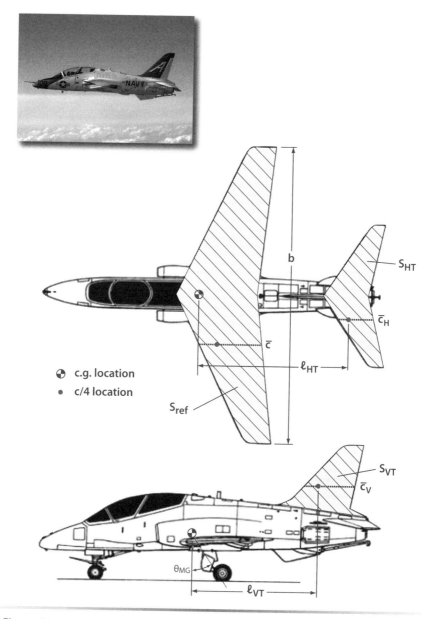

Figure 11.1 Illustration of reference geometry for tail sizing (inset is a T-45).

be realistic so that total aircraft drag can be determined. At this point historical trends determine the vertical tail area S_{VT}.

A convenient parameter to compare across classes of aircraft is the vertical tail volume coefficient:

$$C_{VT} = \frac{\ell_{VT}\, S_{VT}}{b\, S_{ref}} \tag{11.1}$$

where b is the wing span, S_{ref} is the reference wing area, and ℓ_{VT} is the distance between the initial estimate of the c.g. and the quarter chord of the vertical tail mac. S_{VT} is the exposed side view area of the vertical tail as shown in Fig. 11.1.

Tables 11.1–11.7 show the C_{VT} for classes of existing aircraft. Table 11.8 lists typical volume coefficient values for preliminary tail sizing. The designer would select an appropriate value of C_{VT} for a similar class of aircraft and solve Eq. (11.1) for the S_{VT}.

11.3 Sizing the Horizontal Tail (Aft Tailplane)

The horizontal tail provides longitudinal stability and control. The horizontal tail (both aft tailplane and canard) may be sized by one or more of the following conditions (discussed in Chapter 23) [1,2]:

1. **Static longitudinal stability.** The static longitudinal stability derivative C_{m_α} should be negative at all flight speeds as it represents the tendency of the aircraft to resist moving away from equilibrium flight. However, it cannot be too negative as the lift on the tail to trim the aircraft would

Table 11.1 Tail Volume Coefficients for Light Reciprocating–Propeller Aircraft

Aircraft	No. Engines	C_{HT}	C_{VT}
Cessna Skywagon 207	1	0.92	0.046
Cessna Cardinal	1	0.60	0.038
Cessna Skylane	1	0.71	0.047
Piper Cherokee	1	0.61	0.037
Bellanca Skyrocket	1	0.61	0.037
Grumman Tiger	1	0.76	0.024
Cessna 310	2	0.95	0.063
Cessna 402	2	1.07	0.08
Cessna 414	2	0.93	0.071
Piper PA-31	2	0.84	0.056
Piper Chieftain	2	0.72	0.055
Piper Cheyenne I	2	0.85	0.045
Beech Duchess	2	0.67	0.053
Beech Duke B60	2	0.64	0.060

Table 11.2   Tail Volume Coefficients for Turbofan (TF) and Turboprop (TP) Business Aircraft

Aircraft	Engines	C_{HT}	C_{VT}
Beech 1900	Turboprop	1.33	0.076
Beech B200	Turboprop	0.91	0.065
Cessna Conquest	Turboprop	0.91	0.071
DeHavilland DHC-6	Turboprop	0.91	0.077
DeHavilland DHC-7	Turboprop	1.11	0.076
DeHavilland DHC-8	Turboprop	1.47	0.121
BAE 31	Turboprop	1.22	0.120
Dassault Falcon 10/20/50	Turbofan	0.68	0.063
Cessna Citation 500	Turbofan	0.73	0.081
Cessna Citation II	Turbofan	0.64	0.062
Cessna citation III	Turbofan	0.99	0.086
Learjet 24	Turbofan	0.67	0.077
Learjet 35	Turbofan	0.65	0.066
Learjet 55	Turbofan	0.76	0.086
BAE 125	Turbofan	0.72	0.061

Table 11.3   Tail Volume Coefficients for TF and TP Transports

Aircraft	Engines	C_{HT}	C_{VT}
Lockheed C-130E	Turboprop	0.94	0.053
Lockheed C-5A	Turbofan	0.62	0.079
Lockheed L-1011	Turbofan	0.83	0.055
Boeing 727-200	Turbofan	0.82	0.11
Boeing 737-200	Turbofan	1.28	0.10
Boeing 737-300	Turbofan	1.35	0.10
Boeing 747-200	Turbofan	0.74	0.079
Boeing 757-200	Turbofan	1.15	0.086
Boeing 767-200	Turbofan	0.94	0.067
DC-9/S 80	Turbofan	0.96	0.062
DC-10-30	Turbofan	0.90	0.060
Airbus A-300	Turbofan	1.12	0.094
Airbus A-310	Turbofan	1.09	0.098
BAE 146-200	Turbofan	1.48	0.12

Table 11.4   Tail Volume Coefficients
TF and TP Military Trainers

Aircraft	Engines	C_{HT}	C_{VT}
T-34	Turboprop	0.76	0.048
Aero L-39	Turbojet	0.58	0.083
Alphajet	Turbojet	0.43	0.084
Aermacchi MB-339	Turbojet	0.52	0.043
BAE Hawk/T-45	Turbojet	0.61	0.059
Cessna T-37	Turbojet	0.68	0.041

produce a large trim drag. Business jets and commercial transports have C_{m_α} values of -0.7 to -1.5 per radian.

2. **Maneuverability.** Because fighter aircraft have to *maneuver* (move away from an equilibrium condition) their C_{m_α} values would be closer to 0 (neutral stability) or even positive (unstable—this requires a stability augmentation system as discussed in Chapter 23), which makes them maneuverable.

3. **Landing and takeoff.** The horizontal tail must be powerful enough (large enough) to rotate the aircraft about the main gear at V_{TO} to α_{TO} for takeoff. Also, the horizontal tail must be large enough to rotate the aircraft and trim it at low speed to $C_L = 0.8C_{L_{max}}$ for landing approach.

4. **Low trim drag.** The size of the horizontal tail should be such that the drag due to the trim load on the tail at cruise is less than 10% of the total aircraft drag. Otherwise the trim loads are too large and the associated trim drag degrades the aircraft's performance.

Table 11.5   Tail Volume Coefficients for
Supersonic Transport and Bomber Aircraft

Aircraft	C_{HT}	C_{VT}	C_C
Rockwell XB-70	0	0.034	0.10
Tu-144	0	0.081	0
Tu-22M	1.11	0.087	0
Tu-22	0.44	0.059	0
Concorde	0	0.08	0
Rockwell B-1B	0.8	0.039	0
Convair B-58A	0	0.057	0
North American F-108	0	0.045	0.11

Table 11.6 Tail Volume Coefficients for Fighter Aircraft

Aircraft	C_{HT}	C_{VT}
Convair F-106	0	0.075
Grumman A-6A	0.41	0.069
Grumman F-14A	0.46	0.06
North American F-86	0.203	0.0475
North American F-100	0.36	0.0584
Northrop F-5E	0.4	0.098
McDonnell Douglas F-4E	0.26	0.054
McDonnell Douglas F-15	0.2	0.098
General Dynamics F-111A	1.28	0.064
General Dynamics FB-111	0.75	0.054
General Dynamics F-16	0.3	0.094
Cessna A-37B	0.68	0.041
MIG-21	0.214	0.08
MIG-23	—	0.06
MIG-25	0.36	0.1
SU-7	0.4	0.1
Viggen	0	0.0834

Later the horizontal tail (aft tailplane or canard) will be sized to the preceding criteria. For now, however, use the historical trends of horizontal tail volume coefficients, which are defined as follows:

$$C_{HT} = \frac{\ell_{HT}\, S_{HT}}{\bar{c}\, S_{ref}} \qquad (11.2)$$

where $\bar{c}$ is the wing mean aerodynamic chord, ℓ_{HT} is the distance from the initial estimate of the c.g. location to the quarter-chord of the horizontal tail mac, S_{ref} is the wing reference area, and SHT is the total planform area

Table 11.7 Tail Volume Coefficients for Intelligence, Surveillance, and Reconnaissance Aircraft

Aircraft	C_{HT}	C_{VT}
Lockheed Martin U-2S	0.34	0.014
Northrop Global Hawk	0.32	0.0186
Boeing Condor	0.53	0.012

Table 11.8   Typical Values of Volume Coefficients for Preliminary Tail Sizing

Aircraft	C_{HT}	C_{VT}
Sailplane [3]	0.53	0.022
ISR	0.34	0.014
General aviation (one-engine propeller)	0.7	0.032
General aviation (two-engine propeller)	0.76	0.06
Business aircraft (two-engine)	0.91	0.09
Commercial jet transports	1.0	0.083
Military jet trainer	0.6	0.06
Jet fighter (all speeds)	0.5	0.076

of the horizontal tail (includes the aft fuselage carryover as shown in Fig. 11.1).

Tables 11.1–11.7 show the C_{HT} for various classes of existing aircraft. Table 11.8 lists typical volume coefficient values for preliminary tail sizing. The designer would select an appropriate value of C_{HT} for a similar class of aircraft and solve Eq. (11.2) for S_{HT}.

11.4 Horizontal Tail (Canard)

For the canard configuration, the wing aerodynamic center (a.c.) is behind the aircraft center of gravity and thus the wing is stabilizing (i.e., contributes a negative C_{m_α}). The canard "pulls" the configuration a.c. (neutral point) forward; therefore, because the canard is destabilizing, it is not for stability but is for control. The contribution of the canard is statically destabilizing. However, the destabilizing nature of the canard can be helpful for supersonic speeds as it will help offset the aft movement of the a.c. from the wing. The canard S_C is sized by the same criteria as discussed for the aft tailplane in Section 11.3.

The canard volume coefficient is defined as

$$C_C = \frac{\ell_C\, S_C}{\bar{c}\, S_{ref}} \tag{11.3}$$

where ℓ_C is the distance from the c.g. to the mac of the canard and S_C is the exposed top view area of the canard. A value of $C_C = 0.10$–0.11 is suggested for preliminary canard sizing.

11.5 Tailless

For aircraft without a horizontal tail (i.e., F-106 and B-58) $C_{HT} = 0$. Previously suggested values for C_{VT} are appropriate for sizing the vertical tail. Here the wing a.c. is behind the c.g. so that the wing is stabilizing.

The longitudinal control must come from the wing. The wing will have trailing edge surfaces (positive and negative deflection flaps) that can give the wing positive or negative camber to control the pitching moment.

11.6 Vertical Location of the Aft Horizontal Tail

The aft horizontal tail is initially located along the X axis based upon an initial estimate of the fuselage length. Then the horizontal tail is sized using historical tail volume coefficients. Final horizontal location will be established when the length of the propulsion system is known (Chapters 15 and 16). A refined weight estimate (Chapter 20) is performed along with the stability and control analysis discussed in Chapter 21. However, the vertical location of the horizontal tail is unknown. It is not essential that this vertical location is known at this time, unless the designer is anxious to close out the design.

Horizontal tails are located vertically for several reasons. One reason is to reduce their being blanked by the wake from the wing as it stalls, which results in a loss of pitch control as discussed in Section 21.7, Fig. 21.15. Another reason is to move them out of the hot gas exhaust of aft fuselage mounted engines (see Fig. 8.7). Also, of course, the overall appearance of the aircraft is an important consideration.

11.7 Horizontal Location of the Vertical Tail

A spin is a rotation about the Z axis with the aircraft in a 30–60 deg nose-down attitude from the horizontal. The rudder on the vertical tail is the main recovery surface for an aircraft in a spin. The procedure is to stop the spin by applying opposite rudder and then recover from the steep dive using the horizontal tail. For high-maneuvering aircraft (i.e., fighters and trainers) it is important that the rudder be located so that it is not blanked by the horizontal tail. The analysis is graphical, consisting of drawing two lines, one at 30 deg from the horizontal tail trailing edge and the other at 60 deg from the leading edge and positioning the rudder outside of this region. The F-15, F-16, F-18 and T-46 are good spin recovery aircraft.

References

[1] Roskam, J., *Airplane Design*, Pt. II, Roskam Aviation and Engineering Co., Ottawa, KS 66067, 1989. [Available via www.darcorp.com (accessed 31 Oct. 2009).]

[2] McCormick, B., *Aerodynamics, Aeronautics, and Flight Mechanics*, Wiley, New York, 1995.

[3] Thomas, F., *Fundamentals of Sailplane Design*, College Park Press, College Park, MD, 1999.

Chapter 12 Designing for Survivability (Stealth)

- Vulnerability
- Susceptibility (Signature)
- RF/Radar (RCS)
- Shaping
- Absorption & Cancellation
- IR/Infrared
- Visual
- Acoustic

"Father of Stealth" Ben Rich with the F-117A stealth fighter. See "Father of Stealth Wins Big" in Section 12.4.

What is my management style?
I manage with charisma.
Ben Rich

12.1 Putting Things in Perspective

All aircraft should be designed with survivability in mind. Commercial and private aircraft can be at risk to terrorists armed with inexpensive, easy-to-use surface-to-air missiles (SAMs). These SAMs are the man-portable missiles, with infrared (IR) sensors, called MANPADS. Military aircraft often encounter enemy MANPAD units and the more sophisticated integrated air defenses (IADs) with large, expensive SAMs and air-to-air interceptors.

When designing for survivability one should consider a hierarchy of actions. Hardening the aircraft or reducing the aircraft signature is not the first action to consider. This hierarchy consists of the following actions in order:

- **Mission planning.** Plan the mission to avoid the threat. Select the time of day and conditions to minimize the effectiveness of the threat.
- **Plan the mission profile.** Select conditions of speed, altitude, terrain following or terrain avoidance (TF/TA), and so on.
- **Use electronic countermeasures (ECM).** Integrate onboard and off-board ECMs into the aircraft such as flares, chaff, towed decoys, radio-frequency (RF) and IR missile warning systems, and countermeasure electronics (spoofers, jammers, etc.).
- **Defeat the end game.** Using onboard RF/IR missile warning systems a properly timed maneuver can cause the missile to miss.
- **Design-in survivability features.** These survivability features will involve weight, vehicle shape, and cost. These features will impose a lifetime performance penalty on the aircraft and should be considered as a last resort for survivability.

Survivability has two parts: susceptibility and vulnerability. *Susceptibility* is the probability that the aircraft will be detected, tracked, and fired upon:

$$P_{KS} = P_D \times P_T \times P_{FIRE}$$

Vulnerability is the probability that once a missile or bullet is fired it will fuse or hit the aircraft and, if it hits the aircraft, the probability that it will kill the aircraft:

F-117 Shoot Down: During the 1999 Kosovo air campaign the Yugoslav air defenses were able to locate track, and shoot down an F-117 by intercepting its electronic emissions. The Yugoslavs' technique was crude and involved a lot of luck, but it worked, resulting in the only F-117 ever to be shot down. Russi bought the remains of the F-117 from Yugoslavia and used data from component testing to design improved SAMs. Given the faceted design and 1970s coating technology, it's an open question how much useful information was actually obtained [1].

$$P_{KV} = P_{F/H} \times P_{H/K}$$

Then the probability of a kill is $P_K = P_{KS} \times P_{KV}$ and the probability of survival is $P_S = 1 - P_K$.

Reducing susceptibility and reducing vulnerability are two different strategies for high survivability and are shown in Fig. 12.1 for the A-10A and F-117A.

> The A-10A (nicknamed the Warthog) was developed in the early 1970s as a close air support aircraft for low-intensity conflicts. It was designed for low-speed maneuvering and killing tanks with its GAU-8 seven-barrel 30-mm cannon. It also carried a vast array of air-to-surface missiles.

12.2 Designing for Reduced Vulnerability

Designing for *reduced vulnerability* is the strategy of letting the aircraft "take a licking but keep on ticking." The vulnerability reduction concepts are as follows:

- **Critical component redundancy with separation.** Try to have redundancy (usually two) in the critical components and to separate them so that a missile warhead will not take both of them out. Examples are multiple engines, flight control computers, control surfaces, and fuel pumps.
- **Critical component location.** Locate the critical components so that they are not damaged by failure of another component, such as an engine fire or a thrown blade from a damaged compressor or turbine section.
- **Passive damage suppression.** Design critical structure to be damage tolerant or fail safe. Filling the fuel tanks with foam will minimize the voids in the tank and limit the fuel–air mixture leading to an explosion.
- **Active damage suppression.** Filling the fuel tank with an inert gas (such as nitrogen or HALON) as the fuel is consumed will limit the buildup of a fuel–air mixture and greatly reduce the probability of an explosion. The addition of fire detection and suppression systems, especially in the engine bay, has saved many aircraft.
- **Critical component shielding.** Try to shield critical components by deliberate shielding or by locating them so they are shielded by other components. Examples are the titanium "bathtub" surrounding the pilot and locating the two turbofan engines above the wing, providing shielding from ground fire, for the A-10A.
- **Critical component elimination.** If possible, eliminate critical components and replace their function some other way. An example is replacing the pilot on UAVs.

Low Vulnerability (A-10A)

Low Susceptibility (F-117A)

Figure 12.1 Different strategies for high survivability.

These vulnerability concepts are shown in Fig. 12.2 and discussed in great detail in [2]. Figure 12.3 shows that the concepts worked for the A-10A (but they made the pilot nervous).

12.3 Designing for Reduced Susceptibility

Reducing an aircraft's susceptibility starts with reducing the ability of an enemy defense system sensor to detect the presence of the aircraft. The enemy defense system sensors are of five types:

- **RF/Radar.** The radar operates by transmitting RF energy, which reflects off the target aircraft and back to an RF sensor. This RF sensor is usually located at the transmitter. This reflected RF energy represents a radar cross section (RCS) return, or signature, to the sensor. This is the only active defense system. The RF frequencies of interest are 100 MHz to 16 GHz, which covers the Worldwide RF air defense systems. The lower frequency (VHF and UHF band) systems provide long-range detection, the midrange frequencies (L, S, and C bands) are for the tracking radars, and the high-frequency X band with its fine-resolution capability is for the fire control radars.
- **IR/Infrared.** The infrared system operates by sensing the emitted or reflected IR energy from the target aircraft minus the IR energy of the background. This IR signature is a contrast relative to the background and can be negative or positive. This is a passive system represented by a thermal-sensitive sensor in the 1- to 12-micron wavelength bands.
- **Visual.** The visual system operates by sensing the emitted or reflected visual energy from the target aircraft minus the visual energy of the background. An example of emitted visual energy would be a head lamp. The visual signature is a contrast relative to the background and can appear as a glint or a black hole. This is a passive system represented by electro-optical sensors and the human eyeball. The visible band is the 0.37–0.75 micron part of the electromagnetic spectrum.
- **Acoustic.** The acoustic system operates by sensing the emitted or reflected acoustic energy from the target aircraft minus the acoustic energy of the background. The acoustic energy of the target rarely contains any reflected energy. The acoustic signature of the target is a contrast relative to the background. This is a passive system represented by microphones or the human ear.
- **Electronic intelligence and signals intelligence (ELINT/SIGINT).** Aircraft will often emit electronic signals that can be used to locate

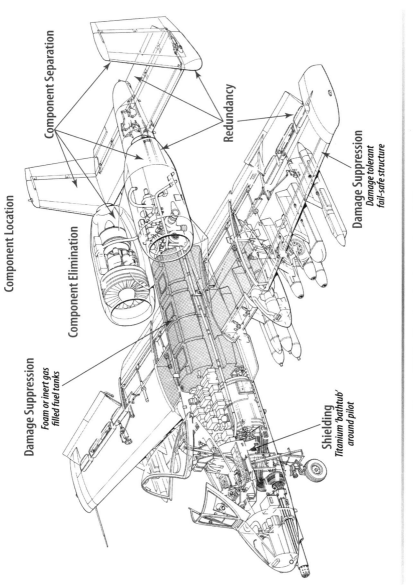

Figure 12.2   Design for low vulnerability.

Figure 12.3   A-10 can tolerate severe damage and still survive. It has excellent vulnerability.

them. The intelligence community uses these electronic signals to locate ground targets and to gather electronic information on enemy weapon systems, both in development and fielded. The United States uses airborne platforms such as the EP-3, EC-130, and U-2 to gather this electronic intelligence.

This chapter discusses design features that reduce the RF, IR, visual, and acoustic signatures of aircraft.

12.4 Radar Cross Section (RCS) Signatures

An aircraft's RCS is measured relative to the radar return from a metal sphere with a cross section of one square meter. The common unit for RCS is decibels relative to a one square meter reference cross section (dBsm):

$$\text{RCS in dBsm} = 10\log_{10}\left(\text{RCS in square meters}\right) \quad (12.1)$$

Figure 12.4 (from [3]) shows the RCS of typical aircraft.

12.4.1 Radar Scattering Phenomena

As the electromagnetic (EM) field generated by a radar washes over a target, the RF energy is reflected. This reflected energy is received by RF sensors as the target RCS. The strategy for reducing the target RCS is as follows [4–6]:

- **Shaping.** If the target surface is properly shaped, the reflected energy will not be received by the threat radar. This strategy controls the

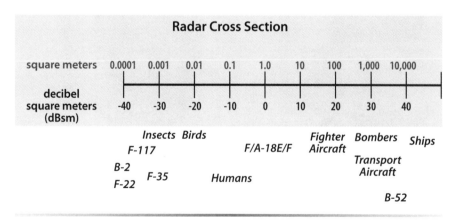

Figure 12.4 Definition of RCS used to assess the level of stealth (data from [3]).

direction of the reflected energy and works well provided the threat RF sensor is collocated with the radar. If the RF sensor is located somewhere else, the target vehicle RCS could actually be enhanced (note that this is the theory behind a bistatic radar defense).

• **Absorption.** If the EM energy interacts with high-resistance iron particles on the target surface, the energy is converted to thermal energy by ohmic heating. Ohmic heating is the temperature rise in a conducting material with electrical resistance when a current flows.

• **Cancellation.** If the target surface can reflect part of the EM energy with a 180-deg phase change, the EM energy will be canceled. Typically this is done passively with coatings on the target surface that have a judiciously chosen thickness to cause the cancellation. The thickness of the absorber depends upon the frequency or wavelength of the incident energy. The cancellation can also be done actively by an onboard electronic system that senses the time and direction of the incident EM energy and then transmits energy of equal strength and opposite phase to cancel the incident energy.

In the late 1970s, the Skunk Works was preparing for its first wind tunnel test of a stealth aircraft design by taking routine pictures of the model with a Polaroid camera. The photographer complained to the test engineers that all of the pictures were out of focus and that something was wrong with the Polaroid camera he was using. After a brief silence someone realized what the problem was. The cameras used a sonar-focusing device that depended on return reflections to adjust the focus length. Because stealth aircraft are designed to not directly return any impinging waves, the camera could not focus on it. It was clear to those in the room that stealth from shaping was going to work. Have Blue and the F-117A designs followed and the rest is history.

The EM energy is reflected at the target vehicle by the three scattering mechanisms shown in Fig. 12.5 [7,8].

• *Specular reflections* result when a radar wave is directly reflected from an object, similar to a flashlight shining on a mirror. The specular reflection angle is equal to the incidence angle of the radar wave. A normal surface reflects the specular energy right back to the radar, whereas an angled surface reflects the energy away from the radar. The specular reflection has a main lobe and side lobes as shown in Fig. 12.6. Specular reflections are controlled by shaping the aircraft to reflect the energy away from the radar.

• *Diffraction* occurs when the EM energy encounters a sudden discontinuity in surface slope or change in electrical impedance (material change). Everyday examples of diffraction are rainbows and reflections from glass prisms. Diffraction scattering is reduced by avoiding surface discontinuities (Fig. 12.5), cancellation, and absorption.

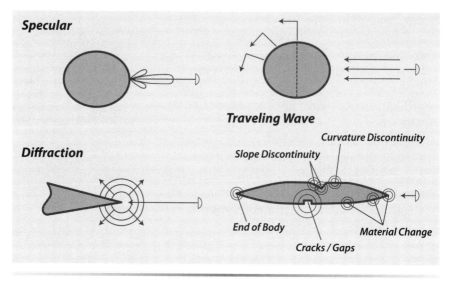

Figure 12.5   Electromagnetic (EM) scattering mechanisms.

- *Traveling waves* (or surface waves) occur as the EM field washes over the target and sets up electrical currents (induced by the incident EM energy) in the conducting surface of the aircraft. Traveling waves, like diffraction, will scatter when they encounter surface discontinuities. Traveling wave scattering is reduced in the same ways as diffraction.

The RCS of an aircraft is the vector sum of all reflected energy from all scattering sources and is dependant upon the orientation of the aircraft to

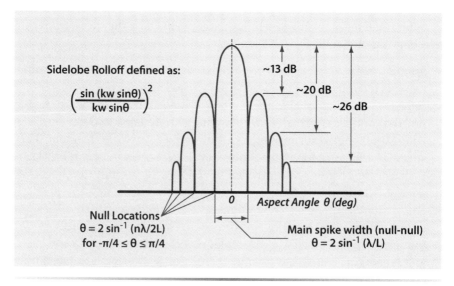

Figure 12.6   Electromagnetic wave backscatter geometry.

the radar and upon the wavelength and polarization of the EM wave. The wavelength of the EM wave is an important parameter in RCS reduction. The wavelength (λ) depends on the frequency (f) of the EM wave as follows:

$$\lambda \left(\text{in inches} \right) = 11.8/f \qquad (12.2)$$

where the frequency is in gigahertz. For example, the 170-MHz Tall King long-range detection radar has a wavelength of 66 inches and the 10-GHz Flap Lid fire control radar has a wavelength of a little over one inch. Every scattering source on the aircraft (i.e., wing leading edge length, vertical tail height, inlet lip radius, outer mold line (OML), gaps and cracks, skin surface imperfections, etc.) has a characteristic dimension L and a scattering size in wavelengths of L/λ. The primary scattering mechanisms (specular, diffraction, and traveling wave) vary depending upon the scattering source size in wavelengths as shown in Fig. 12.7. Figure 12.7 also shows that the technique for reducing the scattering mechanism varies depending upon the source size in wavelengths. Notice from Fig. 12.6 that the width of the specular main lobe, the null locations, and the sidelobe rolloff are all dependant on the parameter L/λ.

Father of Stealth Wins Big

Ben Rich took over the Skunk Works from Kelly Johnson in early 1975 and maneuvered it, over Kelly Johnson's objection, into the DARPA XST program. By October Lockheed and Northrop were locked in a "winner take all" competition on the USAF's radar test range at White Sands, New Mexico. Each company built an RCS model of their "Have Blue" design.

Ben had ball bearings made with the same RCS as Lockheed's design. Ben then prowled the halls of the Pentagon, rolling a ball bearing across the desks of various generals and announcing "General, here's your new fighter airplane." The generals' eyes would bug out of their heads. Northrop yelled foul and Ben stopped approaching anyone not cleared into the DARPA program. Thanks to the creative genius of several Skunk Works engineers and Ben's rapport with the customer, the Skunk Works won the program with a design that later became the F-117A stealth fighter.

Ben was the chief Skunk until his retirement in 1991. Ben's management style was very different from Kelly's. Kelly ruled the Skunk Works by his bad temper ... Ben ruled by his bad jokes. Ben was a schmoozer, a glad hander, a cheer leader. He was well-liked and respected by his customers and his employees. Ben became the chief spokesman for the pursuit of stealth technology within the DOD and rightly earned the title of the "Father of Stealth."

	HF VHF UHF	Microwave	MMW
	L/λ < 3 ◄──┼──► L/λ > 3		L/λ ≫ 3
Primary Scattering Sources	Diffraction Traveling Waves Resonances	Specular Reflection Apertures Details	Details Diffuse Scattering
Scattering Reduction Techniques	Radar Absorbers Nulling Techniques	Shaping Radar Absorbers Shielding	Radar Absorbers Tolerances
Design Approaches	High-Power Computation with Optimization Experiment	Simple Computer Models Experiment	Experiment

Figure 12.7   Scattering for aircraft-size targets.

With this dependence on L/λ one would expect a different polar RCS pattern for the case of a small L/λ aircraft vs a large L/λ. This is indeed the case, as shown in Fig. 12.8 for a diamond-shaped metal (no coatings) aircraft. For the case of $L/\lambda < 3$ shaping is not very effective although it does steer the diffraction scattering and the RCS pattern is a blob with no distinct spikes. The RCS reduction design would feature absorption and cancellation. For the $L/\lambda > 3$ case the pattern is characterized by very distinct and narrow spikes and a fuzz ball (surface detail scattering). Here the RCS reduction design would feature shaping as well as absorption.

The history of current U.S. aircraft stealthy design is shown in Fig. 12.9. The basics of stealth have been known since the 1950s but the analytical techniques were experimental because computer power was lacking. In the 1970s computer power had increased to where the RCS of faceted vehicles could be estimated. This second-generation stealth capability led to the DARPA XST (Experimental Stealth Technology)/UASF Have Blue demonstrator program shown in Fig. 12.10. Even before the Have Blue flight test was completed, the U.S. Air Force ordered 59 F-117As. In the 1980s the United States entered the third generation of stealth due to increasing computer power that allowed the RCS of curved surface configurations to be analyzed. This led to the Northrop B-2 stealth bomber (the aircraft proved to be very expensive and only 21 were produced), the AGM-129 Advanced Cruise Missile (over 400 produced with nuclear warheads), the F-22, and the AGM-158 JASSM. The F-35 Joint Strike Fighter represents the fourth generation of stealth. Figure 12.11 shows a gallery of U.S. low-signature (stealthy) demonstrator vehicles and production aircraft.

a)

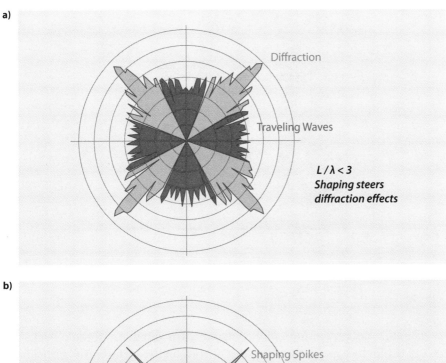

b)

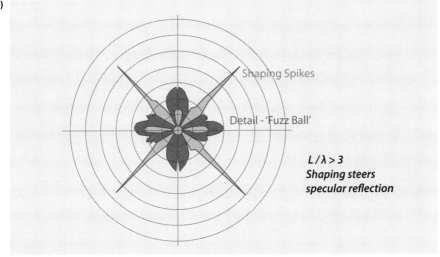

Figure 12.8   RCS pattern shaping and details for **a)** VHF and UHF, and **b)** microwave.

12.4.2 Vehicle Shaping

The RCS design starts by establishing a smooth conducting ground plane completely around the aircraft. The ground plane keeps the EM energy from penetrating into the interior of the aircraft and reflecting off all the structure and subsystems. Silver paint is a popular treatment with conductive films and fabrics used over gaps, cracks, and fasteners. The glass canopy and sensor lenses are made conducting by painting them with a thin film of indium tin oxide (ITO).

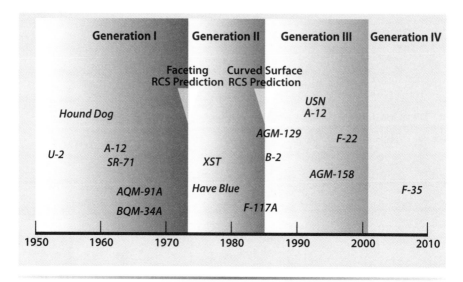

Figure 12.9 Definition of stealth generations.

The RCS design continues by controlling the direction of the reflected energy through shaping, then coatings are put on the surface to absorb and cancel the energy. Here it is assumed that there is intelligence information on how and where the threat radar is deployed and how many. It is necessary to know if the threat is above, below, co-altitude, in front, off to the side, or behind when it is encountered so that the reflected energy can be deflected away from the receiver.

Vehicle shaping works best for the case of an $L/\lambda > 3$ target. This is because the reflected energy spikes are narrow and well defined. The sweep of the wing and tail leading edge (LE) and trailing edge (TE) establishes the basic spike structure for the target vehicle to defeat the threat. Then the

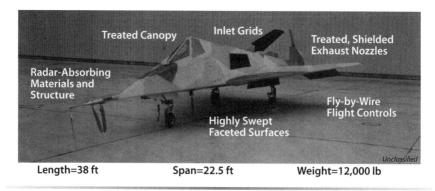

Figure 12.10 Have Blue low-observable technology demonstrator.

Figure 12.11   Gallery of low-observable aircraft.

scattering spikes from all other sources are aligned with this basic spike structure. This strategy is shown in Figs. 12.12 and 12.13. The F-117 and B-2 are termed four-spike designs, whereas the YF-22 and YF-23 are six-spike designs due to the fuselage side spikes. Notice how the inlet and nozzle apertures, and all the gaps and cracks from the bomb bay doors, landing gear doors, and control surface hinge lines are swept to line up with the wing or tail spikes. Typically the wing and tails are swept such that their

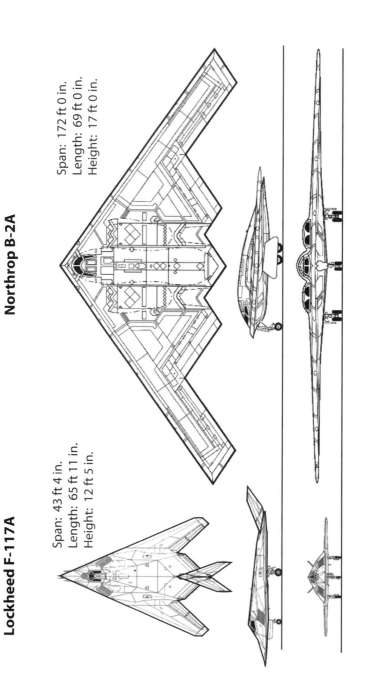

Northrop B-2A

Span: 172 ft 0 in.
Length: 69 ft 0 in.
Height: 17 ft 0 in.

Lockheed F-117A

Span: 43 ft 4 in.
Length: 65 ft 11 in.
Height: 12 ft 5 in.

Figure 12.12 Planform shaping for low observables.

Northrop YF-23   **Lockheed YF-22**

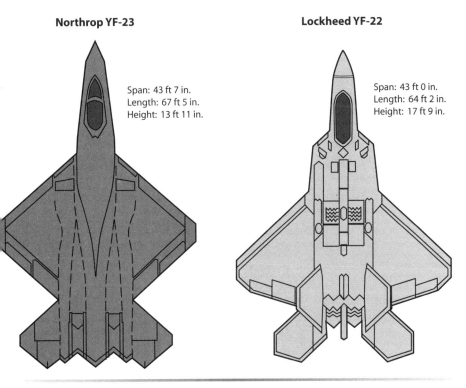

Span: 43 ft 7 in.
Length: 67 ft 5 in.
Height: 13 ft 11 in.

Span: 43 ft 0 in.
Length: 64 ft 2 in.
Height: 17 ft 9 in.

Figure 12.13   Planform shaping for low observables (YF-23 vs YF-22).

LE and TE spikes are away from a threat directly in front of the aircraft (the Tier 3-Minus Dark Star was an exception).

Careful attention must be paid to the shaping of the vehicle surface so that there are not any surface discontinuities to trigger diffraction and travelling-wave scattering (see Fig. 12.5). This means continuous second derivatives (slope change gradient) everywhere. Certainly gaps, cracks, control surface hinge lines, and facet edges do not meet this criterion and must be swept to align their scattering with the basic spike structure of the aircraft.

12.4.3 Absorption and Cancellation

As mentioned earlier the EM energy can be absorbed by ohmic heating and in some cases can-

In 1975 the Lockheed Skunk Works was testing the Have Blue RCS model when range operators abruptly started to get large readings. Visual checks showed a small bird was now sitting on the model. After chasing the bird away measurements were still not consistent with previous data. A closer check of the model revealed that the bird had left droppings on the model. Once these were removed the new data agreed with the old. RCS measurements are very sensitive to many model irregularities.

celled. This action all takes place in the coatings that are put on the surface of the aircraft. These coatings are usually of three types: radar absorbing structure (RAS, sometimes called loaded edge), radar absorbing material (RAM), and resistive sheet. These coatings are shown in Fig. 12.14 applied to a leading edge.

The RAS is the main treatment for low-frequency radars. It is an aerodynamic fairing made of a material that is transparent to the RF energy but can take flight loads. The fairing is reinforced with structural honeycomb core or foam that is impregnated with a resistive liquid (similar to printer's ink) that absorbs the penetrating EM energy through ohmic heating. The depth of the RAS edge should be 0.38λ for maximum cancellation.

RAM is primarily a treatment for high-frequency radars. It absorbs the EM energy and provides cancellation when the thickness is one-quarter wavelength. RAM is very heavy and pretty much limited to high-frequency application because of the quarter-wavelength thickness criterion. The RAM is an iron powder held together in a binder. The iron powder is carbonyl iron (most common but oxidation is a problem), FeSI (excellent corrosion resistance), and FeAl or FeCoV for high-temperature applications. The following are binders for the iron particles:

• **Urethane.** This is the toughest, most common, lowest cost, fast curing, user friendly binder, and it adheres to most materials. It is comercially available as a paste and a spray. It has a good temperature range of $-65°F$ to $250°F$.

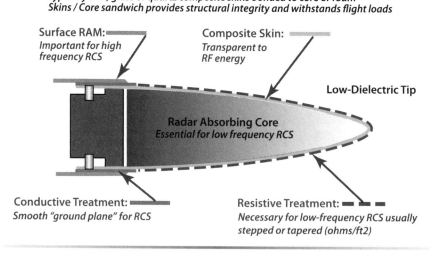

Figure 12.14 Low-observable materials selection and implementation—radar absorbing structure (RAS) edge construction.

- **Silicone.** The aircraft industry has over 40 years' experience with silicone RAM. It is available in sheet, paste, and spray forms. It has a very good temperature range of −65°F to 250°F. The disadvantage of silicon RAM is that nothing bonds to it except silicone.
- **Ceramic.** Ceramic has an excellent temperature range of 600°F to 2000°F and is used in nozzle applications. It is made by loading thin ceramic tiles with the iron powder. It is also used in brick form for RAS.

The resistive sheet absorbs the surface currents traveling along the target vehicle surface through ohmic heating. It is also known as edge card or R-card and is available comercially as a thin decal or appliqué (resistive ink), resistive mat (resistive fibers suspended in a resin soluble sheet), metalized film (sputter or vapor deposit Nichrome or Nickel on Kapton film) or fabric (glass fabric with Nickel treatment).

The weights of these treatments are as follows:

- High-frequency RAM (quarter-wavelength thick), 0.6 lb/ft^2
- Resistive sheet, 0.05 lb/ft^2
- Low-frequency edge RAS—carbon loaded foam/core
  - VHF, 24-inch edge, 6 lb/ft
  - UHF, 12-inch edge, 2 lb/ft

12.4.4 Inlet and Nozzle RCS Design

If not properly designed, the inlet or nozzle can drive the RCS of the entire vehicle in the front or rear sector. This is because the normal reflection from the compressor face or turbine blades bounces right back to the radar. A popular design trick is to block the line of sight (LOS) into the inlet or nozzle by giving the duct an "S" or serpentine shape. This causes the EM energy entering the duct to bounce off the duct walls, reflect off the compressor or turbine face, and then bounce off the duct walls again as it exits. If the duct walls are coated with RAM, each bounce reduces the reflected energy back to the radar.

Another popular design trick is to block the energy entering the inlet or nozzle cavity by a physical phenomenon called aperture cutoff. If the inlet or nozzle cavity dimension (normal to the polarization) is less than $\lambda/2$ the EM wave cannot enter the cavity. This is why your car AM radio (530–1600 kHz, $\lambda \sim 1000$ ft) will not work in a 60 ft diameter tunnel, but your FM radio (88–108 MHz, $\lambda \sim 10$ ft) continues to work. The cavity appears as a black hole. The inlet or nozzle aperture is then swept to reflect the energy away from the radar. This design trick was used on the F-117A inlet by putting a grid into the inlet. The grid had a cell size of 0.6 in., which kept the EM energy of all frequencies below 10 GHz from entering. The inlet aperture was highly swept so that the inlet contributed very little to the F-117A overall RCS. The disadvantage for the F-117 was a higher than normal inlet total pressure loss at cruise speed and ice buildup on the grid. This latter problem was solved by having a wiper blade sweep the ice off of the grid.

The F-117 flew at altitudes of 28,000 ft so that most threats were below it or co-altitude. The F-117 nozzle featured a high-AR, two-dimensional nozzle that provided aperture cutoff for low and middle frequencies. In addition the nozzle lower surface ramp blocked the LOS into the nozzle cavity for all co-altitude threats and below. Ceramic tile RAM was applied to the duct, and ceramic brick was used for the loaded edges.

Inlet and nozzle RCS design is summarized on Fig. 12.15. The front–rear frame referred to in Fig. 12.15 is a device that looks like a potato chip that prevents a normal reflection off of the compressor or turbine blades. Reference [9] is an excellent article on inlet design.

12.4.5 RCS Design Summary

Figure 12.16 shows a typical RCS design. The overall configuration should have as few spikes as possible (the minimum is a three-spike delta configuration) and their directions should be away from the threat sensors (usually located at the radars). All the scattering sources should align their individual spikes with the basic spike structure. Edges should be RAS for low-frequency threats and RAM-coated for high-frequency threats. Resistive sheet should be applied to reduce the travelling-wave scattering. All gaps, cracks, and hinge lines should be filled, treated, and swept. Inlets and nozzles should have LOS blockage to the compressor and turbine blades either by aperture cutoff, serpentine-shaped ducts, or front/rear frame. All ducts should be coated with RAM. The fuselage side slopes should direct

Line of Sight (LOS) blockage
  Low frequency–cutoff frequency
    Dimension < λ/2 normal to polarization
  Serpentine ducts
    Need length/diameter (L/D) ~ 2.5–3.5; for nozzles
    Need L/D ~ 4–6 for inlets
  Front and Rear Frames (approx. one diameter in length)
  Nozzle ramp angle
Absorb reflected energy
  MagRAM on duct walls
  Nozzle ducts need high-temperature RAM
Sweep inlet and nozzle lips

Figure 12.15 Inlet and nozzle design guidelines for low-RCS configuration.

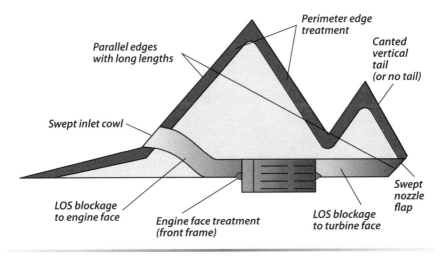

Figure 12.16   Typical low-observable design features (planview).

the reflected energy away from the threat and the side shape should have continuous second derivatives.

12.5 Infrared

Infrared and visual are contrast signatures. This means that they are observed relative to their background:

$$\text{Contrast} = E_T + E_R - \text{Background} \qquad (12.3)$$

where

E_T = target emissions
E_R = emissions due to the reflections from the sun, earth, and sky (clouds)

For Contrast > 0: the target is brighter than the background and appears as a glint (need to reduce E).

For Contrast < 0: the target is dimmer than background and appears as a black hole (need to add E).

For Infrared:

$$I_{\text{Contrast}} = I_E + I_R - \text{Background} \quad (\text{W/sr}) \qquad (12.4)$$

For Visual:

$$V_{\text{Contrast}} = V_E + V_R - \text{Background} \quad (\text{lm/sr}) \qquad (12.5)$$

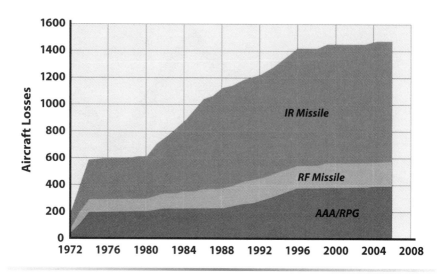

Figure 12.17 Combat aircraft losses (1972–2006).

IR signature reduction is a hard problem because the IR threats are passive (you do not know where a threat is) and the background varies with time of day, orientation, and weather. In addition most (over 60%) of the aircraft kills since 1972 have been from IR missiles (see Fig. 12.17). IR SAMs are the weapon of choice for downing aircraft by terrorist elements because they are user friendly, require minimum maintenance, and are much cheaper than RF SAMs.

12.5.1 Introduction to Infrared (IR 101)

The IR radiation sources are shown in Fig. 12.18. The IR signature is determined as follows:

$$I_{\text{Contrast}} = I_E + I_R - \text{Background} \quad (\text{W/sr}) \qquad (12.4)$$

The aircraft emissions are

$$I_E = \sigma \, \varepsilon f T^4 A_p \qquad (12.6)$$

where

σ = Stephan–Boltzmann constant = 0.481×10^{-12} Btu/ft^2·s·°R
ε = Emissivity of emitting surface
f = Distribution of IR energy in band of interest (i.e., SWIR, MWIR, or LWIR)
T = Absolute temperature of emitter in °R
A_p = Projected area

The

$$I_E = I_{\text{Hot Parts}} + I_{\text{Plume}} + I_{\text{Airframe}} \qquad (12.7)$$

The following are some representative temperatures [10]:

- Turbine blades (usually cooled), 2300°F
- Nozzle exit (turbojet), 1800°F
- Nozzle exit (turbofan), 1100°F
- Plume (turbojet), 1000°F
- Plume (turbofan), 500°F
- Airframe aero heating at Mach 0.85, 122°F
- Airframe aero heating at Mach 3.2
  - LE stagnation point, 800°F
  - Surface, 550°F

The reflected IR energy is

$$\text{IR} = \left[\left(1 - \varepsilon\right)\big/\pi \right] \sum \left(E_{\text{sun}} F_{\text{sun}} + E_{\text{sky}} F_{\text{sky}} + E_{\text{earth}} F_{\text{earth}} \right). \qquad (12.8)$$

Notice that the IR reflectance = $(1 - \varepsilon)$, which poses a dilemma when the designer wants to select an IR paint that will reduce emissions and reflections at the same time.

The IR sensor bands are shown in Fig. 12.18:

- Fire control: SWIR (near IR) 1–3 microns and MWIR (middle IR) 3–6 microns
- Detection, IRST (IR search and track): LWIR (far IR) 6–12 microns

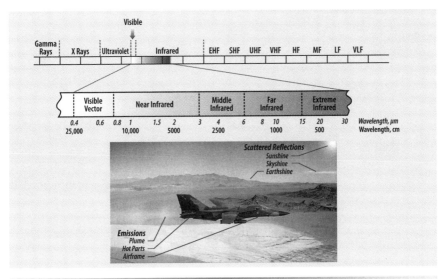

Figure 12.18 Aircraft infrared radiation sources.

12.5.2 IR Design

The rule for reducing the hot-parts emissions is hide what you cannot cool and then coat it with low-emissivity paint. Figure 12.19 shows some concepts for blocking the LOS to the aircraft hot parts. The A-10A very carefully located the twin vertical tails so that they blocked the LOS into the engine cavities at most tail-on angles. The 2-D nozzle and lower surface nozzle ramp on the F-117A effectively shield the exhaust hot parts from co-altitude and below look angles. Changing from a turbojet to a medium bypass ratio (i.e., 1–2) turbofan reduces the engine hot parts and plume emissions significantly. Low-emittance (0.2) paints are available commercially.

The design rule for reducing the exhaust plume emissions is to use a high-bypass turbofan engine if possible and then promote aggressive mixing of the plume with the ambient air. The exhaust mixers on commercial transports to reduce noise do a very good job of reducing the plume temperature. Ejector nozzles also promote plume mixing.

Airframe emissions due to aerodynamic heating can be controlled by flying slower and using low-emissivity coatings. Most of the time the airframe emissions are small compared with the hot parts.

Once the aircraft is deployed its IR emissions are pretty much fixed. However, the aircraft can be repainted from time to time with different ε

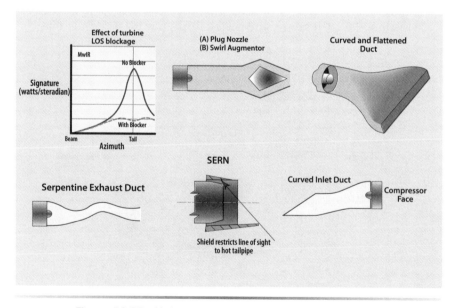

Figure 12.19 Hot parts blocking—IR design guidelines.

paints. From then on it is managing the reflections and background to drive the contrast to zero.

The strategy for managing the IR reflections is to tailor the mission in terms of time of day, background, and weather and in some cases to change the reflectance by changing the emissivity.

12.6 Visual Signature

The visual signature is

$$V_{\text{Contrast}} = V_E + V_R - \text{Background} \quad (\text{lm/sr}) \tag{12.5}$$

The V_E is usually zero; however, sometimes illumination can be added (positive V_E using a head lamp, for example) but illumination can never be taken away (negative V_E).

The visual sensor is usually a human eyeball. Thus, the detection ranges are small—typically 5–8 n mile.

Once again the strategy needs to be to plan the mission and mission profile to avoid or minimize the threat before designing-in performance penalty features. For example, the visual signature can be eliminated by shielding the target from the sensor with terrain or by flying above 5 n mile and not contrailing. The visual signature can be minimized by flying at night.

If the target needs to fly at low altitudes and within visual range of the human eyeball, then the tradeoff between reflectance $(1 - \varepsilon)$ and background must be considered. Figure 12.20 shows how the target reflectance needs to vary with a daytime background to reduce the detection range for a C-130-sized target. For a clear full-moon night background the same C-130 would need a reflectance of about 0.85 to fill in the black hole contrast.

12.7 Acoustic Signature

The acoustic sensor is the human ear and sometimes a microphone. The acoustic signature is a contrast between the emitted noise and the background:

For Acoustic:

$$A_{\text{Contrast}} = A_E + A_R - \text{Background} \quad (\text{EPNdB}) \tag{12.9}$$

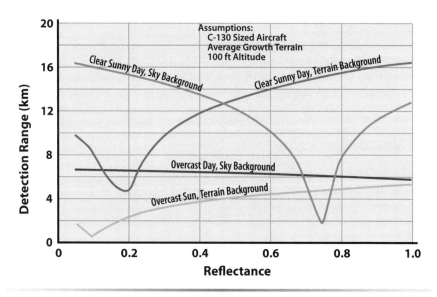

Figure 12.20 Daytime visual detection ranges.

where EPNdB is the effective perceived noise level in decibels and the re-flected noise term A_R is usually zero.

Here a negative contrast (the background is more noisy than the target) is a good thing. If you are a special forces team, the best place to land your aircraft is in the middle of a noisy mall. The good news is that the locals will never hear you. The bad news is that you will probably be seen, with the result being the same as if they heard you.

Acoustic energy is absorbed by buildings, walls (e.g., the noise barriers between residential areas and freeways), humidity, and trees.

Figure 12.21 shows the noise source characteristics for an aircraft. Notice that the main sources of noise are the airframe (aircraft in a dirty

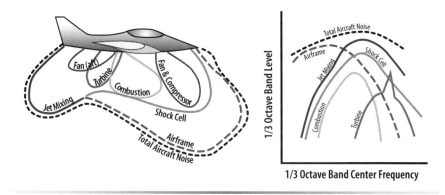

Figure 12.21 Noise source characteristics.

configuration with gear and flaps down) and the jet mixing. The reader is urged to return to Section 4.7 for more discussion of aircraft noise and its suppression. Another aircraft noise source is the sonic boom at speeds greater than Mach 1.0 (see Section 4.6). The main noise source for helicopters is the "slapping" of the rotor blades.

There are design features that can reduce the acoustic signature, such as nozzle noise suppressors, but the most effective approach is mission planning and tactics:

- Avoid acoustic sensors (human ears)
- Power down or slow down when possible
- Carefully select the background and environment

12.8 Case Study—AGM-129A Advanced Cruise Missile

In 1977 the U.S. Air Force was convinced (from the Have Blue program) that stealth could greatly increase the survivability of their strategic cruise missile fleet and issued the requirements for the Advanced Cruise Missile (ACM). The requirement was for a low-signature air-launched cruise missile that could deliver a nuclear weapon (W-80) against a high-value strategic target from a distance of 1900 n mile. Industry went to work, with Lockheed pursuing a medium-altitude design and Boeing a low-altitude design. Having started the Have Blue program with their XST program, DARPA was emotionally involved with the stealth technology and started their own ACM program called Teal Dawn. The first author was an Air Force colonel at DARPA and became the Teal Dawn program manager. Teal Dawn was selected for the ACM and entered development by the U.S. Air Force, with GD Convair as the contractor, in 1983. The ACM entered operation in 1991 with over 460 cruise missiles produced. The AGM-129A ACM is shown in Fig. 12.22.

Example 5.3 (Low-Altitude, Subsonic Cruise Missile) in Chapter 5 is essentially the AGM-129A ACM. The ACM requirement was pretty clear except for the mission profile: high altitude or low altitude. After much discussion, including intense interaction with the Defense Science Board, it was decided that Teal Dawn would fly a low-altitude TF/TA profile at Mach 0.7 similar to the AGM -86 ALCM and the AGM-109 Tomahawk. This meant that the RCS design against the low-frequency detection radars (Tall King and Spoon Rest) would be made easier because most of the flight would be below the radar horizon. The threats would be the short-range, high-frequency SAMs defending the high-value targets (located co-altitude and head-on) and the airborne interceptors (AIs, located above). The signature requirements were as follows:

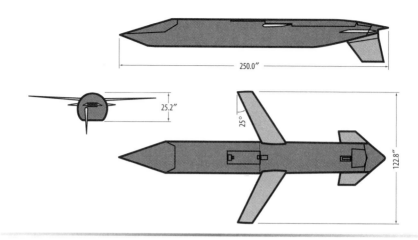

Figure 12.22 Three-view of Convair AGM-129A ACM.

• Very low RCS in the ± 20 deg front sector (X-band)
• No side or rear sector RCS requirement
• Low IR signature (top sector for the lookdown–shootdown AIs)

The TF/TA flight profile compounded the AI RF and IR detection problem because the ACM was operating in ground clutter, and the dense air (and possible clouds) increased the IR transmission losses. It was concluded that the TF/TA flight profile resulted in more relaxed signature requirements than a medium-altitude profile.

The RCS design was a sharp chisel nose shape and a flush inlet on the bottom of the fuselage. The wings were swept forward 25 deg for a more favorable packaging of the wing deployment mechanism. The wing LE spikes reflected off the fuselage and outside of the ±20 deg front sector. The missile was treated with high-frequency RAM. The IR design was to use a Williams F 112 turbofan (derivative of the F107 turbofan) for a cool exhaust plume. The plume was cooled further using an aspect ratio 4, 2-D nozzle that enhanced the plume mixing with the ambient air. The exhaust cavity was shielded from RF and IR sensors by an upper surface nozzle ramp. Finally the missile was painted with a high-emissivity paint to reduce the sunshine and cloudshine IR reflections (because the ground background is dark).

An AGM-129A has never been launched in anger!

References

[1] Rich, B. R., *Skunk Works*, Little, Brown, Toronto, 1994.
[2] Ball, R. E., *The Fundamentals of Aircraft Combat Survivability Analysis and Design*, AIAA Education Series, AIAA, Reston, VA, 1985.

[3] Fulghum, D. A., "Stealth Retains Value, but Its Monopoly Wanes," *Aviation Week and Space Technology*, 5 Feb. 2001, pp. 53–57.

[4] Barrie, D., "LO and Behold," *Aviation Week and Space Technology*, 11 Aug. 2003, pp. 50–53.

[5] Whitford, R., "Designing for Stealth in Fighter Aircraft (Stealth from the Aircraft Designer's Viewpoint," Paper 965540, 1996 World Aviation Congress, 21–24 Oct., Los Angeles, CA (sponsored by AIAA and SAE).

[6] Lynch, D., "How the Skunk Works Fielded Stealth," *Air Force Magazine*, Nov.1992, pp. 22–28.

[7] Piccirillo, A. C., "The Have Blue Technology Demonstrator and Radar Cross Section Reduction," Paper 965538, 1996 World Aviation Congress, 21–24 Oct., Los Angeles, CA (sponsored by AIAA and SAE).

[8] Aronstein, D. C., "The Development and Application of Aircraft RCS Prediction Methodology," Paper 965539, 1996 World Aviation Congress, 21–24 Oct., Los Angeles, CA (sponsored by AIAA and SAE).

[9] Fulghum, D. A., "Stealth Engine Advances Revealed in JSF Designs," *Aviation Week and Space Technology*, 19 March 2001, pp. 53–57.

[10] Varney, G. E., "IR Signature Measurement Techniques and Simulation Methods for Aircraft Survivability." Paper 79-1186, AIAA–Society of Automotive Engineers–American Society of Mechanical Engineers 15th Joint Propulsion Conf., 18–20 June 1979, Las Vegas, NV.

Chapter 13 Estimating Wing–Body Aerodynamics

- Three-Dimensional Lift Curve Slope
- Inviscid Drag-Due-to-Lift (Induced)
- Viscous Drag-Due-to-Lift
- Skin Friction Drag
- Wing Sweep Effects
- Drag Divergence Mach Number
- Canopy & Boattail Drag
- Vehicle Aerodynamics

After 25 years of service with the U.S. Air Force, the F-117 (lower) was retired to the Tonapah Nevada test range. With its second-generation stealth, 64 F-117s were produced from 1982 to 1991. The F-22 (upper), with its third-generation stealth, replaced the F-117. The F-117 received the Collier trophy in 1989, as did the F-22 in 2006.

If the aerodynamic estimates appear "too good to be true," they probably are. Always check estimates with real data.

At this point the design has matured to an aircraft wing–body–tail configuration. Next an aerodynamic analysis is performed to get a refined estimate of the lift and drag (and later the stability derivatives) to determine baseline takeoff and fuel weights. If the design does not close (i.e., the sum of the payload fraction plus the fuel fraction plus the empty-weight fraction does not equal 1.0), then go back to Chapter 5 and start over.

If this were an industry study, the design would be handed to the aerodynamics group and they would start modeling to input to computational fluid dynamics (CFD) codes. However, because this is more of a nonindustry–academic study, we employ well-respected rapid methods developed in the 1970s by the Air Force Flight Dynamics Laboratory [1], the National Advisory Committee for Aeronautics (NACA; now the National Aeronautics and Space Administration [NASA]), and others. The aerodynamic derivatives covered in this chapter are C_{L_α}, K, K', K'', and C_{D_0}.

Aircraft Big and Small

It is worth mentioning that the aero analysis and performance methods for a 6-ft wing span, 8-lb radio-controled (R/C) model airplane are identical to those for a full-scale airplane such as a Cessna 172 (36-ft wing span, 2300 lb). Thus, the methods discussed in this text are applicable to aircraft big and small. The only differences between the R/C model and the full-scale airplane are the wing loading, the Reynolds number, and the moments of inertia.

The R/C model wing loading is one to two orders of magnitude less than a full-scale airplane (because of the "square–cube law"; see Appendix I). R/C models typically have wing loadings of 1–3 lb/ft² whereas full-scale airplanes have greater than 10 (Cessna 172 has 12.6 lb/ft²). The impact is lower stall speeds and shorter takeoff and landing distances.

The R/C model will typically have Reynolds numbers less than 500,000, which gives the wing a predominately laminar boundary layer. Full-scale airplanes have Reynolds number greater than one million and have turbulent boundary layer wings. The impact is that the full-scale airplanes have higher maximum lift coefficients due to the turbulent boundary layer delaying flow separation over the wing better than the laminar boundary layer. The R/C models and the full-scale airplanes are in a Reynolds number region where the drag coefficients are about the same.

The R/C model will have much smaller moments of inertia than the full-scale airplane. The impact is that the time-to-double-amplitude (t_2) from a disturbance will be much shorter for the R/C model because $t_2 = f(\text{moment of inertia})^{1/2}$. R/C pilots will have their hands full with a neutral or unstable model.

13.1 Linear Lift Curve Slope

The subsonic C_L vs α curve for low aspect ratio wings (AR < 4) has a linear and a nonlinear region as shown in Fig. 13.1 (see the C_{L_α} for AR = 2 in Figs. H.2 and H.3).

The wing lift coefficient is given by the expression

$$C_L = \left(C_{L_\alpha}\right)\left(\alpha - \alpha_{0L}\right) + C_1\alpha^2 \qquad (13.1)$$

where C_{L_α} is the linear lift curve slope and C_1 is the nonlinear lift factor. The value of C_1 is determined from Fig. 2.13. The value of C_1 can be assumed zero for $M_\infty > 1$ flight. Performance calculations of cruise range, climbout, and acceleration to cruise Mach seldom require angles-of-attack in the nonlinear range. However, the nonlinear lift may be important for landing and takeoff.

13.1.1 Subsonic

The subsonic linear lift curve slope C_{L_α} per radian is given by Eq. (2.13) of Chapter 2:

$$\frac{\mathrm{d}C_L}{\mathrm{d}\alpha} = C_{L_\alpha} = \frac{2\pi\,\mathrm{AR}}{2 + \sqrt{4 + \mathrm{AR}^2\beta^2\left(1 + \left[(\tan^2 \Delta t/c)/\beta^2\right]\right)}} \qquad (13.2)$$

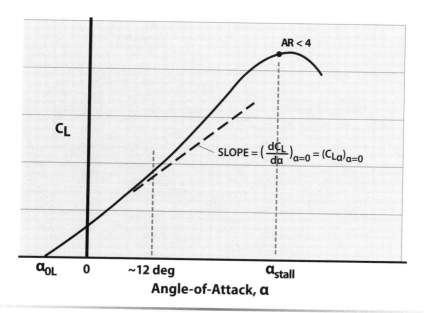

Figure 13.1 Wing C_L-vs-α curve showing nonlinear behavior for low-AR wings.

where

$$\beta = \sqrt{1 - M_\infty^2}$$

$\Delta t/c$ = sweep of maximum thickness line

Equation (13.2) is the C_{L_α} of the wing only and is therefore based upon the exposed wing planform area *Se.*

13.1.2 Supersonic

The method for estimating the wing supersonic lift curve slope is developed using supersonic linear theory corrected for three-dimensional flow effects. The wing C_{L_α} is determined using the charts in Fig. 13.2, where

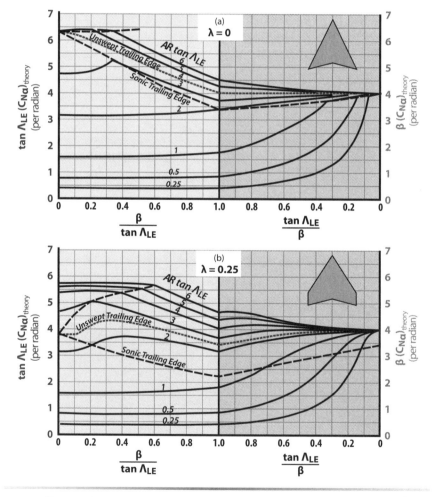

Figure 13.2 Theoretical wing lift curve slope (data from [2,3]).

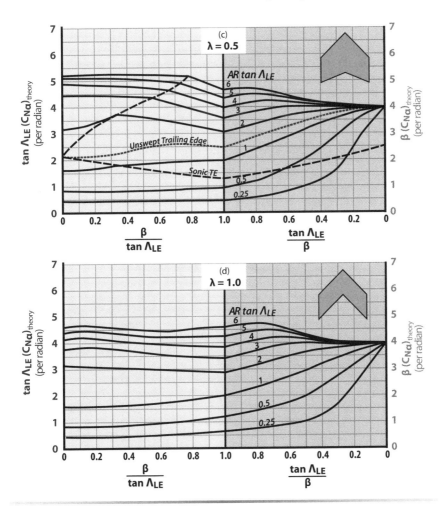

Figure 13.2 (continued) Theoretical wing lift curve slope (data from [2,3]).

C_N is the normal force coefficient slope and is equal to C_L for small to moderate angles-of-attack. In Fig. 13.2, $\beta = \sqrt{M_\infty^2 - 1}$, AR is the wing aspect ratio, and lambda (λ) is the taper ratio.

Here again C_{L_α} is based upon exposed wing planform area.

13.1.3 Transonic

There is no well-defined method for estimating transonic C_{L_α}. Reference [1] reports an empirical method that works reasonably well. The method is too complicated and cumbersome to present here, so an alternate method is suggested.

The C_{L_α} vs Mach number behavior will be as shown in Fig. 13.3a. Use the subsonic method up to about Mach 0.9 and extend the supersonic method down to about Mach 1.3. Then fair in a curve between Mach = 0.9 and 1.3 similar to the curves shown in Fig. 13.3a.

13.1.4 Wing-Body C_{L_α}

The lift characteristics of a wing and a body do not add directly to give the wing–body lift. Rather, there are interference effects of one component

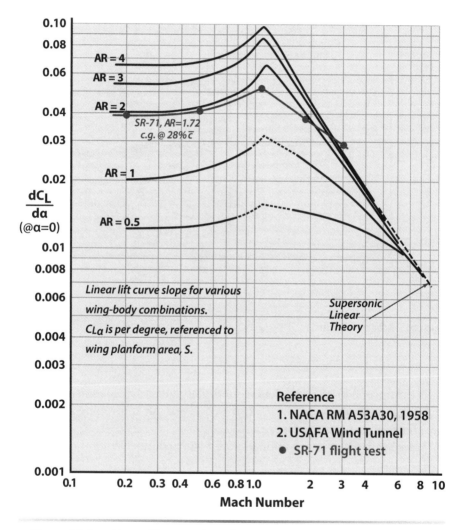

Figure 13.3a Linear lift curve slope for various wing–body combinations (C_{L_α} based on total planform area).

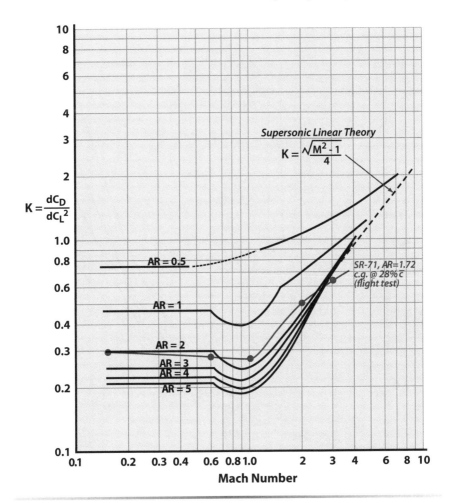

Figure 13.3b Drag-due-to-lift factor (based on total planform area) for uncambered wing–body combinations with delta planforms and LE radius of 0.045% chord.

on the other that make the wing–body lift greater than the sum of the individual components [4]. A method that gives good results for the wing–body linear lift curve slope is

$$\left(C_{L\alpha}\right)_{\text{WB}} = F\left(C_{L\alpha}\right)_{W} \tag{13.3}$$

where $(C_{L\alpha})_W$ is the linear lift curve slope (based upon the exposed wing area) of the wing and F is the wing–body lift interference factor shown in

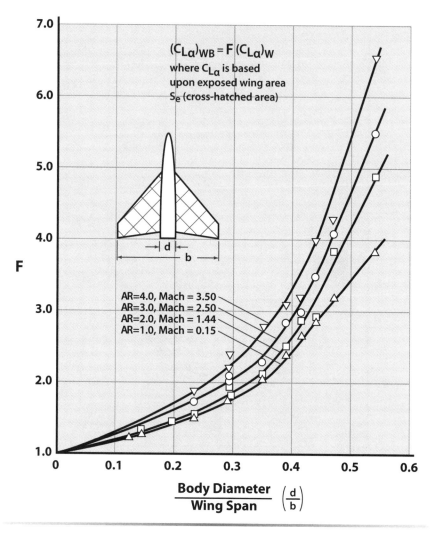

Figure 13.4 Wing–body interference factor (data from [5–7]).

Fig. 13.4. The $(C_{L_\alpha})_{WB}$ is the wing–body lift curve slope and is referenced to the exposed wing planform area S_e.

The final curve of wing–body C_{L_α} vs Mach number can be compared with the experimental data shown in Fig. 13.3a. Notice that the $(C_{L_\alpha})_{WB}$ presented in Fig. 13.3a are referenced to the total wing planform area S_W. The aircraft aerodynamic derivatives can be referenced to either S_e or S_W (see Fig. 7.1), but they must all be referenced to the same reference area. The total wing planform area S_W is more conventional and is recommended.

13.2 Drag-Due-to-Lift

The total drag coefficient for a wing–body combination is expressed as

$$C_D = \left(C_{D0}\right)_{\text{wing}} + \left(C_{D0}\right)_{\text{body}} + \Delta C_{D0} + C_{DL} \qquad (13.4)$$

where ΔC_{D0} is the zero-lift drag coefficient due to miscellaneous protuberances (canopy, pitot tube, etc.) and C_{DL} is the drag coefficient due to lift. Estimating the wing–body C_{DL} is difficult as discussed below; [3] calls it more of an art than a science. The wing–body C_{DL} is primarily due to the wing so that it is safe to assume

$$\text{wing–body } C_{DL} \approx \text{wing } C_{DL}$$

The methods for C_{DL} that follow use wing geometry primarily but represent the entire wing–body C_{DL} referenced to S_W.

13.2.1 Subsonic

In subsonic flow the total drag coefficient for the wing is expressed as

$$C_D = C_{D\text{min}} + K'C_L^2 + K''\left(C_L - C_{l\text{min}}\right)^2 \qquad (13.5)$$

The terms containing K' and K'' are collectively called the *drag-due-to-lift*. This parabolic behavior of C_D with C_L is shown in Figs. 2.16 and 2.17.

The K' term in Eq. (13.5) is the *inviscid* drag-due-to-lift called the *induced drag*. This drag results from the vortices trailing off a finite wing inducing a downwash at the wing aerodynamic center. The K'' term is the *viscous* drag-due-to-lift caused by flow separation and increased skin friction. This drag results from the viscous nature of the fluid causing the separation point on the upper surface to move forward from the trailing edge as the wing rotates to higher angles-of-attack and the region of adverse pressure gradient spreads. There is also an increase in skin friction occurring in the leading edge region due to the local supervelocities associated with increasing lift. The $C_{l\text{min}}$ is the lift coefficient for minimum drag coefficient C_d. For cambered airfoils, $C_{l\text{min}} \neq 0$ and is approximately equal to the C_l for $\alpha = 0$. For symmetric airfoils, $C_{l\text{min}} = 0$ and Eq. (13.5) is expressed as

$$C_D = C_{D0} + KC_L^2 \qquad (13.6)$$

where $K = K' + K'' = dC_D/dC_L^2$ and is called the drag-due-to-lift factor. The variation of K with Mach number is shown in Fig. 13.3b for low-AR wing-bodies. The SR-71 is certainly a low AR aircraft (AR = 1.72) and its K from flight test data agrees well with Fig. 13.3b. Figure G.9 shows the subsonic

$K = 1/\pi$ AR e for many real aircraft (symmetric and cambered). Figure G.9 is empirical and discussed in Appendix G. The C_{D_0} is called the zero-lift drag coefficient. It should be pointed out that $C_{D_0} \approx C_{D_{\min}}$ for wings with cambered airfoils, and the terms C_{D_0} and $C_{D_{\min}}$ are often used interchangeably. This text will use the term C_{D_0} to mean both C_{D_0} (for wings with symmetric airfoils) and $C_{D_{\min}}$ (for wings with cambered airfoils). This is not done to confuse the reader but rather in keeping with convention.

Equations (13.5) and (13.6), which display the parabolic behavior of C_D with C_L, are valid only up through moderate values of C_L. At a C_L called the *break* C_L, C_{LB}, the drag coefficient ceases to be parabolic with C_L as shown in Fig. 2.17. As the C_L increases past C_{LB} the drag-due-to-lift increases sharply from that expected from a parabolic behavior. The flow phenomenon involved here is not too well understood. However, it is connected with the onset of trailing edge separation spreading rapidly over the upper surface and/or the onset of leading edge separation spreading rapidly over the upper surface with no reattachment [8]. For C_Ls above C_{LB} the expression for total drag coefficient is expressed as

$$C_D = C_{D_0} + K'C_L^2 + K''\left(C_L - C_{l_{\min}}\right)^2 + \Delta C_{DB} \tag{13.7}$$

where ΔC_{D_B} is the drag deviation from a parabolic behavior (see Fig. 2.17).

The prediction method for ΔC_{D_B} is complicated and will not be presented here. The method is presented in [8].

The viscous drag-due-to-lift factor K'' is dependant primarily on LE radius, camber, and Re, and secondarily on taper ratio for sharp-edged airfoils. Determining K'' is difficult as it is viscous-dominated. Reference [9] offers a method shown on Fig. 13.6. Fig. 13.6 is independent of camber and Re, and it tends to overestimate K''; however, for symmetric or low camber airfoils, it offers a rapid estimate of K''.

A better method (and the one recommended) is to determine

$$K'' = \Delta\left(C_d - C_{d\min}\right)\Big/\Delta\left(C_l - C_{l\min}\right)^2$$

directly from airfoil polar data by plotting $\Delta(C_d - C_{d\min})$ vs $\Delta(C_l - C_{l\min})^2$ and determining the slope (see Section F.4).

The induced drag-due-to-lift factor K' is given as

$$K' = 1/\pi \text{ AR } e \tag{13.8}$$

Where e is called the wing efficiency factor and corrects the finite wing theory result (see Chapter 2) for taper ratio, sweep and body effects on the span loading. The e factor is best determined from CFD using a vortex lattice method.

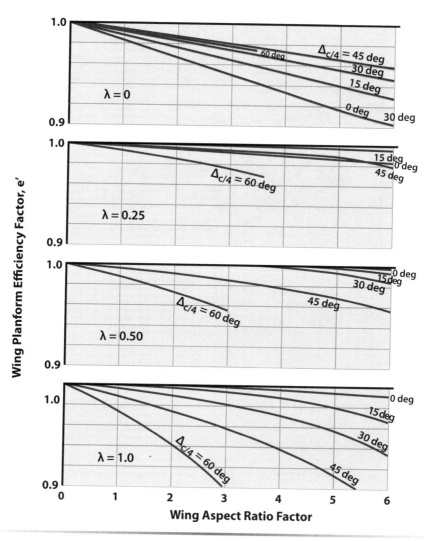

Figure 13.5 Weissinger wing planform efficiency factor (data from [1]).

The e factor can also be determined from

$$e = e'\left[1 - (d/b)^2\right]$$ (13.9)

where d/b is the ratio of body diameter to wing span (see Fig. 13.4). The e' factor has been formulated by Weissinger in [10] and is presented in Fig 13.5. Figure 13.5 was developed for fighter type aircraft and tends to overestimate the e for large aspect ratio configurations.

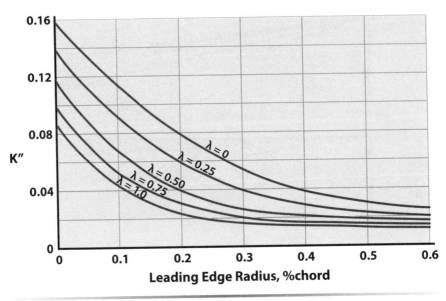

Figure 13.6   Viscous drag-due-to-lift factor K'' (data from [9]); LE radius for NACA airfoils shown on Fig. F.2.

An alternate method (and the one recommended) is to estimate K from Fig. G.9 and determine K' from $K' = K - K''$ as discussed in Section G.2.

13.2.2 Supersonic

The supersonic drag-due-to-lift is developed from supersonic linear theory (Chapter 2). For wings with supersonic leading edges the drag-due-to-lift factor K is given by

$$K = \frac{1}{C_{L_\alpha}} \tag{13.10}$$

where C_{L_α} is the wing–body lift curve slope (per radian) referenced to S_{ref}. Using the $C_{L_\alpha} = 1.6$ per radian value for the SR-71 at Mach $= 3.0$ from Fig. 13.3a gives $K = 0.62$, which agrees well with the flight test data of Fig. 13.3b.

For wings with subsonic leading edges, the drag-due-to-lift is less than that given by Eq. (13.10) because of the suction of the leading edge. Thus, the general expression for supersonic drag-due-to-lift factor K is

$$K = \frac{1}{C_{L_\alpha}} - \Delta N \tag{13.11}$$

where ΔN is the leading edge suction parameter. The ΔN parameter is determined from [8] as

$$\Delta N = \left(\frac{\Delta N}{\Delta N_{M=1.0}} \right) \left(\Delta N_{M=1.0} \right) \qquad (13.12)$$

where $(\Delta N / \Delta N_{M=1.0})$ is obtained from Fig. 13.7 and

$$\Delta N_{M=1.0} = \frac{1}{\left(C_{L\alpha} \right)_{M=1.0}} - \left(K' + K'' \right)$$

The K' and K'' are the subsonic inviscid and viscous drag-due-to-lift factors already determined. The term $(C_{L\alpha})_{M=1.0}$ is the wing–body lift curve slope at Mach = 1.0 (from Fig. 13.3a).

13.2.3 Transonic

There is no reliable method for estimating the transonic drag-due-to-lift factor. It is suggested that a curve be faired between the subsonic and supersonic K curves similar to the experimental data curves presented in Fig. 13.3b.

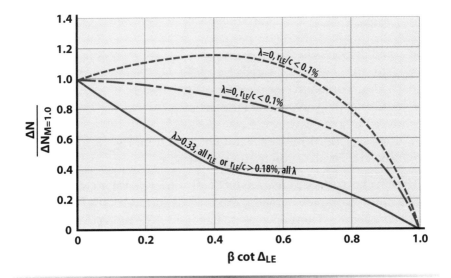

Figure 13.7 Values of LE suction parameter at supersonic speeds (data from [8]).

13.3 Zero-Lift Drag Coefficient

The total drag coefficient for a wing–body combination is given by Eq. (13.4) as

$$C_D = \left(C_{D0}\right)_{\text{wing}} + \left(C_{D0}\right)_{\text{body}} + \Delta C_{D0} + C_{DL} \qquad (13.4)$$

where the C_{D0}s for the wing and the body are determined separately and then added together. Equation (13.4) implies that the individual drag coefficient terms are all referenced to the same reference area S_{ref}. This S_{ref} can be S_e or S_{ref} but must be the same for all.

The methods for predicting the fuselage and wing C_{D0} will be discussed separately. The wing methods are limited to wings with straight leading edges. For nonstraight wings, such as a double delta (Swedish SAAB-35, Draken) or an ogee (Anglo-French Concorde SST), the methods presented in [1] or [9] should be used.

13.3.1 Wing: Subsonic

The subsonic wing C_{D0} is primarily skin friction. The expression for $(C_{D0})_W$ based upon the reference area S_{ref} is given by

$$\left(C_{D0}\right)_W = C_f \left[1 + L\left(\frac{t}{c}\right) + 100\left(\frac{t}{c}\right)^4 \right] R \frac{S_{\text{wet}}}{S_{\text{ref}}} \qquad (13.13)$$

where

L = airfoil thickness location parameter
L = 1.2 for maximum t/c located at $x \geq 0.3c$
L = 2.0 for maximum t/c located at $x < 0.3c$
t/c = maximum thickness ratio of the airfoil
S_{wet} = wetted area of the wing ($2S_e$)
R = lifting surface correlation factor obtained from Fig. 13.8
C_f = turbulent flat plate skin friction coefficient

The effect of surface roughness on the skin friction values is determined using a cutoff Reynolds number. The type of surface is selected and the roughness height is determined from Table 13.1. The ratio ℓ/k is computed and the cutoff Reynolds number, Re_ℓ, determined from Fig. 13.9. The ℓ is the mean aerodynamic chord $\bar{c}$ of the wing (see Fig. 7.1). The wing flight Reynolds number, $Re_e = \rho \bar{c} \, V/\mu$, based upon $\bar{c}$ is determined along a typical subsonic trajectory (see Chapter 4). Then the smaller of the two Reynolds numbers, Re_e or Re_ℓ, is used to determine the C_f from Fig. 2.6.

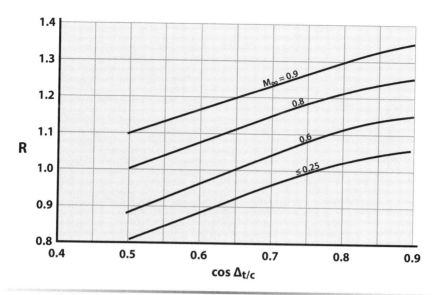

Figure 13.8 Lifting surface correlation factor for wing subsonic C_{D_0}.

13.3.2 Wing: Transonic

The transonic regime for the wing begins at M_{CR} but the drag rise is delayed slightly until the divergence Mach number, M_D. The divergence Mach number is defined as that Mach number where $(\partial C_{D_0}/\partial M) = 0.1$. The transonic wing C_{D_0} is expressed as

$$\left(C_{D_0}\right)_W = C_{Df} + C_{DW} = C_f\left[1+L\left(\frac{t}{c}\right)\right]\frac{S_{\text{wet}}}{S_{\text{ref}}}+C_{DW}$$

Table 13.1 Roughness Height Values
(in Equivalent Sand Roughness)

Type of Surface	k (in.)
Aerodynamically smooth	0
Polished metal or wood	$0.02–0.08 \times 10^{-3}$
Natural sheet metal	0.16×10^{-3}
Smooth matte paint, carefully applied	0.25×10^{-3}
Standard camouflage paint, average	0.40×10^{-3} application
Camouflage paint, mass-production	1.20×10^{-3} spray
Dip-galvanized metal surface	6×10^{-3}
Natural surface of cast iron	10×10^{-3}

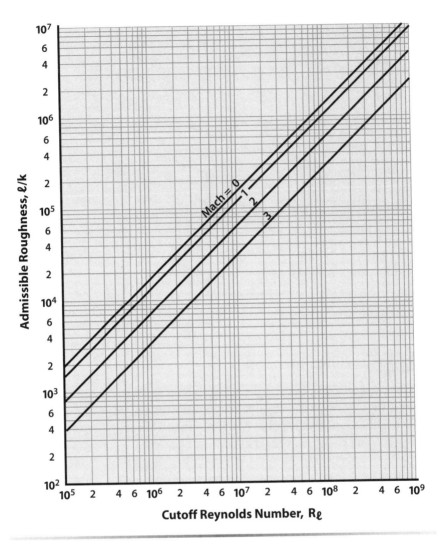

Figure 13.9 Cutoff Reynolds number (data from [1]).

The skin friction drag C_{D_f} is assumed to be a constant value throughout the transonic region. The value for C_{D_f} is the value at Mach = 0.6.

The task of constructing the wing transonic C_{D_W} curve is one of correcting experimental data for sweep, aspect ratio, and t/c using the von Kármán similarity laws for transonic flow. The transonic C_{D_W} curve for unswept wings is shown in Fig. 13.10. Table 13.2 presents useful values for t/c.

The Mach number for drag divergence, M_D, of the unswept wing is obtained by locating the point on the C_{D_W} vs Mach curve where the slope is 0.1. The values of peak C_{D_W}, Mach number for peak C_{D_W}, and M_D are corrected for sweep as follows:

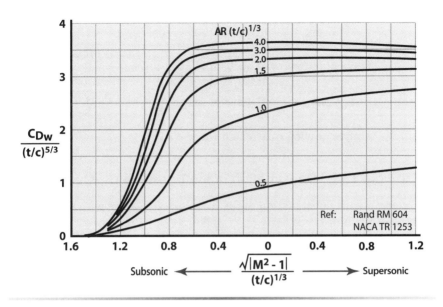

Figure 13.10   Transonic zero-lift wing wave drag for unswept wings.

Table 13.2   Unswept Wings (Values for t/c: Wave Drag)

t/c	$(t/c)^{1/3}$	$(t/c)^{5/3}$
0.12	0.493	0.0293
0.11	0.479	0.0254
0.10	0.464	0.0217
0.09	0.448	0.0181
0.08	0.431	0.0148
0.07	0.412	0.0118
0.06	0.392	0.0092
0.05	0.368	0.0068
0.04	0.342	0.00468
0.03	0.311	0.00292
0.02	0.271	0.00147

$$\text{Swept } M_D = \left[\text{Unswept } M_D\right] / \left(\cos\Lambda_{c/4}\right)^{0.5} \tag{13.14}$$

$$\text{Swept } C_{DW_{\text{peak}}} = \left[\text{Unswept } C_{DW_{\text{peak}}}\right] / \left(\cos\Lambda_{c/4}\right)^{2.5} \tag{13.15}$$

$$\text{Swept } M_{CDW_{\text{peak}}} = \left[\text{Unswept } M_{CDW_{\text{peak}}}\right] / \left(\cos\Lambda_{c/4}\right)^{0.5} \tag{13.16}$$

where $\Lambda_{c/4}$ = angle of wing quarter-chord.

Example 13.1   Construction of the Transonic C_{D_0} Curve

Determine the construction of the transonic C_{D_0} curve with sweep $c/4 =$ 45 deg:

Delta wing with AR	3
t/c	0.03
C_{D_f}	0.006 at Mach 0.6
$(t/c)^{1/3}$	0.311
$(t/c)^{5/3}$	0.00292 (Table 13.2)
AR $(t/c)^{1/3}$	0.933
Unswept $C_{DW_{\text{peak}}}$	0.0082 (from Fig. 13.10)
Unswept $M_{CDW_{\text{peak}}}$	1.09 (from Fig. 13.10)
Swept $C_{DW_{\text{peak}}}$	$(0.0082)(0.42) = 0.00344$
Swept $M_{CD_{\text{peak}}}$	$1.09/0.841 = 1.3$

The construction of the wing transonic C_{D_0} curve is shown on Fig. 13.11. The unswept M_D is located by finding the point where the slope is 0.1. The swept wing M_D is determined by Eq. (13.14). The swept wing C_D curve is then faired in as shown in Fig. 13.11.

13.3.3 Wing: Supersonic

The supersonic wing C_{D_0} based upon S_{ref} is given by

$$\left(C_{D_0}\right)_W = C_{D_f} + C_{DW}$$

The wing supersonic skin friction is expressed as

$$C_{D_f} = C_f \frac{S_{\text{wet}}}{S_{\text{ref}}} \tag{13.17}$$

where $C_f = (C_{f_c}/C_{f_i})C_{f_i}$. The ratio C_{f_c}/C_{f_i} is obtained from Fig. 13.12 and C_{f_i} is determined the same way as for subsonic flow using cutoff and flight Reynolds number comparison.

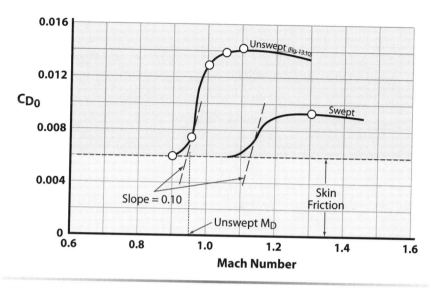

Figure 13.11 Construction of transonic wing C_{D_0} for AR = 3 delta wing with $t/c = 0.03$.

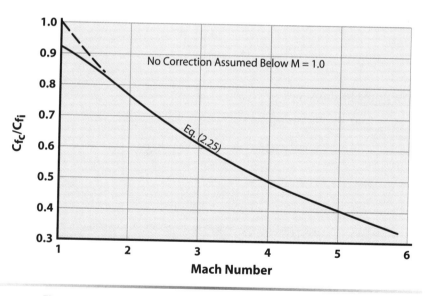

Figure 13.12 Compressibility effect on turbulent skin friction.

The method for predicting the wing supersonic wave drag coefficient is developed from supersonic linear theory. Wings with round leading edges will exhibit a detached bow wave, accompanied by an additional wave drag term due to leading edge bluntness.

For wings with sharp-nosed airfoil sections and the following:

1. Supersonic leading edge ($\beta \cot \Lambda_{LE} \geq 1$), use

$$C_{DW} = \frac{B}{\beta}\left(\frac{t}{c}\right)^2 \frac{S_e}{S_{ref}}$$ (13.18)

2. Subsonic leading edge ($\beta \cot \Lambda_{LE} < 1$), use

$$C_{DW} = B \cot \Delta_{LE} \left(\frac{t}{c}\right)^2 \frac{S_e}{S_{ref}}$$ (13.19)

where B is a constant factor for a given sharp-nosed airfoil. B factors for sharp-nosed airfoils are presented in Table 13.3.

For wings with round-nosed airfoil sections and the following:

1. Supersonic leading edge ($\beta \cot \Lambda_{LE} \geq 1$), use

$$C_{DW} = C_{DLE} + \frac{16}{3\beta}\left(\frac{t}{c}\right)^2 \frac{S_e}{S_{ref}}$$ (13.20)

2. Subsonic leading edge ($\beta \cot \Lambda_{LE} < 1$), use

$$C_{DW} = C_{DLE} + \frac{16}{3}\cot \Delta_{LE}\left(\frac{t}{c}\right)^2 \frac{S_e}{S_{ref}}$$ (13.21)

Table 13.3 B Factor for Sharp-Nosed Airfoils

Basic Wing Airfoil Section	B	Section
Biconvex	$\dfrac{16}{3}$	
Double wedge	$\dfrac{c/x_t}{1-x_t/c}$	
Hexagonal	$\dfrac{c(c-x_2)}{x_1 x_3}$	

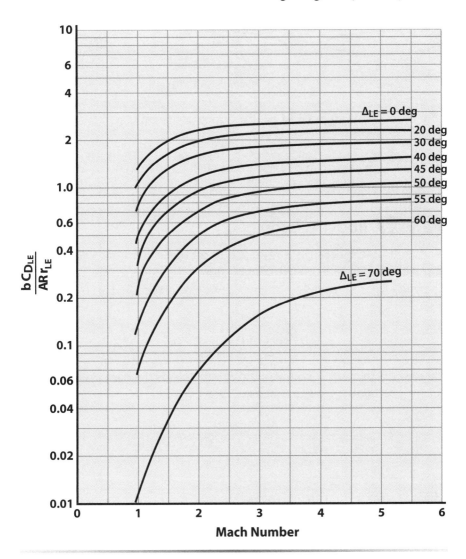

Figure 13.13 Supersonic round LE bluntness drag coefficient (data from [12]).

where the leading edge bluntness term $C_{D_{LE}}$ is determined from Fig. 13.13. In Fig. 13.13 b is the wing span in feet and r_{LE} is the radius of the leading edge at the mean aerodynamic chord in feet.

Sometimes the C_{D_0} values determined in the transonic and supersonic regimes do not match so that it is difficult to fair a smooth curve through all the points. This is usually because the transonic method does not account for leading edge radius. In this event, average the data point values

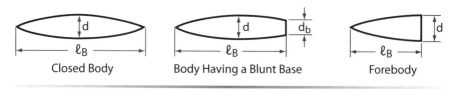

Figure 13.14   Body shapes and geometry.

until a smooth curve can be drawn. The peak C_{D_0} should occur at the Mach number given by Eq. (13.16).

13.3.4 Body: Subsonic

At subsonic speeds the body C_{D_0} of smooth slender bodies is primarily skin friction (Fig. 8.11). Figure 13.14 shows the body shapes considered. The body C_{D_0} referenced to the maximum cross-sectional area S_B is given as

$$\left(C_{D_0}\right)_B = \left(C_{D_f}\right)_B + C_{D_b}$$

where C_{D_f} is the skin friction coefficient and C_{D_b} is the base pressure coefficient. The body C_{D_f} is expressed as [13]

$$\left(C_{D_f}\right)_B = C_f \left[1 + \frac{60}{\left(\ell_B / d\right)^3} + 0.0025\left(\frac{\ell_B}{d}\right)\right]\frac{S_S}{S_B} \qquad (13.22)$$

where S_s is the wetted area of the body surface and ℓ_B/d is the body fineness ratio (see Fig. 13.14).

For noncircular bodies, the equivalent diameter should be used:

$$d_{\text{equiv}} = \sqrt{S_S / 0.7854}$$

The C_f is the turbulent skin friction coefficient and is determined in the same manner as the wing subsonic skin friction. The reference length is the body length ℓ_B.

The base pressure coefficient is expressed in [14] as

$$C_{D_b} = 0.029\left(d_b / d\right)^3 \Big/ \sqrt{\left(C_{D_f}\right)_B} \qquad (13.23)$$

The designer should avoid blunt-base bodies if at all possible because the C_{D_b} term can become quite large. If a jet engine exhaust completely fills the base region, then the base drag is zero.

13.3.5 Body: Transonic

The transonic body C_{D_0} is given as

$$\left(C_{D_0}\right)_B = C_{D_f} + C_{D_p} + C_{D_b} + C_{D_W} \tag{13.24}$$

The $C_{D_f} = C_f S_s/S_B$ is the skin friction drag coefficient, where C_f is the turbulent skin friction coefficient at Mach $= 0.6$. This value is assumed to be constant throughout the transonic region.

The pressure drag coefficient C_{D_p} is evaluated at Mach $= 0.6$ by

$$C_{D_p} = \left(C_f\right)_{M=0.6}\left[\frac{60}{\left(\ell_B/d\right)^3} + 0.0025\left(\frac{\ell_B}{d}\right)\right]\frac{S_S}{S_B} \tag{13.25}$$

The base drag term C_{D_b} is evaluated using

$$C_{D_b} = -C_{p_b}\left(\frac{d_b}{d}\right)^2 \tag{13.26}$$

where the base pressure coefficient C_{p_b} is obtained from the three-dimensional curve in Fig. 2.27.

The wave drag coefficient C_{D_W} is obtained from Fig. 13.15 (data from [15]).

The body transonic C_{D_0} curve is constructed by adding the four drag terms of Eq. (13.24). The divergence Mach number for bodies having fineness ratios of 4 and greater is about 0.95.

13.3.6 Body: Supersonic

The supersonic body C_{D_0} method presented in this section is taken from [11], which contains an excellent summary of the various supersonic theories compared with experimental data. The method presented here is restricted to nonblunt closed-nosed bodies of revolution. If the body is open nosed (such as the fuselage of the F-100 or MIG-21) or has significant nose bluntness, the method of [1] should be used.

The body supersonic C_{D_0} referenced to the maximum cross-sectional area S_B is expressed as

$$\left(C_{D_0}\right)_B = C_f\frac{S_S}{S_B} + C_{D_{N2}} + C_{D_A} + C_{D_{A(NC)}} + C_{D_b} \tag{13.27}$$

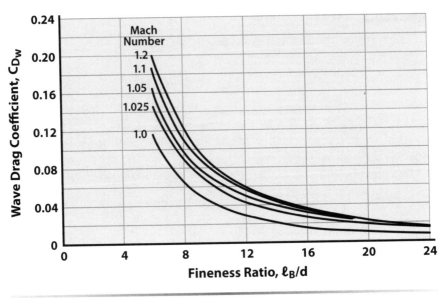

Figure 13.15 Wave drag for parabolic-type fuselage (data from [15]).

where the terms are defined as:

C_f = compressible turbulent skin friction determined in the same fashion as the supersonic wing skin friction

S_s = body wetted area

$C_{D_{AN2}}$ = interference drag coefficient acting on the afterbody due to the center body (cylindrical section) and the nose, obtained from Figs. 13.16 and 13.17

$C_{D_{N2}}$ = nose wave drag obtained from Figs. 13.18, 13.19, and 13.20, where f_N is nose fineness ratio ℓ_N/d (see Fig. 13.16)

C_{D_A} = body afterbody wave drag obtained from Figs. 13.21 and 13.22, where f_A is the afterbody fineness ratio ℓ_A/d (see Fig. 13.16)

C_{D_b} = base drag term given by Eq. (13.26); C_{p_b} is obtained from Fig. 2.27

13.3.7 Miscellaneous Drag Items

The designer should not neglect the drag of miscellaneous items such as external stores, the canopy, and other protuberances. The drag for these items is best obtained from experiment. Figure 13.23 shows the approximate C_{D_0} for a one-man canopy, typical protuberances (such as the pitot tube, antenna mounts, gun ports), and nozzle boattail. The nozzle-boattail approximate C_{D_0} shown in Fig. 13.23 is for a fuselage-mounted jet engine with a gentle afterbody taper down to the exhaust nozzle. This approximate

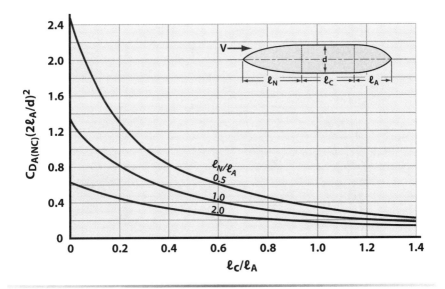

Figure 13.16   Interference drag for pointed bodies with parallel center section.

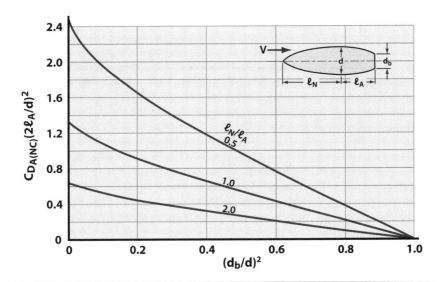

Figure 13.17   Interference drag of truncated afterbodies behind pointed forebodies with no parallel center section.

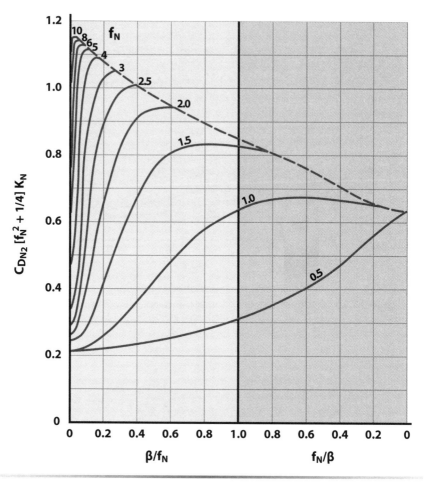

Figure 13.18   Supersonic pressure drag of ogive noses.

C_{D_0} of the nozzle boattail would replace the afterbody and base drag terms mentioned earlier.

Figure 13.24 presents the approximate C_D for external stores.

The data in Figs. 13.23 and 13.24 is from [16] and is referenced to a wing area of 280 ft². Thus, the data must be corrected for the appropriate S_{ref}. For example, the canopy drag coefficient at Mach = 1.0 would be

$$\text{Canopy } \Delta C_{D_0} = (0.004)\frac{280}{S_{ref}}$$

The ΔC_{D_0} for a landing gear can be seen in Fig. 10.4.

13.3.8 Wing–Body C_{D_0}

The problem of estimating the wing–body combination C_{D_0} is one of properly accounting for the mutual interference effects of one component on the other. The problem is extremely complicated and requires a fairly accurate picture of the flow field interactions. This information is not available at this point in the design game. Correction studies have been

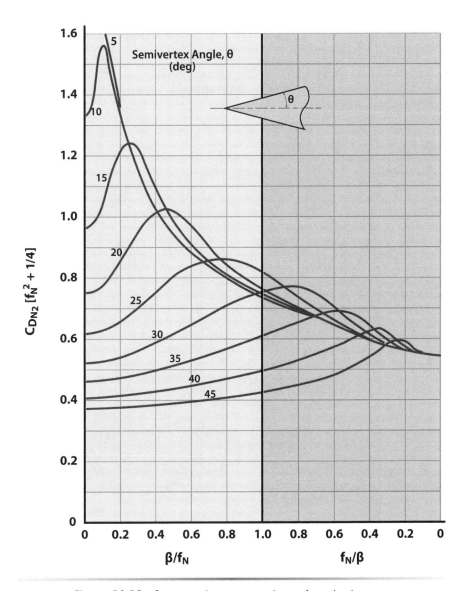

Figure 13.19 Supersonic pressure drag of conical noses.

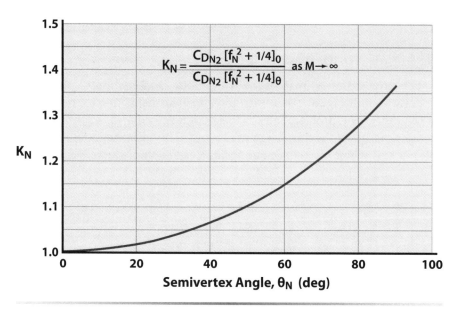

$$K_N = \frac{C_{DN2}\,[f_N^2 + 1/4]_0}{C_{DN2}\,[f_N^2 + 1/4]_\theta} \quad \text{as } M \to \infty$$

Figure 13.20 Correlation factor for pressure drag of ogive noses.

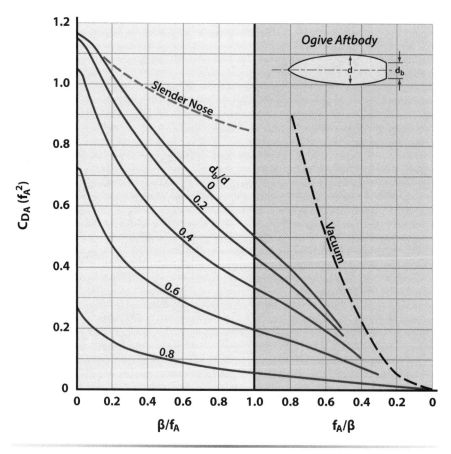

Ogive Aftbody

Slender Nose

d_b/d
0
0.2
0.4
0.6
0.8

Vacuum

$C_{DA}\,(f_A^2)$

β/f_A

f_A/β

Figure 13.21 Supersonic pressure drag of ogive boattails (data from [11]).

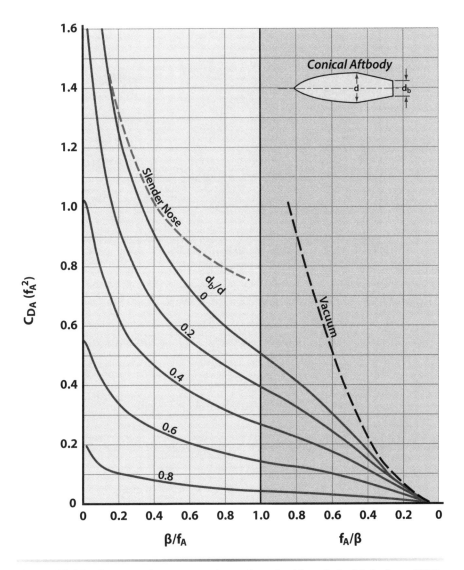

Figure 13.22   Supersonic pressure drag of conical boattails (data from [11]).

conducted on wing–body interference. The results [9] indicate that the wing–body interference effects amount to about ± 5% for subsonic flow with the generous use of fillets. It is hard to argue at this point that the C_{D_0} of the components is accurate to within 5%. Thus, the wing–body subsonic C_{D_0} will be assumed to be simply the sum of the components,

$$\left(C_{D0}\right)_B = C_f \frac{S_S}{S_B} + C_{DN2} + C_{DA} + C_{DA(\text{NC})} + C_{Db} \qquad (13.27)$$

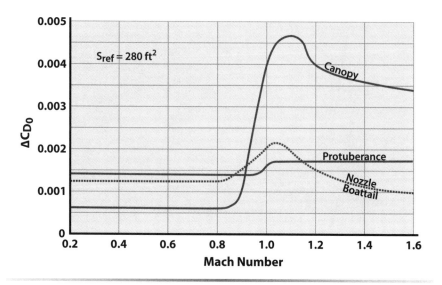

Figure 13.23   Incremental drag for miscellaneous items (data from [16]).

based upon the maximum cross-sectional area S_B. Then the wing-body C_{D_0} referenced to S_{ref} is

$$\left(C_{D_0}\right)_{WB} = \left(C_{D_0}\right)_B \frac{S_B}{S_{\text{Ref}}} + \left(C_{D_0}\right)_W + \Delta C_{D_0} \qquad (13.28)$$

where ΔC_{D_0} is any miscellaneous drag items referenced to S_{ref}.

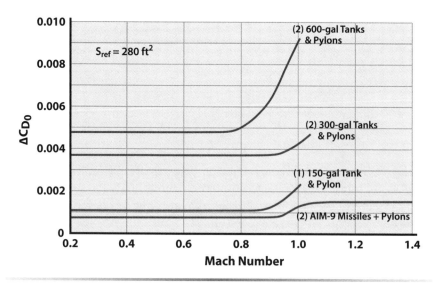

Figure 13.24   Incremental drag for external stores (data from [16]).

For transonic and supersonic flow the interference effects can be significant. The interference drag is usually positive for configurations not specifically contoured to reduce this drag component. However, for area-ruled configurations, this interference drag can be negative. It is recommended that aircraft designed for transonic and supersonic flight be area-ruled. Area-ruling is discussed in Chapter 8.

13.4 Combined Vehicle Aerodynamics

The complete aircraft aerodynamics can now be estimated. First tail surfaces (t/c, planform, symmetrical section) are designed based upon the preliminary estimates for tail size (from Chapter 11), then their individual aerodynamics are estimated and then combined with the wing–body aerodynamics. A popular trick at this point is to assume the tail surfaces to be miniature wings and nacelles to be miniature fuselages so that their aerodynamics are similar. This might appear as cheating, but it is appropriate for the first design loop. Our complete aircraft aerodynamics can be estimated as follows:

$$\left(C_{L_\alpha}\right)_{\text{a/c}} = \left(C_{L_\alpha}\right)_{\text{W/B}} \qquad (13.29a)$$

$$\left(C_{\ell_\alpha}\right)_{\text{a/c}} = \left(C_{\ell_\alpha}\right)_{\text{W/B}} \qquad (13.29b)$$

$$K_{\text{a/c}} = K_{\text{W/B}} \qquad (13.30)$$

$$\left(C_{D0}\right)_{\text{a/c}} = \left(C_{D0}\right)_{\text{W/B}} + \left(C_{D0}\right)_{\text{wing}} \frac{S_{\text{VT}} + S_{\text{HT}}}{S_{\text{ref}}} + \left(C_{D0}\right)_{\text{fuse}} S_{\text{nacelle}}/S_{\text{ref}} \qquad (13.31)$$

Remember to do a "sanity check" on your aerodynamic estimates. Compare your results with real aircraft data such as those found in Appendix G.

References

[1] Ellison, D. E., "USAF Stability and Control Handbook (DATCOM)," U.S. Air Force Flight Dynamics Laboratory, AFFDL/FDCC, Wright–Patterson AFB, OH, Aug. 1968.

[2] Jones, R. T., "Properties of Low-Aspect-Ratio Pointed Wings at Speeds Below and Above the Speed of Sound," NACA TR-835, 1946.

[3] Mirels, H., "Aerodynamics of Slender Wings and Wing–Body Combinations Having Swept Trailing Edges," NACA TN-3105, 1954.

[4] Pitts, W. C., Nielsen, J. N., and Kaattari, P. J., "Lift and Center of Pressure of Wing–Body–Tail Combinations at Subsonic, Transonic and Supersonic Speeds," NACA Rept. 1307, 1959.

[5] Maher, R. J., and Bores, J. H., "Low Aspect Ratio Wing–Body Combination Lift Curve Slope Determination at Subsonic Speeds," Aero 350 Rept., U.S. Air Force Academy, CO, May 1970, pp. 13–36.

[6] Sanchez, F., "Lift Curve Slope Interference Factor for Low Aspect Ratio Wing–Body Combinations," Aero 499 Rept., U.S. Air Force Academy, CO, May 1971.

[7] Nicolai, L. M., and Sanchez, F., "Correlation of Wing–Body Combination Lift Data," *Journal of Aircraft*, Vol. 10, No. 2, Oct. 1973, pp. 126–128.

[8] Simon, W. E., Ely, W. L., Niedling, L. G., and Voda, J. J., "Prediction of Aircraft Drag Due to Lift," U.S. Air Force Flight Dynamics Laboratory, AFFDL-TR-71-84, Wright–Patterson AFB, OH, June 1971.

[9] Benepe, D. B., Kouri, B. G., Webb, J. B., "Aerodynamic Characteristics of Non-Straight Taper Wings," U.S. Air Force Flight Dynamics Laboratory, AFFDL-TR-66-73, Wright–Patterson AFB, OH, 1966.

[10] Furlong, G. C., and McHugh, J. G., "A Summary and Analysis of the Low-Speed Longitudinal Characteristics of Swept Wings at High Reynolds Number," NACA RM L52D16, Aug. 1952.

[11] Morris, D. N., "A Summary of the Supersonic Pressure Drag of Bodies of Revolution," *Journal of the Aeronautical Sciences*, Vol. 28, No.7, July 1961, pp. 516–521.

[12] Crosthwait, E. L., "Drag of Two-Dimensional Cylindrical Leading Edges," General Dynamics, F/W Rept. AIM No. 50, 1966.

[13] Blakeslee, D. J., Johnson, R. P., and Skavdahl, H., "A General Representation of the Subsonic Lift–Drag Relation for an Arbitrary Airplane Configuration," RAND RM 1593, 1955.

[14] Hoerner, S. F., *Fluid-Dynamic Drag*, published by the author, 1958.

[15] Gollos, W. W., "Transonic and Supersonic Pressure Drag for a Family of Parabolic Type Fuselages at Zero Angle of Attack," RAND Rept. RM 982, 1952.

[16] Smith, C. W., "Aerospace Handbook," General Dynamics, Fort Worth, TX, FZA-381, March 1976.

Chapter 14

Propulsion System Fundamentals

- Propeller Systems (Reciprocal & Turbine)
- Turbine Engine Fundamentals
- Electric Aircraft System
- Solar Aircraft System
- Ramjet
- Rocket Engines
- Sir Frank Whittle & Hans von Ohain

Sir Frank Whittle (left) and Dr. Hans von Ohain (right) changed the propulsion world of aviation by inventing the jet turbine engine independently of each other. Their story is told at the end of this chapter.

For any isolated system not in equilibrium the entropy will increase until that system attains equilibrium.
Second law of thermodynamics

14.1 Introduction

The second law of thermodynamics traces its origin to the French physicist Sadi Carnot (1796–1832). In 1824, he published "Reflections on the Motive Power of Fire," which presented the view that motive power (work) is due to the fall of fire (caloric heat) from a hot to a cold body (working substance). Simply stated it is an expression of the universal law of increasing entropy.

The primary purpose of all aircraft propulsion devices is to impart a change in momentum to a mass of fluid. The fluid may be air, air and combustion products, or combustion products only. In the case of a watercraft the fluid would be water. Newton's second law states that the force or thrust produced on a system is equal to the change in momentum of the system in unit time. This fundamental principle is shown in Fig. 14.1 for a stream tube of air [1]. The entrance conditions are denoted by the freestream symbol a and the exit conditions denoted by e. The mass flow rate of air through the stream tube is $\rho A V$ and has units of slugs per second (slug/s). The stream tube boundaries are the fluid streamlines. The force or *net thrust* acting on the stream tube system is given by

$$T_n = (\dot{m}_{\text{air}} + \dot{m}_{\text{fuel}})V_e - \dot{m}_{\text{air}}V_a + P_e A_e - P_a A_a \qquad (14.1)$$

Notice that there may be a difference in the pressure and area at the entrance and exit such that a small pressure force would act on the system. Because the mass flow rate of the fuel added to the system is very small compared with the mass flow rate of the air, Eq. (14.1) is usually written

$$T_n = \dot{m}_{\text{air}}(V_e - V_a) + P_e A_e - P_a A_a \qquad (14.2)$$

The principal types of propulsion devices accelerating the flow inside the stream tube are listed next and are shown qualitatively in Figs. 4.6 and 4.7:

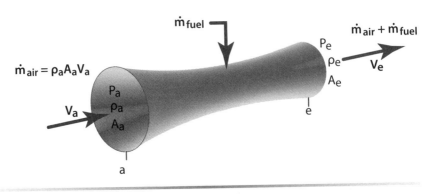

Figure 14.1 Momentum change on a fluid system.

- **Propellers.** Propellers are driven by reciprocating piston engines, gas turbines (turboprops), or electric motors. A propeller operates by producing a relatively small change in velocity of a relatively large mass of air. Propellers are limited to tip speeds less than sonic due to the formation of shocks and thus have a practical speed limit less than 500 kt (Mach = 0.75).
- **Gas turbines.** In the forms of simple jets (turbojets), afterburning turbojets, and turbofans, gas turbines accelerate a small mass of air to a large velocity change and can operate supersonically to about Mach 3.5.
- **Ramjets** (both subsonic and supersonic combustion)
- **Pulsejets**
- **Rockets.** Rockets carry their own oxidizer ($m_{air} = 0$) and thus accelerate the very small (relative to turbines) combusted propellant products to very high velocities. The thrust equation for rockets becomes

$$T = \dot{m}_{cp} V_e + A_e \left(P_e - P_a \right) \tag{14.3}$$

where $\dot{m}_{cp}$ is the mass flow per unit time of combusted propellant products.

14.2 Operation of Propeller Systems

The analysis and design of propellers is discussed in Chapter 17. This subsection will discuss the engines that drive the propeller. The engine provides a *thrust power available* equal to TV, which may be taken as the propeller output. The power input to the propeller from the engine shaft is the *engine brake horsepower*; thus, the *propeller efficiency* is

$$\eta_p = \left(\text{propeller thrust power} \right) / \left(\text{engine shaft brake horsepower} \right)$$

In flight, the propeller accelerates a large mass of air rearward to a velocity only slightly greater than the flight speed, exhibiting efficiencies at normal flight speeds of between 85% and 90%. The lost horsepower appears mainly as unrecoverable kinetic energy of air in the slip stream.

The *horsepower required* for an aircraft to fly at a speed V is

$$\text{hp}_{Req} = DV / 550 \eta_p \tag{14.4a}$$

where the 1/550 converts foot-pounds per second (ft·lb/s) to horsepower.

Using the equation for one-g drag $D = W/(L/D)$ gives the useful equation

$$\text{hp}_{Req} = WV / (L/D) 550 \eta_p \tag{14.4b}$$

14.2.1 Reciprocating Piston Engines

The aircraft *reciprocating piston engine* uses the well-known four-cycle Otto cycle [2]. An aircraft piston engine is similar to an automobile engine with a few differences. First, engine weight [given in horsepower per pound (hp/lb)] is a major performance parameter. Most aircraft engines are air cooled for this very reason. Second, reliability is very important because a malfunction at any altitude is a serious situation. The current piston engines are well developed to give high performance (hp/lb), low *brake specific fuel consumption* (BSFC) in pounds of fuel per hour per brake horsepower [(lb of fuel/h)/bhp], and high reliability.

Current piston and turboprop engines are shown in Fig. 14.2. The hp/lb for the current piston engines varies from about 0.6 for the small engines (less than 600 hp) to almost 1.0 for the larger engines. The BSFC for all the piston engines in Fig. 14.2 at sea level static (SLS) conditions varies from approximately 0.5 for the smaller engines (less than 400 bhp) to 0.42 for the larger engines. Most engines have a major overhaul recommended at 2000 hours. The engines have two spark plugs on each cylinder fired independently from engine-driven magnetos.

The power output from a piston engine depends primarily on two parameters: the engine rpm and the absolute pressure in the intake manifold. Maximum power is typically at 2800 rpm and SLS conditions of 59°F and 14.7 psia (30 in. of Hg).

Table 14.1 presents the specifications for the Lycoming 0–360-A aircraft engine. The 0–360-A (shown in Fig. 14.3) is in the Piper Cherokee 180 and represents a very typical general aviation piston engine. Notice that it is designed to cruise at 65%–75% of maximum power, which is a range of 2200–2450 rpm. The maximum throttle performance degradation with altitude is linear from 700 ft (180 hp at 2700 rpm and 28 in. of Hg) to 21,000 ft (76 hp at 2700 rpm). Cruise power is linear with altitude also.

A useful expression (from [1]) for the power loss (reduction in brake horsepower, Bhp) with altitude is

$$\text{Bhp} = \text{Bhp}_{\text{SL}} \left(\frac{\rho}{\rho_{\text{SL}}} - \frac{1 - \dfrac{\rho}{\rho_{\text{SL}}}}{7.75} \right) \quad (14.5)$$

Piston engines are sometimes supercharged to increase sea-level power for air racing or to increase the operating altitude. *Supercharging* involves compressing the air entering the intake manifold by means of a compressor. In earlier piston engines, this compressor was driven by a gear train from the engine crankshaft. The more modern supercharged engines employ a turbine-driven compressor powered by the engine's exhaust and

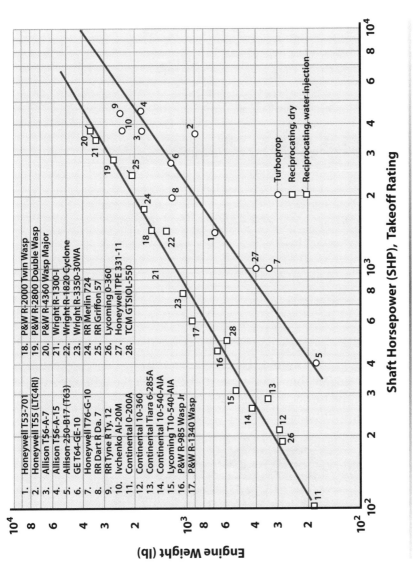

Figure 14.2   Weight vs shaft horsepower (SHP), for turboprop and piston aircraft engines.

Table 14.1 Lycoming 0–360-A Aircraft Engine Specifications and Description (data from [2])

Type: Four-Cylinder, Direct Drive, Horizontally Opposed, Wet-Sump, Air-Cooled Engine	
Weight, pounds	282
Bore, inches	5.125
Stroke, inches	4.375
Displacement, cubic inches	361
Compression ratio	8.5:1
Cylinder head temperature, max. °F	500
Cylinder base temperature, max. °F	325
Fuel: aviation grade, octane	100–130
Performance, hp	
Takeoff rating at SLS, hp	185 at 2900
Max. rated at 700 ft (28 in. of Hg), hp	180 at 2700
Max. rated at 7000 ft, hp	143 at 2700
Max. rated at 21,000 ft, hp	76 at 2700
Cruise rpm at 7000 ft, hp	135 at 2450
Cruise rpm at 21,000 ft, hp	74 at 2450
Cruise rpm at 7000 ft, hp	126 at 2200
Cruise rpm at 21,000 ft, hp	70 at 2200
Cruise BSFC, lb/bhp·h	0.47

are called *turbochargers*. The advantage of the turbocharger over the gear-driven supercharger is twofold. First, the compressor does not extract power from the engine, but uses exhaust energy that would normally be wasted. Second, the turbocharger is able to provide sea level rated power up to much higher altitudes than the gear-driven type.

Figure 14.3 Lycoming 0-360A aircraft engine.

An example of a turbocharger designed for high-altitude operation is the one on the Boeing/DARPA Condor. The Condor was designed to fly at 65,000 ft, where the freestream air pressure is 1/18 that of sea level. The Condor used two Continental GTSIOL-300 piston engines (175 hp, six cylinders, reduction gearing, spark ignition, fuel injected, liquid cooled) each weighing 289 lb. There were two stages of turbocharging, each boosting the pressure 4.2:1 and cooling the air. Each turbocharger weighed 560 lb. Each engine drove an 81-lb, three-bladed, 16-ft propeller geared down 3:1 from the 2700 rpm engine speed. The propeller efficiency was reported as 90%.

The HAARP aircraft (from Section 5.8) uses a three-stage turbocharger to boost the pressure of the air going into its piston engines to 93:1 for operation at 100,000 ft. The HAARP turbocharger is designed for 108:1, giving it a little margin to operate past 100,000 ft. The turbocharger was designed by Teledyne Continental (TCM) and is shown schematically in Fig. J.1.

The practical limit for pressure boost across a turbocharger stage is about 5:1 for current compressor design and materials. Thus, the Condor needed a two-stage and HAARP a three-stage turbocharger. The temperature of the air is increased through each compressor stage and needs to be cooled before going into the next stage. The cooling requirement for one HAARP engine and turbocharger is as follows:

- Engine coolant, heat load 3380 Btu/min
- Engine oil, heat load 1450 Btu/min
- Turbocharger intercoolers, heat load 9900 Btu/min
- Generator and gearbox, heat load 600 Btu/min

The cooling system for these items comprises ram air-cooled heat exchangers located in the leading edge of the wing that weigh 1147 lb total for the two sides. The ram drag for the cooling system is estimated to be equal to 25% of the aircraft $C_{D_{min}}$. This greatly reduces the HAARP maximum L/D from 39 for a clean aircraft to the 27 reported in the example of Section 5.8.

Figure J.2 shows typical weights of the turbochargers, intercoolers, heat exchangers, and ducting as a function of maximum horsepower and altitude.

The engine for HAARP will be sized and selected in Chapter 18 (Section 18.10).

14.2.2 Turboprop Engines

The thermodynamics of the turboprop engine will be discussed in detail in the next section. This section discusses its characteristics as a propeller system.

The performance (hp/lb) of current turboprop engines is shown in Fig. 14.2. Turboprops are lighter than an equivalent piston engine with hp/lb of approximately 2.2–2.4 for all engines. The shaft on a turbine engine typically rotates at 10,000 rpm, a speed much too high for propeller operation. In most cases, the weights shown in Fig. 14.2 for the turboprop includes the weight of the reduction gearing required for a propeller speed of approximately 2000–2700 rpm. The BSFC is about 25% higher for turboprops than for a piston engine.

In a turboprop engine most of the power is extracted as shaft power to drive the propeller. However, there is a residual energy that is expanded through the nozzle as jet thrust T_J, which is not included in the listed shaft horsepower (SHP).

To account for the power produced by this jet thrust an *equivalent shaft horsepower* (ESHP) has been devised to account for the total power output of the engine. Using Eq. (14.4) the jet thrust is converted to a *thrust horsepower* by

$$\text{THP} = T_J V / (0.8)(550) \tag{14.6}$$

where the 0.8 accounts for a conventionally assumed 80% propeller efficiency. With this expression the ESHP may be written

$$\text{ESHP} = \text{SHP} + T_J V / (0.8)(550) \tag{14.7}$$

Notice that this relationship does not account for thrust horsepower under static conditions where $V = 0$. For such cases (and for $V < 100$ kt) another convention has been adopted to equate a given thrust level per horsepower. Some European turboprop companies use 2.6 lb of thrust per horsepower, but the usual equivalence is 2.5 lb of thrust equals one horsepower. Thus, for $V < 100$ kt,

$$\text{ESHP} = \text{SHP} + T_J / 2.5 \tag{14.8}$$

For example, the Honeywell (formerly Garrett) TPE 331–11 is rated statically at 1000 SHP and 1045 ESHP. This engine therefore produces a static thrust from the turbine exhaust of approximately 113 lb.

14.2.3 Electric Motors

Electric motors are simple and reliable (design life of 30,000 h when operated at ~60% rated power). They have a specific power of approx 0.27 hp/lb (0.2 kW/lb). Electric motors get their power from onboard auxiliary power units (APU; either piston or turboshaft engines driving

Table 14.2 Electric Aircraft System Data (2010)

Characteristic	Electric Motor	Solar Cell	Fuel Cell	Batteries
Specific energy (kW·h/lb)	0.2[a]	NA	0.89[b,c]	0.27[c,d]
Design life	30,000 h	[e]	NA	300[f]
Efficiency (%)[g]	97	28	55	90
Installed weight (lb/ft²)	NA	0.1	NA	NA

[a]Weight includes motor, controller, and propeller. Increase weight by 25% for installation.
[b]H_2/O_2 regenerative fuel cell using proton exchange membrane technology.
[c]Specific power based on discharge time.
[d]Li–S batteries are projected to increase to 0.336 kWh/lb by 2015.
[e]Solar cells degrade about 1.5% of power output per year.
[f]300 full-depth discharges in 2010. Decreasing the discharge to 50% would increase number of recharges to approximately 1000.
[g]Efficiency is energy out per energy in. Solar cell efficiency is projected to increase to 32% and fuel cell efficiency to 65% by 2015.

electric generators), batteries, fuel cells, or solar cells (photovoltaic cells that convert incident solar energy into electricity).

For missions having several day–night cycles the electric aircraft would need to be a solar-powered vehicle. It would collect solar energy from the sun during the day and convert it to electricity through the photovoltaic action of solar cells. It would need to store energy in batteries or fuel cells to power the vehicle during the night. The solar cells would then recharge the batteries or fuel cells for the next nighttime operation by collecting excess power during the day. Theoretically this cycle could go on forever; however, the batteries and fuel cells have finite recharging limits and their performance degrades over time [3]. Table 14.2 contains data on electric motors, solar cells, batteries, and fuel cells.

14.3 Operation of Turbine Systems

The *turbine engine* or *turbojet engine*, shown schematically in Fig. 14.4a, operates in a similar fashion to the other aircraft propulsion devices. Air is brought in the inlet and slowed down to approximately Mach = 0.4 at the face of the compressor. The air mass is compressed and pressure is built up (increasing pressure energy of fluid) as the air goes through the compressors with little change in velocity. The air is mixed with fuel in the combustor section, ignited, and burned, increasing the thermal energy of the air–fuel fluid mixture. The heated fluid expands in the turbine section, driving the turbines, which in turn powers the compressor section. The fluid is further expanded through the nozzle section to a high velocity (conversion of pressure and thermal energy into kinetic energy), thus increasing the

Turbojet with Afterburner

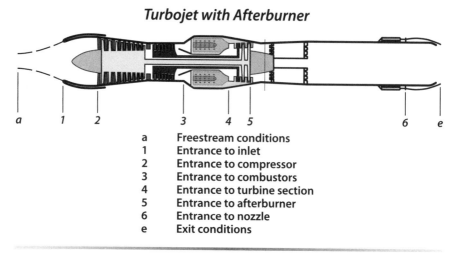

a	Freestream conditions
1	Entrance to inlet
2	Entrance to compressor
3	Entrance to combustors
4	Entrance to turbine section
5	Entrance to afterburner
6	Entrance to nozzle
e	Exit conditions

Figure 14.4a   Schematic of typical turbojet with afterburner.

momentum of the fluid and producing a thrust. Figure 14.4b shows the internal pressure variations inside a typical turbojet engine.

The efficiency of the turbine engine as a propulsion device depends on many factors. One of the major factors is the *compression ratio* of the air through the compressor [overall pressure ratio (OPR)], which is a function of the number of compression stages and their stage efficiencies. The effi-

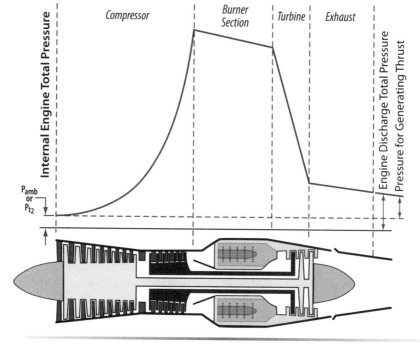

Figure 14.4b   Typical turbojet engine, internal pressure variation.

ciency of the compressor and turbine stages depends upon the blade geometry (number and shape), the ratio of blade length to hub, and the ratio of blade length to tip clearance. The operating temperature of the combustor and turbine determines the amount of thermal energy in the gas available for power extraction and expansion to jet velocity.

The *net thrust* produced by a turbojet engine is given by [from Eq. (14.2) for $A_a = A_e$]

$$T_n = \dot{m}_{\text{air}}(V_e - V_a) + (P_e - P_a)A_a \tag{14.9}$$

where

$\dot{m}$ = mass flow of air, in slugs per second
V = velocity of air, in feet per second
P = static pressure, in pounds per square foot

and the subscripts correspond to the station locations of Fig. 14.4a. Notice that the mass flow of the fuel is not included in the $\dot{m}$ term of Eq. (14.9). This is because the fuel flow is small compared to the air flow; also the weight of air leakage through the engine can be assumed to be approximately equivalent to the weight of the fuel consumed.

The *gross thrust* is defined as the product of the mass flow rate in the jet exhaust and the velocity attained by the jet after expanding to freestream static pressure

$$T_g = \dot{m}_{\text{air}}V_e$$

And the term $\dot{m}_{\text{air}}V_a$ is called the *ram drag*. For static operation, T_g and T_n are equal.

To enable an accurate comparison to be made between turbine engines, fuel consumption is reduced to a common denominator, applicable to all types and sizes of turbine engines. The term used is *thrust specific fuel consumption* (TSFC) and is expressed as

$$C = \text{TSFC} = W_f / T_n \tag{14.10}$$

where W_f is the fuel weight flow in pounds per hour and T_n is the net thrust in pounds.

Frequently, a turbojet engine is equipped with an afterburner for increased thrust. Roughly, about 25% of the air entering the compressor and passing through the engine is used for combustion. Only this amount of air is required to attain the maximum temperature that can be tolerated by the metal parts. The balance of the air is needed primarily for cooling purposes. Essentially, an *afterburner* is simply a huge stovepipe, attached to the rear of the engine, through which all of the exhaust gases must pass.

Fuel is injected into the forward section of the afterburner and is ignited. Combustion is possible because 75% of the air that originally entered the engine still remains unburned. The result is, in effect, a tremendous blow-torch, which increases the total thrust produced by the engine by approximately 50%, or more. Although the total fuel consumption increases almost two-and-a-half times, the net result is profitable for short bursts of aircraft speed, climb, or acceleration. A turbojet aircraft with an afterburner can reach a given altitude with the use of less fuel by climbing rapidly in "afterburning" than by climbing much more slowly in "nonafterburning." The weight and noise of an afterburner, which is used only occasionally on long flights, precludes the device being employed in present-day, transport-type aircraft.

The turbine engines shown in Fig. 14.5 are termed *two spool*. The shaft from the first stage of the turbine is hollow and drives the high-pressure stage of the compressor (called the *high spool*). The power shaft from the aft stages of the turbine runs through the hollow high spool shaft and drives the low-pressure stage of the compressor (called the *low spool*).

14.3.1 Turboprop

The *turboprop* (sometimes called a *turboshaft*) is essentially a turbojet designed to drive a propeller. The turboprop is shown schematically in Fig. 14.5 and uses the basic gas generator section of a turbine engine. The propeller operates from the same shaft as the low-spool compressor through reduction gearing. The hot gases are nearly fully expanded in the turbine first stage, which develops considerably more shaft power than required to drive the low-spool compressor and accessories. The excess power is used to drive a conventional propeller equipped with a speed-regulated pitch control. The remainder of the hot gases are expanded through the nozzle, providing a jet thrust as discussed in Section 14.2.2. This engine retains the advantage of having a light weight and low frontal area, together with the ease of installation that goes with turbojet engine design. In addition, it has a high efficiency at relatively low speeds. However, present propeller design limits the use of this type of powerplant to speeds below 500 kt (see Fig. 4.5 in Section 14.3.2).

14.3.2 Turbofan

The *turbofan* version of an aircraft gas turbine engine is shown in Fig. 14.5 and is the same as the turboprop, the geared propeller being replaced by a duct-enclosed fan driven at engine speed. One fundamental operational difference between the turbofan and the turboprop is that the airflow through the fan of the turbofan is unaffected by airspeed of the aircraft.

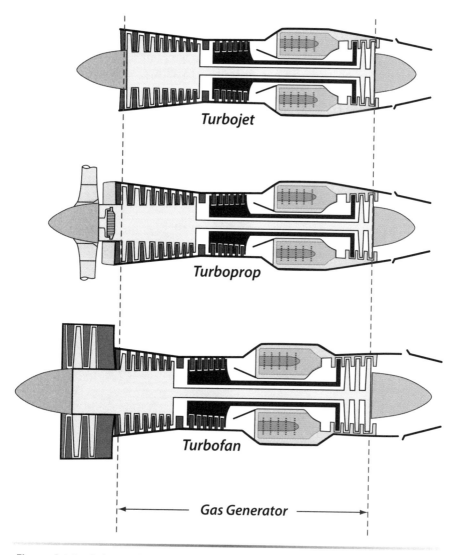

Figure 14.5 Schematic of typical turbojet, turboprop, and turbofan engines showing basic gas generator core.

This eliminates the loss in operational efficiency at high airspeeds, which limited the air-speed capability of a turboprop engine. Also the total air-flow through the turbofan engine is much less than that through the propeller of a turboprop. In the turbofan engine, 30% to 60% of the propulsive force is produced by the fan.

The *bypass ratio* (BPR) for a turbofan is defined as the ratio of the airflow through the fan to the airflow through the gas turbine core. Some modern turbofan engines have bypass ratios as high as BPR = 10 (see Table J.1).

The turbofan engine offers several advantages over a turbojet, such as better takeoff thrust for the same-weight engine and lower TSFCs at high-subsonic speeds (see Fig. 14.6). This advantage comes about because the turbofan can accelerate a higher airflow $\dot{m}_a$ to a lower jet velocity, giving a higher propulsive efficiency (see [1]) than a turbojet of equivalent weight and fuel flow. Figure 14.6 shows the influence of bypass ratio on the sea level static TSFC for current turbine engine technology. The turbofan's advantage decreases at high-subsonic and all supersonic speeds due to the higher drag associated with the larger frontal area.

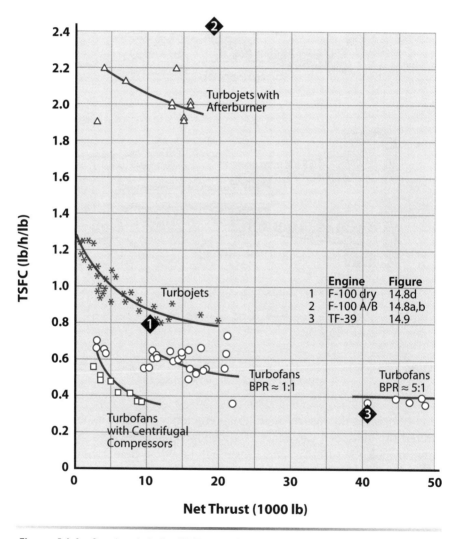

Figure 14.6   Sea level static (SLS) specific fuel consumption for turbojet and turbofan engines.

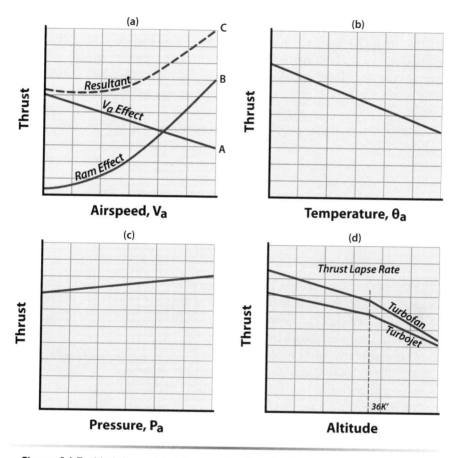

Figure 14.7 Variation of turbine engine thrust with airspeed, temperature, pressure, and altitude.

14.3.3 Factors Affecting Thrust and TSFC

As the aircraft increases its speed, the velocity of the air entering the engine, V_a, increases. The nozzle is usually close to a choked condition (V_e near the speed of sound, see [1]) such that V_e is relatively constant. Thus, the $(V_e - V_a)$ term in Eq. (14.9) decreases with increasing airspeed and the result is a decrease in thrust as shown by curve (A) in Fig. 14.7a. However, as V_a increases, the airflow into the engine, $\dot{m}_a$ = density × velocity × capture area, increases due to ram effect and the result is an increase in thrust as shown by curve (B) in Fig. 14.7a. The overall result of increasing airspeed is a combination of these two effects and is shown as curve (C) in Fig. 14.7a.

The ram effect is important, particularly in high-speed aircraft, because eventually, when airspeed becomes high enough, the ram effect will produce a significant overall increase in engine thrust. At Mach numbers greater than 3.0 the ram effect can replace the compressor sections of turbine engines, resulting in a ramjet engine. At subsonic speeds the ram effect is not very large and does not greatly affect engine thrust. At supersonic speeds, however, ram can be a major factor in determining how much thrust an engine will produce.

The most significant variable in the thrust equation is mass airflow, $\dot{m}_a$. Because

$$\dot{m}_a = \rho V A$$

and

$$\rho = p / R' \theta$$

from the perfect gas relation (R' is the characteristic gas constant) it can be observed that an increase in temperature will result in a decrease in thrust as shown in Fig. 14.7b. Similarly, an increase in pressure p will give an increase in thrust as shown on Fig. 14.7c.

As the aircraft climbs in altitude, the temperature decreases until at the tropopause (36,000 ft) it remains constant (see Appendix B). The pressure decreases steadily with increasing altitude. The result of climbing in altitude is an interplay between the pressure and temperature variations giving a decreasing thrust (called the *thrust lapse rate*) as shown in Fig. 14.7d. The lapse rate is greater for a turbofan than a turbojet. The variation of thrust with altitude can be approximated by

$$T_n = (T_n)_{SL} (p / p_{SL}) (\theta_{SL} / \theta) \qquad (14.11)$$

The TSFC for a turbine engine is given by Eq. (14.10), where the fuel flow to the engine is dependent on the throttle position. Thus, for a constant throttle setting (i.e., military continuous power) the TSFC varies with thrust.

The optimum altitude for subsonic cruise is that altitude where the TSFC is a minimum for a cruise power setting. For a turbine-powered aircraft cruising near Mach 0.8 the cruise power setting is around 80%–100% of normal rated thrust. The best altitude for cruise under these conditions is around 36,000 ft [4].

14.3.4 Turbine Engine Data

Appendix J contains information on the current stable of turbojet, turboprop, and turbofan engines. The engine characteristics in Table J.1 do not reflect thrust losses and weights associated with installing the engines into aircraft. Turbine engine corrections for installation into aircraft are discussed in Chapter 16. Reference [5] is an excellent source of turbine engine data. It is published annually and is kept up to date.

Table 14.3 presents the characteristics for the Pratt and Whitney F-100-PW-100 afterburning turbofan. Figure 14.8 presents the variation of thrust and TSFC with altitude, airspeed, and throttle setting for the engine. The thrust shown in Fig. 14.8 is the installed thrust, which is the net thrust T_n from the basic engine corrected for inlet and nozzle losses, airflow bleed, and turbine power extraction. Figure 14.8g presents the mass airflow $\dot{m}_a$ required for the F-100 in afterburner and military power.

Table 14.4 presents the characteristics for the General Electric TF-39-GE-1 turbofan. The installed engine data are shown in Figure 14.9 and are appropriate to a podded nacelle installation similar to that of the C-5A [6]. Figure 14.10 shows the GE CF-6 engine, which is a popular commercial engine. CF-6 engines were produced in many models providing power from DC-10s during the 1980s to today's Boeing 747/767/777.

Table 14.3 Pratt and Whitney F-100-PW-100 Afterburning

Turbofan Characteristics	
Sea level static thrust	23,000 lb (uninstalled)
Sea level static TSFC	2.248
Bare engine weight	2737 lb
Sea level static airflow, $\dot{m}_a$	217 lbm/s
Engine length (including nozzle)	190 in.
Maximum diameter	44 in.
Compressor face diameter	40 in.
Bypass ratio	0.71
Miscellaneous: Accessory Equipment Weight	
Fuel system	433 lb
Engine controls	22 lb
Starting system	28 lb

The installed engine data of Fig. 14.8 reflects the following propulsion unit corrections:

1. Power extraction of 70 hp to drive electric generators and auxiliary equipment. This 70 hp is at all power settings and flight conditions.
2. Normal shock inlet pressure recovery.
3. Nozzle corrections at moderate pressure ratios.
4. High-pressure bleed air extracted from compressor for operating environmental control system.

The bleed airflow rate is 0.4 lb/s.

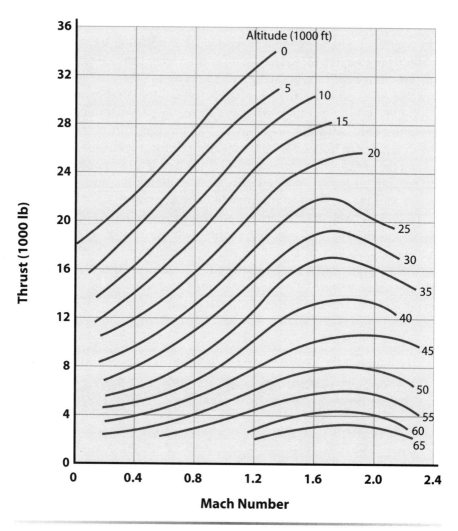

Figure 14.8a   F-100 installed thrust, maximum afterburning.

Table 14.4   General Electric TF-39-GE-1
Turbofan Characteristics

Sea level static thrust (uninstalled)	41,100 lb
Sea level static TSFC	0.315
Sea level static airflow, $\dot{m}_a$	1541 lb/s
Bare engine weight	7026 in.
Engine length	271 in.
Engine diameter (maximum)	100 in.
Bypass ratio	8
Overall pressure ratio	26

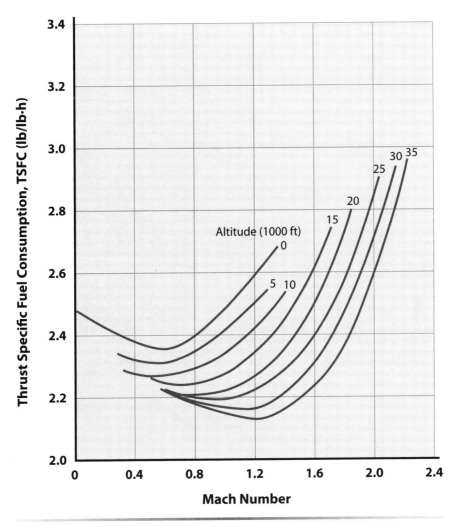

Figure 14.8b F-100 TSFC for maximum afterburning (low altitudes).

14.4 Ramjet Engine Operation

The *ramjet* operates on essentially the same gas cycle as the turbine. The ramjet is a very simple device and is shown schematically in Fig. 4.6. However, all the compression portion of the cycle occurs at the inlet and in the diffuser where the incoming air velocity is decreased producing a rise in static pressure. Fuel is burned and the mixture expanded to ambient through a nozzle. The ramjet is compared with other propulsion devices in Fig. 4.7.

At forward speeds of Mach ≤1.0 the ramjet is prohibitively expensive to operate because its low combustion efficiency results in a TSFC greater

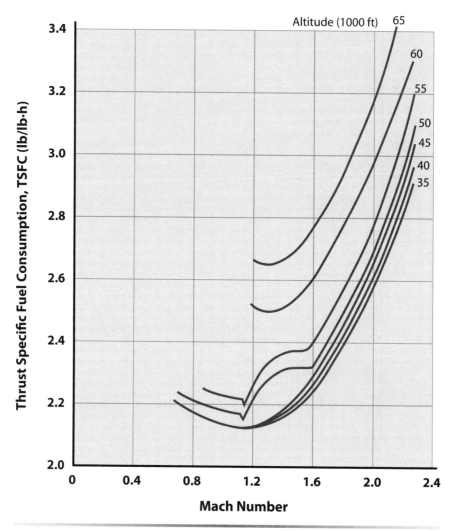

Figure 14.8c   F-100 TSFC for maximum afterburning (high altitudes).

than 6.0 (see Fig. 4.7c). At supersonic speeds a normal shock is located just ahead of the inlet. A normal shock compression, although not ideal is a practical substitute for a compressor. At Mach = 2.0 the shock compression ratio is about 4.5 and the ramjet TSFC is competitive with an afterburning turbojet. Above Mach = 2 the ramjet starts rivaling the dry turbojet. Thus, at Mach ≥2 the normal shock compression is an acceptable substitute for a mechanical compressor making the ramjet a very light and simple machine. Because no turbine is present the usable temperature limits are considerably higher than for a turbojet.

One of the major problems connected with the ramjet is the issue of flame stability. The high speed of the air through the duct tends to blow out the combustion. The art of the ramjet is in the design of a flame holder that will stabilize the combustion but produce minimum resistance to the flow.

14.5 Rocket Operation

All of the propulsion devices considered thus far depend upon atmospheric air and, to some extent, forward speed for their operation. Rockets,

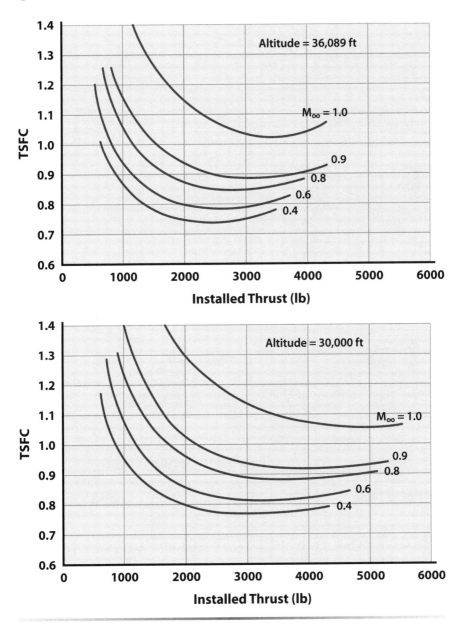

Figure 14.8d F-100 TSFC for partial power settings (nonafterburning).

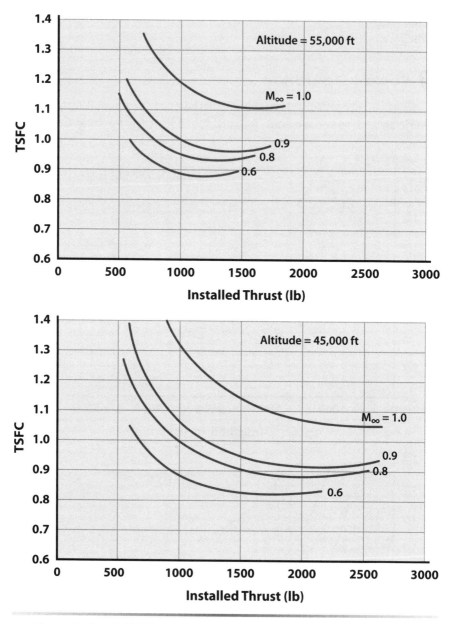

Figure 14.8e F-100 TSFC for partial power settings (nonafterburning).

however, are independent of atmospheric air or forward speed. The atmospheric independence provides an advantage in that the rocket offers the only method of developing thrust outside of the earth's atmosphere. However, this independence is also a disadvantage in that all the mass creating the thrust must be carried in the rocket. Note that all of the propulsion

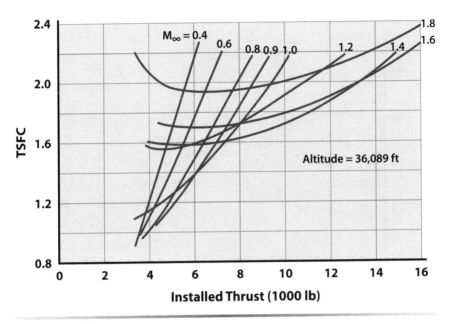

Figure 14.8f F-100 TSFC for partial afterburning.

devices discussed earlier carried only their fuel and that most of the mass accelerated rearward for thrust consisted of the ambient air.

The thrust of a rocket is expressed as

$$T = \dot{m}_{CP} V_e + A_e \left(P_e - P_a \right) \tag{14.3}$$

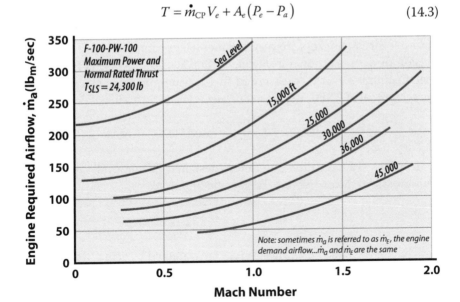

Figure 14.8g Required mass flow for PW F-100 turbofan engine at maximum power.

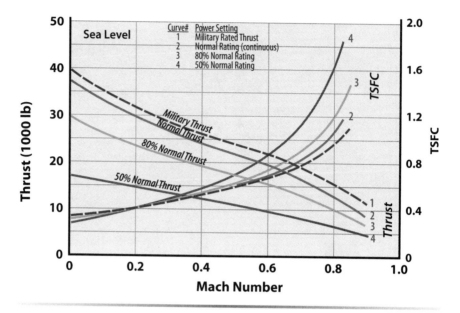

Figure 14.9a Installed thrust and TSFC for TF-39 turbofan engine (see Table 14.4).

The exhaust velocity depends on the composition of the propellant, the design of the exhaust nozzle, and the ambient conditions. The thrust specific fuel consumption of a rocket is TSFC = propellant weight flow (lb) per hour per thrust. Rockets are very fuel inefficient compared to all other propulsion devices, as shown in Fig. 4.7c. The World of Rockets likes to use specific impulse I_{SP} as a measure of fuel consumption. Specific impulse is the reciprocal of TSFC, expressed in seconds and written as

$$I_{SP} = T / g\dot{m}_{CP} = 3600 / TSFC \tag{14.12}$$

In space (vacuum) for a perfectly designed nozzle (full expansion to zero pressure) the expression for specific impulse is

$$I_{SP} = V_e / g = V_e / 32.17 \tag{14.13}$$

The highest specific impulse values are obtained by using hydrogen as a fuel and burning it with either oxygen or fluorine. At sea level with a combustion chamber operating at 500 psia the specific values are

$$\text{Hydrogen} + \text{Fluorine} \rightarrow I_{SP} = 375 \text{ seconds}$$
$$\text{Hydrogen} + \text{Oxygen} \rightarrow I_{SP} = 362 \text{ seconds}$$

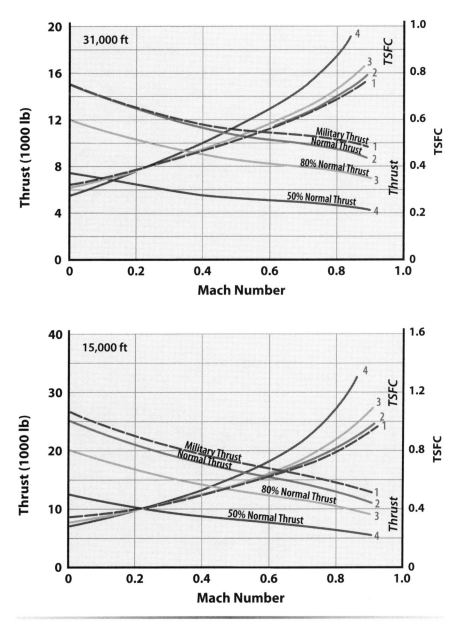

Figure 14.9b Installed thrust and TSFC for TF-39 turbofan engine (see Table 14.4).

The combinations of hydrogen and oxygen or fluorine are difficult to handle so that modern rockets use more modest fuel–oxidizer combinations, including solid propellants. The current space rockets have specific impulses at sea level of 200–300 seconds.

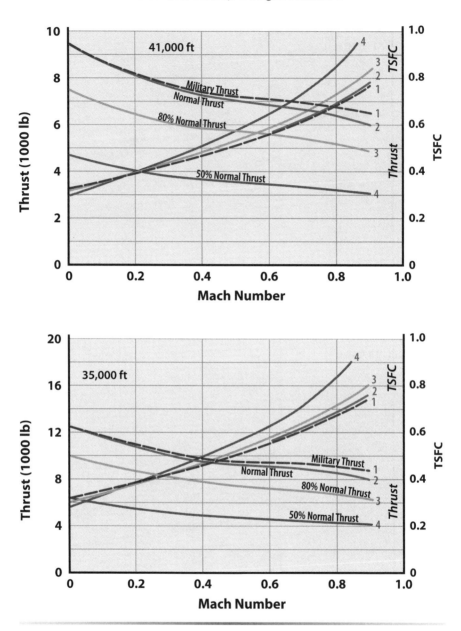

Figure 14.9c   Installed thrust and TSFC for TF-39 turbofan engine (see Table 14.4).

Whittle and von Ohain Change Aviation

Hans von Ohain and Frank Whittle developed the jet turbine engine about the same time in the 1930s but completely independent of one another—Ohain in Germany and Whittle in England. Whittle did his graduate work at

Figure 14.10 CF-6 engine based on the TF-39 design.

Cranwell College as an RAF Flight Officer. His field of study was a new type of gas turbine and he was granted a patent in 1930. His RAF duties and lack of money prevented any serious development of the jet engine until 1937. The British Air Ministry was slow in recognizing the potential of the jet turbine but finally contracted with Whittle for an engine and with Gloster Aircraft for a jet engine powered aircraft in 1939. The Gloster E.28/39 flew on 15 May 1941 with Whittle's jet engine.

Hans von Ohain did his graduate work at the University of Göttingen and received a doctorate in physics and a patent for his jet engine concept in November 1935. Unlike Whittle, Hans was a man of means and hired a mechanic to build a working model of his concept. Ernst Heinkel (Heinkel Aircraft Co.) was impressed with the model and hired Hans in March 1936 to develop a jet turbine engine. A prototype jet engine was developed and ran successfully on hydrogen gas in March 1937. Heinkel was pleased with von Ohain's success and commissioned him to develop a flightworthy, kerosene-fueled engine to power the Heinkel He-178 aircraft shown in Fig. 14.11. Hans developed a jet engine with 992 lb of thrust; it flew in the He-178 on 27 August 1939 and changed the world forever.

After World War II Whittle's engine was produced by several companies in England and the United States. Frank Whittle was knighted by King George VI of England in August 1948. He eventually immigrated to the United States, where he became a research professor at the U.S. Naval Academy in Annapolis, Maryland. He died on 9 August 1996.

(continued)

After World War II Hans came to the United States as part of Operation Paper Clip. He was assigned to Wright–Patterson AFB, Dayton, Ohio, as a propulsion consultant, then as Chief Scientist to the Propulsion Laboratory, and finally to the Aeronautical Research Laboratory. He retired from government service in 1979 and continued as a consultant to the University of Dayton Research Institute. He and his family settled in nicely in midwestern suburbia, living in Centerville (south of Dayton).

I met Hans in 1972 while on active duty at Wright–Patterson AFB, and our friendship flourished until his death in 1998. He was very gracious with his advice and was an annual visitor to my aircraft design short course in Dayton from 1975 to the mid-1990s. Hans and Sir Frank changed the aviation world with their invention of the jet engine. Hans was a technical giant, a true gentleman, and very humble.

Leland Nicolai

Figure 14.11 Heinkel He-178 aircraft designed for the first jet engine.

References

[1] Hill, P. G., and Peterson, C. R., *Mechanics and Thermodynamics of Propulsion*, Addison Wesley, Reading, MA, 1965.

[2] McCormick, B., *Aerodynamic, Aeronautics and Flight Mechanics*, Wiley, New York, 1995.

[3] Mitlisky, F., Weisberg, A., and Myers, B., "Regenerative Fuel Cells," Lawrence Livermore National Laboratory, UCRL-JC-134540, June 1999, paper prepared for the U.S. Department of Energy Hydrogen Program 1999 Annual Review Meeting, Lakewood, CO, 4–6 May 1999.

[4] "The Aircraft Gas Turbine Engine and Its Operation, Pratt and Whitney Operating Instruction 200," United Aircraft Corp., East Hartford, CT, Nov. 1960.

[5] *Aviation Week and Space Technology, Annual Aerospace Source Book*, McGraw-Hill, New York (published annually in January).

[6] Smith, P. R., "C-5A Aerodynamic Substantiating Data Based on Flight Test," Rept. LGIC22-1-1, Lockheed-Georgia Co., Lockheed Aircraft Co., Marietta, GA, 16 Aug. 1971.

Chapter 15 Turbine Engine Inlet Design

- Pitot or Normal Shock Inlet
- External Compression Inlet
- Mixed Compression Inlet
- SR-71 Inlet Operation
- Mass Flow Ratio
- Total Pressure Recovery
- Inlet Examples

The SR-71 mixed compression inlet (left) and the F-22 external compression inlet (right). The SR-71 inlet design and operation is discussed in detail in this chapter. See Example 15.1 for an external compression inlet design similar to the F-22.

The inlet messes up an otherwise clean design!

An aerodynamicist

15.1 Introduction

T he primary purpose of an inlet is to supply the correct quantity and quality of air to the compressor of the engine. The correct mass flow of air must be delivered to the compressor face at a Mach number of about 0.4. The mass flow must also be delivered with an acceptable velocity distribution across the engine face and with minimum loss in the total energy content of the air. In addition, the inlet is required to do this at all flight conditions and at least weight, cost, and drag. The installation of a turbine engine on an aircraft is a most challenging task.

A typical subsonic turbine engine installation consists of a high-compression engine with a short fixed inlet and probably a variable convergent nozzle. The supersonic installation, on the other hand, requires a powerplant with a sophisticated variable geometry inlet having its own automatic control system and a fully variable convergent–divergent (C–D) nozzle in order to extract the full performance from the engine throughout the speed range. A typical subsonic and supersonic installation is shown in Fig. 15.1. Notice that the supersonic inlet is more than two engine diameters in length as opposed to one diameter for the subsonic installation. The complications of the supersonic inlet and its influence on weight will be discussed in this chapter.

Subsonic Installation

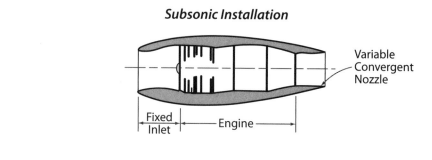

Supersonic Installation

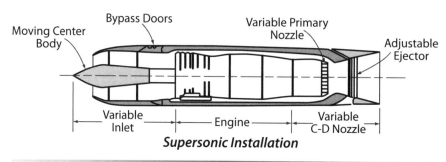

Figure 15.1 Typical subsonic and supersonic powerplant installations.

The performance of the inlet is related to the following four characteristics:

- Total pressure recovery
- Quality of airflow—distortion and turbulence
- Drag
- Weight and cost

The overall value of an inlet must always be determined by simultaneously evaluating all four characteristics because the gain in one is often achieved at the expense of another. It should be kept in mind that the most serious aspect of the engine–inlet problem is concerned with off-design operation; none of the first three characteristics should deteriorate rapidly under conditions of varying power settings and angles-of-attack. As a result, in actual vehicles many compromises have to be made to achieve an acceptable performance throughout the variations in flight Mach number, angles-of-attack, and sideslip as well as variations in the properties of the atmosphere.

15.2 Pressure Recovery and Inlet Types

A supersonic airflow entering an inlet is decelerated through a shock wave or series of shock waves to a subsonic value. The flow is further decelerated in the subsonic diffuser (the diverging section of the inlet between the throat and the compressor face) to a value of about Mach = 0.4 at the compressor face.

The *total pressure recovery* of the inlet is defined as the ratio of the total pressure at the compressor face to that of the freestream:.

$$\eta_R = \text{total pressure recovery} = P_{0_c} / P_{0_\infty}$$

The total pressure recovery of the inlet is an important measure of the inlet performance. It is desired to recover as much of the total pressure at the compressor face as possible (high value of η_R) because the total pressure of the freestream represents the available mechanical energy of the flow that can be converted into a static pressure increase as the flow is decelerated. A large static pressure is desirable at the compressor face because then the compressor section of the turbine engine does not have to be as large in order to compress the flow to the required pressure for combustion. Total pressure is lost due to the viscous dissipation (friction) in the shock waves, the boundary layer, and separated flow regions. The gradual deceleration of a supersonic stream to subsonic through a series of oblique (or conical) shocks prior to the final normal shock is less costly in terms of total pressure loss than a rapid deceleration through a normal shock. The maximum total pressure recovery for different numbers of shocks in an optimum shock wave system is shown in Fig. 15.2. The curve for the single shock is from the normal shock data of Appendix D.

Inlets are of three types, characterized by their shock wave system: pitot or normal shock; external compression; and mixed compression. These three inlet types are shown in Fig. 15.3. The simplest type is the pitot inlet, with the supersonic compression being achieved through a normal shock and further compression carried out in the subsonic diffuser (Fig. 15.3a). The pitot inlet is simple, short, lightweight, and low cost. It gives tolerable total pressure recoveries up to about Mach 1.6. For aircraft having top speed requirements up to about Mach = 1.6, such as the F-16 and F-18 (and needless to say all subsonic aircraft), the pitot inlet is the best arrangement.

For speeds above Mach = 1.6, the flow needs to be decelerated gradually through one or more oblique shocks before the final deceleration through the normal shock. The external compression inlet, Fig. 15.3b, accomplishes the flow compression external to the inlet throat. The external ramp (with flow deflection angle θ_R) can be variable to put the oblique shock on the cowl lip for a variety of different Mach numbers. This "shock-on-lip" is called the *design condition* and will be discussed later. The desired operation is with the normal shock located at the inlet throat. This inlet provides tolerable pressure recoveries up to about Mach = 2.5.

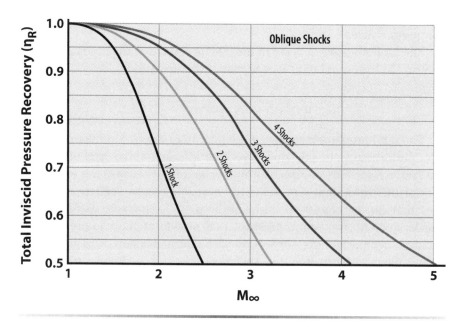

Figure 15.2a Maximum inviscid total pressure recovery—optimum oblique shock system.

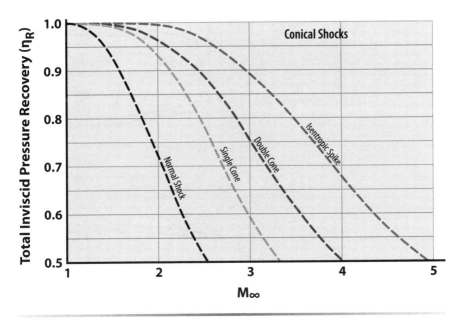

Figure 15.2b Maximum inviscid total pressure recovery—optimum conical shock system.

At freestream Mach numbers above Mach = 2.5 the inlet must provide a multiple shock system and would be a *mixed compression inlet* (Fig. 15.3c). Here again the external ramp can be a series of ramps (or cones) providing a series of external oblique shocks. The shock system continues into the supersonic diffuser, with the normal shock located in the subsonic diffuser. The location of the normal shock is dependent upon the backpressure at the compressor face. The ideal location for the normal shock is just slightly downstream of the throat to minimize the strength of the normal shock and the total pressure loss across it. However, this position is very sensitive to the backpressure. Any perturbation downstream can cause the normal shock to "pop out" of the diffuser and move to a position forward of the inlet lip, "unstarting" the inlet. The mixed compression inlets usually have bypass doors (see Fig. 15.6, Section 15.2.1) in the subsonic diffuser to control the backpressure and thereby the location of the normal shock. These vents can also be used to bypass the excess air in the inlet that cannot be accommodated by the engine. If this excess air is not bypassed it must be spilled ahead of the inlet, causing the mixed compression inlet to unstart.

The mixed compression inlet, sometimes called an *internal contraction inlet*, must have a variable geometry feature to obtain peak performance. The compression ramps must be able to collapse (fold down), giving a ratio of throat area A_T to cowl area A_c of about 0.8 in order to "swallow" the

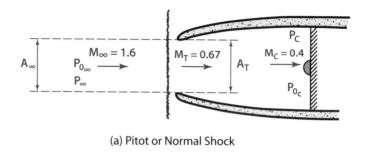

(a) Pitot or Normal Shock

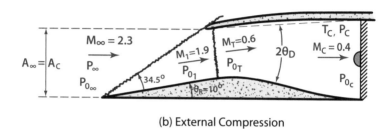

(b) External Compression

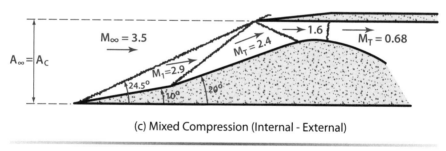

(c) Mixed Compression (Internal - External)

Figure 15.3 Types of inlets operating at supersonic "design" Mach numbers.

normal shock and locate it in the subsonic diffuser. Once the inlet is started, the throat area is decreased to an A_T/A_c of about 0.4 or less (dependent on Mach number, see [1]), which locates the normal shock just downstream of the throat for minimum total pressure loss.

Inlets can be *two-dimensional* with compression ramps as shown in Figs. 15.3b and 15.3c or *axisymmetric* with conical centerbodies as shown in Fig. 15.6 (Section 15.2.1). Axisymmetric inlets have a slight advantage over two-dimensional inlets in terms of weight and pressure recovery. Round ducts can usually be made lighter than rectangular ducts to take the large internal pressures. Also, the total pressure loss across a conical shock is less than across an oblique shock for the same upstream Mach number and flow deflection angle.

The determination of the total pressure recovery for an inlet is accomplished by examining the shock wave system and subsonic diffuser separately. Each shock is considered independently, with its characteristics determined by the flow deflection angle and upstream Mach number. The characteristics for oblique and conical shocks are presented in Appendix E and for normal shocks in Appendix D. The total pressure recovery for the shock wave system is the product of the individual total pressure ratios across each shock. Appendix Figs. E.6 and E.7 present the pressure recovery for cone inlets and Appendix Figs. E.8 and E.9 for ramp inlets.

The total pressure loss in a subsonic diffuser is dependent upon the diffuser geometry, throat Mach number, and the presence of a normal shock ahead of the diffuser entrance. Figure 15.4 shows an empirically determined [2] diffuser loss coefficient, ε, as a function of throat Mach number M_T and the ratio of diffuser length to throat height, L_D/H_T. The presence of a normal shock ahead of the diffuser entrance aggravates the boundary layer growth and tendency for the flow to separate, resulting in an increased diffuser loss coefficient. Figure 15.4 indicates that the designer should avoid short and long subsonic diffusers. Short diffusers, L_D/H_T of 4 or less, tend to cause flow separation, and long diffusers result in large friction losses. Long diffusers are heavy and should be avoided for that reason also. The ratio of the total pressure at the compressor face to that at the diffuser entrance, P_{0_c}/P_{0_T} is determined from Fig. 15.5.

15.2.1 SR-71 Mixed Compression Inlet Operation

The SR-71, shown on the Chapter 2 cover page, has two Pratt and Whitney J58 afterburning turbojet engines (34,000 lb SLS thrust). At this point the reader would be well served to review the SR-71 Case Study in Volume 2. The operating speed and altitude (Mach = 3.2 at 85,000 ft) of the SR-71 dictated a variable-geometry, mixed compression inlet. Figure 15.6 shows the axisymmetric mixed compression inlet on the SR-71. Figure 15.6b (from [3]) shows the position of the centerbody spike and bypass doors to locate the shock structure at four Mach numbers. This inlet design achieved a total pressure recovery of 78% at maximum speed and altitude.

The inlet control system operates to supply a flow of air, at correct pressure and velocity, to the engines throughout the flight envelope. The system includes the centerbody spikes, which are translated fore and aft to capture and retain the normal shock, and forward bypass doors, which operate to assist the spikes in positioning the normal shock. The system is normally

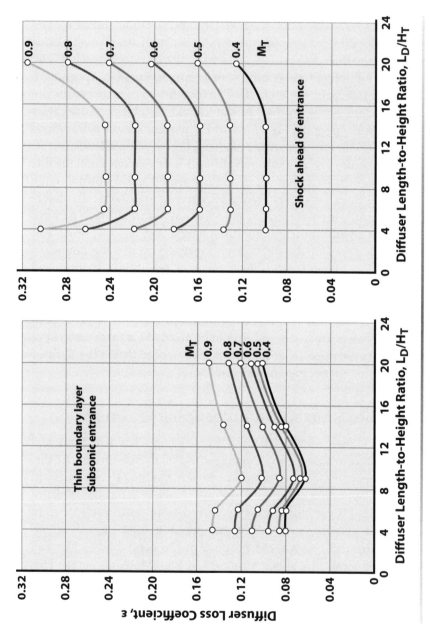

Figure 15.4   Effect of diffuser length on diffuser loss coefficient (data from [2]).

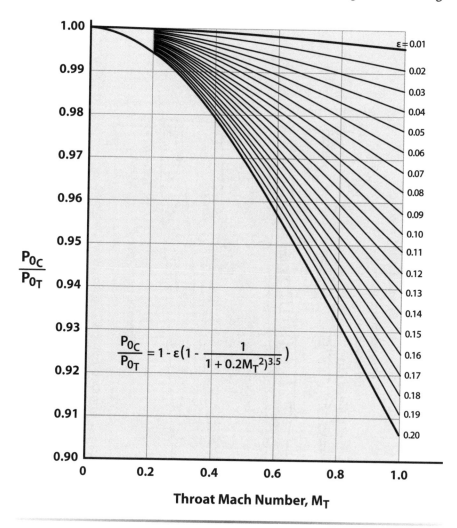

Figure 15.5 Total pressure recovery loss factors for subsonic diffuser.

operated in the automatic mode; however, it can be manually controlled by the pilot.

In operation the spikes are moved forward and aft in the inlet duct as a function of Mach number, varying the size of the inlet throat area and position of the conical shockwaves and the single normal shock. The forward bypass doors are modulated to control inlet duct static pressure and therefore fine tune the location of the normal shock in the inlet throat. The doors operate to prevent excessive duct air pressure.

Operation of the spike and bypass doors, and the resulting airflow patterns, is shown in Fig. 15.6b. At altitudes below 30,000 ft and speeds less

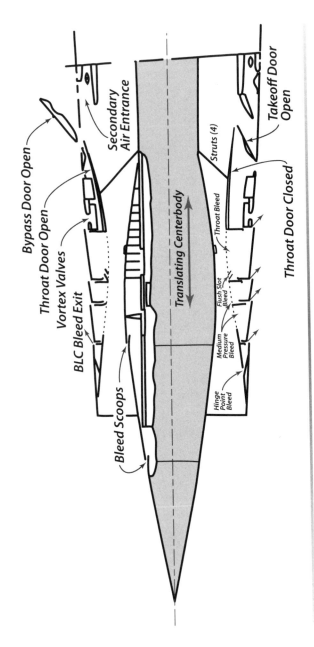

Figure 15.6a Typical supersonic mixed compression axisymmetric inlet.

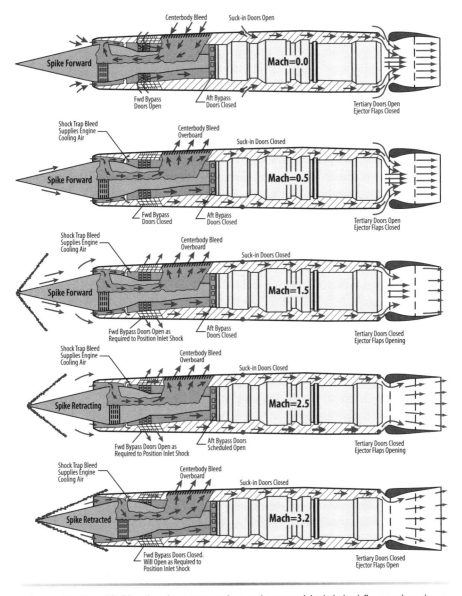

Figure 15.6b   SR-71 mixed compression axisymmetric inlet airflows at various Mach numbers.

than Mach = 1.4, the spike is locked fully forward. At altitudes above 30,000 ft the spike begins to move aft as the speed increases above Mach 1.6. The spike is automatically scheduled aft as a function of Mach number biased by angle-of-attack, sideslip, and vertical acceleration. Aft movement of the spike properly positions the conical and normal shocks relative to the inlet and increases the inlet contraction ratio (inlet area versus throat area). At Mach = 3.2, the spike has moved 26 inches aft of its full forward position and the captured stream tube area has increased 112%, from 8.7 to 18.5 ft^2 , and the throat area has closed down 54% ,from 7.7 to 4.16 ft^2. A peripheral "shock trap" bleed slot, around the inside of the inlet duct just forward of the throat, removes duct boundary layer air, which is ducted aft and exhausted through the ejector nozzle. Spike boundary layer air is removed at a porous bleed section in the spike surface at its maximum diameter. This air is ducted through the spike body and supporting struts to be exhausted overboard through nacelle louvers.

The forward bypass doors consist of two concentric, annular bands, located just aft of the inlet throat. The outer band is rotated slightly about the stationary inner band so that the rectangular openings in the two bands are shifted from doors fully open to fully closed. The doors are fully open when the landing gear is extended and fully closed when the gear is retracted. The doors remain closed until Mach = 1.5 is reached above 30,000 ft, at which point they modulate open or closed as a function of inlet pressure and a pressure ratio schedule to keep the normal shock near the inlet throat. The schedule is followed by comparing the ratio of internal duct pressures with external pressures sensed by pitot probes on the nacelle exterior surface. Any difference between the sensed pressure ratio and the pressure ratio schedule is used as a signal to drive the doors more open or closed.

Inlet unstart is a sudden pressure change in the inlet that results in the normal shock moving forward from its controlled position just aft of the throat—or even expelled from the inlet entirely. This condition reduces the affected engine thrust significantly, causing an asymmetric thrust condition and a violent yaw. Pilots are often slammed against the side of the cockpit, causing their helmets to break.

An unstart causes a sudden decrease in the static pressure near the compressor face. The pressure change initiates an auto restart procedure that moves the spike forward and opens the forward bypass doors in about three seconds. The forward movement of the spike reduces the inlet throat contraction and opening the bypass doors reduces the backpressure, thereby accelerating the airflow and returning the normal shock to its desired position in the inlet throat. The spike and forward bypass doors slowly return to their scheduled positions over a period of 10 seconds. At speeds above Mach = 2.3, the restart operation is applied to both inlets even if a problem is sensed on only one side. This reduces the yaw due to the asymmetric thrust and also prevents a sympathetic unstart of the other inlet which ofttimes occurs during a severe yaw condition. Restart crosstie is not in effect below Mach = 2.3, allowing independent inlet restart operation at slower speed.

The SR-71 mixed compression inlet is truly spectacular—and the fact that it was developed in 1961 before the use of modern computers makes it even more so.

15.3 Capture-Area Ratio or Mass-Flow Ratio (Supersonic Flow)

In Fig. 15.7 the area A_c is the cross sectional area of the inlet and is called the *cowl capture area*. The area A_∞ is the cross-sectional area of the freestream tube of air entering the inlet. The cross-sectional areas A_{∞_I} and A_{∞_E} are defined on Fig. 15.7 and are equal to A_∞ under different conditions. When the engine demand for air, $\dot{m}_E$, and the inlet supply of air, $\dot{m}_I$, are equal, $A_{\infty_I} = A_{\infty_E}$, the engine and inlet are said to be *matched*. When the engine and inlet are not matched, the excess air can either spill around the

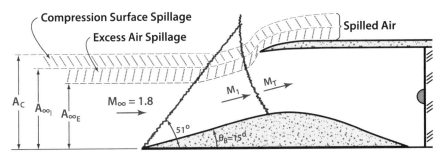

a. Excess Air Spilled (Subcritical Operation)

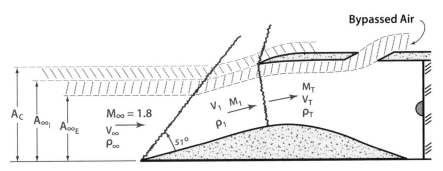

b. Excess Air Spilled (Critical Operation)

From Continuity: $\dot{m}$ = constant

For b: $\rho_\infty V_\infty A_\infty = \rho_1 V_1 A_T = \rho_T V_T A_T$

Engine demand: $\dot{m}_E = \rho_\infty V_\infty A_{\infty E} = \dot{m}_a$

Inlet supply: $\dot{m}_I = \rho_\infty V_\infty A_{\infty I}$

Excess Air: $\dot{m}_X = \dot{m}_I = \dot{m}_E$

Figure 15.7 Two schemes for inlet–engine flow matching.

cowl lips, expelling the normal shock from the inlet throat as shown in Fig. 15.7a, or the excess air can be bypassed through bypass doors in the subsonic diffuser as shown in Fig. 15.7b. When the excess air is spilled, $A_\infty = A_{\infty_E}$ and when the excess air is bypassed, $A_\infty = A_{\infty_I}$. These definitions of A_{∞_I} and A_{∞_E} will be useful when discussing spillage drag and bypass drag.

The *capture area ratio* is defined as A_∞/A_c. The ratio of the mass of air that enters the inlet to the maximum mass that could enter is called the *mass flow ratio*,

$$\frac{\dot{m}_\infty}{\dot{m}_c} = \frac{\rho_\infty V_\infty A_\infty}{\rho_\infty V_\infty A_c} = \frac{A_\infty}{A_c} \tag{15.1}$$

which is the same as the capture area ratio.

As the inlet mass flow ratio changes, the normal shock position and total pressure recovery ratio change, which is shown in Fig. 15.8. The inlet is designed to operate with the oblique shock crossing the lip of the inlet at the required mass flow of the engine (called the design mass flow) as shown in Figs. 15.3b, 15.3c, and 15.8b. At this condition (point B in Fig. 15.8) the normal shock is at the throat, giving a maximum value of mass flow and pressure recovery. This is referred to as the *critical condition*. As the engine demand for air becomes less (i.e., throttling the engine back for cruise) the normal shock is expelled forward to allow the excess air to be spilled over the outside of the lip (assuming no bypass facility). A portion of the air

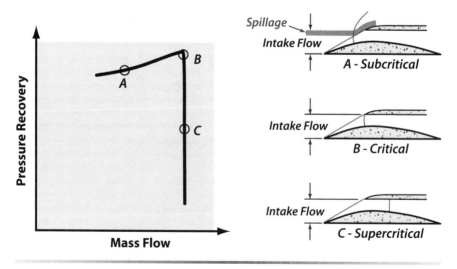

Figure 15.8   Mass flow–pressure recovery characteristic (data from [4]).

entering the inlet in this condition passes only through the single shock formed by the intersection of the normal and oblique shocks. As a result, this air enters the diffuser at a lower total pressure than the air that passes through the oblique and normal shocks. Therefore, there is a lower pressure recovery at this point. This condition is called *subcritical operation* and is point A on Fig. 15.8.

If the inlet is operating critically and the engine suddenly demands more airflow, the backpressure in the inlet is decreased and the normal shock moves back into the subsonic diffuser, becoming stronger as the Mach number in front of it increases. The mass flow cannot increase because the inlet is choked and the result is a reduction in pressure recovery to bring the engine airflow demand down to the inlet capacity. Because the engine cannot get all the airflow it needs, it is said to be *starved* and the inlet operation is termed *supercritical* (point C on Fig. 15.8).

15.4 Variable-Geometry Inlets

The *design Mach number* M_D is that flight vehicle speed that is critical in terms of mission performance. It might be the cruise speed, weapon delivery speed, or a maximum speed. The inlet is designed for this design Mach number to give high η_R, shock-on-lip operation, and $A_{\infty I}$ matched to the engine demand so that spillage and bypass drag due to excess air are minimal. The performance of a fixed-geometry inlet deteriorates rapidly at Mach numbers other than M_D. For flight speeds above M_D, the engine demand airflow is usually greater than the supply airflow and the engine is starved. For flight speeds below M_D the shock is off the cowl lip ($A_\infty < A_c$), giving rise to compression surface spillage of air (Fig. 15.7) and a resulting additive drag. Also, when the flight speed is off the design Mach the inlet throat area may not be sized properly for the incoming airflow.

Variable geometry can provide a resolution to some of these problems but at the expense of increased inlet complexity, weight, and cost. The designer must weigh the pros and cons of the variable-geometry inlet and decide the best compromise.

The inlet might incorporate a variable angle compression ramp or centerbody to keep the shock-on-the-lip at off-design Mach numbers. The ramp or centerbody can also translate to keep the shock-on-the-lip as well as providing a variable throat area. The inlet capture area A_C might also be varied through a hinged cowl lip to provide a better match of engine demand airflow with supply at a variety of flight speeds.

As an engine will accept only a certain amount of air, the excess air must be diverted to the freestream as efficiently as possible. This requires still further variation in geometry. One solution is to increase the compression ramp angle, θ_R, which moves the shock off the cowl lip, diverting the excess air over the lip of the cowl. This results in an additive drag (sometimes called *compression surface spillage* or *critical spillage drag*) but still lower losses than if the shock were left on the lip and the excess air permitted to back up in the inlet and spill around the lip (Fig. 15.7a). Another solution is to provide bypass vents or doors in the subsonic diffuser to bypass the excess air into the freestream (Fig. 15.7b). This approach will produce a bypass drag due to the pressure drag on the spill vents and the momentum change of the bypassed air between the inlet and the spill vent exit. Normally, facilities for both methods of getting rid of the excess air should be provided and compromise settings chosen that will result in a minimum drag for various flight Mach numbers. These two engine–inlet matching schemes are shown in Fig. 15.7.

The variable throat area feature of mixed compression inlets has already been discussed in Section 15.2.1. This variable throat area is necessary in order to swallow the normal shock in the case of an inlet unstart and then position the normal shock for best pressure recovery. The bypass doors are the main control over the normal shock location once the mixed compression inlet has started.

At low speed, most inlets do not have enough cowl area A_C to provide the required engine airflow. Auxiliary doors or suck-in doors are located in the subsonic diffuser to provide additional air during takeoff (see Fig. 15.6).

Figure 15.9 illustrates a two-dimensional variable-geometry inlet, designed for Mach = 2.2, in operation at different flight conditions. It should be clear to the reader that the mechanism details are a real design challenge.

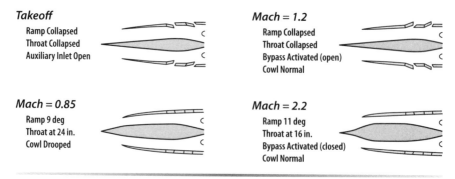

Takeoff	Mach = 1.2
Ramp Collapsed	Ramp Collapsed
Throat Collapsed	Throat Collapsed
Auxiliary Inlet Open	Bypass Activated (open)
	Cowl Normal

Mach = 0.85	Mach = 2.2
Ramp 9 deg	Ramp 11 deg
Throat at 24 in.	Throat at 16 in.
Cowl Drooped	Bypass Activated (closed)
	Cowl Normal

Figure 15.9   Mach 2.2 variable-geometry mixed-compression inlet operating at different flight conditions.

15.5 Quality of the Airflow—Distortion and Turbulence

Another characteristic of inlet performance is the quality of the airflow delivered to the engine compressor. It is important that the distortion and turbulence of the flow at the compression face be minimal, otherwise compressor stall and even engine flameout can result. The elements of distortion are swirl and uneven spatial distribution of total pressure, velocity, and temperature. *Turbulence* is a dynamic characteristic and results in a time variation of the distortion pattern. A poor velocity distribution at the compressor face can cause the compressor blades to pass through alternating high- and low-speed regions, which may cause vibration and possible blade failure. Local velocity variations may be interpreted as local variations in angle-of-attack of the compressor face flow. This variation in α along the blade (i.e., radially) may be sufficient to cause the blade to stall, thereby stalling other blades and surging the engine. An often used measure of the flow quality or flow distortion is given by the distortion parameter.

Figure 15.10 shows a comparison of average and instantaneous recovery maps with K_D for the F-111 at Mach = 0.9, 30,000 ft, and off-design spike position.

The main sources of distortion and turbulence are:

- Flow-field nonuniformity
- Ingestion of low-energy air
- Inlet shock system pressure gradients
- Shock/boundary layer interaction
- Cowl lip separation
- Duct pressure losses and flow separation
- Secondary duct flows

The location of the inlet on the vehicle is often determined by considerations of flow quality. The inlet should not be located in a region of separated or vortical (swirl) flow. Also, the vehicle boundary layer should not be permitted to interact with the inlet. Ingestion of the vehicle boundary layer can aggravate the inlet boundary layer and cause early separation. The inlet should be located out of the vehicle boundary layer by a boundary layer diverter as shown in Fig. 15.13b (in Section 15.7) for the Concorde, XB-70, and F-18. Also notice in this figure that the F-35 has a diverterless inlet.

The inlet boundary layer itself should be removed by boundary layer bleed. The inlet shock wave system interacting with the boundary layer can cause flow separation, resulting in greatly reduced total pressure recovery as well as poor quality flow at the compressor face. The amount of boundary layer that should be removed from within the inlet for satisfactory operation or performance is a function of the inlet type, shape, and shock

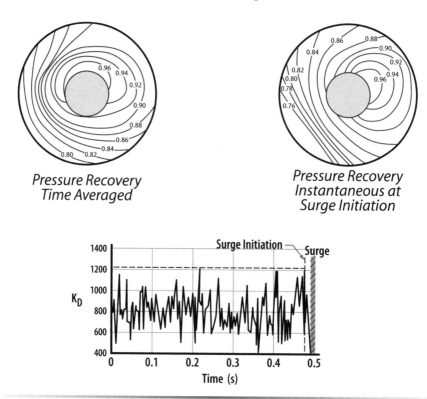

Pressure Recovery Time Averaged

Pressure Recovery Instantaneous at Surge Initiation

Figure 15.10 Comparison of average and instantaneous recovery maps with K_D. F-111 flight conditions: Mach = 0.9 at 30,000 ft and off-design spike position.

wave system; the sensitivity of the engine to the flow distortion and turbulence at the compressor face and the relative sensitivity of vehicle performance to inlet pressure recovery and bleed drag.

External-compression inlet types normally require less bleed for satisfactory operation than mixed-compression inlets because they have fewer shocks interacting with the boundary layer and shorter compression surfaces on which the boundary layer is formed. Two-dimensional inlets usually require more bleed than axisymmetric inlets because of the tendency of the boundary layer to accumulate on the sidewalls and in the corners, and because of the greater amount of surface area usually present in the two-dimensional inlets. A recommended amount of boundary layer bleed is shown in Fig. 15.11. Boundary layer bleed ports should be located in the throat area and in the area of any sharp bends in the subsonic diffuser (see Figs. 15.6 and 15.14, Section 15.8). Angles of bends in ducts should not exceed 15 deg when the duct requires a high Mach number (e.g., 0.85) near the throat [5].

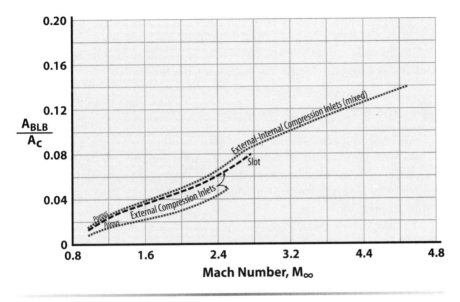

Figure 15.11   Recommended boundary layer bleed (data from [2]).

Podded engines, such as are used on the Boeing 747 and Lockheed C5, are good inlet designs from the standpoint of inlet flow quality. The inlet operates in undisturbed flow and the engine–airframe interactions are minimal. In addition, the podded engine offers good maintainability features such as easy access.

Low-slung inlets will initiate strong inlet vortices normal to the ground plane and with enough energy to scatter debris and to ingest objects. Where possible, the inlet lip should be more than two inlet diameters from the ground [5]. The locating of inlets in trail of the landing gear should be avoided to prevent picking up debris kicked up by the landing wheels.

15.6 Weight and Cost

Although high performance is desired from the inlet, it must be balanced by tolerable weight and cost. For example, an aircraft designed to operate at speeds up to Mach = 1.6 could have a normal shock inlet or a variable ramp external compression inlet. The designer must trade off the simple, low-weight, low-cost normal shock inlet with its shock $\eta_R = 0.9$ at Mach = 1.6 against the more complicated, heavier, and costlier external compression inlet with its shock $\eta_R \sim 0.97$.

The question to be addressed is whether the improved total pressure recovery is worth the added weight and cost. It is not an easy question and one that should not be answered until all the tradeoff information is available. A nacelle weight comparison study reported in [6] showed that a normal shock inlet, designed for low-subsonic flight, weighed 13% of the basic engine weight whereas a mixed compression inlet designed for Mach = 2.7 flight weighed 43% of the basic engine weight. The inlet weights are determined using the weight equations of Chapter 20.

15.7 Inlet Sizing and Design

This section discusses the general sizing and design of inlets and then follows with an example of a Mach = 2.3 inlet for the PW-F-100 engine. Reference [5] is an excellent report on the design of inlets and is recommended very highly to the reader. Many of the ideas of [5] are incorporated in this chapter.

The inlet should be sized to provide enough air to the engine at all flight conditions. It is rare that the inlet can provide exactly the right amount of air at all flight conditions, thus critical flight conditions are selected and the inlet is designed for these "design" conditions. There may be one or more design Mach numbers, M_D.

The engine demand cross-sectional area A_{1_E} is determined for different Mach–altitude conditions using

$$A_{\infty_E} = \frac{\dot{m}_E + \dot{m}_S}{32.17 \rho_\infty V_\infty} \tag{15.2}$$

where $\dot{m}_E$ is the engine airflow in pounds mass per second (lbm/s) (a function of Mach, altitude, and power setting, see Fig. 14.8g) and $\dot{m}_s$ is the secondary airflow required for engine oil cooling, ejector nozzle cooling, and so on. Typical values for secondary airflows are given in Table 15.1. The maximum value of A_{∞_E} is increased by the amount recommended for boundary layer bleed from Fig. 15.11 and the result set equal to the inlet capture area A_c. If the maximum A_{∞_E} occurs at takeoff, a lower value should be selected for A_c and the inlet provided with auxiliary takeoff doors, otherwise the inlet will be oversized for other parts of the mission resulting in

Table 15.1 Typical Secondary Airflows (data from [5])

	$\dot{m}_s/\dot{m}_E$
Engine oil cooling	0–0.01
Engine nacelle cooling	0–0.04
Ejector nozzle secondary air	0.04–0.20
Hydraulic system cooling	0–0.01
Vehicle environmental control	0.02–0.05

large amounts of excess air. Usually the $A_{\infty E}$ for cruise or maximum speed is selected for A_c.

Figure 15.16 (in Section 15.8) shows a typical *engine demand capture area ratio*, $A_{\infty E}/A_c$, for a supersonic aircraft. The inlet is now designed to give a capture area ratio of 1 at the flight condition of the selected A_c. This Mach number is termed the *design Mach number M_D*.

The type of inlet selected should be based upon the M_D. Figure 15.12 offers some good rules of thumb for inlet selection that are based primarily on tolerable total pressure recovery.

The inlet is now designed to match the *inlet supply capture area ratio*, $A_{\infty I}/A_c$, as closely as possible to the engine demand capture area ratio.

The inlet for subsonic and transonic aircraft will probably be the pitot inlet because of its simplicity and low weight and cost. Generally speaking, the inlets for these applications are sized for cruise altitude and Mach number and require very little (or no) variable geometry, bypass, boundary layer bleed, or control complexities to provide satisfactory operation [6]. They are usually characterized by generously rounded cowl lips and are either podded or blended into the fuselage in such a way that no appreciable low-energy air or vortex flows are likely to be ingested. Some designs have incorporated blow-in doors for low-speed, low-altitude flight, but when safety of flight is a prime consideration, the inlet cowl is usually sized

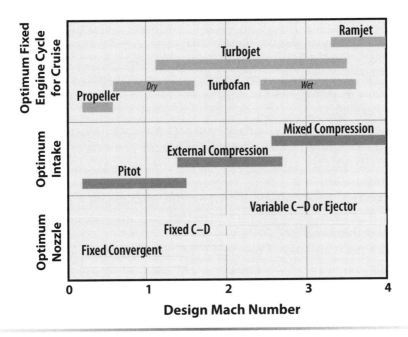

Figure 15.12 Effect of design Mach number on propulsion systems (data from [4]).

sufficiently large to avoid the added complexity of variable geometry. The pitot inlet gives $A_{\infty l}/A_c = 1$ for Mach > 1.0 and $A_{\infty l}/A_c > 1$ for all subsonic Mach numbers. At Mach numbers < 1.10 spillage is a better way of getting rid of excess inlet airflow because it has a lower drag penalty than bypassing at these Mach numbers. Above Mach = 1.10 the designer should examine the tradeoff between spillage and bypass drag and perhaps provide facilities for both.

At supersonic Mach numbers above Mach = 1.6 the inlets should be of the external compression or mixed compression type. For external compression and mixed compression inlets the designer has many decisions to make. Although the A_C is fixed by the airflow demand at M_D, there still remain the questions of two-dimensional vs axisymmetric, single vs multiple compression surfaces, compression surface angles, and fixed vs variable geometry. Axisymmetric inlets are slightly more efficient than two-dimensional inlets in terms of total pressure recovery and weight. However, if there is very much variable geometry, that is, translating or variable-angle compression surfaces, variable cowl area, and so on, the two-dimensional inlet could be less complicated and lighter. Multiple compression surfaces complicate the inlet design and always compound the off-design operation; however, high total pressure recovery at speeds greater than Mach = 2.3 require a multiple shock wave system. Appendix Figs. E.6, E.7, E.8, and E.9 can be used to select the cone or ramp angles to give desired total pressure recovery schedules. The fixed vs variable geometry question depends a lot on the matching of the supply to the demand airflow. These inlets should have boundary layer bleed control according to the schedule in Fig. 15.11 and the inlet supply must account for this. These inlets should also have provision for bypass because bypass is used not only for airflow matching, but for reduction of spillage drag and internal shock control as well.

Once the compression surfaces and angles are selected, the inlet supply capture area ratio $A_{\infty l}/A_c$ as a function of Mach number is determined and compared with the engine demand capture area ratio $A_{\infty E}/A_c$ (see Fig. 15.16 in Section 15.8).

This comparison readily illustrates the spillage and bypass requirements. The designer might choose to fine tune his inlet at this point to minimize the excess air schedule.

The throat area should be checked to insure that it is adequate to pass the supply air at all flight conditions. The *mass flow parameter* (MFP) of Appendix C is a useful parameter to make this quick check.

The subsonic diffuser must decelerate the flow to a Mach number ~ 0.4 at the compressor face with minimum distortion and turbulence. This means a diverging duct with gentle bends. The ratio of the area at the diffuser exit to the throat area is given by

$$\frac{A_{\text{exit}}}{A_T} = \frac{\left(A/A*\right)_{\text{exit}}}{\left(A/A*\right)_T} \tag{15.3}$$

where both $A/A*$ are functions of the Mach numbers at the diffuser exit and throat, and they are determined from Appendix C. Usually A_T and M_T are fixed and A_{exit} is the area of the engine compressor face so that Eq. (15.3) can be used to find the Mach number at the compressor face.

The subsonic diffuser length L_D is usually determined by constraints imposed by acceptable locations for the inlet and engine on the vehicle. The necessity for locating the inlet in a favorable flow field and the engine at a good position for exhaust discharge or for favorable balancing of the vehicle will usually dictate the duct length. Figure 15.4 can be used to help select the subsonic diffuser length for good pressure recovery. A good rule of thumb is to keep the diffuser overall included expansion angle $2\theta_D$ (see Fig. 15.3b) less than 10 deg [5]. There should be a short section of zero slope of one to three throat radii leading into the compressor to permit the flow to stabilize and even out the discharge velocity profile.

Figure 15.13 shows a few of the many inlet designs available to the propulsion engineer. Each aircraft has its own set of operating characteristics and the inlets are tailored to fit these "personalities." For fixed-geometry subsonic operation it is hard to beat the axisymmetric pitot inlet for weight and performance. Most subsonic aircraft use the axisymmetric inlet in one form or another.

The variable-geometry features of the two-dimensional inlet are less complicated than on an axisymmetric inlet and have led to its selection on such supersonic aircraft as the Anglo-French Concorde, Soviet TU-144, B-1, MIG-23, RA5C, and F-15. Underwing location on the Concorde (Fig. 15.13b is a typical installation), the B-1 (Fig. 7.15), and the F-15 provides precompression and permits smaller capture areas than if the inlet were exposed to the freestream. Flow deflection by the wing also reduces the inflow angles at angle-of-attack.

Axisymmetric inlets have the advantage of efficient structural shape for low duct weight and also the lowest wetted area per unit of flow area. However, they require a cone-shaped spike, which presents certain problems. Translating the spike fore and aft to keep the shock-on-lip and collapsing the spike to provide a variable throat area represent design nightmares. The A-12 and SR-71 represent one successful circular inlet. The SR-71 has the inlet canted inboard and pointed down a few degrees to better align the spike with the local flow at cruise angles-of-attack.

The F-16 is not required to operate much past Mach $= 1.6$ and thus uses the simple, lightweight normal shock inlet mounted in a chin fashion.

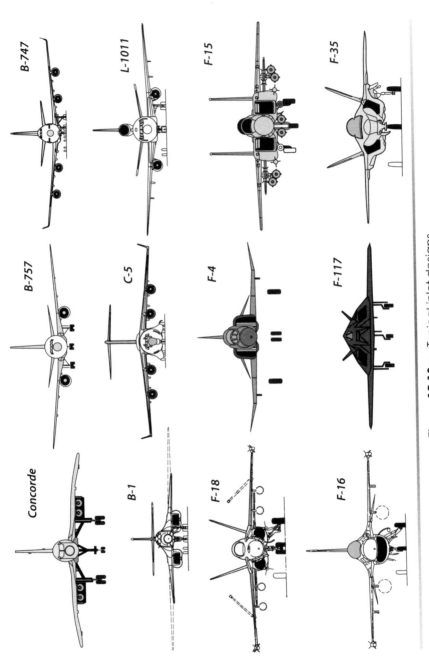

Figure 15.13a Typical inlet designs.

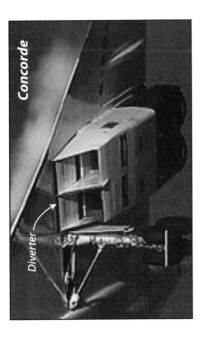

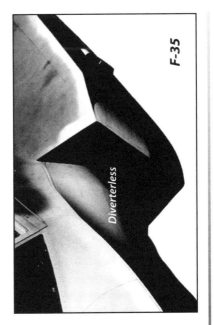

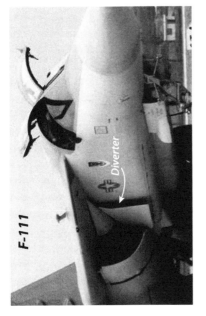

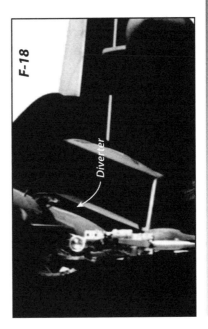

Figure 15.13b Inlet design examples.

Half-round inlets fit very nicely along the fuselage side and have been used on several Mach 2 class aircraft in the past. The F-104 uses this inlet design with a fixed half-round spike. The Mirage 3G uses this configuration located well forward on the fuselage. Also, with long, generous ducts, the system has been operating very well with early versions of the TF-30 engine that were a problem for the F-111.

Quarter-round inlets located in the wing–fuselage armpit, as on the F-111 (Fig. 15.13b), have very low external surface area and also low duct weight because of the short distance to the engine. In addition, the location offers some precompression from the wing shock and allows a smaller design capture area. On the F-111, a splitter plate is used to remove boundary layer air built up on the long forward fuselage. A pie-shaped subinlet is used to remove splitter plate boundary layer. Another pie inlet is used to remove boundary layer from the wing glove. In addition, bleed holes are used on the cone surface. Variable geometry includes both a variable second cone angle (diameter) and translation of the whole cone (fore and aft). As in all cone-type inlets, the cone cannot collapse enough to provide large subsonic flow area. This, plus the fact that the capture area was sized small initially, required the F-111 to employ large suck-in doors at low speed. Downstream of the throat the short subsonic diffuser makes an appreciable turn inboard to the engine. Due to its location, the inlet is sensitive to angle-of-attack as the boundary layer builds up between the fuselage and wing glove. The F-111 was plagued with engine–inlet problems during its flight test and motivated a great deal of activity in inlet distortion and turbulence research during the mid 1960s.

Many versions of the "D" inlet have been used successfully on Mach 2 class aircraft. The J-79-powered F-4 uses a "D" inlet with variable-geometry compression ramps. A slight downward tilt provides better flow alignment at angles of attack.

Example 15.1 External Compression Inlet for M_D = 2.3

The sizing and design of a two-dimensional external compression inlet for the PW-F-100 turbofan engine is demonstrated. The design Mach number will be Mach = 2.3 at 30,000 ft. This example is representative of the inlet on the F-15.

Table 15.2 presents the total pressure recovery data for the Mach 2.3 external compression inlet shown in Fig. 15.14. Notice that this two-dimensional inlet has a detached normal shock from Mach 1.0 to 1.5. Notice also that this inlet design has a short subsonic diffuser $(L_D/H_T = 4.9)$ and the resulting diffuser losses are as large as the losses across the shock wave system at most supersonic Mach numbers. This was a design compromise to give a lightweight (short) and reasonably efficient inlet at the design Mach number.

Table 15.2 Total Pressure Recovery Data for Inlet in Fig. 15.14

M_∞	M_T	P_{0_I}/P_{0_∞} (a)	P_{0_T}/P_{0_I} (b)	P_{0_C}/P_{0_T} (c)	P_{0_C}/P_{0_∞}
0.4	0.4			0.99	0.99
0.6	0.6			0.977	0.977
0.8	0.8			0.955	0.955
1.0	1.0		1.0	0.925	0.925
1.2	0.84		0.993	0.905	0.90
1.4	0.74		0.958	0.927	0.89
1.6	0.91	0.972	0.999	0.875	0.85
1.8	0.81	0.958	0.986	0.91	0.86
2.0	0.72	0.953	0.942	0.933	0.84
2.3	0.63	0.942	0.84	0.957	0.76

[a]Across oblique shock.
[b]Across normal shock.
[c]Across subsonic diffuser.

The assumed trajectory and required engine and secondary air-flows (from Fig. 14.8g) are shown in Table 15.3. Using Eq. (15.2) the A_{∞_E} is determined. The required engine capture area A_{∞_E} at $M_D = 2.3$ is 5.92 ft². The recommended bleed from Fig. 15.11 is 4%, which makes the design cowl area $A_c = 6.2$ ft². The demand capture area

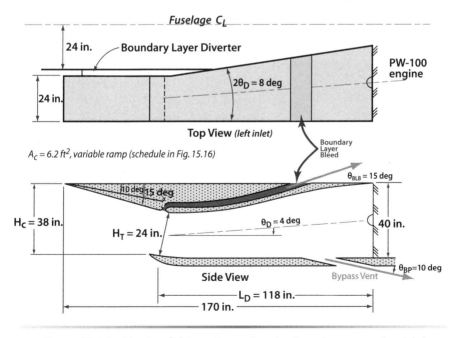

Figure 15.14 Mach = 2.3 two-dimensional external compression inlet.

Table 15.3 Demand Capture Area for PW-F-100 Engine

Mach	Altitude (1000 ft)	$\dot{m}_E + \dot{m}_s$ (lbm/s) [a]	$A_{\infty E}$ (ft^2)	$A_{\infty E}/A_c$ [b]
0.25	2	205	10.9	1.76
0.5	4	220	5.9	0.95
0.75	5	245	4.53	0.73
1.0	30	130	4.56	0.74
1.2	30	154	4.53	0.734
1.4	30	183	4.62	0.749
1.6	30	217	4.76	0.77
1.8	30	257	5.01	0.81
2.0	30	300	5.26	0.853
2.3	30	388	5.92	0.96

[a] $\dot{m}_E$ and $\dot{m}_s$ at maximum power from Fig. 14.8g and Table 15.1.
[b] Selected $A_c = 6.2$ ft^2.

ratio A_∞/A_c is plotted in Fig. 15.16 later in this section. The inlet will have auxiliary takeoff doors to augment the inlet area at takeoff.

The philosophy behind the inlet design, shown in Fig. 15.14, is a lightweight, low-cost inlet giving reasonably good efficiency at a maximum speed of Mach = 2.3. The primary mission of the aircraft is air superiority (similar to the F-15 Eagle), thus it is important that the inlet and engine be well matched in the transonic combat arena (Mach = 0.7–1.2 and 10,000–30,000 ft). The aircraft is designed around two PW-F-100 engines. The two-dimensional, single-ramp, external compression inlet shown in Fig. 15.14 was selected. The single ramp angle of 15 deg at Mach = 2.3 is a little on the high side for good pressure recovery; however, it gives a short inlet for shock-on-lip operation at this Mach number. The pressure recovery characteristics are shown in Table 15.2 and Fig. 15.15. The inlet features a variable ramp according to the schedule shown in Fig. 15.16. This ramp schedule was selected to keep the excess air during maximum power operation to a minimum. For Mach > 1.1 the excess air will be bypassed. Notice that the Mach = 2.3 pressure recovery is 75%, which is comparable with the 78% recovery for the SR-71 at Mach = 3.2.

The inlet supply capture area ratio is shown in Fig. 15.16 and provides a good match for the engine demand during maximum power operation. The maximum bypass requirement (for maximum power operation) above Mach = 1.1 is only 4% and this sized the spill vents

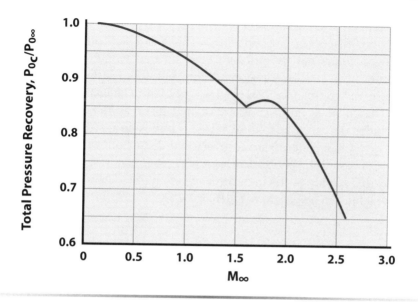

Figure 15.15   Total pressure recovery for inlet of Fig. 15.14.

(see Section 16.8). During cruise at Mach = 0.9, the required thrust is 70% of NRT which means the airflow requirement is approximately 70% [see Eq. (18.2)]. This cruise demand point at Mach = 0.9 is shown in Fig. 15.16 and indicates a spillage of about 27% of the supplied air.

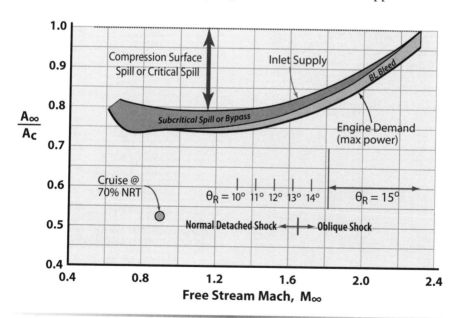

Figure 15.16   Engine–inlet flow matching for inlet of Fig. 15.14 and PW-F-100 engine.

Fortunately, this is a subsonic spill and not too costly in terms of spillage drag (discussed in Chapter 16). This mismatch could have been relieved somewhat by a hinged cowl lip that would decrease the capture area at this cruise condition, but this feature would add weight and complexity (cost) and was not felt to be justified.

The subsonic diffuser was designed to be as short as possible (included diffuser angle $2\theta_D$ of 8 deg for both side and top views) but still give tolerable pressure recovery and distortion levels. The subsonic diffuser total pressure recovery is lower than normal (see Table 15.2) due mainly to the high throat Mach numbers and the low diffuser length to height ratio, $L_D/H_T = 4.9$.

References

[1]   Liepmann, H. W., and Roshko, A., *Elements of Gasdynamics*, Wiley,, New York, 1957.
[2]   Ball, W. H., "Propulsion System Installation Corrections," U.S. Air Force Flight Dynamics Laboratory, AFFDL-TR-72-147, Wright–Patterson AFB, OH, Dec. 1972.
[3]   Urie, D., "Lockheed SR-71, a Supersonic/Hypersonic Research Facility," Lockheed Rept. SR-71–949, Lockheed Advanced Development Co., 1989.
[4]   Henshaw, J. T., *Supersonic Engineering*, Wiley, New York, 1962.
[5]   Crosthwait, E. L., Kennon, I. G., and Roland, H. L., "Preliminary Design Methodology for Air-Induction Systems," Technical Rept. SEG-TR-67-1, Wright–Patterson AFB, OH, Jan. 1967.
[6]   Antonatos, P. P., Surber, L. E., and Stava, D. J. , "Inlet/Airplane Interference and Integration," AGARD Rept. LS-53, NASA, Report Distribution and Storage Unit, Langley Field, VA, May 1972.

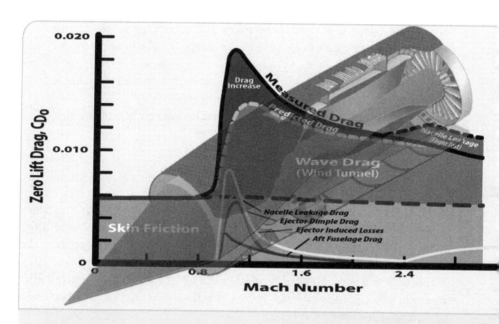

- Total Pressure Recovery
- Additive or Spillage Drag
- Boundary Layer Diverter Drag
- Boundary Layer Bleed Drag
- Exit Flap Drag
- Bypass Drag
- Boattail Drag
- Nozzle Types

Flight test of the SR-71 revealed a higher transonic C_{D_0} than predicted. Further testing showed this to be due to nacelle leakage, ejector-induced losses, ejector dimple drag, and aft fuselage drag. As a result the SR-71 routinely went into a dive maneuver to accelerate past Mach = 1.0.

Those who ignore the mistakes of the past are destined to repeat them.

16.1 Introduction

The engine thrust data provided by the engine manufacturer is termed *uninstalled* thrust. This uninstalled thrust data must now be corrected by the designer for airframe–engine integration effects.

These corrections to the propulsion system data are of three types:

1. **Installed engine thrust corrections.** Effects of inlet recovery and distortion, internal nozzle performance, engine bleed, and power extraction
2. **Inlet drag.** Effects of additive drag, cowl drag, boundary layer bleed drag, bypass drag, and boundary layer diverter drag
3. **Nozzle–afterbody drag.** Effects of nozzle–afterbody interference drag

16.2 Total Pressure Recovery

The *inlet* supplies air to the engine at a certain total pressure recovery schedule, and this schedule is very dependent upon the airframe–inlet design and flight condition. The engine manufacturer determines the thrust of the engine based upon a reference recovery schedule $(P_{0_c}/P_{0\infty})_{\text{Ref}}$. There will be a percentage difference in the net thrust if the designer's inlet does not provide the same ram recovery schedule as that used by the manufacturer. The correction is

$$\text{Percent reduction in thrust} = C_R\left[\left(\frac{P_{0_c}}{P_{0\infty}}\right)_{\text{Ref}} - \left(\frac{P_{0_c}}{P_{0\infty}}\right)_{\text{Design}}\right] \quad (16.1)$$

where C_R is the *ram recovery correction factor* and is a function of engine type, power setting, Mach number, altitude, and temperature conditions (i.e., cold, standard, or hot day). Figure 16.1 shows some typical data for standard day, maximum power operation at 45,000 ft [1]. Figure 16.1 is recommended for values of C_R at this point in the design.

The engine manufacturer's *reference ram recovery* $(P_{0_c}/P_{0\infty})_{\text{Ref}}$ should be reported in the engine data. Sometimes the engine manufacturer will use a constant recovery schedule of 1.0. Engine data supplied for military application will usually be based upon a standard military specification total pressure recovery schedule (MIL-E-5008B Ram Recovery is shown in Fig. 16.2).

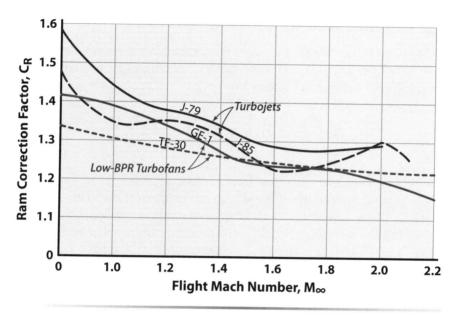

Figure 16.1   Ram correction factor for net thrust (data from [1]).

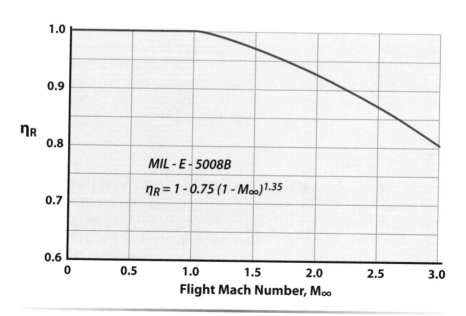

Figure 16.2   Mil-Spec total pressure recovery schedule.

16.3 Engine Bleed Requirements

Airbleed requirements (i.e., environmental control system, anti-icing, boundary layer control, etc.) from the engine compressor will reduce the thrust from the engine. This thrust reduction is estimated from

$$\text{Percent reduction in thrust} = C_B \left(\frac{\dot{m}_B}{\dot{m}_E} \right) \qquad (16.2)$$

where $\dot{m}_B$ is the bleed mass flow from the engine and $\dot{m}_E$ is the engine demand mass flow. The bleed correction factor C_B may be assumed to be equal to 2.0 for design purposes. This bleed requirement is not to be confused with the secondary airflow requirement $\dot{m}_S$ discussed in Section 15.7 (Table 15.1). The engine bleed $\dot{m}_B$ seldom exceeds 5% as it has a significant effect on thrust.

16.4 Inlet Flow Distortion

Inlet flow distortion is actually a velocity distortion, but it has typically been expressed in terms of total pressure variations for the sake of simplicity. The most apparent effect of flow distortion on a turbine engine is a downward shift of the engine surge line [2]. This shift is due primarily to the fact that many of the compressor blades are operating closer to stall in the distorted flow. If the distortion is sufficient to alter the blades' effective angles-of-attack, operating line efficiency will be changed so that the distortion results in a shift along the engine operation line to a lower operating pressure. If surge margin loss due to flow distortion is greater than anticipated, the engine may have to be derated to allow sufficient margin for engine transients. The primary effect of inlet turbulence is to drop the surge line even closer to the operating line [1,2].

16.5 Inlet Drag

Inlet drag is usually defined as all drag associated with the captured streamtube of air and its variations with engine demand and/or aircraft operating conditions. The inlet drag is the responsibility of the propulsion group. The Chief Designer must check the drag bookkeeping to ensure that together the airframe and propulsion groups account for all of the aircraft drag and don't double bookkeep.

The elements of the inlet drag are shown in Fig. 16.3. *Additive drag* is the momentum loss of the streamtube of air defined by the capture area A_c that is diverted around the inlet. Some of this lost momentum may be recovered in lip suction as the diverted flow accelerates over the cowl, cre-

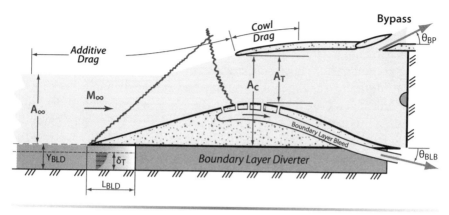

Figure 16.3 Elements of inlet drag.

ating a low-pressure region that acts in the thrust direction. Thus, as is shown later, cowl drag is not really a drag but a thrust force instead. The skin friction drag over the remainder of the inlet external surface, nacelle, or fuselage section is not charged as inlet drag but rather airframe drag, and the airframe group must account for it.

Boundary layer bleed (BLB) *drag* and *bypass drag* are defined as (1) combination of momentum lost by these flows from the time they are taken into the inlet until they exit the aircraft, and (2) the exit door pressure drags. *Boundary layer diverter drag* (usually included in inlet drag, but make sure someone accounts for it) is the momentum lost in the boundary layer that is turned away by the boundary layer diverter.

These drag elements will be discussed in the sections that follow. Methodology will be presented for estimating these inlet drag elements and in all cases the drag coefficients will be referenced to the cowl area A_c.

16.6 Additive (or Spillage) Drag

Additive drag is the momentum loss of the streamtube of air that is diverted around the inlet. For several reasons, not all of the air captured by the cowl (represented by A_c) can enter the inlet and the extra air is diverted over the cowl lip. These several reasons are the presence of a compression surface and the engine airflow demand being less than the inlet supply. The air being diverted around the inlet can be thought of as being "spilled" and thus additive drag is often called *spillage drag*. Additive drag occurs any time $A_\infty/A_c < 1.0$.

Additive or spillage drag is made up of two parts as shown in Fig. 15.16. The first part is called *compression surface spill* or *critical spill* and is due to the physical turning (or deflection) of the flow streamlines by a compression surface [Fig. 15.7b]; the spilled air involved is the difference in airflow between A_c and $A_{\infty p}$. This first part of additive drag is always accompanied by critical or supercritical inlet operation at supersonic Mach numbers.

The second part is called *subcritical spill* and is due to the excess air in the inlet (difference between inlet supply and engine demand) backing up and spilling around the lip as shown in Fig. 15.7a. It should be understood that if all of this excess air in the inlet is bypassed, then the subcritical spillage drag is zero. Thus, the designer can choose to trade off bypass drag for subcritical spillage drag. Subcritical spill is also accompanied by a decrease in pressure recovery as shown in Fig. 15.8.

For capture area ratios A_{∞}/A_c much less than 1, the additive or spillage drag can be appreciable, easily amounting to 20% of the airplane drag. Fortunately in practice this entire penalty is seldom experienced. Proper contouring of the external cowl lip can result in appreciable lip suction effects due to the increased velocities and decreasing pressures on the forward portions of the cowl lip. The magnitude of the lip suction effects can result in the cancellation at subsonic and transonic speeds of up to 80% of the additive drag for subsonic inlets and up to 50% for supersonic inlets [1].

There are several methods to correct the additive drag for cowl lip suction. Reference [3] expresses the corrected additive drag C_{D_A} as

$$C_{DA} = C_{DAdd} - C_{DLS}$$

where C_{DAdd} is the theoretical additive drag and C_{DLS} is the lip-suction term.

The method discussed in this section expresses the corrected additive drag as

$$C_{DA} = C_{DAdd} K_{Add} \qquad (16.3)$$

where K_{Add} is less than 1 and accounts for the cowl effects. The factor K_{Add} is from experimental data and is shown in Fig. 16.4 (data from [1,3,4]). Figure 16.4 represents a first-order correction for the additive drag. Reference [3] presents a more refined method, accounting for more inlet lip geometry details than Fig. 16.4. Figure 16.4 is recommended for conceptual design and [3] for preliminary design.

Figure 16.5 shows the control volume for the theoretical additive drag analysis, where station infinity (∞) is freestream and station 1 is the entrance to the inlet. The angle λ is the angle of the flow velocity through station 1. The same schematic holds for subsonic flow, and the schematic for a pitot inlet is similar.

The A_{∞} is the capture area of the flow streamtube entering the inlet and is expressed as

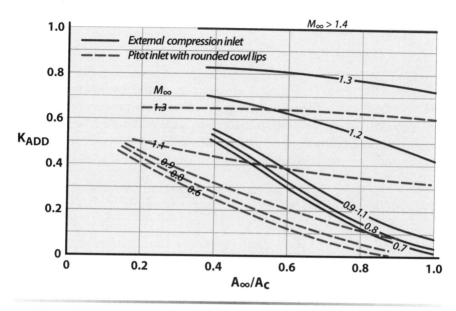

Figure 16.4 Additive drag correction factor for cowl lip suction effects.

$$A_\infty = \frac{\dot{m}_E + \dot{m}_{BP} + \dot{m}_{BLB} + \dot{m}_S}{32.17\ \rho_\infty V_\infty} \tag{16.4}$$

where the mass flows in the numerator (in slugs per second) are the engine demand, bypass, boundary layer bleed, and secondary air, respectively. All other airflow is diverted over the inlet lip thus creating additive drag.

The additive drag is the summation of the pressure forces in the drag direction acting on the control volume surface BC in Fig. 16.5. This drag force, shown as D_A in Fig. 16.5, is expressed as (using gage pressures)

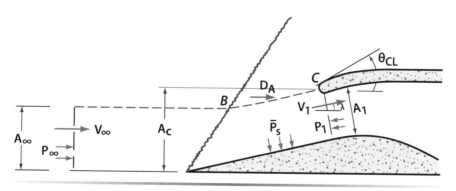

Figure 16.5 Schematic of control volume for additive drag analysis.

$$D_A = -\dot{m}_\infty V_\infty - \left(P_1 - P_\infty\right)A_\infty + \dot{m}_\infty V_1 \cos\lambda + \left(P_1 - P_\infty\right)A_1 \cos\lambda \quad (16.5)$$
$$+\left(\bar{P}_S - P_\infty\right)A_S$$

where $\bar{P}_S$ is the average surface pressure on the ramp or conical centerbody and A_S is the projected area in the drag direction of the ramp or centerbody (equal to zero for a pitot inlet). If there is an angle α between the inlet centerline and the freestream, then D_A is multiplied by $\cos\alpha$.

The additive drag coefficient is

$$
\begin{aligned}
C_{D\text{Add}} &= \frac{D_A}{q_\infty A_c} = \left(\frac{P_\infty}{q_\infty}\right)\left(\frac{A_1}{A_c}\right)\cos\lambda\left[\left(\frac{P_1}{P_\infty}\right)\left(1+\gamma M_1^2\right)-1\right] \\
&\quad -2\left(\frac{A_\infty}{A_c}\right)+\bar{C}_{PS}\left(\frac{A_S}{A_c}\right)
\end{aligned}
\quad (16.6)
$$

where

$$\bar{C}_{PS} = \frac{\bar{P}_S - P_\infty}{q_\infty} = \frac{2}{\gamma M_\infty^2}\left(\frac{\bar{P}_S}{P_\infty}-1\right) \quad (16.7)$$

Everything in Eq. (16.6) is known or can be determined at this point. If the flow is supersonic and the inlet is operating critically or supercritically (normal shock in throat or downstream of throat), then $M_1 > 1$. The M_1 and P_1/P_∞ for two-dimensional ramp inlets are determined straightforwardly from the external shock structure using Fig. E.2. The M_1 and P_1/P_∞ for an axisymmetric inlet with a centerbody is not straightforward due to the nonuniform flow region behind a conical shock. The M_1 behind a conical shock wave is determined as follows. Obtain the cone surface Mach number M_S from Fig. E.5 and the cone shock wave angle β from Fig. E.3. Then determine the value of the Mach number behind an oblique shock having the same shock wave angle β as the conical shock (using Figs. E.1 and E.2). The M_1 behind the conical shock is then estimated to be the average of these two Mach numbers.

$$\left(M_1\right)_{\text{cone}} = 0.5\left(M_s + M_{1\text{wedge}}\right)$$

The pressure ratio P_1/P_∞ for an axisymmetric inlet is determined as follows:

$$\frac{P_1}{P_\infty} = \left(\frac{P_1}{P_{01}}\right)\left(\frac{P_{01}}{P_{0\infty}}\right)\left(\frac{P_{0\infty}}{P_\infty}\right) \quad (16.8)$$

where the static to total pressure ratios are functions of M_1 and M_∞ (from Appendix C) and the $P_{01}/P_{0\infty}$ is determined from

$$\left(\frac{P_{01}}{P_{0\infty}}\right) = \left(\frac{P_{0\text{th}}}{P_{0\infty}}\right)\left(\frac{P_{01}}{P_{0\text{th}}}\right) \tag{16.9}$$

The $P_{0\text{th}}$ is from Figs. E.6 and E.7 and $P_{01}/P_{0\text{th}}$ is the total pressure ratio across a normal shock (from Appendix D) with upstream Mach number M_1.

If $M_\infty > 1$ but the shock is detached or the inlet is operating subcritically, then $M_1 < 1$. The M_1 is found as follows:

$$(A/A^*)_{M_1} = (A/A^*)_{M_\infty}(A_1/A_c)(P_{01}/P_{0\infty})(A_c/A_\infty) \tag{16.10}$$

and M_1 is the Mach number for $(A/A^*)_{M_1}$ as found from Appendix C. The pressure ratio P_1/P_∞ is found using Eq. (16.8), where

$$\left(\frac{P_{01}}{P_{0\infty}}\right) = \left(\frac{P_{0\text{th}}}{P_{0\infty}}\right)$$

For $M_\infty < 1$ the M_1 is found as follows:

$$\left(\frac{A_\infty}{A_1}\right) = \frac{(A/A^*)_{M_\infty}}{(A/A^*)_{M_1}} \tag{16.11a}$$

or

$$\left(\frac{A}{A^*}\right)_{M_1} = \frac{(A/A^*)_{M_\infty}}{(A_\infty/A_1)} \tag{16.11b}$$

where $(A/A^*)_{M_\infty}$ is determined for M_∞ from Appendix C and A_∞/A_1 is known from geometry. The M_1 is then found as a function of $(A/A^*)_{M_1}$ from Appendix C. The pressure ratio is

$$P_1/P_\infty = (P_1/P_{01})/(P_\infty/P_{0\infty}) \tag{16.12}$$

since $P_{01} = P_{0\infty}$.

The surface pressure coefficient (averaged for multiple compression surfaces) $\bar{C}_{P_S}$ is determined from Figs. E.2 and E.10, for a ramp surface, and from Figs. E.4 and E.11, for a conical surface.

The theoretical additive drag coefficient is now determined from Eq. (16.6) and corrected for cowl lip effects using Eq. (16.3) and Fig. 16.4. Reference [3] is recommended for more refined estimates of K_{Add}. The C_D is referenced to the cowl area A_c.

16.7 Boundary Layer Bleed Drag

The methodology presented here (from [4]) estimates the drag produced by the removal and disposal of boundary layer air from the inlet. This air is normally removed to ensure satisfactory stability and uniformity of flow at the diffuser exit for good pressure recovery. Figure 15.11 shows recommended boundary layer bleed levels; however, the designer should choose the bleed mass flow after examining the tradeoff between inlet pressure recovery and boundary layer bleed drag.

The *bleed drag* is composed of two parts: (1) the change of momentum of the bleed air between the bleed system entrance and the exit to the freestream (FS), and (2) the pressure drag on the exit flap door.

The following symbols and definitions apply for the methodology:

M_∞ = freestream Mach numbers

M_E = Mach number at bleed exit

P_{0_E}/P_{0_∞} = total pressure recovery of bleed airflow FS to bleed exit (use Fig. 16.6)

θ_{BLB} = exit of bleed air relative to freestream (15 deg or less is desirable)

A_{BLB}/A_c = boundary layer bleed mass flow ratio (Fig. 15.11)

A_E = bleed nozzle exit area

A_T = bleed duct throat area

A bleed exit discharging at a low θ_{BLB} into a region of low base pressure is desired. This type of exit will provide the highest exit momentum and will reduce base drag at the same time. A convergent discharge nozzle is satisfactory for nozzle pressure ratios up to about 4. At higher pressure ratios, a convergent–divergent nozzle is desired [3].

The methodology presented here will be for a convergent nozzle. The reader should examine [4] for a convergent–divergent nozzle.

The freestream Mach number that will give choked flow in the bleed duct is determined from

$$(M_\infty)_{\text{Ch}} = \left[\frac{6}{\left(P_{0_E}/P_{0_\infty}\right)^{0.286}} - 5 \right]^{1/2} \tag{16.13}$$

where P_{0_E}/P_{0_∞} is determined from Fig. 16.6. Equation (16.13) is plotted in Fig. 16.7.

16.7.1 Choked Flow

If $M_\infty > (M_\infty)_{\text{Ch}}$, the nozzle throat is choked, $M_E = M_T = 1.0$, and $P_E = P_\infty$. The bleed duct throat area is calculated from

$$A_T = \frac{\left(A_{\text{BLB}}/A_c\right)A_c}{\left(A/A^*\right)_{M_\infty}\left(P_{0_E}/P_{0_\infty}\right)} \tag{16.14}$$

where $(A/A^*)_{M_\infty}$ is determined from Appendix C for M_∞. The boundary layer bleed drag coefficient, $C_{D_{\text{BLB}}}$ (referenced to A_c) is given by

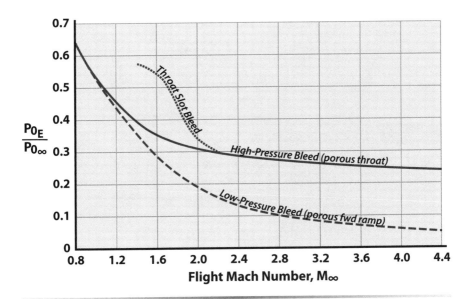

Figure 16.6 Total pressure recovery of bleed airflow (data from [4]).

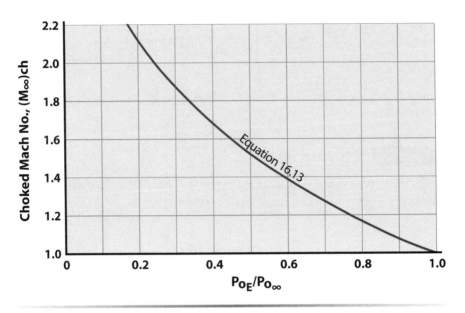

Figure 16.7 Freestream Mach number for choked flow in bleed and bypass ducts.

$$C_{D_{\text{BLB}}} = 2\left(\frac{A_{\text{BLB}}}{A_c}\right)\left(1 - \frac{\cos\theta_{\text{BLB}}}{M_\infty}\sqrt{0.833 + 0.167M_\infty^2}\right.$$

$$\left.\times\left\{1.715 - \left[\frac{0.715}{\left(P_{0_E}/P_{0_\infty}\right)\left(0.833 + 0.167M_\infty^2\right)^{3.5}}\right]\right\}\right) \quad (16.15)$$

16.7.2 Unchoked Flow

If $M_\infty < (M_\infty)_{\text{Ch}}$, the static to total pressure ratio at the duct exit is

$$P_E/P_{0_E} = \left(P_\infty/P_{0_\infty}\right)/\left(P_{0_E}/P_{0_\infty}\right) \quad (16.16)$$

where P_∞/P_{0_∞} is determined from Appendix C for M_∞.

The Mach number at the exit, M_E, is determined from Appendix C, corresponding to P_E/P_{0_E}. The duct throat area is given by

$$A_T = \frac{\left(A_{\text{BLB}}/A_c\right)A_c\left(A/A^*\right)_{M_E}}{\left(A/A^*\right)_{M_\infty}\left(P_{0_E}/P_{0_\infty}\right)} \quad (16.17)$$

where the area ratios A/A^* are determined from Appendix C for Mach numbers M_E and M_∞. The bleed drag coefficient $C_{D_{BLB}}$ (referenced to A_c) is determined from

$$C_{D_{BLB}} = 2\left(\frac{A_{BLB}}{A_c}\right)\left(1-\cos\theta_{BLB}\left[\frac{5}{M_\infty^2}+1\right]^{1/2}\right.$$
$$\left.\times\left[1-\frac{1}{\left(1+0.2M_\infty^2\right)\left(P_{0E}/P_{0\infty}\right)^{0.286}}\right]\right) \qquad (16.18)$$

16.7.3 Exit Flap Drag

If there is a flap-type door over the exit of the boundary layer bleed duct, there will be a pressure drag on the flap. This *exit flap drag* can be omitted at this point but should be included later as the design is refined and fine tuned. The method of [4] is recommended. If the exit is a flush-type, there is no flap drag.

16.8 Bypass Drag

The methodology presented here (from [4]) estimates the drag of the airflow that enters the inlet but bypasses the engine for airflow matching, reduction of additive drag, or internal shock control.

The *bypass drag* is composed of two parts: (1) the change in momentum of the bypass air between the bypass exit and the freestream, and (2) the pressure drag on the bypass exit flap door.

The methodology for a bypass system with a convergent nozzle is identical to that presented in Section 16.7 for the boundary layer bleed drag, with values for the bypass system substituted for the bleed values. The total pressure recovery for the bypass airflow from freestream to bypass exit is approximated by $0.85(P_{0_c}/P_{0\infty})$.

16.9 Boundary Layer Diverter Drag

The *boundary layer diverter* is a splitter plate arrangement that diverts the boundary layer, built up ahead of the inlet, away from the inlet. The reasons for the boundary layer diverter are discussed in Section 15.5. A boundary layer diverter is shown in Fig. 16.3 and locates the inlet out of the upstream boundary layer. The boundary layer diverter height Y_{BLD} should be at least twice the local turbulent boundary layer thickness given by

$$\delta_T = \frac{0.37x}{Re_x^{0.2}}$$ (16.19)

where x = distance from fuselage nose or wing leading edge to boundary layer diverter and the local Reynolds number is

$$Re_x = \frac{\rho_\infty V_\infty x}{\mu_\infty}$$ (16.20)

The boundary layer diverter is usually a compression ramp surface of flow deflection angle θ_{BLD}. The drag of the boundary layer diverter is due primarily to the surface pressure on the compression ramp influenced by the presence of right-angle surfaces. The expression for the boundary layer drag coefficient (referenced to cowl area A_c) is expressed as (from [3])

$$C_{DBLD} = (C_{DBLD}/2\theta_{BLD})(2.6\theta_{BLD})(A_{BLD}/A_c)$$ (16.21)

where the ratio $C_{DBLD}/2\theta_{BLD}$ is obtained from Fig. 16.8, and the distance L in Fig. 16.8 is defined in Fig. 16.3 as L_{BLD}. The θ_{BLD} is the diverter compression ramp angle in degrees and A_{BLD} is the projected surface area of the boundary layer diverter in the flow direction.

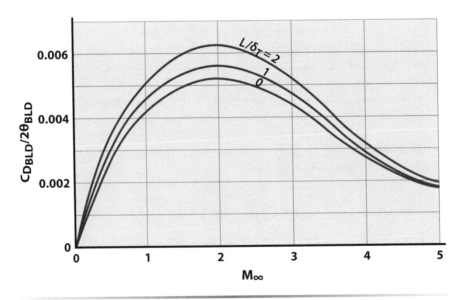

Figure 16.8   Boundary layer diverter drag variation with freestream Mach number.

16.10 Nozzle-Airframe Interference Effects

A jet exhausting from a nozzle has two effects on the surrounding flow field and hence the aircraft. First, the jet acts like a solid body (whose size and shape varies with power setting, nozzle setting, Mach number, and altitude) displacing the external flow; second, it normally entrains mass flow from the external stream. The jet contour affects the pressure distribution on the afterbody and nearby surfaces, which, in subsonic flow, transmits a strong upstream influence. In supersonic flight there is a limited upstream influence because any disturbance can only be propagated upstream through the subsonic part of the boundary layer. The shock system within the jet will continue through the jet boundary and may impinge on nearby surfaces. For aircraft configurations with two or more jet engines the mutual interference becomes even more complex. The influence of the elevated temperatures in the jet exhaust is another interference that must be considered but is not discussed here.

Computation methods available today are either not sufficiently accurate or fail completely to predict the complex afterbody flow field [5]. This is particularly true in subsonic flow incorporating boundary layer separation and strong upstream influences. Therefore, aircraft development relies heavily on wind tunnel tests with simulated jets. The aim of such jet effects testing is to obtain information on critical areas of nozzle–airframe interference. As mentioned in Chapter 1, this configuration fine tuning in the wind tunnel is done during the latter part of the preliminary design phase. At this point in the conceptual design phase only the gross features of the nozzle–airframe configuration will be considered.

The primary parameters influencing the nozzle–airframe interference are the nozzle type, the boattail angle, the base area, the nozzle spacing for multiengine aircraft, and the interfairing length between nozzles. The following subsections discuss each of these parameters briefly.

16.10.1 Nozzle Types

Figure 16.9 shows typical jet pressure ratios for turbojet and turbofan engines versus flight Mach number (the turbojet is the upper limit of the band). Two extreme engine operating conditions are shown. For cruise in the subsonic flight regime the nozzle pressure ratio is low, requiring little or no divergence. For maximum acceleration, that is, full afterburning, the throat area is increased by a factor of about 2 (depending on bypass ratio).

The required nozzle divergence increases gradually with increasing flight speed and reaches a value of $A_e/A_t \sim 2.6$ at nozzle pressure ratios of 14.

Besides cruise and maximum acceleration all intermediate operating conditions are possible (military, partial afterburner). This requires in the

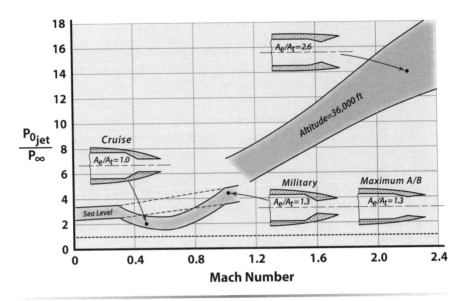

Figure 16.9 Required variation of nozzle geometry (data from [5]).

ideal case a fully variable nozzle with independent variation of throat size and divergence. In many practical cases more simple systems with either purely convergent nozzles or a fixed relation in throat-to-divergence are chosen as a compromise.

Figure 16.10 shows some typical nozzle concepts for afterburning engines. These nozzle types are discussed next (from [5]):

- **Short convergent nozzle.** This concept represents a mechanically simple lightweight nozzle. The major disadvantage from the aerodynamic point of view is the larger base in the closed position.
- **Iris-nozzle.** With the mechanically more complex iris nozzle, annular bases are avoided in all positions. As with the short convergent nozzle, large thrust losses occur at high pressure ratios because no divergence is provided.
- **Plug-nozzle.** The necessary variation in throat area is accomplished by variation of the plug position or geometry. As a consequence, a fixed lightweight shroud can be used. Large cooling air flows, however, are necessary for reheat operation.
- **C–D iris nozzle.** This provides some divergence in the reheat position. The variation in throat size and in divergence is coupled. Thus, the C–D iris is a compromise between the simple iris and a fully variable C–D nozzle.
- **Simple ejector.** This is a frequently chosen nozzle concept. Primary and secondary flaps are mechanically linked. Relatively large secondary airflows are required, associated with drag penalties.

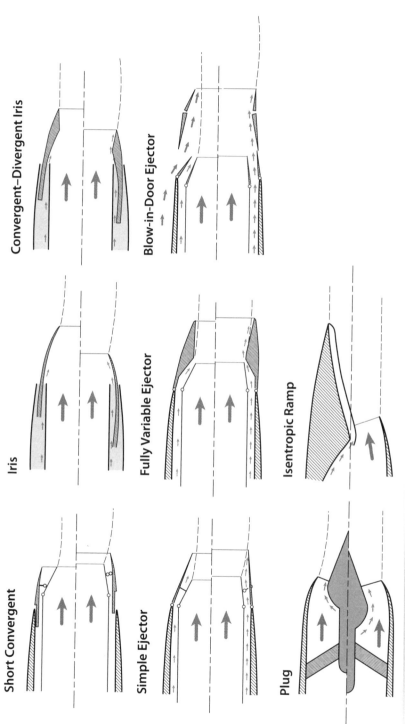

Figure 16.10   Typical nozzle concepts for afterburning engines (upper half of each sketch denotes dry power; lower half is maximum afterburning) [5].

- **Fully variable ejector.** This design yields near-optimum aerodynamic performance: throat area and divergence are independently variable; the required secondary mass flows can be kept low. High weights and complex design are associated with this nozzle concept.
- **Isentropic ramp.** This design is difficult to adapt to varying operating conditions, which normally results in undesirable changes in pitching moment.
- **Blow-in-door ejector.** This nozzle concept provides similarly good performance as the ordinary ejector in the reheat position. In the closed position, large quantities of tertiary air are taken aboard through spring-loaded flaps in order to fill the large annular base of the short primary nozzle. Large quantities of air, however, require careful handling to avoid losses in the sharp turnings of the secondary and tertiary flow passages. Especially this nozzle represents a highly integrated concept with respect to merging of internal and external flows. Peripheral nonuniformities (blockage) of the external flow may cause unfavorable interferences, which is particularly true with closely spaced twin jet installations.

Table 16.1 gives some incremental afterbody drag data for the nozzle concepts of Fig. 16.10. The drag data were taken from several sources and thus there is some scatter in the data.

16.10.2 Boattail Drag

The pressure and skin friction on the afterbody section surrounding the nozzle is called *boattail drag*. The boattail drag coefficient is shown in Fig. 16.11 as a function of Mach number and boattail angle β. For freestream Mach numbers greater than 1.0 the expression for C_D presented in Fig. 16.11 should be used. This C_{D_β} is referenced to the maximum cross-sectional area.

Table 16.1 Incremental Afterbody Drag

Nozzle Type	ΔC_D
Short convergent	0.036–0.042
Blow-in-door ejector	0.025–0.035
Plug	0.015–0.02
Fully variable ejector	0.01–0.02
Iris	0.01–0.02
Ramp	0.01

$M_\infty = 0.8$–0.9
Nozzle pressure ratios = 2.5–3.0
ΔC_D referenced to fuselage maximum cross-sectional area

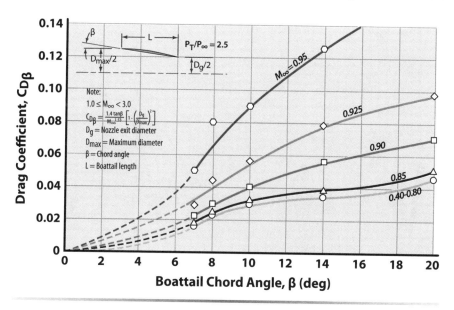

Figure 16.11   Nozzle boattail drag coefficients (data from [4]).

16.10.3 Base Area and Multiengine Installation

The designer should avoid any blunt-based areas as these regions result in large drag increases (Figs. 16.12 and 16.13). Figure 16.13 shows this behavior for blunt and tapered interfairings. Also, blunt-based areas upstream

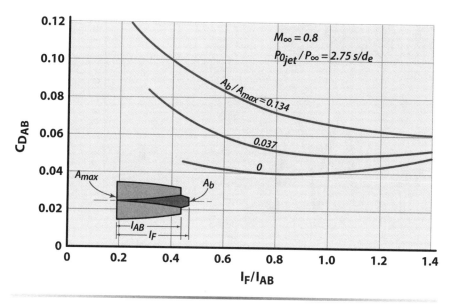

Figure 16.12   Effect of interfairing length on drag for constant-base areas (data from [5]).

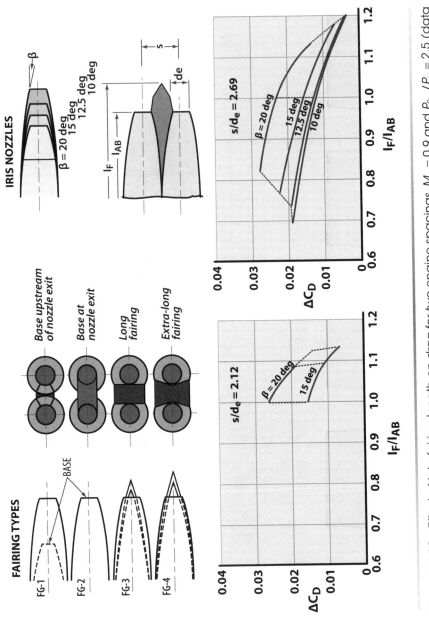

Figure 16.13 Effect of interfairing length on drag for two engine spacings. $M_\infty = 0.9$ and $P_{0jet}/P_\infty = 2.5$ (data from [5]).

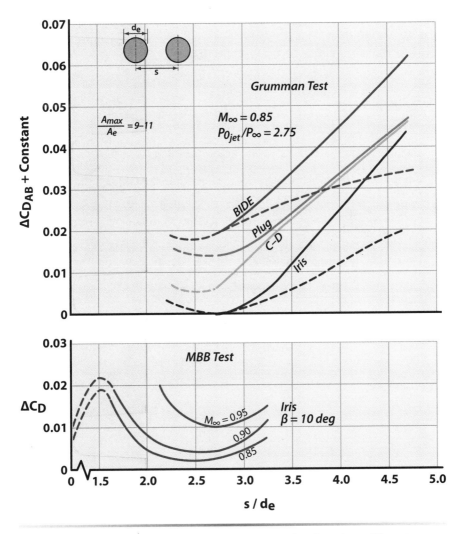

Figure 16.14 Optimization of engine spacing from two different investigations (data from [5]).

of the nozzle exit plane are worse than those downstream as shown in Figs. 16.12 and 16.13 by comparing the ΔC_D for configurations FG-1 and FG-2. The drag due to a blunt base can be estimated using Fig. 2.27.

Fuselage-mounted multiengine installations should have a tapered interfairing between the engine nozzles. Figures 16.12 and 16.13 show the effect of interfairing location and length.

Engine nozzle spacing is a design parameter that needs to be negotiated between the configuration (layout) group and the propulsion group. Figure 16.14 shows the effect of engine spacing on ΔC_D and indicates an optimum s/d_e of about 2.5 at high-subsonic speeds. However, this optimum s/d_e

might aggravate some other feature of the design to the extent that a different s/d_e is warranted. This is another example of the compromise necessary between the design groups that was discussed in Chapter 1.

References

[1] Antonatos, P. P., Surber, L. E., and Stava, D. J., "Inlet/Airplane Interference and Integration," AGARD Rept. LS-53, NASA, Report Distribution and Storage Unit, Langley Field, VA, May 1972.

[2] Zonars, D., "Dynamic Characteristics of Engine Inlets," AGARD Rept. LS-53, May 1972.

[3] Crosthwait, E. L., Kennon, I. G., and Roland, H. L., "Preliminary Design Methodology for Air-Induction Systems," Technical Rept. SEG-TR-67-l, Wright–Patterson AFB, OH, Jan. 1967.

[4] Ball, W. H., "Propulsion System Installation Corrections," U.S. Air Force Flight Dynamics Laboratory, Technical Rept. AFFDL-TR-72-l47, Wright–Patterson AFB, OH, Dec. 1972.

[5] Aulehla, F. and Lotter, K., "Nozzle/Airframe Interference and Integration," AGARD Rept. LS-53, May 1972.

Chapter 17　Propeller Propulsion Systems

Wright Brothers' Propeller

- Why Propellers?
- Theories
- Power Coefficient
- Thrust Coefficient
- Advance Ratio
- Activity Factor
- Propulsive Efficiency
- Vendor Propeller Charts

Backdrop for the propeller used on a Wright Brothers 1908 aircraft. A propeller is best described as a rotating wing with all the complexity of aerodynamics, structures, materials, and control. Design pioneers confronted a situation aptly titled "Propellers and Mystery Are Synonymous" in a 1919 aircraft design textbook.

Propellers and mystery are synonymous.

17.1 Introduction

Despite the great deal of attention and fanfare that has been attached to the design of turbojet and turbofan aircraft since World War II, a large performance regime still exists that can be adequately filled only by propeller-driven aircraft. Certain STOL transport and long-endurance maritime reconnaissance missions lend themselves perfectly to turboprop power schemes, and the high cost of turbine powerplants guarantees that the reciprocating engine will play a major role in general aviation for decades to come. Indeed, one has only to look at the number and dollar value of light aircraft produced annually to realize that the propeller is alive and well in Wichita, Kansas, and in Lock Haven, Pennsylvania. Also, propellers are the only thing that make sense on the slow airships and solar-powered aircraft.

In this chapter we do the following:

- Discuss the theories of propeller performance
- Present a methodology for designing custom propellers
- Discuss the practical use of vendor-supplied propeller charts

17.2 Why Propellers?

The *open propeller*, or *airscrew*, offers an efficient means of propulsion in the low- to medium-subsonic speed range. Just as the turbofan engine is generally more efficient than a turbojet of the same thrust, a propeller–turbine engine combination is more efficient than either of them. The reason can be found from a brief look at Newton's second law in the form that a propulsion engineer would use:

$$T_n = \dot{m}_{air}\left(V_e - V_a\right) \qquad (14.2)$$

where, for this analysis, the pressure term is ignored and the resultant force is the thrust of the powerplant. A propeller achieves a specified level of thrust by giving a relatively small acceleration to a relatively large mass of air, whereas the turbofan and the turbojet each give a correspondingly higher acceleration to a correspondingly smaller mass of air. From energy considerations, the powerplant producing the smallest change in kinetic energy will require the smallest expenditure of fuel; thus, the propeller–shaft-engine powerplant provides the highest efficiency of the methods considered.

In 1908 the Wright brothers were conducting flight trials of their latest Wright Flyer design for the Army. The goal of the next flight was to demonstrate one hour at 40 mph with a passenger on board. Orville was the pilot and Lt. Selfridge was the passenger. Well into the flight one of the wooden propellers broke, sending it into the rigging and causing the aircraft to crash. Orville was badly injured but Lt. Selfridge died on the operating table, making him the first crash victim in a powered aircraft. Selfridge AFB in Michigan is named after him.

As will be shown, this argument breaks down at higher flight speeds, where compressibility effects cause additional and unacceptable blade losses.

The propeller offers an additional advantage to the designer of a multi-engine STOL aircraft by bathing large segments of the wing in its high-dynamic-pressure slipstream. This slipstream produces a significant amount of wing lift independently of any freestream dynamic pressure effects and provides an equivalent increase in wing lift coefficient.

High values of takeoff acceleration can be obtained by optimizing propeller design for static thrust conditions. However, this effort to improve STOL capability can only be made by compromising the cruise performance of the aircraft. As was mentioned in Chapter 10, the utilization of reversible-pitch propellers provides the designer with a high deceleration capability at little or no increase in weight or cost.

17.3 Theory

The analysis of propeller performance can be accomplished using one or more of the following theories: momentum theory, blade element theory, and vortex theory. Each method has its own distinct advantages as well as shortcomings, yet all play an important role in providing an understanding of airscrew performance. The following discussion is intended to convey a general working knowledge of the pertinent theory, but for deeper insight the reader is directed to [1–6].

17.3.1 Momentum Theory

Any aerodynamic propulsive device produces a thrust by imparting a change in momentum flux to a specified mass of air (Newton's second law). The basic momentum theory analyzes the effects of this change in momentum, the work done on the air, and the energy imparted to the air. Certain simplifying assumptions are made about the propeller and its surroundings in the development of this theory that divorce them from the real world, and yet the method remains a useful tool in calculating the maximum theoretical efficiency a propeller can obtain.

The first assumption made by momentum theory is that the propeller is replaced by an infinitesimally thin actuator disk that consists of an infinite number of blades. The disk is held to be uniformly loaded and is thus experiencing uniform flow and imparting a uniform acceleration to the air passing through it.

The actuator disk is further assumed to be surrounded by a sharply delineated streamtube that divides the flow passing through the propeller and the surrounding air. Far upstream and downstream from the disk the walls of the streamtube are parallel, and the static pressure inside the streamtube at these points is equal to the freestream static pressure.

Momentum theory deals with a working fluid (air in this case) that is inviscid and incompressible. As a consequence, the propeller does not impart any rotation to the air, and any profile losses from the blades of the propeller are ignored.

To an observer moving with the actuator, the air far upstream will be moving with the freestream velocity V (Fig. 17.1). This air will be gradually accelerated until, at station 1, the actuator disk $V_1 = V + v$, where v is the induced velocity imparted to the air by the actuator. It can be shown at station 2, far downstream from the actuator disk, that $V_2 = V + 2v$. The net change in velocity through the control volume defined by the streamtube and planes perpendicular to the flow far upstream and far downstream is

$$(V + 2v) - V = 2v \tag{17.1}$$

and, from continuity considerations for an incompressible fluid,

$$A_1 = A = 2A_2 \tag{17.2}$$

For steady flow the mass flux will be constant across every plane of the streamtube that is perpendicular to the flow. Using the actuator disk as a reference plane,

$$\dot{m} = \rho A (V + v) \tag{17.3}$$

The thrust T produced by the actuator disk will be

$$T = \Delta\text{Momentum flux}$$

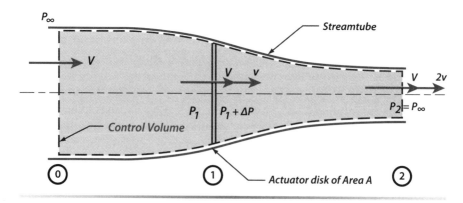

Figure 17.1   Propeller analysis by momentum theory.

$$T = \rho A (V + v) 2v \qquad (17.4)$$

To produce this level of thrust, the actuator (propeller) must supply energy to the slipstream. Because the theory ignores profile and rotational losses, this energy goes only to increasing the kinetic energy of the flow. The power required for this purpose, the induced power P_i will equal the change in kinetic energy flux through the control volume and may be shown to be simply the product of the resultant thrust and the velocity at which the thrust is applied, or

$$P_i = T(V + v) \qquad (17.5)$$

Equation (17.5) indicates that, to minimize induced power requirements at a given thrust level and freestream velocity, the induced velocity must be kept as small as possible. Solving Eq. (17.4) for v (and remembering that $v > 0$ for a propeller) yields

$$V = -\frac{V}{2} + \left(\frac{V^2}{4} + \frac{T}{2\rho A} \right)^{1/2} \qquad (17.6)$$

Two important conclusions may be gleaned from this expression. To minimize V (hence, P_i) at given values of V and T, the quantity T/A, the *disk loading*, must be minimized. Thus, within the limits of the assumptions made, it may be stated that the larger the propeller used to produce a given thrust, the smaller will be the power and energy requirements. The second result is that, for a given thrust as the freestream velocity increases, the induced velocity will decrease. This is not to imply, however, that the induced power requirement will decrease. For a given level of thrust, V will increase faster than v will decrease, and thus the required P_i will increase as freestream velocity increases. In practice, however, the thrust of a propeller will not remain constant with changing velocity, but the power of the engine turning it will, over moderate speed ranges, remain fixed. Because profile and rotational losses are being neglected, P_{avail} will remain constant and thrust will decrease with increasing velocity. This condition may be illustrated by combining Eqs. (17.5) and (17.6) to form an expression for P_i as a function of T and V. Solving for the static condition ($V = 0$ as designated by the 0 subscript) and for the condition of $V \neq 0$, and assuming that P_i is constant for all V, gives

$$\frac{T}{T_0} = \frac{2}{(V/v_0) + \left[(V/v_0)^2 + 4(T/T_0) \right]^{1/2}} \qquad (17.7)$$

Although a general solution of this function $T/T_0 = f(V/V_0)$ is not possible, the approximation

$$\frac{T}{T_0} \approx 1 - 0.32 \frac{V}{v_0} \tag{17.8}$$

will hold for $V/v_0 \ll 1$.

The theoretical power required by the propeller has been defined as the product $T(V + v)$. By defining the useful power output of the propeller as TV, it is possible to form an ideal efficiency

$$\eta_i = \frac{\text{Power Output}\left(\text{Useful Power}\right)}{\text{Power Input}}$$

$$\eta_i = \frac{TV}{T\left(V + v\right)} = \frac{V}{\left(V + v\right)} \tag{17.9}$$

Notice that the static ideal efficiency will be zero, but that η_i will increase with V. This concept of ideal efficiency is misleading for cases where $V/v < 1$. The use of the word "ideal" must again be emphasized as no real-world losses are included in its calculation.

The momentum theory does not provide a means to predict propeller losses due to blade skin friction, rotational motion, or mutual blade interference, nor does it account for any geometry parameters other than disk area. Although it is simple to apply, this theory must be combined with some other analytical tool to be of use to the designer.

17.3.2 Blade Element Theory

An aircraft propeller is nothing more than an airfoil rotating about a translating axis dividing a propeller blade into a number of chordwise strips. It is possible to analyze the performance of the entire propeller by summing the contributions of all segments on all blades of the airscrew. This is essentially what is done by the *blade element theory* (sometimes called *strip theory*).

In Fig. 17.2 a small element of the propeller blade is marked for consideration. This infinitesimal element is dr wide, has chord c, and is located a distance r from the axis of rotation. The entire blade has a radius of R. A cross section of the blade element is shown in Fig. 17.3. The airfoil shape can be clearly seen, and many of the angular and velocity notations are analogous to those used in wing theory.

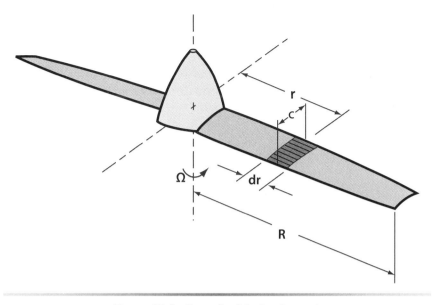

Figure 17.2   Propeller blade element.

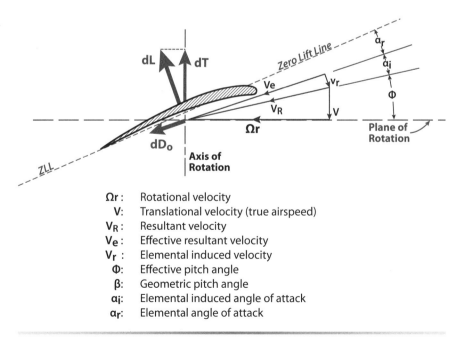

Ωr :	Rotational velocity
V:	Translational velocity (true airspeed)
V_R :	Resultant velocity
V_e :	Effective resultant velocity
V_r :	Elemental induced velocity
Φ:	Effective pitch angle
β:	Geometric pitch angle
α_i:	Elemental induced angle of attack
α_r:	Elemental angle of attack

Figure 17.3   Forces, velocities, and angles for a blade element.

To simplify the development of the blade element theory, it is assumed that each element is subjected to two-dimensional flow only and that each element is independent of its neighbors.

The aerodynamic lift force produced by the elemental lift dL will be perpendicular to the effective velocity V_e and will be inclined from the axis of rotation by the angle

$$\varphi + a_i \approx \tan^{-1}\left(\frac{V + vr}{\Omega r}\right)$$

For freestream velocities up to the mid-subsonic range, it may be assumed that this angle is small, and

$$\sin(\varphi + \alpha_i) \approx \varphi + \alpha_i \quad \text{(in radians)}$$

$$\tan(\varphi + \alpha_i) \approx \varphi + \alpha_i \quad \text{(in radians)}$$

$$\cos(\varphi + \alpha_i) \approx 1$$

Thus, the elemental thrust is

$$\left|dT\right| = \left|dL\right|\cos(\varphi + \alpha_i) \approx \left|dL\right| \tag{17.10}$$

Similarly, the drag force opposing the rotation of the propeller element will consist of a drag component, dD_0, and a component due to the inclination of the lift force, the induced drag dD_i:

$$\left|dD\right| = \left|dD_0\right|\cos(\varphi + \alpha_i) + \left|dL\right|\sin(\varphi + \alpha_i)$$

$$\left|dD\right| \approx \left|dD_0\right| + \left|dL\right|(\varphi + \alpha_i) \tag{17.11}$$

It is now possible to express the thrust produced by a single element as

$$dT \approx dL = \text{dynamic pressure} \times \text{area} \times \text{lift coefficient}$$

$$= \left(\frac{1}{2}\rho V_e^2\right)(cdr)c_\ell \tag{17.12}$$

where c_ℓ is the two-dimensional lift coefficient of the element. To determine the thrust of the propeller one must integrate this expression across the span of the blade and multiply by the number of blades b:

$$T = b\int_0^R 0.5\rho V_e^2 cc_\ell dr \tag{17.13}$$

Practical propeller blades do not run to the axis of rotation because some allowance must be made for a mounting hub and, possibly, a pitch-changing mechanism. For this reason the inner limit of integration, r_i, is usually taken as $0.1R$. Similarly, some accounting must be made for losses caused by a decrease in effectiveness of outboard blade elements which results from the formation of a blade tip vortex. The outer integration limit is usually taken to BR, where the empirically determined tip-loss factor $B \approx 0.96$.

By ignoring compressibility effects, Eq. (17.13) becomes

$$T = 0.5\rho b \int_{r_i}^{BR} V_e^2 c c_\ell \, dr \qquad (17.14)$$

where V_e will vary with r, and c and c_ℓ may or may not be functions of radial position. Generally, $c = c(r)$ is specified but, to calculate the propulsive thrust, one must know $V_e = V_e(r)$ and $c_\ell = c_\ell(r)$.

From Fig. 17.3 it is obvious that

$$V_e \approx \left[\left(V_r + V \right)^2 + \left(\Omega r \right)^2 \right]^{1/2} \qquad (17.15)$$

and the two-dimensional lift coefficient may be expressed as

$$c_\ell = a\alpha_r = a\left[\beta - \left(\varphi + \alpha_i \right) \right] \approx a\left[\beta - \frac{V + vr}{\Omega r} \right] \qquad (17.16)$$

where $a = dc_\ell/d\alpha$. Due to variations in local Mach number across the blade span, a will vary with r. However, with little loss of accuracy, it may be assumed that a is a constant with a value appropriate for the conditions at $r = 0.75R$.

Equations (17.15) and (17.16) still cannot produce the key to solving for the thrust of the propeller until the local induced velocity, v_r, is known at every blade location. An expression for v_r can be obtained by employing simple momentum theory in an elemental approach. Figure 17.4 shows an actuator disk upon which an annulus dr wide and located a distance r from the center has been using the same logic as was used to develop Eq. (17.4), the differential thrust produced by this annulus will be

$$dT = \rho \left(2\pi r \, dr \right)\left(V + v_r \right) 2 v_r \qquad (17.17)$$

From blade element considerations, the thrust generated by this same annulus will be the product of the thrust produced by a single element located a distance r from the axis of rotation [Eq. (17.12)] and the number of blades b. With Eqs. (17.15) and (17.16) this becomes

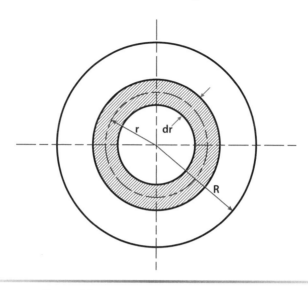

Figure 17.4   Annulus of an actuator disk.

$$dT = \frac{b}{2}\rho\left[(v_r+V)^2+(\Omega r)^2\right]ca\left[\beta-\frac{V+v_r}{\Omega r}\right] \qquad (17.18)$$

By Eqs. (17.17) and (17.18) and solving for v_r:

$$v_r = \left(\frac{V}{2}+\frac{bca\Omega}{16\pi}\right)\left\{-1+\left[1+\frac{2\Omega r\left(\beta-\dfrac{V}{\Omega r}\right)}{\dfrac{4\pi V^2}{bca\Omega}+V+\dfrac{bca\Omega}{16\pi}}\right]^{1/2}\right\} \qquad (17.19)$$

which, within the limitations of the theory, will predict the induced velocity at a radial distance r of a propeller of known physical characteristics that is axially translating at a velocity V.

Theoretically, it would be possible to introduce Eq. (17.19) into (17.18) and integrate the latter expression between appropriate limits to calculate the thrust of a propeller of arbitrary twist distribution. In practice, however, the resulting expression would prove extremely difficult to handle. Satisfactory results can be obtained by dividing the blade into a finite number of stations, calculating v_r and dT at each station, and finally computing r total thrust via graphical integration or some numerical technique such as Simpson's rule.

The calculation of propeller thrust can be greatly simplified by the recognition that, as expressed by Eq. (17.19), the local induced velocity will be constant across the blade if the quantity $2\Omega r[\beta-(V/\Omega r)]$ is also a constant.

(It can be shown [5] that constant v_r across the blade will require the minimum induced power for a given thrust and is thus desirable for reasons other than convenience of computation.) This may be accomplished by providing the blade with ideal twist such that, for any element located at r, the geometric pitch angle is defined by

$$\beta = \frac{\beta_t R}{r} \qquad (17.20)$$

where β_t is the pitch of the tip section. This expression becomes unmanageable for $r \rightarrow 0$ as a result of the *small-angle assumption*

$$(\varphi + \alpha_i)\tan^{-1}\left[(V + v_r)/\Omega r\right]$$

Practically, as $r \rightarrow 0$, $\beta \rightarrow \pi/2$. It must be noted that a unique twist distribution will be ideal only for a limited number of thrust and airspeed combinations. Because $T = f(\beta_t)$ for a given V, varying thrust levels will require variable β_t. However, because $\beta = \pi/2$ at $r = 0$ for all cases, the ideal twist distribution must be optimized for a single thrust–airspeed combination.

The blade element theory furnishes a method for approximating the total power requirements of the propeller by providing insight into the profile losses of the blade. From Fig. 17.3 it can be seen that the power required to rotate the propeller (and thus generate thrust) will be the power needed to overcome the forces in the plane of rotation. For a single infinitesimal element this is

$$dP = \Omega r \, dD_0 \cos(\varphi + \alpha_i) + \Omega r \, dL \cos(\varphi + \alpha_i) \qquad (17.21)$$

The term $dD_0 = (1/2)\rho V_e^2 c c_{d_0} \, dr$ is the profile drag acting on the element, and thus the first series of terms in Eq. (17.21) may be thought of as the *elemental profile power*, whereas the second group is the *elemental induced power*. Then

$$dP = dP_0 + dP_i \qquad (17.22)$$

It must be noted that the induced power requirements are directly associated with the production of propeller thrust, and when the expression for dP_i is integrated across the blade radius provisions must be made for the loss of thrust at the tips. Profile losses, however, are present across the entire exposed radius of the blade. Thus, each of the terms in Eq. (17.22) must be integrated between separate limits:

$$P = 0.5\rho b \left[\int_{r_i}^{R} (\Omega r)^2 V_e c c_{d_0} \, dr + \int_{r_i}^{BR} V_e c (\beta \Omega r - V - v_r)(V + v_r) \, dr \right] \quad (17.23)$$

This general equation is for modern high-speed propellers that employ ideal twist. Also, because most propellers are designed so that each section is operating at a low angle-of-attack, each element will also be functioning in the angle-of-attack region where the two-dimensional, incompressible profile drag coefficient c_{d_0} is approximately constant, and for low-speed application c_{d_0} may be removed from the integral. This last statement is certainly not true for high-speed propellers, however. As shown in Fig. 17.5, the resultant tip speed of a rotating blade is a function of rotational velocity and the true airspeed. At high flight speed and high propeller rpm (necessary for high thrust), the tip Mach number may approach or surpass the critical Mach number (~0.9) of the tip sections, and c_{d_0} will experience a drastic increase as $r \to R$. (For simplicity, skin friction, pressure, and wave drag effects are lumped together in c_{d_0}.)

Equation (17.23) provides a key to understanding the rationale behind the selection of a certain propeller geometry to fulfill given design requirements. For low-to-moderate airspeeds where c_{d_0} will be constant, power requirements may be reduced by minimizing the blade chord toward the tip where dynamic pressure is greatest. However, this high dynamic pressure in the blade tip region is also responsible for the lion's share of the

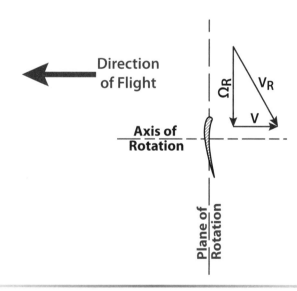

Figure 17.5   Resultant velocity at a propeller tip.

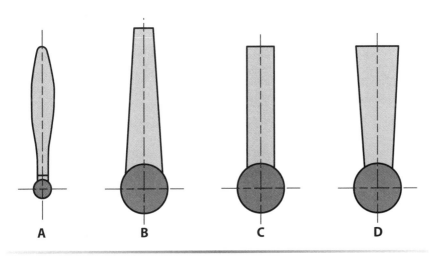

A   B   C   D

Figure 17.6 Typical propeller blade planforms.

resulting thrust, and larger tip chords would be desirable from this stand-point. Some compromise must be reached, and the results are planforms of the type shown in Fig. 17.6. Blade A is a type used on low-speed general aviation craft. It features a circular or elliptical root section developing into an 8%–12% thick section at the outer radii. Operating at rotational tip Mach numbers approaching 0.8, a propeller utilizing this blade can fly at airspeeds up to approximately Mach = 0.4 before compressibility effects begin to be felt. Blade B exhibits a planform designed for use at high-subsonic Mach numbers and features thin sections and reduced chord at the tip. In this way the drag effects of transonic tip conditions can be mini-mized. This class of propeller blade has not found widespread application because the speed range for which it is designed (Mach = 0.6–0.8) can be more efficiently handled by turbofan engines.

A practical blade planform for the middle subsonic range is the "paddle blade" design, blade C, which was used on the original C-130 and Electra aircraft. The wide tip chord of this blade would seem to produce higher compressibility losses, but, as demonstrated in [7], the opposite is true. The blade with a large chord at the tip will be more efficient than a tapered blade producing the same thrust at the same operating conditions because the tip sections of the wider, untapered blade will be operating at a lower c_ℓ and will have a higher critical Mach number. This argument would indicate that an even more efficient design would employ inverse taper as shown by blade D.

Although promising from an aerodynamic viewpoint, this approach has not been accorded wide acceptance because of structural difficulties.

17.3.3 Vortex Theory

Although providing a rapid method for the preliminary calculation of propeller performance, blade element theory does not provide the accuracy needed for detailed design work. Such factors as tip losses, three-dimensional effects, and mutual blade interference cannot be predicted by this method. For example, blade element theory indicates that a linear increase in thrust with no change in efficiency will result from adding blades to a propeller, whereas, in fact, the most efficient propeller consists of a single blade with efficiency decreasing as the number of blades increases.

The third major branch of propeller theory, vortex theory, overcomes many of the limitations of the previous two methods and offers the capability for great accuracy. The equations required to implement this theory satisfactorily, however, necessitate the use of large-capacity, high-speed computers. The details of the vortex theory are beyond the scope of this text and are more the tool of the propeller designer rather than the aircraft designer. The interested reader is referred to [1,3,4,8].

17.4 Preliminary Design

Although the previously discussed theoretical methods for propeller analysis provide convenient and relatively accurate schemes for predicting the performance of airscrews of known design, they would prove to be too cumbersome for preliminary design applications. To establish the propeller design parameters required by the preliminary design process, various semiempirical methods may be employed. Reference [9], for example, provides rapid performance calculations for light aircraft propellers driven by engines of up to 300 horsepower and at flight speeds ranging up to 200 kt. Propellers for larger and faster aircraft may be accurately analyzed through the methods and charts of [10]. The method developed here is based on [11] and is applicable for engine ratings over 300 horsepower and for flight Mach numbers up to 0.8.

The task of identifying the characteristics of a propeller to meet a given set of performance specifications is essentially a two-part problem attempting to relate the horsepower available to the thrust provided by the propeller in the takeoff mode and in the cruise mode. Each of these segments requires independent methodology, and the results must be faired together to provide a continuous picture of thrust output for a selected propeller from brake release up through the limits of the aircraft performance.

At this stage of the design loop the drag characteristics of the airframe should be well established and should include a rough approximation of nacelle drag for the selected number of engines. The major design parameters to be determined at a given flight condition are the propeller diameter and the engine shaft horsepower required for that condition. All other parameters are defined by technology within rather narrow limits. If the propeller diameter is fixed by some structural consideration such as wing location or landing gear limitations, or by some aerodynamic consideration such as the ratio of propeller diameter to wing chord [12,13], then the resulting efficiency will be less than optimum, but the design process will be simplified as the required shaft horsepower can be calculated without iteration.

Certain definitions must be made at this point. As with most aerodynamic quantities, the thrust developed and power required by a propeller are conveniently expressed as nondimensional coefficients in the following forms:

Power Coefficient:

$$C_P = \frac{P}{\rho n^3 D^5} \tag{17.24}$$

Thrust Coefficient:

$$C_T = \frac{T}{\rho n^2 D^4} \tag{17.25}$$

where n is the *propeller rotational velocity* in revolutions per second (rps) and D is the diameter of the airscrew in feet. For generality, a third coefficient is defined to cover the torque Q generated by the propeller:

Torque Coefficient:

$$C_Q = \frac{Q}{n^2 D^5} = \frac{C_p}{2\pi} \tag{17.26}$$

The ideal efficiency η_i has been defined by Eq. (17.9) and should not be confused with another measure of effectiveness, the *propulsive* (or *propeller*) *efficiency*,

$$\eta = \frac{\text{Thrust Power Output}}{\text{Shaft Power Input}} = \frac{TV}{P} \tag{17.27}$$

This expression accounts for profile losses as well as induced losses and may be written as the product of an induced (or ideal) efficiency η_i and a *profile efficiency* η_0. Thus,

$$\eta = \eta_i \eta_0 \tag{17.28}$$

As with the ideal efficiency, the propeller efficiency will be zero under static condition.

Another useful parameter is the *rotational tip speed* of the propeller, V_{tip}, defined as

$$V_{tip} = \Omega R = \pi n D \qquad (17.29)$$

The rotational tip speed has been given close consideration as a design point in recent years because of its importance in determining the operating noise level of the aircraft. Producing an aircraft with acceptable sideline noise levels is a major challenge to the designers of both civil and military STOL aircraft, and the reader is encouraged to consult [10,14–16] for further background on this problem. Suffice it to say that 700–800 ft/s is an upper limit on V_{tip}; in light of the high sound levels created by the C-130 and Electra with their 720 ft/s tip speeds, even lower values might prove to be more realistic starting points.

The ratio of the true airspeed V to the tip speed has proven to be a powerful design variable in that it is related both to efficiency and to the aerodynamic coefficients. This ratio is most often expressed as the *proportional advance ratio*

$$J = V / nD \qquad (17.30)$$

Two more parameters are needed to completely define the propeller and its operational conditions: one is to establish the blade planform and the other to set the sectional lift characteristics. The latter condition was defined in the section on blade element theory as a *two-dimensional lift coefficient* c_ℓ, which could vary across the blade span. In practical propeller designs the sectional camber is defined by the *design lift coefficient* c_{ℓ_d}, and the camber for the entire blade is designated by specifying c_{ℓ_d} at $r = 0.7R$. Generally c_{ℓ_d} at $r = 0.7R$ will vary from 0.4 to 0.6, and minor excursions from the specified value at sections on either side of $r = 0.7R$ will have a negligible effect on the propeller performance.

The blade planform is expressed by the *activity factor* (AF), which represents the rated power absorption capability of all blade elements. Equation (17.23) indicates that the power absorbed by a blade element will be proportional to the area of the element times the cube of the velocity. By assuming $V_e \sim \Omega r$, the power may be expressed as

$$dP \alpha c (\Omega r)^3 dr \qquad (17.31)$$

because at flight velocities $dP_0 \gg dP_i$. This expression has been nondimensionalized with V_{tip} and D to form a function of purely geometric properties and yet one that reflects the relative ability of the blade to absorb power. The activity factor is conventionally defined as

$$AF = \frac{100,000}{16} \int_{0.15}^{1.0} \left(\frac{c}{D}\right)\left(\frac{r}{R}\right)^3 d\left(\frac{r}{R}\right) \qquad (17.32)$$

for a single blade. The propeller AF is simply the blade AF times the number of blades b. Values for blade AF are usually constrained by structural considerations to values between 80 and 180. For example, the blade AF for the C-130 is 162.

The design process may be initiated with either the takeoff condition or the cruise condition, depending on the mission specifications. The designer must realize that the requirements for the two regimes may not be compatible and that a compromise solution most probably will be required. The following discussion assumes that the takeoff performance is not a driving consideration and that the aircraft design is being optimized for cruise.

The general methodology for designing a propeller for cruise flight ($J > 1.4$) is outlined in Fig. 17.7. Again, it is emphasized that only the drag characteristics of the airframe and the approximate specific fuel consumption (SFC) vs power setting and flight condition (speed and altitude) of the class of engine to be used need be known to begin the design procedure. All other parameters may be selected within the limitations previously discussed or calculated from the accompanying charts.

Figure 17.8 permits the computation of the propeller AF and a *basic induced efficiency* η_i'. Because this chart is based on a six-blade propeller and a C_ℓ at $r = 0.7R$ of 0.5, a correction must be made to η_i' to produce the induced efficiency η_i. The value specified for the representative lift coefficient is a reasonable value and may be used with good success for preliminary design purposes. However, changes to c_ℓ at $r = 0.7R$ within the acceptable 0.4–0.6 range will produce a negligible change in η and will produce thrust and power figures still well within the accuracy limitations required in initial aircraft design iterations. The basic induced efficiency should be corrected for number of blades and total activity factor using Fig. 17.9. With a value for $\Delta\eta_i$, the actual induced efficiency becomes

$$\eta_i = \eta_i' + \Delta\eta_i \qquad (17.33)$$

The profile efficiency is obtained from Fig. 17.10 as a function of advance ratio J and flight Mach number. The total efficiency is then calculated using Eq. (17.28). This value of η may be corrected for compressibility effects with the addition of a term $\Delta\eta_c$ from Fig. 17.11 [17].

A word of explanation should be given: the term *shaft horsepower* (SHP) used in this procedure is the engine output that is actually available to turn

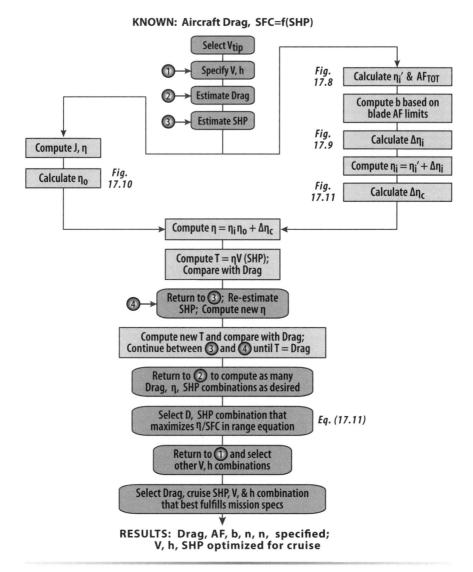

Figure 17.7   Propeller analysis procedure for cruise.

the propeller; it consists of the engine power at the given flight condition corrected for any bleed and auxiliary equipment losses and accounting for the inefficiency of the reduction gear.

This method applies to single-rotation propellers and does account for energy lost in the rotational motion of the slipstream. Approximately 60% of this lost rotational energy may be recovered through the use of dual counter-rotating propellers. The *rotational power expended* (P_R) may be obtained from Fig. 17.12 in the ratio of P_R/R. The induced efficiency of a dual-rotation propeller may then be found using the expression

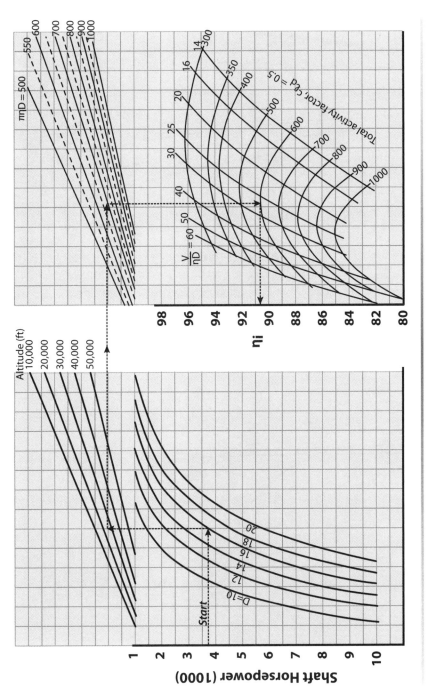

Figure 17.8 Propeller basic induced efficiency for cruise ($b = 6$, $c_{l_d} = 0.5$) (data from [11]).

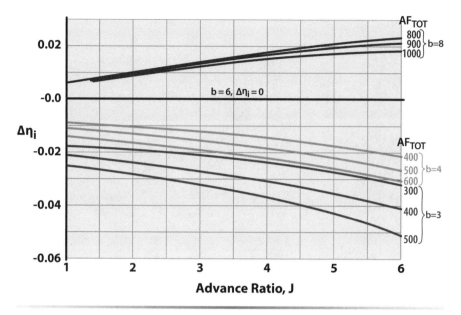

Figure 17.9 Blade number correction to basic induced efficiency.

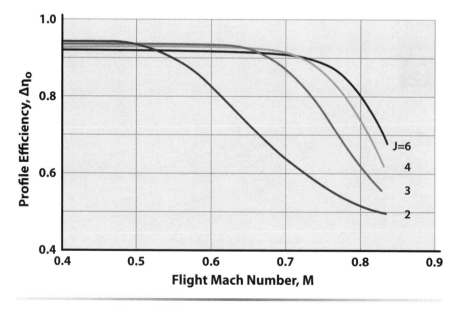

Figure 17.10 Estimated profile efficiency.

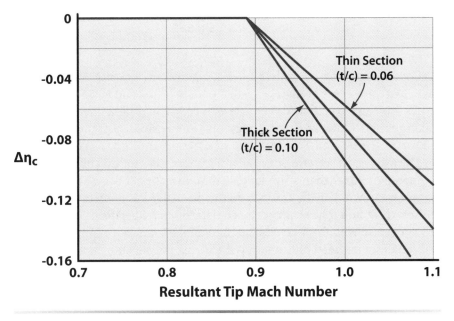

Figure 17.11 Compressibility correction to propeller efficiency.

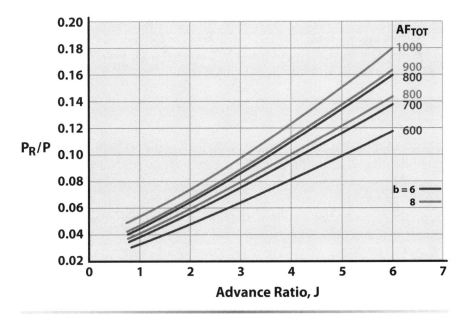

Figure 17.12 Efficiency correction for dual-rotation propellers.

$$\eta_i = \eta_i' + \Delta\eta_i + 0.6\frac{P_R}{P} \qquad (17.34)$$

Counter-rotating propellers offer a significant increase in efficiency, particularly for cases of high propeller activity factor and large numbers of blades (≥ 6); however, they do require an increase in weight due to the associated gearing. Only the Soviet Union made extensive use of this feature on the An-22, Tu-95, and Tu-114.

The process of selecting or evaluating a propeller for takeoff conditions is generally simpler than a similar task for cruise flight because more information is known. For the case where the propeller has been optimized for cruise, AF, D, and n have already been determined. Specification of SFC vs SHP in cruise will, for a given class of engines, establish the takeoff SHP, and the designer need only analyze the thrust produced by the propeller–engine combination for use in calculating the takeoff distance. If the propeller is to be optimized to meet a takeoff specification, a more complex iterative procedure must be utilized to pick the combination of propeller characteristics that will require the least power and, thus, the lowest engine weight.

The methodology for both of the preceding procedures is outlined in Fig. 17.13. In each case the takeoff velocity of the aircraft must be known (Chapter 10). Because the concept of propeller efficiency becomes meaningless for low airspeeds, the takeoff problem is one of finding the relationship between thrust produced and power required (or between C_T and C_P). Figure 17.14 provides this relationship but it is expressed in terms of an *intermediate power coefficient* C_{P_x}, which is not corrected for variable activity factors. The correction to C_P may be obtained from Fig. 17.15, and the actual power coefficient may be computed from

$$C_P = X\left(C_{P_x}\right) \qquad (17.35)$$

As with the calculation of cruise performance, the propeller tip speed is an important parameter for propeller analysis at takeoff. Noise criteria are especially critical during airfield operations, and the designer must make a difficult tradeoff between performance and sideline noise levels. Depending on the type of engine utilized, the tip speed in takeoff need not be the same as that in cruise, but generally

$$\left(V_{tip}\right)_{TO} \geq \left(V_{tip}\right)_{cruise}$$

Once more it must be emphasized that the designer may have to accept a propeller that is optimized for neither takeoff nor cruise to produce an

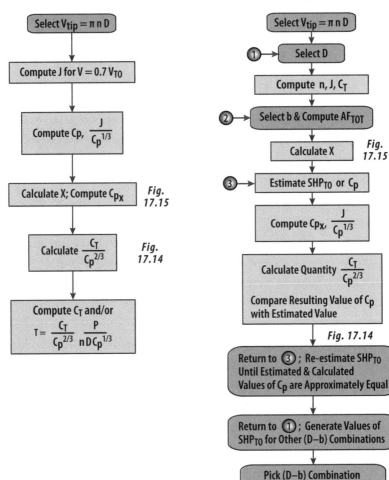

Figure 17.13 Propeller analysis procedure for cruise compared with takeoff.

acceptable performance level for both regimes. A propeller with high cruise efficiency would have blades with low-cambered sections, whereas the optimum blade for takeoff would have a highly cambered section. The solution to this dilemma is to design a propeller with variable-camber blades. Although such a propeller would increase weight and cost somewhat, it does permit very low tip speeds at takeoff to produce desirable noise signatures.

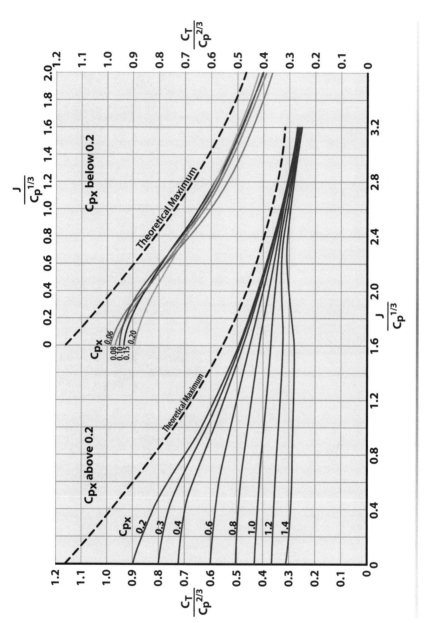

Figure 17.14 Propeller performance chart for takeoff (data from [11]).

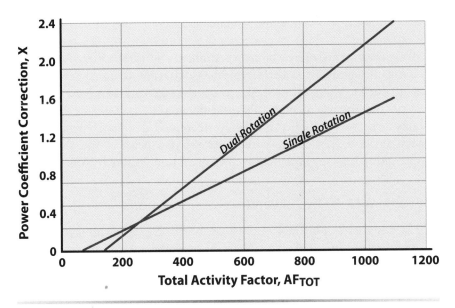

Figure 17.15 Power coefficient adjustment factor.

17.5 Shaft Engine Characteristics

In designing a propeller-driven aircraft, the designer must consider the propeller and its engine together. No discussion of propeller propulsion systems would be complete without some mention of the engine that will turn the airscrew.

Figure 14.2 shows the SHP-to-weight relationships for a spectrum of reciprocating and turboshaft engines. In most cases, the turbine engines include the weight of the reduction gearing required for their application as turboprop engines. The output of these engines is listed in terms of the shaft horsepower being produced to turn the propeller. This is not a complete picture of the capability of the turboprop powerplant because a certain amount of residual jet thrust T_J is also being generated (discussed in Section 14.2.2).

The propeller tip speed is a function of both propeller diameter and shaft speed n; thus, the designer is concerned with the gear ratio between the power turbine and the output shaft. The Allison T56-A-15 (engine data in Appendix J, Fig. J.3) is designed to operate at a constant turbine speed of 13,820 rpm while the propeller shaft turns at a more reasonable 1021 rpm. Powerplant thrust changes are accomplished via simultaneous changes in fuel flow and propeller blade pitch. This same turbine engine could be designed to operate at a different output rpm with the attachment of a reworked gearbox. Each turboprop engine is evolved with a specific propeller in mind, and thus the performance is based on the use of a standard reduction gear.

The performance of a turboprop is similar to a turbojet in that a performance gain is realized with increased velocity due to ram recovery up to about 400 kt, where the propeller losses start to dominate. For a reciprocating engine there is no power increase due to ram recovery; as the velocity increases the propeller efficiency decreases due to compressibility effects, and the drag due to cooling causes the thrust of the propeller–reciprocating-engine combination to drop off rapidly above 200 kt.

Example 17.1   Use of Vendor Propeller Charts

The designer will normally have available both engine and propeller operating charts supplied by the propeller and engine suppliers. This example uses the Piper PA-28–180 Cherokee Archer with a fixed-pitch propeller and the PA-28–200 Cherokee Arrow with a constant-speed, variable-pitch propeller. These two aircraft have the characteristics given in Table 17.1.

The Cherokee Archer at 7000 ft and 2450 rpm has 135 hp available (from Table 14.1). Using Eq. (17.31) the advance ratio at the 122-kt cruise speed is

$$J = V/nD = 206/(40.8)(6.2) = 0.814$$

Table 17.1   Comparison between the Cherokee Archer and Arrow

Aircraft	Cherokee Archer	Cherokee Arrow
Span (ft)	32	32
Wing area (ft^2)	170	170
Aspect ratio	6.02	6.02
TOGW (lb)	2450	2650
W/S (lb/ft^2)	14.4	15.6
Landing gear	Fixed	Retractable
Engine	Lycoming O-360-A	Lycoming IO-360-CIC
Maximum rated hp	185	200
Propeller	Sensenich fixed pitch	Hartzell variable pitch
Propeller diameter (ft)	6.2	6.2
Maximum speed (kt)	129	152
Cruise speed (kt)	122 at 7000 ft, 2450 rpm	143 at 7000 ft, 2450 rpm
Stall speed (kt)	53	57

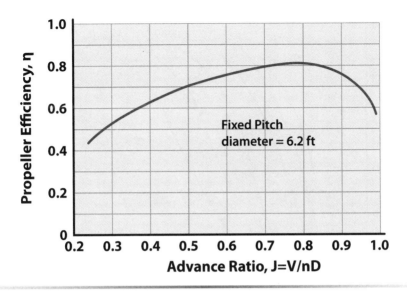

Figure 17.16 Estimated propeller efficiency for the Piper Cherokee Archer PA-28 (courtesy of Sensenich Propeller Manufacturing Co., Inc.).

From Fig. 17.16 the fixed propeller efficiency $\eta = 0.8$. Using Eq. (17.27) the propeller thrust $T = \eta P/V = (0.8)(135)(550)/206 = 288$ lb. If the weight fraction for engine start and climb to 7000 ft is assumed to be 0.97, then the aircraft weight is 2376 lb and the $L/D = 8.26$. The aircraft $C_D = 0.0435$ at the cruise condition.

Using the published brake specific fuel consumption (BSFC) for the Lycoming O-360-A (from Table 14.1) of 0.47 lb/bhp·h, the fuel flow is 63.5 lb/h. For a trip of 400 n mile the Archer would burn about 208 lb of fuel.

For the Cherokee Arrow (Figs. 17.17 and 17.18) at 7000 ft and 2450 rpm the available power from its IO-360-CIC is 150 hp. At its cruise speed of 143 kt the propeller advance ratio is $J = 0.95$. Because the aircraft is equipped with a constant-speed propeller, use the blade-pitch data as shown in Fig. 17.18. Calculating the power coefficient using Eq. (17.24) gives

$$C_P = P\left(\text{ft}\cdot\text{lb}\right)/\rho n^3 D^5 = \left(150\right)\left(550\right)/\left(0.001927\right)\left(67917\right)\left(99161\right) = 0.069$$

Entering Fig. 17.18 with $J = 0.95$ and $C_P = 0.069$ gives a blade-pitch angle $b \sim 27$ deg.

Figure 17.17 shows the estimated propeller efficiency for a variable-pitch propeller. The advantage of a variable-pitch propeller is

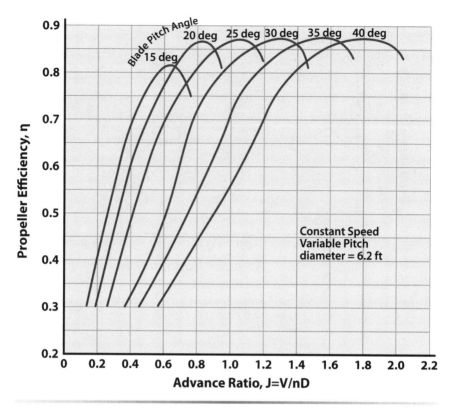

Figure 17.17 Estimated propeller efficiency for the Piper Cherokee Arrow PA-28R (data courtesy of Hartzell Propeller Inc.).

that the blade angle can be adjusted to give a maximum propeller efficiency for different advance ratios and power loadings (horsepower per propeller area). Entering Fig. 17.17 with $J = 0.95$ and $b = 27$ deg gives a propeller efficiency $\eta = 0.85$. The propeller thrust is

$$T = \eta P / V = (0.85)(150)(550)/(241.5) = 290 \text{ lb}$$

Notice that the drag of the Arrow is only 2 lb more than the Archer but it is flying 21 kt faster. The lower drag for the Arrow is due to the retractable gear. The $L/D = 8.85$ for the Arrow and its $C_D = 0.0319$. The Arrow for the same 400–n mile trip would burn 197 lb of gas, or 11 lb less than the Archer, and take almost 30 minutes less time.

At zero forward speed, the efficiency of a propeller is zero by definition, even though its thrust is not zero. In fact, for the same shaft power a variable-pitch propeller will produce the most thrust in zero forward velocity (i.e., its static thrust is greater than the thrust produced in forward flight). Figures 17.19 and 17.20 can be used to esti-

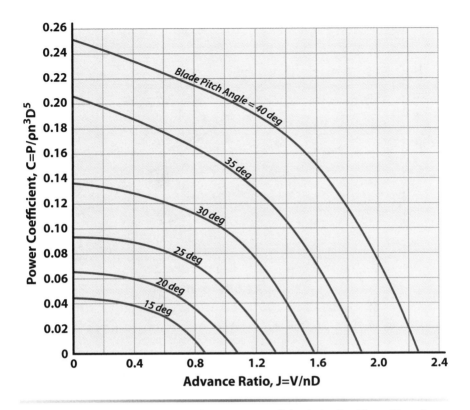

Figure 17.18 Estimated propeller power coefficients for the Piper Cherokee Arrow PA-28R (data courtesy of Hartzell Propeller Inc.).

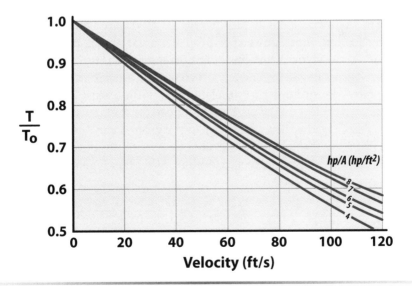

Figure 17.19 Decrease of thrust with velocity for different power loadings (data from [18]).

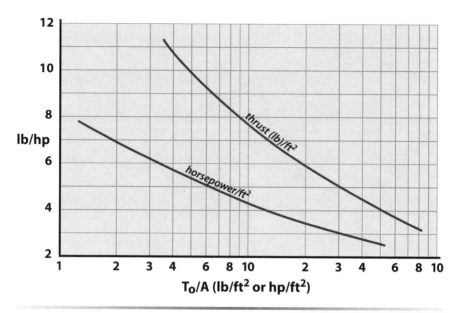

Figure 17.20   Static thrust and power performance of propellers or rotors (data from [18]).

mate the thrust available from a variable-pitch propeller at low forward speeds. The static thrust is first obtained from Fig. 17.20 and then reduced by a factor from Fig. 17.19. These charts apply only to a constant-speed propeller, which allows the engine to develop its rated power regardless of the forward speed. These charts are used to estimate the static thrust for the Cherokee Arrow.

The Arrow at takeoff has 200 hp at 2700 rpm. This gives it a power loading (horsepower per propeller area) of 6.62 hp/ft^2. From Fig. 17.20 the static thrust-level per horsepower (lb/hp) is 4.9, giving a static thrust of 980 lb. The takeoff analysis presented in Chapter 10 calculates the ground run acceleration for the thrust available at $0.7V_{TO}$, where $V_{TO} = 1.2V_{Stall}$. The stall speed for the Arrow is 57 kt, or 96 ft/s. So, use the thrust at 80 ft/s in the ground run analysis. From Fig. 17.19 the thrust at 80 ft/s is about 67.5% of the static thrust, or 662 lb. This is a respectable acceleration T/W of 0.25, giving a takeoff distance of about 1000 ft.

References

[1] McCormick, B. W., Jr., *Aerodynamics of V/STOL Flight*, Academic Press, New York, 1967.
[2] Theodorsen, T., *Theory of Propellers*, McGraw-Hill, New York, 1948.

[3] Dommasch, D. O., *Elements of Propeller and Helicopter Aerodynamics*, Pitman, New York, 1953.

[4] Glauert, H., "Airplane Propellers," *Aerodynamic Theory*, Vol. 4, edited by William F. Durand, Dover, New York, 1963.

[5] Gessow, A. and Myers, G. C., Jr., *Aerodynamics of the Helicopter*, Ungar, New York, 1952.

[6] Stepniewski, W. Z., *Introduction to Helicopter Aerodynamics*, Rotorcraft Publishing Committee, Morton, PA, 1950.

[7] Stack, J., Draley, E. C., Delano, J. B., and Feldman, L., "Investigation of the NACA 4-(3)(08)-045 Two-Blade Propellers at Forward Mach Numbers to 0.725 to Determine the Effects of Compressibility and Solidity on Performance," NACA TR-999, 1950.

[8] Goldstein, S., "On the Vortex Theory of Screw Propellers," *Proceedings of the Royal Society, Series A: Mathematical and Physical Sciences*, Vol. 123, 1929.

[9] Crigler, J. L., and Jaquis, R. E., "Propeller-Efficiency Charts for Light Airplanes," NACA TN 1338, 1947.

[10] "Generalized Method of Propeller Performance Estimation," Hamilton Standard Div., Hamilton Standard Publ. PDB 6101A, United Aircraft Corp., 1963.

[11] Gilman, J., Jr., "Propeller-Performance Charts for Transport Airplanes," NACA TN 2966, 1953.

[12] Kuhn, R. E., and Draper, J. W., "Some Effects of Propeller Operation and Location on the Ability of a Wing with Plain Flaps to Deflect Propeller Slipstreams Downward for Vertical Takeoff," NACA TN 3360, 1955.

[13] Spreeman, K. P., "Investigation of a Semi-Span Tilting-Propeller Configuration and Effects of Wing Chord to Propeller Diameter on Several Small-Chord Tilting-Wing Configurations," NASA TN D-1815, 1963.

[14] Stepniewski, W. Z., and Schmitz, F. H., "Noise Implications for VTOL Development," Society of Automotive Engineers Paper 70–286, 1970.

[15] Hubbard, H. H., "Propeller Noise Charts for Transport Airplanes," NACA TN 2968, 1953.

[16] Rosen, G., and Rohrbach, C., "The Quiet Propeller—A New Potential," AIAA Paper No. 69–1038, 1969.

[17] Perkins. C. D., and Hage, R. E., *Airplane Performance, Stability and Control*, Wiley, New York, 1949.

[18] McCormick, B. W., *Aerodynamics, Aeronautics and Flight Mechanics*, Wiley, New York, 1995.

Chapter 18

Propulsion System Thrust Sizing

- Turbine Engine Thrust Sizing
- Turbine Engine Scaling
- Piston Engine Sizing
- Solar-Power Sizing
- Human-Power Sizing
- Rocket Engine Sizing

Formula One racing aircraft Nemesis rounding the pylon on its way to another win at the Reno National Championship Air Races. (Photograph courtesy of Jon Sharp.)

Small deeds done are better than great deeds planned.

Peter Marshall

18.1 Introduction

At this point in the design game, the size of the fuselage is known, the general configuration has been established, and the first estimate of the aircraft aerodynamics is complete. Now the propulsion unit needs to be sized, that is, select the $(T/W)_{TO}$ so that the aircraft performance can be determined and, in the case of a jet aircraft, the inlet sized and designed.

Sometimes the designer works with existing "off the shelf" engines, such as those reported in Appendix J. The designer would vary the number and type, finding the combination that gives the required vehicle performance at minimum weight, cost, and noise. Other times the designer might work with a nonexisting conceptual engine. In this case, the engine manufacturer would give the designer a "rubber" engine, that is, a paper engine that can be scaled up or down according to scaling laws established by the engine manufacturer. In the case of a jet engine, the designer would vary the engine thrust and perhaps the number of engines to fit the parametric study. The appropriate engine weight, diameter, and length would be determined from the engine scaling information. Occasionally, the designer might examine the influence of engine turbine temperature, overall pressure ratio, fan pressure ratio, and bypass ratio on aircraft design. However, this type of parametric study is usually performed by the engine manufacturer as part of the paper engine design.

The $(T/W)_{TO}$ is usually sized by one or more of the following items:

1. Cruise/Loiter
2. Energy maneuverability (air combat)
3. Acceleration-time and fuel burned during acceleration
4. Takeoff
5. Maximum speed

These criteria are often in conflict with one another. The designer must consider the $(T/W)_{TO}$ for these mission requirements, establish their priorities, and then decide upon an appropriate compromise, after looking at the entire mission. Table 18.1 indicates trends in $(T/W)_{TO}$ based upon current aircraft.

18.2 Turbine Engine Scaling

Assume that the aircraft designer has selected the propulsion type, in terms of bypass ratio, turbine inlet temperature, and pressure ratio, and now desires to scale it up or down to get the proper T/W. The engine man-

Table 18.1 T/W Range for Various Aircraft Types

Dominant Mission Requirement	Range for $(T/W)_{\text{TO}}$ (uninstalled)
Long range	0.2–0.35
Short and intermediate range with moderate field length	0.3–0.45
STOL and utility transport	0.4–0.6
Fighter—close air support	0.4–0.6
Fighter—strike interdiction	0.45–0.7
Fighter—air-to-air	0.8–1.3
Fighter—interceptor	0.55–0.8

ufacturer provides a reference engine (either a paper engine or an existing engine) and the appropriate scaling information. The engine weight, diameter, and length scale according to the mass flow as follows:

$$W_{\text{eng}} = \left(\frac{\dot{m}}{\dot{m}_{\text{REF}}} \right)^{n} \left(W_{\text{eng}} \right)_{\text{REF}} \tag{18.1}$$

where $n = 0.8$–1.3 (usually about 1.0) and $\dot{m}$ is sea level static (SLS) airflow required for the engine. From Chapter 14,

$$T = \dot{m}\left(V_e - V_a \right) + A_e \left(P_e - P_a \right) \tag{14.1}$$

and if constant nozzle velocity V_e is assumed for any thrust size, then

$$\left(\frac{T}{T_{\text{REF}}} \right) = \left(\frac{\dot{m}}{\dot{m}_{\text{REF}}} \right) \tag{18.2}$$

Equation (18.2) is usually a pretty good assumption and facilitates the engine scaling.

The engine diameter d and length ℓ scale as follows:

$$d = \left(\frac{\dot{m}}{\dot{m}_{\text{REF}}} \right)^{1/2} d_{\text{REF}} \tag{18.3}$$

$$\ell = \left(\frac{\dot{m}}{\dot{m}_{\text{REF}}} \right)^{n-(1/2)} \ell_{\text{REF}} \tag{18.4}$$

18.3 Turbine Engines Sized for Cruise Efficiency

The turbine engine is sized by matching the thrust required (drag) during best cruise condition with the thrust available at the power setting for minimum thrust specific fuel consumption (TSFC). This power setting for minimum TSFC varies from engine to engine. For example, the F-100 (Fig. 14.8d) at Mach $= 0.8$ and 36,089 ft has a minimum TSFC at 70% normal rated thrust (NRT) whereas the TF-39 (Fig. 14.9c) at Mach $= 0.8$ and 35,000 ft has a minimum at 100% NRT. The engine sizing should be checked at several points during the cruise, as the aircraft cruise climbs, to find the best sizing compromise.

Example 18.1 Sizing for Optimum Cruise Performance

Consider a four-engine long-range cruise transport using scaled TF-39 engines. Size the engines for optimum cruise performance.

Assume	
W_{TO}	500,000 lb
$(W/S)_{TO}$	120 psf
W_{fuel}/W_{TO}	0.40
S_{ref}	4167 ft^2
W/S at start of cruise	116 psf
Cruise at Mach	0.8
$C_{D_{min}}$	0.018, $d/b = 0.1$
Wing, AR	8.0
Λ	30 deg
λ	0.37
Section	NACA 64$_2$-215 airfoil
Determine	
Cruise	from 31,000 ft to 41,000 ft
From Appendix F:	
α_{0L}	2.6 deg
r_{LE}	1.1% chord
From Section 13.1.1:	
C_{L_α}	0.1 per degree
$C_{L_{min}}$	0.26 (see Fig. F.5)
From Section 13.2.1 and Fig. G.9 ($e = 0.65$):	
K	0.0606
K'	0.0406
K''	0.02
$(L/D)_{max}$	17.3 [from Eq. (3.10b), Chapter 3]
Drag and thrust required (start of cruise):	
q	269.5 psf at Mach $= 0.8$ and 31,000 ft
$C_L = W/q\,S_{ref}$	$116/270 = 0.430$
$C_D = C_{D_{min}} + K'\,C_L^2 + K''(C_L - C_{L_{min}})^2$	$0.018 + 0.0081 = 0.0261$

Cruise L/D	$0.43/0.0261 = 16.48$ [agrees well with result from Eq. (3.29)]
Drag	$C_D q S (0.0261) (269.5)(4167) =$ 29,308 lb
Required thrust	7327 lb (each engine)

An examination of Fig. 14.9b for 31,000 ft indicates that the power setting for the TF-39 that gives the lowest TSFC in continuous operation is 100% NRT. Thus, for 100% NRT: thrust available each engine = 9460 lb. Therefore, use four TF-39 engines that are scaled: scaling factor = 7327/9460 = 0.775.

The scaled engines will have 77.5% of the thrust and airflow (assume V_e to be the same) of the TF-39 engines:

Assume	
Engine weight	$(0.775)(7026) = 5442$ lb
Diameter	$\sqrt{(0.775)(100)} = 88$ inches
Bullet length	$\sqrt{(0.775)(271)} = 238.6$ inches for $n = 1.0$
T_{SLS}	$0.775 (41,100) = 31,853$ lb
$(T/W)_{TO}$	0.255
Check engine sizing for end of cruise	
At end of cruise	$W/S \sim 80$ psf
q	168 psf at Mach = 0.8 and 41,000 ft
C_L	0.477
C_D	$0.018 + 0.010 = 0.028$
Cruise L/D	16.93
Drag	19,568 lb
Thrust required each engine	4892 lb
Thrust available each engine	$(6400)(0.775) = 4960$ lb at 100% NRT (from Fig. 14.9c)

Thus, the 77.5% scaling of the TF-39 provides a good engine match at the beginning and end of cruise.

18.4 Energy Maneuverability (Air-to-Air Combat)

The performance of an aircraft in air combat at a point in velocity–altitude space is indicated by its value of *maximum sustained turn rate* (from Chapter 3):

$$\dot{\psi}_{MS} = \frac{g\sqrt{n_{MS}^2 - 1}}{V} \qquad (3.32)$$

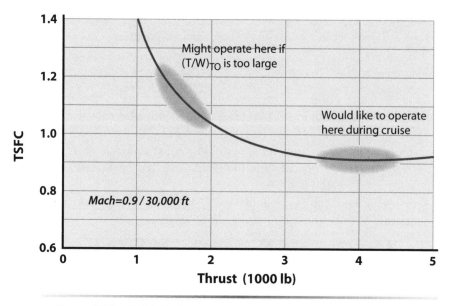

Figure 18.1   F-100 TSFC for partial power setting at Mach = 0.9 and 30,000 ft (see Fig. 14.7).

where n_{MS} is the *maximum sustained load factor*. The n_{MS} can be expressed by

$$n_{MS} = \frac{q}{\sqrt{W/S}} \sqrt{\frac{1}{K}\left[\frac{T}{W}\frac{1}{q} - \frac{C_{D_0}}{W/S}\right]} \tag{18.5}$$

(from Section 3.9). Equation (18.5) indicates that large T/W gives improved combat maneuverability. However, as T/W increases the cruise situation worsens because the engine would have to be throttled back during cruise, which moves away from the minimum TSFC bucket as shown in Fig. 18.1.

Also, as T/W increases so does the propulsion weight. Thus, for a large T/W there would be a lot of extra weight that is used for only a portion of the mission (admittedly the most important part of the mission). Again it is emphasized to examine the entire mission fuel requirement before selecting the $(T/W)_{TO}$.

18.5 Engine Sizing for Acceleration

Acceleration can be examined very simply by looking at the ideal rocket equation

$$\Delta V = I'_{sp} g \ln\left(\frac{W_i}{W_f}\right) \tag{18.6}$$

where I'_{sp} is the effective I_{sp} available for accelerating the vehicle. The effective I_{sp} is defined as

$$I'_{sp} = I_{sp}\left(1 - \frac{D}{T}\right) \qquad (18.7)$$

where D is the drag, T is the thrust available, and $I_{sp} = 3600/\text{TSFC}$ is the *engine specific impulse*. It is clear from Eqs. (18.6) and (18.7) that the thrust must be much larger than the drag; otherwise the I'_{sp} will be small and considerable fuel will be expended during a ΔV acceleration. If ΔV is a very large increment, the I'_{sp} must be an averaged value over the acceleration interval.

The acceleration performance of an aircraft improves as D/T decreases or $(T/W)_{TO}$ is increased. The reason is that the excess thrust (i.e., $T - D$) increases, which decreases acceleration time and fuel burned. Absolute minimum intercept or acceleration time would mean $(T/W)_{TO} \to \infty$. A typical plot of time vs $(T/W)_{TO}$ is shown in Fig. 18.2 and it is observed that after a certain $(T/W)_{TO}$ the curve gets rather flat, resulting in a small improvement for additional $(T/W)_{TO}$. Equations (18.6) and (18.7) indicate that the fuel burned during an acceleration continues to decrease as $(T/W)_{TO}$ increases. It is misleading to look solely at acceleration fuel burned because, as $(T/W)_{TO}$ increases, the weight of the propulsion system increases. Thus, a better quantity to examine is the sum of engine weight plus fuel weight. This is shown plotted in Fig. 18.2 for a conceptual Advanced Manned Interceptor (AMI). Near-minimum acceleration time is certainly important for an interceptor; however, the designer must trade off decreased acceleration time with increasing engine-plus-fuel weight. Figure 18.2 shows the point design for the AMI at a $(T/W)_{TO} = 0.586$ (two reference turbo-ramjet engines), which is a compromise between acceleration time and engine-plus-fuel weight. The AMI is also range dominated and the designer is reminded to examine the cruise efficiency also before making the final selection for $(T/W)_{TO}$.

18.6 Turbine Engine Sizing for Takeoff

If a short takeoff distance is a primary mission requirement, it should be considered in a fair amount of detail at this point because it may size the engines. The takeoff analysis is discussed in Chapter 10. It must be remembered that a short takeoff distance can be achieved using combinations of $(T/W)_{TO}$, $(W/S)_{TO}$, and high-lift devices [see Fig. 6.3 and Equation (6.3)]. Thus, a short takeoff distance need not have a high $(T/W)_{TO}$.

Example 18.2 Turbine Engine Sizing Dilemma

Figure 18.3a shows the mission profile for the Advanced Tactical Fighter (ATF) that became the F-22. The mission profile is very

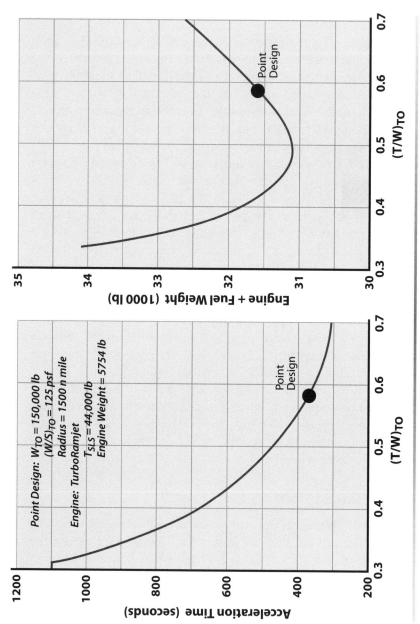

Figure 18.2 Acceleration performance of an Advanced Manned Interceptor (acceleration from 250 ft/s to Mach 4.5 at 75,000 ft).

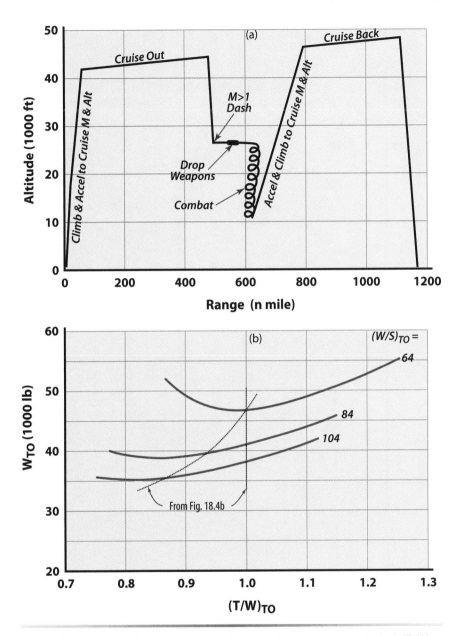

Figure 18.3 Typical tactical fighter mission profile and its associated $(T/W)_{TO}$ variation.

demanding as it calls for a significant supersonic cruise phase, a supersonic dash, several acceleration phases, and air combat. The Air Force was asking for supercruise, supermaneuver, and superstealth— all in the same airplane. This example will bring out the dilemma

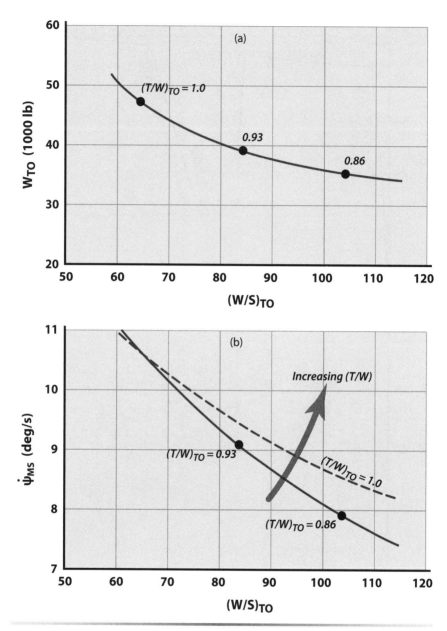

Figure 18.4 $(W/S)_{TO}$ variation for a typical tactical fighter on its basic mission.

facing the designer when selecting the $(W/S)_{TO}$ and $(T/W)_{TO}$ for an aircraft that is driven by several conflicting requirements.

Figures 18.3b and 18.4a show that the W_{TO} decreases for increasing $(W/S)_{TO}$ and decreasing $(T/W)_{TO}$. The decreasing $(W/S)_{TO}$ is understandable from the discussion in Chapter 6 and the fact that the level of air combat is not specified. The best $(T/W)_{TO}$ from Figures

18.3b and 18.4a would be a compromise between the supercruise, supersonic dash, and acceleration requirements. When air combat is considered, Fig. 18.4b, the situation is reversed; the desire to have high $\dot{\psi}_{MS}$ would drive the design to high $(T/W)_{TO}$ and low $(W/S)_{TO}$. The designer has a dilemma and must compromise the design to give tolerable cruise performance, acceleration fuel, and air combat levels.

18.7 Solar Power

The sun is a source of unlimited energy during the day. Every day it bathes the outer edge of the earth's atmosphere with 127 W/ft^2 of solar energy on average. The 127 W/ft^2 is termed the *solar constant*. The amount of solar energy received anywhere on the earth at a point in time depends on the latitude Φ of the surface, the tilt (inclination) of the earth's spin axis as it orbits around the sun, and its position relative to the sun (time of day).

This dependence is shown in Fig. 18.5. The inclination of the earth to the orbital plane varies between +23.5 deg on 21 June and −23.5 deg on 21 December and is the reason the earth has its four seasons. On 21 June the northern hemisphere is getting more solar energy and is enjoying summer while the southern hemisphere is getting less and is having winter. On 21 December the situation reverses. On 21 June the northern hemisphere has its longest day of the year and on 21 December the shortest.

The solar energy received on earth is converted to useful electrical energy by the photovoltaic action of solar cells. The electrical energy per unit area available from a horizontal solar cell of efficiency η_{SC} at an altitude H and solar elevation angle θ is

$$P_{\text{Elect}} = P_{\text{Solar}}\eta_{SC} \sin\theta \tag{18.8}$$

where θ is the elevation angle of the sun above the horizon and $\sin\theta$ accounts for the presented area of the horizontal solar cell. The solar elevation angle is a complicated function of the latitude, inclination angle (time of year), and orientation to the sun (time of day) [1,2]. The best way to determine θ is to go to the National Oceanic and Atmospheric Administration (NOAA) Web site and use their solar position indicator.

The quantity P_{Solar} is the average solar radiation at altitude H and solar elevation angle θ. The earth's *atmospheric mass* (AM) has a significant effect on the value of P_{Solar}. The water and ozone in the atmosphere absorb and scatter the solar radiation: $P_{\text{Solar}} = 127$ W/ft^2 in space (outside the earth's atmosphere at an altitude of approximately 320,000 ft or 53 n mile), whereas $P_{\text{Solar}} = 96.5$ W/ft^2 on the earth's surface and $\theta = 90$ deg, having suffered a 24% energy loss due to atmospheric attenuation. The space condition is termed AM0 and the condition on the earth's surface and $\theta = 90$ deg is AM1.0. Values for P_{Solar} at altitude H and solar elevation angle θ are given

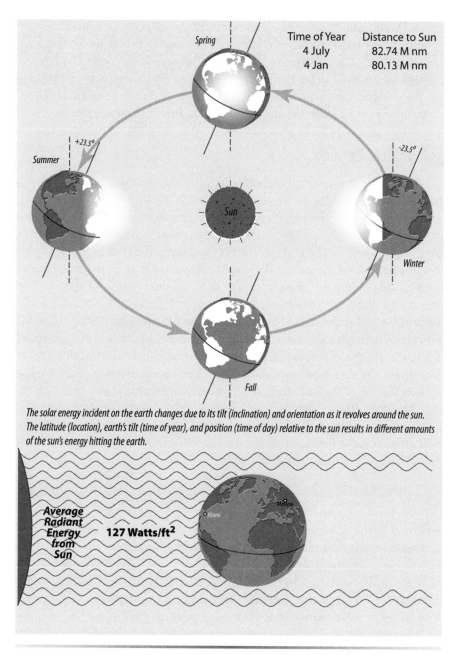

The solar energy incident on the earth changes due to its tilt (inclination) and orientation as it revolves around the sun. The latitude (location), earth's tilt (time of year), and position (time of day) relative to the sun results in different amounts of the sun's energy hitting the earth.

Figure 18.5 Solar energy radiated to earth during the year.

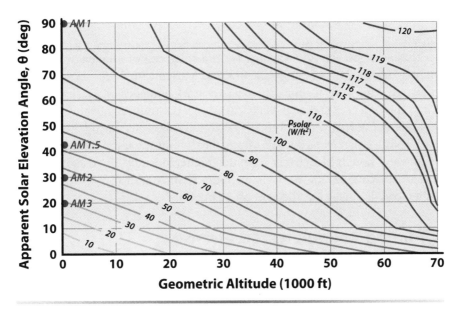

Figure 18.6 Direct clear-sky P_{solar} (W/ft²).

in Fig. 18.6 (essentially, H and θ define the *slant range* through the atmosphere).

Even though the solar energy is limitless and free, it is small when compared with the energy available from burning hydrocarbon fuels (i.e., gasoline or kerosene). Because the power required increases by the cube of the speed and the available solar power is small, the speed of a solar aircraft will always be less than 30 KEAS (knots equivalent airspeed). This can be shown by setting power required = power available.

$$\text{Power required} = (\text{Drag})(\text{Velocity})/(\text{Propulsive efficiency})$$
$$= \left(\tfrac{1}{2}\rho C_D S_{\text{Wing}} V^2\right)(V)/\eta_{\text{Prop}} \tag{6.10}$$

$$\text{Power available} = P_{\text{Solar}}\eta_{\text{SC}} S_{\text{SC}} \sin\theta \tag{18.8}$$

Then assume typical values for the parameters and solve for V as follows:

Altitude = 60,000 ft	$\rho = 0.000224$ slug/ft³
$\theta = 90°$ (optimistic)	$P_{\text{Solar}} = 120$ W/ft² (Fig. 18.6)
$\eta_{\text{SC}} = 31\%$	35% in lab, 31% installed on wing
$\eta_{\text{Prop}} = 0.81$	Motor, propeller and line losses
$S_{\text{Wing}} = S_{\text{SC}}$	Reasonable and it makes the math simpler
$C_D = 0.0356$	Condor at best loiter condition (see Fig. G.11)
Payload and vehicle power = 0	Not realistic but makes the point

V = 177 ft/sec = 105 knots at 60,000 ft ~ 30 KEAS, and this is the best it can do.

The electrical energy available from a solar cell of efficiency 28.7% at Miami, Florida (latitude $\Phi = +25°46'$), and Moscow ($\Phi = +55°45'$) is shown in Fig. 18.7 for 4 July and 5 January over a 24-hour period [1,2]. The area under the curves is the total electrical energy per unit area [watt-hours per square foot ($W \cdot h/ft^2$)] captured by horizontal solar cells during the daylight hours. Notice that Moscow and Miami have about the same total energy on 4 July even though Moscow is at a much higher latitude. The reason for this is that Moscow has more daylight hours than Miami (17 and 14 hours, respectively). This is not the case on 4 January.

18.8 Sizing Solar-Powered Aircraft

It is time to return to the Solar Snooper, introduced in Chapter 6, and close the design shown in Fig. 6.8 by determining if the wing size S_W = 2793 ft^2 from Section 6.7 will provide enough solar-cell area to meet the power required. Then the design can be closed by developing a weight buildup to meet the assumed 4800 lb takeoff gross weight (TOGW).

The requirement for the Solar Snooper is to provide 4 weeks of "24/7" intelligence, surveillance, and reconnaissance (ISR) at 64,000 ft, in midlatitudes during the summer months. The payload is 500 lb, which requires 1 kW of power.

The goal is to design a solar-powered aircraft for operation over Miami and Moscow during the 4-week period starting 7 June and ending 4 July. Trade studies reveal that 4 July is the critical day during the 4-week interval in terms of energy available from the sun (longest distance to sun).

The analysis of Section 6.7 concluded that the total power required to operate at 68 kt at 64,000 ft at C_L = 1.33 was a continuous 24 kW. Refer to the electric aircraft data base shown in Table 14.2 and assume a solar-cell efficiency of 32% (note, these will be multijunction cells and will be expensive). Also, assume that the energy for nighttime operation will be stored in batteries instead of fuel cells (this is an important trade study that needs to be conducted and is left as an exercise for the reader). The baseline round trip efficiency for the batteries is 0.9. However, there are line losses that need to be considered. First there is a transmission efficiency η_{Trans} = 0.98 going in and coming out of the battery. Then there is a power control switch/step efficiency η_{Switch} = 0.90 going in and coming out of the battery. Thus, the total round trip efficiency for the battery storage system is η_{RT} = (0.98)(0.98)(0.9)(0.9)(0.9) = 0.7. The continuous power that needs to be provided to the batteries is 24/0.7 = 34.28 kW for operating at night. The continuous required power loading for the batteries is P_{Req} = (34.28)(1000)/ (2793) = 12.3 W/ft^2, where the solar cell area is assumed to be the wing

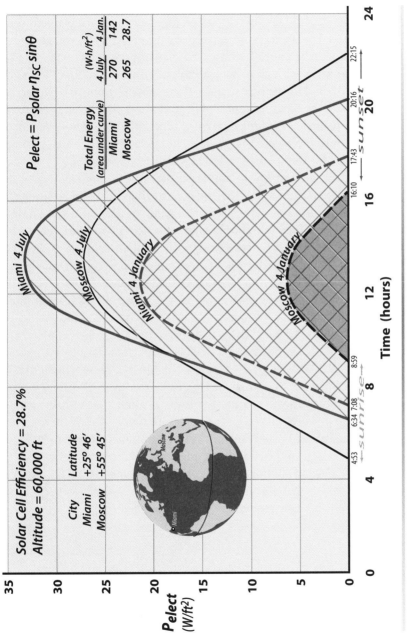

Figure 18.7 Comparison of electrical energy available for Miami and Moscow at 60,000 ft.

surface area. This P_{Req} will be balanced with the available energy collected by the solar cells.

Solar cells are assumed to have a laboratory efficiency of 32%. As mentioned, these are multijunction cells and will be expensive. Each solar cell generates 0.5 volt. The cells are connected together to form a blanket (typically 36 individual cells are connected together, generating 18 volts DC with a blanket packing efficiency of ~95%). The blankets are connected together to form a solar array with an array electronics efficiency of 95%. The solar arrays are glued onto the vehicle surface. If the cells are going to be in service for long periods, the environment will degrade the cell efficiency by about 1.5% per year (called *end-of-life efficiency*). The Solar Snooper cells will be in service for only 4 weeks so that this will not be a concern. Thus, the end-of-life efficiency for the solar cells will be $(32)(0.95)(0.95) = 28.7\%$.

The solar cells are put on the upper surface of the wing. However, the cells should not run right to the leading edge as their heat and contour disturbance will trip the boundary layer to turbulent and limit the laminar extent of the wing. The cells should start at about 15% chord, where the boundary layer thickness is large compared with the thickness of the solar cells. Similarly, there is about 5% of the trailing edge region that is not usable for the solar cells. Common practice initially sizes the horizontal tail to recover the 20% of the wing area lost for the solar cell installation (note, this initial tail sizing can be compared with the tail volume coefficient method of Chapter 11 later). Thus, the horizontal tail is $S_{HT} = 558$ ft^2 and the solar cell area is assumed to be 2793 ft^2. This assumption lets us now balance $P_{req} = 12.3$ W/ft^2 with the power available P_{Elect} shown in Fig. 18.7.

This power balance is shown in Fig. 18.8. The total electrical energy available on 4 July over Miami and Moscow is 270 and 265 W · h/ft^2 respectively. The total power required by the Solar Snooper is 12.3 W/ft^2 continuous over the nighttime period and 8.6 W/ft^2 continuous during the daytime (the difference is due to the round trip efficiency of the batteries).

During the daylight hours of 4 July the aircraft must collect an excess amount of power A_1 that equals the storage power required $(R_1 + R_2)$. From Fig. 18.8 the power sizing results are (shown for Moscow)

Over Miami:

$$A_1 - (R_1 + R_2) = 167 - 154.6 = 12.4 \text{ W} \cdot \text{h/ft}^2 \left(8\% \text{ margin}\right)$$

Over Moscow:

$$A_1 - (R_1 + R_2) = 140 - 127.2 = 12.8 \text{ W} \cdot \text{h/ft}^2 \left(10\% \text{ margin}\right)$$

Because 4 July is the critical day (least total energy available) there will be excess energy collected on all other days in the 4-week surveillance

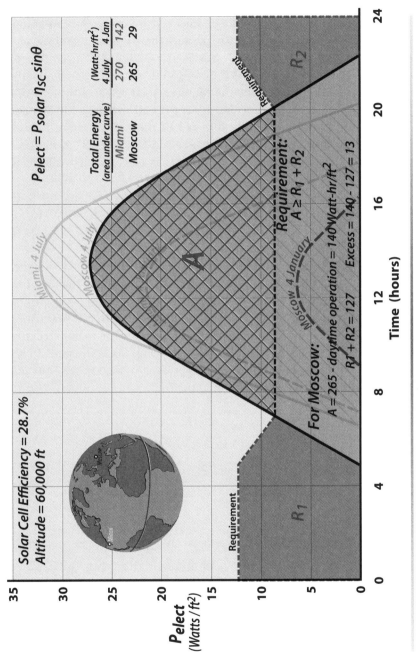

Figure 18.8 Diurnal energy balance example for stationkeeping over Moscow.

period. It is a good idea to carry at least a 10% margin during the conceptual design phase. A power balance closure for Miami and Moscow has been obtained.

The horizontal tail area should be checked by using the tail volume coefficient method of Chapter 11 to make sure there is adequate static pitch stability. From Table 11.8, $C_{HT} = \ell_{HT} S_{HT}/S_W \bar{c} = 0.34$ for ISR aircraft. From Fig. 6.8, $\ell_{HT} = 41$ ft and $\bar{c} = 8$ ft so that $S_{HT} = 185$ ft^2. Observe that S_{HT} is sized not by static pitch stability but by the required solar cell area by a factor of 3. This results in a neutral point that is very far aft. Because the center of gravity should be located slightly forward of the neutral point (a static margin of approximately +5% $\bar{c}$ as discussed in Chapter 22) to minimize trim drag, the location of the center of gravity can be changed by sliding the payload pod fore and aft.

The vertical tail area is determined by static yaw stability using the method of Chapter 11. From Table 11.8, $C_{VT} = \ell_{VT} S_{VT}/S_W b = 0.014$ for ISR aircraft. From Fig. 6.8, $\ell_{VT} = 55$ ft and $b = 317$ ft so that $S_{VT} = 225$ ft^2 for the two verticals.

Now it is time to re-examine the assumed TOGW = 4800 lb. Weight is estimated for electric motors, solar cells, and batteries using the data from Table 14.2. Estimating the weights of the wing, tails, landing gear, payload pod, and booms is a real challenge for a $W/S < 5$ lb/ft^2 aircraft because the historical data base is almost nonexistent. The wing is the major structural component and its weight is estimated using Fig. 20.1. Chapter 19 will discuss this dilemma but it will remain a design weakness. The Solar Snooper weight summary is shown in Table 18.2.

Table 18.2 Solar Snooper Weight Summary

Component	Weight (lb)	Reference
Wing	838	0.30 lb/ft^2 (Fig. 20.1) for $W/S = 1.72$
Pod and booms	341	Sailplane data [3]
Motor, propellers, install	188	0.2 kW/lb + 25% install factor (Table 14.2)
Solar cells	279	0.1 lb/ft^2 (Table 14.2)
Landing gear	96	2% of TOGW (sailplane data)
Tails	235	0.30 lb/ft^2 (Fig. 20.1)
Payload	575	Requirement (500 lb) + 15% for installation
Batteries	1524	34.3 kW for 12 hr at 3.7 lb/kW (Table 14.2)
Battery installation	230	15% installation factor
Avionics, actuators	100	Double that for Helios
Margin	393	8% (should carry at least a 6% margin)
Total	**4800**	

An interesting question at this point is "Is there a Solar Snooper design that could operate over Moscow on 4 January?" It is clear from Fig. 18.8 that the design shown in Fig. 6.8 will not work because the total available electrical energy of 28.7 W·h/ft^2 is nowhere close to the required nighttime $R_1 + R_2 = 127.2$ W·h/ft^2. Thus, the design must change considerably.

If it is assumed that the daytime continuous $P_{Req} = 1$ W/ft^2 over ~7 h (or 7 W·h/ft^2), then the excess electrical energy $A_1 = 28.7 - 7 = 21.7$ W·h/ft^2. The required nighttime energy $R_1 + R_2 = (1)(14)/0.7 = 20$ W·h/ft^2, which gives a positive power balance with an 8.5% margin—so the design is "in the ballpark."

However, the challenge is to decrease the daytime continuous P_{Req} from 8.6 W/ft^2 for the current design to 1.0 W/ft^2 for the new design. Some design changes to consider are the following:

- Decrease the payload and aircraft operation power required to 0.5 kW each.
- Decrease the W/S from 1.72 to 1.0 lb/ft^2. This would increase the wing area (more area for solar cells) and decrease the speed to about 90 ft/s. The increase in wing area would increase the TOGW (heavier wing and more solar cells) and drag, but the overall impact would be a lower propulsion power required.
- Increase the vertical tail area and cover it with solar cells. This would provide more electrical energy especially at the low solar elevation angles over Moscow in the winter. The flight path would have to be tailored to take advantage of the vertical solar cells.
- Finally, the solar cell efficiency could be increased within reason.

It remains as an exercise for the reader to determine if there is a design that closes. It should be obvious to the reader that the design of an aircraft powered by hydrocarbon fuels (i.e., gasoline, diesel, JP-4, etc.) is a much easier challenge than the design of a solar-powered aircraft. This is because the hydrocarbon-powered aircraft:

- Is not expected to have an endurance of more than about 3 days (72 hours).
- The size of the required propulsion unit is independent of latitude, time of year, and time of day.
- The size of the wing is independent of latitude, time of year, and time of day.
- If there is a thrust shortfall, get a bigger engine (do not have to resize the whole aircraft).
- And the list goes on!

18.9 Piston Engine Sizing—HAARP

From the discussion in Chapter 5 (Section 5.8) HAARP is required to fly at 100,000 feet and Mach 0.6 (594.7 ft/s). The design information for sizing the piston engine is as follows:

TOGW	16,000 lb
Weight at start of cruise	14,880 lb
L/D at start of cruise	27
Drag at start of cruise	14,880/27 = 551 lb
High-altitude propeller (designed using ISES code):	
Diameter	24 ft
RPM	528
Efficiency η_P	0.85
Power required at start of cruise	(drag)(speed)/550 η_P = 700 hp

Teledyne Continental (TCM) had provided the engines and developed the two-stage turbochargers for the Boeing Condor. Discussions with TCM centered around their family of geared, liquid-cooled piston engines and their turbocharger experience. Their GTSIOL-550 piston engine was selected. The engine specifications were as follows:

Takeoff–climb power	500 hp
Continuous cruise maximum power	375 at BSFC = 0.42
94% maximum power	350 at BSFC = 0.40
Number of cylinders	6
Weights (total = 581 lb):	
Engine	445 lb
Ignition and plugs	30 lb
Exhaust manifold	12 lb
Starter	19 lb
Gearbox	75 lb

The HAARP configuration shown in Fig. 5.11 is a twin-engine pusher design. The engines are in wing pods. The propeller arrangement consists of an 8 ft diameter, four-blade propeller for takeoff, landing, and climb and a 24 ft diameter, two-blade propeller for high altitude. The small propeller would operate all the time whereas the large propeller would be clutched in at 45,000 ft. The two engines would be operated at maximum power (500 hp) for climb but throttled back to 94% (350 hp) for cruise at Mach=0.6 at 100,000 ft.

The Lockheed Skunk Works submitted a proposal to NASA Dryden Flight Research Center in 1991 to build and operate two HAARP aircraft. NASA declined the offer and instead contracted with Aurora Flight Systems to build the Perseus aircraft (Fig. 18.9). The Perseus B exceeded 60,000 ft in 1998 with a three-stage turbocharged piston engine.

Figure 18.9 The Aurora Flight Systems Perseus UAV developed for NASA high altitude ozone measurements.

A structural analysis and design will be conducted for the HAARP wing in Chapter 19 (Section 19.14).

18.10 Human-Powered Aircraft—Daedalus

The design of a human-powered aircraft starts with the description and performance of the propulsion system. The powerplant in this case is the human engine. Yale University investigated the limits of endurance and the power level of the human powerplant. Their research concluded that an endurance-trained athlete using a specially built recumbent ergometer (essentially a reclined bicycle) could produce a specific power of 3 W/kg (0.00183 hp/lb) for several hours [4,5]. For peak performance the athlete needed preloading with glycogen, controlled temperature, and adequate water supply.

Thus, the available power would be 0.2745 hp for a well-trained 150-lb athlete.

Because the power available is small the design approach is very similar to that of a solar-powered aircraft, such as the Perseus shown in Fig. 18.9 and the Solar Snooper discussed in Example 6.7. The aircraft speed will be low (less than 20 kt) and the wing loading less than 1 lb/ft^2.

Previous human-powered projects have shown that the aircraft is about two-thirds the weight of the pilot. For our 150-lb pilot the aircraft weight would be about 100 lb. For the human-powered aircraft the payload is essentially the pilot so that the total aircraft weight would be approximately 250 lb.

If analysis methods from the early part of the book are used, then specifications can be estimated for the human-powered aircraft. Start by assuming the speed to be 12 kt and the propeller efficiency to be 0.85. The rationale is based upon observations with the Solar Snooper analysis. Then the drag from the power required Eq. (6.10) is 6.25 lb. This means that the aircraft cruise L/D needs to be 40. From the Solar Snooper example the aspect ratio would be ~36.

The altitude will be less than 500 ft above ground level (AGL) because the aircraft would like to take advantage of ground effects. This gives us a Reynolds number per foot $\rho V/v = 127,300$ per ft. If the $W/S = 1$ lb/ft^2, then $S = 150$ ft^2 and the wing span $= 73.5$ ft. The average chord is 2.0 ft and the $Re = 260,000$. The cruise $C_L = W/qS = 1.0/0.48 = 2.1$ for the $W/S = 1.0$ lb/ft^2. This is much too high for current low-Re airfoils [6]. Because the aircraft cannot fly any faster than ~20 ft/s the q is fixed at 0.48 lb/ft^2. A cruise $C_L = 1.0$ is more realistic. Thus, the wing loading needs to decrease to about 0.5 lb/ft^2 and the wing area increase to 300 ft^2. Holding the aspect ratio constant gives a wing span of 104 ft. This is a good trade as it gives us a slightly larger average chord of 2.9 ft and a wing $Re = 368,000$.

The design specifications for our human-powered aircraft are as follows:

Power available	0.2745 hp
Pilot weight	150 lb
Aircraft weight	100 lb
Cruise speed	12 kt
Propeller efficiency	0.85
Cruise L/D	40
Wing aspect ratio	36
Wing span	104 ft
Wing average chord	2.9 ft
Wing area	300 ft^2
Wing Re	368,000

Readers are now requested to read the Daedalus case study in Volume 2. They should recognize the Daedalus specifications as being very similar to the preceding estimates. The case study will add substance and realism to the design analysis in this section.

Figure 18.10 shows the human-powered Daedalus at sunrise, at the start of its 3 h 54 min historic flight across the Sea of Crete on 23 April 1988 [7]. The Daedalus case study in Volume 2 was written by Harold Youngren, the chief engineer on the MIT project.

18.11 Rocket Engine Sizing

Rockets are sized for acceleration and burn-out speed. Acceleration is a function of the T/W of the rocket. A typical T/W is 1.4–2.0 so that the rocket accelerates quickly through the atmosphere in about 140 seconds.

The burn-out speed is determined by the amount of fuel carried by the rocket W_i/W_f as given by Eq. (18.6). The rocket sizing needs to account for the gravity and drag losses as the rocket exits the atmosphere. Gravity

Figure 18.10 Human-powered Daedalus takes off on its historic flight across the Sea of Crete (courtesy of Charles O'Rear).

losses are insignificant for aircraft but significant for rockets as they usually are boosting vertically until outside of the atmosphere. Similarly the drag losses are small for a rocket because they accelerate through the atmosphere quickly. Instead of correcting the rocket I_{sp} for these losses as was done for jet aircraft in Eq. (18.7), the ΔV will be increased to account for drag and gravity losses.

A low earth orbit (LEO) is defined as an orbit outside of the earth's atmosphere where orbital decay of the spacecraft due to drag is not a problem. The edge of the atmosphere is approximately 90 miles up from the earth's surface. The outer limit for a LEO is below the inner Van Allen radiation belt (about 1088 n mile). A geosynchronous orbit (GEO) is an orbit where a spacecraft would appear stationary over a point on the surface of the earth (would have a period of 24 h). A GEO orbital altitude would be 19,468 n mile above the earth's surface.

> The Daedalus was a project undertaken by the MIT Department of Aeronautics and Astronautics in 1985 to recreate the mythical escape of Daedalus from his tower cell on the island of Crete, across the Sea of Crete to the island of Santorini—a distance of almost 65 n mile [7]. According to Greek mythology Daedalus and his son Icarus escaped from their cell by gluing bird feathers onto their bodies with wax. Icarus flew too close to the sun and the wax melted, causing Icarus to plummet to his death. The older and wiser Daedalus stayed a safe distance from the sun and flew to freedom.

Example 18.3   Rocket Sizing for a Low Earth Orbit

Size a rocket to put 1000 lb of payload into a 500,000 ft (82 n mile) LEO. The speed of the rocket at 500,000 ft needs to be 25,638 ft/s tangent to the curve of the earth, balancing the centrifugal and gravitational forces. If the orbital direction is to the east at latitude $\Phi°$, the rocket will get a ΔV boost of 1520 cos Φ ft/s due to the earth's rotation. If launching to the west, the rocket has to increase its ΔV by the amount 1520 cos Φ ft/s.

Assume the following launch conditions:

Launch location	Cape Canaveral, Florida ($\Phi = 26°$ latitude)
Launch direction	east
Earth rotation speed	1366 ft/s (1520 cos Φ)
Gravity losses during boost	3100 ft/s
Drag losses during boost	1200 ft/s
Rocket I_{sp}	330 s (kerosene/O_2)

The ΔV required from the rocket is $\Delta V = 25{,}638 - 1366 + 3100 + 1200 = 28{,}572$ ft/s.

The rocket weight fraction using Eq. (18.6) is

$$W_i / W_f = \exp\left[\Delta V / gI_{sp}\right] = 14.7$$

which means that the payload, structure, and motor comprise 7% of the rocket and the fuel the remaining 93%. For a launch weight of 200,000 lb and a payload of 1000 lb, the fuel weighs 186,000 lb, leaving 13,000 lb for structure and motor. Fortunately, the motors are light (liquid propellant rockets have $T/W_{Motor} \sim 55$, and solid rockets are even better) and the structure is mostly a fuel tank. Using the T/W of 1.4–2.0 gives a rocket thrust of 280,000–400,000 lb.

References

[1] Youngblood, J. W., and Talay, T. A. "Solar Powered Airplane Design for Long Endurance, High Altitude Flight," AIAA Paper AIAA-82-0811, 18 May 1982.

[2] Youngblood, J. W., Talay, T. A., and Pegg, R. J., "Design of Long Endurance Unmanned Airplanes Incorporating Solar and Fuel Cell Propulsion," AIAA Paper AIAA-84-1430, 13 June 1984.

[3] Stender, W., "Sailplane Weight Estimation," OSTIV (International Scientific and Technical Gliding Organization), Elstree-Wassenaar, The Netherlands, 1969.

[4] Wierwille, W. W., "Physiological Measures of Aircrew Workload," *Human Factors*, Vol. 21, No. 5, 1979, pp. 575–593.

[5] Nadel, E. R., "Physiological Adaptations to Aerobic Training," *American Scientist*, Vol. 73, July–Aug. 1985.

[6] Drela, M., "Low Reynolds Number Airfoil Design for the MIT Daedalus Prototype: A Case Study," *Journal of Aircraft*, Vol. 25, No. 6, 1988, pp. 724–732.

[7] Lloyd, P., "Man's Greatest Flight," *Aeromodeller Magazine*, Aug. 1988 (Argus Specialist Publ., London).

Chapter 19 Structures and Materials

by Walter Franklin

Lockheed Martin Fellow in Structures and Materials

- Structural Design Criteria
- Strength vs Buckling Stability
- Finite Element Modeling
- Loads
- Stress Analysis
- Joints
- Material Selection
- Example

This F-117 was one of 64 built in total secrecy by Lockheed in Burbank, California. Completed aircraft were disassembled, put into a C-5A, flown to a secret base at Tonopah, Nevada, and reassembled. At base, they were kept in shelters during the day and flown for training only at night.

One simple test can be worth a whole lot of analysis.

19.1 Introduction

A ircraft structural design and analysis embodies a philosophy that is significantly different from the approach used for many civil engineering structures, such as bridges and buildings. Structural efficiency and minimum weight are of paramount importance for aircraft structure; and taking advantage of the inherent capability of thin-sheet structures to carry substantial load, even in a postbuckled state, is one of the key differences that separates aircraft structural design from other types of structural engineering. Since the Wright brothers' flight in 1903, the aircraft industry has developed a comprehensive body of design and analytical methods, based on extensive structural development testing combined with a wealth of lessons-learned from flight hardware, that make possible airframe structure that is safe, robust, and lightweight.

Aircraft structural engineering combines aspects of design, analysis, and manufacturing; and a basic knowledge in each of these areas is essential to the aircraft structural design process. The engineering disciplines that make up the Structures Group include the following:

1. External loads
2. Stress
3. Flutter and Dynamics
4. Mass properties
5. Materials and Processes
6. Structural testing

Factor of Safety for Aircraft Structural Design

The ultimate factor-of-safety of 1.5 for aircraft structural design was first introduced in the early 1930s. Prior to this time, aircraft were designed to withstand, without failure, a certain load factor which was typically on the order of 6.0 g's. The concept of limit load and ultimate load had not been developed at this time. Since aircraft structure designed in this manner did not show any widespread evidence of permanent yielding or structural failure, it was felt the existing load factor requirements must have included an inherent factor-of-safety. As aircraft speed and performance increased during this time period, it was felt necessary to define this factor-of-safety for future design efforts. The selection of 1.5, although somewhat arbitrary, was based in part on the ratio of ultimate strength to yield strength of the aluminum alloys that were being used at that time. Although a higher factor-of-safety could have been selected, there was also a desire to keep the resulting "limit load" as high as possible to not unduly penalize future aircraft designs.

Professor F. R. Shanley, 1961

The functions and responsibilities of each of these disciplines, and many of the technical challenges that each discipline encounters during the aircraft design and development process, are discussed in the following sections.

19.2 Structural Design Criteria and External Loads

The starting point for the design of airframe structure involves definition of the structural design criteria. The *structural design criteria* are the key parameters, such as design load factors, vehicle weights, speeds and altitudes, design life, factors-of-safety, and other operational considerations, that drive the design of the airframe. Although there are similarities in the structural design criteria among the various types of aircraft, many detailed requirements can vary greatly from one aircraft to another depending on a number of factors, including the agency that will grant flight certification (for example, commercial vs military certification), the particular class of aircraft (for example, fighter vs transport), and other requirements as dictated by the intended operator of the aircraft (for example, U.S. Air Force vs U.S. Navy requirements).

For military aircraft, the MIL-A-8860 series of documents provides a good starting point for defining structural design criteria. The various documents contained in the MIL-A-8860 series are summarized as follows:

1. MIL-A-8860, Aircraft Strength and Rigidity—General Specification
2. MIL-A-8861, Flight Loads
3. MIL-A-8862, Landplane Landing and Flight Handling Loads
4. MIL-A-8863, Ground Loads for Navy Procured Airplanes
5. MIL-A-8864, Water and Handling Loads for Seaplanes
6. MIL-A-8865, Miscellaneous Loads
7. MIL-A-8866, Reliability Requirements, Repeated Loads, and Fatigue
8. MIL-A-8867, Ground Tests
9. MIL-A-8868, Data and Reports

Similar design criteria for commercial and private aircraft are covered under the Federal Aviation Administration (FAA) guidelines contained in the Federal Aviation Regulation (FAR) documents.

Although the MIL-A-8860 documents are a valuable source of commonly used structural design requirements, new military aircraft development programs commonly employ "tailored" design criteria that are unique to the particular aircraft being developed. The "Joint Service Specification Guide—Aircraft Structures" (JSSG-2006) [1] provides a framework for developing such tailored design criteria. JSSG-2006 is one of eight Joint Service Specification Guides that were developed as part of acquisition reform by the U.S. government. These specification guides were developed to provide the aerospace industry with a single, consistent approach for defining design requirements that would be common among the different military services [2]. JSSG-2006 is a comprehensive "fill-in-the-blank" type of template that covers a wide range of structural design requirements such as vehicle weight and center-of-gravity requirements, loading conditions, airframe construction parameters, aeroelastic requirements, material properties, and structural durability requirements. As such, it is an excellent resource to guide the development of a tailored set of design criteria to insure that all pertinent structural requirements are being addressed.

> The Bell X-1, which in 1947 became the first aircraft to break the sound barrier, was designed to a vertical load factor (n_z) of +/− 18 g, which was about 50% higher than the known g capability of any other aircraft being flown at that time.

A key design parameter contained in the structural design criteria is the flight envelope for the aircraft, commonly represented as a *V–n* diagram plotting aircraft speed in KEAS vs vertical load factor n_z, commonly expressed in g's. Figure 19.1 shows a typical *V–n* diagram. Here, KEAS translates to *knots-equivalent airspeed*, which is the true airspeed corrected for the difference in density of the air at altitude compared with sea level (SL), as shown in the following expression:

$$V_e = \sqrt{\sigma}\, V_t$$

where

V_e = equivalent airspeed (this is always written in knots as KEAS)
$\sigma = \rho_{\text{alt}}/\rho_{\text{SL}}$ (air density ratio)
V_t = true airspeed

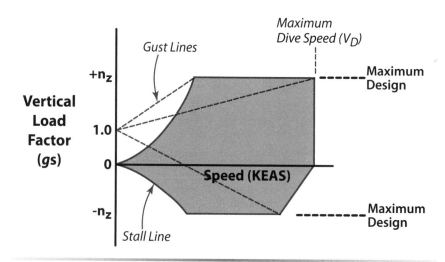

Figure 19.1 Typical *V–n* diagram used for airplane design.

Equivalent airspeed is often the measure of aircraft speed preferred by the Loads Engineer because it represents a speed with a constant dynamic pressure (q), regardless of the aircraft's altitude.

The *vertical load factor* n_Z is of particular interest because it is a key indicator of the critical flight loads that drive the design of the airframe structure, especially the wing structure. MIL-A-8861 provides guidance on the appropriate vertical load factors for different classes of aircraft. Maximum positive vertical load factors, as would be experienced during a pull-up maneuver, typically range from +3.0 g for many transport-type aircraft to +7.5 g, or more, for fighter-type aircraft. Maximum negative vertical load factors, as might occur during a push-over maneuver, commonly range from −1.0 g for transport-type aircraft to −3.0 g for fighter-type aircraft [3].

Gust load factors, which result from the aircraft flying through turbulent air, are also typically included on the $V-n$ diagram in the form of "gust lines." When an aircraft experiences a gust, the effect is an increase or decrease in the angle-of-attack, resulting in a change in lift and, consequently, a change in load factor. The load factor resulting from a gust can be estimated using the following *discrete gust relationship*:

$$n = 1 \pm \frac{K_g C_{L_\alpha} U_e V_e}{498 \, W/S} \tag{19.1}$$

where

C_{L_α} = lift curve slope (per radian) for the complete airplane
U_e = equivalent gust velocity (ft/s)
V_e = equivalent airspeed (KEAS)
W/S = wing loading (lb/ft^2)
K_g = gust alleviation factor = $0.88\mu/(5.3 + \mu)$ (subsonic aircraft)

where

μ = $(2 \, W/S)/(\rho \, \bar{c} C_{L_\alpha} g)$
ρ = air density (slug/ft^3)
$\bar{c}$ = mean aerodynamic chord (ft)
C_{L_α} = lift curve slope (per radian)
g = acceleration due to gravity (ft/s^2)

The *equivalent gust velocity* U_e, input into Eq. (19.1), is defined as a function of both aircraft speed and altitude. Consequently, there is a range of equivalent gust velocities used for aircraft design, as shown in Fig. 19.2. There is an inverse relationship between gust velocity and aircraft speed— as aircraft speed increases the gust velocity used for design decreases. This relationship is representative of customary aircraft operation in which a pilot will reduce speed consistent with the level of turbulence that is encountered [4].

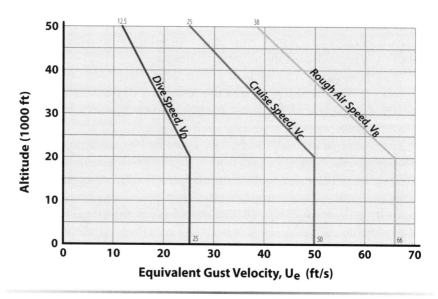

Figure 19.2 Equivalent gust velocity as a function of speed and altitude.

Load factors generated by gust conditions can be more critical than maneuver load factors depending on the speed, altitude, and wing loading (W/S) of the aircraft. In general, aircraft with low wing loading are more susceptible to being designed by gust loads, and gust velocities are typically higher at altitudes less than 20,000 feet. Therefore, an aircraft with a lightly loaded wing that generally flies at lower altitudes is likely to be designed by gust conditions, not flight maneuver conditions. MIL-A-8861 and FAA FAR Part 25 provide additional guidance on defining gust loads for military and commercial aircraft.

The key task of the Loads Engineer is to develop a set of aerodynamic and inertia design loads based on the flight envelope and intended usage of the aircraft. Aerodynamic and inertia loads that are applied to the aircraft are referred to as *external loads* to differentiate them from the *internal loads* that are distributed within the airframe and carried internally by the various structural members. As illustrated in Fig. 19.3, there are a number of tools available to predict the external design loads, ranging from computational fluid dynamics (CFD) and other computational methods such as VORLAX (vortex lattice method), to wind tunnel testing in which the overall forces and moments applied to the vehicle, as well as surface pressure distributions, can be measured. A set of external load conditions must be defined that cover the range of flight weights and center-of-gravity locations for the vehicle, as well as all altitudes, speeds, and possible configurations of flight control surfaces such as ailerons, rudder, flaps, and spoilers. It is common to utilize all types of loads analysis tools in generating a full

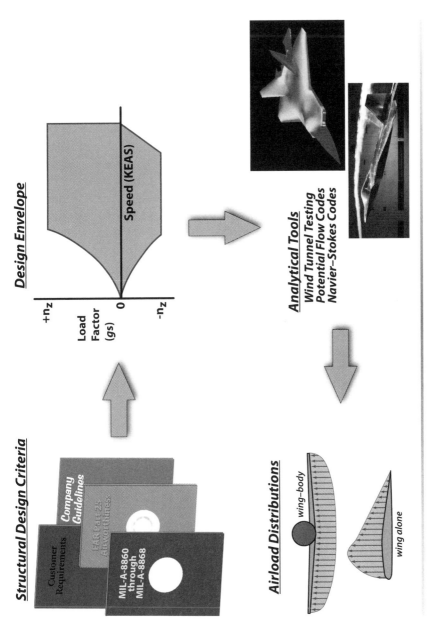

Figure 19.3 External loads definition process.

set of external design loads. Wind tunnel testing is commonly used to define and verify key design points within the flight envelope and it is often complemented by other analytical methods, such as CFD and VORLAX, to populate intermediate design points and off-design conditions.

In addition to the basic aerodynamic and inertia loads resulting from flight maneuvers, there are many other types of load conditions that must be considered in the design of an airframe. As shown in Fig. 19.4, these other types of design loads include landing and taxi loads, cabin and fuel pressures, crash loads, propulsion system loads, control surface loads, loads generated by cargo or stores, and loads associated with various ground handling activities such as jacking, hoisting, or towing of the aircraft. All of these types of design loads may not be applicable to all types of aircraft. Therefore, it is very important to document the pertinent load conditions and criteria to be used in the design of the aircraft in a Structural Design Criteria Document. The Structural Design Criteria Document serves as a single source of requirements that will keep all designers and structural analysts involved in an aircraft development effort working to a consistent set of requirements and focused on the same technical goals and objectives.

Although there are many powerful analytical tools and wind tunnel testing methods available to the Loads Engineer, a key ingredient that should always be used in generating design load conditions is sound reasoning and good judgment based on a thorough understanding of how the aircraft will be flown and operated. Ensuring safety-of-flight must always be of paramount importance. However, there are many second-tier design requirements, not directly related to safety-of-flight, that are often subject to interpretation. Many of these secondary requirements can be tailored to fit the specific mission requirements of the aircraft. An example involves the structural design requirements for low-observable (LO) aircraft. From a structural and signature standpoint, an important design requirement for LO aircraft involves the allowable step, gap, and waviness of the outer skin panels of the vehicle. Steps, gaps, and waviness can result from several sources, such as manufacturing tolerances, but the primary source is often structural deflections caused by in-flight maneuver loads. Meeting the surface smoothness criteria may be critical to the signature of the aircraft, but it is likely not a safety-of-flight issue per se, and satisfying very tight surface smoothness requirements can significantly increase the structural weight of the vehicle by requiring increased thickness of skins or tighter spacing of substructure frames and ribs. Therefore, it may be advantageous to specify a subset of the flight envelope where the surface smoothness requirements must be satisfied for maximum survivability with minimum weight impact. This concept is illustrated in Fig. 19.5 and highlights that the airframe must possess sufficient strength and structural integrity to

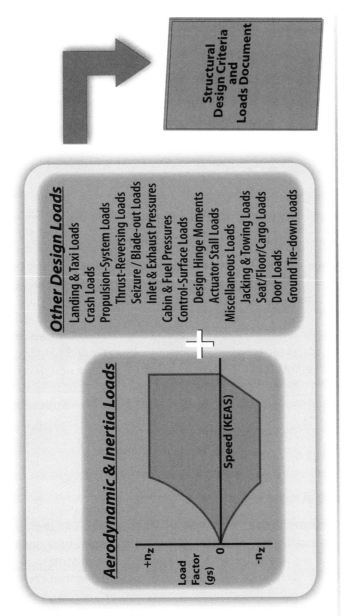

Figure 19.4 Example of design loads contained in Structural Design Criteria Document.

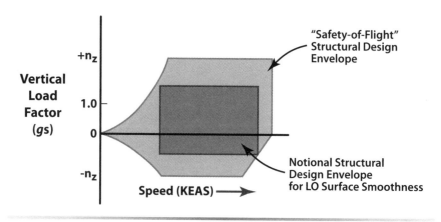

Figure 19.5 *V–n* diagram showing notional LO design envelope as a subset.

provide safe operation throughout the flight envelope, but the airframe weight does not need to be unduly penalized to meet surface smoothness requirements for relatively short duration excursions at the corners of the flight envelope where high-*g* maneuvers are being performed, or for the low-speed portions of the flight envelope where deployment of flaps or wing leading edge (LE) devices might preclude meeting signature requirements regardless of the smoothness of the outer surface of the aircraft.

Although this example is specific to LO vehicles and the issue of surface smoothness, the same basic philosophy can be extended to other types of design requirements for all classes of aircraft. Structural design criteria dictated by safety-of-flight requirements, versus those driven by non-flight-critical considerations, should be evaluated carefully in this manner to avoid a "worst case on worst case" approach that can burden the vehicle with overly conservative design requirements resulting in unnecessary structural weight.

19.3 Stress Analysis

The primary objective of stress analysis is to insure that each structural member of the airframe is properly designed and sized to meet structural requirements with the lowest possible weight. These structural requirements typically include strength and buckling stability, but they can also include stiffness and deflection requirements, as well as durability and damage-tolerance analysis (DaDTA) considerations related to fatigue and crack growth.

Stress is simply force (load) per unit area, and it can result from four basic types of loading: tension, bending, shear, and compression. Figure 19.6 illustrates these four basic types of loading and provides the expres-

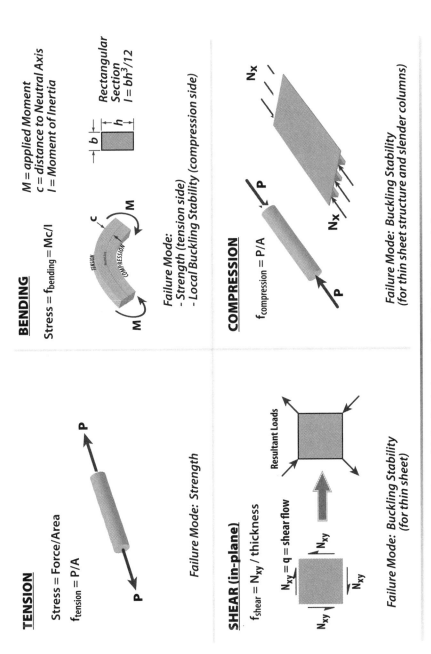

Figure 19.6 Basic types of applied loads and stresses.

sion for calculating the resulting stress for each case. Any single structural member in an airframe will likely be subjected to some combination of these four types of loads for any single design condition. Therefore, the challenge of stress analysis is to understand the interaction of the different types of load, anticipate the potential failure modes of the structural member, design the structural member so that failure does not occur within a specified design envelope, and minimize the structural weight of the component.

The failure mode for tension loading is typically material strength, either tension ultimate strength or tension yield strength. For compression loading the failure mode is usually buckling instability, either global or local buckling instability. Bending results in both tension and compression stress, as shown in Fig. 19.6. Depending on the configuration of the structural member to which the bending moment is applied, the failure mode can be either strength (at the tension side) or local buckling instability (compression side). An in-plane shear load can also be resolved into tension and compression loads as shown in Fig. 19.6. Therefore, for a thin-sheet structure such as a wing or fuselage skin, the failure mode for in-plane shear loading is usually buckling instability caused by the resultant compression loads. The key point is that buckling instability, and not necessarily material strength, can be the governing failure mode for a significant amount of an airframe, especially if constructed of lightweight, thin-sheet structure. As an example, Fig. 19.7 lists the various failure modes, and the percentage of airframe structural weight driven by that failure mode, for the Lockheed Martin S-3A Viking aircraft [5]. This data illustrate that only 30% of the airframe structural weight for this particular aircraft is driven by tension strength, but over 40% of the weight is driven by buckling stability. These characteristics are typical of many other airframe designs.

The concept of limit load vs ultimate load is fundamental to understanding aircraft stress-and-loads analysis. *Limit load* is defined as the maximum load that an airframe will experience anytime during its service life. *Ultimate load* is simply limit load multiplied by a factor-of-safety. The *ultimate factor-of-safety* is typically 1.5 for manned aircraft and 1.25 for unmanned aircraft. However, there are many other factors that may be required depending on the type of structure and type of loading. For example, compartments subjected to internal pressure, such as pressurized passenger cabins or crew compartments, are usually required to be designed to withstand a *proof pressure* that is 1.33 times the maximum attainable pressure and a *burst pressure* that is 2.0 times the maximum attainable pressure. The MIL-8860 series of documents provides guidance on many of these other factors required for structural design.

For strength-critical components, stresses resulting from limit load and ultimate load are compared with the yield and ultimate strength of the par-

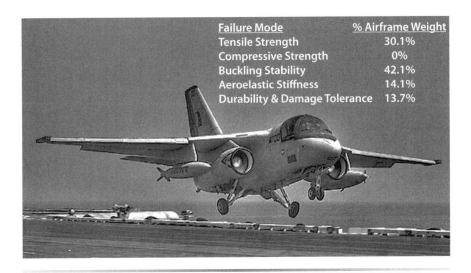

Failure Mode	% Airframe Weight
Tensile Strength	30.1%
Compressive Strength	0%
Buckling Stability	42.1%
Aeroelastic Stiffness	14.1%
Durability & Damage Tolerance	13.7%

Figure 19.7 Airframe structural weight per failure mode for S-3A aircraft.

ticular material from which the component is constructed. Figure 19.8 shows a stress–strain curve for a typical ductile material. The *yield stress* is defined as the point on the stress–strain curve at which permanent deformation starts to occur (also called *plastic deformation*). Structural design criteria for most aircraft state that no detrimental permanent deformation

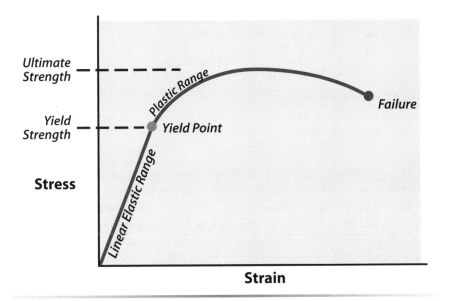

Figure 19.8 Engineering stress–strain curve for a ductile material.

is allowed at, or below, limit load and no failure at, or below, ultimate load [6]. This implies that yielding of the material may be allowed above limit load. However, some yielding may be allowed below limit load provided that the permanent deformation does not interfere with safe operation of the aircraft.

Margin-of-safety is a measure of how much capability a structural component possesses in excess of design requirements. For structural components, margin-of-safety is usually expressed in terms of a material *allowable* (for example, a material strength allowable such as ultimate strength or yield strength) compared against an applied stress:

$$\text{Margin-of-Safety (M.S.)} = \frac{\text{Allowable Stress}}{\text{Applied Stress}} - 1.0$$

Margin-of-safety should not be confused with factor-of-safety; the two quantities serve two distinctly different purposes.

Example 19.1   Margin-of-Safety

Consider a rod with a 1.0-in.^{2} cross section, loaded in tension with 40,000 lb, as shown in Fig. 19.9.

Limit load for this example is 40,000 lb and ultimate load is $1.5 \times 40,000$ lb = 60,000 lb. Based on these applied loads, the tension stress at limit load is calculated to be 40,000 lb/1.0 in.^{2} = 40,000 psi, and the stress at ultimate load is 60,000 lb/1.0 in.^{2} = 60,000 psi. The Structural Design Criteria state that permanent deformation is not allowed below limit load and failure is not allowed below ultimate load. Therefore, the two margins-of-safety are

$$\text{M.S. (Yield Strength)} = \frac{48,000 \text{ psi}}{40,000 \text{ psi}} - 1.0 = +0.20$$

$$\text{M.S. (Ultimate Strength)} = \frac{63,000 \text{ psi}}{60,000 \text{ psi}} - 1.0 = +0.05$$

Yield strength is compared against applied limit stress, and ultimate strength is compared against applied ultimate stress, resulting in margins-of-safety of +20% and +5%. Unless specified otherwise in the Design Criteria, it is permissible to drive all margins as close to zero as possible for minimum weight. Therefore, the cross-sectional area of the rod could be reduced slightly from 1.0 in.^{2} to 0.96 in.^{2}, resulting in a margin of 0% at ultimate and +15% at limit. Additional reduction in cross-sectional area to bring the yield margin closer to zero would result in a negative margin at ultimate load, which is unacceptable in

Cross-Sectional Area = A = 1.0 in.^{2}

$f_{tension} = P/A$

P=40,000 lb

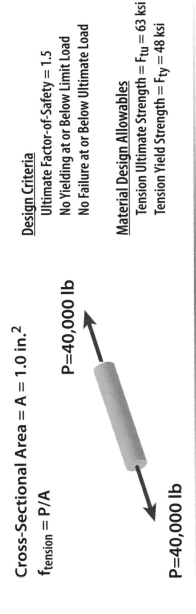

P=40,000 lb

Design Criteria

Ultimate Factor-of-Safety = 1.5
No Yielding at or Below Limit Load
No Failure at or Below Ultimate Load

Material Design Allowables

Tension Ultimate Strength = F_{tu} = 63 ksi
Tension Yield Strength = F_{ty} = 48 ksi

Figure 19.9 Example of design criteria and allowables.

this example. In all cases, however, the ultimate factor-of-safety is unchanged at 1.5.

Example 19.1 illustrates several key points related to stress analysis and sizing of airframe structure: (1) There are usually multiple failure modes, and therefore multiple margins-of-safety, for every structural member. (2) It is desirable to drive the margins-of-safety to zero for minimum weight. (3) It is virtually impossible to drive all margins for all failure modes of a particular structural member to zero at the same time. Therefore, it is important to determine which margins drive the weight of the component and, therefore, warrant the highest priority for being minimized.

Material strength allowables for metallic materials commonly used in the aerospace industry can be found in the government handbook "Metallic Materials Properties Development and Standardization" or MMPDS [7]. Prior to 2004, the MMPDS was known as MIL-HDBK-5 "Metallic Materials and Elements for Aerospace Vehicle Structures." The MMPDS is a source of metallic material and fastener allowables for aluminum, titanium, steel, and high-temperature alloys and is accepted by the Federal Aviation Administration (FAA), all departments and agencies of the Department of Defense (DoD), and the National Aeronautics and Space Administration (NASA).

An example of a typical material allowable data sheet found in the MMPDS is shown in Table 19.1. The headings "A" and "B" near the top of the MMPDS data sheet refer to the statistical basis used in generating the material design allowable. *A-basis allowables* are defined as those for which 99% of the material population is expected to equal or exceed the stated allowable with a 95% confidence level; *B-basis allowables* are defined as those for which 90% of the material population is expected to equal or exceed the allowable with 95% confidence. For most airframe primary and secondary structure, B-Basis allowables are used for design. However, A-basis allowables may be required for a single-load path, safety-of-flight structure, depending on customer requirements and company design policy.

Compared with strength analysis, determination of the buckling stability of a structure can entail a more-involved analysis. There are several forms of buckling instability, such as shear and compression buckling of thin skins or webs, local "crippling" buckling of beam flanges, torsional buckling of open-section columns, and Euler buckling of slender columns loaded in compression, and there are numerous analytical and empirical methods available for addressing each of these types of buckling instability. Unlike yield or ultimate strength, which is an inherent property of a material, the allowable buckling load is dependent on material properties (such as compression modulus E_c or shear modulus G), the geometry of the structural member, and the boundary conditions of the member (usually defined as fixed, simply supported, or free). Also, unlike strength analysis, where there is often a clear "not-to-exceed" strength allowable, it is possible for a structure to experience certain types of buckling and still carry 100% of the required load. Therefore, provided the buckling does not initiate a global instability leading to catastrophic structural failure, it may be permissible to allow buckling below ultimate or limit load. Understanding and taking advantage of this "postbuckled" capability to achieve minimum weight are key features that separate aircraft structural design philosophy from other forms of structural design.

19.4 Finite Element Modeling

Finite element modeling is arguably the most powerful analytical tool available to the stress engineer. The theory of finite element modeling is based on the fundamental mechanics-of-materials relationship:

$$Force = Stiffness \times Displacement$$

which can be expressed in matrix notation as

$$\{F\} = [K] \times \{d\}$$

A *finite element model* (FEM) is a mathematical representation of the airframe structure in terms of a stiffness matrix $[K]$. Once this stiffness matrix is defined, forces $\{F\}$ can be applied to the FEM (commonly in the form of external loads supplied by the Loads Group), and displacements $\{d\}$ can be solved. From displacements the stresses and loads in each individual structural member in the FEM can then be determined.

As with any analytical tool, the results of a FEM are only as accurate as the input data and the fidelity of the model itself. FEM results can be greatly affected by the types of elements used in the model, mesh density, and model boundary conditions. Therefore, it is always good practice to perform a first-order hand analysis of the problem being modeled to provide a sanity check of the FEM results.

Finite element models can range from a detailed model of a fitting to a complete airframe, as shown in Fig. 19.10. A full-vehicle FEM is typically used to determine the load distribution within the airframe and is commonly called an *internal loads model*. An internal loads FEM will include a set of external-loads cases (represented by the $\{F\}$ matrices) that cover all the critical conditions to which the airframe must be designed. These load conditions typically include symmetric and unsymmetric flight maneuvers, internal pressures (such as cabin pressures), propulsion system loads, landing loads, ground handling loads, and any other loading condition that

Table 19.1 Design Mechanical and Physical Properties of Clad 2024 Aluminum Alloy Sheet and Plate

Specification	QQ-A-250/5																			
Form	Flat sheet and plate																			
Temper	T3								T351											
Thickness, in.	0.008–0.009		0.010–0.062		0.063–0.128		0.129–0.249		0.250–0.499		0.500–1.000		1.001–1.500		1.501–2.000		2.001–3.000		3.001–4.000	
Basis	A	B	A	B	A	B	A	B	A	B	A	B	A	B	A	B	A	B	A	B
Mechanical Properties																				
F_{tu}, ksi																				
L	59	60	60	61	62	63	63	64	62	64	61	63	60	62	60	62	58	60	55	57
LT	58	59	59	60	61	62	62	63	62	64	61	63	60	62	60	62	58	60	55	57
ST	—	—	—	—	—	—	—	—	—	—	—	—	—	—	—	—	52	54	49	51
F_{ty}, ksi																				
L	44	45	44	45	45	47	45	47	46	48	45	48	45	48	45	47	44	46	39	41
LT	39	40	39	40	40	42	40	42	40	42	40	42	40	42	40	42	40	42	39	41
ST	—	—	—	—	—	—	—	—	—	—	—	—	—	—	—	—	38	40	38	39
F_{cy}, ksi																				
L	36	37	36	37	37	39	37	39	37	39	37	39	37	39	36	38	35	37	33	35
LT	42	43	42	43	43	45	43	45	43	45	42	45	42	44	42	44	41	43	39	41
ST	—	—	—	—	—	—	—	—	—	—	—	—	—	—	—	—	46	48	44	47
F_{su}, ksi	37	37	37	37	38	39	38	39	39	40	37	38	36	37	35	37	34	35	32	34
F_{bru}, ksi																				
$(e/D) = 1.5$	96	97	97	99	101	102	102	104	94	97	92	95	91	94	91	94	88	91	83	86
$(e/D) = 2.0$	119	121	121	123	125	127	127	129	115	119	113	117	111	115	111	115	107	111	102	106

Specification QQ-A-250/5

Form: Flat sheet and plate

Temper: T3 / T351

Thickness, in.	0.008–0.009		0.010–0.062		0.063–0.128		0.129–0.249		0.250–0.499		0.500–1.000		1.001–1.500		1.501–2.000		2.001–3.000		3.001–4.000	
Basis	A	B	A	B	A	B	A	B	A	B	A	B	A	B	A	B	A	B	A	B

Mechanical Properties

	A	B	A	B	A	B	A	B	A	B	A	B	A	B	A	B	A	B	A	B
F_{bru}, ksi																				
(e/D=1.5)	68	70	68	70	70	73	70	73	69	72	69	72	69	72	69	72	69	72	67	70
(e/D=2.0)	82	84	82	84	84	88	84	88	82	86	82	86	82	86	82	86	82	86	80	84
e, percent (S-basis)																				
LT	10	—	—		15	—	15	—	12		8		7	—	6	—	4	—	4	—
E, 10^3 ksi																				
Primary			10.5										10.7							
Secondary	9.5				10.0								10.2							
E_c, 10^3 ksi																				
Primary			10.7										10.9							
Secondary	9.7				10.2								10.4							
G, 10^3 ksi																				
μ											0.33									

Physical Properties

ω, lb/in³											0.101									
C, K, and α											—									

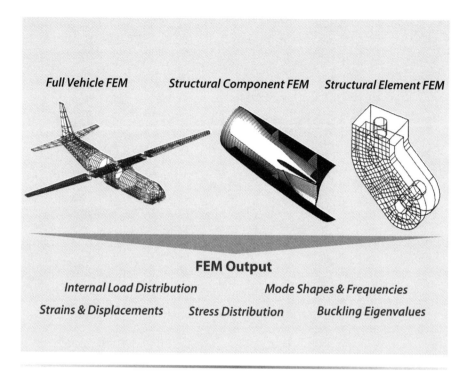

Figure 19.10 Types of structural finite element models.

might drive the structural design of the airframe. The possible range of gross weights and c.g. locations for the aircraft are also commonly included in these FEM runs. Once the internal loads are defined, the stress engineer will use this information to calculate stresses and perform detailed stress analysis and sizing of structure.

19.5 Structural Joints

Sizing of major structural members such as skins, frames, bulkheads, and spars is a major focus of stress analysis, but proper design and analysis of structural joints is also of critical importance to the structural integrity of an airframe. A structure is only as good as its weakest link, and joints can be a common cause of structural failure if not addressed correctly. The majority of airframe structural joints fall into three primary categories: *mechanically fastened* joints, *adhesively bonded* joints, and *welded or brazed* joints. Although it is not uncommon for all three types of joints to be found in any particular aircraft, mechanically fastened and adhesively bonded are usually more prevalent in airframe primary structure.

Potential failure modes for mechanically fastened joints are illustrated in Fig. 19.11. Fastener shear failures can be precluded by selecting a

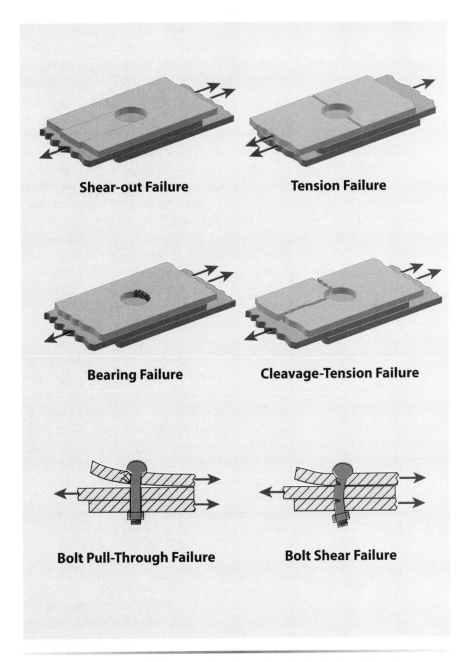

Shear-out Failure   Tension Failure

Bearing Failure   Cleavage-Tension Failure

Bolt Pull-Through Failure   Bolt Shear Failure

Figure 19.11 Failure modes of mechanically fastened joints.

fastener of appropriate diameter and shear strength to carry the required loads. Bearing failures can be precluded by selecting a fastener of appropriate diameter and by maintaining sufficient thickness in the parts being joined together. Design guidelines regarding fastener minimum spacing, minimum edge distance, fastener type (for example, tension head vs shear head fasteners), and minimum sheet thickness for countersunk fasteners help guard against many of the other failure modes shown in Fig. 19.11, but it is the responsibility of the stress engineer to insure that the joint is analyzed and sized properly to preclude all possible failure modes.

Bearing stress results from the shank of the fastener compressing, or "bearing," against the side of the hole as the fastener transmits load from one plate to the next. A *bearing failure* is a local compression-like failure of the plate or skin from this type of loading. It is normally good practice to design a fastened joint to be *bearing-critical*; that is, design the joint so that a bearing failure would occur first compared with the other possible failure modes. Figure 19.12 presents the equation used to calculate bearing stress for a single lap shear joint and shows that, for a given applied load, the two parameters that can be adjusted to determine bearing stress are the diameter of the fastener and the thickness of the parts being joined. A bearing-critical joint is preferred because it provides a degree of fail-safety in that the joint, even if inadvertently loaded beyond intended design levels, will remain intact and capable of transferring some amount of load until a repair can be implemented.

Bonded joints are usually preferred over fastened joints for laminated composite materials, such as graphite–epoxy composites, due to the relatively poor bearing strength of these materials. Poor bearing strength can necessitate a localized increase in the thickness of the parts being joined, as well as an increase in the diameter and number of fasteners. All of these factors tend to add weight. However, use of bonded joints in primary structure places an emphasis on strict adherence to proven manufacturing process specifications and use of nondestructive inspection (NDI) techniques to insure that the bond is structurally reliable.

Stress analysis of bonded joints focuses on the shear stress distribution in the adhesive layer. As shown in Fig. 19.13, this shear stress distribution

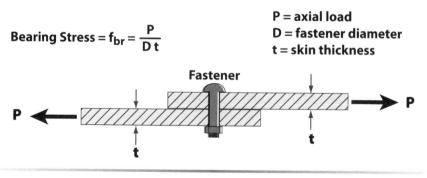

$$\text{Bearing Stress} = f_{br} = \frac{P}{Dt}$$

P = axial load
D = fastener diameter
t = skin thickness

Fastener

P

P

t

t

Figure 19.12 Bearing stress equation for a single lap shear joint.

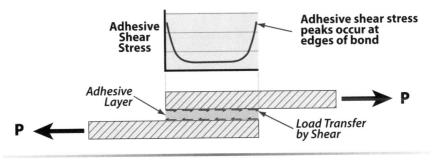

Figure 19.13 Bonded-joint shear stress distribution.

may not be uniform and can peak near each end of the bonded joint. As a result, simply increasing the bond area by increasing the overlap length of the joint may not be sufficient to reduce the peak shear stresses to within the adhesive shear strength allowable. Tailoring the stiffness of the bonded components, usually by tapering each end of the joint or by optimizing the composite layup, is often employed to reduce this peaking of shear stresses and minimize out-of-plane "peel" stresses that can lead to premature bondline failure.

Detailed design and analysis of structural joints is usually not performed in the Conceptual Design phase, but it is important to identify the basic types of joining methods that will be used in construction of the airframe as early as possible. Both bonded and fastened joints represent a source of structural inefficiency; therefore from a weight standpoint it is desirable to minimize the number of joints in the airframe. However, other considerations such as manufacturing constraints on part size, material limitations on maximum billet size, and maintainability considerations related to ease of replacing damaged structural components, may tend to increase the number and influence the location of major structural joints. These types of issues are appropriate topics for structural and manufacturing trade studies during the Conceptual Design phase. In addition, any joint design concepts that might be critical to the structural viability of the airframe design are likely candidates for component-level development testing during Preliminary Design. Planning for these tests, ordering of long-lead materials, and designing of test specimens may need to be accomplished during Conceptual Design in order to meet downstream program schedule milestones.

19.6 Durability and Damage Tolerance

Durability and damage tolerance analysis (DaDTA) addresses issues such as fatigue and other types of structural damage that may be incurred during operation of the aircraft. For a demonstrator aircraft that might have a design life on the order of a hundred flight hours, the influence of fatigue considerations on the airframe design are likely to be minimal. However, for an operational aircraft with a design life of perhaps tens of thousands of flight hours, fatigue considerations can drive the selection of structural design concepts and materials and have a significant impact on structural weight.

Two terms are important in determining the best design approach for satisfying DaDTA requirements: *fail safe* and *safe life*. The goal of the fail safe design philosophy is a structure that, even though damaged to a limited extent, is still capable of carrying a reasonable percentage of its design load to allow an emergency landing or return to base. Complete failure of any single structural member is made safe by providing alternate load paths. However, providing alternate load paths with redundant structure can entail a weight penalty. This weight penalty can be mitigated by applying the fail safe design philosophy to selected areas only, and not the entire airframe structure.

Safe life refers to a design approach that relies heavily on fatigue or crack growth analysis to show the airframe can meet design life requirements. This approach also involves implementation of inspection intervals to ensure that any premature fatigue damage is located and repaired before reaching critical proportions. Replacement of a structural component may be required once the predicted fatigue life is expended, even if the component shows no signs of fatigue damage. A safe life design is often a lighter approach than a fail safe design because it does not rely on redundant structure and multiple load paths. However, the safe life approach can be very analysis-intensive and typically requires detailed definition of the planned operational usage of the aircraft in order to develop the repeated loads spectrum required for crack growth analysis.

19.7 Mass Properties

The universal challenge for all aircraft development programs is achieving vehicle weight and performance while staying within program cost and schedule constraints. Achieving minimum weight structure is always a top priority for the Aircraft Structures engineer, and every decision made during the design process should be balanced with its potential impact on weight.

Several weight-prediction methods are used as a vehicle progresses through the design cycle, as illustrated in Fig. 19.14. During the Conceptual Design phase, the predominant method for predicting weight is based on parametric equations. Chapter 20, "Refined Weight Estimate," contains a detailed discussion of parametric weight-estimating methods for both air-

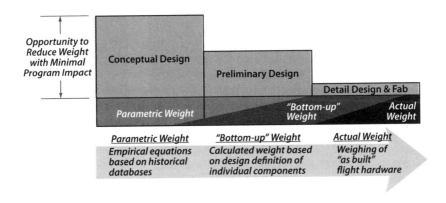

Figure 19.14 Weight-estimating methods utilized throughout the design cycle.

frame structure and subsystems, and provides parametric weight equations for wing, fuselage, and empennage structure as well as propulsion system components, surface controls and hydraulics, avionics, electrical system, and various furnishings such as seats, windows, and cargo-handling provisions.

Because parametric weight equations are based on actual weights of previously developed aircraft, there is a risk these equations may not accurately predict the weight of a new, unconventional configuration that falls well outside the existing database of aircraft. In these cases, it is advisable to validate the parametrically estimated weight by performing a "bottom-up" weight analysis. A *structural bottom-up weight* is composed of calculated weights for each of the structural members (frames, spars, keelsons, ribs, longerons, skins, etc.) based on a sufficient level of design definition for each member, supported by stress sizing. The weight for each of these components is then summed, resulting in a total weight comprising each of the individual pieces, in other words, a total weight derived from the bottom up.

Figure 19.14 also highlights that the opportunity for reducing weight without major impact to program cost and schedule decreases drastically as the design becomes more mature. Design decisions made during the early phases of a program often "lock in" the weight of the final product. A useful philosophy for controlling structural weight during these early design phases is to always strive to approach the airframe design from the "light side"—that is, initiate the design process with the minimum structural sizing (i.e., skin thickness, cap area, number of stiffeners, etc.) deemed necessary to satisfy requirements. As the configuration is matured through the design cycle, any additional structure or increase in sizing must "earn its way" onto the vehicle. This philosophy is in contrast to the approach of starting with a structure that is overdesigned, and then presupposing weight will be reduced as the design matures. In effect, this is approaching weight from the "heavy side" and rarely leads to a true minimum weight design. Once a superfluous capability or design feature finds its way into a design, it can be very difficult to isolate that item during later design stages and reach consensus with all stakeholders that it can be eliminated.

19.8 Flutter and Dynamics

Aeroelasticity refers to the structural response of a flexible airframe when subjected to aerodynamic forces. As illustrated in Fig. 19.15, there are several types of aeroelastic phenomena that can occur, such as flutter, divergence, control reversal, and aero-propulsion-servo-elasticity (APSE). *Flutter* is a dynamic instability of an elastic structure in an airstream; it occurs when the phasing between motion and aerodynamic loading extracts an amount of energy from the airstream that is equal to the energy dissipated by damping within the structure. *Divergence* occurs when the torsional stiffness of a wing or aerosurface is not sufficient to maintain the structure in a statically stable position as the speed of the aircraft increases. *Control reversal* is also related to insufficient torsional stiffness and is characterized by movement that is opposite to the desired direction based on the control input. *APSE* is the coupling of the airframe aeroelastic response with the dynamic characteristics of the flight control and propulsion systems.

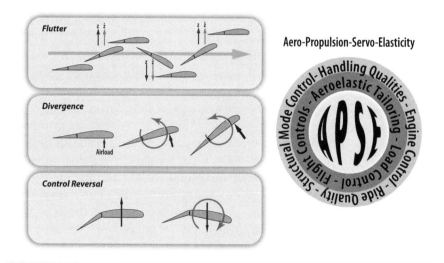

Figure 19.15   Types of aeroelastic behavior.

Detailed evaluation of any of these aeroelastic phenomena requires a considerable amount of structural design definition, particularly the mass and stiffness distribution of the vehicle. As a result, the Conceptual Design phase, where this type of detailed structural definition is often not available, has historically contained a minimal amount of aeroelastic analysis. Not performing detailed aeroelastic analysis during the Conceptual Design phase probably introduces minimal program risk provided the aircraft configuration is similar to previous, flight-proven designs. However, for new aircraft with very unconventional design features, such as extremely thin wings, extremely slender fuselages, nontraditional control surfaces, or placement of large mass items (such as engines) in unconventional locations, it is imperative to perform at least a first-order aeroleastic analysis to provide confidence in the feasibility of the design and insure that any weight penalties for additional structural stiffness are captured. This first-order analysis typically involves evaluating the *EI* (bending stiffness) and *GJ* (torsional stiffness) of the wing, fuselage, and tail structure and may involve a simple "stick model" FEM that represents the wing, fuselage, and tail structure with simple beam elements.

Structural dynamics is concerned with the vibration, shock, and vibroacoustic environment of the vehicle structure and subsystems. As with aeroleastic analysis, there is usually little need to perform a great deal of dynamics analysis in the Conceptual Design phase as long as the vehicle configuration and flight environment are fairly conventional. The vibration and shock environment for the aircraft is usually more of a driver for the design and mounting of subsystem components such as avionics and electrical components. Although primary structure is rarely sized by the vibration or shock environment, high vibroacoustic levels can drive skin thickness and stiffener spacing, such as structure near the exhaust system. For unconventional configurations or operating environments, such as higher than normal acoustic levels resulting from a new propulsion concept, it may be necessary to perform a preliminary structural dynamics evaluation to insure the feasibility of the vehicle configuration and that all associated weight penalties are captured.

19.9 Structural Layout

Major load paths of an airframe are defined by a *structural layout drawing*, sometimes called a *structural "bones" drawing*. As the basic configuration of the vehicle is being developed, it is important to define these load paths to insure that adequate volume is reserved within the vehicle for primary structure such as frames, bulkheads, keelsons, spars, and ribs, and that major design features and subsystems such as engine, landing gear, and inlet-and-exhaust structure are successfully integrated into the design.

The structural layout philosophy used for different aircraft is varied, but there are several recurring themes that can be discerned for the wing and fuselage structure of many aircraft. Lessons-learned and optimization of airframe structure over many years have generated several basic structural layout approaches that have been demonstrated to achieve minimum weight with superior strength and stiffness.

19.9.1 Wing Structure

Wing structure can account for as much as half of the total structural weight of an aircraft. Therefore, selecting the most weight-efficient structural layout for the wing is always a high priority. For most conventional aircraft, two basic types of wing structural layout are most prevalent; the multi-rib wing and the multi-spar wing as shown in Fig. 19.16. The *multi-rib wing* typically features two spars (a forward spar and a rear spar, with a third intermediate spar sometimes present), upper and lower stiffened wing covers (or skins), and numerous ribs that are generally oriented in a chordwise direction. Taken together, this structural system of spars, ribs, and skins is called the *wing box*. The primary function of the spars is to carry vertical shear (P_z) loads (carried in the spar webs) and a percentage of wing spanwise bending (M_x) loads (carried in the spar caps). The wing upper and lower covers react the majority of the spanwise bending loads by carrying tension and compression loads. For example, a wing upbending moment would be reacted by compression loads in the upper skin and tension loads in the lower skin. The wing cover features a thin skin with discrete spanwise stiffeners to provide buckling stability. The primary function of the ribs is to support the wing skins to resist global buckling of these stiffened skins when loaded in compression. Torsional stiffness of the wing is provided by the wing box, which acts as a torque box and carries torsional (M_y) loads as a shear load distributed around the periphery of the box structure.

Multi-rib wings are commonly found in transport-type aircraft that feature relatively high aspect ratio wings with generous thickness (i.e., relatively large wing thickness-to-chord ratios). These types of wings are usually subjected to moderate spanwise bending loads, as might be expected from a design vertical load factor (n_z) of less than 6.0 g.

Multi-spar wings, on the other hand, are commonly found in high-speed or fighter aircraft that feature relatively thin, highly loaded wings. The upper and lower wing skins tend to be thicker than the covers of a multi-rib design and in many cases may not require any discrete stiffening other than the multiple spars. The tight spacing of the multiple spars combined with the thicker skins precludes the need for tightly spaced ribs to resist column buckling of the skins. However, some ribs may be present in

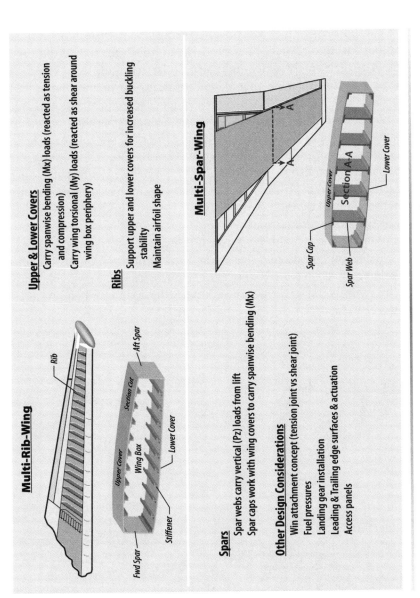

Figure 19.16 Wing structural configurations.

The following text appears within the figure:

Multi-Rib-Wing

Rib
Aft Spar
Section Cut
Upper Cover
Wing Box
Lower Cover
Fwd Spar
Stiffener

Upper & Lower Covers

Carry spanwise bending (Mx) loads (reacted as tension and compression)
Carry wing torsional (My) loads (reacted as shear around wing box periphery)

Ribs

Support upper and lower covers for increased buckling stability
Maintain airfoil shape

Multi-Spar-Wing

Spar Cap
Upper Cover
Section A-A
Spar Web
Lower Cover

Spars

Spar webs carry vertical (Pz) loads from lift
Spar caps work with wing covers to carry spanwise bending (Mx)

Other Design Considerations

Win attachment concept (tension joint vs shear joint)
Fuel pressures
Landing gear installation
Leading & Trailing edge surfaces & actuation
Access panels

the multi-spar design to serve as attachment points for external stores or to provide back-up structure for attachment of leading or trailing edge control surfaces and actuation.

Cutaway drawings of many aircraft illustrate these two popular wing structural layouts, but notable exceptions can be found. Some aircraft feature a combination of the two concepts, using a multi-spar approach for thin outboard wing sections and a multi-rib approach for thicker inboard sections. Also, extremely lightweight aircraft such as sailplanes typically use a wing structural design that is neither multi-rib or multi-spar, but rather a single spar supporting sandwich wing covers requiring few, if any, ribs for buckling stability. Figure 19.17 shows a single spar wing typical of many sailplane designs. Extremely low wing structural weights have been achieved on high-altitude long-endurance (HALE) vehicles by using the tubular spar concept also shown in Fig. 19.17. Although somewhat similar in appearance to a multi-rib design, the wing skins in this concept act only as an aerodynamic covering, with all wing bending and torsional loads carried by the single tubular spar.

This type of wing structural approach is especially attractive for span-loader vehicle configurations such as the AeroVironment Centurion and Helios vehicles (see Chapter 20, Fig. 20.1) that distribute the vehicle mass across the entire wing span, which reduces wing spanwise bending (M_x) moments.

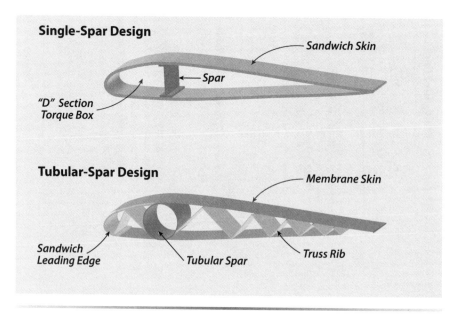

Figure 19.17 Ultralightweight wing structure concepts.

Although the various wing structural concepts presented here offer a good starting point for the design process, the final choice of wing structural layout for a particular aircraft should be supported by trade studies and weight optimization given the specific wing geometry and loads for that particular aircraft. In addition, other design issues related to integration of major subsystems into the wing, such as landing gear, propulsion system, and fuel system, can drive the preferred wing structural concept and must be considered.

19.9.2 Fuselage Structure

Fuselage structure also accounts for a significant percentage of airframe structural weight and can be subject to more demanding subsystem integration challenges than wing structure. This is especially true for densely packed fighter-type aircraft, where integration of inlet, cockpit, engine, internal stores, landing gear, and other subsystems can greatly affect the structural layout options that are available. As with wing structure, there are several recurring themes for fuselage structural layouts that are evident among the many different types of aircraft. Three of these design approaches (skin–stringer, frame–longeron, and sandwich-skin fuselage) are shown in Fig. 19.18.

The *skin–stringer approach* is typical of many commercial airliners. The longitudinal stringers, in conjunction with the skin, react fuselage bending (M_y and M_z) loads. The primary function of the frames is to reduce the column length of the stringers for improved buckling resistance, as well as to maintain the overall shape of the fuselage. Fuselage torsional (M_x) loads are reacted in the skin as shear, and internal cabin pressure loads are primarily carried in the skins in hoop tension (for fuselages with circular cross section).

The *frame–longeron approach* is very similar to the skin–stringer except that the axial-load carrying function of the numerous stringers is consolidated into discrete longerons. This design approach might be preferred for a number of reasons. For one, if the fuselage longitudinal bending loads are relatively low, it may be difficult to design and manufacture weight-efficient stringers due to material minimum gage limitations. Also, if the fuselage design features numerous cutouts for doors or windows, the frame–longeron approach may be advantageous because the longerons can be positioned either above or below the cutouts to provide an uninterrupted axial load path. It is interesting to note that the optimum frame spacing of the frame–longeron approach is typically less than the skin–stringer approach due to the skin between frames being less supported, thus requiring more closely spaced frames to achieve the same buckling load capability.

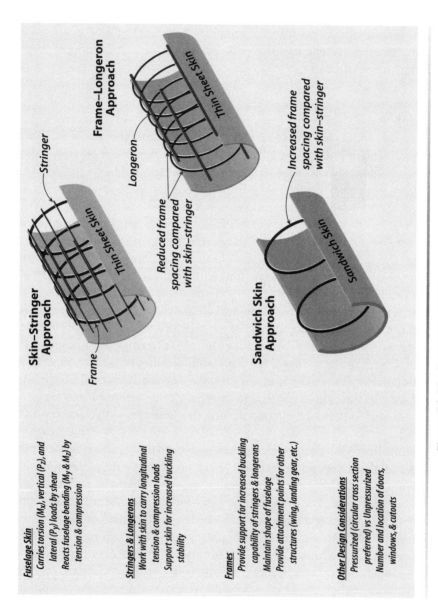

Fuselage Skin
Carries torsion (M_x), vertical (P_z), and lateral (P_y) loads by shear
Reacts fuselage bending (M_y & M_z) by tension & compression

Stringers & Longerons
Work with skin to carry longitudinal tension & compression loads
Support skin for increased buckling stability

Frames
Provide support for increased buckling capability of stringers & longerons
Maintain shape of fuselage
Provide attachment points for other structures (wing, landing gear, etc.)

Other Design Considerations
Pressurized (circular cross section preferred) vs Unpressurized
Number and location of doors, windows, & cutouts

Figure 19.18 Fuselage structural configurations.

The *sandwich-skin approach* effectively eliminates the need for either stringers or longerons by utilizing the inherent compression buckling resistance of the sandwich skins. As might be expected, the optimum frame spacing for this configuration is usually increased compared with the other two concepts.

Selection of which fuselage structural configuration is optimum from a weight standpoint for a particular aircraft depends on a number of factors, including the following: fuselage geometry; number and location of design features such as doors and cutouts; magnitude of the fuselage bending, torsional, and shear loads; and material minimum gage limitations. As with wing structure, the fuselage structural concepts presented can provide a good starting point for analysis and trade studies to determine which particular design approach is preferred.

19.9.3 Structural Design Rules-of-Thumb

Regardless of the specific type of structural approach selected for the wing or fuselage, there are several basic rules-of-thumb that should always be considered:

1. **Keep load paths simple and direct.** Simple load paths can result in a number of benefits. A structural layout with easily understood load paths is easier to design and analyze, often resulting in a lighter weight solution. Also, a simple design is often easier to fabricate and assemble, resulting in decreased manufacturing schedule and cost.

2. **All six components of structural loading must be considered.** All structural members can be subjected to six components of loading, namely axial loads in the three principal axes (P_x, P_y, P_z) and moments about the three axes (M_x, M_y, M_z), as shown in Fig. 19.19. Although it is common for one or two of these loads or moments to dominate the design of any particular structural member, all six components of loading must be considered. For example, design of the root attachment joint for a wing may be dominated by the vertical shear (P_z) load, resulting from lift, and the spanwise (M_x) bending moment. However, the wing joint must also be capable of reacting the other four types of loading resulting from drag (P_x), inboard–outboard loads (P_y), wing torsion (M_y), and fore–aft wing bending (M_z).

3. **A statically determinate structure is usually preferred for minimum weight.** A *statically determinate structure* is one where the reaction forces can be solved directly from the equations of equilibrium, namely, the sum of the forces equals zero and the sum of the moments equals zero. Therefore, for a given set of applied loads, there can be only one set of reaction forces with a determinate structure. Conversely, an *indeter-*

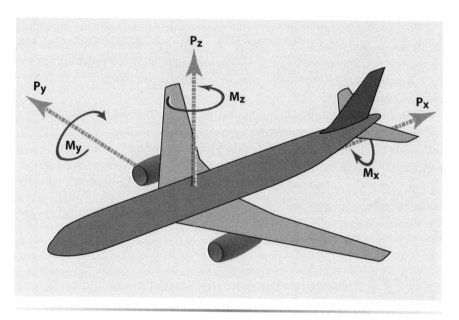

Figure 19.19   Six components of structural loading.

minate structure can have multiple solutions for the reaction forces depending on the relative stiffness of the redundant load paths within the structure. A determinate structure, because it does not include redundant structure, usually represents the minimum amount of material required to carry a specific load and is often a lighter weight solution. Determinate structure is also typically easier to analyze and often easier to build, all leading to lighter weight. However, other design requirements, such as the fail safe requirements discussed in Section 19.6, might dictate that a statically indeterminate design is required.

4. **Each structural component should serve multiple functions.** A key philosophy for achieving minimum weight structure is to require that every major structural member serve multiple functions. For example, the primary function of a wing spar is to carry wing spanwise bending (M_x) and vertical shear (P_z) loads. However, with a well-thought-out structural layout, a main spar can also provide support structure for the main landing gear, serve as a fuel tank wall, and provide attachment points for the engine and external stores.

5. **Subsystems integration requirements must be considered early.** Subsystems integration and accessibility should be considered at the earliest stages of a structural layout, especially for tightly packed vehicles such as fighter aircraft. Location of doors, windows, and nonstructural access panels, as well as integration of major subsystems such as landing gear, engines, inlet and exhaust structure, flight crew stations, and

weapons bays can have a major impact on airframe weight and structural performance if not integrated in an intelligent and synergistic manner.

19.10 Material Selection

From a structures standpoint, one of the most important decisions made during the Conceptual Design phase is selecting the materials from which to build the airframe. Material selection can have far-reaching influence on a number of programmatic issues, including vehicle weight and performance, manufacturing cost and schedule, and the reliability and maintainability of the aircraft in operational service. Key parameters that should be considered in selecting airframe materials are the following:

- Specific strength
- Specific stiffness
- Usage environment
- Fracture toughness
- Manufacturability
- Minimum gage limitations
- Availability

Specific strength and *specific stiffness* are measures of the structural performance of a material per unit weight; specific strength is usually expressed as *ultimate tension strength* (F_{tu}) divided by material density, and specific stiffness is expressed as Young's modulus (E) divided by density. Table 19.2 compares the room-temperature specific values for a number of common airframe materials. Note that laminated composite materials, such as graphite–epoxy, show a wide range of specific strength and stiffness values depending on the specific orientation of the various plies. Table 19.2 also highlights that many metallic materials, while displaying wide variation in density, have very similar specific stiffness properties at room temperature.

In conjunction with specific strength and stiffness, a key discriminator for material selection is the usage environment, specifically, the operational temperatures that the structure will experience. Table 19.2 also lists the approximate maximum usage temperature and it is interesting to note that, for the materials listed, density increases as temperature capability increases. Specific strength and stiffness values for each material will change at different rates as the usage temperature increases, implying that a material that has a clear advantage in specific strength or stiffness at room temperature may not be the best choice at elevated temperature conditions. Therefore, candidate materials must be evaluated and compared throughout the range of expected operational temperatures.

Table 19.2 Comparison of Material Specific Properties and Maximum Use Temperatures

Material	Density (lb/in.^{3})	Specific Ultimate Tension Strength at 70°F (ksi/lb/in.^{3})	Specific Stiffness at 70°F (msi/lb/in.^{3})	Maximum Usage Temperature (°F)
Composite	0.057	368 (quasi-iso layup) 1105 (all 0° layup)	61 (quasi-iso layup) 368 (all 0° layup)	~275
Aluminum (2024)	0.100	630	105	~300
Aluminum (7050)	0.102	745	101	~300
Titanium (6A1-4V)	0.160	812	100	~700
Carbon steel (4130)	0.283	336	102	~800
Stainless steel (301 Full Hard)	0.286	646	91	~1000
Inconel (718 STA)	0.297	606	99	~1200

Fracture toughness, denoted by the symbol K_{IC}, is the measure of the inherent capability of a material to resist crack growth, and it can be a very important material selection parameter for high-usage, long-life aircraft such as commercial airliners and military transports. Very brittle materials, such as ceramics and glass, typically have a very low fracture toughness, whereas more ductile materials have a higher fracture toughness. Material strength properties are often compromised to achieve increased fracture toughness; that is, an improvement in fracture toughness may correspond with a slight reduction in ultimate tension strength. Therefore, the tradeoff between strength and fracture toughness, and the resulting weight impact, is important to understand and quantify. For a fatigue-critical airframe with a design life of tens of thousands of hours, high fracture toughness can be more important than high specific strength.

Manufacturability addresses the ability to fabricate an end product from a particular material, and it is a selection criterion that should not be overlooked in the initial stages of the material selection process. Commonly used metallic materials such as aluminum, titanium, and steel alloys have a variety of manufacturing processes that can be utilized to make a final product starting from the initial sheet, plate, or billet stock. These processes include various forming and machining methods. Likewise, commonly used composite materials, such as graphite–epoxy systems, have a variety of material placement and curing methods. However, all manufacturing methods are not equally applicable to all materials. For example, aluminum alloys are easily cold-formed but many titanium alloys require hot-forming methods. Hot-forming usually involves more complex tooling and, therefore, can have an impact on program manufacturing cost and schedule. Similarly for composites, out-of-autoclave processing using vacuum bag pressure and an oven cure may be perfectly acceptable for some airframe applications. However, if optimum material properties are required, an autoclave cure may be necessary. The increased pressure and temperature of an autoclave cure can drive tooling costs, and autoclave size limitations can restrict the overall dimensions of the part being cured, potentially resulting in additional assembly joints and structural weight. Therefore, it is important to consider the manufacturing impacts during the material selection process to make sure the material selected, and its associated manufacturing process, is compatible with program technical requirements, cost, and schedule.

Minimum gage refers to the minimum thickness to which a material can be produced. For a metallic material, this can be either the minimum thickness of the sheet stock or the minimum thickness that can be achieved by machining. For a laminated composite material, minimum gage can refer to the minimum thickness available for each individual ply. Minimum thickness limitations can have an impact on the structural design approach and weight of airframe structure. For example, if initial stress analysis indicates that a 0.020-in. thickness is required for a metallic skin in a particular region, but the material is only available in a minimum sheet thickness of 0.030 in., then there is an inherent weight inefficiency resulting from incompatibility between the selected material and the selected design concept. Fortunately, there can be possible solutions to this type of scenario, ranging from revising the structural layout so that a 0.030-in. skin is a more optimum solution, to employing chemical milling to reduce the thickness of the sheet stock.

Material availability can be a significant factor, especially for a demonstrator aircraft with a very aggressive development schedule. Many materials, both composite and metallic, can require from several months to well over a year for delivery of quantities sufficient to fabricate a full airframe. Large billets of the less commonly used metallic materials, such as titanium, Inconel, or other high-temperature alloys, can have especially long lead times. Although ordering of materials is rarely performed in the Conceptual Design phase, an early understanding of the lead times involved in procuring materials is important for developing an overall program schedule that progresses from Conceptual Design to Preliminary Design, Detail Design, Vehicle Assembly, and First Flight.

19.11 Composite Materials

Usage of composite materials in military aircraft has seen a steady increase over the past several decades. The benefits offered by composites are many, including reduced weight (see Chapter 20, Section 20.2.3), excel-

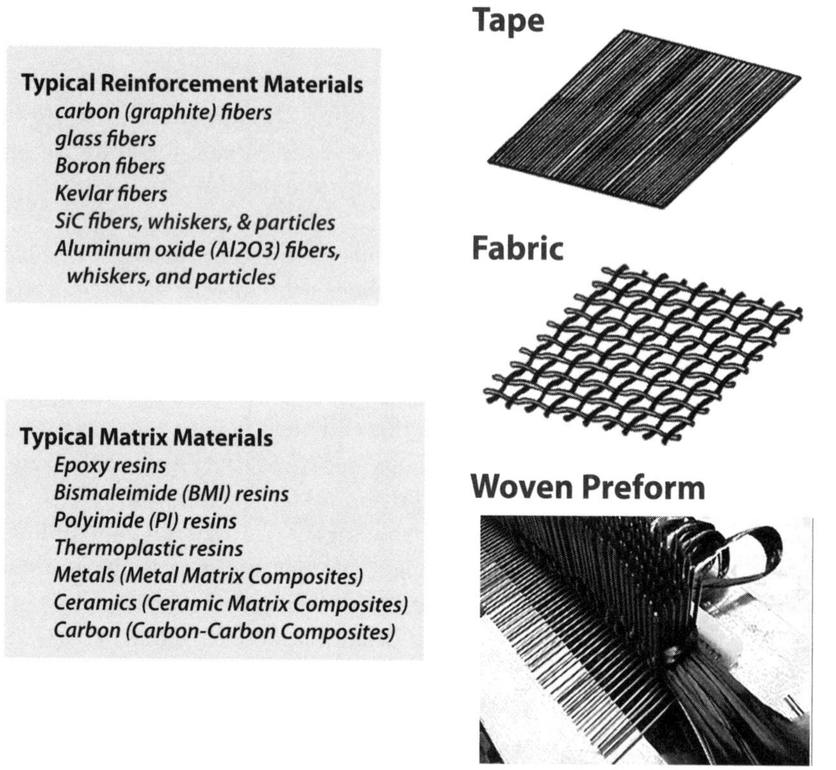

Tape

Typical Reinforcement Materials
carbon (graphite) fibers
glass fibers
Boron fibers
Kevlar fibers
SiC fibers, whiskers, & particles
Aluminum oxide (Al2O3) fibers,
whiskers, and particles

Fabric

Typical Matrix Materials
Epoxy resins
Bismaleimide (BMI) resins
Polyimide (PI) resins
Thermoplastic resins
Metals (Metal Matrix Composites)
Ceramics (Ceramic Matrix Composites)
Carbon (Carbon-Carbon Composites)

Woven Preform

Figure 19.20 Common composite reinforcement and matrix materials.

lent fatigue performance, low coefficient of thermal expansion, corrosion resistance, and the ability to tailor the strength and stiffness properties of the material. Composite materials are composed of a reinforcement material and a matrix, and there are many combinations available, as shown in Fig. 19.20. The reinforcement material provides strength andstiffness to the composite and can be in the form of fibers, whiskers, or particles. The primary function of the matrix is to hold the reinforcement materials in place and distribute loads among the fibers, whiskers, or particles.

Fully realizing the benefits offered by composite materials requires a completely different mind-set for design, analysis, and manufacturing compared with metallic structures. This different way of thinking should be applied early in the design process to avoid the "black aluminum" mentality of designing with composites as if they were simply a different kind of metallic material. As an example, most laminated composites, such as graphite–epoxy tape and fabric laminates, have excellent in-plane strength properties but relatively poor out-of-plane (through-the-thickness) properties. This is in contrast to metallic materials, which are basically isotropic and have comparable properties in all three directions. Therefore, a good composite design should take these types of fundamental characteristics of composites into consideration and enhance the advantages of the material, not accentuate the weaknesses.

Unlike the metallic material design allowables contained in the MMPDS, there is no comprehensive, industry-wide source for composite design allowables. There are several reasons for this. Composite material mechanical properties are very dependant on the specific details of the cure cycle that is utilized (i.e., curing time, temperature, and pressure), and most manufacturers of composite structure have developed unique, and often proprietary, process specifications. Also, new and improved fiber and matrix materials are constantly being developed, and the number of possible combinations of fiber and resin is almost limitless. Therefore, a decision to utilize the "latest & greatest" composite material system may entail an extensive coupon testing program to develop design allowables. Vendor-supplied data may be suitable for early Conceptual Design trade studies, but these data commonly represent "best case" properties and do not include any statistical basis (such as A- or B-basis) or material property knockdowns for environmental exposure and damage tolerance effects. Therefore, vendor data are not normally used for Detail Design unless substantiated by independent tests.

> The Vought XF5U-1 and XF6U-1 aircraft of the mid-1940s featured a sandwich construction called Metalite, which was a balsa wood core with bonded aluminum face sheets. The Metalite panels were formed in molds and cured in a large autoclave, similar to present day composite structures. The Metalite panels minimized the number of ribs or stiffeners required for a lightweight, efficient structure and provided an aerodynamically smooth exterior surface.

19.12 Sandwich Structure

There are many structural design concepts available for integration into airframe structure, but sandwich structure deserves special mention because it can be an extremely weight-efficient and cost-effective method for stiffening a skin or web to achieve increased buckling load capability. *Sandwich structure* is composed of a core material placed between two outer face sheets, as shown in Fig. 19.21. Sandwich core is typically in the form of honeycomb, although foam cores, made of various polymeric materials, and corrugated and truss cores, made of metallic and nonmetallic materials, are also used. Honeycomb core can be fabricated from metallic materials such as aluminum, titanium, and steel, as well as from nonmetallic materials such as Nomex, Fiberglass, and graphite. Similarly, face sheets can be metallic or nonmetallic, with aluminum, titanium, steel, graphite, and Fiberglass being commonly used materials.

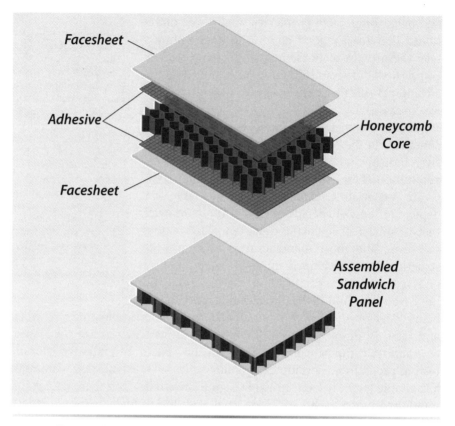

Figure 19.21   Construction of a honeycomb sandwich panel.

The function of the core material is to carry transverse (out-of-plane) shear loads, separate the face sheets for increased moment-of-inertia for reacting bending loads, provide support to the face sheets to prevent buckling, and provide shear continuity between the two face sheets so that the sandwich panel acts as a single structural entity. The primary function of the face sheets is to carry in-plane tension, compression, and shear loads. Panel bending is reacted as tension on one face sheet and compression on the other.

The connection between the face sheets and core is critical to the structural performance of a sandwich panel. In the case of composite or aluminum sandwich panels, this connection is usually made with an adhesive bond by using a film adhesive. For titanium and steel sandwich panels, the connection is usually formed by brazing or welding. When performing trade studies of sandwich panels against other stiffened skin designs, it is important to include the weight of the adhesive or braze material for each face sheet. Although adhesive and braze weights may seem insignificant on a weight per unit area basis, the total adhesive or braze weight can be significant over a large acreage. In addition, the joint concept for attaching the sandwich panel to surrounding structure should be considered in any weight trades. Core ramp-downs, doublers, or core inserts may be required in joint areas, and all of these options have an associated weight penalty that can be significant.

Sandwich construction has excellent stiffness-to-weight characteristics and, therefore, is a very attractive approach for achieving very lightweight airframe structure. However, sandwich construction does have some potential drawbacks that must be understood and addressed for actual design application. Sandwich panels can be subject to moisture entrapment where moisture passes though small pores or microcracks, in either the face sheets or the perimeter of a panel, and accumulates in the honeycomb core cells. Over time, this accumulation of moisture will increase the weight of the panel and can result in significant structural damage if the moisture freezes and expands at altitude, thereby causing the face-sheet–to–core bond to fail. In addition, the face sheets and core of a sandwich panel can be prone to impact damage, especially for lightly loaded sandwich structure where the face sheets can be extremely thin. However, these risks can be mitigated with proper design practices, and the weight reduction advantages often outweigh these potential drawbacks.

19.13 Structural Testing

Although tremendous advances in structural analysis tools, such as finite element modeling, have been made over the last several decades, structural testing is still a very important part of the aircraft design and development process. There are two basic categories of structural testing: development testing, which is focused on generating data required to support detail design and drawing release; and verification testing, which is focused on demonstrating that the as-designed airframe meets structural requirements prior to flight. The bulk of these testing efforts normally occur in the Preliminary and Detail Design phases; however, it may be appropriate to perform "proof-of-concept" testing of new and unproven structural technologies (for example, a new material or innovative design concept) during Conceptual Design, especially if the success of the overall vehicle design hinges on the viability of that new technology. The structural testing philosophy and scope of testing that is envisioned for supporting vehicle development from Conceptual Design to First Flight can influence the airframe structural design approach, minimum margins-of-safety, material selection, and structural weight, as well as overall program cost and schedule. Therefore, it is important to have basic definition of the intended structural test plan and philosophy early in the design process.

Most structural test programs are composed of a series of tests that start at the coupon level, transition to subcomponent- and component-level specimens, and culminate in full-scale structural test articles, as depicted in Fig. 19.22. This progression in scope and complexity of tests is referred to as a *building block* testing approach [8]. Coupon-level testing is commonly focused on material characterization and generation of design allowables. If design allowables for the materials selected already exist in the MMPDS, coupon testing may not be required. However, if a new material is being utilized that has no existing database of design allowables, extensive coupon testing involving hundreds of specimens may be required for generating a full database of A-basis or B-basis design allowables. Subcomponent- and component-level testing typically includes testing of critical structural joints and other key design details. Component-level testing might include testing a section of fuselage or wing structure. In a world with unlimited program schedule and budget, these tests would be performed sequentially and the knowledge gained in each phase of testing would be applied to the next phase. For example, coupon testing would be completed before subcomponent testing is started, and the material data generated by the coupon testing would be used to design the subcomponent test articles. However, programs rarely have the schedule and budget to allow this sequential approach. Compression of the development testing schedule often results in significant overlap of the different levels of testing, with many tests being run in parallel. This places additional emphasis on defining a developing test program that is sufficiently flexible to accommodate test results, both good and bad, as they become available.

Verification testing usually involves static or fatigue testing of a full-scale, flightlike airframe. Figure 19.23 shows a full-scale static test of the Airbus A380 wing. The test article for a full-scale static test can be either an actual flight vehicle or an airframe of identical design to the flight vehicle

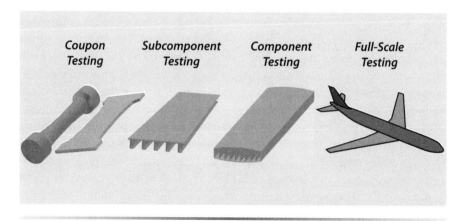

Figure 19.22 Building-block structural testing approach.

Figure 19.23 Airbus A380 full-scale static test (courtesy of Airbus Industrie).

but dedicated for ground testing only. Depending on the size of the aircraft and the scope of testing required, these full-scale tests can represent a substantial cost and schedule investment to the program. Therefore, the verification testing approach should be defined as early as possible in the design cycle, especially if test facilities must be modified or built. For prototype or demonstrator aircraft programs, where perhaps only one or two aircraft are being designed and built, it may not be desirable from a cost standpoint to perform extensive full-scale testing. In these cases, restriction of the flight test envelope or an increased minimum margin-of-safety imposed during the Detail Design phase is sometimes utilized in lieu of extensive full-scale static testing. Increased minimum margins can range from +0.20 to +0.50 instead of the 0.00 margin that is the normal goal for minimum weight structure. Different minimum margins can be required for different parts of the airframe, with the specific values selected dependent on a number of factors, such as the failure mode expected for each component (for example, strength vs stability failure) and the consequence of failure of the component. Any increased margin-of-safety requirement will impact vehicle weight and performance and, therefore, must be defined early in the design cycle. Most important, the overall structural flight certification approach,

The Republic XF-91 Thunderceptor, which first flew in 1949, featured a structurally challenging inverse taper wing in which the chord and thickness of the wing were greater at the tip than at the root. In addition, the entire wing could be tilted to vary the angle of incidence for improved takeoff and landing performance.

whether it includes increased minimum margins-of-safety or extensive structural testing, must be discussed with and agreed to by the customer and the flight certification agency, be consistent with company design policy, and provide a clear path for ensuring a flightworthy and safe design.

Example 19.2   HAARP Wing Structural Analysis

Vehicle Description

Consider the HAARP vehicle shown in Fig. 5.13 and discussed in Sections 5.8, 6.6.1, and 18.10. Vehicle dimensions, weights, and characteristics are as follows:

Wing span = b	269 ft
Wing area = S	2884 ft^2
Wing aspect ratio = AR	25
Wing taper ratio= λ	0.35
t/c	12.2%
Wing structural weight	2708 lb [from Chapter 20, Eq. (20.2)]
Fuel weight in wing tanks	4800 lb (total both tanks)
Fuel tank structural weight + pumps	93 lb (total both tanks, scaled from U2-A, Table I.4)
Payload weight	578 lb (total both sides—located in engine or payload pods)
Heat exchanger weight	1147 lb (total both sides—located in wing LE, Section 14.2.1)
System weights in engine or payload pod (total both sides, Section 18.10):	
Propellers	400 lb (total weight of the two 8-ft and two 24-ft propellers)
Engine + turbocharger + accessories	2803 lb (from Section 18.10 and Fig. J.2)
Pod structural weight	290 lb (total for both sides)
Main landing gear	276 lb (U2-A bicycle gear + pogo, Table I.4)
Wing station for engine or payload pod	21.6 ft from centerline
Wing station for fuel tank	23.75 ft to 44.25 ft from centerline
Wing station for heat exchangers	7.6 ft to 28 ft from centerline
Vehicle takeoff gross weight (TOGW)	16,000 lb
Maximum airspeed	55 KEAS

The example problem will determine the spar cap sizing for the wing encountering a gust at 20,000 ft.

Analysis Approach

Part 1. Calculate the gust positive vertical load factor, $+n_z$, for the HAARP vehicle at the vehicle maximum airspeed and an altitude of

20,000 ft using the discrete gust formula. Assume an equivalent gust velocity of 66 ft/s, a lift curve slope of 0.1 per deg, and a total of 300 lb of fuel burned in reaching altitude.

The discrete gust formula is

$$n = 1 \pm \frac{K_g C_{l\alpha} U_e V_e}{498 \, W/S} \tag{19.1}$$

The gust alleviation factor K_g can be calculated for subsonic aircraft by using the expression

$$K_g = \frac{0.88 \mu}{5.3 + \mu} \tag{19.2}$$

where $\mu = 2(W/S)/\rho cag$.

The vehicle gross weight at altitude is W = 16,000 lb − 300 lb = 15,700 lb. This gives a wing loading W/S = 15,700 lb/2884 ft² = 5.44 lb/ft².

The air density ρ at an altitude of 20,000 ft can be obtained from a standard atmospheric table and is approximately 12.67×10^{-4} slug/ft³.

The acceleration of gravity is g = 32.2 ft/s².

The lift curve slope a is given as 0.1 per degree, which must be expressed as 5.730 per radian for Eq. (19.2).

The mean chord of the wing, c, can be calculated from the wing area S and span b:

$$c = S/b = 2884 \text{ ft}^2/269 \text{ ft} = 10.72 \text{ ft}$$

Plugging these values into Eq. (19.2) for μ gives μ = 4.34. The gust alleviation factor can then be calculated to be K_g = 0.40.

The following values can then be input into the discrete gust formula:

K_g	= 0.40
a	= 5.730 per radian
V_e	= 55 KEAS
U_e	= 66 ft/s
W/S	= 5.44

This gives gust load factors of −2.1 g and +4.1 g. Therefore, the gust positive load factor (resulting in wing upbending) for this example is +4.1 g.

Part 2. Using the positive vertical gust load factor calculated in Part 1, calculate the lift distribution for the wing using Schrenk's approximation [9].

The first step in calculating the lift distribution is to divide the wing into a number of spanwise panels. Although the number and

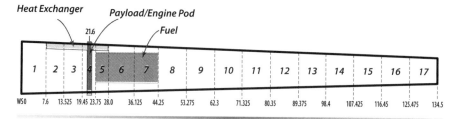

Figure 19.24 HAARP wing panel layout as used for analysis.

size of these panels is somewhat arbitrary, a sufficient number of panels should be used to insure the accuracy of the solution. In addition, subsequent parts of this example problem will involve calculating the inertia relief provided by the weight of the wing structure, fuel, and subsystems. Therefore, it is desirable to divide the wing into panels that match the location of various distributed and concentrated mass items in order to simplify subsequent calculations. For the solution presented here, the HAARP wing has been divided into 17 panels as shown in Fig. 19.24.

The lift distribution applied to the wing can be calculated using *Schrenk's approximation*. This method assumes that the spanwise lift distribution of an untwisted wing or tail is the average of the lift based on the actual trapezoidal wing shape and the lift based on an elliptical wing.

For a trapezoidal wing, lift can be expressed as a function of wing station y by using the following equation:

$$L^{\text{trap}}(y) = \frac{2L}{b(1+\lambda)}\left[1 - \frac{2y}{b}(1-\lambda)\right] \tag{19.3}$$

For an elliptical wing, the expression is

$$L^{\text{ellip}}(y) = \frac{4L}{\pi b}\sqrt{1 - \left[\frac{2y}{b}\right]^2} \tag{19.4}$$

For the trapezoidal wing equation, λ is the wing taper ratio. In both equations, L is the total lift applied to the wing and b is the wing span. For the HAARP wing example under the prescribed vertical gust load,

$$L = (4.1\,g)(15{,}700\text{ lb}) = 64{,}370\text{ lb (total)}$$

$$b = 269\text{ ft}$$

$$\lambda = 0.35$$

The calculations for trapezoidal and elliptical lift as a function of wing station y are shown in the spreadsheet presented in Table 19.3. Because the wing lift distribution is the same for each side, or semi-span of the wing, the calculations are shown only from wing station 0 (WS0, the vehicle centerline) to wing station 134.5 ft (wingtip). In the far right column the total lift per panel, based on the average of the trapezoidal and elliptical lift distributions, is shown. As a check, the lift force for all panels is summed at the bottom of this column and shows a total wing lift of 32,215 lb per side, which is well within 1% of the expected answer of 32,185 lb per side (64,370 lb/2 = 32,185 lb). Figure 19.25 plots the trapezoidal, elliptical, and average lift distributions as a function of wing station.

Part 3. Calculate the distribution of lift minus weight for the gust load calculated in Part 1 using the lift distribution derived in Part 2 and the given wing weights.

Table 19.4 shows the spreadsheet for calculating the weight for each of the 17 wing panels. Column F, the wing unit structural weight (0.939 lb/ft^2), is obtained by dividing the given total wing structural weight (2708 lb) by the wing planform area (2884 ft^2). Column G, the structural weight per panel, is obtained by multiplying this unit structural weight by the planform area of each panel, column E. Column H, the weight per span for each panel, is obtained by dividing each panel structural weight by the panel span.

The heat exchanger weight per span, column I, is obtained by dividing the total heat exchanger weight per side (1147 lb/2 = 573.5 lb/side) by the spanwise length of each heat exchanger (28 ft − 7.6 ft = 20.4 ft), giving a value of 573.5 lb/20.4 ft = 28.113 lb/ft. The spanwise distributed weights for the fuel, column J, and fuel tanks and pumps, column K, are calculated in a similar manner. Columns H, I, J, and K are then added for each panel to give the total distributed weight per panel shown in column L.

Each side of the HAARP wing also contains a number of significant concentrated mass items located at WS 21.6 ft. Specifically, the weight of the payload, payload pod structure, propulsion system (propeller, engine, turbocharger, and accessories), and main landing gear are shown in columns M and N and are summed in column O.

Column P multiples the total 1 g distributed weight per panel of column L by 4.1 g, and column Q multiples the total concentrated weight at WS 21.6 (column O) by 4.1 g. The total 4.1 g distributed and concentrated weights are then summed for each panel in column R; they are totaled at the bottom of the column to serve as an interim check.

Table 19.5 shows the spreadsheet for calculating the (lift-minus-weight) for each of the 17 wing panels. Column F is the average wing

Table 19.3 HAARP Wing Lift Distribution

Panel	Wing Station y (ft)	Panel Span (ft)	Elliptical Lift Distribution		Trapezoidal Lift Distribution		Avg. Lift (Ellip.+Trap.)/2 (lb/ft)	Lift per Panel (lb)
			$[(1-(2y/b)^2)]^{1/2}$	Elliptical Lift (y) at Midpanel (lb/ft)	$1-(2y/b)$ $(1-\lambda)$	Trapezoidal Lift (y) at Midpanel (lb/ft)		
1	0.00 3.80 7.60	7.6	0.9996	304.6	0.9816	348.0	326.3	2479.7
2	10.56 13.53	5.925	0.9969	303.7	0.9490	336.4	320.1	1896.5
3	16.49 19.45	5.925	0.9925	302.4	0.9203	326.3	314.3	1862.4
4	21.60 23.75	4.3	0.9870	300.7	0.8956	317.5	309.1	1329.2
5	25.88 28.00	4.25	0.9813	299.0	0.8750	310.2	304.6	1294.5
6	32.06 36.13	8.125	0.9712	295.9	0.8451	299.6	297.8	2419.2
7	40.19 44.25	8.125	0.9543	290.8	0.8058	285.7	288.2	2341.7

8	48.76 53.28	9.025	0.9320	284.0	0.7643	271.0	277.5	2504.0
9	57.79 62.30	9.025	0.9030	275.1	0.7207	255.5	265.3	2394.4
10	66.81 71.33	9.025	0.8679	264.4	0.6771	240.0	252.2	2276.4
11	75.84 80.35	9.025	0.8259	251.6	0.6335	224.6	238.1	2148.9
12	84.86 89.38	9.025	0.7758	236.4	0.5899	209.1	222.7	2010.3
13	93.89 98.40	9.025	0.7161	218.2	0.5463	193.7	205.9	1858.5
14	102.91 107.43	9.025	0.6439	196.2	0.5027	178.2	187.2	1689.5
15	111.94 116.45	9.025	0.5544	168.9	0.4590	162.7	165.8	1496.5
16	120.96 125.48	9.025	0.4372	133.2	0.4154	147.3	140.2	1265.6
17	129.99 134.50	9.025	0.2569	78.3	0.3718	131.8	105.0	948.0
							TOTAL	32,215.2

Table 19.4 Wing Weight Distribution Spreadsheet

Panel	B Wing Station (y) (ft)	C Panel Span (ft)	D Panel Chord (ft)	E Panel Planform Area (ft²)	F Wing Unit Structural Weight (lb/ft²)	G Wing Structural Wt/Panel (lb)	H Wing Structural Wt/Span (lb/ft)	I Heat Exchanger Wt/Span (lb/ft)	J Fuel Wt/Span (lb/ft)
	0.00								
1	3.80	7.60	15.6	118.5	0.939	111.3	14.64	0.00	0
	7.60								
2	10.56	5.93	15.1	89.3	0.939	83.9	14.15	28.11	0
	13.53								
3	16.49	5.93	14.6	86.6	0.939	81.3	13.73	28.11	0
	19.45								
4	21.60	4.30	14.2	61.2	0.939	57.4	13.36	28.11	0
	23.75								
5	25.88	4.25	13.9	59.1	0.939	55.5	13.05	28.11	109.8
	28.00								
6	32.06	8.13	13.4	109.1	0.939	102.4	12.60	0	109.8
	36.13								
7	40.19	8.13	12.8	104.0	0.939	97.6	12.02	0	109.8
	44.25								
8	48.76	9.03	12.1	109.6	0.939	102.9	11.40	0	0
	53.28								
9	57.79	9.03	11.4	103.3	0.939	97.0	10.75	0	0
	62.30								
10	66.81	9.03	10.8	97.1	0.939	91.1	10.10	0	0
	71.33								
11	75.84	9.03	10.1	90.8	0.939	85.3	9.45	0	0
	80.35								
12	84.86	9.03	9.4	84.6	0.939	79.4	8.80	0	0
	89.38								
13	93.89	9.03	8.7	78.3	0.939	73.3	8.12	0	0
	98.40								
14	102.91	9.03	8.0	72.1	0.939	67.7	7.50	0	0
	107.43								
15	111.94	9.03	7.3	65.8	0.939	61.8	6.85	0	0
	116.45								
16	120.96	9.03	6.6	59.5	0.939	55.9	6.20	0	0
	125.48								
17	129.99	9.03	5.9	53.3	0.939	50.0	5.55	0	0
	134.50								
				1442.0		1353.8			

K Fuel Tank and Pump Wt/Span (lb/ft)	L Total Wing Dist. Wt/Span (lb/ft)	Concentrated Payload and Systems Weight			4.1 g Weight		
		M 1 g POD Systems Weight (lb)	N 1 g POD Payload Weight (lb)	O 1 g POD Payload Weight (lb)	P Total 4.1 g Wing Dist. Wt/Span (lb/ft)	Q 4.1 g Concentrated Weight (lb)	R Total 4.1 g Wing Wt/Panel (lb)
0	14.64	0	0	0	60.0	0	456
0	42.27	0	0	0	173.3	0	1027
0	41.84	0	0	0	171.5	0	1016
0	41.47	1885	289	2174	170.0	8911	9642
2.27	153.19	0	0	0	628.1	0	2669
2.27	124.63	0	0	0	511.0	0	4152
2.27	124.04	0	0	0	508.6	0	4132
0	11.40	0	0	0	46.7	0	422
0	10.75	0	0	0	44.1	0	398
0	10.10	0	0	0	41.4	0	374
0	9.45	0	0	0	38.7	0	350
0	8.80	0	0	0	36.1	0	326
0	8.12	0	0	0	33.3	0	300
0	7.50	0	0	0	30.7	0	277
0	6.85	0	0	0	28.1	0	253
0	6.20	0	0	0	25.4	0	229
0	5.55	0	0	0	22.7	0	205
							26,229

Table 19.5   HAARP Lift-Minus-Weight Distribution Spreadsheet

Panel	B Wing Station (Y) (ft)	C Panel Span (ft)	D Panel Chord (ft)	E Panel Planform Area (ft²)	F 4.1 g Lift (Ellip.+Trap.)/2 (lb/ft)	G Total 4.1 g Wing Distr. Weight/Span (lb/ft)	H 4.1 g Lift Distributed Weight (lb/ft)	I 4.1 g Concentrated Weight (lb)
1	0 3.80 7.60	7.60	15.59	118.49	326.27	60.02	266.25	0
2	10.56 13.53	5.93	15.07	89.30	320.08	173.29	146.79	0
3	16.49 19.45	5.93	14.62	86.61	314.32	171.54	142.79	0
4	21.60 23.75	4.30	14.23	61.17	309.11	170.03	139.08	−8911
5	25.88 28.00	4.25	13.90	59.06	304.59	628.06	−323.48	0
6	32.06 36.13	8.13	13.42	109.05	297.75	510.97	−213.22	0
7	40.19 44.25	8.13	12.80	103.98	288.21	508.57	−220.36	0

8	48.76 53.28	9.03	12.14	109.56	277.46	46.74	230.72	0
9	57.79 62.30	9.03	11.45	103.31	265.31	44.07	221.24	0
10	66.81 71.33	9.03	10.76	97.06	252.23	41.41	210.83	0
11	75.84 80.35	9.03	10.06	90.81	238.11	38.74	199.37	0
12	84.86 89.38	9.03	9.37	84.56	222.75	36.07	186.68	0
13	93.89 98.40	9.03	8.68	78.30	205.92	33.30	172.63	0
14	102.91 107.43	9.03	7.98	72.06	187.20	30.74	156.46	0
15	111.94 116.45	9.03	7.29	65.80	165.82	28.07	137.75	0
16	120.96 125.48	9.03	6.60	59.55	140.23	25.40	114.83	0
17	129.99 134.50	9.03	5.91	53.29	105.04	22.73	82.31	0
			TOTAL = 1442.0					

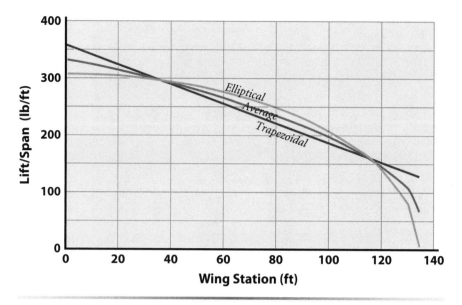

Figure 19.25 Trapezoidal, elliptical, and average wing lift distributions (n_z = +4.1 g).

lift distribution at 4.1 g calculated in Part 2, and column G is the 4.1 g distributed weight that was previously calculated. Column H is obtained by subtracting the lift from the weight for each panel, and column I contains concentrated weight located at WS 21.6.

Figure 19.26 plots the 4.1 g (lift-minus-weight) distribution and 4.1 g concentrated weight located at WS 21.6.

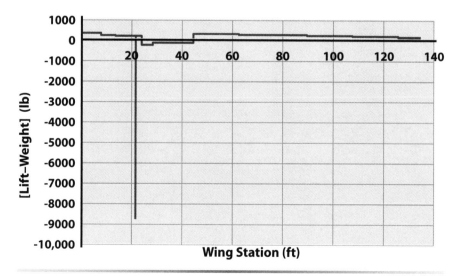

Figure 19.26 HAARP lift-minus-weight distribution.

Part 4. Using the lift-minus-weight distribution calculated in Part 3, calculate the net vertical shear load (P_z) distribution and spanwise bending moment (M_x) distribution for the HAARP wing for the gust condition.

Table 19.6 summarizes the lift-minus-weight distribution and concentrated mass items for the HAARP wing that have been calculated in Parts 1 through 3 of this example problem. Column F presents the net total load for each panel (assumed to act at the panel midpoint), which is derived from multiplying the lift-minus-weight (expressed in pounds per foot of panel span) by the panel span. Notice that the concentrated mass items located at WS 21.6 are handled separately and are not included in the load-per-panel calculations for column F. The vertical shear load (P_z) applied to the wing (column G) is obtained by starting at the wingtip (panel 17) and summing the net load from each panel and progressing toward the wing root (panel 1). This shear load is plotted against wing station in Fig. 19.27 and illustrates the reduction in vertical shear loading resulting from the large mass items (fuel, heat exchanger, propulsion system, payload, etc.) located toward the inner span of the wing. This reduction in wing shear and bending loads due to mass items (both distributed and concentrated mass items) is referred to as *inertia relief.*

The spanwise bending moment (M_x) is obtained by calculating the area under the shear curve. Starting at the wingtip (panel 17) the area under the shear curve is calculated for each panel and summed

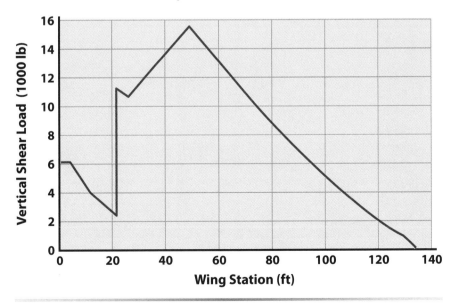

Figure 19.27 HAARP wing vertical shear ($n_z = +4.1\ g$).

Table 19.6   HAARP Wing Vertical Shear Spreadsheet ($n_z = +4.1$ g)

Panel	B Wing Station (y) (ft)	C Panel Span (ft)	D 4.1 g Lift Distr. Weight (lb/ft)	E 4.1 g Concentrated Weight (lb)	F Total 4.1 g Load per Panel (lb)	G Shear Load (lb)
	0					
1	3.80	7.6	266.2	0	2023.5	5,986.4
	7.60					
2	10.56	5.925	146.8	0	869.7	3,962.9
	13.53					
3	16.49	5.925	142.8	0	846.0	3,093.1
	19.45					
4	21.60	4.3	139.1	−8911	598.0	2,247.1
	23.75					
5	25.88	4.25	−323.5	0	−1374.8	10,560.4
	28.00					
6	32.06	8.125	−213.2	0	−1732.4	11,935.2
	36.13					
7	40.19	8.125	−220.4	0	−1790.4	13,667.6
	44.25					
8	48.76	9.025	230.7	0	2082.2	15,458.0
	53.28					
9	57.79	9.025	221.2	0	1996.7	13,375.8
	62.30					
10	66.81	9.025	210.8	0	1902.7	11,379.1
	71.33					
11	75.84	9.025	199.4	0	1799.3	9,476.4
	80.35					
12	84.86	9.025	186.7	0	1684.8	7,677.1
	89.38					
13	93.89	9.025	172.6	0	1558.0	5,992.3
	98.40					
14	102.91	9.025	156.5	0	1412.0	4,434.4
	107.43					
15	111.94	9.025	137.7	0	1243.2	3,022.3
	116.45					
16	120.96	9.025	114.8	0	1036.3	1,779.1
	125.48					
17	129.99	9.025	82.3	0	742.8	742.8
	134.50					

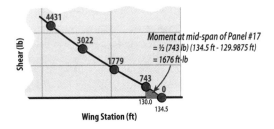

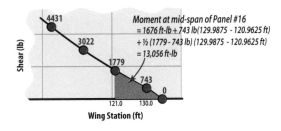

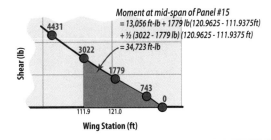

Figure 19.28 Example of method for calculating area under shear curve.

progressively working toward the wing root (panel 1), as illustrated in Fig. 19.28. Table 19.7 summarizes the calculation used to derive the spanwise bending moment based on the area under the shear curve, and Fig. 19.29 is the plot of the spanwise bending moment as a function of wing station.

Part 5. Using the wing moment distribution derived in Part 4, what is the spanwise bending moment at WS 40.2? Assuming the HAARP wing uses a single "I-beam" spar located at maximum t/c of the airfoil that reacts all of the spanwise bending load (i.e., wing skin is ineffective at carrying any load), and assuming that the centroids of the upper and lower spar caps are coincident with the outer surface of the wing skins, what are the spar cap loads at WS 40.2 resulting from the gust condition?

Table 19.7 Spanwise Bending Moment Spreadsheet

Panel	Wing Station (y) (ft)	Panel Span (ft)	Vertical Shear Load (lb)	Delta Spanwise Bending Moment, M_y (ft-lb)	Spanwise Bending Moment, M_y (ft-lb)
	0		5,986	22,748	1,027,053
1	3.80	7.60	5,986	33,641	1,004,305
	7.60				
2	10.56	5.93	3,963	20,903	970,664
	13.53				
3	16.49	5.93	3,093	13,651	949,761
	19.45				
4	21.60	4.30	11,158	46,424	936,110
	23.75				
5	25.88	4.25	10,560	69,596	889,686
	28.00				
6	32.06	8.13	11,935	104,011	820,090
	36.13				
7	40.19	8.13	13,668	124,876	716,079
	44.25				
8	48.76	9.03	15,458	130,113	591,202
	53.28				
9	57.79	9.03	13,376	111,707	461,089
	62.30				
10	66.81	9.03	11,379	94,111	349,383
	71.33				
11	75.84	9.03	9,476	77,405	255,272
	80.35				
12	84.86	9.03	7,677	61,683	177,867
	89.38				
13	93.89	9.03	5,992	47,050	116,183
	98.40				
14	102.91	9.03	4,434	33,648	69,133
	107.43				
15	111.94	9.03	3,022	22,429	35,485
	116.45				
16	120.96	9.03	1,779	11,380	13,056
	125.48				
17	129.99	9.03	743	1,676	1,676
	134.50		0	0	0

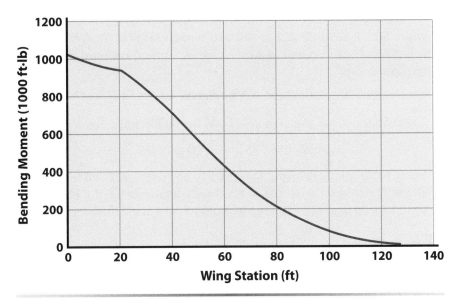

Figure 19.29   Wing bending moment.

The M_x moment at WS 40.2 is 716,079 ft-lb. The wing chord at WS 40.2 is 12.798 ft. Therefore, using a $t/c = 12.2\%$, the spar depth at WS 40.2 = $(0.122)(12.798) = 1.56$ ft.

The wing bending moment is reacted by a couple load in the spar caps as shown in Fig. 19.30. These cap loads can be calculated by dividing the bending moment M by the spar depth d, which at WS 40.2 is 716,079 ft-lb/1.56 ft = ±459,025 lb (tension in the lower cap, and compression in the upper cap for the +4.1 g wing upbending condition).

Part 6. Assuming an ultimate factor-of-safety = 1.5, and assuming that the wing spar is constructed of a material with $F_{ty} = 60$ ksi and $F_{tu} = 75$ ksi, what cross-sectional area is required for the lower spar cap at WS 40.2 when subjected to the gust condition?

Figure 19.30   Wing spanwise bending moment reacted by couple load in the spar caps.

The lower spar is loaded in tension due to this gust condition, with a limit load of 459,025 lb and an ultimate load of $(1.5)(459,025$ lb$)$ = 688,538 lb. The required lower cap area based on tension yield strength is

$$A_{req}\text{-yield} = 459,025 \text{ lb}/60,000 \text{ psi} = 7.7 \text{ in.}^2$$

The required lower cap area based on tension ultimate strength is

$$A_{req}\text{-ult} = 688,538 \text{ lb}/75,000 \text{ psi} = 9.2 \text{ in.}^2$$

Because the spar cap sizing must satisfy both the yield and ultimate strength criteria, the required cross-sectional area for the HAARP wing lower cap at WS 40.2 due to an $n_z = +4.1$ g is 9.2 in.^{2}.

19.14 Summary

The world of aircraft structures involves many diverse technical disciplines related to design, analysis, materials, manufacturing, and testing. There are many options available to the Structures Engineer regarding design concepts, material selection, analysis approach, manufacturing methods, and test verification philosophy; accordingly, it is important to have a rational and objective decision-making process to determine which design options are optimal for satisfying a particular set of vehicle requirements. Although it is sometimes obvious which structural design options are best, there are often multiple paths to achieve the same end result, and most structural design decisions represent a complex balance between weight, risk, cost, and schedule.

References

[1] "Joint Service Specification Guide—Aircraft Structures," JSSG-2006, U.S. Department of Defense, 1998.
[2] "Joint Service Specification Guide—Air Vehicle," JSSG-2001B, U.S. Department of Defense, 2004.
[3] "Aircraft Strength and Rigidity—Flight Loads," MIL-A-8861B, Naval Air Systems Command, 1986.
[4] Niu, M. C. Y., *Airframe Structural Design*, Technical Book Co., Los Angeles, CA, 1988.
[5] Ekvall, J. S., Rhoades, J. E., and Wald, G. G., "Methodology for Evaluating Weight Savings from Basic Material Properties," Design of Fatigue and Fracture Resistant Structures, American Society for Testing and Materials ASTM STP 761, 1982.
[6] "Aircraft Strength and Rigidity—General Specification," MIL-A-8860B, Naval Air Systems Command, 1987.
[7] "Metallic Materials Properties Development and Standardization," MMPDS-04, Federal Aviation Administration, 2008.
[8] "Aircraft Strength and Rigidity—Ground Tests," MIL-A-8867C, Naval Air Systems Command, 1987.
[9] Raymer, D. P., *Aircraft Design: A Conceptual Approach*, AIAA Education Series, AIAA, Washington, DC, 1989.

Chapter 20 / Refined Weight Estimate

- Military, Commercial, & GA WERs
- Adjustment for Advanced Composites
- Low Wing Loading ($W/S < 5$)
- Locating the C.G.
- Estimating Moments of Inertia

Estimating weights is critical in the design of an aircraft. This is especially true for weight-critical aircraft such as the Voyager designed by Burt Rutan (Scaled Composite Inc). Learn more about this aircraft in Section 20.2.4.

Estimating the weight of an airplane or airship involves as much art as it does science.

20.1 Introduction

At this point the designer should do a detailed weight estimate of the aircraft. The original estimate of the aircraft empty weight (from Chapter 5) used the impersonal empty-weight trend curves of Appendix I. As such they were not able to capture the unique and innovative features of the conceptual aircraft. However, they were appropriate for that point in the design cycle when the aircraft information was sparse. Now there is considerable information available on the aircraft and the weights of all the aircraft components can be estimated to get a refined empty weight and center of gravity location. Component weights are determined in a large part through the use of empirical formulations that are conditioned upon the many different geometric properties of the components. A multiple regression analysis is used to determine the best curve-fit expression for the historical data.

The designer must be careful when applying these weight-estimate equations to new designs. If the new aircraft will be considerably different in performance and/or structural design than the aircraft used to develop the weight-estimate equations, then the weight equations might have to be altered. Weight-estimation methods for advanced systems are guarded very closely by aircraft companies as they represent their expertise in the design of advanced systems. The Aeronautical Systems Center at Wright–Patterson AFB, Ohio, served as a clearinghouse for the exchange of weights information in the 1960s and 1970s, hosting weight-prediction workshops. Currently, NAVAIR at Patuxent River, Maryland, serves as the government clearinghouse. The Society of Allied Weight Engineers (SAWE) also promotes the exchange of weight-estimating information.

The weight-estimating equations contained in this chapter have come from many sources. It is recommended that the equations be calibrated case-by-case, by comparing the estimated weights with real aircraft (Appendix I) and other sources (see [1–6]).

As discussed in Appendix I, estimating the empty weight of an aircraft is the most challenging part of the conceptual design process. Historical data are available to make credible weight estimates (estimating weight at the conceptual design level is an art and it will never be a science). Most design groups carry a weight margin through conceptual and preliminary design to account for the uncertainty in the weight estimates and the inevitable and dreaded "weights growth." At the Lockheed Martin Skunk Works the margin on empty weight is 6%.

20.2 Weight-Estimation Methods

20.2.1 Conventional Metal Aircraft—Moderate Subsonic to Supersonic Performance

Credit for the following weight-estimation methods for conventional metal aircraft goes to many sources in the aerospace industry. The weight equations give the component weight in pounds.

20.2.1.1 Structure
Wing

U.S. Air Force (USAF) Fighter Aircraft:

$$
Wt = 3.08 \left(\frac{K_{\mathrm{PIV}} N W_{\mathrm{TO}}}{t/c} \left\{ \left[\tan \Lambda_{\mathrm{LE}} - \frac{2(1-\lambda)}{\mathrm{AR}(1+\lambda)} \right]^2 + 1.0 \right\} \times 10^{-6} \right)^{0.593}
$$
$$
\left[(1+\lambda) \mathrm{AR} \right]^{0.89} S_w^{0.741} \tag{20.1a}
$$

U.S. Navy (USN) Fighter Aircraft:

$$
Wt = 19.29 \left(\frac{K_{\mathrm{PIV}} N W_{\mathrm{TO}}}{t/c} \left\{ \left[\tan \Lambda_{\mathrm{LE}} - \frac{2(1-\lambda)}{\mathrm{AR}(1+\lambda)} \right]^2 + 1.0 \right\} \times 10^{-6} \right)^{0.464}
$$
$$
\left[(1+\lambda) \mathrm{AR} \right]^{0.70} S_w^{0.58} \tag{20.1b}
$$

where

K_{PIV} = wing variable-sweep structural factor
  = 1.00 for fixed wings
  = 1.175 for variable-sweep wings
t/c = maximum thickness ratio
W_{TO} = takeoff weight, in pounds (lb)
Λ_{LE} = leading edge sweep
λ = taper ratio
AR = wing aspect ratio
S_w = wing area, in square feet (ft²)
N = ultimate load factor
  = 13.5 for fighter aircraft (based on a design limit load factor of +9.0 and a margin of safety of 1.5)
  = 4.5 for bomber and transport aircraft (based on a design limit load factor of +3.0)

Subsonic Aircraft (Military and Commercial):

$$\text{Wt} = 0.00428 \left(S_w\right)^{0.48} \frac{\text{AR}^{1.0}\left(M_0\right)^{0.43}\left(W_{\text{TO}}\,N\right)^{0.84}\left(\lambda\right)^{0.14}}{\left(100\,t/c\right)^{0.76}\left(\cos\Lambda_{1/2}\right)^{1.54}} \tag{20.2}$$

where

M_0 = maximum Mach number at sea level
$\Lambda_{1/2}$ = sweep of half-chord
t/c = maximum thickness ratio

This wing weight equation is valid for an M_0 range of 0.4–0.8, a t/c range of 0.08–0.15, and an aspect ratio (AR) range of 4–12.

Horizontal and Vertical Tail
Horizontal Tail:

$$\text{Wt} = 0.0034\,\gamma^{0.915} \tag{20.3a}$$

where

γ = $(W_{\text{TO}}N)^{0.813}(S_{\text{HT}})^{0.584}(b_{\text{HT}}/t_{R_{\text{HT}}})^{0.033}(\bar{c}_{\text{wing}}/L_t)^{0.28}$
N = ultimate load factor
S_{HT} = horizontal tail total planform area (include fuselage carry-through), in square feet (ft^2)
$t_{R_{\text{HT}}}$ = thickness of the horizontal tail at the root, in feet
$\bar{c}_{\text{wing}}$ = mac of the wing, in feet
L_t = tail moment arm, in feet; distance from one-fourth wing mac to one-fourth tail mac. For canard surfaces, use distance from 0.4 wing mac to one-fourth canard mac
b_{HT} = span of horizontal tail, in feet

Vertical Tail:

$$\text{Wt} = 0.19\,\gamma^{1.014} \tag{20.3b}$$

where

γ = $\left(1+h_T/h_V\right)^{0.5}\left(W_{\text{TO}}N\right)^{0.363}\left(S_{\text{VT}}\right)^{1.089} M_0^{0.601}\left(L_t\right)^{-0.726}\left(1+S_r/S_{\text{VT}}\right)^{0.217}$
    $\times\left(\text{AR}_{\text{VT}}\right)^{0.337}\left(1+\lambda_V\right)^{0.363}\left(\cos\Lambda_{\text{VT}}\right)^{-0.484}$

h_T/h_V = ratio of horizontal tail height to vertical tail height. For a "T" tail this ratio is 1.0; for a fuselage-mounted horizontal tail this ratio is 0

S_{VT} = area of vertical tail, in square feet (ft^2)
M_0 = maximum Mach number at sea level
L_t = tail moment arm, in feet; distance from one-fourth wing mac to one-fourth tail mac
S_r = rudder area, in square feet. If unknown, use $S_r/S_V = 0.3$
$\mathrm{AR}_{\mathrm{VT}}$ = aspect ratio of vertical tail
λ_V = taper ratio of vertical tail
Λ_{VT} = sweep of vertical tail quarter-chord

Fuselage

USAF and Commercial:

$$\mathrm{Wt} = 10.43\,(K_{\mathrm{INL}})^{1.42}\,(q \times 10^{-2})^{0.283}\,(W_{\mathrm{TO}} \times 10^{-3})^{0.95}\,(L/H)^{0.71} \quad (20.4)$$

USN:

$$\mathrm{Wt} = 11.03\,(K_{\mathrm{INL}})^{1.23}\,(q \times 10^{-2})^{0.245}\,(W_{\mathrm{TO}} \times 10^{-3})^{0.98}\,(L/H)^{0.61} \quad (20.5)$$

where

q = maximum dynamic pressure, in pounds per square foot ($\mathrm{lb/ft}^2$)
L = fuselage length, in feet (ft)
H = maximum fuselage height, in feet
K_{INL} = 1.25 for inlets on fuselage
= 1.0 for inlets in wing root or elsewhere

Landing Gear

USAF and Commercial:

$$\mathrm{Wt} = 62.21\left(W_{\mathrm{TO}} \times 10^{-3}\right)^{0.84} \quad (20.6)$$

USN:

$$\mathrm{Wt} = 129.1\left(W_{\mathrm{TO}} \times 10^{-3}\right)^{0.66} \quad (20.7)$$

20.2.1.2 Propulsion

Engine

Engine weights should be based upon the engine manufacturer's data and scaling factors. Assume the exhaust, cooling, turbo-supercharger, and lubrication systems weights are included in the engine weight.

Propulsion Subsystems

Propulsion subsystem items are the air induction system, fuel system, engine controls, and starting system.

Air Induction System

The parameters used in determining air induction system weights are defined as follows:

A_i = capture area per inlet, in square feet (ft^2)

N_i = number of inlets, vehicle configuration

L_d = subsonic duct length, per inlet, in feet (ft)

L_r = ramp length forward of throat, per inlet, in feet

K_{GEO} = duct shape factor; use $K = 1.33$ if duct has two or more relatively flat sides; use $K = 1.0$ if duct is round or has one flat side

P_2 = maximum static pressure at engine compressor face, in pounds per square inch absolute (psia)

K_{TE} = temperature correction factor

= 1 for M_D less than 3.0

= $(M_D + 2)/5$ for M_D between 3.0 and 6.0, where design Mach number M_D is the maximum Mach number

K_M = duct material factor, use $K = 1.0$ for $M_D < 1.4$; use $K = 1.5$ for $M_D \geq 1.4$

Duct Provisions:

$$Wt = 0.32 \left(N_i \right) \left(L_d \right) \left(A_i \right)^{0.65} \left(P2 \right)^{0.6} \tag{20.8}$$

This equation accounts for the duct support structure and should only be used for internal installations. Duct provisions are normally included with the weight, but they have been separated out for this discussion to complete the total air-induction system weight.

Internal Duct Weight:

$$Wt = 1.735 \left[\left(N_i \right) \left(L_d \right) \left(A_i \right)^{0.5} \left(P_2 \right) \left(K_{GEO} \right) \left(K_M \right) \right]^{0.7331} \tag{20.9}$$

This equation accounts for the duct structure from the inlet lip to the engine compressor face, and it should only be used for internal engine installations.

Variable-Geometry Ramps, Actuators, and Control Weights:

$$Wt = 4.079 \left[\left(N_i \right) \left(L_r \right) \left(A_i \right)^{0.5} \left(K_{TE} \right) \right]^{1.201} \tag{20.10}$$

This equation should only be used for internal installations. Variable-geometry ramps are normally used with rectangular inlets.

Half-Round Fixed Spike Weight:

$$W_{HFS} = 12.53 \left(N_i \right)\left(A_i \right)$$ (20.11)

This equation should only be used for internal installations.

Full-Round Translating Spike Weight:

$$Wt = 15.65 \left(N_i \right)\left(A_i \right)$$ (20.12)

This equation can be used for either internal or external (podded) engine installations.

Translating and Expanding Spike Weight:

$$W_{TES} = 51.8 \left(N_i \right)\left(A_i \right)$$ (20.13)

This equation can be used for either internal or external engine installations.

External Turbojet Cowl and Duct Weight:

$$Wt = 3.00 \left(N_i \right)\left[\left(A_i \right)^{0.5} \left(L_d \right)\left(P_2 \right) \right]^{0.731}$$ (20.14)

This equation accounts for the exterior cowl or cover panels, ducting, and substructure such as rings, frames, stiffeners, and longerons, from the inlet lip to the engine compressor face, and should only be used for external engine installations.

External Turbofan Cowl and Duct Weight:

$$W_{DTF} = 7.435 \left(N_i \right)\left[\left(L_d \right)\left(A_i \right)^{0.5} \left(P_2 \right) \right]^{0.731}$$ (20.15)

This equation accounts for cowl panels, substructure, and the basic engine duct and the fan duct, and it should only be used for external engine installations.

Fuel System

Self-Sealing Bladder Cells:

$$Wt = 41.6 \left[\left(F_{GW} + F_{GF} \right) \times 10^{-2} \right]^{0.818}$$ (20.16)

where F_{GW} = total wing fuel in gallons and F_{GF} = total fuselage fuel in gallons.

Non-Self-Sealing Bladder Cells:

$$\text{Wt} = 23.10 \left[\left(F_{\text{GW}} + F_{\text{GF}} \right) \times 10^{-2} \right]^{0.758} \tag{20.17}$$

Fuel System Bladder Cell Backing and Supports (Both Self-Sealing and Non-Self-Sealing):

$$\text{Wt} = 7.91 \left[\left(F_{\text{GW}} + F_{\text{GF}} \right) \times 10^{-2} \right]^{0.854} \tag{20.18}$$

In-Flight Refuel System:

$$\text{Wt} = 13.64 \left[\left(F_{\text{GW}} + F_{\text{GF}} \right) \times 10^{-2} \right]^{0.392} \tag{20.19}$$

Dump-and-Drain System:

$$\text{Wt} = 7.38 \left[\left(F_{\text{GW}} + F_{\text{GF}} \right) \times 10^{-2} \right]^{0.458} \tag{20.20}$$

C.G. Control System (Transfer Pumps and Monitor):

$$\text{Wt} = 28.38 \left[\left(F_{\text{GW}} + F_{\text{GF}} \right) \times 10^{-2} \right]^{0.442} \tag{20.21}$$

Engine Controls

Body- or Wing-Root-Mounted Jet:

$$\text{Wt} = K_{\text{ECO}} \left(L_f N_E \right)^{0.792} \tag{20.22}$$

where

L_f = fuselage length, in feet (ft)
N_E = number of engines (per airplane)
K_{ECO} = engine control engine-type coefficient
         = 0.686, nonafterburning engines
         = 1.080, afterburning (A/B) engines

Wing-Mounted Turbojet and Turbofan:

$$\text{Wt} = 88.46 \left[\left(L_f + b \right) N_E \times 10^{-2} \right]^{0.294} \tag{20.23}$$

Wing-Mounted Turboprop:

$$\text{Wt} = 56.84 \left[\left(L_f + b \right) N_E \times 10^{-2} \right]^{0.514} \tag{20.24}$$

Wing-Mounted Reciprocating:

$$\mathrm{Wt} = 60.27 \left[\left(L_f + b \right) N_E \times 10^{-2} \right]^{0.724} \tag{20.25}$$

where b = wing span, in feet.

Starting Systems
One or Two Jet Engines—Cartridge and Pneumatic:

$$\mathrm{Wt} = 9.33 \left(N_E W_{\mathrm{ENG}} \times 10^{-3} \right)^{1.078} \tag{20.26}$$

where N_E = number of engines per airplane, and W_{ENG} = engine weight, in pounds per engine.

One or Two Jet Engines—Electrical:

$$\mathrm{Wt} = 38.93 \left(N_E W_{\mathrm{ENG}} \times 10^{-3} \right)^{0.918} \tag{20.27}$$

Four or More Jet Engines—Pneumatic:

$$\mathrm{Wt} = 49.19 \left(N_E W_{\mathrm{ENG}} \times 10^{-3} \right)^{0.541} \tag{20.28}$$

Turboprop Engines—Pneumatic:

$$\mathrm{Wt} = 12.05 \left(N_E W_{\mathrm{ENG}} \times 10^{-3} \right)^{1.458} \tag{20.29}$$

Reciprocating Engines—Electric:

$$\mathrm{Wt} = 50.38 \left(N_E W_{\mathrm{ENG}} \times 10^{-3} \right)^{0.459} \tag{20.30}$$

Propeller Systems
Propellers:

$$\mathrm{Wt} = K_p N_p \left(N_{\mathrm{BL}} \right)^{0.391} \left(d_p \times \mathrm{HP} \times 10^{-3} \right)^{0.782} \tag{20.31}$$

where

N_p = number of propellers per airplane
N_{BL} = number of blades per propeller
d_p = propeller diameter, in feet per propeller
HP = rated engine shaft horsepower
K_p = propeller-engine coefficient
    = 24.00 for turboprop above 1500 shaft horsepower
    = 31.92 for reciprocating engine at all horsepower and turboprop below
      1500 shaft horsepower

Propeller Controls—Turboprop Engines:

$$\text{Wt} = 0.322 \left(N_{\text{BL}} \right)^{0.589} \left(N_p d_p \ \text{HP} \times 10^{-3} \right)^{1.178} \tag{20.32}$$

Propeller Controls—Reciprocating Engines:

$$\text{Wt} = 4.552 \left(N_{\text{BL}} \right)^{0.379} \left(N_p d_p \ \text{HP} \times 10^{-3} \right)^{0.759} \tag{20.33}$$

20.2.1.3 Surface Controls Plus Hydraulics and Pneumatics

Fighters—USAF:

$$\text{Wt} = K_{\text{SC}} \left(W_{\text{TO}} \times 10^{-3} \right)^{0.581} \tag{20.34}$$

where

K_{SC} = surface control coefficient
 = 106.10 for elevon without horizontal tail
 = 138.18 for horizontal tail
 = 167.48 for variable-sweep wing

Fighter and Attack—USN:

$$\text{Wt} = 23.77 \left(W_{\text{TO}} \times 10^{-3} \right)^{1.10} \tag{20.35}$$

Executive and Commercial Passenger Transports:

$$\text{Wt} = 56.01 \left(W_{\text{TO}} \times q \times 10^{-5} \right)^{0.576} \tag{20.36}$$

where q = maximum dynamic pressure, in pounds per square foot (lb/ft^2).

Commercial and Military Cargo–Troop Transports:

$$\text{Wt} = 15.96 \left(W_{\text{TO}} \times q \times 10^{-5} \right)^{0.815} \tag{20.37}$$

Bombers:

$$\text{Wt} = 1.049 \left(S_{\text{TOT}} \times q \times 10^{-3} \right)^{1.21} \tag{20.38}$$

where S_{TOT} = total surface control area, in square feet (ft^2).

20.2.1.4 Instruments

Flight Instrument Indicators

$$\text{Wt} = N_{\text{PIL}}\left[15.0 + 0.032\left(W_{\text{TO}}\times10^{-3}\right)\right] \tag{20.39}$$

where N_{PIL} = number of pilots.

Engine Instrument Indicators

Turbine Engines:

$$\text{Wt} = N_E\left[4.80 + 0.006\left(W_{\text{TO}}\times10^{-3}\right)\right] \tag{20.40}$$

where N_E = number of engines.

Reciprocating Engines:

$$\text{Wt} = N_E\left[7.40 + 0.046\left(W_{\text{TO}}\times10^{-3}\right)\right] \tag{20.41}$$

Miscellaneous Indicators:

$$\text{Wt} = 0.15\left(W_{\text{TO}}\times10^{-3}\right) \tag{20.42}$$

20.2.1.5 Electrical System

The weight prediction relationships are expressed in terms of the total weight of the fuel system plus the total weight of the electronics system, the prime users of electrical power on most aircraft.

USAF Fighters:

$$\text{Wt} = 426.17\left[\left(W_{\text{FS}}\times W_{\text{TRON}}\right)\times10^{-3}\right]^{0.510} \tag{20.43}$$

where

W_{FS} = weight of fuel system, in pounds (lb)
W_{TRON} = weight of electronics system, in pounds (lb)

USN Fighters and Attack:

$$\text{Wt} = 346.98\left[\left(W_{\text{FS}}\times W_{\text{TRON}}\right)\times10^{-3}\right]^{0.509} \tag{20.44}$$

Bombers:

$$\text{Wt} = 185.46\left[\left(W_{\text{FS}}\times W_{\text{TRON}}\right)\times10^{-3}\right]^{1.286} \tag{20.45}$$

Transports:

$$Wt = 1162.66 \left[\left(W_{FS} \times W_{TRON} \right) \times 10^{-3} \right]^{0.506} \qquad (20.46)$$

20.2.1.6 Furnishings
Fighter and Attack Aircraft
Ejection Seats:

$$Wt = 22.89 \left(N_{CR} \times q \times 10^{-2} \right)^{0.743} \qquad (20.47)$$

where

N_{CR} = number of crew
q = maximum dynamic pressure, in pounds per square foot (lb/ft^2)

Miscellaneous and Emergency Equipment:

$$Wt = 106.61 \left(N_{CR} W_{TO} \times 10^{-5} \right)^{0.585} \qquad (20.48)$$

Bomber and Observation Aircraft
Seats:

$$\textit{Fixed}: Wt = 83.23 \left(N_{CR} \right)^{0.726} \qquad (20.49)$$

$$\textit{Ejection}: Wt = K_{SEA} \left(N_{CR} \right)^{1.20} \qquad (20.50)$$

where

K_{SEA} = ejection seat coefficient
= 149.12 with survival kit
= 99.54 without survival kit

Oxygen System:

$$Wt = 16.89 \left(N_{CR} \right)^{1.494} \qquad (20.51)$$

Crew Bunks:

$$Wt = 12.18 \left(N_{BU} \right)^{1.085} \qquad (20.52)$$

where N_{BU} = number of crew bunks

Transport Aircraft
Flight Deck Seats—Executive and Commercial:

$$Wt = 54.99 \left(N_{FDS} \right) \qquad (20.53)$$

where N_{FDS} = number of flight deck stations

Passenger Seats—Executive and Commercial:

$$Wt = 32.03 \left(N_{PASS} \right) \tag{20.54}$$

where N_{PASS} = number of passengers.

Troop Seats—Troop Transports:

$$Wt = 11.17 \left(N_{TRO} \right) \tag{20.55}$$

where N_{TRO} = number of troops.

Lavatories and Water Provisions—Executive and Commerical:

$$Wt = K_{LAV} \left(N_{PASS} \right)^{1.33} \tag{20.56}$$

where

K_{LAV} = 3.90 for executive
     = 1.11 for long-range commercial passenger
     = 0.31 for short-range commercial passenger

Lavatories and Water Provisions—Military Transport:

$$Wt = 1.11 \left(N_{PASS} \right)^{1.33} \tag{20.57}$$

Food Provisions—Executive and Commercial:

$$Wt = K_{BUF} \left(N_{PASS} \right)^{1.12} \tag{20.58}$$

where

K_{BUF} = 5.68 for long-range (707, 990, 737, 747, 757, 767, 777, etc.)
     = 1.02 for short-range (340, 202, Citation, Learjet, King Air, Jetstream, etc.)

Oxygen System:

$$Wt = 7.00 \left(N_{CR} + N_{PASS} + N_{ATT} \right)^{0.702} \tag{20.59}$$

where N_{ATT} = number of attendants

Cabin Windows—Executive and Commercial:

$$Wt = 109.33 \left[N_{PASS} \left(1 + P_C \right) \times 10^{-2} \right]^{0.505} \tag{20.60}$$

where P_C = ultimate cabin pressure, in pounds per square inch (lb/in²)

Baggage and Cargo Handling Provisions:

$$\text{Wt} = K_{\text{CBC}} \left(N_{\text{PASS}} \right)^{1.456} \tag{20.61}$$

where

K_{CBC} = 0.0646 without preload provisions
    = 0.316 with preload provisions

Miscellaneous Furnishings and Equipment

Executive and Commercial:

$$\text{Wt} = 0.771 \left(W_{\text{TO}} \times 10^{-3} \right) \tag{20.62}$$

Military Passenger:

$$\text{Wt} = 0.771 \left(W_{\text{TO}} \times 10^{-3} \right) \tag{20.63}$$

Military Troop–Cargo:

$$\text{Wt} = 0.618 \left(W_{\text{TO}} \times 10^{-3} \right)^{0.839} \tag{20.64}$$

20.2.1.7   Air Conditioning and Anti-Icing

Fighters

High Subsonic and Supersonic:

$$\text{Wt} = 201.66 \left[\left(W_{\text{TRON}} + 200 \, N_{\text{CR}} \right) \times 10^{-3} \right]^{0.735} \tag{20.65}$$

where

W_{TRON} = weight of electronics system, in pounds (lb)
N_{CR} = number of crew

Subsonic (Below Approximately M = 0.50):

$$\text{Wt} = K_{\text{ACAI}} \left[\left(W_{\text{TRON}} + 200 \, N_{\text{CR}} \right) \times 10^{-3} \right]^{0.538} \tag{20.66}$$

where

K_{ACAI} = air conditioning and anti-icing coefficient
    = 108.64, no wing or tail anti-icing
    = 212.00, wing and tail anti-icing

Bombers and Military Troop–Cargo–Passenger Transports

$$\text{Wt} = K_{\text{ACAI}} \left[V_{\text{PR}} \times 10^{-2} \right]^{0.242} \tag{20.67}$$

where

V_{PR} = pressurized or occupied volume, in cubic feet (ft^3)
K_{ACAI} = air conditioning and anti-ice coefficient
= 887.25, bomber and military transport with wing and tail anti-icing
= 748.15, bomber and military transport without wing and tail anti-icing, supersonic to $M = 2.50$
= 610.56, bomber and military transport without wing or tail anti-icing, subsonic

Executive and Commercial Passenger–Cargo Transports

$$Wt = 469.30 \left[V_{PR} \left(N_{CR} + N_{ATT} + N_{PASS} \right) \times 10^{-4} \right]^{0.419} \tag{20.68}$$

where

V_{PR} = pressurized or occupied volume, in cubic feet (ft^3)
N_{ATT} = number of attendants
N_{PASS} = number of passengers

20.2.1.8 Electronics (Avionics)

Usually requirements will specify the avionics gear for the aircraft. The weight of the avionics can then be determined using Table 8.6 of Chapter 8 or manufacturer information on the particular electronics equipment.

If the electronics gear is not specified, estimates of the weight can be made using the statistical methods of Table 8.7.

20.2.1.9 Landing Retardation Devices

The weight of landing retardation devices (brakes, thrust reversers, and drag chutes) can be determined using the information in Chapter 10. The designer should examine the engine information to see if the thrust reverser is included in the basic engine weight.

20.2.2 Conventional Metal Aircraft— Light Utility Aircraft

The weight equations of Section 20.2.1 predict unrealistic component weights for light utility aircraft such as those reported in Table I.3 in Appendix I. The following equations are recommended for the low-to-moderate performance (up to about 300 kt) light utility aircraft. The weight equations give the component weight in pounds.

20.2.2.1   Structure
Wing

$$Wt = 96.948\left[\left(\frac{W_{TO}\,N}{10^5}\right)^{0.65}\left(\frac{AR}{\cos\Lambda_{1/4}}\right)^{0.57}\left(\frac{S_w}{100}\right)^{0.61}\left(\frac{1+\lambda}{2t/c}\right)^{0.36}\left(1+\frac{V_e}{500}\right)^{0.5}\right]^{0.993}$$

$$(20.69)$$

where

W_{TO} = takeoff weight, in pounds (lb)
N = ultimate load factor ($1.5 \times$ limit load factor)
AR = wing aspect ratio
$\Lambda_{1/4}$ = wing quarter-chord sweep
S_w = wing area in square feet (ft²)
λ = wing taper ratio
t/c = maximum wing thickness ratio
V_e = equivalent maximum airspeed at sea level, in knots

Fuselage

$$Wt = 200\left[\left(\frac{W_{TO}\,N}{10^5}\right)^{0.286}\left(\frac{L}{10}\right)^{0.857}\left(\frac{W+D}{10}\right)\left(\frac{V_e}{100}\right)^{0.338}\right]^{1.1}$$

$$(20.70)$$

where

L = fuselage length, in feet
W = fuselage maximum width, in feet
D = fuselage maximum depth, in feet

Horizontal Tail

$$Wt = 127\left[\left(\frac{W_{TO}\,N}{10^5}\right)^{0.87}\left(\frac{S_H}{100}\right)^{1.2}\left(\frac{\ell_T}{10}\right)^{0.483}\left(\frac{b_H}{t_{HR}}\right)^{0.5}\right]^{0.458}$$

$$(20.71)$$

where

S_H = horizontal tail area, in square feet (ft²)
ℓ_T = distance from wing one-fourth mac to tail one-fourth mac
b_H = horizontal tail span, in feet
t_{HR} = horizontal tail maximum root thickness, in inches

Vertical Tail

$$\text{Wt} = 98.5 \left[\left(\frac{W_{\text{TO}} N}{10^5} \right)^{0.87} \left(\frac{S_V}{100} \right)^{1.2} \left(\frac{b_V}{t_{\text{VR}}} \right)^{0.5} \right] \tag{20.72}$$

where

S_V = vertical tail area, in square feet (ft²)
b_V = vertical tail span, in feet
t_{VR} = vertical tail maximum root thickness, in inches

Landing Gear

$$\text{Wt} = 0.054 \left(L_{\text{LG}} \right)^{0.501} \left(W_{\text{LAND}} N_{\text{LAND}} \right)^{0.684} \tag{20.73}$$

where

L_{LG} = length of main landing gear strut, in inches
W_{LAND} = landing weight (if unknown, use W_{TO} minus 60% fuel)
N_{LAND} = ultimate load factor at W_{LAND}

20.2.2.2 Propulsion
Total Installed Propulsion Unit Weight Less Fuel System
This includes mounting and air induction weight:

$$\text{Wt} = 2.575 \left(W_{\text{ENG}} \right)^{0.922} N_E \tag{20.74}$$

where

W_{ENG} = bare engine weight
N_E = number of engines

Fuel System
This includes fuel pumps, lines, and tanks:

$$\text{Wt} = 2.49 \left(F_G \right)^{0.6} \left(\frac{1}{1 + \text{Int}} \right)^{0.3} \left(N_T \right)^{0.2} \left(N_E \right)^{0.13} \tag{20.75}$$

where

F_G = total fuel, in gallons
Int = percentage of fuel tanks that are integral
N_T = number of separate fuel tanks

20.2.2.3   Surface Controls

For powered surface control systems, use

$$Wt = 1.08 \left(W_{TO}\right)^{0.7} \tag{20.76}$$

For unpowered surface control systems, use

$$Wt = 1.066 \left(W_{TO}\right)^{0.626} \tag{20.77}$$

20.2.2.4   Electrical System

The weight-prediction relationships are expressed in terms of the total weight of the fuel system and the electronics system, the primary users of electrical power on the aircraft:

$$Wt = 12.57 \left(\frac{W_{FS} + W_{TRON}}{1000}\right)^{0.51} \tag{20.78}$$

where

W_{FS} = fuel system weight, in pounds, Eq. (20.75)
W_{TRON} = weight of installed electronics, in pounds, Eq. (20.81)

20.2.2.5   Furnishings

The weight expression for the crew seats is

$$Wt = 34.5\left(N_{CR}\right)\left(q\right)^{0.25} \tag{20.79}$$

where

N_{CR} = number of crew
q = maximum dynamic pressure, in pounds per square foot (lb/ft²)

The weight of the passenger seats is determined from Eq. (20.54) and a weight allowance for miscellaneous furnishings from Eq. (20.62). If the aircraft is pressurized, an additional weight allowance should be considered using Eq. (20.60).

20.2.2.6   Air Conditioning and Anti-Icing

If the aircraft has air conditioning and anti-icing, the following expression can be used to estimate the weight of this equipment:

$$Wt = 0.265 \left(W_{TO}\right)^{0.52} \left(N_{CR} + N_{PASS}\right)^{0.68} \left(W_{TRON}\right)^{0.17} \left(M_E\right)^{0.08} \tag{20.80}$$

where

N_{PASS} = number of passengers
N_{CR} = number of crew
W_{TRON} = weight of installed electronics in pounds, see Eq. (20.81)
M_E = equivalent maximum Mach number at sea level

20.2.2.7   Electronics (Avionics)

The total installed weight of the avionics equipment is

$$W_{\text{TRON}} = 2.117 \left(W_{\text{AU}} \right)^{0.933} \tag{20.81}$$

where W_{AU} = bare avionics equipment weight (uninstalled)

20.2.3   Advanced-Composites Aircraft

The high strength-to-weight and stiffness-to-weight ratios associated with advanced composite materials can significantly reduce aircraft structural weight. The blending of high-strength fibers such as graphite, boron, Kevlar 49, and glass in epoxy, polyimide, or metallic matrices (as discussed in Chapter 19) offers new opportunities for the creative structural engineer to tailor the material to exploit innovative structural designs. The full potential of advanced composites in realizing structural weight reductions (and airframe cost reductions) has not been demonstrated yet; however, there is no question that it will be significant. The designer should review the discussion in Appendix I.

The results of many advanced composites development programs and aircraft conceptual studies indicate that the material can decrease the weight of primary and secondary structural elements by about 25% and 40%, respectively. The conceptual complete aircraft studies indicate that an aircraft should not be 100% composite materials because there are many places where it is more cost effective to use metals, honeycomb, and other materials. Some of the places where it is not cost effective or practical to use advanced composites are canopies, tires, seats, seals, mechanisms, radomes, latches, hinges, and clamps. The optimum composite utilization appears to be about 55% in terms of most cost and weight effectiveness.

Based upon a composite utilization by weight of about 55%, the following methodology is recommended for estimating the aircraft component weights, at this point in the conceptual design:

1. Estimate the weight of the component using the metal weight equations of Sections 20.2.1 or 20.2.2.
2. Reduce the metal weights by the following amounts:
   - Wing, 20%
   - Tail, 25%
   - Fuselage, fighter, 10%
   - Fuselage, transport, 25%
   - Secondary (flaps, slats, access panels, etc.), 40%
   - Landing gear, 8%
   - Air induction, 30%

20.2.4 Low Wing Loading Aircraft

Aircraft in this class are characterized by $W/S \leq 2$ lb/ft^2 and are powered by solar or human energy. Solar or human energy is puny compared to the more traditional sources of energy for aircraft (i.e., turbine, piston, and rocket). Because power required is dependent on speed, this aircraft class will typically have flight speeds less than 30 KEAS. Human-powered aircraft cruise at speeds of 16 KEAS or less (current distance record was established by the MIT Daedalus at 74 miles in 3 hr 54 min, in 1988) and solar-powered at 23 KEAS (typical maximum speed for Helios at 1000 ft), both flying at wing loadings of between 0.6 and 1.0 lb/ft^2. The Solar Snooper solar-powered aircraft of Sections 6.7 and 18.9 had a $W/S = 1.86$ lb/ft^2.

Estimating weights for this class of aircraft is very challenging because the historical data base is almost nonexistant. The large data base for sailplanes is not much help because their wing loadings range from about 5 to 12 lb/ft^2. Sailplanes are designed for aggressive maneuvering as they chase a thermal and high speed for wind penetration. The structural criteria and design of an aircraft with $W/S = 1$ lb/ft^2 are very much different than for one with $W/S = 10$ lb/ft^2. Solar-powered aircraft would have design limit load factors of 2.0 whereas sailplanes would have a limit of 6.0. The patent for the AeroVironment Centurion contains a good discussion of structural design for this class of aircraft [7]. The wing weight (weight per wing area) for existing aircraft and sailplanes with wing loadings ranging between 0.6 and 40 lb/ft^2 is shown in Fig. 20.1.

The construction materials for the wings shown in Fig. 20.1 vary significantly as the wing loading decreases. The U-2A, U-2S, and Boeing Condor wings are conventional built-up metal structures. The Lockheed Martin Tier III-Minus Darkstar wing was made from graphite composites with an aluminum carry-through spar. The Scaled Composites Voyager wing is also made from high-strength composites. Figure 20.1 shows two wing loadings for the Voyager (26.8 to 7.5 lb/ft^2) because its 72% fuel fraction is uncommonly large. The Lockheed Martin Polecat features sandwich structures, water-jet cut ribs and keel, and simplified sandwich skins using LTM45 carbon fiber prepreg. The sailplane group is constructed from Fiberglass and graphite composites. The Helios, Centurion, and Daedalus wings use a hollow carbon tube spar with polystyrene sheet ribs with leading edge and trailing edge members wrapped in a clear plastic Mylar skin covering that is one-half mil thick.

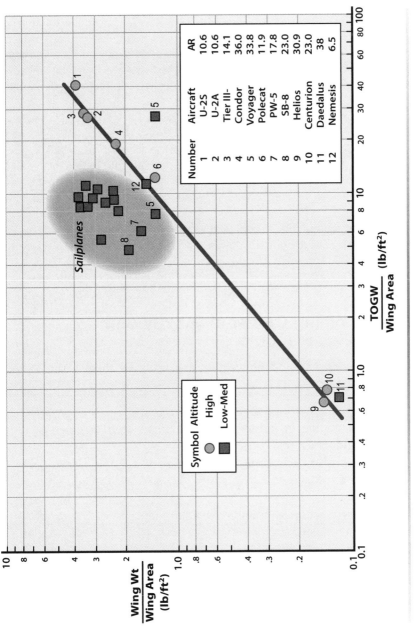

Figure 20.1   Wing weight vs wing loading for various high-AR, low wing loading aircraft.

Weights Rule!

Estimating weights is critical in the design of an aircraft (remember the weights rule). This is especially true for weight-critical aircraft such as the Voyager, designed by Burt Rutan (Scaled Composite Inc). It had to have a fuel fraction of 72 percent in order to fly 22,912 nm around the world nonstop. The Voyager, piloted by Dick Rutan (Burt's brother) and Jeana Yeager, took off from Edwards AFB in California on December 14, 1986 and returned 9 days later—making this unique aircraft the first to complete the first nonstop, nonfueled flight around the world.

The airframe, largely made of fiberglass, carbon fiber, and Kevlar, weighed 939 lb when empty, which gave it a weight fraction of 9.7 percent. The aircraft weighed 9695 lb at takeoff, 2250 lb empty, 534 lb for payload (including the pilots) and 7010 lb of fuel. The Voyager had a wing span of 110 ft, 8 inches and an aspect ratio of 33.8, giving the aircraft a maximum L/D of 27. The takeoff wing loading was 26.8 psf and 7.5 psf at landing (with 3 gallons of fuel remaining in the tanks).

The team behind the Voyager's flight, including designer Rutan, were awarded the Collier Trophy for their record-breaking flight. The Voyager is hanging in the Smithsonian Air and Space Museum in Washington, D.C.

20.3 Determining Center of Gravity and Moments of Inertia

The following component weights are summed to give the aircraft empty weight:

- Structure (wing, fuselage, tail, and landing gear)
- Propulsion (engine, inlet, fuel system, starting system, engine
- controls, and thrust reversers)
- Surface controls plus hydraulics and pneumatics
- Instruments
- Electrical system
- Furnishings (ejection seats and crew equipment)
- Air conditioning and anti-icing
- Electronics (avionics)
- Miscellaneous (drag chutes, etc.)

Adding the fuel weight and fixed weights (crew and payload) gives the aircraft takeoff weight.

Next, it is important to determine the location for each of the components to determine the center of gravity (c.g.) of the aircraft. Many of the component weight locations will be self-evident, such as the pilot, fire control system, and landing gear. Other components, such as fuel cells,

Table 20.1 Weight and Moment Summary

Component	Weight (lb)	Distance from Aircraft Nose (ft)	Moment (ft·lb)
Fuselage			
Wing			
Main gear			
Vertical tail			
Horizontal tail			
etc.			
	ΣWt	Total moment =	ΣM

navigation equipment, bombs, and baggage, can be shifted around to a certain extent to influence the c.g. location.

A weight and moment summary in tabular form is shown in Table 20.1. This serves to provide the designer with a refined estimate of the aircraft c.g. as a function of component placement. Chapter 23 discusses the desired c.g. location to give good flying qualities.

The longitudinal position of the c.g. may now be determined as

$$X_{c.g.} = \text{Total Moment}/\Sigma Wt$$

This center of gravity should be expressed as distance from the nose of the aircraft and percentage of the mean aerodynamic chord. The designer should determine an X location of c.g. for both a full and an empty aircraft as shown in Fig. 23.3 (the c.g. envelope). Figure 23.3 shows the most forward and most aft c.g. locations. It would be embarrassing to have a tricycle-gear aircraft fall on its tail in one of these extreme loading conditions.

The aircraft body axes are defined according to Fig. 2.2 or 21.1. The aircraft moments of inertia are defined as follows:

$$I_{xx} = \int \left(y^2 + z^2 \right) dm$$

$$I_{yy} = \int \left(x^2 + z^2 \right) dm$$

$$I_{zz} = \int \left(x^2 + y^2 \right) dm$$

where dm is an incremental mass element of aircraft. The products of inertia are defined as

$$I_{xy} = \int \left(xy \right) dm$$

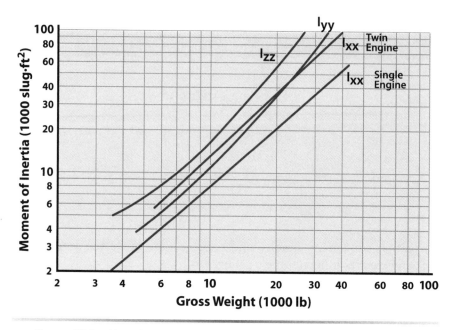

Figure 20.2 Aircraft moments of inertia as a function of gross weight.

and so on. The moments of inertia can be estimated at this point in the design process using Figure 20.2, which is based upon historical data from many existing aircraft.

References

[1] Torenbeek, E., *Synthesis of Subsonic Aircraft Design*, Delft Univ. Press, The Netherlands, 1976.

[2] Roskam, J., *Aircraft Design*, Pt. 5, Roskam Aviation and Engineering Corp., Ottawa, KS, 1985. [Available via www.darcorp.com (accessed 31 Oct. 2009).]

[3] Raymer, D. P., *Aircraft Design: A Conceptual Approach*, AIAA Education Series, AIAA, Reston, VA, 2006.

[4] Thomas, F., *Fundamentals of Sailplane Design*, College Park Press, College Park, MD, 1999.

[5] Stender, W., "Sailplane Weight Estimation," OSTIV (International Scientific and Technical Gliding Organization), Elstree-Wassenaar, The Netherlands, 1969.

[6] Adams, D. F., "High Performance Composite Material Airframe Weight and Cost Estimating Relations," *Journal of Aircraft*, Vol. 11, No. 12, Dec. 1974.

[7] Hibbs, B. D., Lissaman, P. B. S., Morgan, W. R., and Radkey, R. L., "Solar Rechargeable Unmanned Aircraft," U.S. Patent No. 5,810,284, 22 Sept. 1998 (patent for the AeroVironment Centurion).

- Static & Dynamic Stability Modes
- Federal Regulations
- Static S&C Considerations
- Static Longitudinal S&C
- Static Lateral S&C
- Static Directional S&C

The Sopwith Camel was statically unstable (as was the Wright Flyer), giving it a quickness in maneuvering. It was a wonderful machine in combat (it had more aerial victories than any other allied WWI airplane), but it killed many a pilot who was not paying close attention to its deadly lack of stability.

Take your hand off the stick and it would rear right up with a terrific jerk and stand on its tail.
Pilot report on the Sopwith Camel

21.1 Introduction

The discussion in the next three chapters assumes some familiarity with aircraft static *stability and control* (S&C). If not, review this subject in outside texts such as [1–4] for a fundamental understanding of static stability and control. Reference [3] is especially recommended as a companion to this text as it has a very complete discussion of aircraft stability and control.

The axis system for the S&C discussion is shown in Fig. 21.1. An aircraft is in equilibrium if the summation of moments about the three axes is zero. The aircraft is said to be *stable* if it returns to equilibrium about the pitch, roll, and yaw axes when disturbed. The aircraft has *static stability* if it "tends" to return to equilibrium by itself. In other words, the resulting forces and moments from the disturbance push the aircraft toward its original equilibrium state. This static stability will fight the disturbance, making it difficult to move away from the equilibrium condition. Thus, a high degree of static stability will make it hard to maneuver the aircraft. The aircraft has *dynamic stability* if the actual motion of the unsteady forces and moments returns the aircraft (eventually) to its original equilibrium condition.

Figure 21.2 shows an aircraft system disturbed in pitch. In Fig. 21.2a the aircraft has neutral stability and remains at whatever α the disturbance produces. Figure 21.2b shows an unstable system because the tendency of the system is to diverge. In Fig. 21.2c the aircraft has static stability with very high damping, giving it dynamic stability as well. The aircraft slowly returns to its original α without any overshoot. Figure 21.2d shows a more

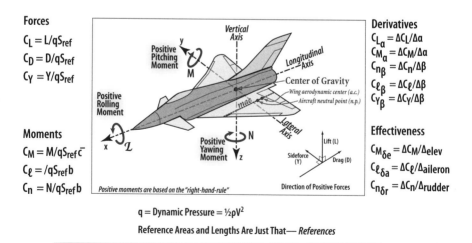

Figure 21.1 Major nondimensional aerodynamic parameters and sign convention.

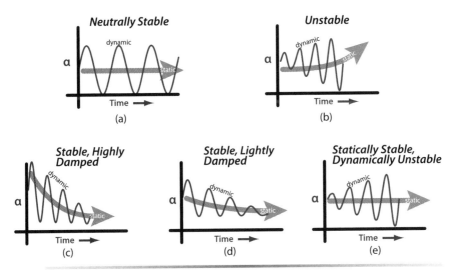

Figure 21.2 Static and dynamic stability about the pitch axis.

typical aircraft response. The aircraft returns to its original state, but experiences overshoot with a converging oscillation. This is acceptable behavior, provided the time to converge is reasonable. In Fig. 21.2e the restoring forces and moments are in the right direction so the aircraft is statically stable. However, the restoring forces and moments are high and the damping is low, so the aircraft overshoots the original equilibrium condition. These restoring forces and moments then push the nose back up, overshooting again, but with increasing amplitude. The pitch oscillations continue to increase in amplitude until the system diverges into an uncontrollable flight mode. It should be obvious that static stability is a necessary but not sufficient condition for dynamic stability.

The degree of dynamic instability is the "time to double amplitude (t_2)" for the system. If the t_2 is large compared with the reaction of the control system, then the aircraft would have acceptable flying qualities. The 1903 Wright Flyer had a t_2 of about 30 seconds for the pitch axis, which permitted Wilbur and Orville to arrest the divergent motion with a pitch control input and fly the aircraft safely. Most aircraft have an unstable lateral mode, the *spiral divergence*. This divergence mode is so slow that the pilot has ample time to make the minor roll correction needed to prevent it.

The strategy of modern flight control systems is to design for low static stability (in fact near neutral stability) and then augment the stability of the aircraft by electronic systems. The *stability augmentation system* (SAS) consists of sensors (rate gyros and accelerometers to sense the movement away from equilibrium), computers (to analyze the aircraft motion and determine the correct control input to counter the aircraft motion), and servos (to input the control deflection to the control surface). These active controls control the aircraft's rapid perturbations from an equilibrium position and make possible the practicality of the unstable aircraft.

A dynamically unstable aircraft would be a very maneuverable aircraft. This was indeed the case for the WWI-era Sopwith Camel (a 1917 British biplane, pictured on the first page of this chapter) as recorded by V. M. Yeates in his "Winged Victory":

> But it was just this instability that gave [Sopwith] Camels their good qualities of quickness in manoeuvre. A stable machine had a predilection for normal flying positions and this had to be overcome every time you wanted to do anything, whereas a Camel had to be held in flying position all the time, and was out of it in a flash. It was nose light, having a rotary engine weighing next to nothing per horsepower, and was rigged tail heavy so that you had to be holding her down all the time. Take your hand off the stick and it would rear right up with a terrific jerk and stand on its tail. Moreover, only having dihedral on the bottom plane gave a Camel a very characteristic elevation. You could tell one five miles off. ... With these unorthodox features, a Camel was a wonderful machine in a scrap. If only it had been 50% faster! There was the rub. A Camel could neither catch anything except by surprise, nor hurry away from an awkward situation, and seldom had the option of accepting or declining combat. But what of it? You couldn't have everything.

21.2 Federal Regulations

Aircraft operating in the United States must conform to regulations. Commercial aircraft must follow the following federally mandated regulations:

1. **FAR 23.** Airworthiness standards for small airplanes in the normal, utility, and acrobatic categories that have passenger seating of nine seats or fewer

   Paragraph 23.171—The airplane must be longitudinally, directionally, and laterally stable. In addition, the airplane must show suitable stability and control "feel" (static stability) in any condition normally encountered in service.

2. **FAR 25.** Airworthiness standards for transport category airplanes

   Paragraph 25.171—The airplane must be longitudinally, directionally, and laterally stable. In addition, the airplane must show suitable stability and control "feel" (static stability) in any condition normally encountered in service.

Military aircraft must follow the following specifications and standards:

1. **MIL-F-8785C (1980).** Flying Qualities of Piloted Airplanes. (Inactive 1996 for new design and no longer used)

   Although inactive the document contains much good design data, theories, and information on aircraft handling and flying qualities [5].

2. **MIL-HDBK-1797 (1997).** Flying Qualities of Piloted Aircraft
   This document replaced MIL-F-8785C and MIL-STD-1797. It contains
   requirements for qualitative and quantitative flying qualities for all
   military aircraft, latest theories, and information relating to pilot
   opinion. In addition to requirements for handling qualities, it also
   specifies other requirements that an aircraft must meet, such as
   operational missions, external stores, configurations, and flight
   envelopes. This handbook also applies to piloted transatmospheric
   flight when flight depends upon aerodynamic lift and/or air-
   breathing propulsion systems.
3. **MIL-F-9490.** Flight Control Systems—Design, Installation and Test for
   Piloted Aircraft
4. **MIL-F-1873.** Flight Control Systems—Design, Installation and Test for
   Aircraft
5. **MIL-C-18244.** Control and Stabilization Systems, Automatic for
   Piloted Aircraft
6. **MIL-F-83300.** Flying Qualities of Piloted V/STOL Aircraft
7. **MIL-H-850.** Flying Qualities of Military Rotorcraft

All of these documents require dynamically stable aircraft—either
inherently stable (passive) or augmented with an SAS.

21.3 Static Stability and Control Considerations

The purpose of the next three chapters is to size and design the aircraft
control surfaces and to determine the trim drags. The criteria and methodology presented will be based upon static considerations only. Dynamic
stability and control analysis is usually reserved for the preliminary design
phase because it requires information about the design that is not available
during the conceptual design phase. For example, the moments of inertia
introduced in Chapter 20 require knowledge of the aircraft mass distribution on all three axes. These are design details that are not generally known
at this point.

Static stability and control considerations will permit the designer to
assess the configuration layout and balance of his design and size the surfaces for adequate stability and control margins. The dynamic analysis in
the preliminary design phase will fine tune the configuration.

The discussions of longitudinal, directional, and lateral motion will
center about the body axes shown in Fig. 21.1.

The mean aerodynamic chord of a wing, denoted by $\bar{c}$ or mac, represents an average chord that, when multiplied by the average section
moment coefficient, dynamic pressure, and reference wing area, gives the
moment for the entire wing. The mac for wings of constant taper and sweep
is given by

$$\bar{c} = \frac{2}{3} c_r \left[\frac{1 + \lambda + \lambda^2}{1 + \lambda} \right]$$

where c_r is the root chord and λ is the wing taper ratio.

The aerodynamic center (a.c.) is that point on an aircraft, wing, or airfoil section about which the pitching moment is independent of angle-of-attack. The aerodynamic center is the most convenient place to locate the lift, drag, and moment of an aircraft wing or airfoil section. This is obvious from stability considerations because $dC_{m_{a.c.}}/d\alpha = 0$ and it is one less term to worry about.

For most aircraft, the body contributes a small amount of lift compared with the wing, resulting in the total aircraft a.c. location being very close to the wing a.c. This is not the case for missiles (where the body is large relative to the wing) and contribution of the body pressure distribution must be considered in locating the missile a.c.

The theoretical position of the aerodynamic center on the mean aerodynamic chord is presented in Fig. 21.3. Notice that at low speed the a.c. is approximately at the quarter-chord and moves aft for supersonic flight. Figures H.8 and H.9 of Appendix H present experimental data on the a.c. location as a function of Mach number for many different low aspect ratio (AR) wing–body combinations.

21.4 Static Longitudinal Stability and Control

The forces and moments acting on an aircraft are shown in Fig. 21.4. The lift and drag are by definition always perpendicular and parallel to V_∞, respectively. It is, therefore, inconvenient to use these forces to obtain moments because their moment arms relative to the center of gravity vary with angle-of-attack α. For this reason, all forces are resolved into normal, N, and chordwise, C, forces whose axes remain fixed with the aircraft and whose arms are, therefore, constant:

$$N = L \cos \alpha + D \sin \alpha$$

$$C = D \cos \alpha - L \sin \alpha$$

For small α, $N \approx L$ and $C \approx D$. For this discussion consider α to be small and use L and D in the development of stability and trim equations.

The moments are summed about the center of gravity for each aircraft. The horizontal tail or canard is usually a symmetric section so that $M_{a.c._c} = 0$ and $M_{a.c._T} = 0$ for $\delta_e = 0$. In Fig. 21.4a, b, and c moments have been neglected due to the fuselage and engine nacelles.

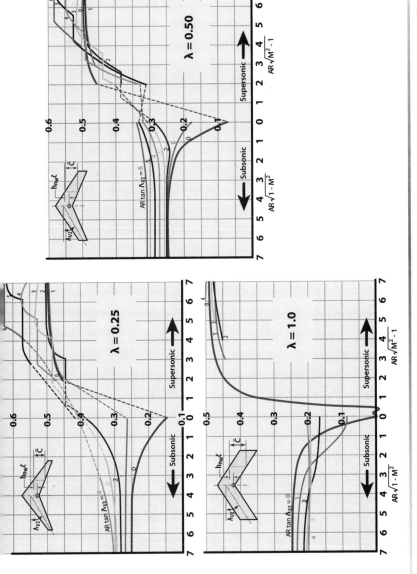

Figure 21.3 Theoretical chordwise position of the aerodynamic center (data from [6]).

a)

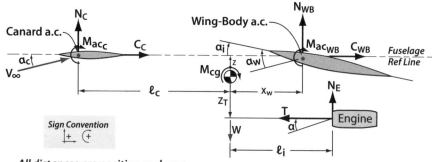

Sign Convention

$\overset{+}{\underset{+}{\llcorner}}$ ⊕

All distances are positive as shown
For small α it is assumed that
N = normal force = L = $C_L q S_{ref}$
C = chord force = D = $C_D q S_{ref}$

$\bar{V}_C$ = Canard Volume Coefficient = $\dfrac{\ell_c S_c}{\bar{c} S_{ref}}$

b)

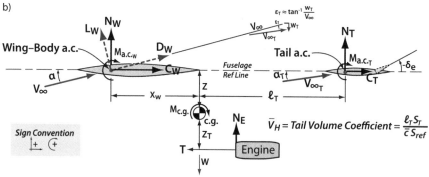

Sign Convention

$\overset{+}{\underset{+}{\llcorner}}$ ⊕

$\bar{V}_H$ = Tail Volume Coefficient = $\dfrac{\ell_T S_T}{\bar{c} S_{ref}}$

Note: Wing and tail chord lines are parallel to fuselage reference line

c)

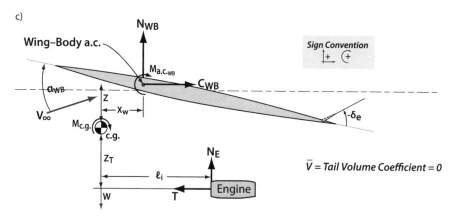

Sign Convention

$\overset{+}{\underset{+}{\llcorner}}$ ⊕

$\bar{V}$ = Tail Volume Coefficient = 0

All distances are positive as shown
ε_T = tail downwash angle due to wing downwash w_T at the tail a.c.
For small α we can assume
N = normal force = L = $C_L q S_{ref}$
C = chord force = D = $C_D q S_{ref}$

Figure 21.4 Forces and moments acting on **a)** an aircraft with a conard tail, **b)** an aircraft with an aft tail, and **c)** a tailless aircraft.

The moment or trim equations are usually placed in coefficient form by dividing through by $(q_\infty S_{ref} \bar{c})$, where q_∞ is the dynamic pressure $(\frac{1}{2}\rho V^2)$, S_{ref} is the wing reference area, and $\bar{c}$ is the mac of the wing. Now, replace the wing by its mac and locate the lift, drag, and moment of the wing at the aerodynamic center a.c.

Trim equations for the three aircraft types shown in Fig. 21.4a–c are as follows:

Aft Tail:

$$C_{Mc.g.} = C_L \frac{x_w}{\bar{c}} + C_D \frac{z}{\bar{c}} + C_{Ma.c.w} + \frac{Tz_T}{q_\infty S_{ref}\bar{c}} - c_{LT}\bar{V}_H \eta_T - C_{Mc.g.inlet} \quad (21.1)$$

Canard:

$$C_{Mc.g.} = -C_L \frac{x_w}{\bar{c}} + C_D \frac{z}{\bar{c}} + C_{Ma.c.w} + \frac{Tz_T}{q_\infty S_{ref}\bar{c}} + c_{LC}\bar{V}_C - C_{Mc.g.inlet} \quad (21.2)$$

Tailless:

$$C_{Mc.g.} = -C_L \frac{x_w}{\bar{c}} + C_D \frac{z}{\bar{c}} + C_{Ma.c.w} + \frac{Tz_T}{q_\infty S_{ref}\bar{c}} - C_{Mc.g.inlet} \quad (21.3)$$

In these equations the moment due to the aft tail or canard drag is much smaller than the wing counterpart and was neglected. The term $q_{\infty T}/q_\infty$ is called the *tail efficiency factor* η_T and comes about because of the influence of the wing on the freestream velocity striking the tail. The wing induces a downwash w_T (due to trailing vortices) at the a.c. of the aft tail (see Fig. 21.4b). Notice that $\eta_T = 1.0$ for the canard. The term $(C_{Mc.g.})$ inlet comes about because of the momentum change in turning the air into the inlet. At small α, this term can be neglected.

Often, $z \ll c$ so that the wing–body drag moment can be neglected. Also, if z_T is small, the thrust term is negligible.

The aircraft must be able to set the trim equation equal to zero for any attitude or flight condition and all thrust levels. The wing primarily establishes the load factor for the aircraft and the aft tail or canard balances the aircraft. If the aircraft is tailless, the moments about the aircraft c.g. are balanced by changing the wing camber (flap deflection), which changes the moment about the wing a.c. The horizontal tail (aft tail and canard) is movable, either all movable (all flying aft tail or canard) or a portion (called the *elevator* on an aft tail) is movable, so that L_T or L_c can be changed independently of the aircraft angle-of-attack. In this way, the horizontal tail can cause the aircraft to rotate from one equilibrium (trimmed) condition to another (i.e., change angle-of-attack α).

If the horizontal aft tail is an all-flying-tail (such as the B-52, 727, and L1011), then the expression for C_{L_T} is

$$C_{L_T} = m_T \left(\alpha_T + \alpha_{cs} \right) = m_T \left[\left(1 - d\varepsilon / d\alpha \right)\alpha + \alpha_{cs} \right]$$

where α_{cs} is the deflection angle that the pilot initiates by moving the control stick, and $d\varepsilon/d\alpha$ is the change in downwash angle ε for a change in α. The m_T is the horizontal tail lift curve slope, $(C_{L_\alpha})_T$.

If the horizontal tail is a stationary stabilizer–movable elevator arrangement, then the expression for C_{L_T} is

$$C_{L_T} = m_T \left[\left(1 - d\varepsilon / d\alpha \right)\alpha - \alpha_{0L_T} \right]$$

where α_{0L_T} is the tail angle for zero lift (see Fig. 2.1a or Table F.1) and is dependent upon the elevator deflection δ_e (note, same as for a flapped airfoil).

If the trim equations (21.1), (21.2), and (21.3) are differentiated with respect to α, the results are as follows:

Aft Tail:

$$\frac{dC_{Mc.g.}}{d\alpha} = C_{M\alpha} = m_w \frac{x_w}{\bar{c}} - m_T \left(1 - \frac{d\varepsilon}{d\alpha} \right) \bar{V}_H \eta_T + C_{M\alpha I} \tag{21.4}$$

Canard:

$$C_{M\alpha} = \bar{V}_c m_c - \frac{x_w}{\bar{c}} m_w + C_{M\alpha I} \tag{21.5}$$

Tailless:

$$C_{M\alpha} = -\frac{x_w}{\bar{c}} m_w + C_{M\alpha I} \tag{21.6}$$

where $m_w = (C_{L_\alpha})_w$ is the wing–body lift curve slope, $m_c = (\dot{C}_{L_\alpha})_c$ is the canard lift curve slope (based upon canard surface area, S_c), $m_T = (C_{L_\alpha})_T$ is the aft tail lift curve slope (based upon aft tail area, S_T) and $(C_{M\alpha})_I$ is the change in inlet moment due to α. In Eqs. (21.4), (21.5), and (21.6) the term due to the wing–body has been neglected. The thrust term disappears because the thrust is not (at least to a first-order approximation) a function of α.

The criterion for static stability in an aircraft is that its value of C_{M_α} be negative. This means that if an aircraft with $C_{M_\alpha} < 0$ is in equilibrium (trimmed) at a positive α and suddenly α is increased (e.g., wind gust), the aircraft will generate a negative moment to push the nose down toward the original equilibrium α.

The inlet term in the trim and stability equation comes about because of the moment generated about the center of gravity when the freestream air is turned at the inlet lip into the engine. The force diagram is shown schematically in Fig. 21.5.

The inlet force N_E can be expressed as

$$N_E = \dot{m}_0 \Delta V = \dot{m}_0 V_\infty \tan \beta \approx \dot{m}_0 V_\infty \beta$$

where $\dot{m}_0$ is the mass flow of air into the inlet in slugs per second and β is the flow turning angle in radians shown in Fig. 21.5. The moment about the center of gravity is $\ell_i N_E$ and is positive for the aircraft in Fig. 21.5. The moment and stability contribution from the inlet is finally expressed as

$$C_{M_{\text{c.g.inlet}}} = \frac{N_E \ell_i}{q_\infty S_{\text{ref}} \overline{c}} \approx \frac{2\dot{m}_0 \beta \, \ell_i}{\rho V_\infty S_{\text{ref}} \overline{c}} \tag{21.7}$$

$$\left(\frac{dC_M}{d\alpha}\right)_I = C_{M_{\alpha\text{inlet}}} \approx \frac{2\dot{m}_0 \beta \, \ell_i}{\rho V_\infty S_{\text{ref}} \overline{c}} \frac{d\beta}{d\alpha} \quad (\text{per radian}) \tag{21.8}$$

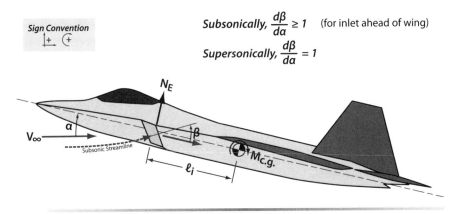

Subsonically, $\dfrac{d\beta}{d\alpha} \geq 1$ (for inlet ahead of wing)

Supersonically, $\dfrac{d\beta}{d\alpha} = 1$

Figure 21.5 Schematic of inlet force on aircraft.

Figure 21.6 Lockheed Jetstar with inlet behind the wing.

where

$\dot{m}_0$ = mass flow rate of air accepted by the inlet (slug/s)
$\bar{c}$ = mean aerodynamic chord (ft)
ρ = air density (slug/ft^3)
V_∞ = freestream velocity (ft/s)
S_{ref} = wing area (ft^2)
ℓ_i = distance of the inlet face ahead of the aircraft c.g. (ft)
$d\beta/d\alpha$ = change in flow direction into the inlet due to upwash of the wing

Note the following:

1. If the inlet is under the wing as for the F-18, the wing turns the airflow into the inlet and there is no inlet moment. For this inlet location use β = 0 and $d\beta/d\alpha$ = 0.

2. For inlets behind the wing trailing edge (such as the JetStar in Fig. 21.6) $d\beta/d\alpha$, may be analyzed as

$$\left(1 - \frac{d\varepsilon}{d\alpha}\right)\frac{x_i}{\ell_h}$$

where x_i is the distance from the wing trailing edge to the inlet and ℓ_h is the length from the wing trailing edge to the horizontal tail mean aerodynamic chord.

3. For inlets ahead of the wing leading edge $d\beta/d\alpha$ = 1 for supersonic flight and may be determined from Fig. 21.7 for subsonic speeds.

The tail downwash term for an aft-tail aircraft, $(1 - d\varepsilon/d\alpha)$, depends largely on the location of the tail with respect to the wing and the position of the horizontal tail with respect to the wing wake. If the horizontal tail is positioned so that it lies either close to or inside the wing wake, large changes in downwash occur, as well as reduced tail efficiency and unpleas-

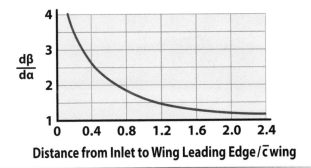

Figure 21.7   Change in flow direction into the inlet due to upwash of the wing (data from [7]).

ant tail buffeting. In the usual design the horizontal tail is kept high enough (or low enough) to avoid the wing wake at all lift coefficients. If this is done, a simplified empirical method, developed from [8], is available to estimate the change of downwash with α at the aft horizontal tail. The method is shown in Fig. 21.8. If increased accuracy is desired, the methods in [9] should be used.

21.5 Static Lateral Stability and Control

The lateral motion for an aircraft is the rolling motion about the fuselage centerline. This lateral motion is shown in Fig. 21.9 with the rolling moment $\mathscr{L}$ being defined as positive for the right wing down. The rolling moment coefficient is C_ℓ (forgive the confusion with section lift coefficient) and is defined as

$$C_\ell = \frac{\mathscr{L}}{qS_{\text{ref}}b} \tag{21.9}$$

where b is the aircraft wing span.

The static lateral stability derivative is

$$\frac{dC_\ell}{d\beta} = C_{\ell\beta}$$

which gives the change of rolling moment coefficient with respect to sideslip angle β. A negative $C_{\ell\beta}$ will cause the right wing to come up for a positive sideslip and is the requirement for lateral static stability. Static lateral stability by itself does not guarantee dynamic lateral stability, but it is a necessary condition. The stability derivative $C_{\ell\beta}$ is influenced by the wing, the vertical stabilizer, and the wing–fuselage interaction. $C_{\ell\beta}$ can be expressed as

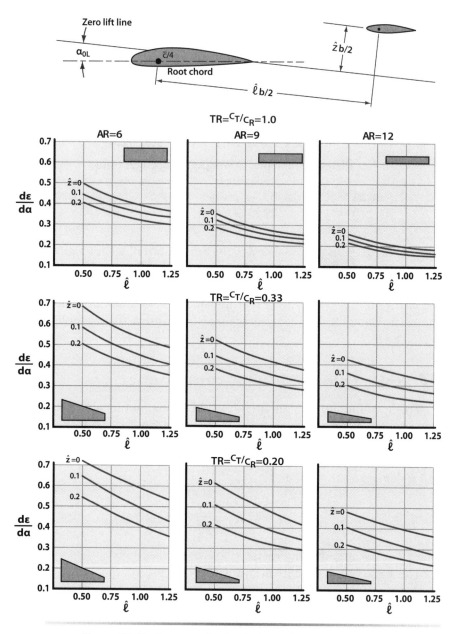

Figure 21.8 Downwash charts for various taper ratios (TR)
(data from [8]).

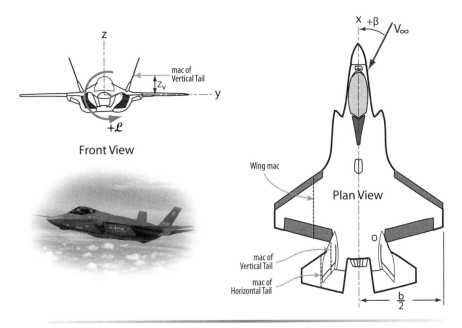

Figure 21.9 Lateral motion of an aircraft and notation for lateral analysis.

$$C_{\ell\beta} = C_{\ell\beta_{\text{wing}}} + C_{\ell\beta_{\text{vertical stabilizer}}} + C_{\ell\beta_{\text{wing-fuselage}}} \qquad (21.10)$$

Now consider the contribution of each component separately.

First, the wing contribution $C_{\ell\beta_{\text{wing}}}$ has three components: the basic wing planform, the sweepback, and the dihedral,

$$C_{\ell\beta_{\text{wing}}} = C_{\ell\beta_{\text{basic}}} + C_{\ell\beta\Delta} + C_{\ell\beta\Gamma} \qquad (21.11)$$

The wing contribution due to the basic wing and sweepback is presented on Fig. 21.10. Notice that the contribution is negative (i.e., stabilizing) and dependent upon the flight C_L. Extrapolate the data on Fig. 21.10 for a delta wing (i.e., $\lambda = 0$).

The wing contribution due to dihedral is stablizing for positive dihedral and its use is the most common way of controlling lateral stability. The expression for the dihedral contribution is given as (from [9])

$$\left(C_{\ell\beta}\right)_{\Gamma} = -0.25 C_{L\alpha}\Gamma\left[\frac{2(1+2\lambda)}{3(1+\lambda)}\right] \qquad (21.12)$$

where the lift curve slope $C_{L\alpha}$ is per radian and Γ is in radians.

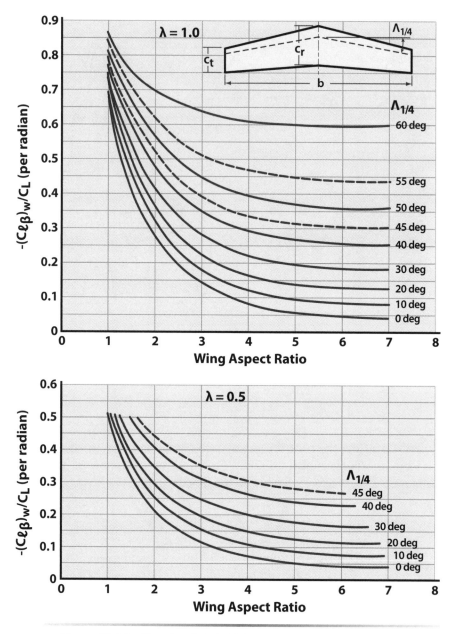

Figure 21.10 The C_{l_β} of straight tapered wings with zero dihedral.

In actual practice, the dihedral angle is usually not set from analytical considerations, because of the large errors involved. Most designers set the wing dihedral only after careful analysis of wind tunnel test data, in which the effects of angle-of-attack, power, and flap settings are carefully analyzed.

Consideration might also be given to using [9, Section 5.1.2.1–1] in determining $C_{\ell\beta\text{wing}}$. This reference combines both the dihedral and sweep-back effects to provide an empirical method for obtaining the desired stability derivative.

Second, $C_{\ell\beta\text{wing-fus}}$ is obtained empirically from [2] and is found to be a function of wing vertical placement on the fuselage:

$$\text{High wing } C_{\ell\beta\text{wing-fus}} \approx -0.0344 \,/\,\text{rad}$$

$$\text{Middle wing } C_{\ell\beta\text{wing-fus}} \approx 0 \qquad (21.13)$$

$$\text{Low wing } C_{\ell\beta\text{wing-fus}} \approx +0.0458 /\text{rad}$$

Third, $(C_{\ell\beta})_{\text{vertical stabilizer}}$: the force on a conventional vertical stabilizer, which is generated as an aircraft sideslips, provides a restoring moment by acting through a moment arm to the aircraft c.g. projection. The opposite is of course true for the ventral fin because it is destabilizing. One may estimate this contribution as

$$\left(C_{\ell\beta}\right)_{\text{VT}} = -C_{L\alpha\text{VT}}\left(1+\frac{d\sigma}{d\beta}\right)\frac{q_{\text{VT}}}{q}\frac{S_{\text{VT}}}{S_{\text{ref}}}\frac{z_v}{b} \qquad (21.14)$$

Terms in Eq. (21.14) are defined as follows:

$(C_{L\alpha})_{\text{VT}}$ = lift curve slope of vertical stabilizer, which is based on an effective aspect ratio that is 1.55 times the actual ratio and is based on the vertical stabilizer area
S_{VT} = planform area of the vertical stabilizer
S_{ref} = wing planform area
z_v = distance from mean aerodynamic chord of vertical stabilizer to aircraft vertical c.g. projection (see Fig. 21.9)

The quantity

$$\left(1+\frac{d\sigma}{d\beta}\right)\frac{q_{\text{VT}}}{q}$$

is a difficult parameter to determine. Reference [9, Section 5.4] presents what appears to be the best analytical method for finding this term:

$$\left(1+\frac{d\sigma}{d\beta}\right)\frac{q_{\text{VT}}}{q} = 0.724 + \frac{3.06\left(S'_{\text{VT}}/S_{\text{ref}}\right)}{1+\cos\Lambda_{c/4}} + 0.4\frac{z_w}{d} + 0.009\,\text{AR} \quad (21.15)$$

where

S'_{VT} = vertical stabilizer area with this area extended to fuselage centerline

z_w = distance along aircraft z axis from the wing root chord to the fuselage centerline

d = maximum fuselage depth

AR = wing aspect ratio

The maximum effect of the vertical stabilizer is greatest as Mach number approaches unity because $(C_{L\alpha})_{VT}$ increases toward that speed condition.

Too large a lateral stability aggravates the condition of Dutch roll (a dynamic lateral response) and does not lend itself to a desirable flight condition. A first approximation to determining the desired amount of lateral stability is suggested as

$$C_{\ell\beta} = -C_{n\beta} \text{ at Mach} = 1.0 \qquad (21.16)$$

Too large a value of $C_{\ell\beta}$ will also result in slow reaction from ailerons and/or spoilers in trying to roll the aircraft.

The lateral control of the aircraft is achieved using ailerons and/or spoilers. As the ailerons deflect, the aircraft begins to roll about the fuselage centerline. If the ailerons remain deflected, the roll rate will increase until the rolling moment due to aileron deflection is balanced by the damping in roll moment. This steady state roll rate condition is given by

$$C_\ell = 0 = C_{\ell_p}\left(\frac{Pb}{2V}\right) + C_{\ell_{\delta_a}}\delta_a \qquad (21.17a)$$

where

C_{ℓ_p} = $dC_\ell/d(Pb/2V)$ is the damping in roll coefficient (determined from Fig. 21.11)

$C_{\ell_{\delta_a}}$ = $dC_\ell/d\delta_a$ is the aileron control power derivative

P = roll rate in radians per second

δ_a = aileron deflection

Rearranging Eq. (21.17a) to solve for the roll rate yields

$$P = -\frac{2V}{b}\frac{C_{\ell_{\delta_a}}}{C_{\ell_p}}\delta_a \qquad (21.17b)$$

The flying qualities in military specification MIL-HDBK-1797 suggest a 90-deg roll in one second for fighter aircraft (discussed further in Section 23.5).

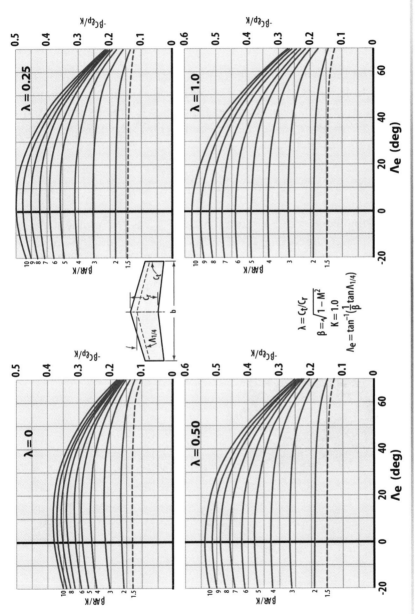

Figure 21.11 C_{l_p} for straight wings (data from [6]).

The $C_{l_{\delta_a}}$ depends upon the amount of aileron area or spoiler area and their locations. Reference [9, Section 6.2.1] or [3] is recommended for determining $C_{l_{\delta_a}}$.

21.6 Static Directional (Weathercock) Stability and Control

The *directional motion* of an aircraft is a rotation about the vertical axis of the aircraft. Figure 21.12 shows a schematic of the forces on an aircraft for directional motion. The directional moment is denoted by N and is positive for the right wing back (clockwise motion). The moment N about the c.g. is (from Fig. 21.12)

$$N = \ell_f L_f + \ell_{VT} L_{VT} + N_{power} + N_{wing} \tag{21.18}$$

where L_f is the side force on the fuselage, N_{power} is the moment due to asymmetric power effects (Fig. 21.12 shows this as a one-engine-out condition), and N_{wing} is the moment due to the wing. If ailerons are deflected, there is a differential lift and drag on each wing and hence an additional moment.

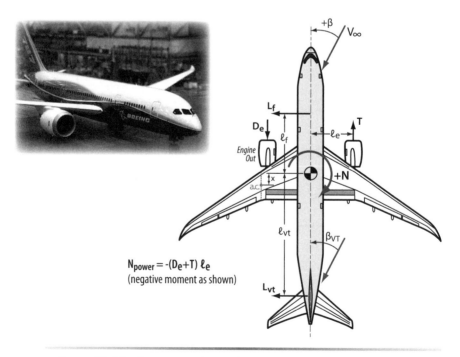

Figure 21.12 Forces on aircraft for directional motion (photograph courtesy of The Boeing Company).

The *directional moment coefficient* is

$$C_n = \frac{N}{q_\infty S_{ref} b}$$

$$C_n = -\frac{\ell_f L_f}{q_\infty S_{ref} b} + \frac{\ell_{VT} S_{VT}}{b S_{ref}} \frac{q_{VT}}{q_\infty} C_{L\alpha VT} \beta_{VT} + \frac{N_{power}}{q_\infty S_{ref} b} + \frac{N_{wing}}{q_\infty S_{ref} b} \quad (21.19)$$

where $\beta_{VT} = (1 + d\sigma/d\beta)\beta$ and accounts for the fuselage sidewash on the vertical tail.

The directional stability derivative is expressed as

$$\frac{dC_n}{d\beta} = C_{n\beta} = C_{n\beta fus} + C_{n\beta wing} + \bar{V}_{VT} C_{L\alpha VT} \left(1 + \frac{d\sigma}{d\beta}\right) \frac{q_{VT}}{q_\infty} \quad (21.20)$$

where $\bar{V}_{VT} = (\ell_{VT} S_{VT}/b S_{ref})$ is the vertical tail volume coefficient. Power effects are usually not dependent on sideslip angle β so that $C_{n\beta power}$ is neglected.

The directional stability derivative $C_{n\beta}$ must be positive for static directional stability. A $C_{n\beta} > 0$ will insure that moments will be generated, for a positive sideslip to rotate the aircraft so that β is reduced.

The vertical tail contribution is stabilizing for vertical tails (or ventral fins) aft of the center of gravity:

$$C_{n\beta VT} = \bar{V}_{VT} C_{L\alpha VT} \left(1 + d\sigma/d\beta\right)\left(q_{VT}/q\right) \quad (21.21)$$

where $C_{L\alpha VT}$ is the lift curve slope of the vertical tail based upon the vertical tail planform area and an effective aspect ratio 1.55 times that of the geometric aspect ratio (the fuselage acts as a large tip plate). The term

$$\left(1 + \frac{d\sigma}{d\beta}\right)\frac{q_{VT}}{q}$$

is determined from Eq. (21.15).

The $C_{n\beta}$ wing is due to the asymmetrical drag and lift distributions on the different wing panels undergoing sideslip. Wing sweep adds to the weathercock stability of the aircraft. An expression for the wing subsonic contribution is (from [9])

$$C_{n\beta wing} = C_L^2 \left[\frac{1}{4\pi AR} - \frac{\tan\Lambda_{c/4}}{\pi AR\left(AR + 4\cos\Lambda_{c/4}\right)} \right.$$

$$\left. \left(\cos\Lambda_{c/4} - \frac{AR}{2} - \frac{AR^2}{8\cos\Lambda_{c/4}} + \frac{6x}{\bar{c}}\frac{\sin\Lambda_{c/4}}{AR}\right)\right] \quad (21.22)$$

per radian, where x is the distance (positive rearward) from the aircraft c.g. to the wing aerodynamic center.

The fuselage at a sideslip angle β behaves like a lifting body. The sideslip derivative of the fuselage is usually destabilizing because the fuselage n.p. is usually ahead of the vehicle c.g. and the effect is very significant. The fuselage yawing moment is easier to calculate than the pitching moment due to $\Delta\beta$. The reason is that longitudinally the lift on the fuselage is very much affected by the wing (upwash and downwash) but directionally it can be assumed that the wing has very little effect. The several references used to find a general formula providing a first-order estimate of this contribution have led to the following form:

$$C_{n\beta\text{fuselage}} = -1.3 \frac{\text{Vol}}{S_{\text{ref}}b} \frac{h}{w} \tag{21.23}$$

per radian, where

Vol $\quad$ = fuselage volume
(h/w) = ratio of mean fuselage depth to mean fuselage width
b $\quad\quad$ = wing span
S_{ref} $\quad$ = wing planform area

The desirable level of directional stability in terms of $C_{n\beta}$ is very difficult to express in general terms. Chapter 23 lists some desired values for $C_{n\beta}$ that have been shown to give pleasant flying qualities. The vertical tail area is sized such that Eq. (21.20) gives desired values for $C_{n\beta}$. The rudder is sized to meet certain low-speed directional control criteria.

The contributions of the wing and fuselage are essentially independent of Mach number. However, the tail $C_{n\beta}$ increases then decreases with increasing Mach number due to the variation in $C_{L_{\alpha VT}}$. Because the wing contributes little stability, the vertical tail is the main component offsetting the destabilizing contribution of the fuselage. Because $C_{L_{\alpha VT}}$ can decrease by a factor of 3 from subsonic to Mach = 3, the static directional stability decreases at high Mach. Some vehicles need extra vertical surfaces at high Mach numbers to give adequate directional stability. The XB-70 did this by folding its wingtips downward (see Fig. 21.13).

The requirements for adequate directional control (discussed in Chapter 23) are that the rudder be powerful enough to hold $\beta=0$ for a one-engine-out (asymmetric power) flight condition at $1.2V_{\text{TO}}$, hold a straight ground path landing and takeoff in a crosswind up to $0.2V_{\text{TO}}$ and overcome the adverse yaw associated with abrupt aileron rolls at V_{TO}.

The asymmetric power condition would be (from Fig. 21.12)

$$C_n = 0 = -\frac{(T+D_e)}{q_\infty S_{\text{ref}}b} + C_{n\delta_r}\delta_r \tag{21.24}$$

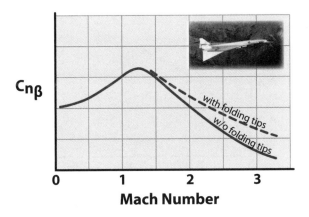

$C_{n\beta}$

0   1   2   3

Mach Number

Figure 21.13 Typical $C_{n\beta}$ variation with Mach number.

where $C_{n\delta_r} = dC_n/d\delta_r$ is the rudder control power and δ_r is the rudder deflection angle.

The crosswind condition is

$$C_n = 0 = C_{n\beta}\beta + C_{n\delta_r}\delta r \tag{21.25}$$

where $\beta = 11.5$ deg for a $0.2V_{TO}$ crosswind. In both cases maximum rudder deflection is ±20 deg.

The rudder control power can be estimated from

$$C_{n\delta_r} \approx 0.9 C_{L\alpha VT}\bar{V}_{VT}\tau \tag{21.26}$$

where $\tau = d\alpha_{0L}/d\delta_r$ and is shown in Fig. 21.14.

The required rudder area S_R for adequate directional control is determined by solving Eqs. (21.24), (21.25), and (21.26) for the maximum value of τ and then going to Fig. 21.14.

21.7 Aft Tail Location for Reduced Pitch-Up

Pitch-up is the longitudinal instability at high lift that results in an aircraft having a positive pitching moment as the wing begins to stall. It is due to the forward shift of the wing center of pressure as the wingtip region stalls and/or the blanking of the aft horizontal tail by the separated wing wake. It is a very undesirable phenomenon as the aircraft tends to pitch up violently with disastrous results at high subsonic speeds. Many of the fighter aircraft (such as the McDonnell F-101 Voodoo) had horns, buzzers,

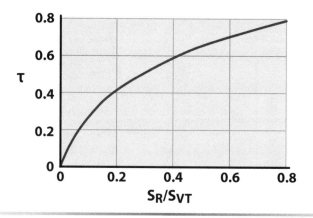

Figure 21.14 Rudder effectiveness chart (from data in Fig. 9.10).

or stick shakers that would warn the pilot about entry into the wing-stall region.

Sometimes the aircraft will pitch up gradually to a high angle-of-attack and the horizontal tail will lose the control power to push the nose down. This situation is termed *pitch hang-up*. If the wing is stalled, it can lead to an unrecoverable deep stall. One design solution is to locate the horizontal tail outside of the wing wake by mounting it on top of the vertical tail in a "T-tail" arrangement. In 1963 the aviation transport world was troubled by several accidents of the newly produced "T-tail" aircraft, including the BAC-111 and the Trident. These accidents were the result of an unrecoverable deep stall, which is a condition marked by softening of the horizontal tail control power at poststall angles-of-attack. Transport aircraft do not normally get to these high angles-of-attack so it is not a problem. However, STOL transports need special features such as stick shakers, stall limiters, and T-tails to stay out of trouble. The Lockheed C-5 had all three features and flew trouble free for over three decades. In 2005 the C-5 was re-engined with the more powerful CF6-80C2 turbofans. Because the engine thrust line was below the c.g. (giving a pitch-up moment) the nose-down pitch authority was reduced, giving the possibility of an unrecoverable deep stall. The solution was to bias the stall limiter to lower angles-of-attack and train the crews in aggressive stall recovery techniques. The C-17 has similar features to those of the C-5. The initial F-16As had a pitch hang-up problem above 35-deg angle-of-attack, and the horizontal tail area was increased on subsequent aircraft.

Reference [10] presents some design guidelines for aircraft planforms and aft tail location for minimizing the possibility of pitch-up at high subsonic speeds. Figure 21.15a presents a boundary for AR–sweep combinations for tailless aircraft. Aspect ratio and sweep combinations in region I (to the right of the boundary) display a tendency to pitch-up at high C_L. Notice that the more modern air-to-air fighters all have region II wings and the high subsonic transports all have region I wings.

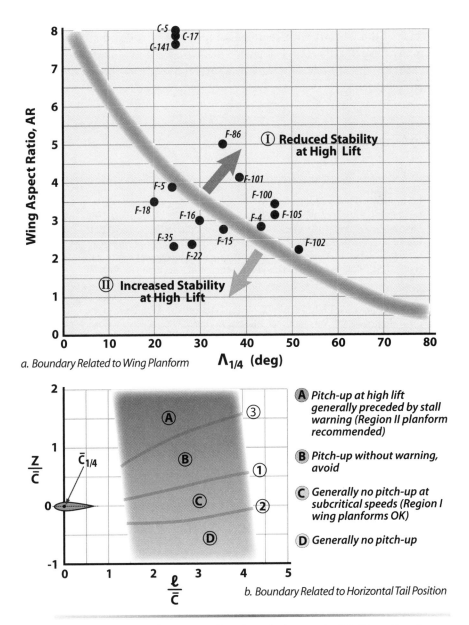

a. Boundary Related to Wing Planform

b. Boundary Related to Horizontal Tail Position

Figure 21.15 Guidelines for wing design and aft tail location for minimal pitch-up at high-subsonic speeds (data from [10]).

Figure 21.15b presents design information on aft tail location for reducing pitch-up tendencies at high-subsonic speeds. The point is to locate the tail out of the high-C_L wing wake so that the tail continues to be effective in providing longitudinal control. Figure 21.8 can be used to locate the wing downwash (wake) behind the wing. Region B of Fig. 21.15b is to be avoided. Region A, although not recommended for aircraft, is permissible provided a region II planform (from Fig. 21.15a) is used.

Most of the aircraft with region I wings have horns and stick shakers, and many have T-tails. The F-4 with a clear region I wing has dihedral on the wing tips and anhedral on the horizontal tail to locate the tail outside of the wing wake.

References

[1] McCormick, B., *Aerodynamics, Aeronautics and Flight Mechanics*, Wiley, New York, 1995.

[2] Etkin, B., *Dynamics of Atmospheric Flight*, Wiley, New York, 1972.

[3] Roskam, J., *Flight Dynamics of Rigid and Elastic Airplanes*, Univ. of Kansas, Lawrence, KS, 1972. [Available via www.darcorp.com (accessed 31 Oct. 2009).]

[4] Roskam, J., *Airplane Design, Part VII: Determination of Stability, Control and Performance Characteristics: FAR and Military Requirements*, Univ. of Kansas, Lawrence, KS, 1988. [Available via www.darcorp.com (accessed 31 Oct. 2009).]

[5] Chalk, C. R., "Background Information and User Guide for MIL-F-8785, Military Specification—Flying Qualities of Piloted Airplanes," U.S. Air Force Flight Dynamics Laboratory, AFFDL-TR-69-72, Wright–Patterson AFB, Dayton, OH, Aug. 1969.

[6] Data sheets, Royal Aeronautical Society, London, UK.

[7] Multhopp, R., "Aerodynamics of the Fuselage," NACA TM-1036, 1942.

[8] Silverstein, A., and Katzoff, S., "Design Charts for Predicting Downwash Angles and Wake Characteristics Behind Plain and Flapped Wings," NACA TR-648, 1939.

[9] Ellison, D. E., "USAF Stability and Control Handbook (DATCOM)," U.S. Air Force Flight Dynamics Laboratory, AFFDL/FDCC, Wright–Patterson AFB, Dayton, OH, Aug. 1968.

[10] Spreemann, K. P., "Design Guide for Pitch-up Evaluation and Investigation at High Subsonic Speeds of Possible Limitations Due to Wing–Aspect-Ratio Variations," NASA TM-X-26, NASA Langley Research Center, Langley, VA, Aug. 1959.

- Neutral Point
- Static Margin
- Aft Trim C_L
- Canard Trim C_L
- Tailless Aircraft Control
- Pitch Damping Coefficient
- Tail & Canard Control Power

An A-10A Thunderbolt II in a tight turn rolling in on a target. Notice the TE up deflection in the horizontal tail. The A-10A's survivability strategy was low vulnerability as it could "take a lickin' and keep on tickin'" (see Figs. 12.2 and 12.3).

Ofttimes the most successful test is the one that failed. Most learning comes from failure. Do not fear failure but keep the cost of failure low.

22.1 Neutral Point and Static Margin

T he aircraft *stick fixed neutral point* is defined as that center of gravity position where $C_{M_\alpha} = 0$. It is determined by setting Eq. (21.4), (21.5), or (21.6) of Chapter 21 equal to zero and solving for X_w. The X_w then gives the neutral point location relative to the wing–body aerodynamic center (see Fig. 21.4a, b, and c). Notice that the neutral point for a tailless aircraft is at the wing–body aerodynamic center location when the inlet stability term is zero. With the center of gravity at the neutral point the aircraft is neutrally stable. The stick fixed neutral point is essentially the total aircraft aerodynamic center.

A convenient way to remember the location of the *neutral point* (n.p.) relative to the wing–body *aerodynamic center* (a.c.) is that the n.p. and a.c. are coincident for a tailless aircraft; then, as you add an aft tail or canard, the n.p. moves in the direction of the horizontal control surface. Thus, the n.p. is behind the a.c. for an aft tail and ahead of the a.c. for a canard.

> Canard is the French word for hoax. When French airplane designers first saw pictures of the Wright brothers' airplane they thought it was a hoax because they knew that the forward control surface makes the airplane unstable and unflyable. The Wrights knew this was unstable but they also knew the pilot could easily control it. Forward control surfaces have been referred to as canards ever since.

The *static margin* (SM) is defined as follows:

$$SM = \frac{X_{\text{n.p.}} - X_{\text{c.g.}}}{\bar{c}} \qquad (22.1)$$

where $X_{\text{n.p.}}$ and $X_{\text{c.g.}}$ are the locations of the neutral point and aircraft center of gravity, respectively. When the neutral point is ahead of the center of gravity the SM is negative and the aircraft is statically unstable. The condition of negative SM is termed *relaxed static margin* and means that, if the aircraft is perturbed from equilibrium, the moments generated will tend to rotate the aircraft further from equilibrium. The aircraft would be extremely sensitive as it would have the tendency to maneuver away from equilibrium. The pilot would have a very maneuverable aircraft and would have to be controlling it all the time; he could not relax for a second. The Sopwith Camel of Chapter 21 was such an aircraft.

The static margin and the longitudinal stability derivative are related as follows:

$$C_{M\alpha} = -\text{SM}\, C_{L\alpha\text{WB}} \qquad (22.2)$$

where $C_{L\alpha\text{WB}}$ is the wing–body linear lift curve slope, m_W.

22.2 **Aft Tail Deflection to Trim $n = 1$ Flight**

For the aft tail aircraft set the trim equation [Eq. (21.1) of Chapter 21] equal to zero:

$$0 = C_L \frac{x_w}{\bar{c}} + C_D \frac{z}{\bar{c}} + C_{Ma.c.w} + \frac{Tz_T}{q_\infty S_{ref}\bar{c}} - C_{LT}\bar{V}_H\eta_T - C_{Mc.g.inlet} \qquad (22.3)$$

where the C_L is the required wing–body C_L for $n = 1$. Equation (22.3) is solved for the trim load coefficient C_{LT} of the tail, where C_{LT} is referenced to tail area S_T.

For statically stable aircraft (positive SM), this C_{LT} is usually negative, that is, the tail trim load is downward. Recalling that

$$n = \frac{\text{Lift}}{\text{Weight}} = \frac{q(C_L S_{ref} + \eta_T C_{LT} S_T)}{W} \qquad (22.4)$$

the C_L of the wing–body will have to be increased to counter the down load on the tail in order to cruise at $n = 1$. Equation (22.3) may have to be iterated several times with different values of C_L to satisfy Eq. (22.4).

The *trim drag* for $n = 1$ flight is expressed as

$$D_{trim} = \eta_T q_\infty S_T K_T C_{LT}^2 \qquad (22.5)$$

The trim drag is only the drag-due-to-lift of the tail because the zero-lift tail drag coefficient is included in the total aircraft C_{D_0}. If the trim is too large, that is, greater than 10% of the aircraft drag, the designer should take steps to reduce the value of C_{LT}. This can be accomplished by the following:

1. Moving the c.g. aft (closer to the neutral point)
2. Increasing the tail volume coefficient $\bar{V}_H$ by increasing S_T and/ or ℓ_T (see Fig. 21.2), both of which will also move the c.g. aft
3. Increasing $C_{L\alpha T}$ by increasing tail aspect ratio (AR)

Figure 22.1 shows a typical variation of C_{LT} with wing–body C_L for different c.g. locations. As the c.g. moves aft and the aircraft becomes less stable and then unstable, the tail trim load reverses from a down load to an up load. Figure 22.1 is for a composite Light Weight Fighter (LWF) at Mach $= 0.9$ and 30,000 ft. Figure 22.2 shows the behavior of the total aircraft cruise drag, wing–body C_L, and aft tail trim coefficient C_{LT} for different c.g. locations. The advantage of negative static margin is clearly evident in Fig. 22.2 as the total aircraft drag decreases to a minimum at an SM of about -8%. As the SM is decreased past -8% the aircraft drag starts to increase

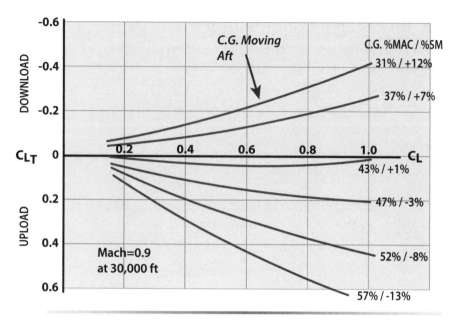

Figure 22.1 Variation of aft tail trim C_{L_T} with aircraft C_L and c.g. (composite LWF).

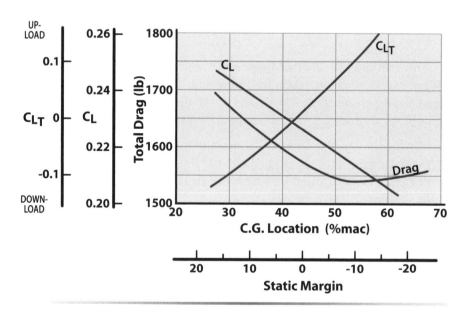

Figure 22.2 Variation of aircraft drag, C_L, and C_{L_T} for LWF at $n = 1$ (cruise), Mach = 0.9, and 45,000 ft.

because the trim drag is increasing faster than the wing–body drag-due-to-lift is decreasing, that is,

$$\text{Drag} = qS_{\text{ref}}\left(C_{D_o} + K\,C_L^2\right) + D_{\text{trim}} \tag{22.6}$$

Normally, aircraft fly at an SM of about +3% to +10%. Figure 22.2 indicates a cruise drag reduction of about 3% by relaxing the SM from +5% to –8%. The overall payoff depends upon the comparison of the decreased aircraft weight due to reduced cruise fuel through relaxed SM with the cost and weight increase of the stability augmentation system. In the case of the F-16 there was a payoff for the relaxed SM and it flies at a –6% SM during its Mach $= 0.9$ cruise.

The tail deflection depends upon the type of aft tail being used. As discussed in Chapter 21 the aft tail can be a stabilizer–elevator arrangement or an all flying tail. If the aft tail is an all flying tail, the expression for C_{L_T} is

$$C_{L_T} = \left(\frac{dC_L}{d\alpha}\right)_T (\alpha_T + \alpha_{\text{cs}}) = \left(\frac{dC_L}{d\alpha}\right)_T \left[\left(1 - \frac{d\varepsilon}{d\alpha}\right)\alpha + \alpha_{\text{cs}}\right] \tag{22.7}$$

where α is the aircraft angle-of-attack and α_{cs} is the deflection angle for the tail. The term $(1 - d\varepsilon/d\alpha)$ is the downwash term and can be determined from Fig. 21.8. The all flying tail is a miniature wing, and the quantity $|\alpha_T + \alpha_{\text{cs}}|$ should not exceed the stall angle for the section.

If the horizontal tail is a stabilizer–elevator arrangement, the expression for C_{L_T} is

$$C_{L_T} = \left(\frac{dC_L}{d\alpha}\right)_T \left[\left(1 - \frac{d\varepsilon}{d\alpha}\right)\alpha + \alpha_{0L}\right] = \left(\frac{dC_L}{d\alpha}\right)_T \left[\left(1 - \frac{d\varepsilon}{d\alpha}\right)\alpha + \frac{d\alpha_{0L}}{d\delta_e}\delta_e\right] \tag{22.8}$$

where α_{0L} is the zero-lift angle for the aft tail and is similar to a flapped wing. The term $d\alpha_{0L}/d\delta_e$ is the same as the τ of Fig. 21.14, Chapter 21. When using Fig. 21.14 for $d\alpha_{0L}/d\delta_e$, replace S_R/S_{VT} by S_e/S_T, where S_e is the elevator area and S_T is the total horizontal tail area.

Equation (21.1) can also be expressed as

$$C_M = C_{M0} + C_{M\alpha}\alpha + C_{M\delta}\left(\delta_e, \alpha_{\text{cs}}\right) \tag{22.9}$$

where $C_{M\delta}$ is called the *horizontal tail power* and, for an all flying tail,

$$C_{M\delta} = -\bar{V}_H \eta_T\, C_{L\alpha_T} \tag{22.10a}$$

and, for an elevator–stabilizer arrangement,

$$C_{M\delta} = -\bar{V}_H \eta_T C_{L\alpha T} \frac{d\alpha_{0L}}{d\delta_e}$$ (22.10b)

and C_{M0}, the moment coefficient at zero α and control deflection, is given by

$$C_{M0} = C_{Ma.c.w} + C_{D0} \frac{z}{\bar{c}} \frac{Tz_T}{q_\infty S_{ref} \bar{c}}$$ (22.11)

Using Eqs. (22.9) and (22.2), the expression for the control deflection for trimming the aircraft is

$$\alpha_{cs}, \delta_e = \frac{-C_{M0} + (SM)C_L}{C_{M\delta}}$$ (22.12)

Figure 22.3 shows a typical variation of the control surface deflection required to trim the aircraft. Notice that the most forward c.g. location is fixed by the maximum control surface deflection (usually ±20 deg). Figure 21.4b shows the sign convention for elevator deflection.

The static longitudinal stability changes with increasing Mach number. The most pronounced Mach number effect is the rearward shift of the wing–body aerodynamic center to about the 50% mac location for supersonic flight. This results in an aftward shift of the neutral point and

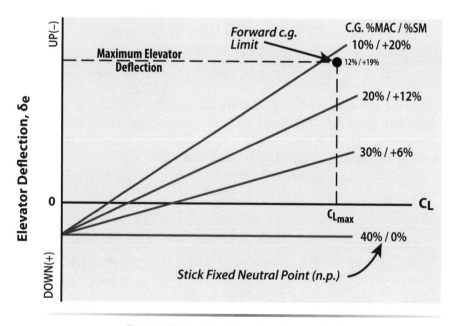

Figure 22.3 Trim C_L vs elevator deflection.

a resulting increase in the SM and aircraft stability. The designer should check the trim drag at this increased stability condition. Sometimes the c.g. is shifted aft by fuel transfer during supersonic flight to reduce the trim drag (see Chapter 23). Another effect is the increase of $(C_{L_\alpha})_T$ at transonic speeds and then its decrease at supersonic speeds (see Fig. 13.3a). This changes the horizontal tail effectiveness and will make the aircraft more stable in the transonic regime and less stable supersonically. At supersonic speeds there is also a decrease in the downwash at the tail.

22.3 Canard Deflection for Trim at $n = 1$

For the canard, set the trim equation [Eq. (21.4a)] equal to zero:

$$0 = C_{M\text{a.c.}w} - \frac{x_w}{\bar{c}} C_L + \frac{z}{\bar{c}} C_D + C_{L_c}\bar{V}_c + \frac{Tz_T}{q_\infty S_{\text{ref}}\bar{c}} - C_{Mc\text{-g-inlet}} \qquad (22.13)$$

Equation (22.13) is solved for the canard lift coefficient C_{L_c} and the resulting trim drag is

$$D_{\text{trim}} = q_\infty S_c K_c C_{L_c}^2 \qquad (22.14)$$

Because the aircraft α for cruise is usually small assume that nonlinear lift is negligible and

$$C_{L_c} = \left(\frac{dC_L}{d\alpha}\right)_c (\alpha - \alpha_{0L} + \alpha_c) = \left(\frac{dC_L}{d\alpha}\right)_c + (\alpha + \alpha_c) \qquad (22.15)$$

The canard is usually symmetric so that $\alpha_{0L} = 0$.
The trim equation can be expressed in a form similar to Eq. (22.9) as

$$C_M = C_{M0} + C_{M\alpha}\alpha + C_{M\alpha_c}\alpha_c \qquad (22.16)$$

where $C_{M\alpha_c}$ is the canard control power

$$C_{M\alpha_c} = \bar{V}_c C_{L\alpha_c} \qquad (22.17)$$

and

$$C_{M0} = C_{M\text{a.c.}w} + C_{D0}\frac{z}{\bar{c}} + \frac{Tz_T}{q_\infty S_{\text{ref}}\bar{c}} \qquad (22.18)$$

The canard deflection for trimming the aircraft is expressed as [from Eqs. (22.18) and (22.2)]

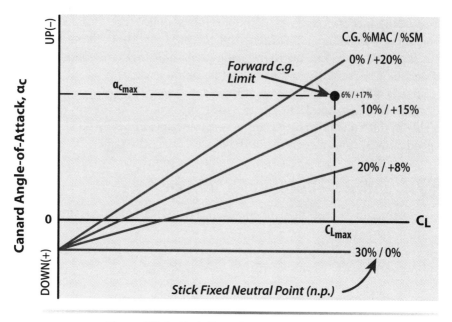

Figure 22.4 Variation of canard angle-of-attack with wing–body C_L and c.g. location.

$$\alpha_c = \frac{-C_{M0} + (\mathrm{SM})C_L}{C_{M\alpha_c}} \tag{22.19}$$

Equation (22.19) indicates that the canard trim load is up for positive SM and reverses to a down trim load as the c.g. moves aft and the SM decreases to zero and then becomes negative. This behavior is shown on Fig. 22.4. Thus, the canard acts opposite to an aft tail. A down trim load works against the wing–body lift and is undesirable, so that relaxed static margin is not as attractive for a wing–canard arrangement as it is for a wing–aft tail configuration.

22.4 Control of a Tailless Aircraft at $n = 1$

As mentioned earlier, the neutral point for a tailless aircraft is located at the wing–body aerodynamic center. Thus, the tailless aircraft must have the center of gravity ahead of the wing–body a.c. for static stability. This is shown on Fig. 22.5. Because the tailless aircraft does not have any horizontal control surfaces (aft tail or canard), the moment to trim the aircraft must come from the wing moment about the a.c. $C_{M_{\alpha c}}$, C_M about the a.c., is changed by deflecting the wing flaps (called *elevons*) up and down, effecting a positive or negative camber change. Figure 22.5 illustrates this. The upsweep of the trailing edge (called *reflex camber*) produces a positive $C_{M_{\alpha c}}$ and balances the aircraft at positive C_L.

For the tailless aircraft set the trim equation (neglecting the moment due to wing drag and inlet) equal to zero:

$$0 = C_{M_{a.c._w}} - \frac{x_w}{\overline{c}} C_L + \frac{Tz_T}{q_\infty S_{ref} \overline{c}} \tag{22.20}$$

As the elevon is deflected the wing lift coefficient C_L changes and $C_{M_{a.c._w}}$ changes. Equation (22.20) can be rewritten as

$$0 = \left(\frac{dC_{M_{a.c.}}}{d\delta_e}\right)\delta_e - \frac{x_w}{\overline{c}}\left[\left(\frac{dC_L}{d\alpha}\right)_w \alpha + \left(\frac{dC_L}{d\delta_e}\right)\delta_e\right] + \frac{Tz_T}{q_\infty S_{ref}\overline{c}}$$

and the elevon deflection is expressed as

$$\delta_e = \frac{(x_w/\overline{c})(dC_L/d\alpha)_w \alpha - \left(Tz_T / q_\infty S_{ref}\overline{c}\right)}{-(x_w/\overline{c})C_{L\delta} + C_{M\delta}} \tag{22.21}$$

The denominator is referred to as the *elevon control power* and [1, Sections 6.1.1–6.1.5; 2], can be used to estimate $C_{L\delta}$ and $C_{M\delta}$.

The trim drag for a tailless aircraft is much different than for an aft tail or canard aircraft. A tailless aircraft trims itself by changing the camber of

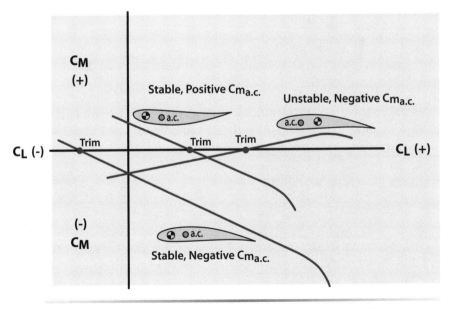

Figure 22.5 Control of a tailless aircraft by camber change.

the wing. Subsonically, this results in a small change in the wing separation drag and can be determined using the method of Section 9.5.

Supersonically the elevon deflection results in a change in the wave drag. From linear theory

$$C_{Dw} = \frac{4}{\sqrt{M^2 - 1}} \left[\alpha^2 + \left(\frac{\overline{dh}}{dx} \right)^2 + \overline{\alpha_c^2} \right]$$

and by deflecting the elevon α and α_c^2 are also changed The result is

$$\frac{\left(C_{Dwave}\right)_{\delta e}}{\left(C_{Dwave}\right)_{\delta e = 0}} = \left[1 - \frac{c_e}{c}\right] + \frac{c_e}{c} \frac{\left(\alpha + \delta_e\right)^2}{\alpha^2} \qquad (22.22)$$

where c_e is the elevon chord and c is the wing chord (including elevon).

22.5 Aft Tail Deflection for Maneuvering Flight— Pull-Up Maneuver

In a pull-up or loop maneuver the weight of the aircraft always opposes the lift vector. The aft tail deflection, δ_e or α_{cs}, for a pull-up maneuver is greater than $n - 1$ times the trim deflection for $n = 1$ flight because of the inertial loading and pitch damping of the aircraft.

The increase in δ_e or α_{cs} for a pull-up maneuver of n gs is given by

$$\left(\Delta\alpha_{cs}, \Delta\delta_e\right)_{\text{maneuver}} = -\frac{\left[-\text{SM} + \left(\rho S_{\text{ref}}\bar{c} / 4m\right)C_{Mq}\right](n-1)C_{Ln=1}}{C_{M\delta}} \qquad (22.23)$$

where m is the mass of the aircraft in slugs, ρ is the density in slugs per cubic foot (slug/ft³), and $C_{L_n} = 1$ is the C_L required for $n = 1$ flight. The C_{Mq} is the pitch damping derivative and is determined from

$$C_{Mq} = \frac{dC_M}{d\left(q\bar{c} / 2V\right)} = -2.2\eta_T \bar{V}_H C_{L\alpha T} \frac{\ell_T}{\bar{c}} \qquad (22.24)$$

The $C_{M\delta}$ for Eq. (22.23) is determined from Eq. (22.10a) for an all flying tail and Eq. (22.10b) for an elevator–stabilizer arrangement. The total aft tail deflection is determined by adding the value from Eq. (22.23) to the value for $n = 1$ from Eq. (22.12). The C_{LT} is determined from Eq. (22.7) or (22.8) and the trim from Eq. (22.5).

22.6 Canard Deflection for Maneuvering Flight—Pull-Up Maneuver

The additional canard deflection to pull n gs may be found from

$$\left(\Delta\alpha_c\right)_{\text{maneuver}} = -\frac{\left[-\text{SM} + \left(\rho S_{\text{ref}}\bar{c}/4m\right)C_{M_q}\right](n-1)C_{L_{n=1}}}{\bar{V}_c C_{L_{\alpha c}}} \tag{22.25}$$

where the denominator is the canard control power $C_{M_{\delta_c}}$ [Eq. (22.17)] and the pitch damping term

$$C_{M_q} = -2.2\bar{V}_c C_{L_{\alpha c}}\frac{\ell_c}{\bar{c}}$$

22.7 Elevon Deflection for a Tailless Aircraft in Maneuvering Flight—Pull-Up Maneuver

For the tailless aircraft the damping in pitch is very small compared with an aircraft with an aft tail or canard. As a first approximation set $C_{M_q} = 0$. This makes the maneuver point coincide with the neutral point, which coincides with the aerodynamic center.

Thus,

$$\left(\Delta\delta_e\right)_{\text{maneuver}} = -\frac{\text{SM}(n-1)C_{L_{n=1}}}{-(x_w/\bar{c})C_{L_\delta} + C_{M_\delta}} \tag{22.26a}$$

$$\Delta\delta_{e\,(\text{maneuver})} = (n-1)\delta_{e_{n=1}} \tag{22.26b}$$

or

$$\Delta\delta_{e\,(\text{total})} = n\delta_{e_{n=1}} \tag{22.26c}$$

where $\delta_{e_{n=1}}$ is the elevon deflection for $n = 1$ flight as given by Eq. (22.21).

22.8 Control Deflection for Level Turn Maneuvering Flight

The previous discussion of maneuvering flight (i.e., Sections 22.5, 22.6, and 22.7) was for a pull-up or loop maneuver where the weight of the aircraft always opposed the lift vector. For a level turn only a component of the lift is equal to the weight (see Section 3.7) and thus the control deflection equations are slightly different from those for a loop.

The increased control deflection for a level turn is given by the following:
Aft Tail:

$$\left(\Delta\alpha_{cs}, \Delta\delta_e\right)_{\text{levelturn}} = -\frac{\left[-\text{SM}(n-1) + C_{M_q}\left(\rho S_{\text{ref}}\bar{c}/4m\right)\left(n-(1/n)\right)\right]C_{L_{n=1}}}{C_{M_\delta}}$$

(22.27)

where $n = 1/\cos\varphi$ for a level turn, φ is the bank angle (from Fig. 3.11), C_{M_q} is given by Eq. (22.24), and C_{M_δ} is given by Eq. (22.10).
Canard:

$$\left(\Delta\alpha_c\right)_{\text{levelturn}} = -\frac{\left[-\text{SM}(n-1) + C_{M_q}\left(\rho S_{\text{ref}}\bar{c}/4m\right)\left(n-(1/n)\right)\right]C_{L_{n=1}}}{\bar{V}_c C_{L_{\alpha_c}}}$$

(22.28)

where

$$C_{M_q} = -2.2\bar{V}_c C_{L_{\alpha_c}}\frac{\ell_c}{\bar{c}}$$

Tailless: Because the damping in pitch is negligible, the control deflection for a tailless aircraft in a level turn is the same as Eq. (22.26).

References

[1] Ellison, D. E., "USAF Stability and Control Handbook (DATCOM)," U.S. Air Force Flight Dynamics Laboratory, AFFDL/FDCC, Wright–Patterson AFB, OH, Aug. 1968.
[2] Roskam, J., *Flight Dynamics of Rigid and Elastic Airplanes*, Univ. of Kansas, Lawrence, KS, 1972. [Available via www.darcorp.com (accessed 31 Oct. 2009).]

Chapter 23 / Control Surface Sizing Criteria

- Typical Values for S&C
- Recommended C_{M_α} Values
- Lateral Control Spin Parameter
- Recommended C_{n_β} Values
- C_{n_β} Dynamic
- C.G. Travel
- Aileron Sizing Criteria

Two Canada geese coming in for a STOL landing. Mother Nature (the ultimate designer) did a superb job of sizing the geese's control surfaces for this high angle-of-attack, low-speed, low-power approach condition. It is too bad she didn't write a design text.

If it looks good, it flies good.
Clarence "Kelly" Johnson

23.1 Government Regulations Require Static Stability

The Federal Aviation Regulations, Parts 23 and 25 [1] are more abbreviated and qualitative than MIL-HDBK-1797 with regard to stability and control requirements and handling qualities. The FARs require the aircraft to be stable in the longitudinal, directional, and lateral modes. The civilian regulations in their current form do not contain guidelines that are sufficiently detailed for use in design.

Both MIL-HDBK-1797 and FAR Parts 23 and 25 require that the elevator fixed neutral point be aft of the center of gravity for all loading conditions for aft tail configurations. This insures that

$$C_{M_\alpha} < 0$$

for all c.g. positions. It is interesting to note that the British Civil Airworthiness Requirements [2, Paragraph 2.1] specify a maximum allowable negative stick fixed static margin of −0.05.

Both MIL-HDBK-1797 and FAR Parts 23 and 25 require static directional stability, that is,

$$C_{n_\beta} > 0$$

in terms of characteristics involving rudder position and rudder force.

Roll damping is an important handling-qualities parameter, particularly in roll maneuvers and in Dutch roll. The government regulations do not specify minimum values for the roll damping derivative C_{ℓ_p} directly. However, to meet the roll-handling requirements of [3,4] it is necessary that

$$C_{\ell_p} < 0$$

Similarly, the government regulations do not give minimum values for the pitch-damping derivative C_{M_q}. However, to meet the short-period damping requirement of [5] it is necessary that

$$C_{M_q} < 0$$

MIL-HDBK-1797 requires that left aileron force be generated for left sideslip. For conventional control arrangements this can be shown to imply that the lateral stability parameter

$$C_{\ell_\beta} < 0$$

Table 23.1   Typical Aircraft Stability and Control Data

	Learjet	B-727	T-38	Cherokee 180 Archer
Takeoff weight (lb)	13,500	152,000	11,250	2450
Empty weight (lb)	7,252	88,000	7,370	1390
Wing area (ft^2)	232	1,560	170	156
Span (ft)	34.2	106	25.25	30
Aspect ratio	5.02	7.2	3.75	5.71
Wing sweep $c/4$ (deg)	0	35	24	0
mac (ft)	7.04	15	7.73	5.25
Vertical tail area (ft^2)	37.4	15,356	7.73	11.4
Horizontal tail area (ft^2)	54.0	376	59.0	25
Derivatives at Mach = 0.8				At Mach = 0.19
C.G. location (% mac)	32	25	19	19
C_{M_0}	0.06	−0.042	—	−0.07
$C_{L_{\delta_e}}$ (per radian)	0.6	0.4	—	—
C_{M_α} (per radian)	−0.7	−1.5	−0.16	−0.32
$C_{M_{\delta_e}}$ (per radian)	−1.6	−1.3	−0.13	−1.1
$C_{L_{\delta_a}}$ (per radian)	0.015	—	—	—
$C_{L_{\delta_r}}$ at $C_L = 1.1$ (per radian)	0.007	0.017	—	
C_{n_β} (per radian)	0.12	0.08	0.28	0.092
$C_{n_{\delta_r}}$ (per radian)	−0.074	−0.098	—	−0.06
C_{M_q} (per radian)	−14	−24	−8.4	−12
C_{ℓ_p} (per radian)	−0.5	−0.30	−0.35	−0.47
C_{ℓ_β} (per radian)	−0.1	−0.13	−0.075	−0.107
I_{xx} (10^4) (slug·ft^2)	3	92	1.48	0.107
I_{yy} (10^4) (slug·ft^2)	1.9	300	2.82	0.1249
I_{zz} (10^4) (slug·ft^2)	5	380	2.9	0.2312

This condition is also necessary to keep the spiral mode from being divergent.

Over the years, all of the aircraft companies have developed their own flying-qualities criteria to supplement the government regulations. These criteria are reflected in the stability and control characteristics of the current inventory of aircraft. Table 23.1 lists the static stability and control characteristics of a representative small general aviation aircraft, FAR 23 business jet, FAR 25 transport, and a fighter-class aircraft.

23.2   Center of Gravity Location

The designer should not leave the c.g. location to chance, but rather force its location by judicious placement of components, including the wing. The c.g. is the most important element in the stability and control of an aircraft. It should be located so that the tail size (if it has a tail) is not unduly large and the trim drags are acceptable.

Current regulations require a statically stable aircraft, $C_{M\alpha} < 0$. Figure 23.1 shows some recommended $C_{M\alpha}$ values for general aviation, FAR 23 business jet, FAR 25 transport, and fighter aircraft. Transport aircraft should be more stable than fighters because of the comfort of the passengers. Fighter aircraft on the other hand need lower values of $C_{M\alpha}$ because of their maneuverability requirements. A good rule of thumb is an SM of +4% to +7% for transport aircraft and neutral to +3% for fighter aircraft. Larger static margins lead to trim drags that become intolerable. The recommended curves of $C_{M\alpha}$ on Fig. 23.1 are based on this range of SM and the expression

$$C_{M\alpha} = -(\text{SM})\left(C_{L\alpha}\right)_{\text{WB}} \qquad (23.1)$$

At this point the designer has a wing–body configuration and thus the wing–body a.c. location can be determined (Fig. 21.3). The usual aft tail or canard will move the n.p. about 5% mac behind or ahead of the a.c., respectively (Table 23.2). If an SM of +5% mac is desired, c.g. locations can be determined using Eq. (22.1).

$$\text{SM} = \left(x_{\text{n.p.}} - x_{\text{c.g.}}\right)/\overline{c} \qquad (22.1)$$

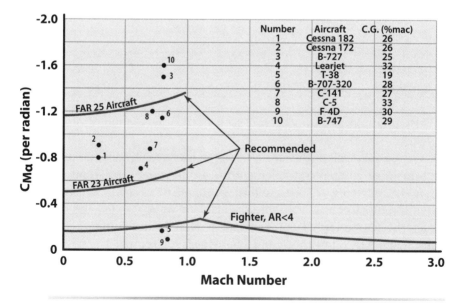

Number	Aircraft	C.G. (%mac)
1	Cessna 182	26
2	Cessna 172	26
3	B-727	25
4	Learjet	32
5	T-38	19
6	B-707-320	28
7	C-141	27
8	C-5	33
9	F-4D	30
10	B-747	29

Figure 23.1 $C_{M\alpha}$ values for current aircraft.

Table 23.2 Approximate N.P. and C.G. Locations

Subsonic: Assume A.C. at 35% mac		
Type	Approximate N.P. Location (% mac)	Approximate C.G. Location (% mac)
Aft tail	40	35
Tailless	35	30
Canard	30	25
Supersonic: Assume A.C. at 50% mac		
Type	Approximate N.P. Location (% mac)	Approximate C.G. Location (% mac)
Aft tail	55	50
Tailless	50	45
Canard	45	40

The c.g. moves around as fuel is burned, ordnance expended, cargo or passengers unloaded and loaded, and so on. The c.g. travel must be watched very closely as it can be costly in terms of excessive trim drag and aircraft safety-of-flight. The designer must allow for all possible c.g. locations, which may require system events (such as fuel transfer) so that the c.g. shift is within tolerable limits. Usually a c.g. shift of less than 10% mac for subsonic aircraft is tolerable; however, it varies from aircraft to aircraft. If the aircraft is scheduled to spend much of its mission at supersonic speeds, there should be some provision for shifting the c.g. aft (such as fuel sequencing) to follow the a.c. shift and keep the SM at a desired level.

Fuel sequencing or transfer is imperative for most aircraft. Here the fuel is located in fuel tanks distributed throughout the fuselage and wing. Fuel is then taken from these tanks in a definite schedule so that fuel is transferred from tank to tank to keep the c.g. located properly. Fuel is also sequenced so that when the weapons are dropped the c.g. shift is within limits. Figure 23.2 shows an example of this scheduling for the Boeing/McDonnell F-4D. This part of the preliminary design is not easy but it is essential to keep the c.g. shift within acceptable limits; otherwise performance benefits of the aircraft can be negated by excessive trim drags.

Finally, the aircraft should be unloaded completely ahead of the c.g. to insure that the aircraft does not fall back on its tail. Aircraft can, as a last resort, have a placard that dictates the load distribution while it is parked.

The designer should try very hard to get the c.g. close to the predetermined location. Payload, subsystems, and fuel can be shifted around, within fuselage constraints, to locate the c.g. at a desired position. Shifting the wing back and forth has a large effect on c.g. location because the

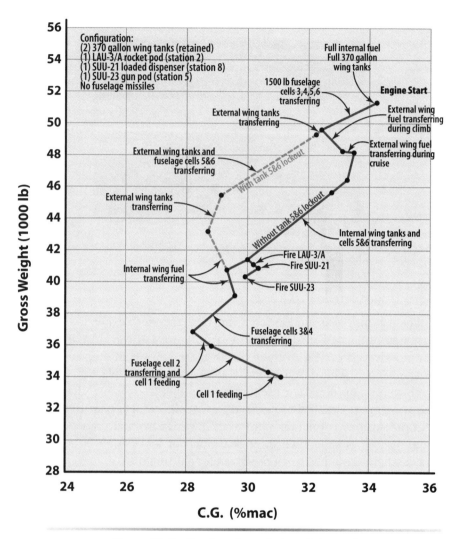

Figure 23.2 Center of gravity travel due to fuel consumption, "potato curve" (F-4D).

mac and a.c. move directly. Shifting the wing should be considered as a last resort because of its effect on inlet location and area-ruling.

23.3 Sizing the Horizontal Surface

The horizontal surface (aft tail or canard) is used for longitudinal stability and control. The designer should recognize at this point that stability and control are two independent functions. The horizontal surface is sized separately for each and the larger of the areas selected.

The designer should locate the horizontal surface on the aircraft and estimate a general planform shape (i.e., aspect ratio, taper ratio, and sweep).

23.3.1 Static Longitudinal Stability

The desired level of stability, C_{M_α}, is determined from Fig. 23.1, and then Eq. (21.4) or (21.5) is used to solve for S_T or S_C. Several Mach numbers should be examined and the largest area for all design conditions is then selected.

23.3.2 Longitudinal Control

The horizontal surface is now sized for adequate longitudinal control. The horizontal control surface can be an elevator (with deflection δ_e), an all flying stabilizer (with control surface angle-of-attack α_{cs}), or an all flying canard (with canard angle-of-attack α_c). The δ_e, α_{cs}, or α_c is usually limited to about ±20 deg.

There are several critical conditions for longitudinal control that should be examined:

1. **Trim drag.** The trim drag during cruise should be less than 10% of the total aircraft drag. Many designers reduce this limit to 5% for range-dominated vehicles.
2. **High-*g* maneuver.** If the aircraft is a fighter, the horizontal control surface should be checked to insure that there is sufficient control power to trim the high-*g* condition.
3. **Takeoff rotation.** The takeoff rotation to climb C_L (see Chapter 10) should be checked. The horizontal control surface must have enough control power to rotate the aircraft about the main landing gear to the takeoff attitude. This problem is shown schematically in Fig. 23.3. Attention paid to the recommended angle between the center of

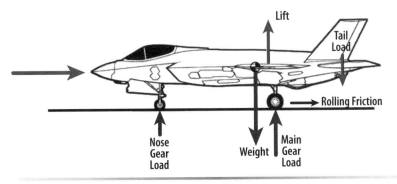

Figure 23.3 Takeoff control schematic.

gravity and the main gear wheel as shown in Table 8.5 will insure that the size of the horizontal tail will be acceptable.

4. **High α, low speed.** The condition of low-speed approach for landing with power at idle, flaps down, and high angle-of-attack is often a critical condition for sizing control surfaces. This condition often determines the most forward c.g. position as shown in Figs. 22.3 and 22.4. Ground effects must be considered as this condition increases the aircraft stability and compounds the control problem.

23.4 Sizing the Vertical Tail

23.4.1 Static Directional Stability

The vertical tail is sized to give adequate static directional stability. Desired values of C_{n_β} are put into Eq. (21.20) and the vertical tail area, S_{VT}, is determined. Figure 23.4 gives some recommended values for C_{n_β}.

23.4.2 Fighter Aircraft Spin Resistance

The degree of susceptibility to spin during hard turns with and without rolling inputs has a significant impact on the dogfighting capability of air superiority aircraft. Reference [5] reports two simple parameters that have been related to spin resistance margin and provide a good approximation of the relative resistance of aircraft to spin departure.

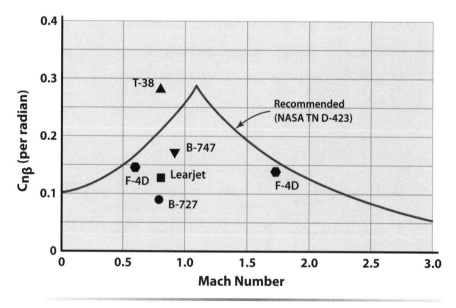

Figure 23.4 Recommended C_{n_β} values.

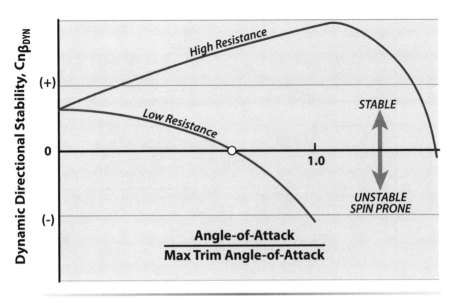

Figure 23.5 Measure of spin resistance (turning without rolling).

The angle-of-attack region of spin susceptibility for nonrolling turning maneuvers without lateral or directional inputs has been correlated with the dynamic directional stability parameter

$$C_{n\beta_{\mathrm{dyn}}} = C_{n\beta} - C_{\ell\beta}\frac{I_{zz}}{I_{xx}}\tan\alpha \qquad (23.2)$$

Figure 20.2 can be used to estimate the moments of inertia I_{xx} and I_{zz}. Unless the $C_{n\beta_{\mathrm{dyn}}}$ is positive throughout the possible operating angle-of-attack range as illustrated in Fig. 23.5, the aircraft will be susceptible to spin in hard nonrolling turns.

For assessing the spin susceptibility in turning maneuvers where lateral control inputs are introduced, it has been found that the angle-of-attack at which the roll reverses due to sideslip opposing the aileron correlates very closely with the region of spin susceptibility of a number of current fighter aircraft. The dominant parameters influencing roll reversal are the yaw due to aileron $C_{n\delta_a}$ and the directional stability $C_{n\beta}$.

High adverse yaw and low directional stability are detrimental. A *lateral control spin parameter* (LCSP) is defined by

$$\mathrm{LCSP} = C_{n\beta} - C_{\ell\beta}\frac{C_{n\delta_a}}{C_{\ell\delta_a}} \qquad (23.3)$$

where $C_{\ell\beta}$ and $C_{\ell\delta_a}$ were introduced in Chapter 21 and $C_{n\delta_a}$ can be estimated from [6] or [7]. Roll reversal occurs at the point where this parameter

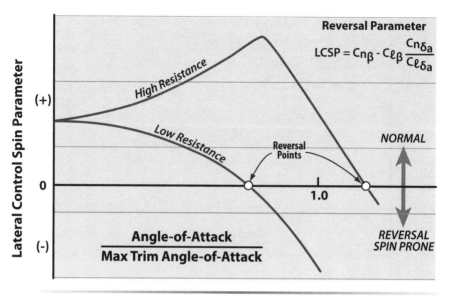

Figure 23.6 Measure of spin resistance (combined turning and rolling maneuvers).

equals zero. Figure 23.6 illustrates the variation of the LCSP vs angle-of-attack normalized to maximum angle.

23.4.3 Static Directional Control Requirements

The requirements on the rudder for adequate static directional control are as follows:

1. **Crosswind landing.** The rudder must be powerful enough to maintain a straight ground path during normal takeoff and landing in 90-deg crosswinds up to velocities of $0.2V_{TO}$. This means adequate rudder power to hold a sideslip of $\beta = 11.5$ deg at approach speeds. The analysis was discussed in Section 21.6.
2. **Antisymmetric power.** The rudder must be powerful enough to hold zero sideslip ($\beta = 0$) in straight flight at all speeds above $1.2V_{stall}$ with gear down, flaps in best setting, thrust on one outboard engine equal to zero (with associated drag), and all other engines developing full thrust. This condition was discussed in Section 21.6.
3. **Adverse yaw.** When an airplane is rolled into a turn, yawing moments are often produced that require rudder deflection to maintain zero sideslip, that is, coordinate the turn. For example, when initiating a roll to the right, aileron deflection may cause yaw to the left. This is termed *adverse yaw* and a rudder deflection is required to eliminate the sideslip. Because adverse yaw will be greatest at high C_L and full deflection of the ailerons, steep turns at low speed may produce a critical requirement for rudder control power.

23.5 Sizing the Ailerons

The *lateral control surface* is the aileron. As discussed in Chapter 21, this lateral control surface has no effect on the lateral stability of the aircraft. The lateral stability derivative is C_{ℓ_β} and is influenced by the wing (independent of the ailerons), the vertical tail, and the wing–fuselage. The regulations require C_{ℓ_β} to be negative. Typical values are given in Table 23.1.

MIL-HDBK-1797 places all aircraft in one of the following classifications, Class I, Class II, Class III, Class IVA, Class IVB, or Class IVC:

Class I

Small, light airplanes such as

- Light utility
- Primary trainer
- Light observation

Class II

Medium-weight, low-to-medium maneuverability airplanes such as

- Heavy utility or search and rescue
- Light or medium transport, cargo, or tanker
- Early warning, electronic countermeasures, or airborne command, control, or communications relay
- Antisubmarine
- Assault transport
- Reconnaissance
- Tactical bomber
- Heavy attack
- Trainer for Class II

Class III

Large, heavy, low-to-medium maneuverability airplanes such as

- Heavy transport, cargo, or tanker
- Heavy bomber
- Patrol, early warning, electronic countermeasures, or airborne command, control, or communications relay
- Trainer for Class III

Class IVA

High-maneuverability airplanes such as

- Fighter–interceptor
- Attack
- Tactical reconnaissance

Table 23.3 MIL-HDBK-1797 Roll Requirements

Class	Roll Performance
I	600 in 1.3 s
II	450 in 1.4 s
III	300 in 1.5 s
IVA	900 in 1.3 s
IVB	900 in 1.0 s
IVB	3600 in 2.8 s
IVC	900 in 1.7 s

- Observation
- Trainer for Class IV

Class IVB

Air-to-air fighter

Class IVC

Air-to-ground fighter with external stores

The ailerons should be sized to provide the roll performance listed in Table 23.3 for the appropriate class of aircraft under consideration. The roll rate P in radians per second is given by Eq. (21.17b):

$$P = -\frac{2V}{b}\frac{C_{\ell_{\delta a}}}{C_{\ell_p}}\delta_a \qquad (21.17b)$$

where V is speed in feet per second, $C_{\ell_{\delta a}}$ is the aileron control power, and C_{ℓ_p} is the roll-damping coefficient.

References

[1] "Airworthiness Standards: Part 23—Normal, Utility and Acrobatic Category Airplanes; Part 25—Transport Category Airplanes," *Federal Aviation Regulation*, Vol. 3, U.S. Department of Transportation, U.S. Government Printing Office, Washington, DC, Dec. 1996.

[2] "British Civil Airworthiness Requirements," Sec. D, Air Registration England, 15 Nov. 1991.

[3] "Military Specification—Flying Qualities of Piloted Aircraft," MIL-F-8785C, Nov. 1980.

[4] "Flying Qualities of Piloted Aircraft," MIL-HDBK-1797, Dec. 1997.

[5] Chambers, J. R., and Anglin, E. L., "Analysis of Lateral-Directional Stability Characteristics of a Twin-Jet Fighter Airplane at High Angles of Attack," NASA TN D-5361, 1969.

[6] Ellison, D. E., "USAF Stability and Control Handbook (DATCOM)," U.S. Air Force Flight Dynamics Laboratory, Wright–Patterson AFB, OH, Aug. 1968.

[7] Roskam, J., *Flight Dynamics of Rigid and Elastic Airplanes*, Univ. of Kansas, Lawrence, KS, 1972. [Available via www.darcorp.com (accessed 31 Oct. 2009).]

Chapter 24 / Life Cycle Cost

- Cost-Estimating Equations
- Economic Escalation Factors
- Design for Reduced O&M
- Cost-Estimating Charts
- UCAV vs Manned Aircraft O&S
- Design for Production

The Boeing 777 transport or "Triple 7" was developed in the 1990s to compete head-on with the Airbus 340 and the MD-11/12. Boeing has delivered over twice as many "Triple 7s" as Airbus has A-340s. (McDonnell Douglas dropped out of this market.) Example 24.1 uses the Triple 7 to estimate development and acquisition cost. (Photograph courtesy of Singapore Airlines.)

A billion here, a billion there—
pretty soon it adds up to real money!
Everett McKinley Dirksen

24.1 Life Cycle Cost

The *life cycle cost* (LCC) of a military aircraft is the total cost to transition the aircraft from "cradle to grave." It includes the following phases (as shown in Fig. 1.16):

- Research
- Development, Test, and Evaluation (DT&E)
- Acquisition (production, ground equipment, initial spares, training aids, etc.)
- Operations and Maintenance (O&M)

The *research phase* involves the basic research and the exploratory and advanced development efforts required to mature those technologies that are essential to the successful operation of the aircraft. This phase can include technology demonstrator aircraft, testbeds, and prototypes. An example of this phase would be the effort expended in researching the integration of the shaft-driven lift fan into the Joint Strike Fighter short takeoff, vertical landing (STOVL) prototype X-35B. This phase is important because, without it, advanced technologies would not find their way onto new aircraft systems. In many cases, a commercial aircraft will build upon technology that was developed for a military aircraft program, thereby reducing the research cost for the commercial program. The research phase is a difficult cost item to estimate because of the uncertainty inherent in a research and technology development program. Also, the research phase is a mixture of contractor funding and government funding.

The *development, test, and evaluation cost* is that cost required to engineer, develop, fabricate, and flight test a number Q_D of aircraft prior to committing to production. The DT&E aircraft might number as few as 2 or as many as 10. The DT&E phase is usually government funded. The cost elements charged to DT&E are as follows:

- Airframe engineering
- Development support
- Flight test aircraft
  - Engine and avionics
  - Manufacturing labor
  - Material and equipment
  - Tooling
  - Quality control
- Flight test operations
- Test facilities
- Profit

The *acquisition cost* includes the cumulative cost of Q_P production aircraft, associated ground equipment (such as starting carts and special equipment for maintenance and operation), initial spares, and training aids (simulators, manuals, etc.). The cost elements charged to production are as follows:

- Engines and avionics
- Manufacturing labor
- Material and equipment
- Airframe engineering (sustaining)
- Tooling
- Quality control
- Manufacturing facilities
- Profit

The *operations costs* comprise the fuel and oil (POL), including storage and delivery, salaries of operating and support personnel, day-to-day (direct) maintenance, depot and overhaul, spares, depreciation of equipment, and indirect costs.

For a military aircraft the breakdown of O&M costs (sometimes called O&S, operations and support) is as follows:

- **Spares.** Initial, replenishment, engines, war reserve material (WRM)
- **Maintenance.** Both on-equipment and off-equipment
- **Management personnel.** System and item managers
- **Training.** Ground and flight training
- **Operations.** Crew, commander, staff, and operations personnel
- **Support.** Base operating support (the care and feeding of all squadron personnel)
- **POL.** Fuel, oil, and lubricants
- **Modifications.** Hardware modifications
- **Munitions and missiles.** Training
- **Personnel.** Permanent change of station (PCS)
- **Attrition**
- **New facilities**

For commercial aircraft the LCC phases are similar to those of a military aircraft except that the research, development, test, and evaluation (RDT&E) phase is all privately funded. This phase ends with the type certification of the aircraft by the Federal Aviation Administration (FAA). As discussed in Chapter 1, this phase is shorter for a commercial aircraft than for a military aircraft.

The acquisition phase for a commercial aircraft is termed the *production phase*.

For commercial operations the cost breakdown is as follows:

- Flight operations—crew, POL, airport fees, insurance
- Station and ground

- Ticketing, sales, and promotion
- Maintenance and overhaul
- Flight equipment depreciation
- Passenger services
- General administration and taxes

The O&M phase costs are the largest element of the LCC because most aircraft are operational for several decades: the commercial aircraft flying 24/7 revenue flights [1] and the military manned aircraft doing peacetime training. Figures 24.1 and 24.2 show the LCC for the B-52 and the F-111, respectively—two military aircraft whose operational life was more than three decades. The B-52 LCC shown in Fig. 24.1 is typical of the military LCC history, where the RDT&E is small relative to the acquisition and operations phases and precedes the acquisition phase with only a small overlap. The F-111, Fig. 24.2, on the other hand, had technical problems during the latter part of the flight test and early operations phase, which resulted in an overlapping RDT&E phase and significant increase in RDT&E cost. The F-111 history was more the exception than the rule, involving several new technologies and occurring during a semiwartime situation. However, it points out the importance of being careful and thorough during the RDT&E phase and committing to production only after completed test and evaluation.

In both the commercial and military O&M phase, direct personnel costs are more than one-third of the costs. This points out the huge impact

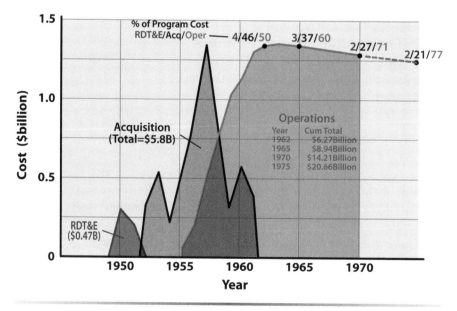

Figure 24.1   B-52 life cycle cost (LCC) (data from [2]).

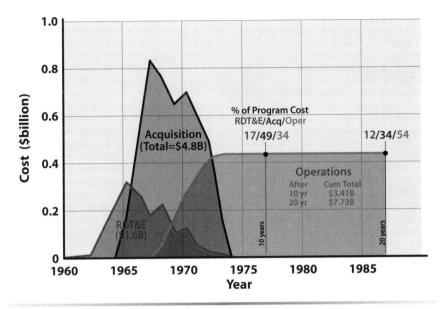

Figure 24.2   F-111 life cycle cost (data from [3]).

that the human element has on aircraft O&M and makes a good argument for military UAVs [this will be examined later in an example comparing squadron O&M costs for the F-16 and an unmanned combat aerial vehicle (UCAV)].

24.2 DT&E and Acquisition–Production Costs

The methodology presented in this section is very preliminary but is adequate for the economic analysis appropriate to this level of design. There are more-refined LCC methods available but they require information that is normally not available at the conceptual design.

Methodology for estimating the research phase costs is not available as this is a very nebulous effort and very much dependent upon the individual aircraft program. The designer should examine the design for the development status of the technologies being used, then talk with the technology community relative to the schedule and funds appropriate to these technologies.

Methodology for estimating costs for the remaining three LCC phases will be discussed in the following sections. In many cases the costs will be estimated in terms of 1998 U.S. dollars. Figure 24.3 can be used to convert the 1998 dollar costs to "now year" costs by multiplying the 1998 dollars by the economic escalation factor [Consumer Price Index (CPI)].

The *cost-estimating relationships* (CERs) for the military DT&E and production phases will be based upon the methodology developed by the

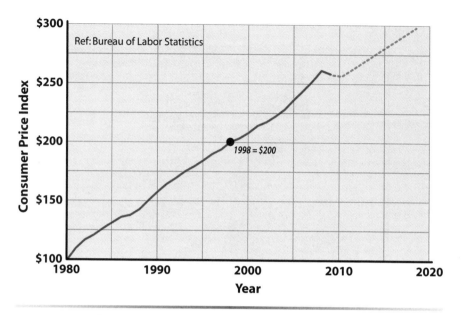

Figure 24.3   Economic escalation factors (CPI).

Rand Corporation in 1986 and presented in [4]. This was preceded by [5] in 1971; it examined 29 aircraft built between 1945 and 1970. Reference [4] added the following 13 more modern aircraft to the data base:

- **Attack.** A-6, A-7, A-10, S-3
- **Fighter.** F-111, F-4, F-14, F-15, F-16, F-18
- **Transport.** C-141, C-5
- **Utility.** T-39

The Rand study of the DT&E and production costs for aircraft built between 1945 and 1986 concluded that the primary characteristics driving these costs were as follows:

1. W, empty weight in pounds (discussed in the next section)
2. S, maximum speed in knots
3. Q, quantity of aircraft produced during DT&E and production

All other aircraft characteristics appeared to be second order.

It is worth bringing to the attention of the reader that the weight influencing the cost of the aircraft is more correctly the weight according to the American Manufacturers Planning Report (the *AMPR weight*), which is the empty weight of the aircraft minus the procured items (such as engines, wheels, instruments, and electrical equipment). To determine the AMPR weight the designer needs a detailed weight summary, which is often not available during the conceptual design phase. Typically, the AMPR weight is approximately 62% of the empty weight. In the cost methodology that follows (from [4]) the always-available empty weight is used and the 62% has been absorbed into the coefficients. Reference [5] used AMPR weight.

The reader will observe that the direct labor hours to produce an item (such as engineering, assembly, or tooling) will decrease as the cumulative number of items produced (Q) increases. The basis for this is that the personnel involved in producing the item get smarter as they produce more items. This improvement is called the learning curve. Early CERs were built upon an 80% learning curve, where the labor hours reduced by 20% every time the quantity produced doubled. Thus, the second-unit labor cost was 80% of that for the first unit, the fourth was 80% of the second, the eighth was 80% of the fourth, and soon. When large quantities of the same item are produced, the rate of improvement with respect to time may be so small as to go unnoticed. For example, if 1000 units have been produced and the production rate is 250 units per year, four years will be required to reach 2000 units: four years to double the quantity and attain a 20% reduction in labor hours. It should be noticed, however, that the 2000th unit would require 8.7% of the labor hours needed for the first unit. Thus, production runs are necessary to drive the unit costs down. If Ford only produced 50 automobiles each year, no one could afford them. It is only through mass production that could put "a car in every garage."

Reference [4] examined the cost–quantity relationship and found it to vary for the different cost elements. Thus, the CERs presented in the next sections will have different values of the cost–quantity curve slope (or exponent for Q) for each cost element. The learning curve is close to 80 for only a few of the cost elements.

24.2.1 Airframe Engineering (DT&E and Production)

The engineering activities involved in the DT&E are as follows:

1. Design studies and integration
2. Engineering for wind tunnel models, mock-ups, and engine tests
3. Test engineering, laboratory work on subsystems and static test items, and development testing
4. Release and maintenance of drawings and specifications
5. Shop and vendor liaison (*)
6. Analysis and incorporation of changes (*)
7. Materials and process specifications (*)
8. Reliability (*)

The starred items (*) are also part of the sustaining engineering effort production. Engineering hours not directly related to airframe design and development are not included here. For example, test engineering, ground handling equipment design and development, mobile training units, and publications are not charged to airframe engineering.

The cumulative total *airframe engineering hours* E can be estimated using the following expression (from [4]):

$$E = 4.86 \ W^{0.777} \ S^{0.894} \ Q^{0.163} \tag{24.1}$$

where

W = empty weight in pounds
S = maximum speed (kt) at best altitude
Q = cumulative quantity produced
   = Q_D for DT&E phase (number of development flight test aircraft)
   = $Q_D + Q_P$ for production phase

The empty weight W is defined as the takeoff gross weight (TOGW) minus the fuel and payload. Said another way, the empty weight of the aircraft is the sum of the (1) airframe structure and canopy, (2) wheels, brakes, and tires, (3) engines and accessories, (4) cooling fluid, (5) rubber or nylon bladder type fuel cells, (6) crew seats and instruments, (7) batteries, electrical power supply, and conversion–conditioning equipment, (8) electronic and avionics equipment, (9) armament and fire-control system, (10) air conditioning units and fluid, (11) onboard power plant unit, and (12) trapped fuel and oil.

Equation (24.1) gives the total engineering hours for either DT&E or production. For the DT&E phase the quantity Q is equal to the number of flight test aircraft Q_D and the engineering hours are just for DT&E. For the production phase, the quantity Q is the total produced (Q_D plus the Q_P production aircraft). The production phase sustaining engineering hours are the hours from Eq. (24.1) minus the DT& E engineering hours (see Example 24.1 in Section 24.2.8).

The hours from Eq. (24.1) are then multiplied by an appropriate engineering dollar rate for the year of interest. This rate includes direct labor, overhead, general and administrative expense, and miscellaneous direct charges. Figure 24.4 presents historical data on average hourly rates, created with data from the U.S. Department of Labor. These labor rates can be estimated from the consumer price index (CPI), which is available from the U.S. Department of Labor, Bureau of Labor Statistics Web site.

24.2.2 Development Support (DT&E)

Development support is defined as the nonrecurring manufacturing effort undertaken in support of engineering during the DT&E phase of an

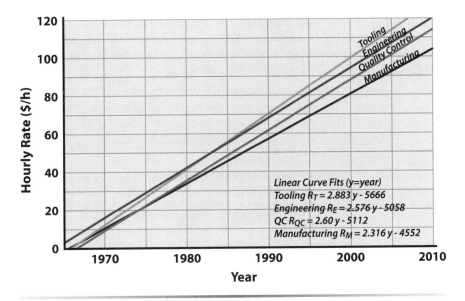

Figure 24.4   Trends in hourly rates in aircraft construction for engineering, tooling, manufacturing, and quality control.

aircraft program. The cost of the development support is the cost of manufacturing labor and material required to produce mock-ups, test parts, static test items, and other items of hardware that are needed for airframe design and development work. The level of this effort is largely dependent upon the extent that new technologies figure into the aircraft program. If the aircraft design involves new and untried concepts, then the development support cost can be high. For example, the KC-135 was largely a derivative of the Boeing 707, and the development support cost was $(1957)37 million, whereas the F-111A incorporated several new and untried technologies and its development support cost was $(1965)178 million.

The development support cost can be estimated using (from [4])

$$D = 66 \ W^{0.63} \ S^{1.3} \tag{24.2}$$

where

D = development support cost in 1998 constant dollars
W = empty weight, in pounds (lb)
S = maximum speed, in knots (kt), at best altitude

24.2.3   Flight Test Operations (DT&E)

The *flight test operations* cost element includes all costs incurred by the aircraft builder to carry out flight tests except the cost of the flight test aircraft. It includes flight test engineering planning and data reduction, manufacturing support, instrumentation, spares, fuel and oil, pilot's pay, facilities rental, and insurance. The flight test establishes the operating envelope of the aircraft, its flying and handling qualities, general airworthiness, initial maintainability features, and compatibility with ground support equipment. Civil and commercial aircraft are establishing the aircraft's compliance with the FARs for airworthiness certification. Military aircraft are demonstrating compliance with government specifications and regulations, such as Air Force Regulation 80–14.

The cost for flight test operations can be estimated using (from [4])

$$F = 1852 \ W^{0.325} \ S^{0.822} \ Q_D^{1.21} \tag{24.3}$$

where F is the flight test operations cost in 1998 constant dollars and W, S, and Q_D are as defined in Eq. (24.1).

24.2.4 Tooling (DT&E and Production)

Tools are the jigs, fixtures, dies, and special equipment used in the fabrication of an aircraft. *Tooling hours* are defined as the hours charged for tool design, tool planning, tool fabrication, production test equipment, checkout of tools, maintenance of tooling, normal changes, and production planning. Tooling hours are dependent upon a new variable called *production rate*. Tools designed for low production rates do not have to be as well engineered as tools for high production rates. Sometimes tools are destroyed during the fabrication process (called *soft tooling*) and have to be rebuilt for each aircraft. Tooling can be as simple as 2 × 4s or as complicated and costly as matched metal dies of stainless steel accurate to one ten-thousandth of an inch (1/10,000 in.).

The tooling hours can be estimated using the following expression (from [4]):

$$T = 5.99 \ W^{0.777} \ S^{0.696} \ Q^{0.263} \tag{24.4}$$

where

T = cumulative tooling hours
Q = cumulative quantity, $Q_D + Q_P$
Q_D = DT&E
Q_P = production

The *total tooling cost* is the tooling hours multiplied by an appropriate tooling hourly rate. Figure 24.4 shows some historical data on average hourly tooling rates.

Equation (24.4) gives the total tooling hours for either DT&E or production. For the DT&E phase tooling hours, the quantity Q is Q_D. For the production phase, the quantity Q is Q_D plus the Q_P production aircraft and the tooling hours are the hours from Eq. (24.4) minus the DT&E tooling hours.

24.2.5 Manufacturing Labor (DT&E and Production)

Manufacturing labor hours include those hours necessary to fabricate, process, and assemble the major structure of an aircraft, and to install purchased parts, government furnished equipment (GFE), and off-site manufactured assemblies (i.e., subcontract components). Airframe structure direct manufacturing man-hours also include effort on those parts that, because of their configuration or weight characteristics, are design controlled for the basic aircraft. These normally represent significant proportions of the airframe weight and manufacturing effort, and they are included regardless of their method of acquisition. Such parts specifically include [4] the following:

1. Actuating hydraulic cylinders
2. Radomes, canopies, and ducts
3. Passenger and crew seats
4. Fixed external tanks

The manufacturing labor hours can be estimated using the expression (from [4])

$$L = 7.37 \ W^{0.82} \ S^{0.484} \ Q^{0.641} \tag{24.5}$$

where L is cumulative total manufacturing hours, and W, S, and Q are defined in Eq. (24.1). The manufacturing labor hours for DT& E and for production are determined separately as discussed in Sections 24.2.1 and 24.2.4. The cumulative manufacturing cost is obtained by multiplying the manufacturing labor hours L by a representative hourly rate. Figure 24.4 gives representative average hourly rates for manufacturing.

24.2.6 Quality Control

Quality control (QC) is the task of inspecting fabricated and purchased parts, subassemblies, and assembled items against material and process standards, drawings, and/or specifications. Quality control is an extremely important activity in the manufacture of aircraft because of their complexity. Government specifications and standards require close inspection of all facets of fabrication. Quality control is closely related to direct manufacturing labor and is considered to be a percentage of the labor hours. The quality control hours can be estimated as follows (from [4]):

$$QC = 0.076 \, L \text{ for cargo and transport aircraft} \qquad (24.6a)$$

$$QC = 0.13 \, L \text{ for all other aircraft} \qquad (24.6b)$$

The total cost for quality control is obtained by multiplying the man-hours from Eq. (24.6) by the representative manufacturing hourly rate.

24.2.7 Manufacturing Material and Equipment (DT&E and Production)

The *material and equipment list* (sometimes called the BOM, bill of materials) includes the raw material, hardware, and purchased parts required for the fabrication and assembly of the airframe. All airframe equipment except engines and avionics are included in this cost item. Specific items in the material and equipment cost are as follows:

1. Raw materials in sheets, plates, bars, rods, and so on
2. Raw castings and forgings
3. Wires, cables, fabrics, tubing, windshield glass and canopies, and so on
4. Fasteners, clamps, bushings, and so on
5. Hydraulic and plumbing fittings, valves, and fixtures
6. Standard electrical products such as motors, transformers, inverters, alternators, voltage regulators, switches, controls, generators, batteries, and auxiliary power units (APUs)
7. Pumps for fuel, hydraulic, water, and so on
8. Environmental systems, air conditioning, and oxygen equipment
9. Crew furnishings, seats, instruments, bunks, and so on
10. Bladder-type fuel tanks

The manufacturing material and equipment costs can be estimated from the following expression [4]:

$$M = 16.39 \, W^{0.921} \, S^{0.621} \, Q^{0.799} \qquad (24.7)$$

where M = cumulative total manufacturing material and equipment cost in 1998 dollars and W, S, and Q are defined in Eq. (24.1). The costs for DT&E

and production are determined separately using development $Q = Q_D$ and production $Q = Q_P$ (see example in Section 24.3).

24.2.8 Engine and Avionics Costs

The engine and avionics will be assumed to be off-the-shelf items so that DT& E costs of these subsystems will not be considered. Only production unit costs are considered.

Costs in 1998 dollars for current turbine engines are shown in Fig. 24.5. Figure 24.5 shows quite a bit of scatter in the data. This scatter is explained by the fact that the engines represent different types, levels of technology, and production quantities. More refined propulsion cost methodology would take these variables into consideration [6]. However, at this point in the design, the data from Fig. 24.5 or the following expression is adequate:

$$P = 2306 \, [0.043 \, T_{\text{SLS}} + 243.3 \, M_{\text{max}} + 0.969 \, T_R - 2228] \qquad (24.8)$$

where

P = production engine unit cost in 1998 dollars
T_{SLS} = sea level maximum thrust in pounds
M_{max} = maximum Mach number
T_R = turbine inlet temperature in degrees absolute (Rankine)

Avionics equipment is so varied that space will not be taken here to list avionics gear and associated costs. The designer is referred to the avionics vendors for prices of selected avionics equipment.

Figure 24.6 presents unit prices [$(1993)] for existing fighters, bombers, transports, bizjets, cruise missiles, and targets. The charts confirm that aircraft are bought by the pound. The prices can be adjusted to reflect any year by ratioing the escalation factors from Fig. 24.3.

Unit prices are not the only thing taken into account when determining an aircraft's selling price. Often the selling price will include initial spares for initial fleet operation, data and publications, and flight training for the pilots and maintenance training for the ground crews. These "extras" can easily be 10% of the selling price.

The cost-estimating relations presented in Sections 24.2.1 through 24.2.8 will be demonstrated by estimating the cost for the Boeing 777-200LR. This will be a good checkout for the CERs as they were developed from a military aircraft data base.

Note the cost numbers in this example: it is a major decision for a company to commit to an $8 billion DT&E cost for a new aircraft line.

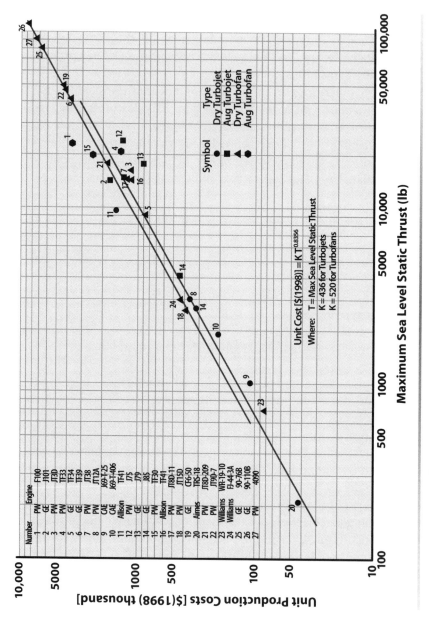

Figure 24.5 Engine production unit costs in 1998 dollars.

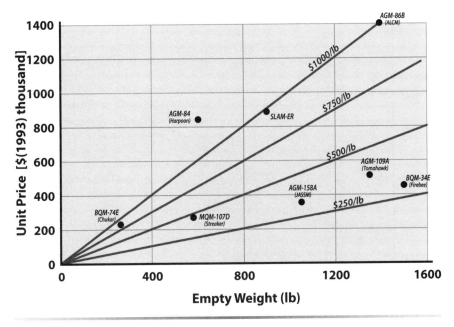

Figure 24.6a Unit prices for target and cruise missiles.

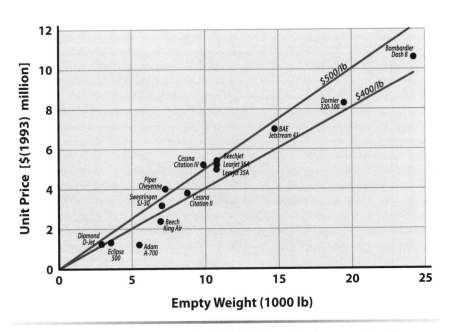

Figure 24.6b Unit prices for light bizjets and turboprop aircraft.

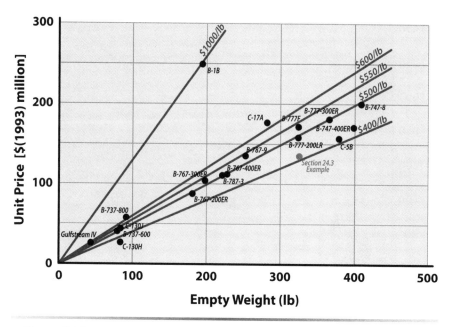

Figure 24.6c   Unit prices for medium and large transports and bombers.

Example 24.1   DT&E and Acquisition Cost of the Boeing 777

Estimate the cost of the Boeing 777 (called the triple seven). It was designed and developed between October 1990 and October 1994. The first flight of the 777-200 was June 1994 and the aircraft became

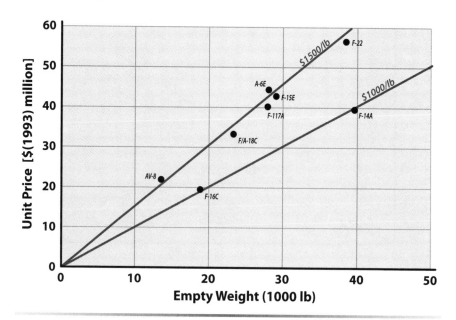

Figure 24.6d   Unit prices for fighter aircraft.

operational with United Airlines in June 1995. The aircraft received its FAA and Joint Aviation Authority (JAA, the European FAA) certificates in April 1995.

The 777-200LR (LR for longer range) became the world's longest unrefueled range commercial airliner when it entered service in 2006. In November 2005, a 777-200LR flew 11,664 n mile on a special flight from Hong Kong to London, a new world record. The aircraft can carry 440 passengers in an economy-class arrangement. The 777 family features a digital fly-by-wire flight control system, a supercritical airfoil on a wing swept 31.6 deg, 9% of the structural weight is composite materials, and the largest landing gear and tires ever used on a commercial aircraft. The aircraft was designed entirely on a computer. All design drawings were created on 3-D CAD software system known as CATIA. The aircraft was entirely "paperless." As of November 2009 Boeing had delivered 816 aircraft in all models.

The information needed for developing the selling price for the 777-200LR is as follows (at this point it is recommended that the reader review the Boeing 777 case study in Volume 2):

Time frame for costing	1998
TOGW	766,000 lb
Empty weight	326,000 lb
Maximum speed	510 kt (Mach 0.89)
Engines	two GE 90–110B (T_{SLS} = 110,000 lb)
Avionics cost	$250,000 (estimated)
Flight test aircraft number	9
Production quantity for costing	500 units
Labor rates, dollars per hour ($/h) for 1998	
Engineering R_E	$88.85
Tooling R_T	$94.23
Manufacturing R_M	$75.37
Quality control R_{QC}	$82.80

Engineering hours	
	$E = 4.86\ W^{0.777}\ S^{0.894}\ Q^{0.163}$ (24.1)
Development	$Q_D = 9$
	$E_D = 35{,}201{,}600$ h
Cost	$3,127,664,790
Production	$Q_P = 500 + 9 = 509$
	$E_P = 67{,}909{,}467 - 35{,}201{,}600 = 32{,}707{,}867$ h
Cost	$2,906,093,988

Development support	
	$D = 66\ W^{0.63}\ S^{1.3}$ (24.2)
	$D = \$666{,}235{,}760$
Flight test operations	
	$F = 1852\ W^{0.325}\ S^{0.822}\ Q_D^{1.21}$ (24.3)
	$F = \$275{,}407{,}260$
Tooling	
	$T = 5.99\ W^{0.777}\ S^{0.696}\ Q^{0.263}$ (24.4)
Development	$Q_D = 9$
	$T_D = 15{,}718{,}000$ h
Cost	$\$1{,}481{,}107{,}000$
Production	$Q_P = 9 + 500 = 509$
	$T_P = 45{,}419{,}128 - 15{,}718{,}000 = 29{,}707{,}000$ h
Cost	$\$2{,}799{,}324{,}000$
Manufacturing labor	
	$L = 7.37\ W^{0.82}\ S^{0.484}\ Q^{0.641}$ (24.5)
Development	$Q_D = 9$
	$L_D = 20{,}437{,}778$ h
Cost	$\$1{,}540{,}395{,}000$
Production	$Q_P = 9 + 500 = 509$
	$L_P = 271{,}521{,}328 - 20{,}437{,}778 = 251{,}062{,}540$ h
Cost	$\$18{,}922{,}583{,}650$
Quality control	
	$QC = 0.13\ L$ (24.6b)
Development	$Q_{CD} = 2{,}657{,}000$ h
Cost	$\$219{,}992{,}000$
Production	$Q_{CP} = 32{,}638{,}000$ h
Cost	$\$2{,}702{,}431{,}000$
Material and equipment, in 1998 dollars [\$(1998)]	
	$M = 16.39\ W^{0.921}\ S^{0.621}\ Q^{0.799}$ (24.7)
Development	$Q_D = 9$
	$M_D = \$544{,}418{,}700$
Production	$Q_P = 500$
	$M_P = \$13{,}682{,}352{,}500$
Engine cost [\$(1998)]	
Engines	Maximum $T_{SLS} = 110{,}000$ lb
Unit cost	$\$8{,}483{,}800$ (from Fig. 24.5)
Development	Assume three per aircraft
Cost	$\$229{,}041{,}000$
Production	two per aircraft
Cost	$\$8{,}483{,}000{,}000$

Avionics cost [$(1998)]: $250,000	
Development	Cost = $2,250,000
Production	Cost = $125,000,000
Total DT&E cost [$(1998)]	
Airframe engineering	$3,127,665,000
Development support	$666,236,000
Flight test aircraft	$4,017,204,000
Engines	$229,041,000
Avionics	$2,250,000
Manufacturing labor	$1,540,395,000
Material and equipment	$544,418,700
Tooling	$1,481,107,000
Quality control	$219,992,000
Flight test operations	$275,407,000
Test facilities	$0
Subtotal	$8,086,512,000

If this were a contracted effort there would be a profit added onto the DT&E cost. Because this is a private company project there is no profit. Assume no special test facilities were built for this program.

This DT&E cost is amortized over some number of units. Assume it to be amortized over 250 units so that the price per aircraft will be increased by $32,346,000.

Total production and unit cost [in $(1998)]	
Engines	$8,483,000,000
Avionics	$125,000,000
Manufacturing labor	$18,922,584,000
Material and equipment	$13,682,353,000
Sustaining engineering	$2,906,094,000
Tooling	$2,199,324,000
Quality control	$2,702,786,000
Manufacturing facilities	0
Subtotal for 500 aircraft	$49,020,786,000

Assume for this discussion that there were no new manufacturing facilities needed for the production of the 777-200LR. This was not the case, as new facilities were built for the 777 family—but this information was not available for this example.

The unit cost is the total production cost divided by 500 aircraft plus the amortized cost: The unit cost = $130.4 million.

The unit price is the unit cost plus the profit on each aircraft. Assuming a 15% profit the unit selling price is $(1998)150 million. Boeing quotes a 2008 selling price for the 777-200LR of $237.5 million. Adjusting our estimated selling price to 2008 dollars (using the economic escalation factors on Fig. 24.3) gives (260/200) = $194million, or a 22% cost difference. In the world of cost estimating a difference of 22% is considered quite close.

24.3 Operations and Maintenance Phase

The O&M costs are based upon a period of operation, usually 10 years. A fleet size and number of flying hours (FH) per year are estimated.

The aircraft operating characteristics are known at this point so that an average fuel flow per hour, in gallons per hour, can be determined. At the time of this writing the fuel prices were in a state of random motion. The designer should obtain current and projected fuel prices from petroleum vendors and then determine the operating fuel costs. The oil and lubricant costs are less than 0.5% of the operating fuel costs and could be neglected in the total POL costs.

Each fleet of aircraft has a crew ratio that varies with the type of aircraft and the utilization rate. Table 24.1 gives information on crew ratios for different aircraft and annual flying rates. Salaries for these personnel are estimated and the aircrew costs determined.

The direct maintenance personnel costs are best determined using the *maintenance-man-hours per flying-hour* (MMH/FH). Table 24.2 gives MMH/FH for current aircraft. This ratio varies with the type of aircraft, the mission or sortie length, the utilization rate, and the years-in-service of the aircraft. The MMH/FH decreases with increased sortie length because the takeoffs and landings are harder on the aircraft than cruising flight. In addition, maintenance cannot be performed on a failed item until the aircraft lands; when the aircraft is flying it continues to accumulate flying hours.

Table 24.1   Crew Ratio for LCC Planning

Aircraft Type	Flying Hours	Ratio
Transport	Less than 1200	1.5
Transport	1200–2400	2.5
Transport	2400–3600	3.5
Bomber	500	1.5
Fighter	500	1.1

Table 24.2   LCC Planning Data

Aircraft	Average Annual FH per Aircraft	MMH/FH	Year
Cessna 150/172		0.3	1974
Cessna Skywagon		0.5	1974
Beech Kingair		1.0	1974
Citation II		3.0	1988
T-37		7.8	1981
T-38	400	10	1981
T-39	600	9.8	1974
T-43	700	10	1974
F-5E	410	17	1981
A-7D	300	25	1974
A-10A	300	13	1984
F-14	314	48	1988
F-15C	302	22	1998
F-16C	346	19	1998
F-18C	360	18	1988
F-4E	302	33	1981
F-105G	316	58	1974
F-111D	280	40	1974
F-117A	—	113	1983 (IOC)
F-117A	—	45	2003
F-22A	316	10.5	2009
B-2A	—	124	1997 (IOC)
B-2A	—	51	2002
B-2A	—	32	2004
C-17	780	24	2005
C-17	780	20	2007
C-17	780	16	2008
C-5B	716	58	2005
C-5B	716	41	2007
C-5B	716	33	2008
C-130E	720	20	1974
C-141B	1080	21	1981
B-52D	424	37	1981
B-52G	516	49	1981
B-58A	430	54	1974
KC-135	377	27	1974
L1011	1870	14.1	1981

(continued)

Aircraft	Average Annual FH per Aircraft	MMH/FH	Year
DC-10-10	2450	11	1981
B727-100	2670	8	1974
B727-200	2800	6.5	1974
B737-200	2200	6.6	1974
B747	3525	14.5	1981
B757	3010	9.1	1998
B767	3010	11.4	1998
B777	3010	10.2	1998
SR-71	260	~400	1981

Data sources:
General Aviation—Cessna and Beech Aircraft
Military—AFM-173-10, 3M data and U.S. Air Force and U.S. Navy maintenance records
Commercial—CAB Form 41, Boeing Airplane Co.
Blackbird—Lockheed SR-71 Researcher's Handbook

Notice the big difference in annual flying hours for the military and commercial. This is because the commercial aircraft are losing money when they are sitting on the ground. Thus, the commercial aircraft average about 14 hours in the air each day for the long-range transports and 10–12 hours for the shorter route aircraft. In the military each pilot needs about 260 flying hours per year to stay proficient. Using the crew ratios from Table 24.1, this gives about 300 h for fighters and 400 h for bombers per year, or about one hour per day.

The *utilization rate* (flying hours per period of time) also affects the MMH/FH. This reduction in man-hour requirement with increased utilization can be explained as follows. Aircraft systems, used daily, normally receive better upkeep and experience fewer failures per flight hour. Also, aircraft that fly frequently are on the ground less time and require maintenance to be accomplished in a limited amount of time. Because of this pressure, maintenance is accomplished more efficiently and frequently by personnel with higher skill level. Maintenance personnel can more easily retain knowledge of failures and maintenance accomplished the day before, hence there is better continuity between maintenance tasks [3].

The MMH/FH are not static but vary with the point in the aircraft's service life. The MMH/FH decreases from the initial deployment to a point where the aircraft is a mature, well-understood member of the fleet and then starts increasing as the aircraft begins to wear out. Two good examples are the F-117A and B-2A. Both aircraft had a very high MMH/FH at initial operational capability (IOC) due largely to the new stealth technologies on the aircraft. As the maintenance crews became familiar with the aircraft and the new low-observable (LO) materials the MMH/FH dropped dramatically.

The data in Table 24.2 are for typical sortie lengths and several years-in-service. The maintenance personnel costs are determined from an estimated MMH/FH, the maintenance personnel hourly rates and the annual flying hours.

24.4 O&M Costs

The elimination of peacetime flying (or at most minimal flying) would result in a large O&M cost saving for a UCAV relative to a manned fighter squadron. The following example is notional and is used to develop values for the various O&M cost elements. The example indicates that the cost savings in annual O&M for the UCAV could be greater than 80%. This O&M cost saving needs to be quantified with a careful and thorough study that examines the peacetime training (of both operators and ground crew) and the design impact of long-term flyable storage.

Example 24.2 Comparison of O&M Costs for Manned and Unmanned Tactical Fighter Squadrons

Manning of a 24-aircraft UCAV squadron is shown in Table 24.3 and compared with a 24-aircraft F-16C squadron. Both squadrons are air-to-ground strike/SEAD (suppression of enemy air defenses) units. The costs shown are annual O&M during peacetime. The number of officers (primarily pilots or remote operators) is about the same for the two squadrons, but the number of enlisted personnel doing maintenance and support is very much less for the UCAV squadron. The UCAVs would be stored in a humidity-controlled, flyable storage facility. To ensure readiness, four UCAVs would be taken out of storage each year and flown for manned aircraft interface and maintenance–support crew training, The total squadron flying hours would be about 140. In contrast, a 24-unit F-16 squadron would fly about 8300 h each year. The composition of the enlisted personnel are 15 ground support crew (6 weapons handlers, 7 vehicle support, 1 chief, and 1 administration), 10 technicians (ground control station and associated avionics maintenance), and 7 administration and support. During wartime the active duty ground support crew would be augmented by 4 reserve crews to support a tempo of four sorties per day for 30 days. During peacetime the active-duty ground support crew would train the 4 reserve crews, support the limited UCAV flying, and maintain the aircraft in flyable storage.

Table 24.3 UCAV Peacetime O&S Cost Compared with an F-16C Squadron [$(2002)Million]

F-16C Annual O&S (per AFI 65–503)		UCAV Annual O&S (per modified AFI 65–503)	
Unit personnel (42 off./307 enl.)	$15.7M	Unit personnel (30 off./32 enl.)	$3.6M
Fuel for 8300 flying hours	5.5M	Fuel for 140 flying hours	0.09M
Base support personnel	10.1M	Base support personnel	1.4M
Depot maintenance	6.7M		
Training and personnel acquisition	5.3M	Training and personnel acquisition	0.69M
Replenish spares	6.6M		
System support and mods.	4.3M	System support and mods.	0.66M
Munitions and missiles	1.2M		
Total	**$55.4M**		**$6.44M**

24.5 Design for Reduced Cost

The designer must recognize and appreciate the fact that he has a powerful influence over the life cycle cost of an aircraft system. The major portion of the LCC is locked in at the conceptual and early preliminary design phases, because it is during these early design phases that the aircraft is taking on its shape and size. Once the design is in preliminary design, details are being fine tuned and all the gross features, good and bad, are already locked in. This argument was presented in Fig. 1.16 and will be developed in the following paragraphs.

The designer influences the RDT&E costs directly by the choice of new technologies to be incorporated in the new design. The selection of new technologies that are not quite mature (i.e., ready for system application) can cause this cost to skyrocket. The F-111 is an example of incorporating new technologies that needed more research before moving them into the DT&E phase. The F 111A cost for development support and engineering was more than any other U.S. Air Force production fighter. The technologies should be fully demonstrated and validated before putting them on a new aircraft.

24.5.1 Design for Production

The key to reducing production costs is to reduce the "touch labor." The designer has more influence over this than any other person. Some design guidelines for reducing production costs are as follows:

1. Minimize the part count; this in turn reduces the tooling, fabrication, and assembly time, which reduces touch labor.
2. Standardize left and right tooling; this is another way to keep the part count down. Examples would be interchangeable right and left ailerons, main landing gears, and horizontal tails.
3. Require structural parts to perform multiple functions. An example would be the main landing gear mounted to the wing carry-through structure.
4. Use large unitary pieces of structure rather then build up the structure from many smaller pieces. This reduces touch labor and is ofttimes the rationale for using composites (large co-cured pieces) rather than metal built-up parts.
5. Minimize complex checkout.
6. Combine engineering and quality testing.
7. Use simple curvature shapes; the use of compound curvature surfaces greatly increases the tooling and fabrication time.
8. Use simple and common parts; use parts that are common to other aircraft such as landing gears, crew furnishings, and equipment.
9. Use state-of-the-art materials and structures design; this means the use of technology demonstrators during the research phase to fully develop and validate materials and structural concepts before committing them to the aircraft.
10. Use proven engines and inlet–nozzle configurations.

The overall design rule is "Keep It Simple."

24.5.2 Design for O&M

The best thing that a designer can do for reduced O&M costs is to design for quick and easy access to everything. This is difficult and it means a far from optimum packaging in the fuselage and wings. However, a slightly larger and roomier fuselage, although weighing more and giving lower performance, will pay for itself in reduced MMH/FH. The MMH/FH is a direct function of accessibility (getting to the faulty or suspicious item), complexity of the system, and ease of component removal. The designer should recognize that

• Avionics equipment is always going to need attention
• Hydraulic systems are going to leak
• Fasteners are going to "unfasten"
• Mechanisms are going to wear out and/or need adjusting

so design for the situation. The location of most of the components and the roominess of the equipment bays are locked-in by the conceptual and early preliminary design.

The early McDonnell F-4s had some communications avionics located beneath the rear seat. Every time the communications gear needed adjusting (which was about every 3 sorties) the rear ejection seat needed to be removed and then replaced. This poor design added 2.3 maintenance hours just to gain access to the avionics equipment [7]. A tightly packed fuselage might be elegant from a design viewpoint but it is a nightmare for the ground crew as they often have to remove good equipment just to gain access to a faulty piece of equipment. A good design rule is to only package equipment "one deep."

The McDonnell F-4 and F-15 are aircraft of similar size and weight. The F-15, designed in the early 1970s, emphasized reduced MMH/FH. The primary design solution was to improve the accessibility of the F-15 over that of the F-4. The result was 570 ft^2 of access doors and panels on the F-15 compared with 55 ft^2 on the F-4. This feature was largely responsible for the MMH/FH being reduced from 33 for the F-4 to 22 for the F-15.

The painful compromise that a designer must make between performance and cost must surely be evident by now. It is paramount that the designer appreciate the importance of cost, otherwise Calvin Coolidge's recommendation of "buy one aircraft and let the aviators take turns flying it" may someday become a reality.

References

[1] "LCC Breakdown," *Aviation Week and Space Technology*, 22 Oct. 2007, p. 23.
[2] Reel, R. E., Totey, C. E., and Johnson, W. L., "Weapon System Support Cost Reduction Study (U)," Aeronautical Systems Div., Deputy for Development Plans, ASD /XR Rept. 72–49, Wright–Patterson AFB, OH, June 1972.
[3] Johnson, W. L., and Reel, R. E., "Maintainability/Reliability Impact on System Support Costs," U.S. Air Force Flight Dynamics Laboratory, AFFDL/PTC, AFFDL-TR-73-152, Wright–Patterson AFB, OH, Dec. 1973.
[4] Hess, R. W., and Romanoff, H. P., "Cost-Estimating Relationships for Aircraft Airframes," Rand Rept. R-3255-AF, Rand Corporation, Santa Monica, CA, Dec. 1987.
[5] Levenson, G. S., Boren, H. E., Tihansky, D. P., and Timson, F., "Cost Estimating Relationships for Aircraft Airframes," Rand Rept. R-761-PR, Rand Corp., Santa Monica, CA, Dec. 1971.
[6] Large, J. P., "Estimating Aircraft Turbine Engine Costs," Rand Corp. Rept. RM-6384/1-PR, Sept. 1970.
[7] "R&M Proof," *Aerospace Daily*, Vol. 135, No. 15, 23 Sept. 1985, p. 113.

Chapter 25 Trade Studies and Sizing

- Trade Studies
- Carpet Plots
- Knothole Plots
- Risk Assessment
- Risk Mitigation
- Kelly's 14 Rules

Kelly Johnson was perhaps the greatest airplane designer of the 20th century. His legendary designs included among many others the P-38, XF-90, F-94, F-104, F-117A, C-130, U-2, SR-71, and D-21 drone. His famous 14 rules of management were the forerunners of "concurrent engineering" and are summarized at the end of this chapter.

Be quick, be quiet, and be on time!
Clarence "Kelly" Johnson

25.1 Introduction

A re we done yet? The answer is No. We have essentially completed one iteration for a baseline configuration as shown in Fig. 25.1. The design may meet the mission requirements, or exceed some and fall short on others. In any event the designer, being very close to the design, has definite feelings on what should be changed to make the design better. The designer is now ready to start another iteration, hoping to make the estimates of aerodynamics, weight, propulsion data, and performance more refined.

There are three major trade studies shown in Fig. 25.1 that the designer needs to conduct:

1. **Design.** Helps the designer select the best combination of design features to meet the measures of merit (MoM)
2. **Mission.** Indicates the sensitivity of the baseline design to changes in the mission requirements
3. **Technology.** Indicates the sensitivity of the baseline design to the selected technologies and forms the basis for the risk assessment

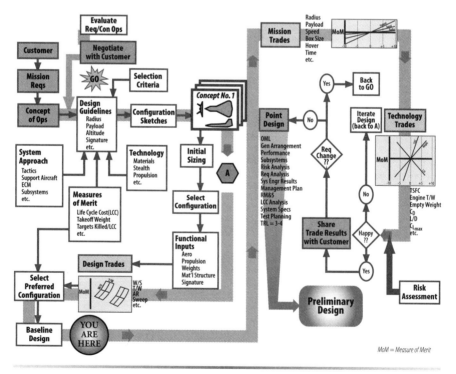

Figure 25.1 What happens after the first design iteration?

The results of the trade studies are extremely important as they indicate the sensitivity of some measure of merit to changes in the design parameters, mission requirements, and technologies. These measures of merit are usually one or more of the following:

- **Takeoff weight.** Indicates the general vehicle size and hence cost and energy requirements
- **Cost.** The total life cycle cost (LCC) over a fixed period such as 10 years; tradeoff between RDT&E, acquisition, and O&M costs
- **Energy.** Total fuel required for mission
- **System effectiveness.** Some parameter that combines performance, cost, and/or energy, such as the following:
  - Return on investment (ROI)
  - Bombs on target per hour per dollar
  - Kill ratio per aircraft dollar
  - Survivability
  - Transport direct operating cost (DOC)
  - Energy effectiveness parameter

25.2 Carpet Plots and Knotholes

The number of variables that might be considered in a tradeoff study may be less than 10 or more than 50. The designer has the difficult task of sorting through all combinations in a systematic fashion to find the best combination. These data should also be used to visually explain why certain design decisions were made so internal and external managers understand why the final design looks the way it does. Sometimes the designer might want to display several parameters on the trade study charts.

Figure 25.2 shows an example of examining the three design variables T/W, W/S, and aspect ratio (AR). The mission requirement calls for a deep strike interdiction fighter with a payload of 4500 lb and a mission radius of 400 n mile. The fighter also has the acceleration requirement of P_S = 700 fps at M = 1.6/35,000 ft and a maximum sustained maneuver load factor of 4.5 g at M = 0.9/20,000 ft. The takeoff gross weight (TOGW) is the measure of merit for this example. The AR is held constant and the T/W and W/S are iterated to give the minimum TOGW vehicle that just meets the mission. The design cycle is then repeated for other values of AR. The minimum TOGW for each AR is then plotted versus AR to find the best AR for the interdiction fighter. Admittedly, the computer must be used to perform the design iterations; however, the designer is in the loop to assess the results and make the final design selection. The computer cannot be asked to select the final design as some measures of merit are qualitative and will often change with time. The designer must be aware of this and

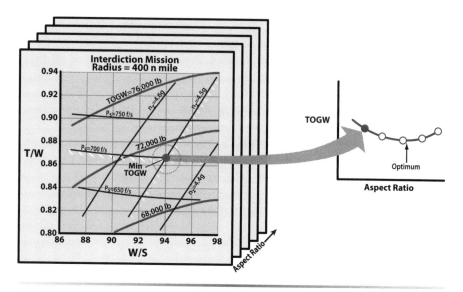

Figure 25.2 Parametric tradeoff showing a three-variable example of wing loading, thrust-to-weight ratio, and aspect ratio.

project into the future as best as he can. Alan Mullaly (Project Manager for the Boeing 777 and later Boeing CEO) said it best: "Planes are made by people not computers [1]."

Sometimes during a parametric trade study, the data on the charts become dense and tend to overlap. This makes interpolation difficult. One way of spreading the data of more than two variables apart for better visibility and still providing direct interpolation is to present the data on *design carpet plots*. A design carpet plot buildup is demonstrated in Fig. 25.3 using a Navy multimission fighter–attack aircraft.

Figure 25.3a shows the relative interdiction mission gross weight required to fly the desired radius with a constant W/S of 100 psf, with T/W varying from 0.35 to 0.45. The relative gross weights required to fly this constant radius interdiction mission at other wing loadings can be presented by shifting the abscissa and plotting a second wing loading on the new shifted scale as shown in Fig. 25.3b. Additional wing loadings can be added in the same manner and points of constant T/W are connected to form a final design carpet plot as shown in Fig. 25.3c, where the abscissa scales have been eliminated. By interpolation between the curves the relative interdiction mission gross weight can be determined for any combination of W/S and T/W. Other information could be presented in Fig. 25.3c by superimposing lines of constant design characteristics. For example, Fig. 25.3d shows lines of constant relative acquisition cost. Other design constraint lines can be added such as approach speed, takeoff field length, and airport noise.

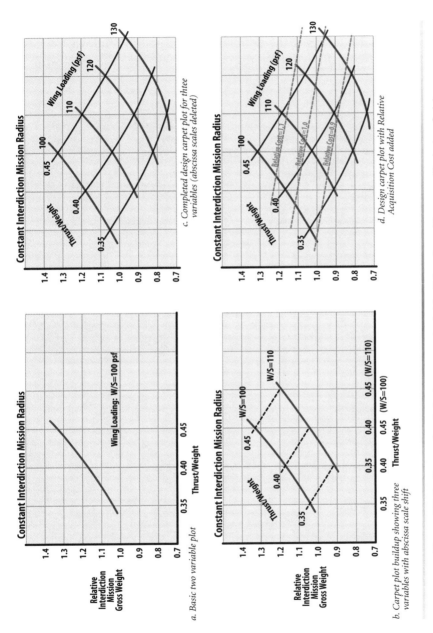

Figure 25.3 Example of design carpet plot buildup for a Navy multimission fighter.

The trade space data just discussed can also be presented in a format that is capable of illustrating an entire trade study on a single plot. This plot is called a *knothole* based on its usual form. Figures 25.4 and 25.5 summarize the entire process used to create a complete "knothole" for a commercial transport. The benefits of putting the data in this format are that it clearly communicates where the optimum design point is and what constraints are preventing the optimum from being selected. The extra work

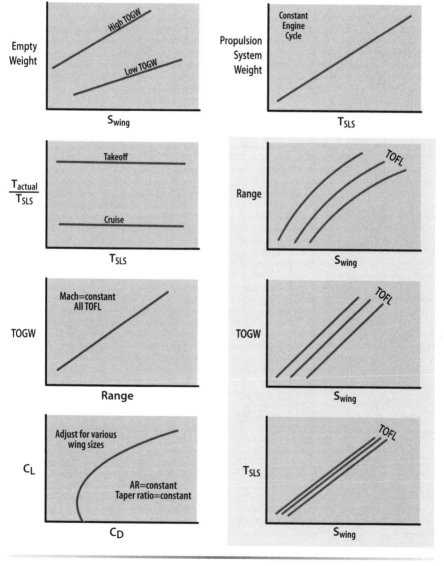

Figure 25.4a Performance trade results used to construct "knotholes."

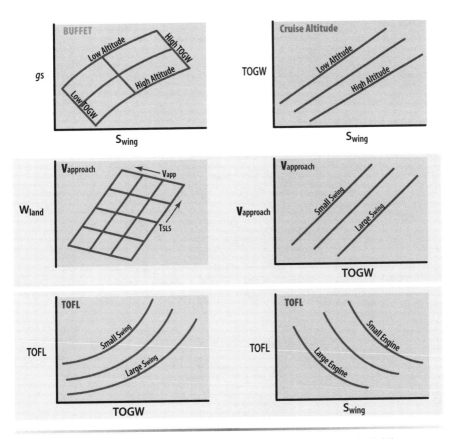

Figure 25.4b Performance results needed to draw constraint lines on "knotholes."

involved to generate this plot is repaid many times over because it distills numerous technical issues into a form that is easily understandable by customers, management, and nontechnical program personnel.

The included example is for defining an optimum large-scale commercial transport that has a size somewhere between an L-1011 and a B-747. Figures 25.4a and 25.4b summarize all of the data that must be accumulated. This data is a mix of experimental, analytical, and sometimes "best guess" information. The data plots are unique to a commercial transport and differ significantly from that of a military fighter or bomber. Here the constraints were noise, buffet, approach speed, and takeoff and landing field lengths. The final knotholes are shown in Fig. 25.5 and present that data in two different ways. One plot holds range constant and shows rings of constant TOGW. The other is for a constant TOGW and shows rings of constant range. This entire study assumed a constant wing aspect ratio of 8.0.

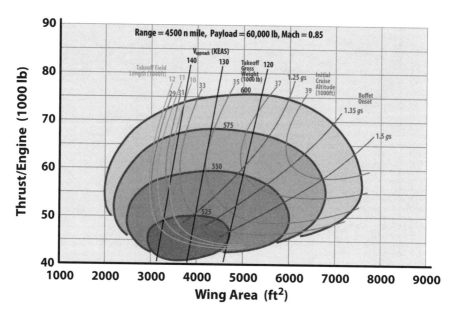

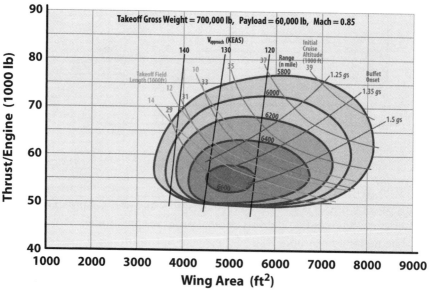

Figure 25.5 Parametric trade study results presented as "knotholes" (commercial transport).

These knotholes used T/W and W/S for the axes, but for studies where the engine is known and the wing planform is also known the axes will become T (thrust) and S (wing area). The process is identical.

Some words of caution when generating knotholes are appropriate at this point. First, knotholes take extra time to generate so they must be useful for communicating outside the technical study group. If the results are only going to be used by technical groups, then the carpet plots in Fig. 25.2 will likely suffice. Second, drawing the entire ring can be difficult particularly on the low wing loading and low thrust/weight edges. Solutions will blow up for very slight changes, making a little "artistic license" necessary for the final shapes. In the end knotholes offer a means to concisely summarize large amounts of carpet-plotted data in an easily understood form.

To illustrate that knotholes can look vastly different depending on the type of air vehicle that is being studied, Figure 25.6 is presented without discussion but represents a small autonomous UAV that has a portable ground control station. It is similar yet very different from the commercial transport knothole in Fig. 25.5.

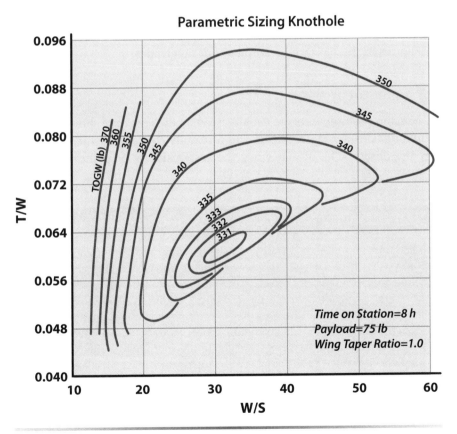

Figure 25.6 Parametric sizing study for Class III UAV.

25.3 Design Trades

The tradeoff information is used to select the proper combination of design features to achieve the most efficient vehicle relative to the measure of merit criteria. Some of the design parameters that are often varied during a parametric study are the following:

- Body shape (fineness ratio, nose shape, cross-sectional area distribution)
- Wing size and wing loading
- Wing shape (sweep, aspect ratio, taper ratio, thickness ratio, variable versus fixed geometry)
- High-lift devices (mechanical vs powered)
- Tail configuration (aft tail, canard, or tailless)
- Stability level (degree of static margin)
- Engine [T/W, number of engines, bypass ratio, fan pressure ratio, overall pressure ratio, turbine temperature, propulsion concept (turboprop, turbofan, turbojet, etc.)]
- Inlet or nozzle (location, type of inlet, type of nozzle)
- Materials (metals vs composites)

Figures 25.3–25.6 show examples of design trades.

25.4 Mission Trades

Mission requirements are usually fixed by the customer; however, they should be considered negotiable and forcefully challenged when they distort the design. The designer has the responsibility of pointing out the

Nontechnical Issues Can Drive Technical Decisions

Sometimes technical decisions are made based on nontechnical events. In 1987 the YF-22 design team (Lockheed, Boeing, and General Dynamics) was conducting design trades to select the best wing (planform, sweep, aspect ratio, and span) for their advanced tactical fighter (ATF) prototype. The clipped diamond planform won out over the swept trapezoidal planform (even though the "trap" wing had more aspect ratio) because it offered more wing area at a lighter structural weight (more root chord). The 48 deg LE wing sweep was a compromise between supersonic drag and subsonic aero performance. The 43-ft wing span was selected (even though it had a low aspect ratio of 2.2) because the width of the door on the TAB-V aircraft shelters was 45 ft. The wing span on the production F-22A was increased to 44.5 ft and sweep decreased to 42 deg to increase the aspect ratio to 2.4.

sensitivity of the aircraft design to the mission requirements. If one mission requirement, such as range, is driving the aircraft design to large takeoff weights (and hence high cost) the designer should advise the customer of this situation. The customer might choose to back off on the performance requirement to bring the cost down to an affordable level. The designer should provide mission requirement tradeoff information to the decision makers to permit the best compromise between performance and cost. Thus, the mission requirements of range, payload, turning performance, field length, and so on are typical candidates for the mission requirements trade study.

This trade is briefly discussed in Chapter 5, where the composite Lightweight Fighter (LWF) was sized.

25.5 Technology Trades

The results of a technology trade are used in two different ways by different groups:

1. The results are used by the program community to form the basis for a risk analysis as it answers two very important questions: What is the consequence (on the MoM) of the technology failing to perform as expected and what is the probability of the technology not performing as expected?
2. The results are used by the technology planners to form the basis for technology investment decisions as they show the payoff for spending research dollars on maturing the technology.

An example of a technology trade is shown in Fig. 25.7. The example is an intelligence, surveillance, and reconnaissance (ISR) aircraft (called SensorCraft) with the mission requirements shown on the chart. SensorCraft uses the off-the-shelf (OTS) engines AE 3007H and the aircraft TOGW is 130,000 lb using state-of-the-art technologies. The technologies are allowed to improve or degrade and the sensitivity of TOGW is determined. The change in the TOGW is the consequence of failure (or success) of the technology. The technology community would be asked to assess the probability of technology failure (or success).

The technology trade study shows that aircraft L/D and engine thrust specific fuel consumption (TSFC) have the most impact on the aircraft TOGW: both have a $\Delta TOGW/\Delta\%$ change $= -1000$. Getting a $+5\%$ improvement in TSFC is probably expensive, otherwise Allison (Rolls Royce) would have done it long ago. Similarly the TSFC is not likely to degrade because the AE 3007H is a mature engine in the Global Hawk. Thus, it is probably unwise to invest dollars in the AE 3007H. Likewise, there is little concern about the TSFC degrading (low probability of failure). On the other hand

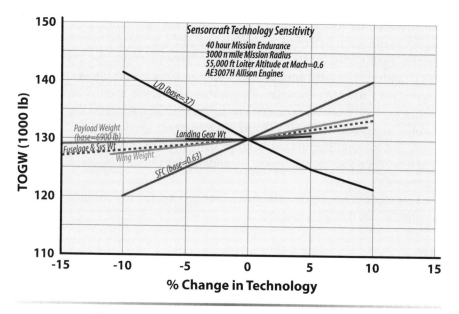

Figure 25.7   ISR aircraft technology trade results.

dollars might be invested in improving the L/D (more laminar flow, winglet, and airfoil research) and be concerned about degraded L/D (high probability of failure) due to losing laminar extent on the wing. The trade study shows that reducing aircraft structural and system weight has less impact on TOGW and the technology community would probably agree that it is harder to achieve.

25.6 Risk Analysis

Risk is an increasingly popular topic in the aerospace industry because there is risk in every decision that is made [2]. Choosing between configurations that have similar performance could prove to be easy if their respective risks were quite different. Folding in risk to conceptual and preliminary design efforts adds another element to consider when making engineering choices. Understanding potential risks early in the program is important but there needs to be an objective means of assigning risk to unfavorable program events. Assigning risk allows customers and management to better understand how the engineering group plans on maturing technologies that are part of a selected design and the priorities of each risk.

Quantifying risk uses relationships from probability and set theory as its mathematical basis (see [3]). Risk is simply defined mathematically as the union of failures and impacts or the probability of occurrence of unfavorable events. A *failure* is an unfavorable event and an *impact* is an unfavorable event that follows a failure; an impact may also be a failure. Risk has two major components. The first is the probability that the item will fail (Pf) (or likelihood that the failure will occur, Lo) and the second is the consequence (impact) of that failure, Cf (or the consequence of occurrence, Co). The terms Pf and Cf are considered to be too negative and so, popular usage has replaced them with their equivalents Lo and Co. These two parameters are mathematically combined (based on set theory) in Eq. (25.3) to yield a single number, generally referred to as the *risk index*, that represents the total risk of that item. Generally, Co is broken into its three parts representing (technical, schedule, and cost). Scoring the three components of Co can be combined using the same set theory mathematics to yield a single value for Co ([see Eq. (25.2)]. Once Co is calculated then Eq. (25.3) can be used to obtain the *Risk Index*. Both equations depend on Eq. (25.1) realistically representing the risk of any system. Table 25.1 shows a sample calculation of the total system risk based on the individual risks of each of its components. Notice how high the total risk index is. Managers and engineers often underestimate the aggregate risk of numerous items when assigning a total system risk:

$$\text{Risk(Overall)} = A \cup B - A \cap B = \text{Lo} \cup \text{Co} - \text{Lo} \cap \text{Co} \qquad (25.1)$$

$$\text{Risk}\left(\text{Consequences, Co}\right) = P\left(C_T\right) \cup P\left(C_S\right) \cup P\left(C_C\right) \qquad (25.2)$$

where

$P(C_T)$ = probability of the consequence from technical failure
$P(C_S)$ = probability of the consequence from schedule failure
$P(C_C)$ = probability of the consequence from cost failure

$$\text{Risk Index}\left(\text{RI}\right) = \text{Lo} + \text{Co} - \text{Lo} \cdot \text{Co} \qquad (25.3)$$

Table 25.1 Sample Risk Calculation

Component No.	Lo	Co	Risk Index
1	0.3	0.2	0.44
2	0.2	0.2	0.36
3	0.4	0.3	0.58
4	0.6	0.1	0.64
5	0.1	0.4	0.45
TOTAL =	0.879	0.758	0.971

Although the mathematics is appealing and objective, many system engineers assess risk in other ways. However, any other mathematical technique is clearly inferior to the set theory approach. Many engineers use a nonmathematical technique that ultimately ends up with risk simply having a value of low, moderate, or high. This approach will be discussed next.

In the non-mathematical approach a risk matrix is used to determine the overall Risk Index value. Although this matrix can vary from user to user, Fig. 25.8 is a good representative of this matrix. The 5 ratings of Lo and Co must have agreed-upon definitions (see Fig. 25.9) so that they are consistent across the program and independent of the person(s) giving the rating. Once Lo and Co (remember Lo and Co are probabilities with values from 0 to 1) ratings have been assigned they are located on the matrix and will end up in the low, moderate, or high category. General rules state that any item that is high risk must have a mitigation plan and a backup plan in case the original plan fails. Any item that has moderate risk must have a mitigation plan. Items that are low risk are just watched to make sure their risk does not change over the course of the program. Some managers will also use the results of risk assessment to allocate resources. Obviously, there is a relationship between the amount of risk (i.e., difficulty) and the cost of mitigating or maturing that risk.

The final result of these risk identification and assessment processes becomes a risk mitigation chart (waterfall chart) that shows the amount of risk reduction (mitigation) as a function of time. Figure 25.10 illustrates the relationship between risk assessment and risk mitigation.

It is a good time to reflect on how individuals assign risk to systems containing numerous components. Which system has more risk, (1) a

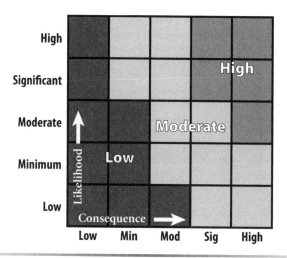

Figure 25.8   Risk assessment matrix.

Risk Assessment Template—Performance

Likelihood of Occurrence—Lo

Low	Minor	Moderate	Significant	High
Off the shelf	Flight Test	Element test complete	Partial test mixed with mostly analysis	Analysis only

Cost — Consequence of Occurrence—Co

Low	Minor	Moderate	Significant	High
Negligible cost impact	Minor cost impact on functional area No cost impact on overall program	Minor cost impact on the program	Significant cost impact on the program	Major cost impact on the program

Schedule

Low	Minor	Moderate	Significant	High
Negligible impact on program success	Could impact noncritical path milestones No impact on program critical path	Minor impact on program critical path milestones Workaround will likely maintain schedule	Moderate impact on program critical path milestones	Major impact on program schedule (>4 month slip)

Technical

Low	Minor	Moderate	Significant	High
Negligible impact on mission performance No impact on program success Can accept degradation	Minor impact on mission performance No impact on program success Can accept degradation	Minor impact on mission performance Minor impact on program success Acceptable workaround available	Degrades mission performance Impacts program success Expedited resolution required	Significantly impacts mission performance Endangers program success Must be fixed prior to aircraft delivery

Figure 25.9 Example of risk assessment template (performance).

system that has one high-risk item and one low-risk item or (2) a system that has one moderate-risk item and two low-risk items? There is no single right answer and answers will vary with individuals. Although the mathematical technique (Eq. 25.1) can consistently calculate relative risks regardless of the number of components, managers will often substitute subjective values for the mathematical values.

25.7 Now We Are Done

There is no set rule on how many iterations and parametric tradeoffs are necessary for a design—it depends upon the skill and thoroughness of the designer and the design team and upon the time and budget available for the conceptual phase. The conceptual phase usually continues until the decision is made to move the most promising design into the preliminary design phase or to terminate the design effort.

25.8 Kelly Johnson's 14 Rules of Management

1. The Skunk Works manager must be delegated practically complete control of his program in all aspects. He should report to a division president or higher.
2. Strong but small project offices must be provided by both the military and industry.

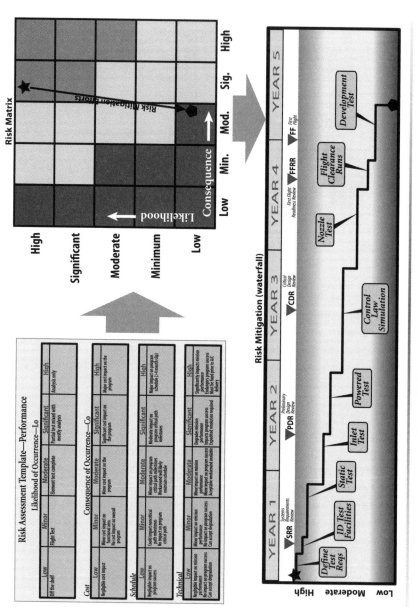

Figure 25.10 Assessing and mitigating risk.

3. The number of people having any connection with the project must be restricted in an almost vicious manner. Use a small number of good people (10% to 25% compared with the so-called normal systems).

4. A very simple drawing and drawing release system with great flexibility for making changes must be provided.

5. There must be a minimum number of reports required, but important work must be recorded thoroughly.

6. There must be a monthly cost review covering not only what has been spent and committed but also projected costs to the conclusion of the program. Don't have the books 90 days late, and don't surprise the customer with sudden overruns.

7. The contractor must be delegated and must assume more than normal responsibility to get good vendor bids for subcontract on the project. Commercial bid procedures are very often better than military ones.

8. The inspection system as currently used by the Skunk Works, which has been approved by both the Air Force and Navy, meets the intent of existing military requirements and should be used on new projects. Push more basic inspection responsibility back to subcontractors and vendors. Don't duplicate so much inspection.

9. The contractor must be delegated the authority to test his final product in flight. He can and must test it in the initial stages. If he doesn't, he rapidly loses his competency to design other vehicles.

10. The specifications applying to the hardware must be agreed to well in advance of contracting. The Skunk Works practice of having a specification section stating clearly which important military specification items will not knowingly be complied with and reasons therefore is highly recommended.

11. Funding a program must be timely so that the contractor doesn't have to keep running to the bank to support government projects.

12. There must be mutual trust between the military project organization and the contractor with very close cooperation and liaison on a day-to-day basis. This cuts down misunderstanding and correspondence to an absolute minimum.

13. Access by outsiders to the project and its personnel must be strictly controlled by appropriate security measures.

14. Because only a few people will be used in engineering and most other areas, ways must be provided to reward good performance by pay not based on the number of personnel supervised.

Kelly Johnson and Lulu Belle

Axis Germany's Messerschmitt Me 262 was the first operational jet fighter, becoming operational during the summer of 1944. It instantly raised the bar for fighter aircraft, having a 100-mph advantage over every WWII Allied fighter.

The US Army Air Force had commissioned Bell aircraft in September 1941 to build a jet fighter—the P-59 Airacomet using a British jet engine with the Whittle design. The YP-59 had its first flight in October 1942, but from the beginning its performance was disappointing. On June 21, 1943 the US Army gave Lockheed a contract to build one prototype of a jet fighter using the British Goblin jet engine. The contract was for $642,000 with a delivery date of November 1943 (180 days).

The project lead was given to a young engineer named Clarence "Kelly" Johnson. The Lockheed Advanced Development Projects (ADP, better known as the Skunk Works) was born. Kelly set up a super-secret operation with about 20 engineers and 80 shop men working 10 hour days, 6 days a week (Sunday was a day of rest ... no matter what). Kelly shaped his 14 rules of management during this mission-critical project.

Four days early on November 17, 1943, the XP-80, dubbed "Lulu Belle," (see the first page of Appendix K) rolled out with the Goblin engine installed and ready for systems check-out. Problems with engine/inlet integration delayed first flight until January 8, 1944. Lulu Belle made two flights that day and reached 490 mph on the second flight—50 mph more than the maximum speed of the fastest Allied aircraft, the P-38. The XP-80 led to the P-80 Shooting Star, which eventually reached 600 mph. Kelly Johnson and his team's implementation of his 14 rules of management led to the Skunk Works' success: Lockheed built 1,715 aircraft for the USAF and Navy.

References

[1] Sabbagh, Karl, "21st Century Jet-The Making of the Boeing 777," MacMillan General Books, London, UK, 1995.

[2] Bernstein, P. L., *Against the Gods, The Remarkable Story of Risk*, Wiley, New York, 1998.

[3] Blanchard, B. S., *System Engineering Management*, 3rd ed., Wiley, New York, 2004.

Appendix A   Conversions

- Unit Conversions
- Temperature Conversions
- Gases and Liquids

USAF/Northrop B-2 Spirit stealth bomber being refueled by a KC-10 tanker. The B-2 was designed to penetrate dense anti-aircraft defenses and deliver both conventional and nuclear weapons. The program has been controversial because of the high unit and O&S costs, and Northrop has only built 21 aircraft to date.

A scientist discovers that which exists.
An engineer creates that which never was.
Theodore von Kármán

A.1 Unit Conversions

A.1.1 Length

Multiply	By	To Obtain
Centimeter (cm)	3.281×10^{-2}	Feet
	3.938×10^{-1}	Inches
	1.000×10^{-5}	Kilometers
	1.000×10^{-2}	Meters
	1.094×10^{-2}	Yards
Foot (ft)	30.48	Centimeters
	12.00	Inches
	3.048×10^{-4}	Kilometers
	3.048×10^{-1}	Meters
	1.894×10^{-4}	Miles
	3.333×10^{-1}	Yards
Inch (in.)	2.540	Centimeters
	8.333×10^{-2}	Feet
	2.540×10^{-2}	Meters
	2.778×10^{-2}	Yards
	1.000×10^{-3}	Miles
Meter (m)	1.000×10^{2}	Centimeters
	3.281	Feet
	39.37	Inches
	1.000×10^{-3}	Kilometers
	6.214×10^{-4}	Miles
	1.094	Yards
Statute mile (mile or mi)	5.280×10^{3}	Feet
	1.609	Kilometers
	1.760×10^{3}	Yards
	0.868976	Nautical miles
Nautical mile (n mile)	6.076×10^{3}	Feet
	1.852×10^{3}	Meters
	1.15078	Miles
Yard (yd)	91.44	Centimeters
	3.000	Feet
	36.00	Inches
	9.144×10^{-1}	Meters
	5.682×10^{-4}	Miles

A.1.2 Area

Multiply	By	To Obtain
Acre	4.356×10^4	Square feet
	4.047×10^3	Square meters
	1.562×10^{-3}	Square miles
Square centimeter (cm²)	1.076×10^{-3}	Square feet
	1.550×10^{-1}	Square inches
	1.000×10^{-4}	Square meters
	1.000×10^2	Square millimeters
Square foot (ft²)	2.296×10^{-5}	Acres
	1.440×10^2	Square inches
	9.290×10^{-2}	Square meters
	3.587×10^{-8}	Square miles
	1.111×10^{-1}	Square yards
Square inch (in.²)	6.4516	Square Centimeters
	6.944×10^{-3}	Square feet
	6.452×10^{-4}	Square meters
Square kilometer (km)	2.471×10^2	Acres
	1.076×10^7	Square feet
	3.861×10^{-1}	Square miles
Square meter (m²)	2.471×10^{-4}	Acres
	1.000×10^4	Square centimeters
	10.76	Square feet
	1.550×10^3	Square inches
	3.861×10^{-7}	Square miles
Square mile	6.40×10^2	Acres
	2.778×10^7	Square feet
	2.590	Square kilometers
	2.590×10^6	Square meters
	3.0976×10^6	Square yards

A.1.3 Volume

Multiply	By	To Obtain
Cubic centimeter (cm³)	3.531×10^{-5}	Cubic feet
	6.1024×10^{-2}	Cubic inches
	1.000×10^{-6}	Cubic meters
	1.308×10^{-6}	Cubic yards
	3.381×10^{-2}	Fluid ounce
Cubic foot (ft³)	2.832×10^{4}	Cubic centimeters
	1.728×10^{3}	Cubic inches
	2.832×10^{-2}	Cubic meters
	28.317	Liters
	7.481	Gallons
Cubic inch (in.³)	16.39	Cubic centimeters
	5.787×10^{-4}	Cubic feet
	1.639×10^{-5}	Cubic meters
Cubic meter (m³)	1.000×10^{6}	Cubic centimeters
	35.31	Cubic feet
	6.102×10^{4}	Cubic inches
	1.308	Cubic yards
Gallon (U.S.) (gal)	1.3368×10^{-1}	Cubic feet
	3.78542	Liters
	3.785×10^{-3}	Cubic meters
	231	Cubic inches
	128	Fluid ounces
	8.000	Pints
	4.000	Quarts
Imperial gallon	2.774×10^{2}	Cubic inches
	1.201	Gallons (U.S.)
	4.546	Liters
Liter	3.532×10^{-2}	Cubic feet
	0.2642	Gallons
	1.000×10^{-3}	Cubic meters
	2.113	Pints
	1.05669	Quarts
	33.8142	Fluid ounces
Pint (U.S.) (pt)	1.671×10^{-2}	Cubic feet
	1.250×10^{-1}	Gallons
	4.732×10^{-1}	Liters
	0.5	Quarts
	28.875	Cubic inches
	16	Fluid ounces
Quart (U.S.) (qt)	3.342×10^{-2}	Cubic feet
	2.500×10^{-1}	Gallons
	9.463×10^{-1}	Liters
	2	Pints

A.1.4 Velocity

Multiply	By	To Obtain
Centimeter per second (cm/s)	3.281×10^{-2}	Feet per second
	3.937×10^{-1}	Inches per second
	1.000×10^{-2}	Meters per second
Foot per second (fps or ft/s)	30.48	Centimeters per second
	1.097	Kilometers per hour
	5.921×10^{-1}	Knots
	3.048×10^{-1}	Meters per second
	6.818×10^{-1}	Miles per hour
Inch per second (ips)	8.333×10^{-2}	Feet per second
	2.540	Centimeters per second
Kilometer per hour (km/h)	9.113×10^{-1}	Feet per second
	5.396×10^{-1}	Knots
	6.214×10^{-1}	Miles per hour
Knot (kt)	1.689	Feet per second
	1.151	Miles per hour
	1.000	Nautical miles per hour
	1.852	Kilometers per hour
Meter per second (m/s)	3.281	Feet per second
	3.600	Kilometers per hour
	1.943	Knots
	2.237	Miles per hour
Mile per hour (mph)	1.467	Feet per second
	1.609	Kilometers per hour
	0.8684	Knots
	0.4470	Meters per second

A.1.5 Acceleration

Feet per second^2 (ft/s^2)	30.48	Centimeters per second^2
	0.6818	Miles per hour-second

A.1.6 Angular Rate and Frequency

Multiply	By	To Obtain
Radians per second (rad/s)	0.1592	Revolutions per second
	9.549	Revolutions per minute
	57.296	Degrees per second
Revolutions per minute (rpm)	0.01667	Revolutions per second
	0.10472	Radians per second
	6	Degrees per second
Cycle per second (cps)	1.000	Hertz
	2π	Radians per second

A.1.7 Mass

Multiply	By	To Obtain
Kilogram (kg)	1.000×10^3	Grams
	6.854×10^{-2}	Slugs
Slug	1.459×10^4	Grams
	14.59	Kilograms

A.1.8 Weight

Multiply	By	To Obtain
Gram (g)	3.528×10^{-2}	Ounces
	2.205×10^{-3}	Pounds
Pound (lb)	4.536×10^2	Grams
	16	Ounces
Short ton	2000	Pounds
	907.185	Kilograms
Metric tonne	2205	Pounds
	1000	Kilograms

A.1.9 Force

Multiply	By	To Obtain
Dyne	1.020×10^{-3}	Grams
	1.000×10^{-5}	Newtons
	2.248×10^{-6}	Pounds
Gram (g)	3.528×10^{-2}	Ounces
	2.205×10^{-3}	Pounds
	9.807×10^2	Dynes
	9.807×10^{-3}	Newtons
Kilogram (kg)	2.205	Pounds
	9.807	Newtons
	70.93	Poundals
Pound (lb)	4.536×10^{-1}	Kilograms
	4.448	Newtons
	32.17	Poundals
Poundal	1.410×10^{-2}	Kilograms
	1.383×10^{-1}	Newtons
	3.108×10^{-2}	Pounds

A.1.10 Pressure

Multiply	By	To Obtain
Atmosphere (atm)	29.92	Inches of mercury (0°C)
	760	Millimeters of mercury (0°C)
	1.0133	Bars
	14.70	Pounds per square inch
	1.01325×10^6	Dynes per centimeter
	1.01325×10^5	Newtons per meter
Bar	9.870×10^{-7}	Atmospheres
	1.000	Dyne per square centimeter
	1.0×10^5	Newtons per square meter
	7.501×10^2	Millimeters of mercury (0°C)
	1.451×10^{-5}	Pounds per square inch
Dyne per square centimeter (dyne/cm^2)	2.952×10^{-5}	Inches of mercury (0°C)
	1.020×10^{-2}	Kilograms per square meter
	7.501×10^{-4}	Millimeters of mercury (0°C)
	1.450×10^{-5}	Pounds per square inch
Inch of mercury (in. Hg)	3.342×10^{-2}	Atmospheres (0°C)
	3.388×10^{-2}	Bars
	3.388×10^3	Dynes per square centimeter
	13.60	Inches of water
	25.40	Millimeters of mercury
	3.388×10^3	Newtons per square meter
	70.73	Pounds per square foot
	4.912×10^{-1}	Pounds per square inch
Inch of water (in. H$_2$O) (4°C)	2.458×10^{-3}	Atmospheres
	7.355×10^{-2}	Inches of mercury
	1.868	Millimeters of mercury
	2.491×10^2	Newtons per square meter
	3.613×10^{-2}	Pounds per square inch
	5.203	Pounds per square foot
Kilogram per square meter (kg/m^2)	9.678×10^{-5}	Atmospheres
	98.07	Bars
	2.896×10^{-3}	Inches of mercury
	9.807	Newtons per square meter
	6.588	Poundals per square foot
	2.048×10^{-1}	Pounds per square foot
	1.422×10^{-3}	Pounds per square inch

(continued)

Multiply	By	To Obtain
Millimeter of mercury (0°C) (torr or mm Hg)	1.333×10^3	Dynes per square centimeter
	3.937×10^{-2}	Inches of mercury
	5.354×10^{-1}	Inches of water
	1.333×10^2	Newtons per square meter
	1.934×10^{-2}	Pounds per square inch
Newton per square meter [pascal (Pa)] (N/m²)	9.869×10^{-6}	Atmospheres
	10	Dynes per square centimeter
	2.953×10^{-4}	Inches of mercury
	1.020×10^{-1}	Kilograms per square meter
	2.089×10^{-2}	Pounds per square foot
	1.450×10^{-4}	Pounds per square inch
Pound per square foot (psf)	4.725×10^{-4}	Atmospheres
	4.788×10^{-4}	Bars
	4.788×10^2	Dynes per square centimeter
	1.414×10^{-2}	Inches of mercury
	4.882	Kilograms per square meter
	47.88	Newtons per square meter
	6.944×10^{-3}	Pounds per square inch
Pound per square inch (psi)	6.804×10^{-2}	Atmospheres
	6.895×10^4	Dynes per square centimeter
	2.036	Inches of mercury
	7.031×10^{-2}	Kilograms per square meter
	6.895×10^3	Newtons per square meter
	1.44×10^2	Pounds per square foot

A.1.11 Density

Multiply	By	To Obtain
Pound per cubic foot (lb/ft³)	5.787×10^{-4}	Pounds per cubic inch
	16.018	Kilograms per cubic meter
	1.6018×10^{-2}	Grams per cubic centimeter

A.1.12 Work and Energy

Multiply	By	To Obtain
British thermal unit (Btu)	2.530×10^2	Calories
	7.783×10^2	Foot pounds
	3.927×10^{-4}	Horsepower hours
	1.055×10^3	Joules
	1.055×10^3	Newton meters
	2.930×10^{-4}	Kilowatt hours
	1.055×10^3	Watt seconds
Foot pound (ft·lb)	1.285×10^{-3}	British thermal units
	5.050×10^{-7}	Horsepower hours
	1.356	Joules
	3.766×10^{-7}	Kilowatt hours
	1.356	Newton meters
Horsepower hour (hp·h)	2.545×10^3	British thermal units
	1.980×10^6	Foot pounds
	2.684×10^6	Joules
	7.457×10^{-1}	Kilowatt hours
Joule	9.486×10^{-4}	British thermal units
	2.389×10^{-1}	Calories
	1.000×10^7	Dyne centimeters (ergs)
	7.376×10^{-1}	Foot pounds
	1.000	Newton meter
	1.000	Watt second
Kilowatt hour (kWh)	3.415×10^3	British thermal units
	2.655×10^6	Foot pounds
	1.341	Horsepower hours
	3.600×10^6	Joules
	3.670×10^5	Kilogram meters
	3.600×10^6	Watt seconds
Dyne centimeter	7.3756×10^{-8}	Foot pounds
	1.000×10^{-7}	Newton meters

A.1.13 Power

Multiply	By	To Obtain
British thermal unit per minute (BTU/min)	3.969×10^6	Calories per second
	12.97	Foot-pounds per second
	2.357×10^{-2}	Horsepower
	17.58	Joules per second
	2.987×10^{-2}	Kilogram meters per second
	17.58	Watts
Foot-pound per second (ft·lb/s)	7.713×10^{-2}	British thermal units per minute
	3.239×10^{-1}	Calories per second
	1.818×10^{-3}	Horsepower
	1.356	Joules per second
	1.383×10^{-1}	Kilogram meters per second
	1.356	Watts
Horsepower (hp)	42.42	British thermal units per minute
	550	Foot-pounds per second
	33,000	Foot-pounds per minute
	7.457×10^2	Joules per second
	76.04	Kilogram-meters per second
	7.457×10^2	Watts
Kilogram-meter per second	33.47	British Thermal Units per minute
	7.233	Foot-pounds per second
Watt (joule per second) (W)	5.689×10^{-2}	British thermal units per minute
	2.388×10^{-1}	Calories per second
	7.376×10^{-1}	Foot-pounds per second
	1.341×10^{-3}	Horsepower
	1.020×10^{-1}	Kilogram-meters per second

A.2 Temperature Conversions

- $T(°C) = (5/9) [T(°F) - 32]$
- $T(°C) = (5/9) [T(°R) - 491.67]$
- $T(°C) = T(°K) - 273.15$
- $T(°F) = (9/5) T(°C) + 32$
- $T(°F) = (9/5) [T(°K) - 273.15] + 32$
- $T(°F) = T(°R) - 459.67$

A.3 Gases and Liquids

A.3.1 Standard Values for Air at Sea Level

- $p_0 = 2116.22$ psi $= 1.01325 \times 10^5$ N/m$^2 = 29.92$ in. Hg $= 760$ mm Hg
- $T_0 = 518.67°$R $= 59.0°$F $= 288.15°$K $= 15.0°$C
- $g_0 = 32.174$ ft/s$^2 = 9.80665$ m/s^2
- $\rho_0 = 0.002377$ slug/ft$^3 = 0.12492$ kg·s^2/m^4
- $\nu_0 = 1.5723 \times 10^{-4}$ ft^2/s $= 1.4607 \times 10^{-5}$ m^2/s
- $\mu_0 = 1.2024 \times 10^{-5}$ lb/ft·s $= 1.7894 \times 10^{-5}$ kg/m·s
- $\mu_0 = 3.737 \times 10^{-7}$ slug/(ft·s)

A.3.2 Specific Weights of Other Gases at One Atmosphere and 0°C

- Carbon dioxide $= 0.12341$ lb/ft^3
- Helium $= 0.01114$ lb/ft^3
- Hydrogen $= 0.005611$ lb/ft^3
- Nitrogen $= 0.07807$ lb/ft^3
- Oxygen $= 0.089212$ lb/ft^3

A.3.3 Specific Weights (Specific Gravity) of Some Liquids at 0°C

- Alcohol (methyl) $= 50.5$ lb/ft^3 (0.810)
- Gasoline $= 44.9$ lb/ft^3 (0.72)
- JP1 $= 49.7$ lb/ft^3 (0.80)
- JP3 $= 48.2$ lb/ft^3 (0.775)
- JP4 $= 49.0$ lb/ft^3 (0.785)
- JP5 $= 51.1$ lb/ft^3 (0.817)
- JP7 $= 48.6–50.3$ lb/ft^3 (0.779–0.806)
- JP8 $= 55.81$ lb/ft^3 (0.894)
- JP10 $= 58.62$ lb/ft^3 (0.939)
- Kerosene $= 51.2$ lb/ft^3 (0.82)
- Sea water $= 63.99$ lb/ft^3 (1.025)
- Water $= 62.43$ lb/ft^3 (1.000)

Appendix B Atmospheric Data

- 1976 U.S. Standard Atmosphere
- MIL-210A Atmospheric Data
- MIL-210A 10% & 1% Probability Atmospheric Data
- MIL-210A Navy Non-Standard Atmospheric Data

T-46A flight test aircraft, built by Fairchild Republic for the Next Generation Trainer (NGT) to replace the Cessna T-37B "Tweeti Bird" used for undergraduate USAF pilot training. Despite winning the USAF competition, Fairchild Republic closed its doors in 1987. A case study of the T-46A appears in Volume II.

To raise new questions, new possibilities, to regard old problems from a new angle requires creative imagination and marks real advances in science.

Albert Einstein

The U.S. military has definitions for atmospheres that are different from a standard day, whether "standard day" is defined by MIL-210A or the 1976 U.S. Standard Atmosphere (Table B.1). Tables B.2–B.4 summarize the temperature profiles of those nonstandard atmospheres as defined by the U.S. Air Force and the U.S. Navy. The standard day temperature profile is shown on each plot (Figs. B.1 and B.2) for reference. The 1% and 10% hot and cold days represent temperature profiles that would occur only 1% or 10% of the time, respectively. The unique temperature profiles defined by the U.S. Navy are also included for completeness. All temperature profiles are plotted, and the plotted data are included in the tables.

Table B.1 1976 U.S. Standard Atmosphere

Alt.	Press.	P/P_0	Density	ρ/ρ_0	Temp.	Temp.	T/T_0	Viscosity	V_{sonic}
(ft)	(lb/ft²)	(δ)	(slug/ft³)	(σ)	°R	°F	(θ)	(slug/ft-s)	(ft/s)
0	2116.2	1.0000	0.002377	1.0000	518.7	59.00	1.0000	3.737E-07	1116.4
1,000	2040.9	0.9644	0.002308	0.9711	515.1	55.43	0.9931	3.717E-07	1112.6
2,000	1967.7	0.9298	0.002241	0.9428	511.5	51.87	0.9862	3.697E-07	1108.7
3,000	1896.6	0.8962	0.002175	0.9151	508.0	48.30	0.9794	3.677E-07	1104.9
4,000	1827.7	0.8637	0.002111	0.8881	504.4	44.74	0.9725	3.657E-07	1101.0
5,000	1760.8	0.8320	0.002048	0.8617	500.8	41.17	0.9656	3.637E-07	1097.1
6,000	1695.9	0.8014	0.001987	0.8359	497.3	37.60	0.9587	3.616E-07	1093.2
7,000	1632.9	0.7716	0.001927	0.8106	493.7	34.04	0.9519	3.596E-07	1089.3
8,000	1571.9	0.7428	0.001868	0.7860	490.1	30.47	0.9450	3.575E-07	1085.3
9,000	1512.7	0.7148	0.001811	0.7620	486.6	26.90	0.9381	3.555E-07	1081.4
10,000	1455.3	0.6877	0.001755	0.7385	483.0	23.34	0.9312	3.534E-07	1077.4
11,000	1399.7	0.6614	0.001701	0.7156	479.4	19.77	0.9244	3.513E-07	1073.4
12,000	1345.9	0.6360	0.001648	0.6932	475.9	16.21	0.9175	3.493E-07	1069.4
13,000	1293.7	0.6113	0.001596	0.6713	472.3	12.64	0.9106	3.472E-07	1065.4
14,000	1243.2	0.5875	0.001545	0.6500	468.7	9.07	0.9037	3.451E-07	1061.4
15,000	1194.3	0.5643	0.001496	0.6292	465.2	5.51	0.8969	3.430E-07	1057.3
16,000	1146.9	0.5420	0.001447	0.6090	461.6	1.94	0.8900	3.409E-07	1053.3
17,000	1101.1	0.5203	0.001400	0.5892	458.0	-1.62	0.8831	3.388E-07	1049.2
18,000	1056.8	0.4994	0.001355	0.5699	454.5	-5.19	0.8762	3.367E-07	1045.1
19,000	1013.9	0.4791	0.001310	0.5511	450.9	-8.76	0.8694	3.345E-07	1041.0
20,000	972.5	0.4595	0.001266	0.5328	447.3	-12.32	0.8625	3.324E-07	1036.8
21,000	932.4	0.4406	0.001224	0.5150	443.8	-15.89	0.8556	3.303E-07	1032.7
22,000	893.7	0.4223	0.001183	0.4976	440.2	-19.46	0.8487	3.281E-07	1028.6
23,000	856.3	0.4046	0.001142	0.4807	436.6	-23.02	0.8419	3.259E-07	1024.4
24,000	820.2	0.3876	0.001103	0.4642	433.1	-26.59	0.8350	3.238E-07	1020.2

Alt.	Press.	P/P_0	Density	ρ/ρ_0	Temp.	Temp.	T/T_0	Viscosity	V_{sonic}
(ft)	(lb/ft²)	(δ)	(slug/ft³)	(σ)	°R	°F	(θ)	(slug/ft·s)	(ft/s)
25,000	785.3	0.3711	0.001065	0.4481	429.5	−30.15	0.8281	3.216E−07	1016.0
26,000	751.64	0.3552	0.001028	0.4325	425.95	−33.72	0.8212	3.194E−07	1011.7
27,000	719.15	0.3398	0.000992	0.4173	422.38	−37.29	0.8144	3.172E−07	1007.5
28,000	687.81	0.3250	0.000957	0.4025	418.82	−40.85	0.8075	3.150E−07	1003.2
29,000	657.58	0.3107	0.000923	0.3881	415.25	−44.42	0.8006	3.128E−07	999.0
30,000	628.43	0.2970	0.000889	0.3741	411.69	−47.98	0.7937	3.106E−07	994.7
31,000	600.35	0.2837	0.000857	0.3605	408.12	−51.55	0.7869	3.084E−07	990.3
32,000	573.28	0.2709	0.000826	0.3473	404.55	−55.12	0.7800	3.061E−07	986.0
33,000	547.21	0.2586	0.000795	0.3345	400.99	−58.68	0.7731	3.039E−07	981.7
34,000	522.12	0.2467	0.000765	0.3220	397.42	−62.25	0.7662	3.016E−07	977.3
35,000	497.96	0.2353	0.000737	0.3099	393.85	−65.82	0.7594	2.994E−07	972.9
36,000	474.71	0.2243	0.000709	0.2981	390.29	−69.38	0.7525	2.971E−07	968.5
37,000	452.43	0.2138	0.000676	0.2843	389.97	−69.70	0.7519	2.969E−07	968.1
38,000	431.20	0.2038	0.000644	0.2710	389.97	−69.70	0.7519	2.969E−07	968.1
39,000	410.97	0.1942	0.000614	0.2583	389.97	−69.70	0.7519	2.969E−07	968.1
40,000	391.68	0.1851	0.000585	0.2462	389.97	−69.70	0.7519	2.969E−07	968.1
41,000	373.30	0.1764	0.000558	0.2346	389.97	−69.70	0.7519	2.969E−07	968.1
42,000	355.78	0.1681	0.000531	0.2236	389.97	−69.70	0.7519	2.969E−07	968.1
43,000	339.09	0.1602	0.000507	0.2131	389.97	−69.70	0.7519	2.969E−07	968.1
44,000	323.17	0.1527	0.000483	0.2031	389.97	−69.70	0.7519	2.969E−07	968.1
45,000	308.01	0.1455	0.000460	0.1936	389.97	−69.70	0.7519	2.969E−07	968.1
46,000	293.56	0.1387	0.000439	0.1845	389.97	−69.70	0.7519	2.969E−07	968.1
47,000	279.78	0.1322	0.000418	0.1758	389.97	−69.70	0.7519	2.969E−07	968.1
48,000	266.65	0.1260	0.000398	0.1676	389.97	−69.70	0.7519	2.969E−07	968.1
49,000	254.14	0.1201	0.000380	0.1597	389.97	−69.70	0.7519	2.969E−07	968.1
50,000	242.21	0.1145	0.000362	0.1522	389.97	−69.70	0.7519	2.969E−07	968.1
51,000	230.85	0.1091	0.000345	0.1451	389.97	−69.70	0.7519	2.969E−07	968.1
52,000	220.01	0.1040	0.000329	0.1383	389.97	−69.70	0.7519	2.969E−07	968.1
53,000	209.69	0.0991	0.000313	0.1318	389.97	−69.70	0.7519	2.969E−07	968.1
54,000	199.85	0.0944	0.000299	0.1256	389.97	−69.70	0.7519	2.969E−07	968.1
55,000	190.47	0.0900	0.000285	0.1197	389.97	−69.70	0.7519	2.969E−07	968.1
56,000	181.53	0.0858	0.000271	0.1141	389.97	−69.70	0.7519	2.969E−07	968.1
57,000	173.01	0.0818	0.000258	0.1087	389.97	−69.70	0.7519	2.969E−07	968.1
58,000	164.90	0.0779	0.000246	0.1036	389.97	−69.70	0.7519	2.969E−07	968.1
59,000	157.16	0.0743	0.000235	0.0988	389.97	−69.70	0.7519	2.969E−07	968.1
60,000	149.78	0.0708	0.000224	0.0941	389.97	−69.70	0.7519	2.969E−07	968.1
61,000	142.75	0.0675	0.000213	0.0897	389.97	−69.70	0.7519	2.969E−07	968.1
62,000	136.05	0.0643	0.000203	0.0855	389.97	−69.70	0.7519	2.969E−07	968.1
63,000	129.67	0.0613	0.000194	0.0815	389.97	−69.70	0.7519	2.969E−07	968.1

(continued)

Alt. (ft)	Press. (lb/ft^2)	P/P_0 (δ)	Density (slug/ft^3)	ρ/ρ_0 (σ)	Temp. °R	Temp. °F	T/T_0 (θ)	Viscosity (slug/ft·s)	V_{sonic} (ft/s)
64,000	123.59	0.0584	0.000185	0.0777	389.97	−69.70	0.7519	2.969E−07	968.1
65,000	117.79	0.0557	0.000176	0.0740	389.97	−69.70	0.7519	2.969E−07	968.1
66,000	112.26	0.0530	0.000168	0.0705	390.18	−69.49	0.7523	2.970E−07	968.3
67,000	107.00	0.0506	0.000160	0.0671	390.73	−68.94	0.7533	2.974E−07	969.0
68,000	101.99	0.0482	0.000152	0.0639	391.28	−68.39	0.7544	2.977E−07	969.7
69,000	97.22	0.0459	0.000145	0.0608	391.83	−67.84	0.7554	2.981E−07	970.4
70,000	92.68	0.0438	0.000138	0.0579	392.37	−67.30	0.7565	2.984E−07	971.1
71,000	88.36	0.0418	0.000131	0.0551	392.92	−66.75	0.7576	2.988E−07	971.7
72,000	84.25	0.0398	0.000125	0.0525	393.47	−66.20	0.7586	2.991E−07	972.4
73,000	80.33	0.0380	0.000119	0.0500	394.02	−65.65	0.7597	2.995E−07	973.1
74,000	76.60	0.0362	0.000113	0.0476	394.57	−65.10	0.7607	2.998E−07	973.8
75,000	73.05	0.0345	0.000108	0.0453	395.12	−64.55	0.7618	3.002E−07	974.4
76,000	69.67	0.0329	0.000103	0.0432	395.67	−64.00	0.7628	3.005E−07	975.1
77,000	66.45	0.0314	0.000098	0.0411	396.22	−63.45	0.7639	3.009E−07	975.8
78,000	63.38	0.0300	0.000093	0.0392	396.76	−62.91	0.7650	3.012E−07	976.5
79,000	60.46	0.0286	0.000089	0.0373	397.31	−62.36	0.7660	3.016E−07	977.1
80,000	57.67	0.0273	0.000084	0.0355	397.86	−61.81	0.7671	3.019E−07	977.8
81,000	55.02	0.0260	0.000080	0.0338	398.41	−61.26	0.7681	3.023E−07	978.5
82,000	52.50	0.0248	0.000077	0.0322	398.96	−60.71	0.7692	3.026E−07	979.2
83,000	50.09	0.0237	0.000073	0.0307	399.51	−60.16	0.7703	3.030E−07	979.8
84,000	47.79	0.0226	0.000070	0.0293	400.06	−59.61	0.7713	3.033E−07	980.5
85,000	45.61	0.0216	0.000066	0.0279	400.60	−59.07	0.7724	3.037E−07	981.2
86,000	43.52	0.0206	0.000063	0.0266	401.15	−58.52	0.7734	3.040E−07	981.9
87,000	41.54	0.0196	0.000060	0.0253	401.70	−57.97	0.7745	3.043E−07	982.5
88,000	39.65	0.0187	0.000057	0.0242	402.25	−57.42	0.7755	3.047E−07	983.2
89,000	37.84	0.0179	0.000055	0.0230	402.80	−56.87	0.7766	3.050E−07	983.9
90,000	36.12	0.0171	0.000052	0.0219	403.35	−56.32	0.7777	3.054E−07	984.5
91,000	34.48	0.0163	0.000050	0.0209	403.90	−55.77	0.7787	3.057E−07	985.2
92,000	32.92	0.0156	0.000047	0.0200	404.44	−55.23	0.7798	3.061E−07	985.9
93,000	31.43	0.0149	0.000045	0.0190	404.99	−54.68	0.7808	3.064E−07	986.5
94,000	30.01	0.0142	0.000043	0.0181	405.54	−54.13	0.7819	3.068E−07	987.2
95,000	28.66	0.0135	0.000041	0.0173	406.09	−53.58	0.7829	3.071E−07	987.9
96,000	27.36	0.0129	0.000039	0.0165	406.64	−53.03	0.7840	3.074E−07	988.5
97,000	26.13	0.0123	0.000037	0.0157	407.19	−52.48	0.7851	3.078E−07	989.2
98,000	24.96	0.0118	0.000036	0.0150	407.74	−51.93	0.7861	3.081E−07	989.9
99,000	23.84	0.0113	0.000034	0.0143	408.29	−51.38	0.7872	3.085E−07	990.5
100,000	22.77	0.0108	0.000032	0.0136	408.83	−50.84	0.7882	3.088E−07	991.2

Unit conversions: m = 0.3048(ft); N/m^2 = 47.88(lb/ft^2); kg·s^2/m^4 = 0.01903(slug/ft^3); Temp°K = 5/9(Temp°R); q/M^2 = 0.7P; Temp°C = 5/9(Temp°R−491.67); kg/m·s = 0.02088(slug/ft·s); m/s = 0.3048(ft/s); knots = 0.5921(ft/s)

Table B.2 MIL-210A Nonstandard Atmospheres

Cold Day		Polar Day		STANDARD DAY		Tropical Day		Hot Day	
Alt. (ft)	T (°R)	Alt. (ft)	T (°R)	Alt. (ft)	T (°R)	Alt. (ft)	T (°R)	Alt. (ft)	T (°R)
0	399.7	0	444.0	0	518.7	0	549.5	0	562.7
3,311	444.7	3,243	453.9	5,000	500.8	5,000	530.1	5,000	543.4
5,000	444.7	5,000	453.0	10,000	483.0	10,000	510.7	10,000	523.6
10,000	444.7	9,882	450.3	15,000	465.2	15,000	491.3	15,000	504.6
10,744	444.7	10,000	450.0	20,000	447.3	20,000	472.0	20,000	485.2
15,000	430.6	15,000	435.9	25,000	429.5	25,000	452.7	25,000	466.4
20,000	413.6	20,000	421.7	30,000	411.7	30,000	433.4	30,000	447.4
25,000	395.8	25,000	407.4	35,000	393.9	35,000	414.1	35,000	429.6
30,000	376.9	30,000	393.0	36,089	390.0	40,000	395.1	39,400	414.7
30,715	374.7	30,065	392.7	40,000	390.0	45,000	376.9	39,500	414.7
35,000	374.7	35,000	391.4	45,000	390.0	50,000	359.6	40,000	414.9
40,000	374.7	40,000	390.1	50,000	390.0	53,595	347.7	45,000	417.1
42,377	374.7	45,000	388.8	55,000	390.0	55,000	350.7	50,000	419.5
45,000	361.1	50,000	387.5	60,000	390.0	60,000	361.7	50,400	419.5
50,000	336.8	55,000	386.2	65,000	390.0	65,000	373.0	50,500	419.5
50,583	334.7	60,000	385.0	65,617	390.0	69,620	383.7	55,000	420.6
55,000	334.7	65,000	383.7	70,000	392.4	70,000	384.2	60,000	421.5
60,000	334.7	70,000	382.4	75,000	395.1	75,000	390.7	65,000	422.5
61,087	334.7	75,000	381.1	80,000	397.9	80,000	397.4	70,000	425.0
65,000	346.5	80,000	379.8	85,000	400.6	85,000	404.2	75,000	428.5
70,000	359.2	85,000	378.7	90,000	403.3	90,000	410.9	80,000	432.0
73,055	365.7	86,092	378.3	95,000	406.1	95,000	417.6	85,000	435.9
75,000	365.2	90,000	378.3	100,000	408.8	100,000	424.3	90,000	439.9
80,000	363.8	95,000	378.3					95,000	443.8
85,000	361.9	100,000	378.3					100,000	448.1
90,000	360.0								
95,000	358.0								
100,000	355.8								

Table B.3 MIL-210B Nonstandard Days

Alt. (ft)	COLDB (1%) T (°R)	COLDB (10%) T (°R)	HOTB (10%) T (°R)	HOTB (1%) T (°R)
0	381.7	394.7	572.7	579.7
5,000	403.7	419.7	550.7	553.7
10,000	409.7	422.7	527.7	533.7
15,000	401.7	411.7	509.7	517.7
20,000	389.7	395.7	495.7	503.7
25,000	375.7	381.7	479.7	487.7
30,000	363.7	373.7	465.7	474.7
35,000	359.7	369.7	450.7	463.7
40,000	359.7	369.7	436.7	449.7
45,000	357.7	363.7	427.7	437.7
50,000	347.7	355.7	421.7	427.7
60,000	334.7	343.7	421.7	427.7
70,000	339.7	343.7	421.7	427.7

Table B.4   MIL-210B Navy Nonstandard Days

	Coastal		COLDC		HOTC
Alt. (ft)	T (°R)	Alt. (ft)	T (°R)	Alt. (ft)	T (°R)
0	572.7	0	445.7	0	545.7
1,000	563.7	1,000	447.7	1,000	549.7
2,000	555.7	2,000	447.7	2,000	549.7
2,500	551.7	2,500	447.7	2,500	549.7
5,000	538.7	5,000	442.7	5,000	538.7
7,500	527.7	7,500	437.7	7,500	527.7
10,000	518.7	10,000	431.7	10,000	518.7
12,500	509.7	12,500	426.7	12,500	509.7
15,000	500.7	15,000	421.7	15,000	500.7
17,500	491.7	17,500	415.7	17,500	491.7
20,000	483.7	20,000	409.7	20,000	483.7
22,500	474.7	22,500	403.7	22,500	474.7
25,000	465.7	25,000	397.7	25,000	465.7
27,500	456.7	27,500	391.7	27,500	456.7
30,000	447.7	30,000	384.7	30,000	447.7
32,500	439.7	32,500	377.7	32,500	439.7
35,000	429.7	35,000	374.7	35,000	429.7
37,500	421.7	37,500	372.7	37,500	421.7
39,500	415.7	40,000	368.7	39,500	415.7
40,000	416.7	42,500	365.7	40,000	416.7
42,500	419.7	45,000	360.7	42,500	419.7
45,000	421.7	47,500	355.7	45,000	421.7
47,500	421.7	50,000	349.7	47,500	421.7
50,000	421.7			50,000	421.7
52,500	421.7			52,500	421.7

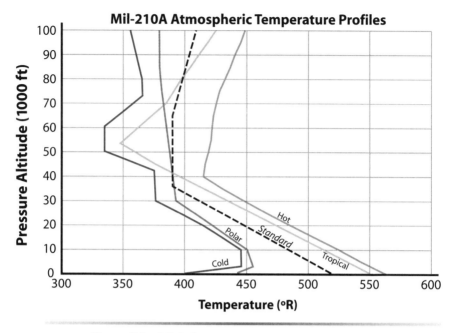

Figure B.1 Mil-210A atmospheric temperature profiles.

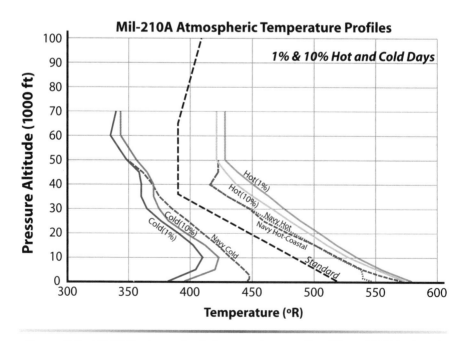

Figure B.2 Mil-210A atmospheric temperature profiles 1% and 10% hot and cold days.

Appendix C

Isentropic Compressible Flow Functions

- One-Dimensional Isentropic Compressible Flow
- Perfect Gas
- Constant Specific Heat
- Constant Molecular Weight

Lockheed Martin F-35B Lightning II is the U.S. Marine Corps variant preparing for a vertical landing. The F-35 Joint Strike Fighter is a single-seat, single-engine, stealthy, multi-role fighter that can perform close air support, tactical bombing, and air defense missions. See story at end of Chapter 9 and case study in Volume II.

Engineering is done with numbers.
Analysis without numbers is only an opinion.
Akin's First Law of Spacecraft Design

C.1 One-Dimensional Isentropic Compressible Flow Functions for a Perfect Gas with Constant Specific Heat and Molecular Weight, $\gamma = 1.4$

The following relations are tabulated in Table C.1 as a function of Mach number:

Temperature ratio

$$\frac{T}{T_0} = \left[1 + \frac{\gamma - 1}{2} M^2\right]^{-1}$$

Pressure ratio

$$\frac{P}{P_0} = \left[1 + \frac{\gamma - 1}{2} M^2\right]^{-\gamma/(\gamma-1)}$$

Nozzle design expansion ratio

$$P_{\bar{n}} = P_c / P_a$$

where

$P_a = P_e$
P_a = atmospheric pressure
P_e = exit pressure
P_c = chamber pressure
$P_{\bar{n}} = P_0/P$ = optimum expansion for a nozzle

Area ratio

$$\frac{A}{A^*} = \frac{1}{M}\left[\frac{2}{\gamma+1}\left(1 + \frac{\gamma-1}{2} M^2\right)\right]^{(\gamma+1)/[2(\gamma-1)]}$$

Mass flow parameter (MFP)

$$\frac{\dot{m}\sqrt{T_0}}{AP_0} = \frac{\sqrt{\left(\dfrac{2}{\gamma+1}\right)^{(\gamma+1)/(\gamma-1)} \dfrac{\gamma \bar{M} g}{\bar{R}}}}{\dfrac{1}{M}\left[\dfrac{2}{\gamma+1}\left(1 + \dfrac{\gamma-1}{2} M^2\right)\right]^{(\gamma+1)/[2(\gamma-1)]}}$$

where

$\dot{m}$ = in pound-mass per second (lbm/s)
T_0 = in degrees Rankine (°R)
A = in square feet (ft²)
P_0 = in pound-force per square foot (lbf/ft²)
γ = specific heat ratio
$\overline{M}$ = molecular weight
g = acceleration due to gravity
$\overline{R}$ = 1545.43 (ft·lbf)/(lb·mole °R)
$\overline{M}$ = 28.97 lb·mole (air)

The mass flow parameter (MFP) that is tabulated in Table C.1 is for air; also, MFP at $M = 1$ is

$$\frac{\dot{m}\sqrt{T_0}}{A^*P_0} = 0.532 \frac{\text{lbm}\sqrt{°\text{R}}}{\text{lbf} \cdot \text{s}}$$

Table C.1 One-Dimensional Isentropic Compressible Flow Functions for a Perfect Gas with Constant Specific Heat and Molecular Weight, $\gamma = 1.4$

Mach No.	A/A^*	$P_{\bar{n}}$	P/P_0	T/T_0	MFP
0.00	∞	1.00000	1.00000	1.00000	0.00000
0.02	28.942	1.00028	0.99972	0.99992	0.01837
0.04	14.481	1.00112	0.99888	0.99968	0.03672
0.06	9.6659	1.00252	0.99748	0.99928	0.05501
0.08	7.2616	1.00449	0.99553	0.99872	0.07323
0.10	5.8218	1.00702	0.99303	0.99800	0.09134
0.12	4.8643	1.01012	0.98998	0.99713	0.10932
0.14	4.1824	1.01379	0.98640	0.99610	0.12714
0.16	3.6727	1.01803	0.98228	0.99491	0.14479
0.18	3.2779	1.02286	0.97765	0.99356	0.16223
0.20	2.9635	1.02828	0.97250	0.99206	0.17944
0.22	2.7076	1.03429	0.96685	0.99041	0.19640
0.24	2.4956	1.04090	0.96070	0.98861	0.21309
0.26	2.3173	1.04813	0.95408	0.98666	0.22948
0.28	2.1656	1.05596	0.94700	0.98456	0.24556
0.30	2.0351	1.06443	0.93947	0.98232	0.26130
0.32	1.9219	1.07353	0.93150	0.97993	0.27670
0.34	1.8229	1.08329	0.92312	0.97740	0.29172
0.36	1.7358	1.09370	0.91433	0.97473	0.30636
0.38	1.6587	1.10478	0.90516	0.97193	0.32059
0.40	1.5901	1.11655	0.89561	0.96899	0.33442
0.42	1.5289	1.12902	0.88572	0.96592	0.34781
0.44	1.4740	1.14221	0.87550	0.96272	0.36076
0.46	1.4246	1.15612	0.86496	0.95940	0.37327
0.48	1.3801	1.17078	0.85413	0.95595	0.38531
0.50	1.3398	1.18621	0.84302	0.95238	0.39689
0.52	1.3034	1.20242	0.83165	0.94869	0.40799
0.54	1.2703	1.21944	0.82005	0.94489	0.41861
0.56	1.2403	1.23727	0.80823	0.94098	0.42874
0.58	1.2130	1.25596	0.79621	0.93696	0.43839
0.60	1.1882	1.27550	0.78400	0.93284	0.44754
0.62	1.1656	1.29594	0.77164	0.92861	0.45620
0.64	1.1451	1.31729	0.75913	0.92428	0.46437
0.66	1.1265	1.33959	0.74650	0.91986	0.47204
0.68	1.1097	1.36285	0.73376	0.91535	0.47922
0.70	1.0944	1.38710	0.72093	0.91075	0.48591

Mach No.	A/A^*	$P_{\bar{n}}$	P/P_0	T/T_0	MFP
0.72	1.0806	1.41238	0.70803	0.90606	0.49212
0.74	1.0681	1.43871	0.69507	0.90129	0.49784
0.76	1.0570	1.46612	0.68207	0.89644	0.50309
0.78	1.0471	1.49466	0.66905	0.89152	0.50787
0.80	1.0382	1.52434	0.65602	0.88652	0.51219
0.82	1.0305	1.55521	0.64300	0.88146	0.51605
0.84	1.0237	1.58730	0.63000	0.87633	0.51946
0.86	1.0179	1.62066	0.61703	0.87114	0.52243
0.88	1.0129	1.65531	0.60412	0.86589	0.52497
0.90	1.0089	1.69130	0.59126	0.86059	0.52710
0.92	1.0056	1.72868	0.57848	0.85523	0.52881
0.94	1.0031	1.76749	0.56578	0.84982	0.53012
0.96	1.0014	1.80776	0.55317	0.84437	0.53104
0.98	1.0003	1.84956	0.54067	0.83887	0.53159
1.00	1.0000	1.89293	0.52828	0.83333	0.53177
1.02	1.0003	1.93792	0.51602	0.82776	0.53159
1.04	1.0013	1.98457	0.50389	0.82215	0.53108
1.06	1.0029	2.03296	0.49189	0.81651	0.53023
1.08	1.0051	2.08313	0.48005	0.81085	0.52906
1.10	1.0079	2.13514	0.46835	0.80515	0.52759
1.12	1.0113	2.18905	0.45682	0.79944	0.52582
1.14	1.0153	2.24492	0.44545	0.79370	0.52377
1.16	1.0198	2.30282	0.43425	0.78795	0.52145
1.18	1.0248	2.36281	0.42322	0.78218	0.51888
1.20	1.0304	2.42497	0.41238	0.77640	0.51606
1.22	1.0366	2.48935	0.40171	0.77061	0.51301
1.24	1.0432	2.55605	0.39123	0.76481	0.50973
1.26	1.0504	2.62513	0.38093	0.75900	0.50625
1.28	1.0581	2.69666	0.37083	0.75319	0.50257
1.30	1.0663	2.77074	0.36091	0.74738	0.49870
1.32	1.0750	2.84745	0.35119	0.74158	0.49466
1.34	1.0842	2.92686	0.34166	0.73577	0.49045
1.36	1.0940	3.00908	0.33233	0.72997	0.48609
1.38	1.1042	3.09418	0.32319	0.72418	0.48159
1.40	1.1149	3.18227	0.31424	0.71839	0.47695
1.42	1.1262	3.27345	0.30549	0.71262	0.47220
1.44	1.1379	3.36780	0.29693	0.70685	0.46732
1.46	1.1501	3.46545	0.28856	0.70110	0.46235

(continued)

Mach No.	A/A^*	$P_{\bar{\eta}}$	P/P_0	T/T_0	MFP
1.48	1.1629	3.56649	0.28039	0.69537	0.45728
1.50	1.1762	3.67103	0.27240	0.68966	0.45212
1.52	1.1899	3.77919	0.26461	0.68396	0.44689
1.54	1.2042	3.89109	0.25700	0.67828	0.44158
1.56	1.2190	4.00684	0.24957	0.67262	0.43622
1.58	1.2344	4.12657	0.24233	0.66699	0.43080
1.60	1.2502	4.25041	0.23527	0.66138	0.42533
1.62	1.2666	4.37849	0.22839	0.65579	0.41983
1.64	1.2836	4.51095	0.22168	0.65023	0.41429
1.66	1.3010	4.64792	0.21515	0.64470	0.40873
1.68	1.3190	4.78955	0.20879	0.63919	0.40315
1.70	1.3376	4.93599	0.20259	0.63371	0.39755
1.72	1.3567	5.08739	0.19656	0.62827	0.39195
1.74	1.3764	5.24391	0.19070	0.62285	0.38634
1.76	1.3967	5.40570	0.18499	0.61747	0.38073
1.78	1.4175	5.57294	0.17944	0.61211	0.37513
1.80	1.4390	5.74580	0.17404	0.60680	0.36954
1.82	1.4610	5.92444	0.16879	0.60151	0.36397
1.84	1.4836	6.10906	0.16369	0.59626	0.35842
1.86	1.5069	6.29984	0.15873	0.59104	0.35289
1.88	1.5308	6.49696	0.15392	0.58586	0.34739
1.90	1.5553	6.70064	0.14924	0.58072	0.34192
1.92	1.5804	6.91106	0.14470	0.57561	0.33648
1.94	1.6062	7.12843	0.14028	0.57054	0.33108
1.96	1.6326	7.35297	0.13600	0.56551	0.32572
1.98	1.6597	7.58490	0.13184	0.56051	0.32040
2.00	1.6875	7.82445	0.12780	0.55556	0.31512
2.02	1.7160	8.07184	0.12389	0.55064	0.30989
2.04	1.7451	8.32731	0.12009	0.54576	0.30471
2.06	1.7750	8.59110	0.11640	0.54091	0.29958
2.08	1.8056	8.86348	0.11282	0.53611	0.29451
2.10	1.8369	9.14468	0.10935	0.53135	0.28949
2.12	1.8690	9.43499	0.10599	0.52663	0.28452
2.14	1.9018	9.73466	0.10273	0.52194	0.27961
2.16	1.9354	10.0440	0.09956	0.51730	0.27475
2.18	1.9698	10.3632	0.09649	0.51269	0.26996
2.20	2.0050	10.6927	0.09352	0.50813	0.26522
2.22	2.0409	11.0327	0.09064	0.50361	0.26055
2.24	2.0777	11.3836	0.08785	0.49912	0.25594

Mach No.	A/A^*	$P_{\overline{n}}$	P/P_0	T/T_0	MFP
2.26	2.1153	11.7455	0.08514	0.49468	0.25139
2.28	2.1538	12.1190	0.08251	0.49027	0.24690
2.30	2.1931	12.5043	0.07997	0.48591	0.24247
2.32	2.2333	12.9017	0.07751	0.48158	0.23811
2.34	2.2744	13.3116	0.07512	0.47730	0.23381
2.36	2.3164	13.7344	0.07281	0.47305	0.22957
2.38	2.3593	14.1704	0.07057	0.46885	0.22539
2.40	2.4031	14.6200	0.06840	0.46468	0.22128
2.42	2.4479	15.0836	0.06630	0.46056	0.21724
2.44	2.4936	15.5616	0.06426	0.45647	0.21325
2.46	2.5403	16.0544	0.06229	0.45242	0.20933
2.48	2.5880	16.5623	0.06038	0.44841	0.20547
2.50	2.6367	17.0859	0.05853	0.44444	0.20168
2.52	2.6865	17.6256	0.05674	0.44051	0.19794
2.54	2.7372	18.1818	0.05500	0.43662	0.19427
2.56	2.7891	18.7549	0.05332	0.43277	0.19066
2.58	2.8420	19.3455	0.05169	0.42895	0.18711
2.60	2.8960	19.9540	0.05012	0.42517	0.18362
2.62	2.9511	20.5809	0.04859	0.42143	0.18019
2.64	3.0073	21.2268	0.04711	0.41772	0.17682
2.66	3.0647	21.8920	0.04568	0.41406	0.17351
2.68	3.1233	22.5772	0.04429	0.41043	0.17026
2.70	3.1830	23.2829	0.04295	0.40683	0.16706
2.72	3.2440	24.0096	0.04165	0.40328	0.16393
2.74	3.3061	24.7579	0.04039	0.39976	0.16084
2.76	3.3695	25.5284	0.03917	0.39627	0.15782
2.78	3.4342	26.3217	0.03799	0.39282	0.15485
2.80	3.5001	27.1383	0.03685	0.38941	0.15193
2.82	3.5674	27.9789	0.03574	0.38603	0.14906
2.84	3.6359	28.8441	0.03467	0.38268	0.14625
2.86	3.7058	29.7346	0.03363	0.37937	0.14349
2.88	3.7771	30.6511	0.03263	0.37610	0.14079
2.90	3.8498	31.5941	0.03165	0.37286	0.13813
2.92	3.9238	32.5644	0.03071	0.36965	0.13552
2.94	3.9993	33.5627	0.02980	0.36647	0.13296
2.96	4.0763	34.5897	0.02891	0.36333	0.13046
2.98	4.1547	35.6461	0.02805	0.36022	0.12799
3.00	4.2346	36.7327	0.02722	0.35714	0.12558

(continued)

Mach No.	A/A^*	$P_{\overline{n}}$	P/P_0	T/T_0	MFP
3.20	5.1210	49.4370	0.02023	0.32808	0.10384
3.40	6.1837	66.1175	0.01512	0.30193	0.08600
3.60	7.4501	87.8369	0.01138	0.27840	0.07138
3.80	8.9506	115.889	0.00863	0.25720	0.05941
4.00	10.719	151.835	0.00659	0.23810	0.04961
4.20	12.792	197.548	0.00506	0.22085	0.04157
4.40	15.210	255.256	0.00392	0.20525	0.03496
4.60	18.018	327.595	0.00305	0.19113	0.02951
4.80	21.264	417.665	0.00239	0.17832	0.02501
5.00	25.000	529.090	0.00189	0.16667	0.02127
5.20	29.283	666.084	0.00150	0.15605	0.01816
5.40	34.175	833.523	0.00120	0.14637	0.01556
5.60	39.740	1037.02	0.00096	0.13751	0.01338
5.80	46.050	1283.02	0.00078	0.12940	0.01155
6.00	53.180	1578.88	0.00063	0.12195	0.01000
6.20	61.210	1932.94	0.00052	0.11510	0.00869
6.40	70.227	2354.70	0.00042	0.10879	0.00757
6.60	80.323	2854.83	0.00035	0.10297	0.00662
6.80	91.594	3445.38	0.00029	0.09758	0.00581
7.00	104.14	4139.84	0.00024	0.09259	0.00511
7.20	118.08	4953.30	0.00020	0.08797	0.00450
7.40	133.52	5902.58	0.00017	0.08367	0.00398
7.60	150.58	7006.41	0.00014	0.07967	0.00353
7.80	169.40	8285.51	0.00012	0.07594	0.00314
8.00	190.11	9762.85	0.00010	0.07246	0.00280
8.20	212.85	11463.8	0.00009	0.06921	0.00250
8.40	237.76	13416.1	0.00007	0.06617	0.00224
8.60	265.01	15650.6	0.00006	0.06332	0.00201
8.80	294.77	18200.7	0.00005	0.06065	0.00180
9.00	327.19	21103.3	0.00005	0.05814	0.00163
9.20	362.46	24398.4	0.00004	0.05578	0.00147
9.40	400.78	28129.9	0.00004	0.05356	0.00133
9.60	442.32	32345.3	0.00003	0.05146	0.00120
9.80	487.30	37096.4	0.00003	0.04949	0.00109
10.00	535.94	42439.2	0.00002	0.04762	0.00099

Appendix D Normal Shock Functions for a Perfect Gas

- One-Dimensional Normal Shock Flow
- Perfect Gas
- Constant Specific Heat
- Constant Molecular Weight

Schlerien photograph of the normal shock wave ahead of a blunt-nosed bullet. A pitot inlet (Chapter 15) going supersonic would have a similar shock structure. The total pressure loss across the normal shock makes the pitot inlet a poor design for supersonic flight.

A bad design with a good presentation is doomed eventually. A good design with a bad presentation is doomed immediately.

Akin's 20th Law of Spacecraft Design

D.1 One-Dimensional Normal Shock Functions for a Perfect Gas with Constant Specific Heat and Molecular Weight, $\gamma = 1.4$

For a one-dimensional normal shock as shown in Fig. D.1, the following relations are tabulated as functions of the upstream Mach number:

Downstream Mach number:

$$M_2 = \sqrt{\frac{M_1^2 - \left[2/(1-\gamma)\right]}{\left[2\gamma/(1-\gamma)\right]M_1^2 - 1}}$$

Stagnation pressure ratio:

$$\frac{P_{0_2}}{P_{0_1}} = \left[\frac{\left[(\gamma+1)/2\right]M_1^2}{1+\left[(\gamma-1)/2\right]M_1^2}\right]^{\gamma/(\gamma-1)} \left[\frac{2\gamma}{\gamma+1}M_1^2 - \frac{\gamma-1}{\gamma+1}\right]^{1/(1-\gamma)}$$

Pressure ratio:

$$\frac{P_2}{P_1} = \frac{2\gamma}{\gamma+1}M_1^2 - \frac{\gamma-1}{\gamma+1}$$

Density ratio:

$$\frac{\rho_2}{\rho_1} = \frac{P_2}{P_1}\frac{T_1}{T_2}$$

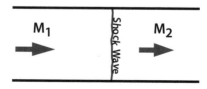

Figure D.1. One dimensional normal shock nomenclature

Temperature ratio:

$$\frac{T_2}{T_1} = \frac{\left(1 + \dfrac{\gamma-1}{2}M_1^2\right)\left(\dfrac{2\gamma}{\gamma-1}M_1^2 - 1\right)}{\dfrac{(\gamma+1)^2}{2(\gamma-1)}M_1^2}$$

Table D.1   One-Dimensional Normal Shock Functions for a Perfect Gas with Constant Specific Heat and Molecular Weight, $\gamma = 1.4$

M_1	M_2	(P_{02}/P_{01})	P_2/P_1	ρ_2/ρ_1	T_2/T_1
1.00	1.00000	1.00000	1.0000	1.00000	1.00000
1.02	0.98052	0.99999	1.0471	1.03344	1.01325
1.04	0.96203	0.99992	1.0952	1.06709	1.02634
1.06	0.94445	0.99975	1.1442	1.10092	1.03931
1.08	0.92771	0.99943	1.1941	1.13492	1.05217
1.10	0.91177	0.99893	1.2450	1.16908	1.06494
1.12	0.89656	0.99821	1.2968	1.20338	1.07763
1.14	0.88204	0.99726	1.3495	1.23779	1.09027
1.16	0.86816	0.99605	1.4032	1.27231	1.10287
1.18	0.85488	0.99457	1.4578	1.30693	1.11544
1.20	0.84217	0.99280	1.5133	1.34161	1.12799
1.22	0.82999	0.99073	1.5698	1.37636	1.14054
1.24	0.81830	0.98836	1.6272	1.41116	1.15309
1.26	0.80709	0.98568	1.6855	1.44599	1.16566
1.28	0.79631	0.98268	1.7448	1.48084	1.17825
1.30	0.78596	0.97937	1.8050	1.51570	1.19087
1.32	0.77600	0.97575	1.8661	1.55055	1.20353
1.34	0.76641	0.97182	1.9282	1.58538	1.21624
1.36	0.75718	0.96758	1.9912	1.62018	1.22900
1.38	0.74829	0.96304	2.0551	1.65494	1.24181
1.40	0.73971	0.95819	2.1200	1.68966	1.25469
1.42	0.73144	0.95306	2.1858	1.72430	1.26764
1.44	0.72345	0.94765	2.2525	1.75888	1.28066
1.46	0.71574	0.94196	2.3202	1.79337	1.29377
1.48	0.70829	0.93600	2.3888	1.82777	1.30695
1.50	0.70109	0.92979	2.4583	1.86207	1.32022
1.52	0.69413	0.92332	2.5288	1.89626	1.33357
1.54	0.68739	0.91662	2.6002	1.93033	1.34703
1.56	0.68087	0.90970	2.6725	1.96427	1.36057

(continued)

M_1	M_2	(P_{02}/P_{01})	P_2/P_1	ρ_2/ρ_1	T_2/T_1
1.58	0.67455	0.90255	2.7458	1.99808	1.37422
1.60	0.66844	0.89520	2.8200	2.03175	1.38797
1.62	0.66251	0.88765	2.8951	2.06526	1.40182
1.64	0.65677	0.87992	2.9712	2.09863	1.41578
1.66	0.65119	0.87201	3.0482	2.13183	1.42985
1.68	0.64579	0.86394	3.1261	2.16486	1.44403
1.70	0.64054	0.85572	3.2050	2.19772	1.45833
1.72	0.63545	0.84736	3.2848	2.23040	1.47274
1.74	0.63051	0.83886	3.3655	2.26289	1.48727
1.76	0.62570	0.83024	3.4472	2.29520	1.50192
1.78	0.62104	0.82151	3.5298	2.32731	1.51669
1.80	0.61650	0.81268	3.6133	2.35922	1.53158
1.82	0.61209	0.80376	3.6978	2.39093	1.54659
1.84	0.60780	0.79476	3.7832	2.42244	1.56173
1.86	0.60363	0.78569	3.8695	2.45373	1.57700
1.88	0.59957	0.77655	3.9568	2.48481	1.59239
1.90	0.59562	0.76736	4.0450	2.51568	1.60792
1.92	0.59177	0.75812	4.1341	2.54633	1.62357
1.94	0.58802	0.74884	4.2242	2.57675	1.63935
1.96	0.58437	0.73954	4.3152	2.60695	1.65527
1.98	0.58082	0.73021	4.4071	2.63692	1.67132
2.00	0.57735	0.72087	4.5000	2.66667	1.68750
2.02	0.57397	0.71153	4.5938	2.69618	1.70382
2.04	0.57068	0.70218	4.6885	2.72546	1.72027
2.06	0.56747	0.69284	4.7842	2.75451	1.73686
2.08	0.56433	0.68351	4.8808	2.78332	1.75359
2.10	0.56128	0.67420	4.9783	2.81190	1.77045
2.12	0.55829	0.66492	5.0768	2.84024	1.78745
2.14	0.55538	0.65567	5.1762	2.86835	1.80459
2.16	0.55254	0.64645	5.2765	2.89621	1.82188
2.18	0.54977	0.63727	5.3778	2.92383	1.83930
2.20	0.54706	0.62814	5.4800	2.95122	1.85686
2.22	0.54441	0.61905	5.5831	2.97837	1.87456
2.24	0.54182	0.61002	5.6872	3.00527	1.89241
2.26	0.53930	0.60105	5.7922	3.03194	1.91040
2.28	0.53683	0.59214	5.8981	3.05836	1.92853
2.30	0.53441	0.58329	6.0050	3.08455	1.94680
2.32	0.53205	0.57452	6.1128	3.11049	1.96522
2.34	0.52974	0.56581	6.2215	3.13620	1.98378

M_1	M_2	(P_{02}/P_{01})	P_2/P_1	ρ_2/ρ_1	T_2/T_1
2.36	0.52749	0.55718	6.3312	3.16167	2.00249
2.38	0.52528	0.54862	6.4418	3.18690	2.02134
2.40	0.52312	0.54014	6.5533	3.21190	2.04033
2.42	0.52100	0.53175	6.6658	3.23665	2.05947
2.44	0.51894	0.52344	6.7792	3.26117	2.07876
2.46	0.51691	0.51521	6.8935	3.28546	2.09819
2.48	0.51493	0.50707	7.0088	3.30951	2.11777
2.50	0.51299	0.49901	7.1250	3.33333	2.13750
2.52	0.51109	0.49105	7.2421	3.35692	2.15737
2.54	0.50923	0.48318	7.3602	3.38028	2.17739
2.56	0.50741	0.47540	7.4792	3.40341	2.19756
2.58	0.50562	0.46772	7.5991	3.42631	2.21788
2.60	0.50387	0.46012	7.7200	3.44898	2.23834
2.62	0.50216	0.45263	7.8418	3.47143	2.25896
2.64	0.50048	0.44522	7.9645	3.49365	2.27972
2.66	0.49883	0.43792	8.0882	3.51565	2.30063
2.68	0.49722	0.43070	8.2128	3.53743	2.32168
2.70	0.49563	0.42359	8.3383	3.55899	2.34289
2.72	0.49408	0.41657	8.4648	3.58033	2.36425
2.74	0.49256	0.40965	8.5922	3.60146	2.38576
2.76	0.49107	0.40283	8.7205	3.62237	2.40741
2.78	0.48960	0.39610	8.8498	3.64307	2.42922
2.80	0.48817	0.38946	8.9800	3.66355	2.45117
2.82	0.48676	0.38293	9.1111	3.68383	2.47328
2.84	0.48538	0.37649	9.2432	3.70389	2.49554
2.86	0.48402	0.37014	9.3762	3.72375	2.51794
2.88	0.48269	0.36389	9.5101	3.74341	2.54050
2.90	0.48138	0.35773	9.6450	3.76286	2.56321
2.92	0.48010	0.35167	9.7808	3.78211	2.58607
2.94	0.47884	0.34570	9.9175	3.80117	2.60908
2.96	0.47760	0.33982	10.055	3.82002	2.63224
2.98	0.47638	0.33404	10.194	3.83868	2.65555
3.00	0.47519	0.32834	10.333	3.85714	2.67901
3.20	0.46435	0.27623	11.780	4.03150	2.92199
3.40	0.45520	0.23223	13.320	4.18841	3.18021
3.60	0.44741	0.19531	14.953	4.32962	3.45373
3.80	0.44073	0.16447	16.680	4.45679	3.74260
4.00	0.43496	0.13876	18.500	4.57143	4.04688

(continued)

M_1	M_2	(P_{02}/P_{01})	P_2/P_1	ρ_2/ρ_1	T_2/T_1
4.20	0.42994	0.11733	20.413	4.67491	4.36657
4.40	0.42554	0.09948	22.420	4.76847	4.70171
4.60	0.42168	0.08459	24.520	4.85321	5.05233
4.80	0.41826	0.07214	26.713	4.93010	5.41842
5.00	0.41523	0.06172	29.000	5.00000	5.80000
5.50	0.40897	0.04236	35.125	5.14894	6.82180
6.00	0.40416	0.02965	41.833	5.26829	7.94059
6.50	0.40038	0.02115	49.125	5.36508	9.15643
7.00	0.39736	0.01535	57.000	5.44444	10.46939
7.50	0.39491	0.01133	65.458	5.51020	11.87948
8.00	0.39289	0.00849	74.500	5.56522	13.38672
8.50	0.39121	0.00645	84.125	5.61165	14.99113
9.00	0.38980	0.00496	94.333	5.65116	16.69273
9.50	0.38860	0.00387	105.125	5.68504	18.49152
10.00	0.38758	0.00304	116.500	5.71429	20.3875

Appendix E

Plane Oblique and Conical Shocks

- Plane Oblique & Conical Shocks
- Single- & Double-Ramp Inlet Data
- Single- & Double-Cone Inlet Data

Schlerien photograph of an oblique or conical shock attached to a sharp-nosed bullet. An external compression inlet (Chapter 15) going supersonic would have a similar shock structure. The total pressure loss across an oblique or conical shock is much less than across a normal shock.

Failure is not fatal,
but failure to change might be.

John Wooden

E.1 Perfect Gas with Constant Specific Heat and Molecular Weight, $\gamma = 1.4$

E.1.1 Oblique Shock

For a two-dimensional oblique shock as shown in Fig. E.1, the following property ratios apply across the shock:

$$\frac{P_2}{P_1} = f_1\left(M_1 \sin \beta\right)$$

$$\frac{P_2}{\rho_1} = f_2\left(M_1 \sin \beta\right)$$

$$\frac{T_2}{T_1} = f_3\left(M_1 \sin \beta\right)$$

$$\frac{P_{02}}{P_{01}} = f_4\left(M_1 \sin \beta\right)$$

$$M_2 \sin\left(\beta - \theta\right) = f_5\left(M_1 \sin \beta\right)$$

Because an oblique shock wave acts as a normal shock to the flow perpendicular to the shock, the normal shock relations can be applied to oblique shocks. The normal shock relations apply if M_1 and M_2 are replaced by their normal components $M_1 \sin \beta$ and $M_2 \sin(\beta - \theta)$. The relation between M_1 upstream Mach, number, β shock-wave angle, and θ flow-deflection angle for an oblique shock is

$$\frac{1}{M_1^2} = \sin^2 \beta - \frac{\gamma + 1}{2} \frac{\sin \beta \sin \theta}{\cos\left(\beta - \theta\right)}$$

which is plotted in Fig. E.2. Figure E.3 shows the pressure ratio P_2/P_1 across the plane oblique shock plotted as a function of θ and upstream Mach number M_1. Figure E.3 also shows M_2 behind the shock.

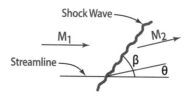

Figure E.1 Oblique shock nomenclature.

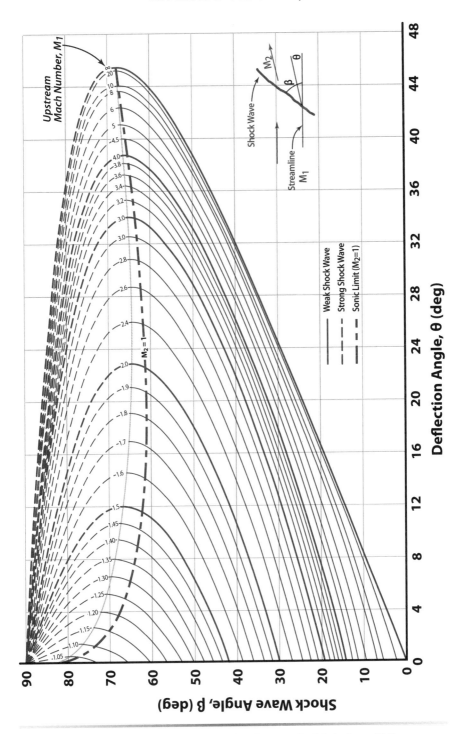

Figure E.2   Oblique shock deflection angle (data from [1]).

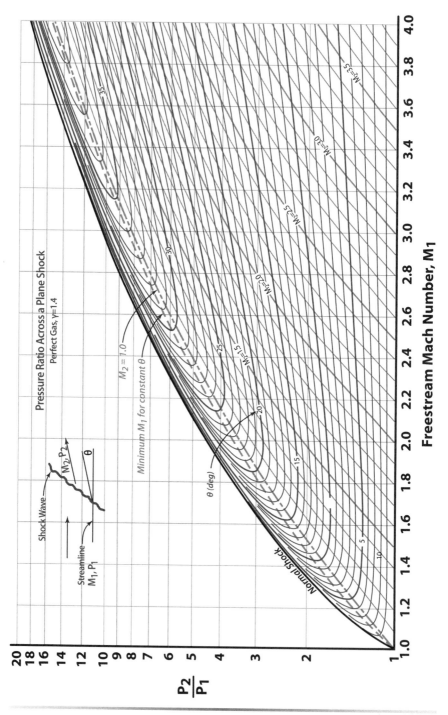

Figure E.3 Oblique shock pressure ratio (data from [1]).

E.1.2 Conical Shock

Consider a cone immersed in a supersonic stream. If the cone is sufficiently sharp, a conical shock will be attached to the vertex as shown in Fig. E.4.

The flow field for a cone is not as simple as for the two-dimensional oblique shock. In the latter case the flow region between the oblique shock and the compression surface is uniform. In the three-dimensional case of the cone, however, the flow region between the shock and the cone surface is not uniform.

Flow conditions are constant along rays emanating from the vertex. Between the shock and the cone, the conditions vary from one ray to another, and thus along the streamlines that cross them. Due to the three-dimensional effect, the compression by a cone is much weaker than for a two-dimensional wedge; therefore, shock detachment occurs at a much lower Mach number for a cone than for a wedge.

Plots of shock-wave angle β and surface pressure coefficient are presented in Figs. E.5 and E.6, respectively.

Mach number on the cone surface behind the shock is plotted in Fig. E.7. The inviscid total pressure recovery for various inlets is plotted in Figs. E.8–E.11.

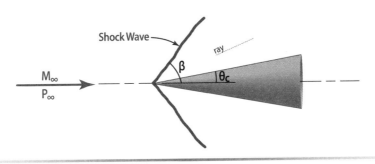

Figure E.4 Conical shock nomenclature.

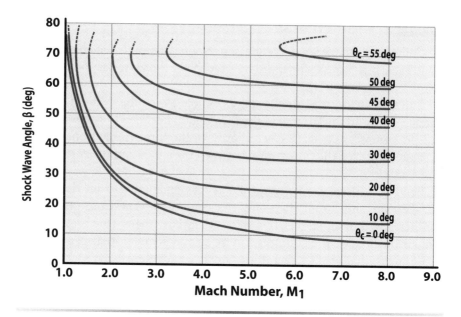

Figure E.5   Conical shock wave angle vs Mach number for various cone angles (data from [1]).

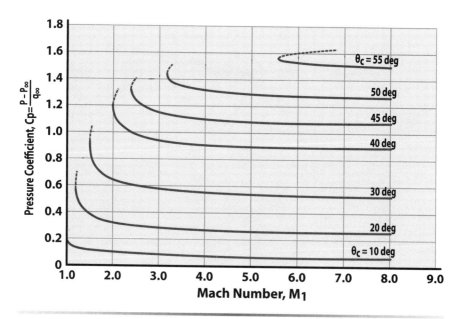

Figure E.6   Pressure coefficient on cone surface vs Mach number for various cone angles (data from [1]).

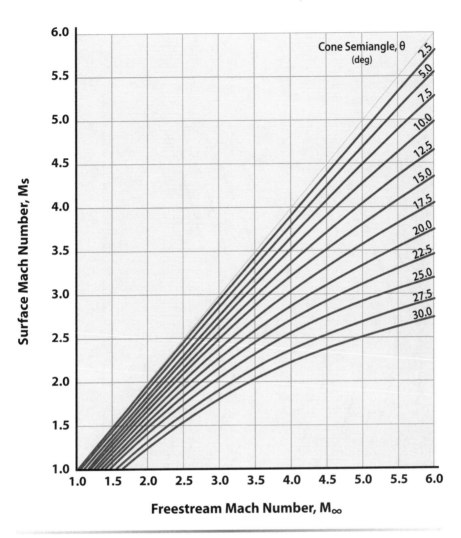

Figure E.7 Mach number on cone surface behind conical shock.

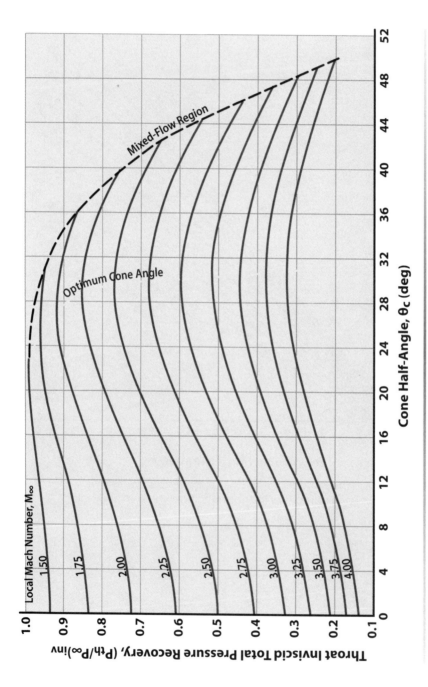

Figure E.8 Inviscid total pressure recovery for single-cone inlets.

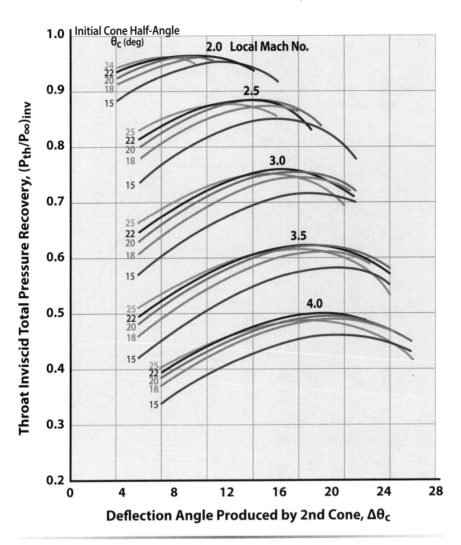

Figure E.9   Inviscid total pressure recovery for double-cone inlets.

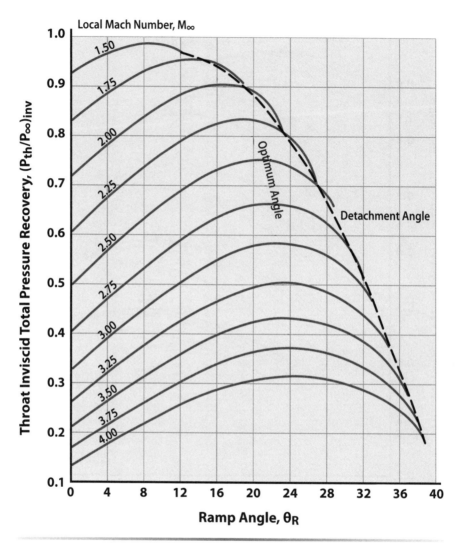

Figure E.10   Inviscid total pressure recovery for single-ramp inlets.

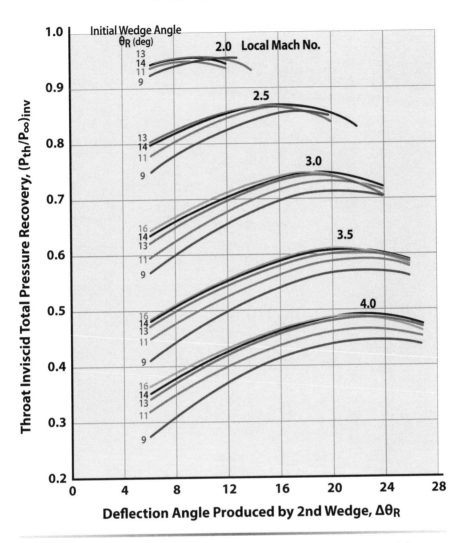

Figure E.11 Inviscid total pressure recovery for double-ramp inlets.

Reference

[1] "Equations, Tables, and Charts for Compressible Flow," NACA TR 1135, 1953.

Appendix F NACA Airfoil Nomenclature and Data

- NACA Airfoils
- Effect of Location of Maximum Thickness on Airfoil Performance
- Effect of Airfoil Maximum Thickness on Airfoil Performance
- Effect of Airfoil Camber on Drag Polar
- Airfoils on Various Aircraft

Flow streamlines representing the Kutta Condition.

People who hate their work are slaves, no matter how much they make.

Jim Clemmer

F.1 Introduction

The difference between aircraft and land machines like pick-up trucks is that aircraft can climb into the air and fly around. The feature that permits aircraft to fly is the wing, specifically its cross-sectional shape—referred to as the *airfoil*. The airfoil is carefully shaped so that the air flowing over the top surface of the wing goes further than the air flowing along the bottom surface. Air molecules traveling over the top and bottom wing surfaces arrive at the wing trailing edge at precisely the same time.

The flow along the bottom is prevented from turning around the trailing edge and flowing upstream by the pointed shape of the trailing edge (called the *Kutta Condition*). Because the flow on the upper surface must travel further, it has to go faster to arrive at the same time. As it goes faster, the static pressure in the flow decreases (known as *Bernoulli's Theorem*). Because the speed of the flow over the upper surface is greater than the flow over the lower surface, there is a lower surface pressure on the top than on the bottom. The difference between these two pressures is called *lift*.

A barn door at an angle of attack in an airstream produces lift by deflecting the airstream downward (remember Newton's second law—a reaction force results from a change in momentum). The problem with the barn door is that the drag produced is as big as the lift produced. The science of aerodynamics is to design airfoils that generate lift without producing a lot of drag. The efficiency of the airfoils is measured by their L/D, which is always much greater than one. This appendix provides aerodynamic data on many popular airfoil shapes.

This appendix will first discuss the airfoils developed by the National Advisory Committee for Aeronautics (NACA) between 1929 and 1945. The NACA airfoils were used by aircraft designers around the world and are still popular today. Next the airfoils used on many of the past and present aircraft will be tabulated in Section F.2 (Table F.2).

F.2 NACA Airfoil Nomenclature and Characteristics

The National Advisory Committee for Aeronautics was created, by Act of Congress approved 3 March 1915, for the supervision and direction of the scientific study of the problems of flight. The NACA had its headquarters in Washington, DC, but conducted most of its experimental research at three laboratories: the Langley Aeronautical Laboratory (Langley Field, Virginia), the Ames Aeronautical Laboratory (Moffett Field, California), and the Lewis Flight Propulsion Laboratory (Cleveland, Ohio). The NACA was redesignated the National Aeronautics and Space Administration

(NASA) in 1958, and their mission was broadened to include the study of problems of space flight as well as flight within the atmosphere.

In 1929, the NACA started a systematic investigation of airfoils. They developed families of airfoils and examined their aerodynamic characteristics. The most popular families of airfoils are the four-digit, the five-digit and the series-6 sections. The meanings of these designations are illustrated by the following examples, using the airfoil terminology of Fig. F.1.

NACA 4415

4 = maximum camber of the mean line is $0.04c$
4 = position of the maximum camber is at $0.4c$
15 = maximum thickness is $0.15c$

NACA 23012

2 = maximum camber of the mean line is approximately $0.02c$ (design lift coefficient is 0.15 times the first digit for this series)
30 = position of the maximum camber is at $0.30/2 = 0.15c$
12 = maximum thickness is $0.12c$

NACA 65₃-421

6 = series designation
5 = minimum pressure is at $0.5c$
3 = drag coefficient is near its minimum value over a range of lift coefficients of 0.3 above and below the design lift coefficient
4 = design lift coefficient is 0.4
21 = maximum thickness is $0.21c$

NACA Report No. 824 (1945) presents a summary of the airfoil data gathered during the extensive systematic investigation begun in 1929. Much of this airfoil data is published in [1]. Airfoil data for the more popular airfoils are presented in Table F.1.

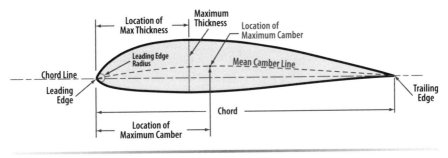

Figure F.1 Airfoil terminology.

Table F.1 Low-Speed Data on Airfoil Sections, $Re = 9 \times 10^6$ [1]

Airfoil	α_{0L} (deg)	Cm_0	C_{ℓ_α} (per deg)	a.c.	$\alpha_{c_{\ell_{max}}}$ (per deg)	$C_{\ell_{max}}$	α^{*a} (deg)
0006	0	0	0.108	0.250	9.0	0.92	9.0
0009	0	0	0.109	0.250	13.4	1.32	11.4
1408	−0.8	−0.023	0.109	0.250	14.0	1.35	10.0
1410	−1.0	−0.020	0.108	0.247	14.3	1.50	11.0
1412	−1.1	−0.025	0.108	0.252	15.2	1.58	12.0
2412	−2.0	−0.047	0.105	0.247	16.8	1.68	9.5
2415	−2.0	−0.049	0.106	0.246	16.4	1.63	10.0
2418	−2.3	−0.050	0.103	0.241	14.0	1.47	10.0
2421	−1.8	−0.040	0.103	0.241	16.0	1.47	8.0
2424	−1.8	−0.040	0.098	0.231	16.0	1.29	8.4
4412	−3.8	−0.093	0.105	0.247	14.0	1.67	7.5
4415	−4.3	−0.093	0.105	0.245	15.0	1.64	8.0
4418	−3.8	−0.088	0.105	0.242	14.0	1.53	7.2
4421	−3.8	−0.085	0.103	0.238	16.0	1.47	6.0
4424	−3.8	−0.082	0.100	0.239	16.0	1.38	4.8
23012	−1.4	−0.014	0.107	0.247	18.0	1.79	12.0
23015	−1.0	−0.007	0.107	0.243	18.0	1.72	10.0
23018	−1.2	−0.005	0.104	0.243	16.0	1.60	11.8
23021	−1.2	0	0.103	0.238	15.0	1.50	10.3
23024	−0.8	0	0.097	0.231	15.0	1.40	9.7
63-006	0	0.005	0.112	0.258	10.0	0.87	7.7
-009	0	0	0.111	0.258	11.0	1.15	10.7
63-206	−1.9	−0.037	0.112	0.254	10.5	1.06	6.0
-209	−1.4	−0.032	0.110	0.262	12.0	1.4	10.8
-210	−1.2	−0.035	0.113	0.261	14.5	1.56	9.6
63$_1$-012	0	0	0.116	0.265	14.0	1.45	12.8
-212	−2.0	−0.035	0.114	0.263	14.5	1.63	11.4
-412	−2.8	−0.075	0.117	0.271	15.	0	1.77
63$_2$-015	0	0	0.117	0.271	14.5	1.47	11.0
-215	−1.0	−0.030	0.116	0.267	15.0	1.60	8.8
-415	−2.8	−0.069	0.118	0.262	15.0	1.68	10.0
-615	−3.6	−0.108	0.117	0.266	15.0	1.67	8.6
63$_3$-018	0	0	0.118	0.271	15.5	1.54	11.2
-218	−1.4	−0.033	0.118	0.271	14.5	1.85	8.0
-418	−2.7	−0.064	0.118	0.272	16.0	1.57	7.0
-618	−3.8	−0.097	0.118	0.267	16.0	1.59	4.2

Airfoil	α_{0L} (deg)	Cm_0	C_{ℓ_α} (per deg)	a.c.	$\alpha_{c_{\ell_{max}}}$ (per deg)	$C_{\ell_{max}}$	α^{*a} (deg)
63_4-021	0	0	0.118	0.273	17.0	1.38	9.0
-221	-1.5	-0.035	0.118	0.269	15.0	1.44	9.2
-421	-2.8	-0.062	0.120	0.275	16.0	1.48	6.7
63,4-420	-2.2	-0.059	0.109	0.265	14.0	1.42	7.6
63,4-420 $_{a=3}$	-2.4	-0.037	0.111	0.265	16.0	1.35	6.0
63(420)-422	-3.2	-0.065	0.112	0.271	14.0	1.36	6.0
63(420)-517	-3.0	-0.084	0.108	0.264	15.0	1.60	8.0
64-006	0	0	0.109	0.256	9.0	0.8	7.2
-009	0	0	0.110	0.262	11.0	1.17	10.0
64-108	0	-0.015	0.110	0.255	10.0	1.1	10.0
-110	-1.0	-0.020	0.110	0.261	13.0	1.4	10.0
64-206	-1.0	-0.040	0.110	0.253	12.0	1.03	8.0
-208	-1.2	-0.039	0.113	0.257	10.5	1.23	8.8
-209	-1.5	-0.040	0.107	0.261	13.0	1.40	8.9
-210	-1.6	-0.040	0.110	0.258	14.0	1.45	10.8
64_1-012	0	0	0.111	0.262	14.5	1.45	11.0
-112	-0.8	-0.017	0.113	0.267	14.0	1.50	12.2
212	-1.3	-0.027	0.113	0.262	15.0	1.55	11.0
-412	-2.6	-0.065	0.112	0.267	15.0	1.67	8.0
64_2-015	0	0	0.112	0.267	15.0	1.48	13.0
-215	-1.6	-0.030	0.112	0.265	15.0	1.57	10.0
-415	-2.8	-0.070	0.115	0.264	15.0	1.65	8.0
64_3-018	0	0.004	0.111	0.266	17.0	1.50	12.0
-218	-1.3	-0.027	0.115	0.271	16.0	1.53	10.0
-418	-2.9	-0.065	0.116	0.273	14.0	1.57	8.0
-618	-3.8	-0.095	0.116	0.273	16.0	1.58	5.6
64_4-021	+0.005	-0.029	0.110	0.274	14.0	1.30	10.3
-221	-1.2	-0.029	0.117	0.271	13.0	1.32	6.8
-421	-2.8	-0.068	0.120	0.276	13.0	1.42	6.4
65-006	0	0	0.105	0.258	12.0	0.92	7.6
-009	0	0	0.107	0.264	11.0	1.08	9.8
65-206	-1.6	-0.031	0.105	0.257	12.0	1.03	6.0
-209	-1.2	-0.031	0.106	0.259	12.0	1.30	10.0
-210	-1.6	-0.034	0.108	0.262	13.0	1.40	9.6
65-410	-2.5	-0.067	0.112	0.262	14.0	1.52	8.0
65_1-012	0	0	0.110	0.261	14.0	1.36	10.0
-212	-1.0	-0.032	0.108	0.261	14.0	1.47	9.4

(continued)

Airfoil	α_{0L} (deg)	Cm_0	C_{ℓ_α} (per deg)	a.c.	$\alpha_{c_{\ell max}}$ (per deg)	$C_{\ell max}$	α^{*a} (deg)
-212 $_{a=6}$	-1.4	-0.033	0.108	0.269	14.0	1.50	9.6
-412	-3.0	-0.070	0.111	0.265	15.5	1.66	10.5
65$_2$-015	0	0	0.111	0.257	15.0	1.42	11.2
-215	-1.2	-0.032	0.112	0.269	15.5	1.53	10.0
-415	-2.6	-0.060	0.111	0.268	16.0	1.61	8.7
-415 $_{a=5}$	-2.6	-0.051	0.111	0.264	20.0	1.60	7.0
65(215)-114	-0.7	-0.019	0.112	0.265	15.0	1.44	10.5
65(216)-415 $_{a=5}$	-3.0	-0.057	0.106	0.267	18.0	1.60	6.0
65,3-018	0	0	0.100	0.262	17.0	1.44	10.0
-418 $_{a=8}$	-3.0	-0.081	0.112	0.266	20.0	1.58	4.4
-618	-4.0	-0.100	0.110	0.273	20.0	1.60	4.9
65$_3$-018	0	0	0.100	0.267	16.0	1.37	10.0
-218	-1.2	-0.030	0.100	0.263	18.0	1.48	8.8
-418	-2.4	-0.059	0.110	0.265	18.0	1.54	4.9
-418 $_{a=5}$	-2.8	-0.055	0.115	0.267	18.0	1.50	6.0
-618	-4.0	-0.102	0.113	0.276	18.0	1.64	5.2
-618 $_{a=5}$	-4.2	-0.078	0.104	0.265	20.0	1.51	5.3
65$_4$-021	0	0	0.112	0.267	18.5	1.40	7.4
-221	-1.3	-0.029	0.115	0.274	20.5	1.46	6.0
-421	-2.8	-0.066	0.116	0.272	22.0	1.56	5.0
-421 $_{a=5}$	-2.8	-0.052	0.116	0.272	20.0	1.43	5.6
65(421)-420	-2.4	-0.061	0.116	0.276	20.0	1.52	4.7
66-006	0	0	0.100	0.252	9.0	0.80	6.5
-009	0	0	0.103	0.259	10.0	1.05	10.0
66-206	-1.6	-0.038	0.108	0.257	10.5	1.00	7.0
-209	-1.0	-0.034	0.107	0.257	11.0	1.17	9.0
-210	-1.3	-0.035	0.110	0.261	11.0	1.27	10.0
66$_1$-012	0	0	0.106	0.258	14.0	1.25	11.2
-212	-1.2	-0.032	0.102	0.259	15.0	1.46	11.6
66$_2$-015	0	0.005	0.105	0.265	15.5	1.35	12.0
-215	-1.3	-0.031	0.106	0.200	16.0	1.50	11.4
-415	-2.6	-0.069	0.106	0.260	17.0	1.60	10.0
66(215)-016	0	0	0.105	0.260	14.0	1.33	10.0
-216	-2.0	-0.044	0.114	0.262	16.0	1.55	8.8
-216 $_{a=6}$	-1.2	-0.030	0.100	0.257	16.0	1.46	7.0
-416	-2.6	-0.068	0.100	0.265	18.0	1.60	4.0
63A010	0	0.005	0.105	0.254	13.0	1.20	10.0

Airfoil	α_{0L} (deg)	Cm_0	C_{ℓ_α} (per deg)	a.c.	$\alpha_{C_{\ell max}}$ (per deg)	$C_{\ell max}$	α^{*a} (deg)
63A210	−1.5	−0.040	0.103	0.257	14.0	1.43	10.0
64A010	0	0	0.110	0.253	12.0	1.23	10.0
64A210	−1.5	−0.040	0.105	0.251	13.0	1.44	10.0
64A410	−3.0	−0.080	0.100	0.254	15.0	1.61	10.0
64₁A212	−2.0	−0.040	0.100	0.252	14.0	1.54	11.0
64₂A215	−2.0	−0.040	0.095	0.252	15.0	1.50	12.0

$^a\alpha^*$ = angle of attack at which lift curve ceases to be linear (incipient stall).
Note: C_{M0} is about the aerodynamic center (a.c.).

Data for the airfoils used on the world's aircraft, past and present, are available from the following website: www.ae.uiuc.edu/m-selig/ads/aircraft.html. Table F.2 contains airfoil data on some of the more popular U.S. aircraft. The geometry and performance effects of some NACA airfoils are plotted in Figs. F.2–F.5. Figures F.3, F.4, and F.5 are discussed in Sections 7.3, 7.2, and 7.5 respectively.

Table F.2   Aircraft Airfoils

Company	Aircraft	Airfoil (Root)
AAI	RQ-2 Pioneer	NACA 4415
AAI	Shadow 200/400	NACA 4415
AerMacchi	MB. 339	NACA 64A114
AerMacchi	MC. 202/205	NACA 23018
AerMacchi	MC. 72	Biconvex
AerMacchi	SF-260	NACA 64-212
Aero Commander	Shrike 500/600/720	NACA 23012
Aero	L-139/39/59	NACA 64A012
AeroVironment	Gossamer Condor	Lissaman 7769
AeroVironment	Helios	Selig S6078
AeroVironment	Pathfinder	Liebeck LA2573A
Airbus	A300B	15%
Avro	Vulcan	NACA 0010 mod
Avro	CF-105 Arrow	NACA 0003.5 mod
Bede	BD-4/6/7	NACA 64-415
Bede	BD-5	NACA 64-212
Beech	Bonanza/Lightning	NACA 23016.5
Beech	King Air	NACA 23016.5
Beech	Queen Air	NACA 23018

(continued)

Company	Aircraft	Airfoil (Root)
Bell	P-39 Aircobra	NACA 0015
Bell	P-59 Aircomet	NACA 66-014
Bell	P-63 Kingcobra	NACA 66-116
Bell	X-5	NACA 64A011
Bellanca	Citabria	NACA 4412
Boeing	B-17	NACA 0018
Boeing	B-29	Boeing 117 (22%)
Boeing	707/727/737/747/757	Boeing airfoils (12%–15%)
Boeing	Condor	Liebeck LD-17A
Boeing	C-17	DLBA 142
Boeing	F-15 Eagle	NACA 64A006.6
Boeing	F-18	NACA 65A005 mod
Breguet	941	NACA 63A416
BAE	AV-8 Harrier	Hawker 10%
Cessna	150/152/172/180/182/206	NACA 2412
Cessna	208 Caravan	NACA 23017
Cessna	310/L-27/U-3	NACA 2412
Cessna	T-37	NACA 23018
Cessna	Citation I/II/V	NACA 23014
Convair	B-58 Hustler	NACA 0003.46
Convair	F-102	Delta Dagger
Convair	F-106 Delta Dart	NACA 0004-65 mod
DeHavilland	DH-106 Comet	NACA 63A116 mod
Douglas	A-4 Skyhawk	NACA 0008-1.1-25
Douglas	DC-3	NACA 2215
Douglas	DC-4	NACA 23016
Douglas	DC-6	NACA 23016
Douglas	DC-8/DC-9/DC-10	DSMA
Douglas	D-558-II Skyrocket	NACA 63-010
Fairchild	A-10 Thunderbolt II	NACA6716
Ford	Trimotor	Goettingen 386
General Atomics	RQ-1 Predator	A
General Atomics	RQ-9	Predator B
General Dynamics	F-111	NACA 64-210.68
Grumman	SA-16/HU-16 Albatross	NACA 23017
Grumman	E-2 Hawkeye	NACA 63A216
Grumman	A-6 Intruder	NACA 64A009 mod
Grumman	F-14 Tomcat	NACA 64A209.65 mod
Grumman	F9F Cougar	NACA 64A010

Company	Aircraft	Airfoil (Root)
Gulfstream	GII/GIII/GIV	NACA 0012 mod
Hawker	Hurricane	Clark YH (19%)
Hawker	Typhoon	NACA 2219
Hughes	H-1 (Long Wing)	NACA 23016.5
Hughes	H-1 (short wing)	NACA 2418
Hughes	H-4 Hercules	NACA 63(420)-321
Learjet	23–60	NACA 64A109
Lesher	(Univ of Mich)Teal	NACA 63A615
Lockheed	P-80 Shooting Star	NACA 65-213
Lockheed	P-38 Lightning	NACA 23016
Lockheed	F-94	NACA 65-213
Lockheed	P-3 Orion/Electra	NACA 0014-1.1
Lockheed	U-2A/R/S	NACA 64A409
Lockheed	C-130	NACA 64A318
Lockheed	L-1011 Tristar	Lockheed airfoil (12.4%)
Lockheed	F-117 Nighthawk	3 flats upper, 2 lower
Lockheed Martin	F-16	NACA 64A204
Lockheed Martin	Polecat	LM airfoil
McDonnell Douglas	F-4 Phantom	NACA 0006.4-64 mod
McDonnell	F-101 Voodoo	NACA 65A007 mod
Messerschmitt	Bf 109	NACA 2R1 14.2
Messerschmitt	Bf 110/161/162	NACA 2R1 18.5
Messerschmitt	Me 209	NACA 2R1 16
Messerschmitt	Me 263	Me 1.8 25 14-1.1-30
Mooney	(All)	NACA 63-215
North American	P-51B/C/D Mustang	NACA 45-100
North American	B-25 Mitchell	NACA 23017
North American	F-86F	NACA 0009-64 mod
North American	F-100C/D	NACA 64A007
North American	T-39 Sabreliner (−40/60)	NACA 64A212
Northrop	B-2 Spirit	Modified supercritical
Northrop	F-5 Tiger	NACA 65A004.8
Northrop	F-20 Tigershark	NACA 65A004.8
Northrop	T-38 Talon	NACA 65A004.8
Northrop	Tacit Blue	Clark Y mod
Northrop Grumman	RQ-4A Global Hawk	NASA LRN 1015
Pilatus	PC-12	NASA LS(1)-0417
Pilatus	PC-6 Turbo Porter	NACA 64-514

(continued)

Company	Aircraft	Airfoil (Root)
Pilatus	PC-7 Turbo Trainer	NACA 64A415
Pilatus	PC-8B Twin Porter	NACA 64-514
Pilatus	PC-9	PIL15M825
Piper	J-3 Cub	USA 35B
Piper	L-14	USA 35B
Piper	PA-22 Tripacer	USA 35B
Piper	PA-24 Comanche	NACA64A215
Piper	PA-28 Cherokee	NACA 65-415
Piper	PA-31 Cheyenne	NACA 63A415
Piper	PA-32 Saratoga	NACA 65-415
Piper	PA-34 Seneca	NACA 65-415
Piper	PA-38 Tomahawk	NASA GA(W)-1
Piper	PA-40 Arapaho	NACA 64A215
Piper	PA-46 Malibu	NACA 23015
Piper	PA-48 Enforcer	NACA 45-100
Pitts	S-1C/D	NACA M-6
Pitts	S-1E	Symmetrical
Republic	F-84F Thunderstreak	NACA 64A010
Republic	F-105 Thunderchief	NACA 65A005.5
Republic	P-47 Thunderbolt	Seversky S-3
Rutan	VariViggen	Roncz R1145MS
Rutan	Proteus	NACA 4414
Rutan	Varieze	NACA 4414
Rutan	Pond Racer	Roncz
Rutan	Raptor	Roncz RQW17B
Rutan	Global Flyer	Roncz
Rutan	Space Ship One	Rutan
Rutan	White Knight	Hatfield
Ryan	AQM-34 Firebee	NACA 10% hybrid
Ryan	XV-5A	NACA 0012-64
Ryan	Supersonic Firebee	Symmetrical 3%
Ryan	Spirit of St. Louis	Clark Y
Ryan	Navion	NACA 4415R
Ryan	RQ-4 Global Hawk	NASA LRN 1015
Schempp-Hirth	Nimbus II	Wortmann FX 67-K-170
Schempp-Hirth	Nimbus 3	Goettingen 681
Schleicher	ASW 20C	Wortmann FX 62-131 (14.4)
Schleicher	ASW 22	Wortmann FX S-02-196
Schleicher	ASW 23	Wortmann FX 61-168

Company	Aircraft	Airfoil (Root)
Schleicher	ASW 24	Delft DU89-158
Schleicher	ASW 28	Delft DU99-146
Schleicher	Ka-6 Rhonsegler	NACA 63-618
Slingsby	T.4 Falcon	Goettingen 535 mod
Slingsby	T.41 Skylark 2	NACA 63-620
Slingsby	T.67 Firefly	NACA 23015
Stinson	Reliant	Clark Y
Stoddard-Hamilton	Glasair	NASA GA(W)-2
Stoddard-Hamilton	Glastar	NASA GA(W)-2
Supermarine	Spitfire	NACA 2213
Swearingen	Queen Air 800	NACA 23018
Swearingen	Merlin III/IV	NACA 65A215

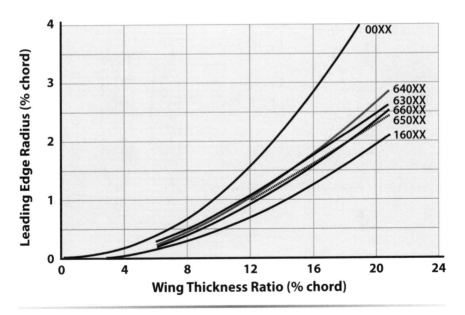

Figure F.2   NACA Airfoil Geometry.

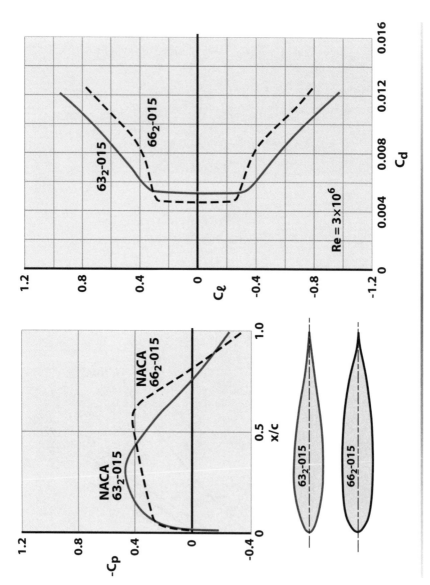

Figure F.3 Effect of Location of Maximum Thickness (Minimum Pressure) on Airfoil Performance.

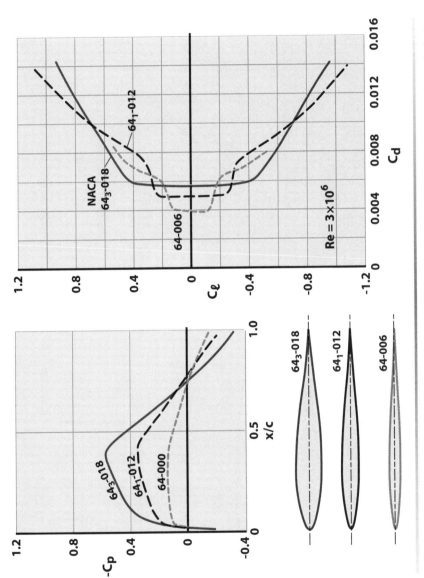

Figure F.4 Effect of Airfoil Thickness on Airfoil Performance (Pressure Distribution and Drag Polar).

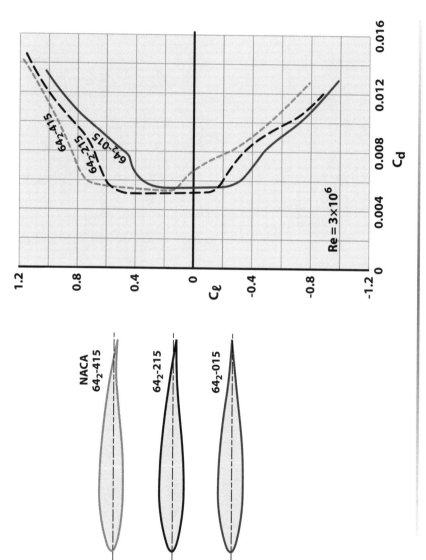

Figure F.5 Effect of Airfoil Camber on Drag Polar.

F.3 Selecting an Airfoil

Some considerations for selecting the airfoil are as follows:

Dominant Aircraft Feature	Airfoil Feature
Low supersonic drag	Symmetric, small LE radius and $t/c < 5\%$ because $C_{Dwave} = f[\text{camber}, (t/c)^2]$
Laminar flow wing	Location of max $t/c > 45\%$ chord
Low C_{Dmin}	Laminar flow wing & low max t/c
Operate at high C_L	High camber (or TE flap)
High max C_L	Large LE radius (or LE slot/slat)
Low viscous drag due to lift	Large LE radius
Low inviscid drag due to lift (induced)	Nothing
Low wing weight	Large t/c
Large wing volume	Large t/c
Tailless (low C_{mo})	Location of max $t/c < 30\%$ chord & low camber
High subsonic critical Mach Number	Low t/c

Example F.1   Lockheed Martin Polecat

In 2004 the Lockheed Martin Skunk Works built a small ISR aircraft on company funds called Polecat (Fig. F.6). The Polecat specifications are as follows:

Mission	High altitude ISR
Speed	Mach 0.5
Propulsion	2 FJ-44-3E (Appendix J)
TOGW	9000 lb
Empty weight	6384 lb
Payload + fuel	2500 lb
Wing AR	12
Wing area	679 ft^2
Wing sweep	23 deg

Because Polecat was a high-altitude ISR, tailless aircraft the airfoil needed to have the following attributes:

Moderate to high operating C_L	Aft camber for $C_{Lmin} > 0$
Tailless	Near zero C_{mo} at operating $C_L = 0.7$
Wing volume for fuel	$t/c > 12\%$ but OK for Mach 0.5
Able to trim out negative C_{mo}	Reflex TE (inverse camber)

The airfoil selected was the JW 1416 shown in Fig. 2.4. Notice that the airfoil is 16% thick, front loaded (max thickness at 30%) giving a

Figure F.6 Lockheed Martin Polecat unmanned aircraft.

near zero C_{mo} at $C_l = 0.7$, modest camber with reflex at the TE. The airfoil is not a laminar flow section as the adverse pressure gradient starts about 30% chord.

In the early 1950s Cessna Aircraft was developing a family of light aircraft for the general aviation public (that is, C 150, C 152, C 172, C 177, and C 180). They wanted an airfoil that was well understood, good t/c for light wing structure, docile stall characteristics, modest camber for good section maximum lift, and large r_{LE} for low viscous drag-due-to-lift factor K''. They chose the NACA 2412. The section characteristics are listed in Table F.1 and shown in Fig. F.7. The section maximum lift coefficient of 1.5 gave a respectable wing $C_{L_{max}}$ of 1.6 with a half-span plain flap. The $K'' = \Delta(C_d - C_{d_0})/(C_l - C_{l_{min}})^2 = 0.0047$ at $Re = 2.9 \times 10^6$.

In the early 1980s DARPA commissioned the Boeing Company to build a long endurance, propeller, ISR aircraft called Condor (discussed in Appendix G). In order to keep the wing small at 60,000 ft and 200 kt, Boeing wanted a highly cambered airfoil with a $C_{l_{min}} > 1.0$. Boeing wanted a high $(L/D)_{max}$ which meant max $t/c > 35\%$ chord for laminar flow, high aspect ratio for low K', and large r_{LE} for low K''. The airfoil LD-17A was developed by Dr. Robert Liebeck of Boeing and is shown in Fig. F.7. Notice that the airfoil is highly cambered with $\alpha_{OL} = -8$ deg and $C_{l_{max}} = 1.7$. Notice also that $C_{l_{min}} = 1.2$ and the airfoil drag polar is not symmetrical about that design C_l. The viscous

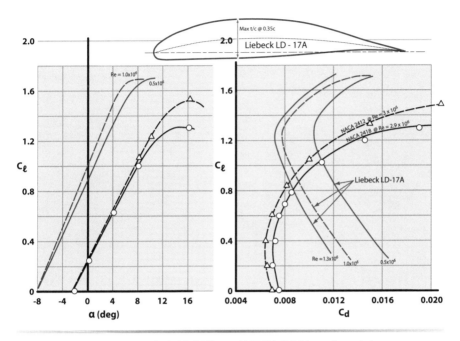

Figure F.7 Airfoils LD-17A and NACA 24XX section data.

drag-due-to-lift factor K''' is different above and below $C_{l_{min}} = 1.2$ as shown here:

Condition	Re	K'''
Below $C_{l_{min}}$	5×10^5	0.0074
Below $C_{l_{min}}$	1×10^6	0.0066
Below $C_{l_{min}}$	1.3×10^6	0.006
Above $C_{l_{min}}$	5×10^5	0.0175
Above $C_{l_{min}}$	1×10^6	0.0193
Above $C_{l_{min}}$	1.3×10^6	0.021

The aerodynamics for the Condor is discussed in Appendix G.2.

Reference

[1] Abbott, I. H., and von Doenhoff, A. E., *Theory of Wing Sections*, Dover, New York, 1959.

Appendix G Aerodynamic Data of Real Aircraft

- Subsonic & Supersonic Drag Polars
- C_{D_0} & $C_{D_{min}}$ vs Mach Number
- Maximum L/D Correlation Curve
- Subsonic C_{D_0} Correlation Curve
- Wing Efficiency Factor Correlation
- Boeing/DARPA Condor ISR Aircraft

Cessna 172 Skyhawk, a four-seat, single-engine, high-wing, unpressurized general aviation aircraft. First flown in 1955, it is still in production. Over 43,000 have been built, more than any other aircraft. The number of variants have been relatively few, a testament to the original design. A case study appears in Volume II.

Few things are created and perfected at the same time.

G.1 How To Use Aerodynamic Data

This appendix contains aerodynamic data on real aircraft. There is drag polar data on propeller aircraft, turbine transports and military cargo aircraft, fighter aircraft, and intelligence, surveillance, and reconnaissance (ISR) aircraft (Figs. G.1–G.9). All aero coefficients are referenced to the aircraft reference area (S_{ref}), which is the total planform wing area. These data are to be used by the designer as a starting point for initial sizing and then as a sanity check on the aero estimates.

When using these data the designer must understand the mission of the aircraft because the mission drives the configuration, which drives the resulting aerodynamics. When designing an apple, use the data from other apples. It is interesting to observe that the aero data for wing–body–tail

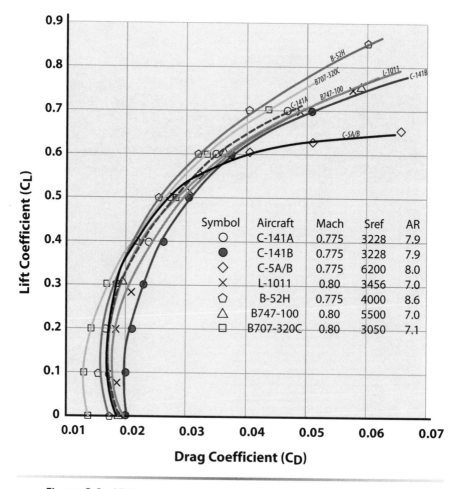

Symbol	Aircraft	Mach	S_{ref}	AR
O	C-141A	0.775	3228	7.9
●	C-141B	0.775	3228	7.9
◇	C-5A/B	0.775	6200	8.0
✕	L-1011	0.80	3456	7.0
⬠	B-52H	0.775	4000	8.6
△	B747-100	0.80	5500	7.0
☐	B707-320C	0.80	3050	7.1

Figure G.1 High-subsonic drag polars for large transport aircraft—flight test data.

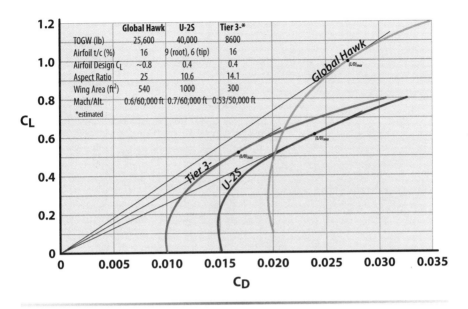

Figure G.2 Drag polars for ISR aircraft.

transport and fighter configurations collapse into a relatively tight band for the subsonic drag polars (see Figs. G.1, G.4, and G.6) and $C_{D_{min}}$ vs Mach number (see Fig. G.5). However the $C_{D_{min}}$ data for swept-wing tailless configurations fall well outside this band (see the F-106 and B-70 in Fig. G.5).

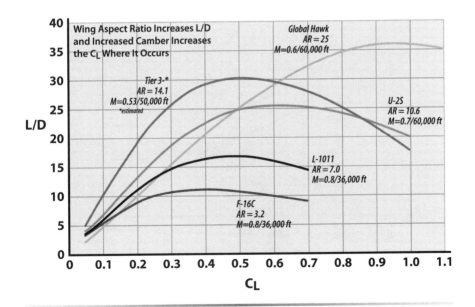

Figure G.3 L/D vs C_l for various aircraft.

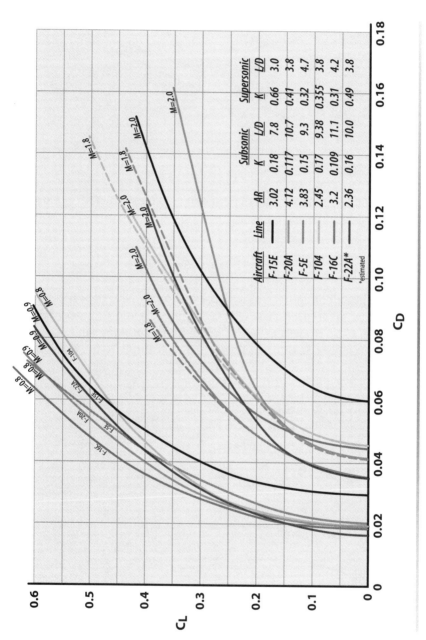

Figure G.4 Drag polars for fighter aircraft.

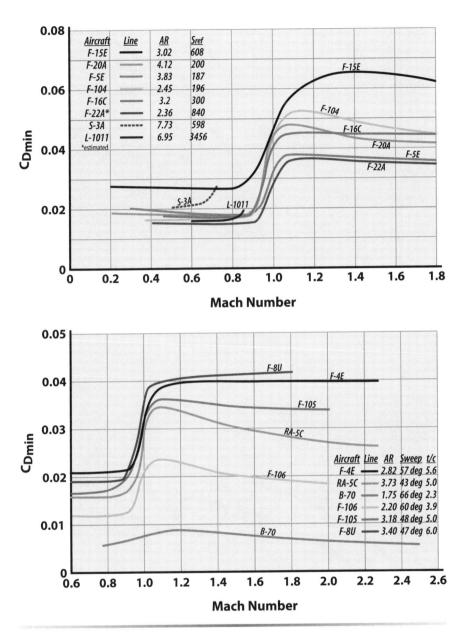

Figure G.5 $C_{D_{min}}$ vs Mach number (flight test data).

Figure G.3 shows a useful way to present aircraft aero data. Plotting L/D versus C_L shows the maximum L/D and the operating C_L region. This C_L is used to select the airfoil and to size the wing.

As discussed in Section 2.21 the subsonic C_{D_0} or $C_{D_{min}}$ is a function of the aircraft wetted area divided by the reference area. This is because the

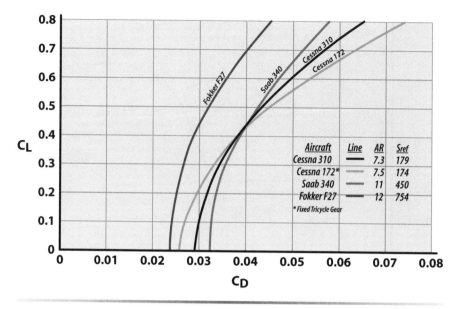

Figure G.6   Drag polars for general aviation and small transport propeller aircraft.

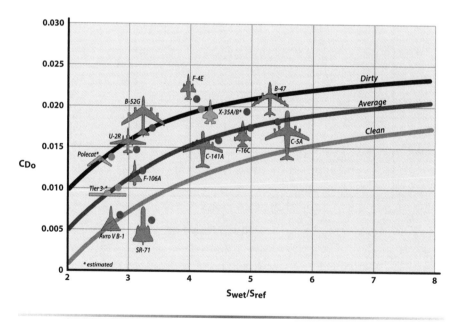

Figure G.7   Correlation of subsonic C_{D_0} with S_{wet}/S_{ref}.

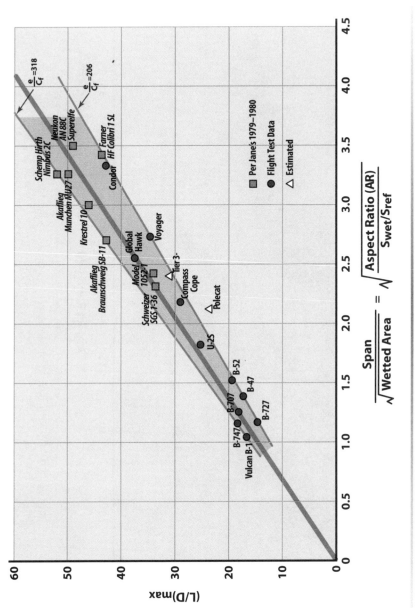

Figure G.8 Maximum lift-to-drag correlation curve.

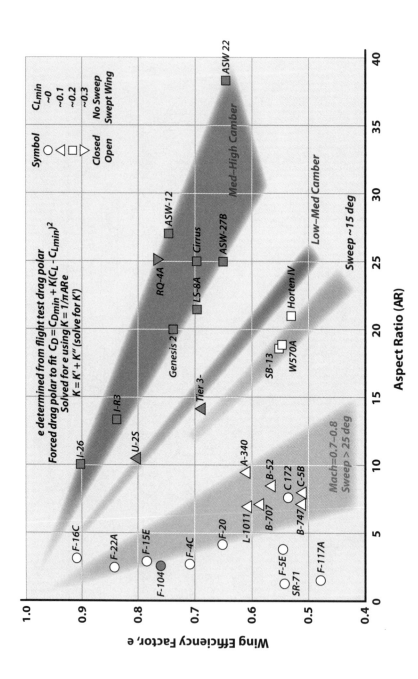

Figure G.9 Subsonic wing efficiency (e) vs aspect ratio for various aircraft.

subsonic C_{D_0} or $C_{D_{min}}$ is primarily skin friction drag. This correlation is shown in Fig. G.7. The difference between the "clean" and "dirty" trend lines is primarily the excrescence/protuberance drag. The U-2R and B-52G have many external antennas. The F-4E has external weapons, targeting pods, and their associated pylons.

Figure G.8 is a very useful chart for checking the $(L/D)_{max}$ of your design. As developed in Section 2.21 the $(L/D)_{max} \sim (\text{span})/(S_{wet})^{1/2}$. This correlation is shown in Fig. G.8 for operational aircraft, demonstrator aircraft, and sailplanes with moderate to high aspect ratio (AR). The trend lines of e/C_f are the wing efficiency factor divided by the skin friction factor (Fig. 2.4). All aircraft need high $(L/D)_{max}$ but it is essential for sailplanes in world-class competition. Sailplanes fall on the upper trend line because they have carefully tailored, highly efficient, high-AR wings and laminar boundary layers (operating $Re < 5 \times 10^5$). Most operational and demonstrator aircraft have good wing designs but turbulent boundary layers.

Figure G.9 is useful in obtaining a preliminary estimate of the wing inviscid factor K'. The chart is *empirical* and assumes that the aircraft drag polar can be expressed as

$$C_D = C_{D_{min}} + K\left(C_L - C_{L_{min}}\right)^2$$

The K in the drag-due-to-lift term is made up of the viscous drag-due-to-lift K'' (due to the viscous separation as seen in the airfoil drag polar) and the inviscid drag-due-to-lift K' (due to the influence of the trailing vortices on the wing a.c.); $K' = 1/\pi ARe'$ from finite wing theory, where e' depends on the spanwise lift distribution and is typically from 0.8 to 0.9; K'' typically ranges from 0.003 to 0.03 and is obtained from airfoil data (see Fig. 7.8 and Section F.4). The K' and K'' are combined as K. The aircraft flight test drag polar is plotted as $C_{D_{min}}$ vs $(C_L - C_{L_{min}})^2$. The slope K is then determined. The wing e in Fig. G.9 is determined from $K = 1/\pi ARe$.

G.2 How to Estimate Subsonic Drag Polar

Example G.1 Boeing Condor

The Condor was a high-altitude, long-endurance demonstrator aircraft built by Boeing in the 1980s for DARPA. It featured an AR = 36.6 wing with a Liebeck LD-17A airfoil, giving it an impressive maximum L/D of 42 (Fig. G.8). It had two 175-hp internal combustion engines with 16 ft diameter propellers and two stages of turbocharging. The Condor had a wing span of 200 ft, a length of 66 ft, a TOGW of 20,300 lb (empty weight of 7880 lb), and 11,080 lb of fuel. It demonstrated a flight of 58 h at 67,000 ft in 1989 and is seen in flight in the photograph (Fig. G.10).

Figure G.10   Condor (courtesy of The Boeing Company).

The subsonic drag polar for the Condor operating at 60,000 ft and 200 kt will be estimated. The Reynolds number per foot is 0.3 million and the reference wing area is 1107 ft^2. The method discussed in Section 13.3 is used assuming the horizontal and vertical tails to be small wings and the engine pod to be a small fuselage. The data for determining $C_{D_{min}}$ is as follows:

Component	Wetted Area (ft^2)	Ref L (ft)	t/c or l/d	$\Delta C_{D_{min}}$
Wing	2214	6.5	0.17	0.007
Fuselage	946	66	18	0.00242
Horiz. Tail	198	4	0.10	0.00025
Vert. Tail	142	6.6	0.10	0.00046
Eng. Pods (2)	531	20.8	6.5	0.002
Total	4031			0.0123

Add 20% to account for cooling drag (from Reference 1 in Chapter 9).

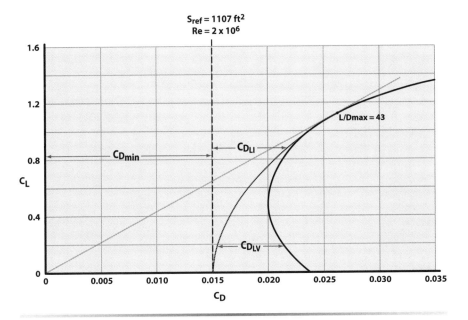

Figure G.11 Condor drag polar at 60,000 ft/200 kt.

Condor $C_{D_{min}}$	0.015
From Section F.4	$K'' = 0.006$ for $C_L < C_{l_{min}} = 1.2$
From Fig. G.9	Assume $e = 0.6$, $K = 0.0145$
Since $K = K' + K''$	$K' = 0.0085$

The expression for

$$C_D = 0.015 + 0.0085C_L^2 + 0.006(C_L - 1.2)^2 \qquad \text{(A.1)}$$

The drag polar is shown on Fig. G.11. Notice that the partition of inviscid and viscous drag-due-to-lift is quite a bit different from that shown on Fig. 2.16. This is because the LD-17A has such a large $C_{l_{min}}$. Notice also that the $C_{D_{min}}$ in equation (A-1) is very much different than the $C_{D_{min}}$ for the drag polar. This is because the $C_{D_{min}}$ buildup for Eq. (A-1) is mostly skin friction and very little pressure drag due to viscous separation. A sanity check reveals an $(L/D)_{max} = 43$, which checks with that reported for the Condor.

References

[1] Hoerner, Sighard F., *Fluid Dynamic Drag*, Midland Park, NJ, 1993.

Appendix H Aerodynamics of Wing–Body Combinations

- Data from NACA RM A53A30, Jan. 1958
  - Lift curve slope, C_{D_0} and a.c.
  - Mach Number 0 to 2.0
  - $Re = 1$ M to 10 M
  - Sears–Haack body
  - AR = 2, 3, 4 and t/c = 3%, 5%, 8%
  - Delta and rectangular planforms
  - Symmetric airfoil with sharp & round LE

The Lockheed Martin F-16 Fighting Falcon (Chapter 5) was designed as a lightweight dogfighter with numerous innovations, including frameless, bubble canopy for better visibility; side-mounted control stick to ease control under high g-forces; and reclined seat (16 deg) to reduce the effect of the g-forces on the pilot.

Knowledge is free at the library.
Bring your own container.

his appendix contains excerpts of the data reported in NACA RM A53A30 [1]. The report is an excellent compilation of aerodynamic data on low aspect radio wing–body configurations.

The fuselage used was a Sears–Haack body with a fineness ratio of about 12. Its shape is expressed by the equation for radius as

$$r = r_o \left[1 - \left(1 - 2x/\ell \right)^2 \right]^{3/4}$$

where ℓ is the body length and r_o is the maximum body radius at $\ell/2$.

The wings for the study were for aspect ratios (AR) 2, 3, and 4 and various planform shapes (delta, rectangular, straight with taper, and swept with taper). The airfoil sections were symmetric with thickness ratios of 3%, 5%, and 8%. A typical section is shown in Fig. H.1.

The Reynolds numbers for the wind tunnel tests were between 10^6 and 10^7. The largest ratio of body diameter to wing span, d/b, was 0.18 so that body lift contribution was small relative to the wing.

The experimental data in this appendix are presented so that the designer can develop a feel for the magnitude of the aerodynamic coefficients. The data represent typical drag, lift, and aerodynamic center (a.c.) behavior.

Figures H.2–H.9 present the lift coefficient, drag coefficient, and aerodynamic center location for various low-AR wing–body configurations at subsonic and supersonic speeds. The $C_{D_{\min}}$ shown in Figs. H.4–H.6 is the same as C_{D_0}. The lift curve slope data, C_{L_α}, and the drag-due-to-lift factor K for the wing–body combinations are presented in Figs. 13.3a and 13.3b, respectively, of Chapter 13.

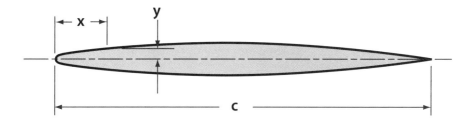

x	y
Percent c	Percent c
0.0	0.0
1.25	0.333
2.5	0.468
5.0	0.653
7.5	0.790
10.0	0.900
15.0	1.071
20.0	1.200
30.0	1.375
40.0	1.469
50.0	1.500
60.0	1.440
70.0	1.260
80.0	0.960
85.0	0.765
90.0	0.540
95.0	0.285
100.0	0.0
LE Radius: 0.045% c	

Figure H.1   Coordinates of 3% thick round-nose airfoil section.

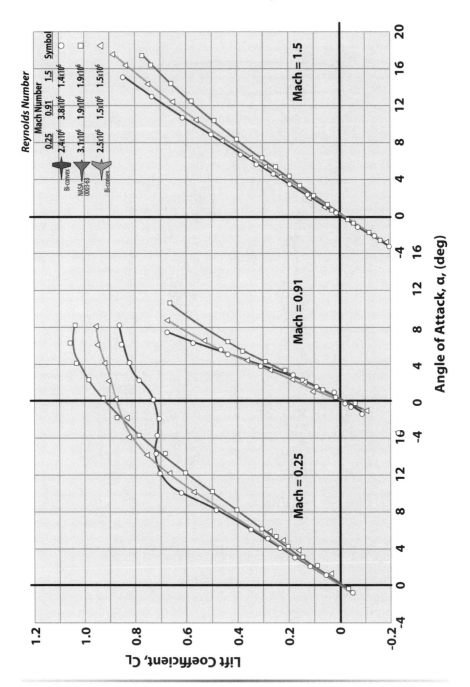

Figure H.2   Variation of lift coefficient with angle-of-attack for plane wings with AR = 3, 3% thick, and different planforms.

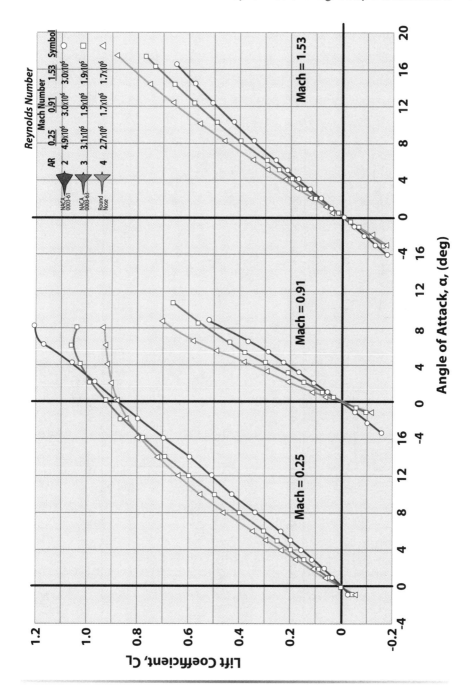

Figure H.3 Variation of lift coefficient with angle-of-attack for plane triangular wings 3% thick, with various aspect ratios.

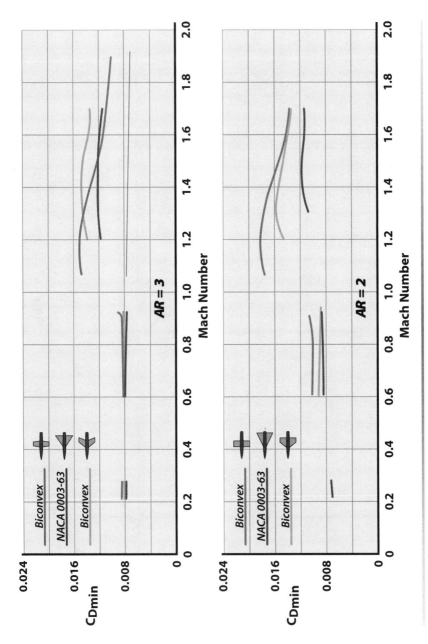

Figure H.4   Minimum drag coefficient for plane wings with 3% thickness and having different planforms.

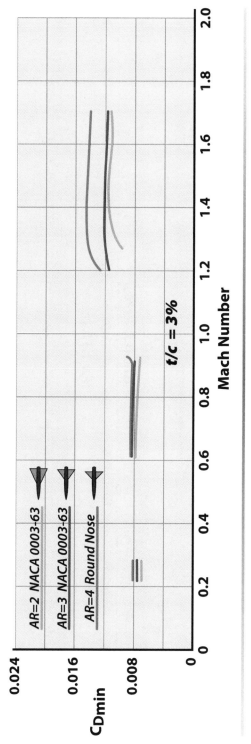

Figure H.5 Minimum drag coefficient of plane triangular wings with 3% thickness.

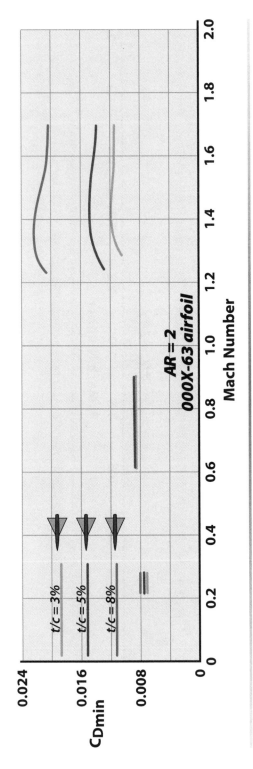

Figure H.6 Minimum drag coefficient of plane triangular wings with AR = 2 and having NACA 000X-63 sections.

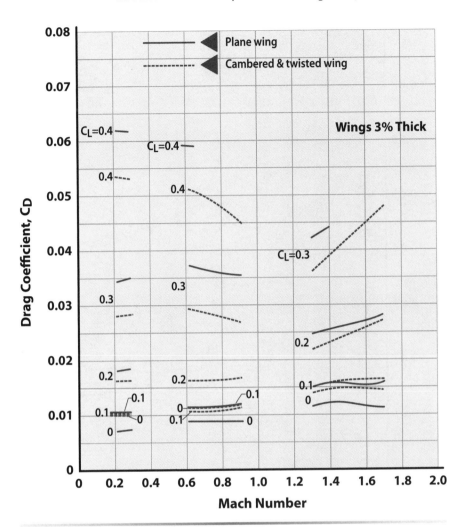

Figure H.7a Drag characteristics for triangular wings 3% thick with AR = 2 that are plane, twisted, and cambered.

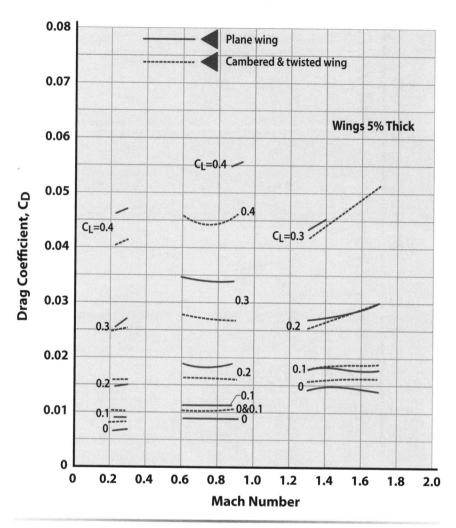

Figure H.7b   Drag characteristics for triangular wings 5% thick with AR = 2 that are plane, twisted, and cambered.

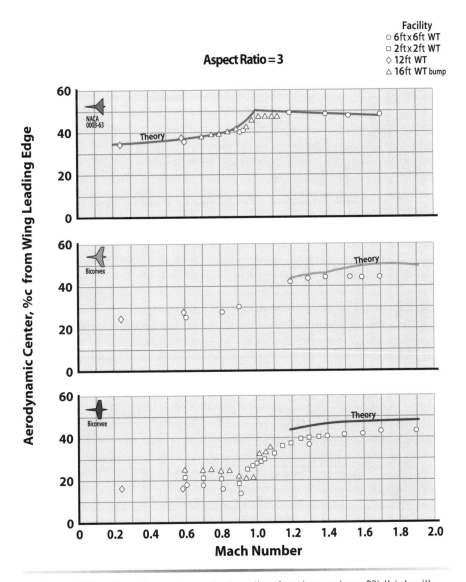

Figure H.8a Aerodynamic center location for plane wings 3% thick with different planforms.

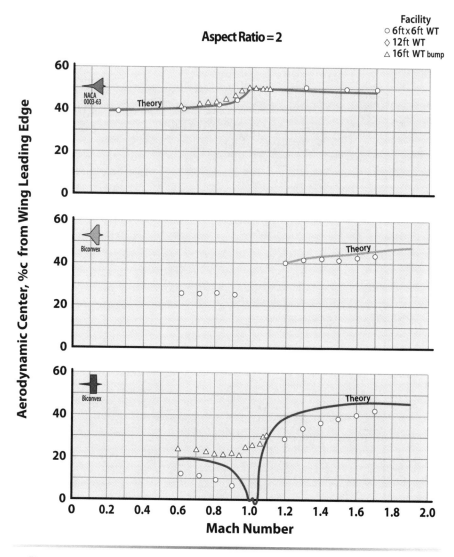

Figure H.8b Aerodynamic center location for plane wings 3% thick with different planforms.

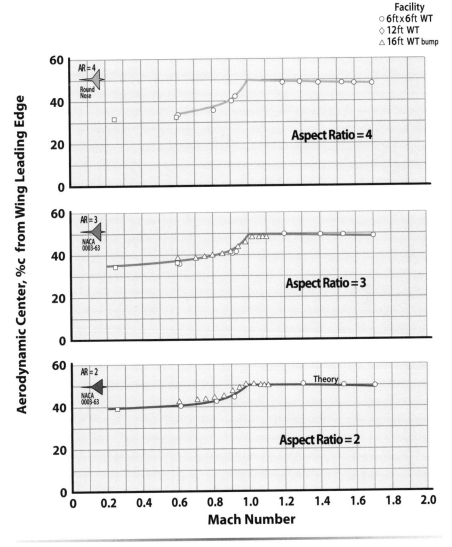

Figure H.9 Aerodynamic center location for plane triangular wings 3% thick.

Reference

[1] Hall, C. F., "Lift, Drag and Pitching Moment of Low-Aspect-Ratio Wings at Subsonic and Supersonic Speeds," NACA RM A53A30, Jan. 1958.

Appendix I Aircraft Weights Data

- Square–Cube Law
- Weight Summaries & Empty-Weight Trend Lines for
  - Fighter & Bomber/Transport Aircraft
  - FAR 23 & Trainer Aircraft
  - ISR & UAV Aircraft
  - Air-Launch Cruise Missiles & Targets
  - Sailplanes

The Eta is the world's highest performing glider. It is a German–Italian open-class, two-seater, self-launching sailplane featuring a 63-hp engine located on top of the wing at the trailing edge that is cranked into position for launch and then retracted. The wing aspect ratio is 51.3 with a reported L/D of 72. (Photograph courtesy of Reiner Kickert.)

If you don't get what you want, then you didn't seriously want it, or you tried to negotiate over its price.

I.1 Introduction

Estimating aircraft empty weight for initial sizing is the most difficult part of the conceptual design process. We rely on historical data to make the weight estimating at the conceptual design level at least an art (it will never be a science). We gather as much weights data as we can on real aircraft and develop weight-estimating relationships (WERs) using regression analysis or plot the data on log–log paper and develop trend line equations. Chapter 20 contains WERs for the aircraft components. This appendix contains weight summaries of actual aircraft (Tables I.1–I.5), current sailplanes (Table I.6), and empty-weight trend lines (Figs. I.1–I.8). The empty weight of a new design is determined by using the tables and figures for aircraft similar to the new design. Notice that the tables and figures are for conventional metal structure (sailplanes and most UAVs being exceptions). A way to adjust these weights data for advanced composites is discussed later in this appendix (Section I.3) and in Chapter 20.

I.2 Empty-Weight Trends for Conventional Metal Structure

The empty weight for fighter aircraft with conventional metal structure is shown in Fig. I.1. There was so much scatter in the data that three trend lines were developed. In all cases the equation of the trend lines was a constant times $(TOGW)^{0.947}$. The first trend line was for fighter aircraft having wing loadings less than about 70 psf or developmental–prototype aircraft. These aircraft tend to have large wings or nonoptimized structure and test instrumentation. The second trend line was for a general class of fighter having wing loadings greater than 70 psf and thrust-to-weight ratios less than 0.9. The third trend line was for air-to-ground fighters designed for external stores.

Figure I.2 shows the empty-weight trend for bombers and transport aircraft with conventional metal structure. Here the data group much more tightly around the trend line than in the fighter case. Those aircraft not falling near the trend line can be explained in most cases. For example, the B-58A carried a large fuel/weapon pod beneath the aircraft and took off at half fuel. It was refueled after takeoff to full fuel and the pod was dropped over the target. This strategy greatly reduced the landing-gear loads and structural weight criteria. Notice that the trend line equation is the same as for the fighter middle trend line.

Figure I.3 shows the empty-weight trend line for light civil aircraft with conventional metal structure. The data display a tight grouping around the trend line, which has the same equation as the bombers and transports.

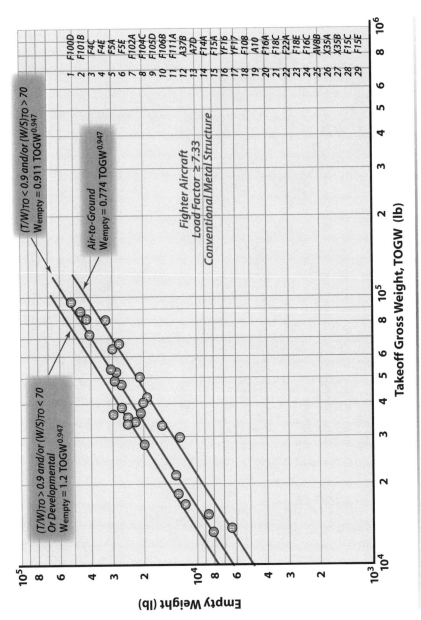

Figure I.1   Trend of empty weight with takeoff gross weight—fighter aircraft.

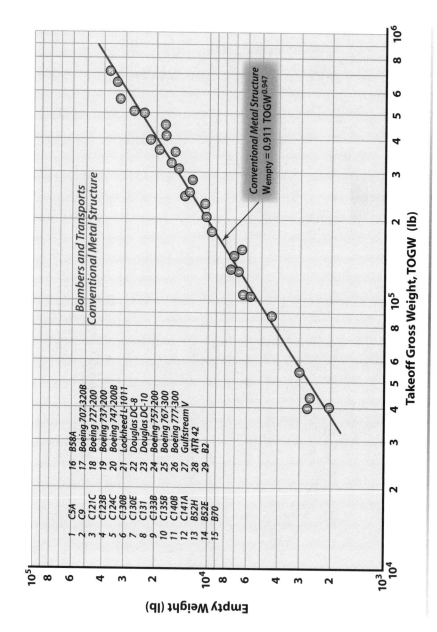

Figure I.2 Trend of empty weight with takeoff gross weight—bombers and transports.

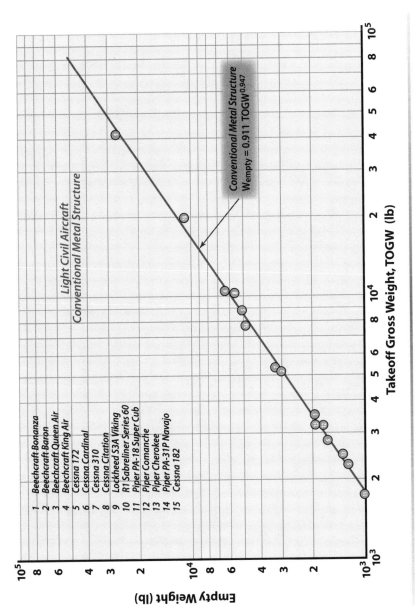

Figure I.3   Trend of empty weight with takeoff gross weight—light civil aircraft.

Table I.1 Jet Transport Aircraft Weights Summary (Weights in Pounds)

Aircraft	McDonnell Douglas		Boeing			Airbus
	MD-80	DC-10-30	737-200	727-100	747-100	A-300
Wing	15,560	58,859	10,613	17,764	86,402	44,131
Empennage	3,320	14,676	2,718	4,133	11,850	5,941
Fuselage	16,150	47,270	12,108	17,681	71,845	35,820
Nacelle	5,340	9,127	1,392	3,870	10,031	7,039
Landing gear	5,340	25,761	4,354	7,211	31,427	13,611
Nose gear	550	1,832	—	—	—	—
Main gear	4,790	23,929	—	—	—	—
Structure total	42,490	155,693	31,185	50,659	211,555	106,542
Engine	8,820	26,163	6,217	9,325	34,120	16,825
Nozzle system and $T_{reverser}$	1,540	6,916	1,007	1,744	6,452	4,001
Air induction system[a]	0	0	0	0	0	0
Fuel system	640	4,308	575	1,143	2,322	1,257
Propulsion install	—	—	378	250	802	814
Propulsion total	11,000	37,387	8,177	12,462	43,696	22,897
Avionics and instruments	2,130	4,274	625	756	1,909	377
Surface controls	2,540	6,010	2,348	2,996	6,982	5,808
Hydraulic system	540	2,587	873	1,418	4,471	3,701

[a]Engines in pods, weight included in nacelle.

Aircraft	McDonnell Douglas		Boeing				Airbus
	MD-80	DC-10-30	737-200	727-100	747-100		A-300
Pneumatic system	1,720	5,912	—	—	—		—
Electrical system	—	—	1,066	2,142	3,348		4,923
Electronics	—	—	956	2,142	3,348		4,923
Auxiliary power units (APU)	840	1,643	836	1,591	4,429		1,726
Oxygen system	220	256	—	—	—		—
Environmental control system (ECS)	1,580	2,723	1,416	1,976	3,969		3,642
Anti-icing system	550	471	—	—	—		—
Furnishings	8,450	34,124	6,643	10,257	37,245		13,161
Miscellaneous	3,650	16,274	124	85	421		732
Equipment total	25,460	76,194	14,887	21,281	63,062		35,053
Empty weight (lb)	78,950	269,274	60,210	88,300	353,398		168,805
Fuel	39,362	247,034	34,718	48,353	331,675		76,512
Oil	—	—	—	—	—		—
Payload (lb)	43,950	98,726	34,790	29,700	140,000		69,865
TOGW (lb)	140,000	555,000	115,500	160,000	710,000		302,000
Wing span (ft)	107.7	165.3	94.0	108.0	195.7		147.3
Wing area, S (ft^2)	1,270	3958	980	1,700	5,500		2,799
Horizontal tail area, S_h (ft^2)	314	1,338	321	376	1,470		748
Vertical tail area, S_v (ft^2)	168	605	233	356	830		487

Table I.2 Business Jet Aircraft Weights Summary (Weights in Pounds)

Aircraft Item[a]	Lockheed Jetstar	Gates Learjet28	Cessna CitationII	HawkerSid-deley HS125	Gulfstream GII
Wing	2,827	1,939	1,288	1,968	6,372
Empennage	879	361	295	608	1,965
Fuselage	3,491	1,624	1,069	1,628	5,944
Nacelle	792	214	220	(infuse)	1,239
Landing gear	1,061	584	465	659	2,011
Nose and main	—	102/482	87/378	—	321/1690
Structure total	9,050	4,722	3,337	4,863	17,531
Engine	609	792	1,100	—	6,570
Nozzle system and $T_{reverser}$	—	—	15	—	—
Air induction system	135	0	26	—	—
Fuel system	360	237	189	—	316
Propulsion install	230	255	105	—	—
Propulsion total	2,475	1,284	1,435	—	6,866
Avionics and instruments	153	383	87	—	1,715
Surface controls	768	275	203	—	1,021
Hydraulic system	262	114	96	217	959
Pneumatic system	—	—	—	—	—

Aircraft	Lockheed	Gates	Cessna	HawkerSid-deley	Gulfstream
Item[a]	Jetstar	Learjet28	CitationII	HS125	GII
Electrical system	973	603	340	—	1,682
Electronics	868	—	313	—	—
APU	—	—	—	—	258
Oxygen system	—	26	—	—	140
ECS	510	285	264	—	927
Anti-ice system	—	162	98	—	—
Furnishings	1,521	768	800	—	4,501
Miscellaneous	10	—	50	—	—
Equipment total	5,065	2,605	2,251	—	11,203
Empty weight (lb)	16,590	8,611	7,023	12,260	35,620
Fuel	11,229	4,684	5,009	9,193	23,300
Oil	204	177	143	—	—
Payload (lb)	2,100	1,962	1,500	1,905	5,380
TOGW (lb)	30,680	15,000	13,500	23,300	64,800
Wing span (ft)	55.0	43.8	52.3	47.0	68.8
Wing area, S (ft^2)	521	265	279	353	794
Horizontal tail area, S_H (ft^2)	149	54	70	100	182
Vertical tail area, S_V (ft^2)	110	37	51	52	155

[a]Abbreviations: APU, auxiliary power units; ECS, environmental control system; TOGW, takeoff gross weight.

Table I.3   General Aviation Aircraft Weights Summary (Weights in Pounds)

Aircraft	Cessna					Beech	
Item[a]	150	172	182	210J	310C	QueenAir	TwinBonanza
Wing	216	226	235	335	453	670	656
Empennage	36	57	62	86	118	153	156
Fuselage	231	353	400	408	319	601	495
Nacelle	22	27	34	28	129	285	261
Landing gear	104	111	132	191	263	444	447
Structure total	**609**	**774**	**863**	**1048**	**1282**	**2153**	**2015**
Engine	197	254	417	450	852	1008	1008
Air induction system	2	1	1	7	7	27	27
Fuel system	17	21	26	24	76	137	137
Nozzle	0	0	0	0	0	0	0
Propeller	22	33	64	64	162	258	258
Installation	28	36	37	36	153	180	180
Propulsion total	**267**	**345**	**545**	**581**	**1250**	**1610**	**1610**
Avionics and instruments	3	4	6	18	46	72	89

Aircraft	Cessna					Beech	
Item[a]	150	172	182	210J	310C	QueenAir	TwinBonanza
Surface controls	31	31	36	48	66	132	120
Electrical system	43	38	43	57	121	166	184
Hydraulics	0	0	0	51	0	0	0
ECS	1	1	1	10	46	90	81
Anti-ice	—	—	—	—	—	—	—
Furnishings	33	85	87	130	154	438	333
Misc	0	0	0	20	65	5	7
Equipment total	**102**	**159**	**173**	**335**	**498**	**903**	**814**
Empty weight (lb)	946	1243	1545	1964	3032	4701	4459
Fuel	156	252	390	464	612	1380	1380
Oil	11	15	22	24	45	60	60
Payload (lb)	398	702	715	693	1186	1287	1311
TOGW (lb)	1500	2200	2650	3400	4830	7368	7150
Wing span (ft)	33.3	36.0	36.0	36.8	37.0	50.3	45.3
Wing area, S (ft^2)	160	175	175	176	175	277	277
Horizontal tail area, S_H (ft^2)	28.5	34.6	34.1	38.6	54.3	79.3	79.3
Vertical tail area, S_V (ft^2)	14.1	18.4	18.4	17.2	25.9	30.8	30.8

[a]Abbreviations: ECS, environmental control system; TOGW, takeoff gross weight.

Table I.4 ISR Aircraft Weights Summary (Weights in Pounds)

Aircraft	Lockheed				Boeing	N/G
Item[a]	U-2A	U-2S	Darkstar	SR-71	Condor[b]	GlobalHawk
No. of Engines	1	1	1	2	2	1
Engine model	PWA J-57-37	GE F118-100	Williams FJ44-2D	PWA J-58	TCM GTSIOL	Allison AE3007H
Wing	2,034	4,585	1,111	14,054	2,519	—
Empennage	320	674	0	1,503	253	—
Fuselage	1,410	2,719	847	6,911	823	—
Landing gear	263	596	391	3,486	243	—
Structure total	4,027	8,573	2,349	25,954	3,838	—
Engine	4076	4,814	448	12,966	1,121	—
Nozzle	204	350	104	1,565	—	—
Air induction	164	235	57	4,941	—	—
Fuel system	311	556	161	1,700	235	—
Propulsion install	111	57	42	481	833[a]	—
Propulsion total	4,865	6059	812	21,653	2,189	—
Avionics and instruments	57	137	417	372	530	—
Surface controls	362	497	287	1,682	295	—
Hydraulic system	66	138	111	1,222	—	—
Pneumatic system	—	—	—	—	—	—
Electrical system	290	1,060	384	898	319	—
Electronics	166	478	—	1,427	—	—

Aircraft	Lockheed				Boeing	N/G
Item[a]	U-2A	U-2S	Darkstar	SR-71	Condor[b]	GlobalHawk
APU	0	0	0	—	0	—
ECS/Anti-ice	135	159	187	1,325	157	—
Oxygen system	61	—	0	75	0	—
Furnishings	82	287	0	520	0	—
Miscellaneous	21	127	31	200	550	—
Equipment total	1,240	2,883	1,394	7,646	1,851	—
Empty weight	10,180	17,616	4,554	55,253	7,878	9,200
Fuel	5,810	17,677	2,846	80,650	11,077	13,015
Oil and trapped fuel	146	57	76	644	221	427
Payload	518	2,714	1,123	3,411	1000	2,300
Crew	0	0	0	570	—	—
Miscellaneous, communications gear	0	0	0	0	0	658
TOGW	17,000	40,000	8,600	140,750	20,300	25,600
Wingspan	80	103	69	56.7	200	116
Wing area, S	600	1000	300	1,605	1,107	540
Horizontal tail Area, S_H	—	145	0	0	110	56
Vertical tail area, S_V	—	55	0	301	71	69
C_H	—	0.34	0	0	0.53	0.32
C_V	—	0.014	0	—	0.012	0.0186

Abbreviations: APU, auxiliary power units; ECS, environmental control system; TOGW, takeoff gross weight.
[b]Condor propulsion install weight includes the propeller installation, cooling system for engine and two-stage turbochargers, lubricating system, and engine controls. Engine weight includes the weight of the two Teledyne Continental 175-hp liquid-cooled engines and the two-stage turbochargers (25:1).

Table I.5a Fighter Aircraft Weights Summary (Weights in Pounds)

Type	NAA F-100F	McD F-101B	McD RF-101	GD F-102A	Republic F-105B	GD F-106A	NAA F-107A	NAA F-86H	Vought F8U-3	McD F4H	Grumman F11F	Grumman F9F-5	Grumman A2F(A6)
No. of engines	1	2	2	1	1	1	1	1	1	2	1	1	2
Weight Item (lb)													
Wing group	3896	3507	3680	3000	3409	3302	3787	2702	4128	4343	2180	2294	4733
Empennage group	979	812	837	535	965	693	1130	329	1045	853	669	404	819
Fuselage group	4032	3901	3955	3409	5870	4401	4792	2035	3850	4042	3269	1779	3538
Engine section	104	99	103	39	106	39	260	42	92	125	47	0	64
Landing gear group	1509	1592	1596	1056	1848	1232	1410	989	949	1735	907	728	2343
Nose gear													
Main gear													
Outrigger gear													
Structure total	10520	9911	10171	8039	12198	9667	11379	6097	10064	11098	7072	5205	11497
Engine(s)	5121	10800	9676	4993	6187	5816	6100	3646	6010	6940	3489	2008	4010
Air induction system	504	729	638	693	524	975	833	167	673	1037	159	225	61
Fuel system	761	1226	1412	394	608	777	983	845	849	953	463	529	936
Propulsion system	414	892	599	278	406	503	368	340	338	106	192	116	632
Power plant total	6800	13647	12325	6358	7725	8071	8284	4998	7870	9036	4303	2878	5639
Avionics + instrumentation	303	318	204	141	227	190	288	111	191	166	118	82	133
Surface controls	1076	772	780	413	1311	445	1454	358	1425	919	760	345	932
Hydraulic system	157	433	359	318	449	431	150	339	150	441	166	267	170
Pneumatic system													
Electrical system	568	825	819	594	700	606	447	476	439	502	459	458	695
Electronics	496	2222	629	2001	737	2743	382	230	840	1386	439	292	2652
Armament	794	228	36	589	719	626	1006	828	376	446	358	416	323
Air conditioning system	435	270	362	259	168	407	390	205	329	341	76	85	164
Anti-icing system													
Furnishings	427	480	242	227	243	290	282	182	210	321	166	144	476
Auxiliary gear	77	84	91	78	92	69	42	12	183		131	51	
APU					224								
Photographic system													

Type	NAA F-100F	McD F-101B	McD RF-101	GD F-102A	Republic F-105B	GD F-106A	NAA F-107A	NAA F-86H	Vought F8U-3	McD F4H	Grumman F11F	Grumman F9F-5	Grumman A2F(A6)
Ballast													
Manufacturing variation													
Fixed equipment total	4333	5632	3522	4620	4870	5807	4441	2741	4143	4522	2673	2140	5545
W(oil) + W(tfo)	166	223	223	216	198	303	143	57	196	131	72	0	195
Max. fuel capacity	7729	8892	9782	7053	7540	8476	11050	3660	14306	13410	6663	7160	8764
Payload (max. fuel)	250	1881	704	1241	757	1374	2560	420	1197	1500	340	0	2000
Expendable payload													
Fixed payload													
Flight design GW, lb	29391	39800	37000	25500	31392	30590	29524	19012	30578	34851	17500	14900	34815
Structure/GW	0.358	0.249	0.275	0.315	0.389	0.316	0.385	0.321	0.329	0.318	0.404	0.349	0.330
Power plant/GW	0.231	0.343	0.333	0.249	0.246	0.264	0.281	0.263	0.257	0.259	0.246	0.193	0.162
Fixed equipment/GW	0.147	0.142	0.095	0.181	0.155	0.190	0.150	0.144	0.135	0.130	0.153	0.144	0.159
Empty weight/GW	0.737	0.733	0.724	0.750	0.797	0.767	0.816	0.728	0.722	0.707	0.771	0.686	0.651
Wing group/GW	0.133	0.088	0.099	0.118	0.109	0.108	0.128	0.142	0.135	0.125	0.125	0.154	0.136
Empennage group/GW	0.033	0.020	0.023	0.021	0.031	0.023	0.038	0.017	0.034	0.024	0.038	0.027	0.024
Fuselage group/GW	0.137	0.098	0.107	0.134	0.187	0.144	0.162	0.107	0.126	0.116	0.187	0.119	0.102
Engine section/GW	0.004	0.002	0.003	0.002	0.003	0.001	0.009	0.002	0.003	0.004	0.003	0.000	0.002
Land. gear group/GW	0.051	0.040	0.043	0.041	0.059	0.040	0.048	0.052	0.031	0.050	0.052	0.049	0.067
Take-off gross weight, lb	30638	41288	37723	28137	34081	33888	39405	18908	38528	40217	21233	17500	34815
Empty weight, W_E, lb	21653	29190	26774	19130	25022	23448	24104	13836	22092	24656	13485	10223	22680
Wing group/S, psf	9.7	9.5	10.0	4.3	8.9	4.7	9.6	8.6	8.9	8.2	8.5	9.2	9.1
Empennage group/S_{emp}, psf	6.3	5.1	5.2	5.6	5.2	6.6	6.4	4.1	7.2	5.2	5.8	3.5	4.4
Ultimate load factor, gs	7.33	10.2	11	10.5	13	10.5	13	11	9.6		9.8	11.3	
Surface areas, ft²													
Wing, S	400	368	368	698	385	698	395	313	462	530	255	250	520
Horizontal tail, S_H	98.9	75.1	75.1	0	96.5	0	93.3	47.2	67.2	96.2	65.5	48	120
Vertical tail, S_V	55.6	84.9	84.9	95.1	88.1	105	83.8	32.2	78.6	67.5	50.3	66	68.4
Empennage area, S_{emp}	155	160	160	95.1	185	105	177	79.4	146	164	116	114	188

Table I.5b Fighter Aircraft Weights Summary

Type	McD F3H-2	NAA A3J	Vought F7U-1	McD F-4E	Boeing F-15C	Boeing F/A-18A	Boeing AV-8B	LM F/A-16A	LM F/A-16C	Boeing F/A-18E	LM F/A-22A
No. of engines	2	2	1	2	2	2	1	1	1	2	2
Weight item (lb)											
Wing group	4314	5072	3583	5226	3642	3798	1443	1829	2145	4799	4586
Empennage group	576	1358	726	969	1104	945	372	719	898	1526	2580
Fuselage group	3551	6851	937	5050	6245	4685	2060	3325	3923	6054	12049
Engine section	93	80	0	166	102	143	141	322	449	175	231
Landing gear group	1458	2173	1181	1944	1393	1992	1011	881	1116	2623	1467
Nose gear				377	264	626	334	166	200	743	253
Main gear				1567	1129	1366	400	627	832	1880	1214
Outrigger gear							277				
Structure Total	9992	15534	6427	13355	12486	11563	5027	7076	8531	15177	20913
Engine(s)	4960	7260	2790	7697	6091	4294	3815	3026	3945	5218	10286
Air induction system	614	767	690	1318	1464	423	236	298	351	682	1426
Fuel system	1262	979	1080	1932	1128	1002	542	361	385	1253	1206
Propulsion system	70	353	937	312	522	558	444	357	412	412	594
Power plant total	6906	9359	5497	11259	9205	6277	5037	4042	5093	7565	13512
Avionics + instrumentation	145	210	108	270	151	94	80	103	107	63	303
Surface controls	1067	1845	482	1167	810	1067	698	727	839	1242	1327
Hydraulic system	474	275	317	543	433	364	176	311	310	447	764
Pneumatic system											
Electrical system	535	821	371	542	607	547	424	457	791	799	1207
Electronics	984	2239	328	2227	1787	1538	697	1054	1683	1979	1924
Armament	662	45	367	641	627	387	152	593	608	319	327
Air conditioning system	101	424	79	406	685	593	218	255	321	888	1290
Anti-icing system						21		3	5	21	4
Furnishings	218	676	279	611	294	317	298	309	334	380	456
Auxiliary gear	253		128	412	119	189		63	71	276	203

Type	F3H-2	A3J	F7U-1	F-4E	F-15C	F/A-18A	AV-8B	F/A-16A	F/A-16C	F/A-18E	F/A-22A
APU								165	170	232	384
Photographic system					24	36				188	
Ballast					318						
Manufacturing variation				57	-97	-19	-16	-9		78	114
Fixed equipment total	4439	6535	2459	6876	5758	5134	2727	4031	5239	6912	8303
$W(oil) + W(tfo)$	147	320	97					176	186	476	579
Max. fuel capacity	9789	19074	5826	12058	13455	17592	7759	13732	13606	21093	34439
Payload (max. fuel)	216	1885	502								
Expendable payload				2193	2571	5453	4271	338	338	6045	2472
Fixed payload				0	0	2231	832	1856	1930	3461	7731
Flight design gross weight, lb	26000	46028	19310	37500	37400	32357	22950	22500	26910	43900	54313
Structure/GW	0.384	0.337	0.333	0.356	0.334	0.357	0.219	0.314	0.317	0.346	0.385
Power plant/GW	0.266	0.203	0.285	0.300	0.246	0.194	0.219	0.180	0.189	0.172	0.249
Fixed equipment/GW	0.171	0.142	0.127	0.183	0.154	0.159	0.119	0.179	0.195	0.157	0.153
Empty weight/GW	0.818	0.679	0.746	0.840	0.733	0.710	0.557	0.673	0.701	0.675	0.787
Wing group/GW	0.166	0.110	0.186	0.139	0.097	0.117	0.063	0.081	0.080	0.109	0.084
Empennage group/GW	0.022	0.030	0.038	0.026	0.030	0.029	0.016	0.032	0.033	0.035	0.048
Fuselage group/GW	0.137	0.149	0.049	0.135	0.167	0.145	0.090	0.148	0.146	0.138	0.222
Engine section/GW	0.004	0.002	0.000	0.004	0.003	0.004	0.006	0.014	0.017	0.004	0.004
Land. gear group/GW	0.056	0.047	0.061	0.052	0.037	0.062	0.044	0.039	0.041	0.060	0.027
Takeoff gross weight, TOGW, lb	32037	53658	21638	58000	68000	51900	29750	35400	42300	66000	84200
Empty weight, W_E, lb	21272	31246	14397	31514	27425	22974	12791	15149	18863	29654	42728
Wing group/S, psf	8.4	7.2	7.1	9.5	6.1	9.5	6.3	6.1	7.2	9.6	5.5
Empennage group/S_{emp}, psf	4.5	3.4	8.3	5.8	4.7	4.9	5.0	6.9	7.6	6.4	8.2
Ultimate load factor, gs	11.25			9.75	11	11.25	10.5	13.5	13.5	11.25	13.5
Surface areas, ft^2											
Wing, S	516	700	507	548	599	400	230	300	300	500	840
Horizontal tail, S_H	82.5	304	0	100	111	88.1	48.5	49	63.7	120	136
Vertical tail, S_V	45.4	101	88	67.5	125	104	26.6	54.75	54.75	120	178
Empennage area, S_{emp}	128	405	88	168	236	192	75.1	103.75	118.45	240	314

Table I.6 Sailplane Weight (data from [1])

Sailplane Item	Nimbus II	Nimbus 4T	ASW 17	ASW 22BL	ASW 12	LS-9	SB-12	SB-8	PW-5	Eta
Span (ft)	67	86.6	66	86.6	60	59	49	59	44	101.4
Aspect ratio	28.6	39.2	27	41.8	25.8	27.7	22.5	23	17.8	51.3
Wing area (ft²)	155	192	160	179	140	126	110	152	109	200
Wing MAC (ft)	2.3	2.08	2.32	1.9	2.28	2.065	2.22	2.43	2.26	2.25
Wing taper ratio	0.36	0.23	0.43	0.25	0.34	0.25	0.43	0.46	0.2	0.35
Wing wt (lb)	506	686	532	678	427	484	257	284	185	814
Fuselage/vert wt (lb)	264	436	279	304	257	319	205	167	220	532
Horiz tail (lb)	13	18	24	18	18	15	15	13	13	18
Empty wt (lb)	783	1140	889	1000	702	818	474	464	418	1364
Max wt (lb)ᵃ	1276	1760	1342	1650	946	1155	990	735	660	1870
Water ballast (lb)	295	422	255	452	46	139	318	73	44	506
Water ballast/empty wt	0.376	0.37	0.287	0.45	0.07	0.17	0.67	0.16	0.11	0.37
Empty wt/max wt	0.614	0.648	0.662	0.61	0.74	0.71	0.48	0.63	0.633	0.73
Max wt/wing area (psf)	8.23	9.17	8.39	9.22	6.76	9.17	9	4.8	6.05	9.35
Wing wt/wing area (psf)	3.26	3.57	3.33	3.79	3.05	3.84	2.34	1.86	1.7	4.07
Horizontal Tail										
Area (ft²)	10.98	14.64	15.39	13.67	10.76	10.76	12.4	14	14.85	14.1
Moment arm (ft)	15.1	16.1	13.85	15.65	13.7	13.45	12.3	15.26	12.47	20.44
C_{HT}ᵇ	0.465	0.59	0.57	0.63	0.46	0.56	0.62	0.58	0.75	0.64
Vertical Tail										
Area (ft²)	13.35	15.93	14.53	18.41	12.92	10.87	9.58	14	7.2	21.64
Moment arm (ft)	14.63	15.75	14.53	15.13	14.11	12.8	12.24	15.49	11	19.88
C_{VT}ᵇ	0.019	0.015	0.021	0.018	0.022	0.0187	0.022	0.024	0.0165	0.0212

ᵃMax weight = empty weight + 198 lb pilot + water ballast.
ᵇTail volume coefficient (see Chapter 11).

Figure I.4 shows the empty weight for trainer aircraft with conventional metal structure. The data group relatively tightly around the trend line, which has a slightly increased slope over the trend line of air-to-ground fighters. The data show several proposal (paper) empty weights. Data points 1, 4, and 11 were designed for the U.S. Air Force next-generation trainer (NGT) procurement in the early 1980s to replace the T-37B (data point 3). The Fairchild T-46A won the competition, and two aircraft were built and flight tested. The T-46A program was canceled in 1986 due to budget cuts, and the T-37B marches on.

Figures I.6 and I.7 show empty-weight data for subsonic unmanned vehicles. Shown are UAVs (production and prototype), air-launch cruise missiles, and aerial targets. The data points represent metal and composite structures, turbine and propeller designs, differing missions, and three orders of magnitude in size and weight. A tight grouping of the data was forced by having multiple trend lines for the different missions and propulsion systems. The BQM-34A Firebee aerial target data point falls well outside the trend line in Fig. I.7. The only explanation is that the BQM-34A was designed and built in the early 1950s by Ryan Aeronautics to be a multipurpose "truck" for the target community and it could "take a licking and keep on ticking." Figure I.7 also shows the reduction in empty weight that can be realized by not having a person onboard or a landing gear. Figure I.8 shows the empty weight trend line for wood, metal, and composite sailplanes.

I.3 Adjustment Factors for Advanced Composite Structure

The question arises as to what adjustment should be made to conventional metal structure data to account for the use of advanced composites. This is a difficult question because we have so few composite aircraft and they are predominately designed to reduce cost and not weight. Most of the composite aircraft that are flying are in the small UAV category. This is because the safety-of-flight issues for manned composite aircraft have not been resolved. The turboprop business aircraft Beech Starship was built in the early 1980s out of advanced composites. It was certified by the FAA in 1986. The FAA requirements for manned safety of flight resulted in the Starship weighing more than an equivalent metal airplane (empty weight of 7693 lb and TOGW of 12,500 lb).

Most UAVs are made of composites for reasons of easier manufacture (cost) rather than lighter weight. The AGM-158 JASSM (stealthy air-launch cruise missile) was designed in the mid-1990s out of advanced composites for reasons of lower cost. The JASSM did have an empty weight about 8% less than a metal structure. However, the JASSM being a bomb, there was

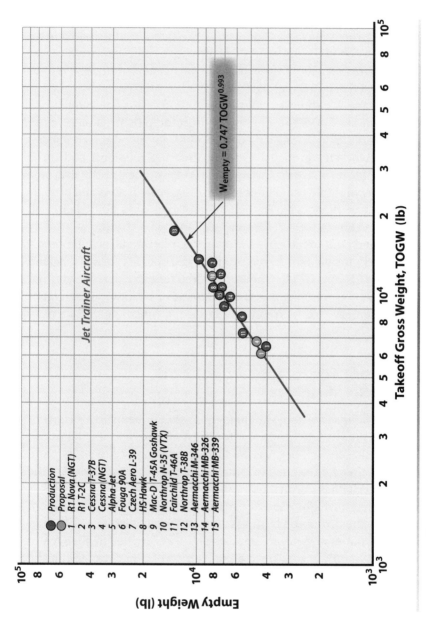

Figure I.4 Trend of empty weight with takeoff gross weight—jet trainer aircraft.

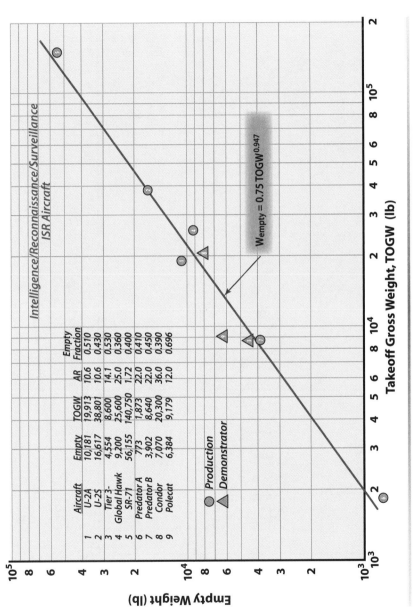

Figure I.5 Trend of empty weight with takeoff gross weight—ISR aircraft.

The following data appears within the figure:

Intelligence/Reconnaissance/Surveillance ISR Aircraft

	Aircraft	Empty	TOGW	AR	Empty Fraction
1	U-2A	10,181	19,913	10.6	0.510
2	U-2S	16,617	38,801	10.6	0.430
3	Tier 3-	4,554	8,600	14.1	0.530
4	Global Hawk	9,200	25,600	25.0	0.360
5	SR-71	56,155	140,750	1.72	0.400
6	Predator A	773	1,873	22.0	0.410
7	Predator B	3,902	8,640	22.0	0.450
8	Condor	7,070	20,300	36.0	0.390
9	Polecat	6,384	9,179	12.0	0.696

Production
Demonstrator

$W_{empty} = 0.75 \, TOGW^{0.947}$

Empty Weight (lb)

Takeoff Gross Weight, TOGW (lb)

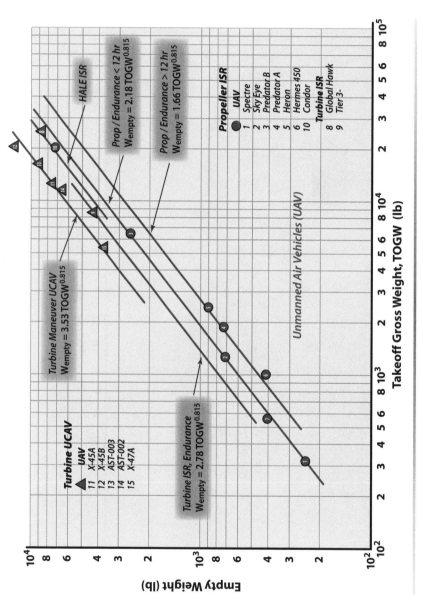

Figure I.6 Trend of empty weight with takeoff gross weight—UAVs.

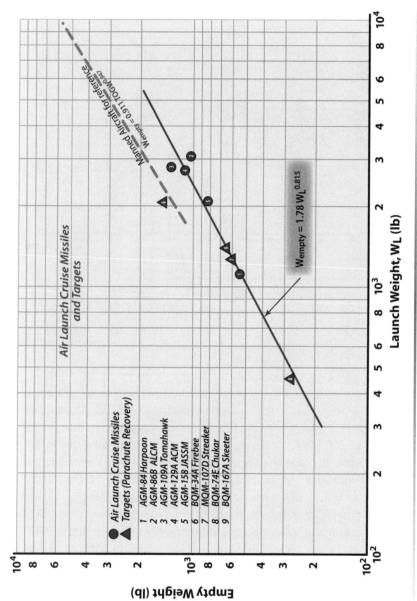

Figure I.7   Trend of empty weight with takeoff gross weight—air-launch cruise missiles and targets.

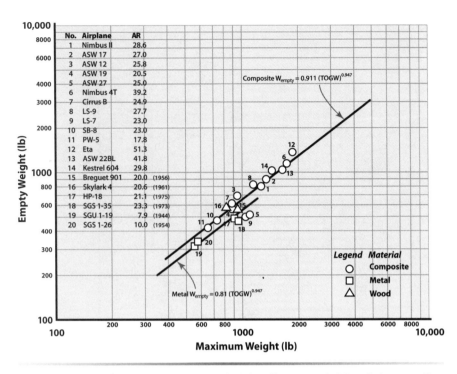

Figure I.8 Trend of empty weight with takeoff gross weight–sailplanes, with water ballast and a 198-lb pilot (data from [1]).

never the motivation to make it lighter than the 2150 lb pylon weight limit. The USAF BQM-167 Skeeter aerial target (produced by CEI, Composite Engineering, Inc.) was designed and fabricated almost entirely from composite material in 2002, but it falls right on the metal trend line of Fig. I.7. The same holds for the AGM-129A ACM. The Boeing 787 Dreamliner is 50% composite material and is expected to have about a 20% airframe weight reduction relative to a conventional metal structure design.

Advanced composite structure R&D has shown that wings and tails designed using composite materials could be made about 20% lighter than conventional metal structures. Fuselage weight reduction for fighter aircraft is only about 10% because of all the cutouts (canopy, access panels, buried engines, weapon and landing-gear bay doors, etc.). Transport aircraft would have a higher fuselage weight reduction (perhaps as much as 20%) due to simpler configurations and podded engines. Research on composite landing gears has shown about an 8% weight reduction. Thus, the empty weights shown in Figures I.1–I.5 (manned aircraft with conventional metal structure) could be reduced by approximately 16% for a manned aircraft aggressively designed for reduced weight using advanced composites.

I.4 Development of the Sailplane and Weight Trends

Figure I.8 is interesting in that the equation for the composite sailplane empty weight is the same as the equation for transports, bombers, general aviation, and multipurpose fighters (see Figs. I.1–I.3 and Table 5.1). The reason is that although sailplanes do not carry fuel they do carry water for ballast. Water ballast is used in high-performance sailplanes to vary the wing loading to meet different environmental conditions and mission phases. High wing loading is used to transit rapidly to an area of high thermal activity and low wing loading for thermalling and landing. While sailplanes seldom fly at the full water ballast condition, their position on Fig. I.8 is for the maximum weight with full water. Typical water ballast weight to empty weight ratios for the current composite sailplanes is less than 0.45 and empty weight over maximum weight is greater than 0.6.

The features that locate sailplane empty weights above the composite trend line in Fig. I.8 are high aspect ratios (A > 30) and low water ballast over empty weight ratios (less than 0.25). The ASW 27 and LS-7 (points 5 and 9) are located quite a bit below the composite trend line in Fig. I.8 because they were designed with modest aspect ratios but large water ballast weight over empty weight ratios of 0.74 and 0.71 respectively.

In the first half of the 20th century sailplanes were made of wood and fabric [1]. These aircraft produced impressive results but were limited in their performance. Wing aspect ratios were limited to about 20 and the surface quality would not meet the requirements set forth by the NACA laminar airfoils (NACA 64 series) in the 1940s. Figure I.8 shows the empty weight trend line for the wooden sailplanes coinciding with the composite trend line, but the wooden sailplanes carried little to no water ballast.

The sailplane community started using metals after WWII and were able to fabricate higher aspect ratio wings up to the high 20s; however, it was still difficult to fabricate wings with the necessary surface quality to take full advantage of the laminar airfoils. The empty weight trend line for the metal sailplanes in Fig. I.8 is below the composite trend line as the designers designed in large amounts of water ballast. Metal construction enjoyed limited popularity due to the emergence of fiberglass composites in the 1950s and carbon fiber composites in the 1970s.

The fiberglass composites provided the surface quality necessary for the laminar airfoils to take full advantage of their potential. The carbon fiber composites permitted the design of higher aspect ratio wings (A > 30). The current standard is the Eta (pictured on the first page of this appendix) with an aspect ratio of 51.3 and demonstrated maximum L/D of 72.

I.5 Square–Cube Law

Galileo in 1638 advanced the idea of the "square–cube law": "When an object undergoes a proportional increase in size, its new volume is proportional to the cube of the multiplier and its new surface area is proportional to the square of the multiplier."

When applied to aircraft, the square–cube law is used as follows: As the size of an aircraft changes by a scale factor (SF), the wing area (lift capability) changes by the square of the SF and the weight changes by the cube of the SF.

As an example, a half-scale (SF = 1/2) demonstrator of a full-scale aircraft would weigh $(1/2)^3 = 1/8$ of the full-scale aircraft and would

• Have a wing area one-fourth that of the full-scale aircraft
• Have a wing loading half that of the full-scale aircraft
• Require an engine one-fourth the size of the full-scale aircraft

However, for buoyant systems (airships) this law is reversed (as discussed in Volume II)! As the size of a buoyant system changes by a scale factor SF its lift capability changes as the cube of the scale factor (volume, i.e., more or less lifting gas) and the weight changes as the square of the scale factor (surface area, i.e., more or less envelope weight).

References

[1] Thomas, F., *Fundamentals of Sailplane Design*, College Park Press, College Park, MD, 1999.

Appendix J Propulsion Data

- Allison T56-A-15, 5000-hp Turboprop
- WRI FJ 44 Family of Turbofans
- F 119 Class Afterburning Turbofan
- AE 3007H, 8000-lb Class Turbofan
- Honeywell TFE 731 Family of Turbofans
- GE 90 98,000-lb Class Turbofan

GE 90 high-bypass turbofan engine, which has an overall pressure ratio of 40, a bypass ratio over 8, and SLS thrust levels of 76,000 to 115,000 lb. It is the largest and highest thrust engine as of 2010; its thrust to engine weight of over 6 raises the bar throughout the engine community.

Given enough thrust,
you can make anything fly.
Ben Rich

J.1 Design Tasks and Propulsion Data

Very early in the conceptual design phase the designer will need information on the propulsion system to conduct the following design tasks:

- Flight envelope (Chapter 4), need engine operating envelope (Fig. 4.6)
- Estimate TOGW (Chapter 5), need engine fuel consumption (cruise, combat)
- Estimate wing loading (Chapter 6), need engine thrust (cruise, takeoff, combat, ceiling)
- Estimate fuselage length and c.g., need engine dimensions and weight
- Estimate takeoff distance (Chapter 10), need engine takeoff thrust
- Design inlet (Chapter 15), need turbine engine airflow
- Thrust sizing (Chapter 18), need engine thrust

J.2 Gas Turbine Engines

An excellent place to start for gas turbine engines is to review the summary data contained in Table J.1. This table can be kept current by adding new engine data from *Aviation Week and Space Technology* magazine, which publishes new engine data once each year (in January). The turbine engine data can be scaled up and down in thrust using the scaling information contained in Chapter 18.

J.3 Piston, Turboprop, and Turbofan Engines

Figure 14.2 contains a summary review of turboprop and piston aircraft engines. Chapter 14 discusses turbochargers for piston engines. Figure J.1 is the three-stage turbocharger for the HAARP aircraft example and is discussed in Section 14.2.1. Figure J.2 shows typical weights of the turbochargers, intercoolers, heat exchangers, and ducting as a function of maximum horsepower and altitude. Propulsion system information is contained in this appendix and elsewhere in the book:

- **F-100-PW-100** (afterburning turbofan), Table 14.3 and Fig. 14.8
- **TF-39-GE-1** [high-bypass-ratio (BPR) turbofan], Table 14.4 and Fig. 14.9
- **Lycoming 0–360-A** (185-hp piston engine), Table 14.1
- **Allison T56-A-15** (5000-hp turboprop), Fig. J.3
- **WRI FJ 44 family** (2400-lb–class turbofan), Table J.2
- **F 119 class** (afterburning turbofan), Table J.3
- **AE 3007H** (8000-lb–class turbofan), Table J.4 and Fig. J.4
- **Honeywell TFE 731** (4000-lb–class turbofan), Table J.5 and Fig. J.5
- **GE 90 Turbofan** (98,000-lb–class turbofan), Table J.6

	Takeoff Conditions												
	Thrust		SFC		FPR	OPR	BPR	Wa	Weight	Max Dia	Fan Dia	Length	Max T/W	Application/Comments
Manufacture/Model	Dry	A/B	Dry	A/B										
	(SLS-lb)		(lb/s/lb)					(lbm/s)	(lb)	(in.)	(in.)	(in.)	(lb/lb)	
Honeywell (ex-Allied-Signal)														
F109-GA-100	Dry 1330		0.396			20.7	5.0		439	29.8	23.5	44.5	3.030	Squalus
TFE731-20	Dry 3641		0.440		1.54	14.3	3.7	123.0	836	39.4	34.2	59.7	4.355	Learjet 45
JETEC	Dry 4000		≤0.8			25+	2.5	55.0			19.0			
TFE731-40	Dry 4248		0.463			22.0			885	39.4	28.2	51.0	4.800	Falcon 50EX, IAI Astra APX
TFE731-5	Dry 4304		0.484			14.6	3.5	143.0	852	42.5	33.8	65.5	5.052	BAe 125, C101
TFE731-5A	Dry 4500		0.469			14.0	3.3	139.7	988	42.5	33.8	72.0	4.555	Falcon 900EX
TFE731-1042	Dry 4585		0.698											
TFE731-60	Dry 4999		0.409			22.0			988	42.5	30.7	72.0	5.060	Falcon 900EX
ATF3	Dry 5316		0.511							33.9		103.2		
ATF3-3	Dry 5440		0.506			22.8			1118	33.9		103.2	4.866	Falcon 2000, Guardian
ATF3-6A	Dry 5440		0.503			21.3			1125	33.9		102.3	4.836	Falcon 2000, Guardian
CFE738-1-1B	Dry 5918		0.351		1.60	35.0	5.3	210	1325	47.7	35.5	98.7	4.466	Falcon 2000, Guardian
F124	Dry 6332		0.788							36.0	23.3	66.8		
F124-GA-100	Dry 6400		0.810			19.4	0.5	93.8	1100	36.0	23.3	66.8	5.818	Aero Vodochody L-159 (dry)
F125-GA-100	A/B 6060	9491	0.785	1.98		19.0	0.5	92.60	1360	33.4	23.3	140.2	6.979	ROC IDF (A/B)
AS907	Dry 6619		0.398											
AS907	Dry 6500		0.420			21.0			1364	46.3		92.4	4.765	Bombardier Continental Jet

(continued)

Manufacture/Model		Thrust Dry (SLS-lb)	A/B	SFC Dry (lb/s/lb)	A/B	FPR	OPR	BPR	Wa (lbm/s)	Weight (lb)	Max Dia (in.)	Fan Dia (in.)	Length (in.)	Max T/W (lb/lb)	Application/Comments
AS977 AS907A	Dry	7092 4-6000		0.416			23.0			1364	49.9	34.0	92.4	5.199	BAe Avro RJX Replacement for ALF503 (PW308 competitor)
LF507-1F	Dry	7000		0.397			13.8	5.0	256.0	1385	50	41.7	65.6	5.054	FADEC-equipped Avro RJ transport series
LF507-1H	Dry	7000		0.406			13.8	5.0	256.0	1375	50	41.7	65.6	5.091	Avro RJ transport w/ hydromechanical fuel control
ALF502L	Dry	7500		0.730			13.3	5.2	256.0	1375	50	41.7	65.6	5.455	Challenger, Bae146
Rolls-Royce Allison															
15S		670		1.20				0	8.3	62.1	12.5		21.2	10.789	Missiles, Target, Decoy, UAV applications
150+		575		1.20				0	7.2	41.2	11.6		19.5	13.956	Missiles, Target, Decoy, UAV applications
150 (J104-AD-100)		485		1.24				0	6.4	41.2	11.6		19.5	11.772	Missiles, Target, Decoy, UAV applications
120		274		1.24				0	3.4	23	8.0		16.0	11.913	Missiles, Target, Decoy, UAV applications
AE2100		6000shp		0.34			16.6			1548	24.5		108.0		C-130J, C-27J, IPTN N-250, Saab 2000

Manufacturer/ Model	Thrust (SLS-lb)		SFC (lb/s/lb)		FPR	OPR	BPR	Wa (lbm/s)	Weight (lb)	Max Dia (in.)	Fan Dia (in.)	Length (in.)	Max T/W (lb/lb)	Application/ Comments
	Dry	A/B	Dry	A/B										
AE3007	7150		0.39			24.0	5		1581	43.5	38.5	106.5	4.522	Citation 10, Embraer ERJ-135, ERJ145, TII+
AE3007A	7580		0.39			24.0	5.3	260	1586	43.5	38.5	106.5	4.779	Citation 10, Embraer ERJ-135, ERJ145, TII+
AE3007C	6495		0.39			24.0	5.3	240	1581	43.5	38.5	106.5	4.108	Citation 10, Embraer ERJ-135, ERJ145, TII+
GEAE														
CF700	4501	—	0.651	—	1.60	6.22	2	130	767	33.0	33.0	53.6	5.868	Sabreline 75A, Falcon D/E/F
J85-GE-13	2720	4080	1.03	2.22	—	7.0	—	44	597	17.7	16.1	105.0	6.834	F-5B
J85-GE-17	2700	2850	0.99	—	—	7.0	—	44	400	17.7	16.1	40.4	7.125	Cessna A-37B
J85-21	3500	5000	1.00	2.13	—	8.3	—	53	684	21.0	18.1	112.5	7.310	F-5E
CFE738-1	5918	—	0.371	—	1.60	35.0	5.3	210	1325	47.7	35.5	98.7	4.466	Falcon 2000 (joint w/Allied)
TF34-GE-100	9065	—	0.37	—	1.50	19.8	6.2	333	1440	46.0	46.0	100.0	6.295	Canadair Challenger, Canadair Regional Jet
CF34-3	9330	—	0.35	—	1.50	19.8	6.3	334	1670	44.0	44.0	82.3	5.587	Canadair Challenger, Canadair Regional Jet
CF34-8C1	13790	—	0.37	—	1.80	27.0	5.0	440	2215	52.0	52.0	128.5	6.226	Canadair CRJ-X
F101-GE-102	18473	33127	0.56	2.29	2.36	26.8	1.91	352	4468	57.8	36.1	180.8	7.414	B-1B Lancer

(continued)

Takeoff Conditions

Manufacture/Model	Thrust Dry (SLS-lb)	Thrust A/B (SLS-lb)	SFC Dry (lb/s/lb)	SFC A/B (lb/s/lb)	FPR	OPR	BPR	Wa (lbm/s)	Weight (lb)	Max Dia (in.)	Fan Dia (in.)	Length (in.)	Max T/W (lb/lb)	Application/Comments
F110-GE-100	17265	28992	0.65	1.86	3.23	31.2	0.76	263	3923	46.5	36.1	187.6	7.390	F101 derivative fighter engine, F-16
F110-GE-400	16333	26950	0.67	1.85	3.19	30.4	0.83	270	4494	46.5	36.1	232.2	5.997	F-14B/D Tomcat
F110-GE-129	17595	29474	0.67	1.85	3.30	31.2	0.74	270	3940	46.5	36.1	187.6	7.481	Lockheed Martin F-16C/D, Boeing F-15E
F110-GE-129EFE	17669	33093	0.68	1.90	3.83	36.4	0.68	275	3990	46.5	36.1	185.3	8.294	
F118-GE-100	20187	—	0.66	—	3.44	34.0	0.76	290	3163	46.5	36.1	100.5	6.382	Non-AB F110 derivative for Northrop Grumman B-2
F118-GE-101	15940	—	0.65	—	3.26	32.8	0.72	251	3150	47.0	38.0	100.5	5.060	Lockheed Martin U-2S
F136-GE-100	26090	40490												LM/NG/BAES F-35A
F136-GE-400	26090	40490												LM/NG/BAES F-35C
F136-GE-600	—	40490												LM/NG/BAES F-35B
F404/F1D2	10600	—	0.80	—	4.10	25.0	0.37	145	1730	35.0	27.7	87.0	6.127	Lockheed Martin/USAF F-117A
F404/RM12	12150	18100	0.81	1.79		27.0	0.28	152	2325	35.0	31.0	159.0	7.785	Swedish JAS 39
F404-GE-100D	11000	—	0.80	—	4.30	25.0	0.30	142	1802	35.0	31.0	89.0	6.104	Singapore A-4S, dry version
F404-GE-400	10650	16000	0.80	1.91	4.30	28.8	0.30	146	2185	35.0	28.0	159.0	7.323	F/A-18A/B/C/D
F404-GE-402	11900	17700	0.81	1.79	4.30	28.6	0.26	146	2230	35.0	28.0	159.0	7.937	F/A-18C/D
F404-GE-F2J3	—	18300	—	1.81		27.0		152	2335	35.0	35.0	159.0	7.837	Indian LCA fighter
F414-GE-400	14447	21496	0.82	1.844	4.49	29.7	0.29	168	2512	39.0	31.2	153.4	8.557	F/A-18E/F

Manufacture/Model		Thrust (SLS-lb)		SFC (lb/s/lb)		FPR	OPR	BPR	Wa (lbm/s)	Weight (lb)	Max Dia (in.)	Fan Dia (in.)	Length (in.)	Max T/W (lb/lb)	Application/Comments
		Dry	A/B	Dry	A/B										
GE56		5800	10100	0.6				2.0	118	835	34.3	24.0	58.3	6.946	New engine
GE56		4424	—		—			2.0	90	637	30.0	21.0	52.3	6.946	New engine
GE56		12288	—		—			2.0	250	1907	49.9	34.9	78.7	6.444	New engine
CF6-6D	Dry	40000					24.3	5.72	1303	7896		86.4	177.0	5.066	DC10-10
CF6-6D1	Dry	41500					25.2	5.76	1328	7896		86.4	177.0	5.256	DC10-10
CF6-45A2	Dry	46500					26.3	4.64	1393	8768		86.4	173.0	5.303	DC10-10,A300B2/4
CF6-50C	Dry	51000					29.3	4.26	1450	8721		86.4	173.0	5.848	B747SR
CF6-50E	Dry	52500					30.1	4.24	1470	8490		86.4	173.0	6.184	B747-200,B747F, USAF E-4A
CF6-50C1	Dry	52500					30.1	4.24	1470	8721		86.4	173.0	6.020	DC10-30,A300B2/4, B747-200
CF6-50E1	Dry	52500					30.1	4.24	1470	8490		86.4	173.0	6.184	DC10-30,A300B2/4, B747-200
CF6-50C2	Dry	52500					30.4	4.31	1476	8731		86.4	173.0	6.013	DC10-10,A300B2/4, B747-200
CF6-50E2	Dry	52500					30.4	4.31	1476	8768		86.4	173.0	5.988	DC10-10,A300B2/4, B747-200
CF6-50C2B	Dry	54000					31.1	4.25	1476	8731		86.4	173.0	6.185	DC10-30
CF6-50E2B	Dry	54000					30.9	4.24	1476	8768		86.4	173.0	6.159	B747-200
CF6-50C2-F	Dry	46500					26.3	4.64	1393	8731		86.4	177.0	5.326	B747SP/SR,A310-200,DC10-15
CF6-80A/A1	Dry	48000					28.0	4.66	1435	8420		86.4	157.4	5.701	B767-200,A310-200
CF6-80A2/A3	Dry	50000					29.0	4.59	1460	8420		86.4	157.4	5.938	B767,A310
CF6-80C2-A1	Dry	59000					30.4	5.15	1754	9135		93.0	160.9	6.459	A300-600

(continued)

Takeoff Conditions

Manufacture/Model		Thrust Dry (SLS-lb)	Thrust A/B (SLS-lb)	SFC Dry (lb/s/lb)	SFC A/B	FPR	OPR	BPR	Wa (lbm/s)	Weight (lb)	Max Dia (in.)	Fan Dia (in.)	Length (in.)	Max T/W (lb/lb)	Application/Comments
CF6-80C2-A3	Dry	60200					31.1	5.09	1769	9135		93.0	160.9	6.590	A300-600, A310-300
CF6-80C2-A5	Dry	61500					31.5	5.05	1781	9135		93.0	160.9	6.732	A300-600
CF6-80C2-A2	Dry	53500					27.8	5.15	1677	9135		93.0	160.9	5.857	A310-300
CF6-80C2-B2	Dry	56700					29.3	5.19	1710	9135		93.0	160.9	6.207	B747-200, B747-300
CF6-80C2-B1F	Dry	57900					29.9	5.15		9135		93.0	160.9	6.338	B747-400
CF6-80C2-B2	Dry	52500					27.4	5.31	1650	9135		93.0	160.9	5.747	B767-200, B767-300, B767-200ER
CF6-80C2-B4	Dry	57900					29.9	5.15	1727	9135		93.0	160.9	6.338	B747-400
CF6-80C2-B6	Dry	60800					31.1	5.06	1790	9164		93.0	160.9	6.635	B767-300ER
CF6-80D1F	Dry	61500					31.5	5.05				93.0	160.9		MD-11
CF6-80E1	Dry	67500					32.7	5.20		10323		93.0		6.539	A330
GE90-B4	Dry	87400					39.3	8.40	3037	14185		123.0	192.8	6.161	B777
GE90-90B	Dry	90000					40.0				134.0		204.0		B777-200, B777-200IGW, B777-300
GE90-92B	Dry	92000					40.0				134.0		204.0		B777-200IGW, B777-300
CFM															
CFM56-2	Dry	22000		0.36			24.7	6.00	821	4612	72.0	68.3	95.7	4.770	DC8-70, KC135, E-3/KE-3/E6
CFM56-3	Dry	23500		0.39			22.6	5.00	655	4280	63.0	60.0	93.0	5.491	B737-300, B737-400, B737-500
CFM56-5-A1	Dry	25000		0.33			26.5	6.00	852	4860	72.0	68.3	95.4	5.144	A320
CFM56-5C-2	Dry	31200		0.32			31.5	6.60	1027	5700	76.6	72.3	103.0	5.474	A340
CFM56-5C-3	Dry	32500		0.32			32.6	6.60	1027	5700	76.6	72.3	103.0	5.702	A340
CFM56-5C-4	Dry	34000		0.33			33.9	6.60	1027	5700	76.6	72.3	103.0	5.965	A340

Manufacture/Model		Thrust Dry (SLS-lb)	Thrust A/B (SLS-lb)	SFC Dry (lb/s/lb)	SFC A/B (lb/s/lb)	FPR	OPR	BPR	Wa (lbm/s)	Weight (lb)	Max Dia (in.)	Fan Dia (in.)	Length (in.)	Max T/W (lb/lb)	Application/Comments
CFM56-5B2	Dry	31000		0.35			32.9			5250	72.0	68.3	103.0	5.905	A321
Microturbo															
TRS 18		259		1.22			3.7	0		92.6	11.8		23.6	2.792	A215J Glider, C22J
TRS 18-1		360		1.18			4.7	0		92.6	11.8		23.6	3.885	MU200/Mirach 100-2/100-3/100-5
TRI 40		764						0		99.2	11.0		26.8	7.705	NSM Antiship missile (Norway)
TRI 60-1		787		1.19			3.7	0		143.3	13.5		33.1	5.491	Sea Eagle
TRI 60-2		832		1.26			3.8	0		143.3	13.5		33.1	5.805	RB15 missile, MQM 107B Raytheon
TRI 60-3		944		1.29			3.9	0		143.3	13.5		33.1	6.589	C22 drone
TRI 60-5		990		1.30			4.1	0		143.3	13.5		33.1	6.909	Super MQM & MQM 107D Raytheon, MQM 107E Tracor
TRI 60-20		1214		1.15			6.3	0		143.3	13.7		33.5	8.472	Super MQM Raytheon
TRI 60-30		1214		1.10			6.3	0		143.3	13.7		33.5	8.472	Apache stand-off weapon; British Storm Shadow
TRI 60-6								0							
P&WA															
ADS9778		4000						3.30	75	600			105.0	6.667	
JT15D-5/5A		2900		0.551			12.6	1.8	92	632	27.3	22.7	63.0	4.589	Cessna T-47A
JT15D-5CTP		3190		0.573			13.5			665	27.0	22.7	45.5	4.797	Augusta S211A
JT15D-5D		3045		0.560			13.1			627	27.3	22.7	63.0	4.856	Cessna Citation Ultra

(continued)

Manufacture/Model	Thrust (SLS-lb) Dry	A/B	SFC (lb/s/lb) Dry	A/B	FPR	OPR	BPR	Wa (lbm/s)	Weight (lb)	Max Dia (in.)	Fan Dia (in.)	Length (in.)	Max T/W (lb/lb)	Application/Comments
PW300	2500								830		27.5		3.012	
PW304														
PW305A	4679		0.388			12.9	4.30		993	44.0	30.1	82.0	4.712	Learjet 60
PW305B	5266		0.391			12.9	4.30		993	44.0	30.1	82.0	5.303	Raytheon Hawker 1000
PW306/5 (+5)	6040		0.400						1043	46.0	31.6	68.1		IAI Galaxy
PW306/9	6050		0.414				4.70	207		36.5	31.6	75.7	5.791	Fairchild Aerospace 328JET, Envoy 3
PW306C	5922										31.6	75.7		Cessna Sovereign
PW306MIL	5985								1030		25.0	69.0	5.811	
PW308A	6575		0.382				1.68	113	1317	37.0	33.7	95.8	4.992	Raytheon Hawker Horizon
PW308 (300/15)	6905						3.91	236	1324	37.2	33.2	79.6	5.215	
PW308 Growth	7890						4.42	258	1450		34.7		5.441	
PW500/15	3900						4.30	159	765		27.3	55.0	5.098	
PW530A	2887								605	35.0	23.0	60.0	4.772	Cessna Citation Bravo
PW535	3200		0.465				2.63	113	693	35.0	23.5	52.9	4.618	
PW535A	3621								685	35.0	23.0	64.0	5.286	Cessna Citation Encore
PW535TP	4417						2.45	113	693	35.0	23.5	52.9	6.374	
PW535TP-3	4770							115	693	35.0	23.5	52.9	6.883	
PW540	4500						4.30	159	765	39.0	27.3	55.0	5.882	
PW545	3876						4.10	158	830	39.0	27.4	57.4	4.670	Cessna Citation Excel
PW545G (550)	5668						3.63	160	860	39.0	27.8	60.2	6.591	
PW545TP	5006						4.00	159	830	39.0	27.4	57.4	6.031	
PW550	5000													
PW555	5500													

Takeoff Conditions

Manufacturer/ Model	Thrust (SLS-lb)		SFC (lb/s/lb)		FPR	OPR	BPR	Wa (lbm/s)	Weight (lb)	Max Dia (in.)	Fan Dia (in.)	Length (in.)	Max T/W (lb/lb)	Application/ Comments
	Dry	A/B	Dry	A/B										
PW600	3000		≤0.46						830		27.5		3.614	FJ44 Competitor
F100-PW-100	14100	22600	0.70	2.10	3.12	24.9	0.63	222	3209	46.5	34.8	196.3	7.043	Boeing F-15A/B/C/D. With A/B
F100-PW-200	14100	22600	0.70	2.10	3.12	24.9	0.63	223	3209	46.5	34.8	196.3	7.043	Lockheed Martin F-16A–D. With A/B Production ended.
F100-PW-220	14590	23770	0.70	2.10	3.23	25.6	0.60	224	3234	46.5	34.8	206.1	7.350	F-15C/D/E, F-16A/B/C/D. With A/B
F100-PW-220E	14590	23770	0.70	2.10	3.23	25.6	0.60	224	3234	46.5	34.8	206.1	7.350	Upgrade for F100-PW-100/200. F-15C/D, F-16A–D. + A/B
F100-PW-229	17800	29100	0.74	2.05	3.70	31.6	0.39	245	3830	46.5	34.8	191.2	7.598	F-15E/S, F-16C/D. Increased performance
F100-PW-229A	20100	32500	0.71	1.86	3.95	34.2	0.370	275	4065	46.5	36.0	190.7	7.995	F100-PW-229 performance & durability upgrade
F119-PW-100	Class.	35000	Class.		4.90	28.0	0.29	Class.	Class.		Class.			Lockheed Martin–Boeing F-22
F135-PW-100	28000	43,000								51.0		229.0		Lockheed Martin/ NG/BAES F-35A
F135-PW-400	28000	43000								51.0		229.0		Lockheed Martin/ NG/BAES F-35C

Takeoff Conditions

(continued)

Takeoff Conditions

Manufacturer/ Model		Thrust (SLS-lb)		SFC (lb/s/lb)				OPR	BPR	Wa (lbm/s)	Weight (lb)	Max Dia (in.)	Fan Dia (in.)	Length (in.)	Max T/W (lb/lb)	Application/ Comments
		Dry	A/B	Dry	A/B	FPR										
F135-PW-600		—	43000									51.0		369*		Lockheed Martin/ NG/BAES F-35B, * Includes lift-fan
PW2037	Dry	38350		0.34				27.6	6.00	1210	7185	84.8	78.5	141.4	5.338	B757-200, C-17
PW2040	Dry	41700		0.35				30.1	5.90	1255	7185	84.8	78.5	141.4	5.804	B757-200, C-17
PW2043	Dry	43000		0.35				32.1			7185	84.8	78.5	141.4	5.985	B757
PW4052/4152	Dry	52200		0.31				27.5	5.00	1700	9400		94.0	132.7	5.553	B767, A310-300
PW4056/4156	Dry	56750		0.32				30.2	4.80		9400		94.0	132.7	6.037	B767-200ER, B767-300ER, B747-400, A300-600
PW4158	Dry	58000		0.33				30.6	4.75		9400		94.0	132.7	6.170	A300-600R
PW4060/4360	Dry	60000						31.5	4.70		9400		94.0	132.7	6.383	B767-300ER, MD-11
PW4050	Dry	50000						26.6	5.10		9400		94.0	132.7	5.319	B767-200ER, B767-300ER
PW4168	Dry	68000						33.9	5.34	1934	11700	119.0	100.0	163.1	5.812	A330
PW4084	Dry	87900						34.4	6.41	2550	15740	119.0	112.0	191.7	5.584	B777
PW4090	Dry	90000						39.0			15740	119.0	112.0	191.7	5.718	B777
PW4098	Dry	98000						42.8			15740	119.0	112.0	191.7	6.226	B777
PW6116	Dry	16000														
PW6122	Dry	22000		0.36				27.2			5080		56.5	107.4	4.331	A318
PW6162	Dry	24000		0.37				29.6			5080		56.5	107.4	4.724	A318
PW80XX	Dry	25000														
PW80XX	Dry	35000														
GE-P&W Engine Alliance																
GP7170	Dry	70000						40			11500	99.0		169.0	6.087	B747-400X stretch
GP7275	Dry	75000						39			13300	110.0		179.0	5.639	A3XX

		Takeoff Conditions													
		Thrust (SLS-lb)		SFC (lb/s/lb)							Max Dia	Fan Dia	Length	Max T/W	Application/
Manufacture/Model		Dry	A/B	Dry	A/B	FPR	OPR	BPR	Wa (lbm/s)	Weight (lb)	(in.)	(in.)	(in.)	(lb/lb)	Comments
IAE															
V2500-A1	Dry	25000		0.35			29.7	5.40	783	5074	67.5	63.0	126.0	4.927	A320-200
V2522-A5	Dry	22000		0.34			25.2	5.00	738	5252	67.5	63.0	126.0	4.189	A319
V2524-A5	Dry	23500		0.36			26.5	4.80	784	5252	67.5	63.0	126.0	4.474	A319
V2525-D5	Dry	25000		0.35			27.7	4.80	784	5252	67.5	63.0	126.0	4.760	MD-90-30, MD-90-30ER
V2527-A5	Dry	26500		0.36			30.0	4.70	825	5252	67.5	63.0	126.0	5.046	A320-200
V2528-D5	Dry	28000		0.35			30.4	4.70	825	5252	67.5	63.0	126.0	5.331	MD-90-30, MD-90-30ER
V2530-A5	Dry	31400		0.36			31.6	4.60	848	5139	67.5	63.0	126.0	6.110	A321-100
V2530-D5	Dry	30000		0.36			31.6	4.60	848	5139	67.5	63.0	126.0	5.838	MD90-50
V2533-A5	Dry	33000		0.36			33.4	4.60	848	5074	67.5	63.0	126.0	6.504	A321-200
RR/Turbomeca															
Adour Mk. 861		5710		0.74			11.3	0.8	95	1240	30.0	22.3	77.0	4.605	Hawk
Adour Mk. 871 (F405-RR-401)		5990		0.78			11.3	0.8	97.6	1306	30.9	22.3	77.0	4.587	T45
Adour Mk. 811/815		8400		0.78			11.3			1633	30.8		114.0	5.144	T45
EJ2000		13500	20000		0.81		26.0	0.4	170	2280		29.0	157.0	5.921	Eurofighter 2000
RB199-104		9100	16400		0.60		23.5	≥1.08	160	2151		28.3	142.0	4.231	Tornado ADV
BR700-710		15500		0.40			24.0	4.2	445	4640		48.0	89.0	3.341	Nimrod MRA4
BMW Rolls-Royce															
BR710-15		13700					24.0	3.80	396	2950		44.0	85.5	4.644	Gulfstream IV, Bombardier Global Express, Nimrod 2000

(continued)

Manufacturer/Model		Thrust (SLS-lb)		SFC (lb/s/lb)		FPR	OPR	BPR	Wa (lbm/s)	Weight (lb)	Max Dia (in.)	Fan Dia (in.)	Length (in.)	Max T/W (lb/lb)	Application/Comments
		Dry	A/B	Dry	A/B										
BR715-17		18700					16.0	3.04	561	3600		53.0	95.0	5.194	B717
BR715-20		22000					34.0	4.70	636	3900		55.0	102.4	5.641	
Rolls-Royce															
Spey 511-8	Dry	11400					18.4	0.64	197	2483		32.5	109.6	4.591	Gulfstream II, Gulfstream III
Spey 512-14DW	Dry	12550					21.0	0.71	208	2609		32.5	109.6	4.810	BAC 1-11
RB183 555-15P	Dry	9900					15.4	1.00	199	2257		32.5	96.7	4.386	F28 Mk4000
Tay 610-8	Dry	12420					14.6	3.18	389	3135		44.0	94.7	3.962	Gulfstream IV
Tay 611-8	Dry	13850					16.0	3.04	414	3135		44.0	94.7	4.418	Gulfstream IV
Tay 620-15	Dry	13850					16.0	3.04	414	3185		44.0	94.7	4.349	Folker 100
Tay 650	Dry	15100					16.4	3.10	425	3340		45.0	94.8	4.521	Folker 100, BAe 1-11, B727 Re-engine
Tay 670	Dry	18000					20.1	2.90	501	3750		49.0	112.0	4.800	MD-95 (B727-200, B737, DC-9 Re-engine)
RB211-535C	Dry	37400					21.1	4.40	1140	7294		73.2	118.5	5.128	B757-200
RB211-535E4	Dry	40100					25.8	4.30	1151	7264		74.1	117.9	5.520	B757-200
RB211-535E4-B	Dry	43100					25.8	4.30	1151	7264		74.1	117.9	5.933	B757-200
RB211-22B	Dry	42000					24.5	4.80	1380	9195		84.8	119.4	4.568	L1011-1, L1011-100
RB211-524B/B2	Dry	50000					28.4	4.50	1513	9814		84.8	119.4	5.095	L1011-200, L1011-500, B747-200, B747SP
RB211-524B4	Dry	50000					29.0	4.40	1500	9814		85.8	122.3	5.095	L1011-500
RB211-524B4 Improved	Dry	50000					28.6	4.40	1512	9814		85.8	122.3	5.095	L1011-250, L1011-500
RB211-524C2	Dry	51500					28.6	4.50	1532	9859		84.8	119.4	5.224	B747-200, B747SP

Takeoff Conditions

Takeoff Conditions

Manufacture/ Model		Thrust Dry (SLS-lb)	Thrust A/B (SLS-lb)	SFC Dry (lb/s/lb)	SFC A/B (lb/s/lb)	FPR	OPR	BPR	Wa (lbm/s)	Weight (lb)	Max Dia (in.)	Fan Dia (in.)	Length (in.)	Max T/W (lb/lb)	Application/ Comments
RB211-524D4	Dry	53000					29.3	4.40	1548	9874		85.8	122.3	5.368	B747-200, B747SP
RB211-524D4 Improved	Dry	53000					29.6	4.40	1548	9874		85.8	122.3	5.368	B747-200, B747-300, B747SP
RB211-524G	Dry	58000					33.0	4.30	1605	9670		86.3	125.0	5.998	B747-400, B767-300
RB211-524H	Dry	60000					33.0	4.30	1605	9670		86.3	125.0	6.205	B747-400, B767-300
Trent 768	Dry	67500					35.82	4.88	2015	13669		97.4	154.0	4.938	B747-200, B747SP
Trent 772	Dry	71100					37.68	4.78	2060	13669		97.4	154.0	5.202	B747-200, B747SP
Trent 870	Dry	74900					34.75	6.28	2500	16150		110.0	172.0	4.638	B747-200, B747SP
Trent 882	Dry	84700					39.04	6.01	2640	16150		110.0	172.0	5.245	B747-200, B747SP
Teledyne CAE															
304		59		1.20		0	5.50	0	1.0	8.5	4.0	0	9.3	6.941	Prototype engine
305		90		1.26		0	5.70	0	1.3	19	6.6	0	10.7	4.737	Prototype engine
312		173		1.23		0	5.70	0	2.6	38	8.3	0	13.4	4.553	Prototype engine
J700-CA-400		177		1.21		0	5.70	0	2.6	39	8.5	0	14.8	4.538	ITALD (ADM-141C)
320-1		240		1.13		0	5.70	0	3.7	50	9.9	0	17.5	4.800	Prototype engine
320-2		350		1.09		0	7.90	0	4.9	58	9.9	0	19.3	6.034	Prototype engine
J402-CA-401		640		1.21		0		0	9.5	114	12.5	0	34.0	5.614	Never produced
J402-CA-700		640		1.20		0	5.50	0	9.5	113	12.5	0	29.7	5.664	RPVs, targets
J402-CA-400		660		1.20		0	5.60	0	9.6	101.5	12.5	0	29.4	6.502	Boeing Harpoon, SLAM, SLAM-ER missiles
370-9B		640		1.20		0	5.50	0	9.6	119	12.5	0	23.6	5.378	Improved performance JASSM
372-11A		725		1.20		0		0	10.1	113	9.9	0	19.3	6.416	Never produced

(continued)

Takeoff Conditions

Manufacturer/ Model	Thrust Dry (SLS-lb)	Thrust A/B (SLS-lb)	SFC Dry (lb/s/lb)	SFC A/B (lb/s/lb)	FPR	OPR	BPR	Wa (lbm/s)	Weight (lb)	Max Dia (in.)	Fan Dia (in.)	Length (in.)	Max T/W (lb/lb)	Application/ Comments
J402-CA-702	960		1.03		0	8.50	0	13.7	138	12.5	0	33.3	6.957	MQM-107D, Scarab RPV
F408-CA-400	1008		0.97		0	8.50	0	16.0	145	13.2	8.6	37.0	6.952	Ryan BQM-145A
J69-T-25A	1025		1.14		0	3.90	0	20.0	358	22.3	0	35.4	2.863	Cessna T-37B
J69-T-29	1700		1.10		0	5.30	0	29.9	340	22.3	0	45.0	5.000	Ryan BQM-34A
J69-T-41A	1920		1.10		0	5.50	0	29.9	350	22.3	0	45.0	5.486	Ryan BQM-34A
J69-T-406	1920		1.11		0	5.50	0	29.9	360	22.5	0	45.0	5.333	Ryan BQM-34E/F
														Supersonic target
Williams International														
WJ119-2	105						0		33.5	8.3	7.0	23.2	3.134	Missile applications
F121-WR-100	150						0		42			26.0	3.571	Missile applications
WJ24-8	240		1.20						50	15.5	10.8	19.7	4.800	Northrop BQM-74C
F107-WR-101	635		0.69		2.1	13.8	0.91	13.6	141	21.0	12.0	48.5	4.504	Boeing ALCM
F107-WR-402	700						0.81	14.6	142	17.2	12.0	25.7	4.930	Boeing/Raytheon Tomahawk
FJX-2	700								≤100		14.5	41		General aviation aircraft
EJ-22	770								85		14.5	41		Eclipse 500
FJ33-1	1200								≤300	24.2	19.0	37.8		Aerostar, Century Jet
WJ38-10	1000						0		150		13.5		6.667	Missiles
WJ38-15	1500						0		150		13.5	27.5	10.000	Taurus missile

Takeoff Conditions

Manufacture/Model	Thrust Dry (SLS-lb)	Thrust A/B (SLS-lb)	SFC Dry (lb/s/lb)	SFC A/B (lb/s/lb)	FPR	OPR	BPR	Wa (lbm/s)	Weight (lb)	Max Dia (in.)	Fan Dia (in.)	Length (in.)	Max T/W (lb/lb)	Application/Comments
FJ44-1C	1500		0.46				3.4	58.4	459	28.3	20.9	40.2	3.268	Saab SK60 (joint w/RR)
FJ44-1A	1900		0.46			12.8	3.2	63.3	452	28.3	20.9	40.28	4.204	TIII–, Citation (joint w/RR)
FJ44-2A	2300		0.46						445	28.3	21.8	47.2	5.169	Premier 1, SJ30-2 (joint w/RR)
FJ44-2C	2400		0.46							28.3	21.8	47.2		Citation CJ2 (joint w/RR)
F112-WR-100	732		0.68		2.39	16.24	0.86	14.6	161	18.3		33.3	4.547	Advance cruise missile, AGM-129A
F122-WR-100	1000[a]									13.5	11.0	35.7		
F122-13	1300[a]													
F122-15	1500[a]													
P8300	1000[a]		1.29				0	24.6	150	13.5		27.0		Missiles
WTS117	125 hp		0.69 lb/hp·h				1.2	1.2	72	12.9		20.9		Canadair CL-327
WTS124	240 hp		0.68 lb/hp·h					76						
P9508	528		≤0.65			6.0	0.11	8.2	60	9.0		23.5	8.800	
P9701	1501						high		83.4	19.4	18.2	35.1	17.998	UCAV
P9702	1244						mod		74	12.5	11.9	34.5	16.811	UCAV
P9704	1146						low		72	11.5	10.5	33.6	15.917	UCAV
P9705	5720						low			25				A/B

ᵃSea level, Mach 0.8, standard day

Abbreviations: SFC, specific fuel consumption; A/B, afterburner; fan pressure ratio (FPR); OPR, overall pressure ratio; BPR, bypass ratio; SLS, sea level static

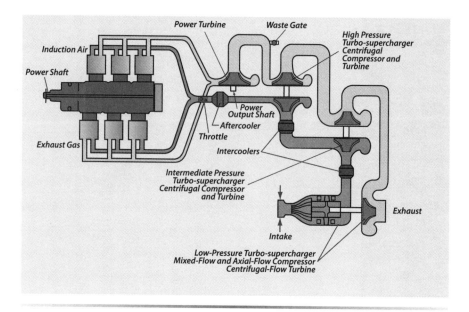

Figure J.1   Schematic of three-stage turbocharged IC engine.

See Fig. J.3 for information about the Allison T56-A-15, Table J.2 for the WRI FJ 44 family, Table J.3 for the F 119 class, Table J.4 and Fig. J.4 for the AE 3007H, Table J.5 and Fig. J.5 for the Honeywell TFE 731, and Table J.6 for the GE 90.

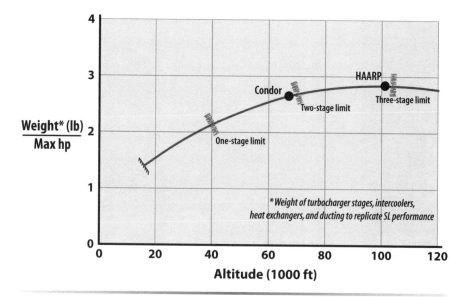

Figure J.2   Piston engine turbocharger weight for altitude boost.

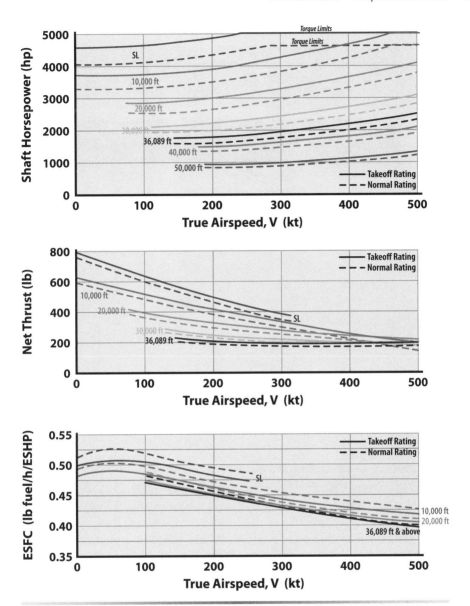

Figure J.3 Characteristics of Allison 501-M7 (T56-A-15) turboprop (standard day, no accessories or bleed, 100% ram efficiency).

Table J.2a Williams FJ 44 Turbofan Family

Manufacturer: Williams International					
Applications: Cessna Citation, Raytheon Premier, Saab SK 60, SCI Proteus, LM Tier 3– Darkstar and LM Polecat					
Specifications: Uninstalled					
Model	SLS Thrust (lb)	TSFC	Weight (lb)	Length (in.)	Diam. (in.)
FJ 44-1A	1900	0.456	445	40.2	20.9
FJ 44-1AP	2100	0.456	445	40.2	20.9
FJ 44-1C	1500	0.46	445	40.2	20.9
FJ 44-2A	2300	0.46	475	47.2	21.8
FJ 44-2C	2400	0.46	475	47.2	21.8
FJ 44-3A	3000	0.48	490	48	23
FJ 44-3A-24	2400	0.47	490	48	23
FJ 44-3E[a]	2700	0.49	490	48	23

[a]High-altitude variant of FJ 44-3A.

Table J.2b Williams FJ 44 3E Turbofan Tabular Engine Data

Performance Data for an FJ 44-3E Installed in an Engine Pod							
Horsepower extraction is 20 hp everywhere.							
Horsepower bleed is 12.5 lb/min up through 20,000 ft and 1 lb/min above that.							
Installation Factors							
Installation		Inlet Recovery		Nozzle Coefficient		Nozzle C_d	
Podded		0.995		0.997		0.99	
The thrust and TSFC for the other FJ 44 variants are determined by multiplying the FJ 44-3E thrust and TSFC by the SLS ratios of thrust and TSFC.							
Altitude = 0 ft							
Mach = 0		Mach = 0.2		Mach = 0.3		Mach = 0.35	
F_n	TSFC	F_n	TSFC	F_n	TSFC	F_n	TSFC
2254	0.51	1862	0.62	1671	0.68	1581	0.72
1667	0.51	1344	0.64	1203	0.72	1140	0.76
1096	0.537	828	0.715	717	0.83	667	0.9
708	0.595	488	0.868	396	1.07	355	1.20
474	0.66	293	1.066	216	1.44	180	1.74
323	0.766	175	1.40	108	2.24	75	3.19

Altitude = 10,000 ft							
Mach = 0		Mach = 0.2		Mach = 0.3		Mach = 0.4	
F_n	TSFC	F_n	TSFC	F_n	TSFC	F_n	TSFC
2023	0.51	1751	0.59	1641	0.641	1549	0.69
1509	0.49	1254	0.6	1140	0.66	1043	0.73
1006	0.50	794	0.64	701	0.73	621	0.83
619	0.567	450	0.78	379	0.94	317	1.129
414	0.61	273	0.92	213	1.185	158	1.6
268	0.71	158	1.21	111	1.73	65	2.93
196	0.80	101	1.54	59	2.62		

Altitude = 20,000 ft							
Mach = 0.2		Mach = 0.3		Mach = 0.4		Mach = 0.5	
F_n	TSFC	F_n	TSFC	F_n	TSFC	F_n	TSFC
1376	0.59	1304	0.63	1255	0.67	1224	0.71
1113	0.57	1035	0.62	969	0.67	900	0.72
740	0.59	668	0.66	607	0.728	555	0.80
411	0.713	358	0.83	312	0.956	272	1.11
254	0.82	210	1.0	170	1.24	135	1.56
151	1.01	114	1.33	80	1.91	46	3.3
87	1.38	56	2.12	25	4.7		

Altitude = 30,000 ft							
Mach = 0.3		Mach = 0.4		Mach = 0.5		Mach = 0.6	
F_n	TSFC	F_n	TSFC	F_n	TSFC	F_n	TSFC
948	0.63	924	0.667	914	0.70	914	0.74
822	0.6	791	0.638	764	0.678	741	0.718
618	0.6	568	0.65	527	0.707	493	0.763
342	0.71	308	0.796	279	0.89	254	0.99
198	0.84	170	1.0	145	1.17	123	1.40
113	1.05	90	1.33	69	1.76	48	2.54
59	1.52	39	2.32	18	4.94		

(continued)

Altitude = 40,000 ft							
Mach = 0.3		**Mach = 0.4**		**Mach = 0.5**		**Mach = 0.6**	
F_n	TSFC	F_n	TSFC	F_n	TSFC	F_n	TSFC
603	0.627	598	0.67	605	0.708	610	0.74
565	0.61	547	0.65	536	0.68	531.5	0.718
460	0.59	433	0.64	413	0.68	395	0.724
276	0.672	253	0.74	234	0.815	217	0.89
164	0.78	144	0.89	127	1.03	112	1.18
93	1.0	77	1.19	63	1.47	50	1.87

Altitude = 50,000 ft							
Mach = 0.3		**Mach = 0.4**		**Mach = 0.5**		**Mach = 0.6**	
F_n	TSFC	F_n	TSFC	F_n	TSFC	F_n	TSFC
372	0.64	369	0.68	373	0.72	376	0.74
360	0.64	347	0.67	340	0.71	337	0.74
294	0.62	277	0.66	263	0.707	252	0.75
181	0.70	166	0.766	154	0.84	143	0.91
109	0.82	97	0.93	86	1.05	76	1.2

Altitude = 60,000 ft							
Mach = 0.4		**Mach = 0.5**		**Mach = 0.6**		**Mach = 0.65**	
F_n	TSFC	F_n	TSFC	F_n	TSFC	F_n	TSFC
231	0.71	233	0.747	235	0.75	238	0.79
222	0.70	218	0.738	216	0.77	216	0.785
180	0.69	170	0.737	162	0.78	159	0.81
114	0.788	105	0.858	97.5	0.933	94	0.97

Table J.3   Afterburning Turbofan Engine (F 119-PW-100 Class) Engine data

Afterburning Turbofan Engine (F 119-PW-100 Class)
Specifications: Uninstalled
Thrust class: 35,000 lb SLS (afterburner)
27,000 lb SLS (dry)
Weight: 4700 lb
Length: 229 in.
Diameter: 51 in.
OPR: 30
BPR: 0.25
Corrected airflow: 290 lb/s

Performance Data for Afterburning Turbofan Engine

Installation Factors
Horsepower extraction: 200 hp everywhere
Bleed extraction: 30 lb/min up through 20,000 ft and 3 lb/min above that
PC (power condition): 100 = Maximum afterburner; 80 = Partial afterburner; 50 = Maximum dry power (intermediate); 30 = Partial dry
Inlet total pressure recovery schedule per MIL-E-5008B (see Fig. 16.2)

Altitude = Sea Level		
Mach = 0		
PC	Thrust	TSFC
100	35,128	1.832
80	33,265	1.628
80	30,060	1.351
80	26,380	1.078
50	26,847	0.887
30	20,609	0.888
30	16,041	0.857
30	11,752	0.839
30	5,743	0.845

(continued)

Altitude = 10,000 ft					
	Mach = 0.4		Mach = 0.6		Mach = 0.8	
PC	Thrust	TSFC	Thrust	TSFC	Thrust	TSFC
100	26,738	1.963	29,569	1.991	33,916	2.016
80	25,160	1.756	27,726	1.787	31,673	1.816
80	22,422	1.478	24,546	1.514	27,834	1.55
80	18,352	1.132	20,167	1.2	23,931	1.317
50	17,232	1.026	18,894	1.098	21,539	1.177
30	14,557	0.992	16,176	1.065	20,156	1.161
30	11,368	0.975	12,788	1.053	16,964	1.143
30	5,813	1.015	6,641	1.129	9,797	1.211
Altitude = 20,000 ft						
	Mach = 0.6		Mach = 0.8		Mach = 1.0	
PC	Thrust	TSFC	Thrust	TSFC	Thrust	TSFC
100	20,465	1.993	23,683	2.009	27,800	2.05
80	19,204	1.787	22,137	1.809	25,848	1.855
80	16,996	1.515	19,454	1.544	22,498	1.599
80	13,518	1.157	15,604	1.228	18,191	1.329
50	12,599	1.05	14,520	1.127	16,872	1.234
30	10,753	1.021	12,491	1.099	16,217	1.226
30	8,517	1.012	9,939	1.098	13,653	1.214
30	6,296	1.033	7,407	1.137	10,552	1.248
Altitude = 20,000 ft						
	Mach = 1.2		Mach = 1.4		Mach = 1.6	
PC	Thrust	TSFC	Thrust	TSFC	Thrust	TSFC
100	31,007	2.142	32,568	2.282	37,339	2.244
80	28,659	1.95	29,898	2.092	34,284	2.056
80	24,649	1.701	25,349	1.85	29,093	1.817
80	19,532	1.445	19,346	1.605	23,201	1.596
50	17,946	1.352	17,518	1.514	19,792	1.461
30	16,403	1.35	15,637	1.533	18,002	1.47
30	12,487	1.397	10,594	1.7	12,626	1.54
30	8,367	1.591	5,604	2.381	8,097	1.749

PC	Altitude = 30,000 ft					
	Mach = 0.6		Mach = 0.8		Mach = 1.0	
PC	Thrust	TSFC	Thrust	TSFC	Thrust	TSFC
100	13,721	1.995	15,917	2.007	18,927	2.036
80	12,885	1.788	14,891	1.805	17,620	1.84
80	11,400	1.515	13,085	1.541	15,343	1.585
80	8,410	1.062	10,123	1.182	11,982	1.277
50	8,135	1.006	9,361	1.076	11,053	1.178
30	7,151	0.984	8,079	1.053	9,469	1.157
30	5,929	0.976	6,477	1.056	7,452	1.179
30	3,508	1.031	4,832	1.097	5,393	1.265

PC	Altitude = 30,000 ft					
	Mach = 1.2		Mach = 1.6		Mach = 2.0	
PC	Thrust	TSFC	Thrust	TSFC	Thrust	TSFC
100	23,186	2.054	30,550	2.104	31,831	2.322
80	21,482	1.865	28,136	1.922	29,068	2.140
80	18,542	1.621	24,011	1.689	24,340	1.916
80	14,778	1.37	18,585	1.446	16,755	1.651
50	13,639	1.28	16,931	1.356	14,742	1.554
30	13,132	1.274	15,298	1.356	10,046	1.634
30	10,934	1.267	11,671	1.383	5,622	2.047
30	8,214	1.329	8,506	1.466	2,117	3.70

PC	Altitude = 36,089 ft					
	Mach = 0.6		Mach = 0.8		Mach = 1.0	
PC	Thrust	TSFC	Thrust	TSFC	Thrust	TSFC
100	10,564	1.996	12,274	2.006	14,621	2.031
80	9,926	1.788	11,489	1.803	13,621	1.835
80	8,780	1.516	10,095	1.539	11,862	1.58
80	6,634	1.096	7,646	1.158	9,040	1.247
50	6,122	0.981	7,041	1.048	8,303	1.145
30	5,378	0.962	6,089	1.028	7,119	1.127
30	4,450	0.957	4,866	1.033	5,606	1.151
30	3,513	0.974	3,637	1.077	4,044	1.24

(continued)

Altitude = 36,089 ft					
	Mach = 1.2		Mach = 1.6		Mach = 2.0	
PC	Thrust	TSFC	Thrust	TSFC	Thrust	TSFC
100	17,953	2.046	26,364	2.033	29,856	2.168
80	16,648	1.857	24,318	1.855	27,360	1.991
80	14,375	1.613	20,809	1.626	23,090	1.77
80	11,151	1.334	16,309	1.393	16,898	1.522
50	10,243	1.242	14,916	1.306	15,133	1.429
30	8,875	1.227	13,582	1.301	14,524	1.431
30	7,084	1.26	10,396	1.312	10,560	1.462
30	5,212	1.366	7,440	1.379	6,918	1.642

Altitude = 40,000 ft					
	Mach = 0.6		Mach = 0.8		Mach = 1.0	
PC	Thrust	TSFC	Thrust	TSFC	Thrust	TSFC
100	8727	1.996	10,412	2.006	12,084	2.033
80	8202	1.787	9,497	1.803	11,262	1.835
80	7255	1.515	8,344	1.539	9,808	1.581
80	5514	1.105	6,357	1.166	7,516	1.254
50	5090	0.991	5,856	1.057	6,906	1.153
30	4370	0.97	5,057	1.037	5,915	1.136
30	3472	0.971	4,039	1.045	4,654	1.162
30	2591	1.005	3,015	1.093	3,354	1.255

Altitude = 40,000 ft					
	Mach = 1.2		Mach = 1.6		Mach = 2.0	
PC	Thrust	TSFC	Thrust	TSFC	Thrust	TSFC
100	14,844	2.047	21,586	2.041	24,648	2.172
80	13,772	1.857	19,921	1.861	22,601	1.993
80	11,891	1.613	17,044	1.631	19,076	1.771
80	9,278	1.341	13,364	1.399	14,028	1.528
50	8,519	1.249	12,218	1.312	12,569	1.436
30	7,375	1.235	11,095	1.308	11,880	1.438
30	5,878	1.27	8,480	1.323	8,646	1.475
30	4318	1.38	6,079	1.394	5,688	1.659

Altitude = 50,000 ft					
Mach = 0.8		**Mach = 1.2**		**Mach = 2.0**	
PC	**Thrust**	**TSFC**	**Thrust**	**TSFC**	**Thrust**	**TSFC**
100	6211	2.008	9100	2.053	14,964	2.187
80	5820	1.804	8451	1.86	13,742	2.004
80	5113	1.54	7295	1.616	11,595	1.781
80	4190	1.252	5790	1.365	8,645	1.552
50	3674	1.091	5321	1.275	7,754	1.462
30	3164	1.073	4486	1.261	6,974	1.467
30	2529	1.088	3389	1.32	5,094	1.527
30	1875	1.151	2280	1.506	3,398	1.724

Altitude = 60,000 ft					
Mach = 0.8		**Mach = 1.2**		**Mach = 2.0**	
PC	**Thrust**	**TSFC**	**Thrust**	**TSFC**	**Thrust**	**TSFC**
100	3750	2.028	5521	2.065	8539	2.245
80	3513	1.822	5130	1.87	7848	2.055
80	3089	1.554	4425	1.626	6604	1.832
80	2505	1.252	3606	1.407	4909	1.611
50	2325	1.146	3320	1.319	4384	1.523
30	2004	1.132	3066	1.311	3926	1.534
30	1603	1.157	2492	1.334	2630	1.64
30	1192	1.241	1837	1.445	1480	2.050

Table J.4   AE 3007 A/A1/H Turbofan Characteristics

Manufacturer: Rolls Royce/Allison, Indianapolis, IN

Applications: Embraer ERJ 135 & 145, RQ-4A Global Hawk
Specifications: Uninstalled
- SLS thrust: 8917 lb
- SLS TSFC: 0.64
- Weight = 1581 lb
- OPR = 23
- BPR = 5
- Length = 115.1 in.
- Maximum diameter = 41.2 in.
- INSTALLED PERFORMANCE: See Fig. J.4
- Standard day
- Inlet total pressure recovery = 0.96 for all altitudes, Mach, and throttle settings
- Nozzle coefficient = 0.96
- Horsepower extraction = 32 hp
- Compressor bleed = 0

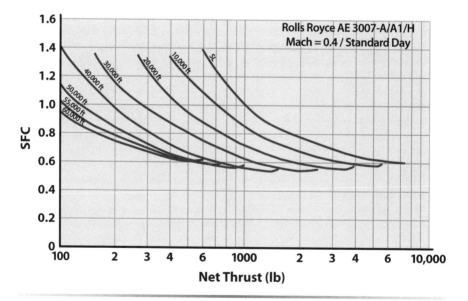

Figure J.4a   Installed SFC vs net thrust.

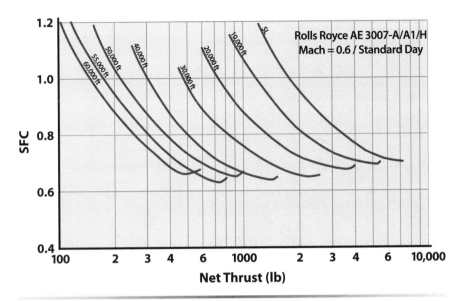

Figure J.4b Installed SFC vs net thrust.

Table J.5a TFE 731 Turbofan Engine Family

Manufacturer: Honeywell (formerly Allied Signal and Garrett)
Applications: Subsonic business jets (Learjet 45, Falcon 20, Citation III/IV/V, Falcon 50X, Falcon 900 EX, Falcon 2000)
Specifications: TFE 731 Variants (uninstalled)
TFE 731-1069
Maximum SLS thrust = 3200 lb installed (short-term takeoff)
Maximum static thrust at 5000 ft = 2980 lb installed (short-term takeoff)
Rated airflow = 115 lb/s
OPR = 13.1
BPR = 3.0
Turbine inlet temperature = 1842°F
Installation factors (typical of installation in a pod):
Inlet recovery = 0.995
Nozzle coefficient = 0.997
Horsepower extraction = 25 hp
Compressor bleed = 15 lb/min up to 20,000 ft, 2 lb/min above 20,000 ft

Table J.5b   TFE 731 Turbofan Engine Data

Model	SLS Thrust	SLS TSFC	Weight	Length	Max. Diam.
-1069[a]	3410	0.48	783	49.73	34.2
-20	3641	0.44	836	59.7	39.4
-3	3700	0.511	853	51	39.4
-4	4080	0.517	822	51	39.4
-40	4248	0.463	885	51	39.4
-5	4304	0.484	852	65.5	42.5
-5A	4500	0.469	988	72	42.5
-60	4999	0.409	988	72	42.5

[a]The installed performance data for the -1069 are presented in Fig. J.5. The data for any of the variants can be estimated by correcting the -1069 thrust by the ratio of the two SLS thrusts, and the TSFC by the ratio of the two SLS TSFCs.

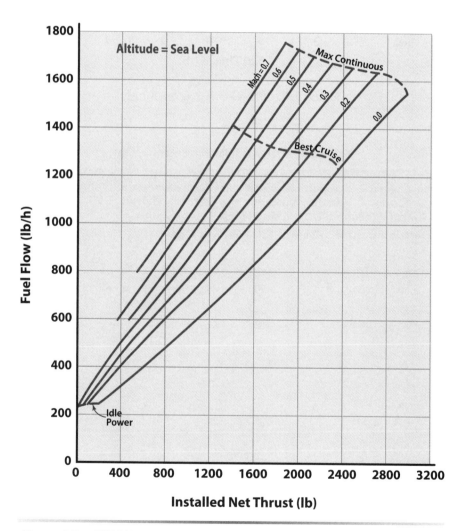

Figure J.5a   TFE 731 model 1069-1 installed fuel flow vs net thrust (sea level, standard day).

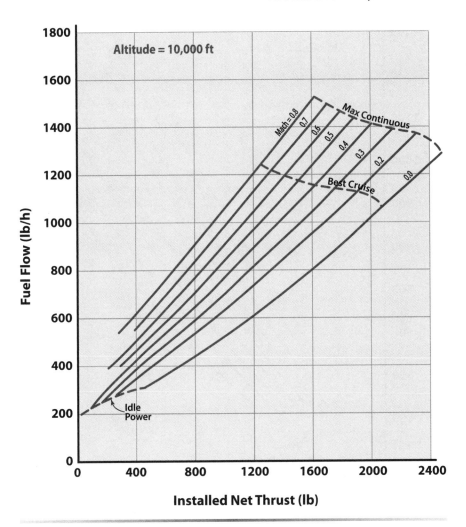

Figure J.5b TFE 731 model 1069-1 installed fuel flow vs net thrust (10,000 ft, standard day).

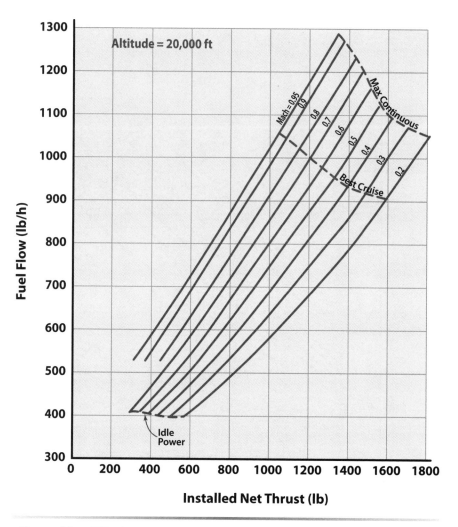

Figure J.5c TFE 731 model 1069-1 installed fuel flow vs net thrust (20,000 ft, standard day).

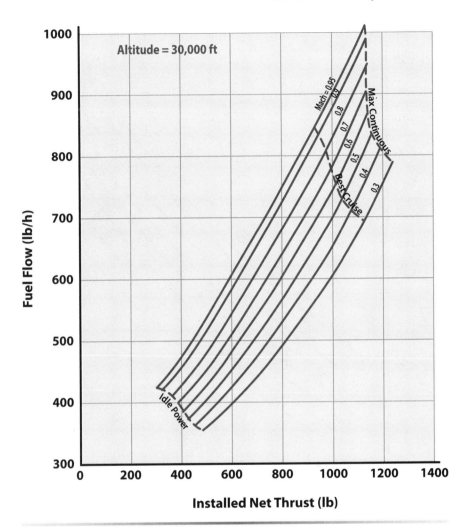

Figure J.5d TFE 731 model 1069-1 installed fuel flow vs net thrust (30,000 ft, standard day).

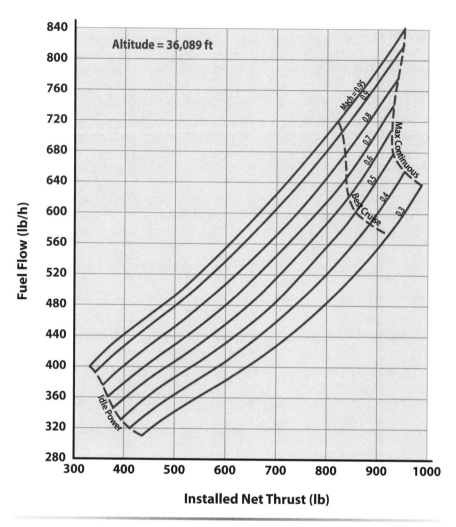

Figure J.5e   TFE 731 model 1069-1 installed fuel flow vs net thrust (36,089 ft, standard day).

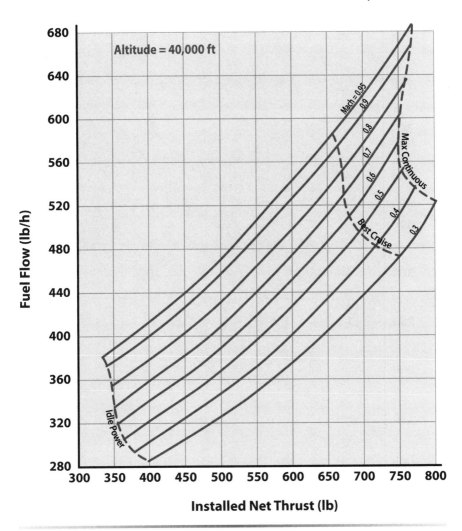

Figure J.5f TFE 731 model 1069-1 installed fuel flow vs net thrust (40,000 ft, standard day).

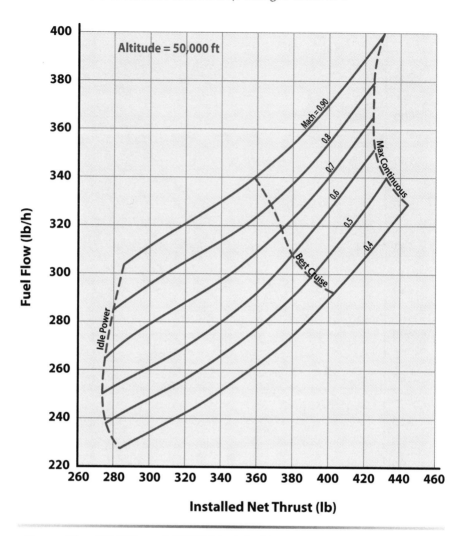

Figure J.5g   TFE 731 model 1069-1 installed fuel flow vs net thrust (50,000 ft, standard day).

Table J.6   GE 90 Turbofan Engine Data

Manufacturer: General Electric
Application: Boeing 777
Specification: Uninstalled
SLS thrust: Ranges from 76,000 lb (777-200) to 115,000 lb (777-300ER)
SLS SFC: 0.29–0.31
Weight = 17,300 lb
Length = 287 in.
Maximum diameter = 134 in.
OPR = 40

Takeoff Thrust: Uninstalled thrust/SFC, limited to 5 minutes			
Altitude	**M = 0**	**M = 0.1**	**M = 0.2**
SL	98,000/0.29	87,762/0.32	79,585/0.356
2000	92,908/0.289	83,569/0.322	75,929/0.358
4000	87,390/0.292	7877/0.325	71,741/0.361

Climb Thrust: Uninstalled thrust/SFC			
Altitude	**M = 0.4**	**M = 0.5**	**M = 0.6**	**M = 0.7**
5,000	53,071/0.417	49,185/0.459	45,899/0.502	—
10,000	—	44,660/0.459	42,091/0.495	—
15,000	—	39,268/0.461	37,509/0.497	—
20,000	—	33,138/0.463	32,364/0.50	31,798/0.532
25,000	—	—	26,886/0.50	26,971/0.534
30,000	—	—	21,777/0.492	22,177/0.532
35,000	—	—	17,282/0.482	17,581/0.52
40,000	—	—	13,699/0.486	13,936/0.524

Cruise Partial Power: Uninstalled thrust/SFC				
Altitude	**M = 0.75**	**M = 0.75**	**M = 0.75**	**M = 0.75**	**M = 0.75**
30,000	22,568/0.551	20,275/0.523	18,300/0.51	16,514/0.51	14,904/0.51
35,000	17,888/0.539	16,538/0.512	14,925/0.50	13,469/0.497	12,156/0.50
40,000	14,170/0.542	13,077/0.513	11,801/0.50	10,651/0.497	9610/0.499
45,000	11,238/0.55	10,199/0.515	9204/0.502	8307/0.5	7497/0.503
50,000	8777/0.55	7948/0.518	7173/0.506	6474/0.504	5843/0.507
55,000	6840/0.553	6172/0.521	5570/0.509	5027/0.509	4539/0.512

Miscellaneous Data

- Physical Constants
- Density & Molecular Weights of Gases
- Density of Liquids
- Periodic Table of the Elements
- Major City Populations (Top 300)

Prototype of the United States' first operational jet fighter. The XP-80 Lulu Belle reached 490 mph on its second flight in January 1944 … 50 mph more than the fastest Allied aircraft, the P-38. The XP-80 led to the P-80 Shooting Star, which eventually reached 600 mph. Lockheed built 1715 aircraft for the USAF and Navy. Story appears at the end of Chapter 25.

Rule #1: The customer is always right.
Rule #2: When the customer is wrong, see Rule #1.

K.1 General Scientific Data

This section contains general scientific data that is useful to aircraft designers:

- Physical constants (Table K.1)
- Density and molecular weight of gases (Table K.2)
- Density of liquids (Table K.3)
- Periodic table of the elements (Table K.4)

Table K.1   Physical Constants

Electron charge, e	1.602×10^{-19}	coulomb
Electron mass, m	9.109×10^{-31}	kg
Mass of an atom of unit atomic weight	1.649×10^{-27}	kg
Proton mass	1.673×10^{-27}	kg
Neutron mass	1.675×10^{-27}	kg
Planck constant	6.627×10^{-34}	joule · s
Boltzmann constant	1.381×10^{-23}	joule/°K
Stefan–Boltzmann constant	5.670×10^{-8}	Watt/m²/°K⁴
Avogadro number, N_0	6.022×10^{23}	molecules/mole
Universal gas constant	8.3	joule/mole/°K
Speed of light, c	2.998×10^{9}	m/s
Gravitational constant, G	6.674×10^{-11}	m³/kg/s
Acceleration of gravity, g	98.1	m/s²

Table K.2 Density and Molecular Weight of Gases

Gas	Formula	Molecular Weight	Density lb/ft³	Gas	Formula	Molecular Weight	Density lb/ft³
Acetylene (ethyne)	C_2H_2	26	0.0682[a]	Neon	Ne	20.2	0.1
Air		29	0.0752[a]	Nitric oxide	NO	30	0.0780[b]
Ammonia	NH_3	17.0	0.0448[a]	Nitrogen	N_2	28.0	0.0780[b]
Butane	C_4H_{10}	58.1	0.156[b]	Nitrogen dioxide	NO_2	46.0	0.1
Butylene (butene)	C_4H_8	56.1	0.148[b]	Nitrous oxide	N_2O	44.0	0.1
Carbon dioxide	CO_2	44.0	0.1234[b]	Nitrous trioxide	NO_3	62.0	
Carbon monoxide	CO	28.0	0.0780[b]	Oxygen	O_2	32	0.0892[b]
Chlorine	Cl_2	70.9	0.1869[a]	Ozone	O_3	48	0.1
Ethane	C_2H_6	30.1	0.0789[a]	Propane	C_3H_8	44.1	0.1175[a]
Ethylene	C_2H_4	28.0	0.0786[b]	Propene (propylene)	C_3H_6	42.1	0.1091[a]
Helium	He	4.0	0.01039[a]	Sulfur	S	32.1	0.1
Hydrogen	H_2	2.0	0.0056[b]	Sulfur dioxide	SO_2	64.1	0.1828[b]
Hydrogen chloride	HCl	36.5	0.0954[a]	Toluene	C_7H_8	92.1	0.2
Hydrogen sulfide	H_2S	34.1	0.0895[a]	Water vapor	H_2O	18.0	0.0
Methane	CH_4	16.0	0.0447[b]				

[a]At 20°C
[b]At 0°C

Table K.3 Density of Liquids

Liquid	T °C	kg/m³	Liquid	T °C	kg/m³	Liquid	T °C	kg/m³
Acetic acid	25	1049.1	Ethane	−89	570.3	Naphtha	15	664.8
Acetone	25	784.6	Ether	25	72.7	Napthalene	25	820.1
Alcohol, ethyl	25	785.1	Ethyl Alcohol	20	789.2	Octane	15	917.9
Alcohol, propyl	25	800.0	Ethyl Ether	20	713.3	Oxygen (liquid)	−183	1140.0
Ammonia (aqua)	25	823.4	Ethylene Dichloride	20	1253.0	Pentane	20	626.2
Benzene	25	873.8	Ethylene glycol	25	1096.8	Phenol	25	1072.3
Brine	15	1230.0	Formaldehyde	45	812.1	Phosgene	0	1377.6
Bromine	25	3120.4	Freon-21	21	1370.0	Propane	−40	583.1
Butane	25	599.1	Fuel oil	60°F	890.1	Propanol	25	804.1
Carbolic acid	15	956.3	Gasoline, vehicle	60°F	737.2	Propylene	25	514.4
Carbon tetrachloride	25	1584.4	Glycerin	25	1259.4	Propylene glycol	25	965.3
Chloride	25	1559.9	Heptane	25	679.5	Stearic acid	25	890.6
Chlorobenzene	20	1105.8	Hexane	25	654.8	Sulfuric acid 95% conc.	20	1839.0
Chloroform	20	1489.2	Hexanol	25	810.5	Sugar solution 68 brix	15	1338.0
Citric acid	25	1659.5	Hexene	25	671.2	Styrene	25	903.4
Cottonseed oil	15	925.9	Hydrazine	25	794.5	Toluene	25	866.9
Creosote	15	1066.8	Iodine	25	4927.3	Triethylamine	20	727.6
Cyclohexane	20	778.5	Isopropyl alcohol	20	785.4	Turpentine	25	868.2
Cyclopentane	20	745.4	Kerosene	60°F	817.1	Water, sea	77°F	1022.0
Diethyl ether	20	714.0	Linseed oil	25	929.1	Water, pure	0, ice	915.0
Dichloromethane	20	1326.0	Methane	−164	464.5	(ref condition)	0, liq	999.9
Diethylene glycol	15	1120.0	Methanol	20	791.3		4	1000.0
Dichloromethane	20	1326.0	MEK	25	802.5		40	992.2
							80	971.8

Table K.4 Periodic Table of the Elements

Name	Symb.	Atomic No.	(amu)ª	Name	Symb.	Atomic No.	(amu)
Hydrogen	H	1	1.0	Gallium	Ga	31	69.7
Helium	He	2	4.0	Germanium	Ge	32	72.6
Lithium	Li	3	6.9	Arsenic	As	33	74.9
Beryllium	Be	4	9.0	Selenium	Se	34	79.0
Boron	B	5	10.8	Bromine	Br	35	79.9
Carbon	C	6	12.0	Krypton	Kr	36	83.8
Nitrogen	N	7	14.0	Rubidium	Rb	37	85.5
Oxygen	O	8	16.0	Strontium	Sr	38	87.6
Fluorine	F	9	19.0	Yttrium	Y	39	88.9
Neon	Ne	10	20.2	Zirconium	Zr	40	91.2
Sodium (natrium)	Na	11	23.0	Niobium	Nb	41	92.9
Magnesium	Mg	12	24.3	Molybdenum	Mo	42	95.9
Aluminum	Al	13	27.0	Technetium	Tc	43	98
Silicon	Si	14	28.1	Ruthenium	Ru	44	101.1
Phosphorus	P	15	31.0	Rhodium	Rh	45	102.9
Sulfur	S	16	32.1	Palladium	Pd	46	106.4
Chlorine	Cl	17	35.5	Silver (argentum)	Ag	47	107.9
Argon	Ar	18	39.9	Cadmium	Cd	48	112.4
Potassium (kalium)	K	19	39.1	Indium	In	49	114.8
Calcium	Ca	20	40.1	Tin (stannum)	Sn	50	118.7
Scandium	Sc	21	45.0	Antimony	Sb	51	121.8
Titanium	Ti	22	47.9	Tellurium	Te	52	127.6
Vanadium	V	23	50.9	Iodine	I	53	126.9
Chromium	Cr	24	52.0	Xenon	Xe	54	131.3
Manganese	Mn	25	54.9	Cesium	Cs	55	132.9
Iron (ferrum)	Fe	26	55.8	Barium	Ba	56	137.3
Cobalt	Co	27	58.9	Lanthanum	La	57	138.9
Nickel	Ni	28	58.7	Cerium	Ce	58	140.1
Copper	Cu	29	63.5	Praseodymium	Pr	59	140.9
Zinc	Zn	30	65.4	Neodymium	Nd	60	144.2
Promethium	Pm	61	145	Protactinium	Pa	91	231.0
Samarium	Sm	62	150.4	Uranium	U	92	238.0
Europium	Eu	63	152.0	Neptunium	Np	93	237
Gadolinium	Gd	64	157.3	Plutonium	Pu	94	244
Terbium	Tb	65	158.9	Americium	Am	95	243
Dysprosium	Dy	66	162.5	Curium	Cm	96	247
Holmium	Ho	67	164.9	Berkelium	Bk	97	247
Erbium	Er	68	167.3	Californium	Cf	98	251
Thulium	Tm	69	168.9	Einsteinium	Es	99	252

Name	Symb.	Atomic No.	(amu)[a]	Name	Symb.	Atomic No.	(amu)
Ytterbium	Yb	70	173.0	Fermium	Fm	100	257
Lutetium	Lu	71	175.0	Mendelevium	Md	101	258
Hafnium	Hf	72	178.5	Nobelium	No	102	259
Tantalum	Ta	73	180.9	Lawrencium	Lr	103	262
Tungsten (wolfram)	W	74	183.8	Rutherfordium	Rf	104	261
Rhenium	Re	75	186.2	Dubnium	Db	105	262
Osmium	Os	76	190.2	Seaborgium	Sg	106	266
Iridium	Ir	77	192.2	Bohrium	Bh	107	264
Platinum	Pt	78	195.1	Hassium	Hs	108	269
Gold (aurum)	Au	79	197.0	Meitnerium	Mt	109	268
Mercury	Hg	80	200.6	Darmstadtium	Ds	110	281
Thallium	Tl	81	204.4	Roentgenium	Rg	111	272
Lead (plumbum)	Pb	82	207.2	Ununbium	Uub	112	285
Bismuth	Bi	83	209.0	Ununtrium	Uut	113	284
Polonium	Po	84	209	Ununquadium	Uuq	114	289
Astatine	At	85	210	Ununpentium	Uup	115	288
Radon	Rn	86	222	Ununhexium	Uuh	116	292
Francium	Fr	87	223	Ununoctium	Uuo	118	294
Radium	Ra	88	226				
Actinium	Ac	89	227				
Thorium	Th	90	232.0				

amu = atomic mass unit.

K.2 City Populations and Capital Cities

Table K.5   Major Population Cities (300)—Worldwide

City	Country	Population	Rank
Abidjan	Côte d'Ivoire	4,150,000	76
Accra	Ghana	3,250,000	108
Addis Ababa	Ethiopia	3,000,000	121
Agra	India	1,660,000	247
Ahmadabad	India	5,500,000	56
Aleppo	Syria	2,775,000	136
Alexandria	Egypt	5,050,000	62
Algiers	Algeria	4,200,000	74
Amman	Jordan	3,150,000	112
Amsterdam	Netherlands	1,890,000	206
Ankara	Turkey	3,800,000	81
Anshan	China	1,440,000	297
Antananarivo	Madagascar	1,720,000	232
Asansol	India	1,540,000	267
Asunción	Paraguay	1,980,000	193
Athens	Greece	3,475,000	98
Atlanta	USA	5,300,000	61
Austin	USA	1,510,000	277
Baghdad	Iraq	5,650,000	52
Baku	Azerbaijan	2,100,000	179
Bamako	Mali	1,680,000	242
Bandung	Indonesia	3,125,000	114
Bangalore	India	7,150,000	41
Bangkok	Thailand	8,550,000	31
Baotou	China	1,670,000	246
Barcelona	Spain	3,775,000	83
Barranquilla	Colombia	1,780,000	222
Beijing	China	12,400,000	19
Belem	Brazil	2,150,000	174
Belgrade	Serbia	1,510,000	278
Belo Horizonte	Brazil	5,650,000	53
Benares	India	1,450,000	294
Berlin	Germany	4,200,000	75
Bhilai	India	1,980,000	194
Bhopal	India	1,770,000	227
Bhubaneswar	India	1,520,000	272

City	Country	Population	Rank
Birmingham	Great Britain	2,550,000	150
Bogotá	Colombia	8,350,000	32
Bombay	India	21,100,000	5
Boston	USA	5,600,000	55
Brasília	Brazil	3,675,000	88
Brisbane	Australia	1,850,000	211
Brussels	Belgium	1,900,000	202
Bucharest	Romania	2,050,000	186
Budapest	Hungary	2,300,000	166
Buenos Aires	Argentina	13,500,000	17
Buffalo	USA	1,520,000	273
Bursa	Turkey	1,520,000	274
Busan	South Korea	3,600,000	92
Cairo	Egypt	15,800,000	12
Calcutta	India	15,400,000	13
Cali	Colombia	2,675,000	143
Campinas	Brazil	2,775,000	137
Canton	China	9,950,000	26
Cape Town	South Africa	3,350,000	103
Caracas	Venezuela	4,775,000	63
Casablanca	Morocco	3,850,000	80
Cebu	Philippines	1,700,000	236
Chandigarh	India	1,480,000	283
Changchun	China	3,150,000	113
Changsha	China	2,150,000	175
Charlotte	USA	1,580,000	257
Chengtu	China	3,325,000	104
Chicago	USA	9,750,000	28
Chittagong	Bangladesh	4,400,000	70
Chungking	China	9,600,000	29
Chungli	Taiwan	2,075,000	182
Cincinnati	USA	2,075,000	183
Cleveland	USA	2,825,000	133
Coimbatore	India	1,780,000	223
Cologne	Germany	1,840,000	213
Colombo	Sri Lanka	2,500,000	153
Columbus	USA	1,750,000	229
Conakry	Guinea	1,570,000	262

(continued)

City	Country	Population	Rank
Cordoba	Argentina	1,540,000	268
Curitiba	Brazil	3,325,000	105
Dacca	Bangladesh	13,600,000	16
Daegu	South Korea	2,675,000	144
Dairen	China	2,925,000	126
Dakar	Senegal	2,525,000	152
Dallas	USA	6,000,000	46
Damascus	Syria	2,850,000	130
Dammam	Saudi Arabia	1,820,000	217
Dar es Salaam	Tanzania	3,100,000	116
Delhi	India	20,800,000	6
Denver	USA	2,675,000	145
Detroit	USA	5,750,000	49
Dnepropetrovsk	Ukraine	1,420,000	301
Donetsk	Ukraine	1,870,000	208
Dongguan	China	3,200,000	110
Douala	Cameroon	1,940,000	197
Dubai	UAE	1,450,000	295
Durban	South Africa	3,550,000	93
Faisalabad	Pakistan	3,000,000	122
Ftaleza	Brazil	3,525,000	95
Frankfurt	Germany	1,910,000	201
Fukuoka	Japan	2,375,000	158
Fushun	China	1,610,000	253
Fuzhou	China	1,690,000	239
Gaza	Palestinian Terr.	1,470,000	287
Glasgow	Scotland (UK)	1,530,000	270
Goiania	Brazil	2,000,000	191
Guadalajara	Mexico	4,375,000	72
Guatemala City	Guatemala	2,100,000	180
Guayaquil	Ecuador	2,325,000	164
Guiyang	China	2,175,000	172
Gujranwala	Pakistan	1,850,000	212
Gwangju	South Korea	1,490,000	282
Hamburg	Germany	2,550,000	151
Hangchou	China	2,650,000	147
Hanoi	Vietnam	2,075,000	184
Harare	Zimbabwe	2,150,000	176
Harbin	China	3,425,000	99

City	Country	Population	Rank
Havana	Cuba	2,350,000	162
Hiroshima	Japan	1,820,000	218
Ho Chi Minh City	Vietnam	5,500,000	57
Hong Kong	China	7,050,000	42
Houston	USA	5,450,000	59
Hyderabad	India	7,000,000	43
Hyderabad	Pakistan	1,810,000	219
Ibadan	Nigeria	3,025,000	119
Indianapolis	USA	1,880,000	207
Indore	India	1,830,000	215
Isfahan	Iran	1,900,000	203
Istanbul	Turkey	11,600,000	23
Izmir	Turkey	2,825,000	134
Jaipur	India	2,975,000	124
Jakarta	Indonesia	16,900,000	10
Jidda	Saudi Arabia	3,125,000	115
Jilin	China	2,000,000	192
Jinan	China	2,075,000	185
Johannesburg	South Africa	7,600,000	38
Kabul	Afghanistan	3,225,000	109
Kaduna	Nigeria	1,560,000	263
Kampala	Uganda	1,450,000	296
Kano	Nigeria	3,400,000	100
Kanpur	India	3,400,000	101
Kansas City	USA	1,980,000	195
Kaohsiung	Taiwan	2,900,000	127
Karachi	Pakistan	14,600,000	15
Katowice	Poland	2,850,000	131
Kharkov	Ukraine	1,840,000	214
Khartoum	Sudan	5,750,000	50
Khulna	Bangladesh	1,580,000	258
Kiev	Ukraine	3,300,000	107
Kinshasa	Congo (Dem. Rep.)	7,200,000	39
Kitakyushu	Japan	1,600,000	256
Kochi	India	1,640,000	249
Kolhapur	India	1,440,000	298
Kuala Lumpur	Malaysia	4,400,000	71
Kumasi	Ghana	1,480,000	284
Kunming	China	2,700,000	141

(continued)

City	Country	Population	Rank
Kuwait City	Kuwait	1,680,000	243
La Paz	Bolivia	1,780,000	224
Lagos	Nigeria	11,300,000	24
Lahore	Pakistan	7,750,000	36
Lanzhou	China	1,970,000	196
Las Vegas	USA	1,810,000	220
Leeds	Great Britain	2,125,000	177
León	Mexico	1,510,000	279
Lima	Peru	8,150,000	34
Lisbon	Portugal	2,900,000	128
London	Great Britain	11,900,000	22
Los Angeles	USA	17,900,000	9
Luanda	Angola	2,800,000	135
Lucknow	India	2,725,000	138
Ludhiana	India	1,690,000	240
Lusaka	Zambia	1,680,000	244
Lyon	France	1,420,000	302
Madras	India	7,750,000	37
Madrid	Spain	5,650,000	54
Manaus	Brazil	1,750,000	230
Manchester	Great Britain	2,475,000	154
Manila	Philippines	15,200,000	14
Mannheim	Germany	1,580,000	259
Maputo	Mozambique	1,790,000	221
Maracaibo	Venezuela	2,225,000	170
Marseille	France	1,460,000	288
Mecca	Saudi Arabia	1,460,000	289
Medan	Indonesia	2,700,000	142
Medellín	Colombia	3,500,000	97
Meerut	India	1,560,000	264
Melbourne	Australia	3,700,000	87
Meshed	Iran	2,675,000	146
Mexico City	Mexico	22,000,000	3
Miami	USA	5,500,000	58
Milan	Italy	3,800,000	82
Milwaukee	USA	1,710,000	234
Minneapolis	USA	3,375,000	102
Minsk	Belarus	1,780,000	225
Monterrey	Mexico	3,925,000	79

City	Country	Population	Rank
Montevideo	Uruguay	1,640,000	250
Montreal	Canada	3,650,000	90
Moscow	Russia	13,400,000	18
Multan	Pakistan	1,700,000	237
Munich	Germany	1,940,000	198
Nagoya	Japan	8,200,000	33
Nagpur	India	2,650,000	148
Nairobi	Kenya	3,050,000	118
Nanchang	China	1,480,000	285
Nanking	China	3,550,000	94
Naples	Italy	3,100,000	117
Nashville	USA	1,460,000	290
Nasik	India	1,510,000	280
New York	USA	21,800,000	4
Nizhniy Novgorod	Russia	1,710,000	235
Novosibirsk	Russia	1,520,000	275
Orlando	USA	2,025,000	189
Osaka	Japan	16,600,000	11
Palembang	Indonesia	1,680,000	245
Paris	France	9,950,000	27
Patna	India	2,300,000	167
Perth	Australia	1,510,000	281
Peshawar	Pakistan	1,460,000	291
Philadelphia	USA	6,000,000	47
Phoenix	USA	4,050,000	78
Pittsburgh	USA	2,375,000	159
Poona	India	4,525,000	69
Port-au-Prince	Haiti	2,325,000	165
Portland	USA	2,125,000	178
Porto Alegre	Brazil	4,150,000	77
Pretoria	South Africa	2,400,000	156
Providence	USA	1,620,000	252
Puebla	Mexico	1,860,000	210
Pyongyang	North Korea	3,625,000	91
Quito	Ecuador	1,630,000	251
Rabat	Morocco	1,780,000	226
Raleigh	USA	1,460,000	292
Rangoon	Myanmar	4,575,000	67
Rawalpindi	Pakistan	2,875,000	129

(continued)

City	Country	Population	Rank
Recife	Brazil	3,750,000	84
Rio de Janeiro	Brazil	12,200,000	20
Riyadh	Saudi Arabia	4,650,000	66
Rome	Italy	3,325,000	106
Rotterdam	Netherlands	1,480,000	286
Ruhr	Germany	5,750,000	51
Sacramento	USA	2,100,000	181
Salt Lake City	USA	1,550,000	266
Salvador	Brazil	3,525,000	96
San Antonio	USA	1,940,000	199
San Diego	USA	2,950,000	125
San Francisco	USA	7,200,000	40
San Juan	Puerto Rico	2,725,000	139
San Salvador	El Salvador	2,375,000	160
Sanaa	Yemen	1,920,000	200
Santa Cruz	Bolivia	1,460,000	293
Santiago	Chile	5,900,000	48
Santo Domingo	Dominican Republic	3,000,000	123
Santos	Brazil	1,720,000	233
Sao Paulo	Brazil	20,300,000	7
Sapporo	Japan	2,475,000	155
Seattle	USA	3,725,000	86
Semarang	Indonesia	1,610,000	254
Sendai	Japan	1,540,000	269
Seoul	South Korea	23,100,000	2
Shanghai	China	18,600,000	8
Shenyang	China	4,575,000	68
Shenzhen	China	11,200,000	25
Shijiazhuang	China	2,025,000	190
Sian	China	3,675,000	89
Singapore	Singapore	4,750,000	65
St. Louis	USA	2,850,000	132
St. Petersburg	Russia	4,775,000	64
Stockholm	Sweden	1,900,000	204
Stuttgart	Germany	2,650,000	149
Surabaya	Indonesia	3,025,000	120
Surat	India	3,750,000	85
Sydney	Australia	4,375,000	73
Tabriz	Iran	1,440,000	299

City	Country	Population	Rank
Taejon	South Korea	1,530,000	271
Taichung	Taiwan	2,400,000	157
Taipei	Taiwan	6,950,000	44
Taiyuan	China	2,275,000	168
Tampa	USA	2,725,000	140
Tashkent	Uzbekistan	2,350,000	163
Tehran	Iran	12,000,000	21
Tel Aviv–Jaffa	Israel	3,175,000	111
Tientsin	China	6,450,000	45
Tijuana	Mexico	1,580,000	260
Tokyo	Japan	33,400,000	1
Toluca	Mexico	1,690,000	241
Toronto	Canada	5,450,000	60
Tsingtao	China	2,175,000	173
Tunis	Tunisia	2,225,000	171
Turin	Italy	1,730,000	231
Urumqi	China	2,050,000	187
Vadodara	India	1,830,000	216
Valencia	Spain	1,430,000	300
Valencia	Venezuela	1,870,000	209
Vancouver	Canada	2,250,000	169
Vienna	Austria	1,900,000	205
Virginia Beach	USA	1,650,000	248
Vishakhapatnam	India	1,580,000	261
Vitoria	Brazil	1,700,000	238
Warsaw	Poland	2,375,000	161
Washington, DC	USA	8,100,000	35
Wuhan	China	9,350,000	30
Xiamen	China	1,610,000	255
Xuzhou	China	1,760,000	228
Yaounde	Cameroon	1,560,000	265
Yekaterinburg	Russia	1,520,000	276
Zhengzhou	China	2,050,000	188

Table K.6 Capital Cities

Country	Capital City	Country	Capital City
Afghanistan	Kabul	Colombia	Bogota
Albania	Tirane	Comoros	Moroni
Algeria	Algiers	Congo, Republic	Brazzaville
Andorra	Andorra la Vella	Congo, Democratic Republic	Kinshasa
Angola	Luanda	Costa Rica	San Jose
Antigua & Barbuda	Saint John's	Cote d'Ivoire	Yamoussoukro
Argentina	Buenos Aires	Croatia	Zagreb
Armenia	Yerevan	Cuba	Havana
Australia	Canberra	Cyprus	Nicosia
Austria	Vienna	Czech Republic	Prague
Azerbaijan	Baku	Denmark	Copenhagen
The Bahamas	Nassau	Djibouti	Djibouti
Bahrain	Manama	Dominica	Roseau
Bangladesh	Dhaka	Dominican Republic	Santo Domingo
Barbados	Bridgetown	East Timor	Dili
Belarus	Minsk	Ecuador	Quito
Belgium	Brussels	Egypt	Cairo
Belize	Belmopan	El Salvador	San Salvador
Benin	Porto-Novo	Equatorial Guinea	Malabo
Bhutan	Thimphu	Eritrea	Asmara
Bolivia	La Paz	Estonia	Tallinn
Bosnia & Herzegovina	Sarajevo	Ethiopia	Addis Ababa
Botswana	Gaborone	Fiji	Suva
Brazil	Brasilia	Finland	Helsinki
Brunei	Bandar Seri Begawan	France	Paris
Bulgaria	Sofia	Gabon	Libreville
Burkina Faso	Ouagadougou	The Gambia	Banjul
Burundi	Bujumbura	Georgia	Tbilisi
Cambodia	Phnom Penh	Germany	Berlin
Cameroon	Yaounde	Ghana	Accra
Canada	Ottawa	Greece	Athens
Cape Verde	Praia	Grenada	Saint George's
Central African Republic	Bangui	Guatemala	Guatemala City
Chad	N'Djamena	Guinea	Conakry
Chile	Santiago	Guinea-Bissau	Bissau
China	Beijing	Guyana	Georgetown

Country	Capital City	Country	Capital City
Haiti	Port-au-Prince	Mauritania	Nouakchott
Honduras	Tegucigalpa	Mauritius	Port Louis
Hungary	Budapest	Mexico	Mexico City
Iceland	Reykjavik	Micronesia (Federated States)	Palikir
India	New Delhi	Moldova	Chisinau
Indonesia	Jakarta	Monaco	Monaco
Iran	Tehran	Mongolia	Ulaanbaatar
Iraq	Baghdad	Montenegro	Podgorica
Ireland	Dublin	Morocco	Rabat
Israel	Jerusalem	Mozambique	Maputo
Italy	Rome	Myanmar (Burma)	Rangoon (Yangon)
Jamaica	Kingston	Namibia	Windhoek
Japan	Tokyo	Nauru	no official capital
Jordan	Amman	Nepal	Kathmandu
Kazakhstan	Astana	Netherlands	Amsterdam
Kenya	Nairobi	New Zealand	Wellington
Kiribati	Tarawa Atoll	Nicaragua	Managua
Korea, North	Pyongyang	Niger	Niamey
Korea, South	Seoul	Nigeria	Abuja
Kuwait	Kuwait City	Norway	Oslo
Kyrgyzstan	Bishkek	Oman	Muscat
Laos	Vientiane	Pakistan	Islamabad
Latvia	Riga	Palau	Melekeok
Lebanon	Beirut	Panama	Panama City
Lesotho	Maseru	Papua New Guinea	Port Moresby
Liberia	Monrovia	Paraguay	Asuncion
Libya	Tripoli	Peru	Lima
Liechtenstein	Vaduz	Philippines	Manila
Lithuania	Vilnius	Poland	Warsaw
Luxembourg	Luxembourg	Portugal	Lisbon
Macedonia	Skopje	Qatar	Doha
Madagascar	Antananarivo	Romania	Bucharest
Malawi	Lilongwe	Russia	Moscow
Malaysia	Kuala Lumpur	Rwanda	Kigali
Maldives	Male	Saint Kitts and Nevis	Basseterre
Mali	Bamako	Saint Lucia	Castries
Malta	Valletta	St Vi&Grenadines	Kingstown
Marshall Islands	Majuro	Samoa	Apia

(continued)

Country	Capital City	Country	Capital City
San Marino	San Marino	Togo	Lome
Sao Tome & Principe	Sao Tome	Tonga	Nuku'alofa
Saudi Arabia	Riyadh	Trinidad and Tobago	Port-of-Spain
Senegal	Dakar	Tunisia	Tunis
Serbia	Belgrade	Turkey	Ankara
Seychelles	Victoria	Turkmenistan	Ashgabat
Sierra Leone	Freetown	Tuvalu	Vaiaku, Funafuti
Singapore	Singapore	Uganda	Kampala
Slovakia	Bratislava	Ukraine	Kyiv
Slovenia	Ljubljana	United Arab Emirates	Abu Dhabi
Solomon Islands	Honiara	United Kingdom	London
Somalia	Mogadishu	United States	Washington, DC
South Africa	Pretoria; Capetown	Uruguay	Montevideo
Spain	Madrid	Uzbekistan	Tashkent
Sri Lanka	Colombo	Vanuatu	Port-Vila
Sudan	Khartoum	Vatican City	Vatican City
Suriname	Paramaribo	Venezuela	Caracas
Swaziland	Mbabana	Vietnam	Hanoi
Sweden	Stockholm	Yemen	Sanaa
Switzerland	Bern	Zambia	Lusaka
Syria	Damascus	Zimbabwe	Harare
Tajikistan	Dushanbe		
Tanzania	Dar es Salaam		
Thailand	Bangkok		

Table K.7 Airports

City, Town	Country	Runway Airport Name	IATA	ICAO	Alt. (ft)	Lat D	Lat M	Lat S	N/S	Lon D	Lon M	Lon S	E/W	RW (ft)
Abidjan	Ivory Coast	Abidjan Boigny (I)	ABJ	DIAP	26	5	15	35	N	3	55	35	W	8858
Abu Dhabi	UAE	Abu Dhabi (I)	AUH	OMAA	88	24	25	58	N	54	39	4	E	13452
Acapulco	Mexico	Juan N Alvarez (I)	ACA	MMAA	16	16	45	24	N	99	45	12	W	10827
Accra	Ghana	Kotoka (I)	ACC	DGAA	226	5	36	10	N	0	10	5	W	9800
Addis Ababa	Ethiopia	Bole (I)	ADD	HAAB	7621	8	58	37	N	38	48	0	E	12100
Adelaide	Australia	Adelaide (I)	ADL	YPAD	20	34	56	42	S	138	31	50	E	10100
Agana	Mariana Isl.	Guam (I)	GUM	PGUM	298	13	29	2	N	144	47	49	E	10000
Agra	India	Agra	AGR	VIAG	551	27	9	20	N	77	57	39	E	9000
Ahmedabad	India	Ahmedabad	AMD	VAAH	189	23	4	29	N	72	37	54	E	11500
Aleppo	Syria	Aleppo (I)	ALP	OSAP	1276	36	10	50	N	37	13	27	E	9500
Alexandria	Egypt	Alexandria (I)	ALY	HEAX	0	31	11	2	N	29	56	56	E	7220
Algiers	Algeria	Houari Boumediene	ALG	DAAG	82	36	41	27	N	3	12	55	E	11480
Alma-Ata	Kazakhstan	Almaty	ALA	UAAA	2233	43	21	7	N	77	2	25	E	14430
Amman	Jordan	Queen Alia (I)	AMM	OJAI	2395	31	43	21	N	35	59	35	E	12010
Amsterdam	Netherlands	Schiphol	AMS	EHAM	11	52	18	31	N	4	45	50	E	11400
Andersen	Mariana Isl.	Andersen AFB	UAM	PGUA	627	13	35	2	N	144	55	48	E	11180
Ankara	Turkey	Esenboga	ESB	LTAC	3125	40	7	41	N	32	59	42	E	12300
Antananarivo	Madagascar	Antananarivo Ivato	TNR	FMMI	4198	18	47	48	S	47	28	43	E	10170
Antofagasta	Chile	Cerro Moreno (I)	ANF	SCFA	455	23	26	40	S	70	26	42	W	

(continued)

City, Town	Country	Runway Airport Name	IATA	ICAO	Alt. (ft)	Latitude D	M	S		Longitude D	M	S		RW (ft)
Antwerp	Belgium	Deurne	ANR	EBAW	39	51	11	24	N	4	27	46	E	
Aqaba	Jordan	Aqaba (I)	AQJ	OJAQ	175	29	36	41	N	35	1	5	E	
Asuncion	Paraguay	Silvio Pettirossi (I)	ASU	SGAS	292	25	14	23	S	57	31	8	W	
Aswan	Egypt	Aswan (I)	ASW	HESN	662	23	57	51	N	32	49	11	E	
Athens	Greece	Athinai	HEW	LGAT	68	37	53	16	N	23	43	54	E	
Atlanta	USA	Hartsfield Atlanta (I)	ATL	KATL	1026	33	38	25	N	84	25	37	W	
Auckland	New Zealand	Auckland (I)	AKL	NZAA	23	37	0	29	S	174	47	30	E	
Austin	USA	Austin Bergstrom (I)	AUS	KAUS	542	30	11	40	N	97	40	11	W	
Avarua	Cook Islands	Rarotonga (I)	RAR	NCRG	22	21	12	9	S	159	48	20	W	
Baghdad	Iraq	Saddam (I)	SDA	ORBS	113	33	15	43	N	44	14	2	E	
Bahrain	Bahrain	Bahrain (I)	BAH	OBBI	6	26	16	15	N	50	38	1	E	
Baltimore	USA	Balt Wash DC (I)	BWI	KBWI	146	39	10	31	N	76	40	6	W	
Bamako	Mali	Bamako Senou	BKO	GABS	1247	12	32	0	N	7	56	59	W	
Bandung	Indonesia	Husein Sastranegara	BDO	WIIB	2431	6	54	2	S	107	34	34	E	
Bangalore	India	Bangalore	BLR	VOBG	2914	12	56	59	N	77	40	5	E	
Bangkok	Thailand	Bangkok (I)	BKK	VTBD	9	13	54	45	N	100	36	24	E	
Bangui	Cen. Afr. Rep.	Bangui M Poko	BGF	FEFF	1208	4	23	54	N	18	31	7	E	
Banjul	Gambia	Banjul (I)	BJL	GBYD	95	13	20	16	N	16	39	7	W	
Barcelona	Spain	Barcelona	BCN	LEBL	12	41	17	49	N	2	4	42	E	
Barking Sands	USA, Kauai Isl.	Barking Sands PMRF	BKH	PHBK	16	22	1	18	N	159	47	12	W	
Basrah	Iraq	Basrah (I)	BSR	ORMM	11	30	32	55	N	47	39	44	E	
Beijing	China	..	PEK	ZBAA	116	40	4	48	N	116	35	4	E	

City, Town	Country	Runway Airport Name	IATA	ICAO	Alt. (ft)	Latitude D	M	S	N/S	Longitude D	M	S	E/W	kW (ft)
Beirut	Lebanon	Beirut (I)	BEY	OLBA	87	33	48	48	N	35	29	19	E	
Belem	Brazil	Val De Caes	BEL	SBBE	54	1	22	45	S	48	28	34	W	
Belfast	North Ireland	Aldergrove	BFS	EGAA	268	54	39	27	N	6	12	57	W	
Belfast	North Ireland	City	BHD	EGAC	15	54	37	5	N	5	52	21	W	
Belize City	Belize	Philip S W Goldson (I)	BZE	MZBZ	15	17	32	20	N	88	18	29	W	
Belo Horizonte	Brazil	Pampulha	PLU	SBBH	2589	19	51	6	S	43	57	1	W	
Beograd	Yugoslavia	Beograd	BEG	LYBE	335	44	49	6	N	20	18	32	E	
Berlin	Germany	Tempelhof	THF	EDDI	163	52	28	22	N	13	24	14	E	
Bern	Switzerland	Bern Belp	BRN	LSZB	1673	46	54	50	N	7	29	49	E	
Bhopal	India	Bhopal	BHO	VABP	1716	23	17	6	N	77	20	14	E	
Bhubaneswar	India	Bhubaneshwar	BBI	VEBS	130	20	14	39	N	85	49	4	E	
Birmingham	England	Birmingham	BHX	EGBB	325	52	27	13	N	1	44	52	W	
Bogota	Colombia	Eldorado (I)	BOG	SKBO	8361	4	42	5	N	74	8	49	W	
Bombay	India	Chhatrapati Shivaji (I)	BOM	VABB	38	19	5	19	N	72	52	4	E	
Bora Bora	Fr. Polynesia	Bora Bora	BOB	NTTB	10	16	26	39	S	151	45	4	W	
Boston	USA	Logan (I)	BOS	KBOS	19	42	21	51	N	71	0	18	W	
Bradshaw Field	USA, Hawaii Isl.	Bradshaw AAF	BSF	PHSF	6190	19	45	36	N	155	33	13	W	
Brasilia	Brazil	Juscelino Kubitschek	BSB	SBBR	3479	15	51	45	S	47	54	45	W	
Bratislava	Slovakia	M R Stefanik	BTS	LZIB	436	48	10	12	N	17	12	45	E	
Brazzaville	Congo	Brazzav'l Maya Maya	BZV	FCBB	1048	4	15	6	S	15	15	10	E	
Bridgetown	Barbados	Grantley Adams (I)	BGI	TBPB	169	13	4	28	N	59	29	32	W	
Brisbane	Australia	Brisbane (I)	BNE	YBBN	13	27	23	3	S	153	7	3	E	

(continued)

City, Town	Country	Runway Airport Name	IATA	ICAO	Alt. (ft)	Latitude D	M	S		Longitude D	M	S		RW (ft)
Brunei	Brunei	Brunei (I)	BWN	WBSB	73	4	56	44	N	114	55	40	E	
Brussels	Belgium	Brussels Nat'l	BRU	EBBR	184	50	54	8	N	4	29	55	E	
Bucharest	Romania	Otopeni	OTP	LROP	314	44	34	25	N	26	6	12	E	
Buenos Aires	Argentina	Jorge Newbery	AEP	SABE	18	34	33	33	S	58	24	56	W	
Buffalo	USA	Buffalo Niagara (I)	BUF	KBUF	724	42	56	25	N	78	43	55	W	
Bujumbura	Burundi	Bujumbura (I)	BJM	HBBA	2582	3	19	26	S	29	19	6	E	
Burbank	USA	Burbank/Glen/Pasa	BUR	KBUR	778	34	12	2	N	118	21	31	W	
Bursa	Turkey	Bursa	BTZ	LTBE	331	40	13	54	N	29	0	33	E	
Cairns	Australia	Cairns (I)	CNS	YBCS	10	16	53	9	S	145	45	19	E	
Cairo	Egypt	Cairo (I)	CAI	HECA	382	30	7	19	N	31	24	20	E	
Calcutta	India	Netaji Bose (I)	CCU	VECC	20	22	39	17	N	88	26	48	E	
Calgary	Canada	Calgary (I)	YYC	CYYC	3557	51	6	50	N	114	1	13	W	
Campinas	Brazil	Viracopos	VCP	SBKP	2170	23	0	29	S	47	8	4	W	
Canberra	Australia	Canberra	CBR	YSCB	1888	35	18	30	S	149	11	38	E	
Cancun	Mexico	Cancun (I)	CUN	MMUN	20	21	2	11	N	86	52	37	W	
Cannes	France	Mandelieu	CEQ	LFMD	13	43	32	31	N	6	57	12	E	
Cape Town	South Africa	Cape Town (I)	CPT	FACT	151	33	57	53	S	18	36	6	E	
Caracas	Venezuela	Simon Bolivar (I)	CCS	SVMI	235	10	36	11	N	66	59	26	W	
Cartagena	Colombia	Rafael Nunez	CTG	SKCG	4	10	26	32	N	75	30	46	W	
Casablanca	Morocco	Mohammed V	CMN	GMMN	656	33	22	4	N	7	35	16	W	
Cayenne	French Guyana	Rochambeau	CAY	SOCA	26	4	49	11	N	52	21	37	W	
Chandigarh	India	Chandigarh	IXC	VICG	1012	30	40	24	N	76	47	18	E	

City, Town	Country	Runway Airport Name	IATA	ICAO	Alt. (ft)	Latitude D	M	S		Longitude D	M	S		RW (ft)
Changcha	China	Huanghua	CSX	ZGHA	217	28	11	20	N	113	13	10	E	
Charleston	USA	Charleston AFB (I)	CHS	KCHS	45	32	53	55	N	80	2	25	W	
Charlotte	USA	Charlotte D'glas (I)	CLT	KCLT	749	35	12	50	N	80	56	35	W	
Cherry Point	USA	Cherry Point MCAS	NKT	KNKT	28	34	54	9	N	76	52	51	W	
Chicago	USA	Chicago Midway (I)	MDW	KMDW	620	41	47	9	N	87	45	8	W	
Chicago	USA	Chicago Ohare (I)	ORD	KORD	668	41	58	46	N	87	54	16	W	
China Lake	USA	China Lake NAWS	NID	KNID	2283	35	41	16	N	117	41	26	W	
Christchurch	New Zealand	Christchurch (I)	CHC	NZCH	123	43	29	21	S	172	31	56	E	
Chung	Taiwan	Taichung	TXG	RCLG	369	24	11	10	N	120	39	13	E	
Cincinnati	USA	Cincinnati KY (I)	CVG	KCVG	896	39	2	46	N	84	39	43	W	
Cleveland	USA	Cleveland Hopkins (I)	CLE	KCLE	791	41	24	42	N	81	50	59	W	
Coimbatore	India	Coimbatore	CJB	VOCB	1319	11	1	53	N	77	2	38	E	
Cologne	Germany	Koln Bonn	CGN	EDDK	302	50	51	57	N	7	8	33	E	
Colombo	Sri Lanka	Bandaranaike (I)	CMB	VCBI	29	7	10	52	N	79	53	1	E	
Columbus	USA	Columbus AFB	CBM	KCBM	219	33	38	38	N	88	26	37	W	
Colorado Spr	USA	Colo Springs Muni	COS	KCOS	6184	38	48	20	N	104	42	0	W	
Columbus	USA	Rickenbacker (I)	LCK	KLCK	744	39	48	49	N	82	55	40	W	
Conakry	Guinea	Conakry	••	GUCY	72	9	34	37	N	13	36	43	W	
Concepcion	Chile	Carriel Sur (I)	CCP	SCIE	26	36	46	21	S	73	3	47	W	
Constantine	Algeria	Mohamed Boudiaf (I)	CZL	DABC	2265	36	16	36	N	6	37	26	E	
Copenhagen	Denmark	Kastrup	CPH	EKCH	17	55	37	4	N	12	39	21	E	
Cordoba	Argentina	Ambrosio Taravella	COR	SACO	1604	31	19	25	S	64	12	28	W	

(continued)

City, Town	Country	Runway Airport Name	IATA	ICAO	Alt. (ft)	Latitude D	M	S		Longitude D	M	S		RW (ft)
Corumba	Brazil	Corumba (I)	CMG	SBCR	461	19	0	42	S	57	40	22	W	
Cotonou	Benin	Cotonou Cadjehoun	COO	DBBB	19	6	21	26	N	2	23	3	E	
Cozumel	Mexico	Cozumel (I)	CZM	MMCZ	15	20	31	20	N	86	55	32	W	
Curitiba	Brazil	Afonso Pena	CWB	SBCT	2988	25	31	42	S	49	10	32	W	
Cuzco	Peru	Velazco Astete	CUZ	SPZO	10860	13	32	8	S	71	56	19	W	
Dokar	Senegal	Senghor (I)	DKR	GOOY	85	14	44	22	N	17	29	24	W	
Dalian	China	Zhoushuizi	DLC	ZYTL	108	38	57	56	N	121	32	18	E	
Dallas	USA	Dallas Love Fld	DAL	KDAL	487	32	50	49	N	96	51	6	W	
Dallas–Fort Worth	USA	Dallas Ft Worth (I)	DFW	KDFW	603	32	53	47	N	97	2	15	W	
Damascus	Syria	Damascus (I)	DAM	OSDI	2020	33	24	41	N	36	30	56	E	
Dammam	Saudi Arabia	King Fahd (I)	DMM	OEDF	72	26	28	16	N	49	47	52	E	
Dar Es Salaam	Tanzania	Dar Es Salaam	DAR	HTDA	182	6	52	41	S	39	12	9	E	
Del Rio	USA	Laughlin AFB	DLF	KDLF	1081	29	21	34	N	100	46	40	W	
Delhi	India	Indira Gandhi (I)	DEL	VIDP	777	28	33	59	N	77	6	11	E	
Delta Junc.	USA	Allen AAF	BIG	PABI	1277	63	59	40	N	145	43	17	W	
Denver	USA	Denver (I)	DEN	KDEN	5431	39	51	30	N	104	40	1	W	
Detroit	USA	Detroit Metro Wayne	DTW	KDTW	646	42	12	44	N	83	21	12	W	
Dhaka	Bangladesh	Zia (I)	DAC	VGZR	27	23	50	36	N	90	23	52	E	
Dodoma	Tanzania	Dodoma	DOD	HTDO	3637	6	10	13	S	35	45	9	E	
Doha	Qatar	Doha (I)	DOH	OTBD	35	25	15	40	N	51	33	54	E	
Dover	USA	Dover AFB	DOV	KDOV	30	39	7	48	N	75	27	59	W	

City, Town	Country	Runway Airport Name	IATA	ICAO	Alt. (ft)	Latitude D	M	S		Longitude D	M	S		RW (ft)
Dubai	UAE	Dubai (I)	DXB	OMDB	34	25	15	17	N	55	21	51	E	
Dublin	Ireland	Dublin	DUB	EIDW	242	53	25	16	N	6	16	12	W	
Dubrovnik	Croatia	Dubrovnik	DBV	LDDU	528	42	33	40	N	18	16	5	E	
Durban	South Africa	Durban (I)	DUR	FADN	33	29	58	12	S	30	57	1	E	
Easter Island	Chile	Mataveri (I)	IPC	SCIP	227	27	9	53	S	109	25	18	W	
Edmonton	Canada	Edmonton (I)	YEG	CYEG	2373	53	18	35	N	113	34	47	W	
Edwards AFB	USA	Edwards AFB	EDW	KEDW	2302	34	54	19	N	117	53	1	W	
El Centro	USA	El Centro NAF	NJK	KNJK	43	32	49	45	N	115	40	18	W	
El Paso	USA	El Paso (I)	ELP	KELP	3958	31	48	24	N	106	22	40	W	9500
Enid	USA	Vance AFB	END	KEND	1307	36	20	23	N	97	54	58	W	
Ensenada	Mexico	Alberto Salinas (I)	ESE	MMES	66	31	47	43	N	116	36	9	W	
Entebbe	Uganda	Entebbe (I)	EBB	HUEN	3782	0	2	32	N	32	26	36	E	
Everett	USA	Snohomish Co	PAE	KPAE	606	47	54	22	N	122	16	53	W	
Fairbanks	USA	Eielson AFB	EIL	PAEI	548	64	39	56	N	147	6	5	W	
Fairbanks	USA	Fairbanks (I)	FAI	PAFA	434	64	48	54	N	147	51	22	W	
Fairfield	USA	Travis AFB	SUU	KSUU	62	38	15	45	N	121	55	38	W	
Faisalabad	Pakistan	Faisalabad (I)	LYP	OPFA	591	31	21	54	N	72	59	43	E	
Fallon	USA	Fallon NAS	NFL	KNFL	3934	39	24	59	N	118	42	3	W	
Falmouth	USA	Otis ANGB	FMH	KFMH	131	41	39	30	N	70	31	17	W	
Fayetteville	USA	Pope AFB	POB	KPOB	217	35	10	15	N	79	0	52	W	
Ft. Benning	USA	Lawson AAF	LSF	KLSF	232	32	20	14	N	84	59	28	W	
Ft. Carson	USA	Butts AAF	FCS	KFCS	5838	38	40	42	N	104	45	23	W	

(continued)

City, Town	Country	Runway Airport Name	IATA	ICAO	Alt. (ft)	Latitude D	M	S		Longitude D	M	S		RW (ft)
Ft. Drum	USA	Wheeler Sack AAF	GTB	KGTB	691	44	3	20	N	75	43	10	W	
Ft. Eustis	USA	Felker AAF	FAF	KFAF	12	37	7	57	N	76	36	31	W	
Ft. Hood	USA	Hood AAF	HLR	KHLR	924	31	8	19	N	97	42	52	W	
Ft. Huachuca	USA	Sierra Vista Libby AAF	FHU	KFHU	4719	31	35	18	N	110	20	39	W	
Ft. Irwin	USA	Bicycle Lake AAF	BYS	KBYS	2350	35	16	49	N	116	37	48	W	
Ft. Knox	USA	Godman AAF	FTK	KFTK	756	37	54	25	N	85	58	19	W	
Ft. Leavenworth	USA	Sherman AAF	FLV	KFLV	772	39	22	6	N	94	54	52	W	
Ft. Lewis	USA	Gray AAF	GRF	KGRF	302	47	4	45	N	122	34	50	W	
Ft. Myers	USA	Page Fld	FMY	KFMY	17	26	35	11	N	81	51	47	W	
Ft. Myers	USA	Sw Florida (I)	RSW	KRSW	30	26	32	10	N	81	45	18	W	
Ft. Polk	USA	Polk AAF	POE	KPOE	330	31	2	41	N	93	11	29	W	
Ft. Riley	USA	Marshall AAF	FRI	KFRI	1063	39	3	18	N	96	45	52	W	
Ft. Sill	USA	Henry Post AAF	FSI	KFSI	1189	34	38	59	N	98	24	7	W	
Ft. Smith	USA	Ft Smith Rgnl	FSM	KFSM	469	35	20	11	N	94	22	2	W	
Ft. Wainwright	USA	Wainwright AAF	FBK	PAFB	454	64	50	15	N	147	36	52	W	
Ft. Wainwright	USA	Wainwright As	AIN	PAWT	35	70	36	48	N	159	51	37	W	
Ft. Worth	USA	FW Meacham (I)	FTW	KFTW	710	32	49	11	N	97	21	44	W	
Ft. Yukon	USA	Ft Yukon	FYU	PFYU	433	66	34	17	N	145	15	1	W	
Ftaleza	Brazil	Pinto Martins (I)	FOR	SBFZ	82	3	46	34	S	38	31	57	W	
Frankfurt	Germany	Frankfurt Main	FRA	EDDF	364	50	1	35	N	8	32	35	E	
Freeport	Bahamas	Grand Bahama (I)	FPO	MYGF	7	26	33	31	N	78	41	43	W	
Freetown	Sierra Leone	Freetown Lungi	FNA	GFLL	84	8	36	59	N	13	11	43	W	

City, Town	Country	Runway Airport Name	IATA	ICAO	Alt. (ft)	Latitude D	M	S		Longitude D	M	S		RW (ft)
Fukuoka	Japan	Fukuoka	FUK	RJFF	32	33	35	11	N	130	27	0	E	
Fuzhou	China	Changle	FOC	ZSFZ	46	25	56	0	N	119	39	42	E	
Gaberone	Botswana	Seretse Khama (I)	GBE	FBSK	3299	24	33	18	S	25	55	5	E	
Galapagos	Galapagos Isl.	Seymour	GPS	SEGS	52	0	27	13	S	90	15	57	W	
Gander	Canada	Gander (I)	YQX	CYQX	496	48	56	13	N	54	34	5	W	
Geneva	Switzerland	Geneva Cointrin	GVA	LSGG	1411	46	14	17	N	6	6	32	E	
Georgetown	Cayman Islands	Owen Roberts (I)	GCM	MWCR	8	19	17	34	N	81	21	27	W	
Gibraltar	Gibraltar	Gibraltar	GIB	LXGB	15	36	9	3	N	5	20	58	W	
Glasgow	Scotland (U.K.)	Glasgow	GLA	EGPF	26	55	52	19	N	4	25	59	W	
Goiania	Brazil	Santa Genoveva	GYN	SBGO	2450	16	37	52	S	49	13	20	W	
Goldsboro	USA	Seymour Johnson AFB	GSB	KGSB	110	35	20	21	N	77	57	38	W	
Goose Bay	Canada	Goose Bay	YYR	CYYR	160	53	19	9	N	60	25	33	W	
Guadalajara	Mexico	Hidalgo (I)	GDL	MMGL	5012	20	31	18	N	103	18	40	W	
Guangzhou	China	Baiyun	CAN	ZGGG	37	23	11	3	N	113	15	57	W	
Guantanamo	Cuba	Mariana Grajales	GAO	MUGT	56	20	5	7	N	75	9	30	W	
Guatemala	Guatemala	La Aurora	GUA	MGGT	4952	14	34	59	N	90	31	39	W	
Guayaquil	Ecuador	Simon Bolivar (I)	GYE	SEGU	15	2	9	28	S	79	53	2	W	
Haifa	Israel	Haifa	HFA	LLHA	28	32	48	40	N	35	2	38	E	
Halifax	Canada	Halifax (I)	YHZ	CYHZ	477	44	52	51	N	63	30	31	W	
Hamburg	Germany	Hamburg	HAM	EDDH	53	53	37	49	N	9	59	17	E	
Hampton	USA	Langley AFB	LFI	KLFI	11	37	4	58	N	76	21	37	W	
Hangzhou	China	Xiaoshan	HGH	ZSHC	23	30	13	42	N	120	25	54	E	

(continued)

City, Town	Country	Runway Airport Name	IATA	ICAO	Alt. (ft)	Latitude			Longitude			RW (ft)
						D	M	S	D	M	S	
Hanoi	Viet Nam	Noibai (I)	HAN	VVNB	39	21	13	18 N	105	48	20 E	
Harare	Zimbabwe	Harare (I)	HRE	FVHA	4887	17	55	54 S	31	5	34 E	
Harbin	China	Taiping	HRB	ZYHB	456	45	37	24 N	126	15	1 E	
Havana	Cuba	Jose Marti (I)	HAV	MUHA	210	22	59	21 N	82	24	33 W	
Helsinki	Finland	Helsinki Vantaa	HEL	EFHK	179	60	19	12 N	24	57	22 E	
Hewandorra	St. Lucia Island	Hewanorra (I)	UVF	TLPL	14	13	43	59 N	60	57	9 W	
Hilo	USA, Hawaii Isl.	Hilo (I)	ITO	PHTO	38	19	43	13 N	155	2	55 W	
Hiroshima	Japan	Hiroshima	HIJ	RJOA	1088	34	26	7 N	132	55	19 E	
Ho Chi Minh City	Viet Nam	Tansonnhat (I)	SGN	VVTS	33	10	49	12 N	106	39	42 E	
Homestead	USA	Homestead ARB	HST	KHST	7	25	29	18 N	80	23	1 W	
Hong Kong	Hong Kong	Hong Kong (I)	HKG	VHHH	28	22	18	32 N	113	54	52 E	
Honolulu	USA, Oahu Isl.	Honolulu (I)	HNL	PHNL	13	21	18	57 N	157	55	36 W	
Hopkinsville	USA	Campbell AAF	HOP	KHOP	573	36	40	6 N	87	29	46 W	
Houston	USA	Bush (I) Houston	IAH	KIAH	97	29	58	49 N	95	20	23 W	
Hunter AAF	USA	Hunter AAF	SVN	KSVN	42	32	0	36 N	81	8	44 W	
Hyderabad	Pakistan	Hyderabad	HDD	OPKD	130	25	19	5 N	68	21	58 E	
Hyderabad	India	Hyderabad	HYD	VOHY	1741	17	27	8 N	78	27	40 E	
Indian Springs	USA	Indian Springs Af Aux	INS	KINS	3133	36	35	13 N	115	40	24 W	
Indianapolis	USA	Indianapolis (I)	IND	KIND	797	39	43	2 N	86	17	39 W	
Indore	India	Ahilyabai Holkar	IDR	VAID	1849	22	43	18 N	75	48	3 E	
Iquique	Chile	Diego Aracena (I)	IQQ	SCDA	157	20	32	6 S	70	10	52 W	
Islamabad	Pakistan	Chaklala	ISB	OPRN	1666	33	36	59 N	73	5	57 E	

City, Town	Country	Runway Airport Name	IATA	ICAO	Alt. (ft)	Latitude D	M	S		Longitude D	M	S		RW (ft)
Istanbul	Turkey	Ataturk	IST	LTBA	163	40	58	36	N	28	49	16	E	
Izmir	Turkey	Adnan Menderes	ADB	LTBJ	412	38	17	32	N	27	9	25	E	
Jacksonville	USA	Jacksonville NAS	NIP	KNIP	22	30	14	9	N	81	40	50	W	
Jaipur	India	Jaipur	JAI	VIJP	1263	26	49	26	N	75	48	35	E	
Jakarta	Indonesia	Soekarno Hatta (I)	CGK	WIII	34	6	7	32	S	106	39	21	E	
Jalalabad	Afghanistan	Jalalabad	JAA	OAJL	1814	34	23	57	N	70	29	58	E	
Jeddah	Saudi Arabia	Abdulaziz (I)	JED	OEJN	48	21	40	46	N	39	9	23	E	
Jerusalem	Israel	Jerusalem/Atarot	JRS	LLJR	759	31	52	0	N	35	13	0	E	
Johannesburg	South Africa	Johannesburg (I)	JNB	FAJS	5558	26	8	21	S	28	14	45	E	
Juneau	USA	Juneau (I)	JNU	PAJN	19	58	21	17	N	134	34	34	W	
Kabul	Afghanistan	Kabul (I)	KBL	OAKB	5878	34	33	57	N	69	12	44	E	
Kadena	Japan	Kadena AB	DNA	RODN	143	26	21	20	N	127	46	3	E	
Kandahar	Afghanistan	Kandahar	KDH	OAKN	3329	31	30	21	N	65	50	52	E	
Kaneohe Bay	USA, Oahu Isl.	Kaneohe Bay MCAF	NGF	PHNG	17	21	26	57	N	157	46	4	W	
Kano	Nigeria	Mallam Aminu (I)	KAN	DNKN	1565	12	2	51	N	8	31	28	E	
Kanpur	India	Kanpur	KNU	VIKA	410	26	26	28	N	80	21	48	E	
Kansas City	USA	Kansas City (I)	MCI	KMCI	1026	39	17	51	N	94	42	50	W	
Kaohsiung	Taiwan	Kaohsiung (I)	KHH	RCKH	31	22	34	31	N	120	21	3	E	
Karachi	Pakistan	Jinnah (I)	KHI	OPKC	100	24	54	23	N	67	9	38	E	
Kathmandu	Nepal	Tribhuvan (I)	KTM	VNKT	4390	27	41	47	N	85	21	33	E	
Keflavik	Iceland	Keflavik NAS	KEF	BIKF	171	63	59	6	N	22	36	20	W	
Key West	USA	Key West NAS	NQX	KNQX	6	24	34	33	N	81	41	20	W	

(continued)

City, Town	Country	Runway Airport Name	IATA	ICAO	Alt. (ft)	Latitude D	M	S		Longitude D	M	S		RW (ft)
Khalid Mil.	Saudi Arabia	Khaled Mil City	HBT	OEKK	1352	27	54	3	N	45	31	41	E	
Khartoum	Sudan	Khartoum	KRT	HSSS	1261	15	35	22	N	32	33	11	E	
Kiev	Russia	Boryspil	KBP	UKBB	427	50	20	42	N	30	53	42	E	
Kigali	Rwanda	Kigali (I)	KGL	HRYR	4892	1	58	7	S	30	8	22	E	
Killeen	USA	Robert Gray AAF	GRK	KGRK	1019	31	4	2	N	97	49	44	W	
Kingston	Jamaica	Norman Manley (I)	KIN	MKJP	10	17	56	8	N	76	47	15	W	
Kinshasa	Zaire	Kinshasa Ndjili (I)	FIH	FZAA	1027	4	23	8	S	15	26	40	E	
Kirkwall	Scotland	Kirkwall	KOI	EGPA	51	58	57	29	N	2	54	18	W	
Kitakyushu	Japan	Kitakyushu	KKJ	RJFR	18	33	50	10	N	130	56	49	E	
Knobnoster	USA	Whiteman AFB	SZL	KSZL	871	38	43	49	N	93	32	52	W	
Kolhapur	India	Kolhapur	KLH	VAKP	1992	16	39	50	N	74	17	17	E	
Krakow	Poland	Balice (I)	KRK	EPKK	791	50	4	39	N	19	47	5	E	
Kuala Lumpur	Malaysia	Kuala Lumpur (I)	KUL	WMKK	69	2	44	44	N	101	42	35	E	
Kunming	China	Wujiaba	KMG	ZPPP	6217	24	59	32	N	102	44	36	E	
Kuwait	Kuwait	Kuwait (I)	KWI	OKBK	206	29	13	36	N	47	58	48	E	
La Paz	Bolivia	El Alto (I)	LPB	SLLP	13313	16	30	47	S	68	11	32	W	
Lahore	Pakistan	Allama Iqbal (I)	LHE	OPLA	712	31	31	17	N	74	24	12	E	
Lakehurst	USA	Lakehurst NAES	NEL	KNEL	103	40	2	0	N	74	21	12	W	
Lanzhou	China	Zhongchuan	ZGC	ZLLL	6388	36	31	0	N	103	37	18	E	
Las Vegas	USA	Mc Carran (I)	LAS	KLAS	2181	36	4	49	N	115	9	8	W	
Las Vegas	USA	Nellis AFB	LSV	KLSV	1868	36	14	10	N	115	2	3	W	
Leeds	England	Leeds Bradford	LBA	EGNM	682	53	51	57	N	1	39	38	W	

City, Town	Country	Runway Airport Name	IATA	ICAO	Alt. (ft)	Latitude D	M	S		Longitude D	M	S		RW (ft)
Lemoore	USA	Lemoore NAS	NLC	KNLC	234	36	19	58	N	119	57	7	W	
Lima	Peru	Jorge Chavez (I)	LIM	SPIM	113	12	1	18	S	77	6	51	W	
Linz	Austria	Horsching (I) AFB	LNZ	LOWL	313	48	14	0	N	14	11	0	E	
Lisbon	Portugal	Lisboa	LIS	LPPT	374	38	46	52	N	9	8	9	W	
Lome	Togo	Lome Tokoin	LFW	DXXX	71	6	9	56	N	1	15	14	E	
Lompoc	USA	Vandenberg AFB	VBG	KVBG	367	34	43	46	N	120	34	36	W	
London	England	Gatwick	LGW	EGKK	196	51	8	53	N	0	11	25	W	
London	England	Heathrow	LHR	EGLL	80	51	28	39	N	0	27	41	W	
Long Beach	USA	Long Beach	LGB	KLGB	60	33	49	3	N	118	9	5	W	
Los Angeles	USA	Los Angeles (I)	LAX	KLAX	126	33	56	33	N	118	24	29	W	
Luanda	Angola	Luanda 4 De Fevereiro	LAD	FNLU	243	8	51	30	S	13	13	52	E	
Lucknow	India	Lucknow	LKO	VILK	402	26	45	38	N	80	53	11	E	
Ludhiana	India	Ludhiana	LUH	VILD	834	30	51	16	N	75	57	4	E	
Lusaka	Zambia	Lusaka (I)	LUN	FLLS	3779	15	19	50	S	28	27	9	E	
Luxemburg	Luxemburg	Findel (I)	LUX	ELLX	1234	49	37	35	N	6	12	41	E	
Lyon	France	Saint Exupery	LYS	LFLL	821	45	43	34	N	5	5	27	E	
Macau	Macau	Macau (I)	MFM	VMMC	20	22	8	58	N	113	35	29	E	
Macon	USA	Robins AFB	WRB	KWRB	294	32	38	24	N	83	35	30	W	
Madras	India	Chennai (I)	MAA	VOMM	53	12	59	39	N	80	10	49	E	
Madrid	Spain	Barajas	MAD	LEMD	2000	40	28	20	N	3	33	39	W	
Mahe	Seychelles	Seychelles (I)	SEZ	FSIA	10	4	40	27	S	55	31	18	E	
Majuro	Marshall Isl.	Marshall Isl (I)	MAJ	PKMJ	6	7	3	53	N	171	16	19	E	

(continued)

City, Town	Country	Runway Airport Name	IATA	ICAO	Alt. (ft)	Latitude D	M	S		Longitude D	M	S		RW (ft)
Malabo	Equat. Guinea	Malabo	SSG	FGSL	76	3	45	19	N	8	42	31	E	
Male	Maldives	Male (I)	MLE	VRMM	6	4	11	30	N	73	31	44	E	
Managua	Nicaragua	Managua (I)	MGA	MNMG	194	12	8	28	N	86	10	5	W	
Manaus	Brazil	Eduardo Gomes (I)	MAO	SBEG	264	3	2	19	S	60	2	59	W	
Manchester	England	Manchester	MAN	EGCC	257	53	21	13	N	2	16	29	W	
Manila	Philippines	Ninoy Aquino (I)	MNL	RPLL	75	14	30	31	N	121	1	10	E	
Mannheim	Germany	Mannheim City	MHG	EDFM	309	49	28	21	N	8	30	49	E	
Maputo	Mozambique	Maputo	MPM	FQMA	145	25	55	15	S	30	34	21	E	
Maracaibo	Venezuela	La Chinita (I)	MAR	SVMC	235	10	33	29	N	71	43	40	W	
Marietta	USA	Dobbins ARB	MGE	KMGE	1068	33	54	55	N	84	30	58	W	
Marseille	France	Provence	MRS	LFML	74	43	26	8	N	5	12	49	E	
Marysville	USA	Beale AFB	BAB	KBAB	113	39	8	9	N	121	26	11	W	
Mazatlan	Mexico	Rafael Buelna (I)	MZT	MMMZ	38	23	9	40	N	106	15	57	W	
Medan	Indonesia	Polonia	MES	WIMM	90	3	33	29	N	98	40	18	E	
Medellin	Colombia	Olaya Herrera	EOH	SKMD	4940	6	13	12	N	75	35	26	W	
Melbourne	Australia	Melbourne (I)	MEL	YMML	434	37	40	24	S	144	50	36	E	
Meridian	USA	Meridian NAS	NMM	KNMM	317	32	33	7	N	88	33	20	W	
Mexico City	Mexico	Benito Juarez (I)	MEX	MMMX	7316	19	26	10	N	99	4	19	W	
Miami	USA	Miami (I)	MIA	KMIA	8	25	47	35	N	80	17	26	W	
Midway	Midway Island	Midway Atoll	MDY	PMDY	13	28	12	5	N	177	22	53	W	
Milan	Italy	Linate	LIN	LIML	353	45	26	43	N	9	16	37	E	
Milton	USA	Whiting Fld NAS N	NSE	KNSE	200	30	43	27	N	87	1	19	W	

City, Town	Country	Runway Airport Name	IATA	ICAO	Alt. (ft)	Latitude D	M	S		Longitude D	M	S	RW
Milwaukee	USA	General Mitchell (I)	MKE	KMKE	723	42	56	50	N	87	53	47	W
Minatitlan	Mexico	Minatitlan	MTT	MMMT	36	18	6	12	N	94	34	50	W
Minneapolis	USA	Minneapolis St Paul	MSP	KMSP	841	44	52	49	N	93	13	0	W
Minot	USA	Minot AFB	MIB	KMIB	1668	48	24	56	N	101	21	27	W
Minsk 2	Russia	Minsk 2	MSQ	UMMS	669	53	52	56	N	28	1	50	E
Miramar	USA	Miramar MCAS	NKX	KNKX	478	32	52	6	N	117	8	33	W
Mogadishu	Somalia	Mogadishu	MGQ	HCMM	31	2	0	49	N	45	18	17	E
Mohanbari	India	Dibrugarh	MOH	VEMN	361	27	29	0	N	95	1	3	E
Monclova	Mexico	Monclova (I)	LOV	MMMV	1854	26	57	20	N	101	28	12	W
Monrovia	Liberia	Roberts (I)	ROB	GLRB	31	6	14	1	N	10	21	44	W
Monterrey	Mexico	Del Norte (I)	NTR	MMAN	1473	25	51	56	N	100	14	14	W
Monterrey	Mexico	Escobedo (I)	MTY	MMMY	1270	25	46	42	N	100	6	24	W
Montevideo	Uruguay	Carrasco (I)	MVD	SUMU	105	34	50	16	S	56	1	49	W
Montgomery	USA	Maxwell AFB	MXF	KMXF	172	32	22	45	N	86	21	45	W
Montreal	Canada	Montreal (I) Dorval	YUL	CYUL	117	45	28	5	N	73	44	29	W
Moscow	Russia	Sheremetyevo	SVO	UUEE	630	55	58	18	N	37	24	54	E
Mt. Pleasant	Falkland Isl.	Mount Pleasant	MPN	EGYP	244	51	49	22	S	58	26	50	W
Mountain Home	USA	Mountain Home AFB	MUO	KMUO	2996	43	2	36	N	115	52	20	W
Mountain View	USA	Moffett Federal Afld	NUQ	KNUQ	32	37	24	54	N	122	2	53	W
Muir	USA	Muir AAF	MUI	KMUI	489	40	26	5	N	76	34	9	W
Multan	Pakistan	Multan (I)	MUX	OPMT	403	30	12	11	N	71	25	8	E
Munich	Germany	Munich	MUC	EDDM	1487	48	21	13	N	11	47	9	E

(continued)

City, Town	Country	Runway Airport Name	IATA	ICAO	Alt. (ft)	Latitude D	M	S		Longitude D	M	S		RW (ft)
Murmansk	Russia	Murmansk	MMK	ULMM	266	68	46	54	N	32	45	2	E	
Muscat	Oman	Seeb (I)	MCT	OOMS	48	23	35	35	N	58	17	4	E	
Nagasaki	Japan	Nagasaki	NGS	RJFU	15	32	55	21	N	129	55	24	E	
Nagoya	Japan	Nagoya	NGO	RJNN	52	35	15	18	N	136	55	28	E	
Nagpur	India	Nagpur	NAG	VANP	1012	21	5	31	N	79	2	49	E	
Nairobi	Kenya	Nairobi Wilson	WIL	HKNW	5536	1	19	18	S	36	48	53	E	
Nanchang	China	Nanchang Airport	KHN	ZSCN	0	28	36	0	N	115	55	0	E	
Nanjing	China	Lu Kou Airport	NKG	ZSNJ	49	31	44	24	N	118	51	36	E	
Naples	Italy	Capodichino	NAP	LIRN	298	40	53	9	N	14	17	26	E	
Nashville	USA	Nashville (I)	BNA	KBNA	599	36	7	28	N	86	40	41	W	
Nasik Road	India	Nasik Road	ISK	VANR	1959	19	57	45	N	73	48	25	E	
Nassau	Bahamas	Nassau (I)	NAS	MYNN	16	25	2	20	N	77	27	58	W	
Nausori	Fiji	Nausori (I)	SUV	NFNA	17	18	2	35	S	178	33	33	E	
N'djamena	Chad	Ndjamena	NDJ	FTTJ	968	12	8	1	N	15	2	2	E	
New Orleans	USA	Armstrong No (I)	MSY	KMSY	4	29	59	36	N	90	15	28	W	
New Orleans	USA	New Orleans NAS JRB	NBG	KNBG	3	29	49	31	N	90	2	6	W	
New York	USA	John F Kennedy (I)	JFK	KJFK	13	40	38	23	N	73	46	44	W	
New York	USA	La Guardia	LGA	KLGA	22	40	46	38	N	73	52	21	W	
Niamey	Niger	Diori Hamani	NIM	DRRN	732	13	28	53	N	2	11	1	E	
Nice	France	Cote D Azur	NCE	LFMN	13	43	39	38	N	7	13	3	E	
Nizhnevartovsk	Russia	Nizhnevartovsk	NJC	USNN	0	60	57	0	N	76	28	0	E	
Norfolk	USA	Norfolk (I)	ORF	KORF	26	36	53	40	N	76	12	4	W	

City, Town	Country	Runway Airport Name	IATA	ICAO	Alt. (ft)	Latitude D	M	S		Longitude D	M	S		RW (ft)
Oakland	USA	Metro Oakland (I)	OAK	KOAK	6	37	43	16	N	122	13	14	W	
Oceana	USA	Oceana NAS	NTU	KNTU	22	36	49	14	N	76	2	0	W	
Odessa	Russia	Odessa	ODS	UKOO	171	46	25	37	N	30	40	41	E	
Ogden	USA	Hill AFB	HIF	KHIF	4789	41	7	26	N	111	58	22	W	
Oklahoma City	USA	Will Rogers World	OKC	KOKC	1295	35	23	35	N	97	36	2	W	
Oklahoma City	USA	Tinker AFB	TIK	KTIK	1291	35	24	53	N	97	23	11	W	
Omaha	USA	Offutt AFB	OFF	KOFF	1052	41	7	6	N	95	54	45	W	
Ontario	USA	Ontario (I)	ONT	KONT	944	34	3	21	N	117	36	4	W	
Oranjestad	Aruba	Reina Beatrix (I)	AUA	TNCA	60	12	30	5	N	70	0	54	W	
Orlando	USA	Orlando (I)	MCO	KMCO	96	28	25	44	N	81	18	57	W	
Osaka	Japan	Osaka (I)	ITM	RJOO	50	34	47	7	N	135	26	17	E	
Oslo	Norway	Oslo Gardermoen	OSL	ENGM	681	60	11	38	N	11	6	1	E	
Ottawa	Canada	Ottawa Cartier (I)	YOW	CYOW	374	45	19	21	N	75	40	9	W	
Ouagadougou	Burkina Faso	Ouagadougou	OUA	DFFD	1037	12	21	11	N	1	30	44	W	
Pago Pago	Samoa	Pago Pago (I)	PPG	NSTU	31	14	19	51	S	170	42	37	W	
Palembang	Indonesia	Mahmud Badaruddin	PLM	WIPP	37	2	53	52	S	104	42	4	E	
Palm Springs	USA	Palm Springs (I)	PSP	KPSP	474	33	49	46	N	116	30	24	W	
Palmdale	USA	Palmdale Flt Test	PMD	KPMD	2543	34	37	45	N	118	5	4	W	
Panama	Panama	Gelabert (I)	PAC	MPMG	31	8	58	24	N	79	33	20	W	
Panama City	USA	Tyndall AFB	PAM	KPAM	18	30	4	11	N	85	34	35	W	
Paphos	Cyprus	Paphos (I)	PFO	LCPH	41	34	43	4	N	32	29	8	E	
Paramaribo	Surinam	Zorg En Hoop	ORG	SMZO	10	5	48	39	N	55	11	26	W	

(continued)

City, Town	Country	Runway Airport Name	IATA	ICAO	Alt. (ft)	Latitude D	Latitude M	Latitude S	Latitude N/S	Longitude D	Longitude M	Longitude S	Longitude E/W	RW (ft)
Paris	France	Le Bourget	LBG	LFPB	218	48	58	10	N	2	26	29	E	
Paris	France	Charles De Gaulle	CDG	LFPG	392	49	0	46	N	2	33	0	E	
Paris	France	Orly	ORY	LFPO	291	48	43	31	N	2	21	34	E	
Paro	Bhutan	Paro	PBH	VQPR	7333	27	24	11	N	89	25	33	E	
Patna	India	Patna	PAT	VEPT	169	25	35	26	N	85	5	16	E	
Pax River	USA	Patuxent River NAS	NHK	KNHK	39	38	17	9	N	76	24	42	W	
Pensacola	USA	Pensacola NAS	NPA	KNPA	28	30	21	9	N	87	19	7	W	
Perth	Australia	Perth (I)	PER	YPPH	67	31	56	25	S	115	58	1	E	
Peshawar	Pakistan	Peshawar	PEW	OPPS	1211	33	59	38	N	71	30	52	E	
Philadelphia	USA	Philadelphia (I)	PHL	KPHL	38	39	52	19	N	75	14	28	W	
Philipsburg	Antilles	Princess Juliana (I)	SXM	TNCM	13	18	2	27	N	63	6	32	W	
Phnom-Penh	Cambodia	Pochentong (I)	PNH	VDPP	40	11	32	47	N	104	50	38	E	
Phoenix	USA	Luke AFB	LUF	KLUF	1085	33	32	6	N	112	22	59	W	
Phoenix	USA	Phoenix Sky Harb (I)	PHX	KPHX	1135	33	26	3	N	112	0	29	W	
Phuket	Thailand	Phuket (I)	HKT	VTSP	82	8	6	47	N	98	19	0	E	
Pittsburgh	USA	Pittsburgh (I)	PIT	KPIT	1204	40	29	29	N	80	13	58	W	
Plaisance	Mauritius	Ramgoolam (I)	MRU	FIMP	186	20	25	48	S	57	41	0	E	
Podgorica	Yugoslavia	Podgorica	TGD	LYPG	121	42	21	33	N	19	15	6	E	
Pohnpei	Micronesia	Pohnpei (I)	PNI	PTPN	10	6	59	6	N	158	12	32	E	
Point Mugu	USA	Point Mugu NAS	NTD	KNTD	12	34	7	13	N	119	7	15	W	
Point Salines	Grenada	Point Salines (I)	GND	TGPY	41	12	0	15	N	61	47	10	W	
Port Moresby	Papua New Guinea	Port Moresby (I)	POM	AYPY	146	9	26	36	S	147	13	12	E	

City, Town	Country	Runway Airport Name	IATA	ICAO	Alt. (ft)	Latitude D	Latitude M	Latitude S	Latitude N/S	Longitude D	Longitude M	Longitude S	RW (ff)
Port-Au-Prince	Haiti	Port Au Prince (I)	PAP	MTPP	121	18	34	48	N	72	17	33	W
Portland	USA	Portland (I)	PDX	KPDX	30	45	35	19	N	122	35	51	W
Porto Alegre	Brazil	Salgado Filho	POA	SBPA	11	29	59	39	S	51	10	17	W
Port-Of-Spain	Trinidad & Tobago	Piarco	POS	TTPP	57	10	35	43	N	61	20	14	W
Prague	Czech Republic	Ruzyne	PRG	LKPR	1247	50	6	3	N	14	15	36	E
Pretoria	South Africa	Wonderboom	PRY	FAWB	4095	25	39	13	S	28	13	27	E
Pristina	Yugoslavia	Pristina	PRN	LYPR	1789	42	34	22	N	21	2	9	E
Puebla	Mexico	Serdan (I)	PBC	MMPB	7380	19	9	29	N	98	22	17	W
Puerto Montt	Chile	El Tepual (I)	PMC	SCTE	294	41	26	19	S	73	5	38	W
Puerto Vallarta	Mexico	Ordaz (I)	PVR	MMPR	20	20	40	48	N	105	15	15	W
Pune	India	Pune	PNQ	VAPO	1942	18	34	55	N	73	55	10	E
Punta Arenas	Chile	Ibanez Del Campo (I)	PUQ	SCCI	139	53	0	10	S	70	51	17	W
Pyongyang	North Korea	Pyongyang / Sunan	FNJ	ZKPY	38	39	2	0	N	125	47	0	E
Quantico	USA	Quantico MCAF	NYG	KNYG	11	38	30	6	N	77	18	19	W
Quebec	Canada	Quebec Jean Lesage (I)	YQB	CYQB	244	46	47	18	N	71	23	51	W
Queenstown	New Zealand	Queenstown	ZQN	NZQN	1171	45	1	16	S	168	44	21	E
Quito	Ecuador	Mariscal Sucre (I)	UIO	SEQU	9228	0	8	28	S	78	29	17	W
Rabat	Morocco	Sale	RBA	GMME	276	34	3	5	N	6	45	5	W
Raleigh-Durham	USA	Raleigh Durham (I)	RDU	KRDU	435	35	52	39	N	78	47	14	W
Ramstein	Germany	Ramstein AB	RMS	ETAR	782	49	26	15	N	7	36	5	E
Rapid City	USA	Ellsworth AFB	RCA	KRCA	3280	44	8	42	N	103	6	12	W
Rawala Kot	Pakistan	Rawalakot	RAZ	OPRT	5500	33	50	57	N	73	47	52	E

(continued)

City, Town	Country	Runway Airport Name	IATA	ICAO	Alt. (ft)	Latitude D	M	S		Longitude D	M	S		RW (ft)
Recife	Brazil	Guararapes	REC	SBRF	33	8	7	35	S	34	55	24	W	
Red River	USA	Grand Forks AFB	RDR	KRDR	911	47	57	39	N	97	24	4	W	
Redstone	USA	Redstone AAF	HUA	KHUA	685	34	40	43	N	86	41	5	W	
Reno	USA	Reno Tahoe (I)	RNO	KRNO	4412	39	29	54	N	119	46	5	W	
Rio De Janeiro	Brazil	Galeao Anton Jobim	GIG	SBGL	28	22	48	32	S	43	14	37	W	
Riverside	USA	March ARB	RIV	KRIV	1535	33	52	50	N	117	15	34	W	
Riyadh	Saudi Arabia	King Khaled (I)	RUH	OERK	2049	24	57	27	N	46	41	55	E	
Robinson	USA	Robinson AAF	RBM	KRBM	587	34	51	0	N	92	18	0	W	
Rome	Italy	Fiumicino	FCO	LIRF	13	41	48	46	N	12	15	11	E	
Roswell	USA	Roswell Air Center	ROW	KROW	3671	33	18	5	N	104	31	50	W	
Rotorua	New Zealand	Rotorua	ROT	NZRO	935	38	6	33	S	176	19	2	E	
Rotterdam	Netherlands	Rotterdam	RTM	EHRD	15	51	57	26	N	4	26	30	E	
Sacramento	USA	Sacramento Mather	MHR	KMHR	96	38	33	14	N	121	17	51	W	
Sacramento	USA	Sacramento (I)	SMF	KSMF	27	38	41	43	N	121	35	26	W	
Salt Lake City	USA	Salt Lake City (I)	SLC	KSLC	4227	40	47	18	N	111	58	39	W	
Salvador	Brazil	Luis Magalhaes	SSA	SBSV	63	12	54	39	S	38	19	51	W	
Salzburg	Austria	Salzburg	SZG	LOWS	1411	47	47	35	N	13	0	15	E	
San Antonio	USA	Randolph AFB	RND	KRND	762	29	31	46	N	98	16	44	W	
San Antonio	USA	San Antonio (I)	SAT	KSAT	809	29	32	1	N	98	28	11	W	
San Antonio	USA	Lackland AFB Kelly	SKF	KSKF	691	29	23	3	N	98	34	51	W	
San Diego	USA	North Island NAS	NZY	KNZY	26	32	41	57	N	117	12	55	W	
San Diego	USA	San Diego Lindb'gh (I)	SAN	KSAN	14	32	44	0	N	117	11	22	W	

City,Town	Country	Runway Airport Name	IATA	ICAO	Alt. (ft)	Latitude D	M	S		Longitude D	M	S		RW (ft)
San Francisco	USA	San Francisco (I)	SFO	KSFO	13	37	37	8	N	122	22	29	W	
San Jose	USA	Mineta San Jose (I)	SJC	KSJC	58	37	21	42	N	121	55	44	W	
San Jose	Costa Rica	Juan Santamaria (I)	SJO	MROC	3021	9	59	37	N	84	12	31	W	
San Juan	Puerto Rico	Marin (I)	SJU	TJSJ	9	18	26	21	N	66	0	6	W	
San Salvador	El Salvador	El Salvador (I)	SAL	MSLP	101	13	26	26	N	89	3	21	W	
Santa Ana	USA	John Wayne / Orange Co	SNA	KSNA	56	33	40	32	N	117	52	5	W	
Santa Cruz	Bolivia	Viru Viru (I)	VVI	SLVR	1224	17	38	41	S	63	8	7	W	
Santa Maria Isl.	Acores	Santa Maria	SMA	LPAZ	308	36	58	17	N	25	10	14	W	
Santiago	Chile	Arturo Benitez (I)	SCL	SCEL	1554	33	23	34	S	70	47	8	W	
Santo Domingo	Dominican Republic	Las Americas (I)	SDQ	MDSD	59	18	25	46	N	69	40	7	W	
Santorini	Greece	Santorini	JTR	LGSR	130	36	24	1	N	25	28	43	E	
Santos	Brazil	Santos Air Base	SSZ	SBST	10	23	55	30	S	46	17	15	W	
Sao Paulo	Brazil	Guarulhos	GRU	SBGR	2459	23	25	56	S	46	28	9	W	
Sao Tome	Sao Tome & Principe	Sao Tome (I)	TMS	FPST	33	0	22	41	N	6	42	43	E	
Sapporo	Japan	Chitose (I) Airport	CTS	RJCC	82	42	46	30	N	141	41	32	E	
Sarajevo	Bosnia-Hercegovina	Sarajevo	SJJ	LQSA	1708	43	49	28	N	18	19	53	E	
Seattle	USA	Seattle Tacoma (I)	SEA	KSEA	429	47	26	56	N	122	18	33	W	
Semarang	Indonesia	Achmad Yani	SRG	WIIS	10	6	58	23	S	110	22	31	E	
Sendai	Japan	Sendai	SDJ	RJSS	15	38	8	22	N	140	55	0	E	
Seoul	South Korea	Gimpo	GMP	RKSS	58	37	33	29	N	126	47	26	E	
Seoul East	South Korea	Seoul Ab	SSN	RKSM	92	37	26	45	N	127	6	50	E	

(continued)

City, Town	Country	Runway Airport Name	IATA	ICAO	Alt. (ft)	Latitude D	M	S		Longitude D	M	S		RW (ft)
Shanghai	China	Hongqiao (I)	SHA	ZSSS	10	31	11	52	N	121	20	10	E	
Shemya	USA	Eareckson AS	SYA	PASY	97	52	42	44	N	174	6	49	E	
Shenzhen	China	Baoan	SZX	ZGSZ	13	22	38	22	N	113	48	44	E	
Shiraz	Iran	Dastghaib (I)	SYZ	OISS	4920	29	32	21	N	52	35	22	E	
Shreveport	USA	Barksdale AFB	BAD	KBAD	166	32	30	6	N	93	39	45	W	
Singapore	Singapore	Singapore Changi	SIN	WSSS	22	1	21	20	N	103	59	14	E	
Sofia	Bulgaria	Sofia	SOF	LBSF	1742	42	41	42	N	23	24	22	E	
Spokane	USA	Spokane (I)	GEG	KGEG	2372	47	37	11	N	117	32	1	W	
Spokane	USA	Fairchild AFB	SKA	KSKA	2462	47	36	54	N	117	39	20	W	
St. Louis	USA	Lambert St Louis (I)	STL	KSTL	604	38	44	51	N	90	21	35	W	
St. Petersburg	Russia	Pulkovo	LED	ULLI	79	59	48	0	N	30	15	54	E	
St. Thomas	Virgin Isl.	Cyril E King	STT	TIST	23	18	20	14	N	64	58	24	W	
Stockholm	Sweden	Arlanda	ARN	ESSA	124	59	39	7	N	17	55	7	E	
Stuttgart	Germany	Stuttgart	STR	EDDS	1276	48	41	23	N	9	13	19	E	
Sumter	USA	Shaw AFB	SSC	KSSC	242	33	58	22	N	80	28	22	W	
Surabaya	Indonesia	Juanda	SUB	WRSJ	9	7	22	47	S	112	47	12	E	
Surat	India	Surat	STV	VASU	50	21	6	54	N	72	44	34	E	
Sydney	Australia	Smith (I) Airport	SYD	YSSY	21	33	56	46	S	151	10	38	E	
Syracuse	USA	Syracuse Hancock (I)	SYR	KSYR	421	43	6	40	N	76	6	22	W	
Tabriz	Iran	Tabriz (I)	TBZ	OITT	4459	38	7	58	N	46	14	5	E	
Tacoma	USA	Mc Chord AFB	TCM	KTCM	323	47	8	15	N	122	28	35	W	
Taipei	Taiwan	Chiang Kai Shek (I)	TPE	RCTP	107	25	4	48	N	121	13	56	E	

City,Town	Country	Runway Airport Name	IATA	ICAO	Alt. (ft)	Latitude D	M	S		Longitude D	M	S		RW (ft)
Tallinn–Ulemiste	Estonia	Tallinn	TLL	EETN	132	59	24	47	N	24	49	58	E	
Tampa	USA	Macdill AFB	MCF	KMCF	14	27	50	57	N	82	31	16	W	
Tampa	USA	Tampa (I)	TPA	KTPA	26	27	58	31	N	82	31	59	W	
Tanger	Morocco	Ibn Batouta	TNG	GMTT	62	35	43	36	N	5	55	0	W	
Tashkent	Uzbekistan	Yuzhny	TAS	UTTT	1414	41	15	26	N	69	16	54	E	
Tbilisi	Georgia	Lochini	TBS	UGGG	1624	41	40	9	N	44	57	17	E	
Tegucigalpa	Honduras	Toncontin (I)	TGU	MHTG	3294	14	3	39	N	87	13	1	W	
Teheran	Iran	Mehrabad (I)	THR	OIII	3962	35	41	21	N	51	18	48	E	
Tel-Aviv	Israel	Ben Gurion	TLV	LLBG	135	32	0	34	N	34	52	36	E	
Tenerife	Canary Islands	Tenerife Norte	TFN	GCXO	2073	28	28	57	N	16	20	29	W	
The Valley	Anguilla Isl.	Wallblake	AXA	TQPF	102	18	12	17	N	63	3	18	W	
Thule	Greenland	Thule Air Base	THU	BGTL	251	76	31	52	N	68	42	11	W	
Tianjin	China	Binhai	TSN	ZBTJ	7	39	7	26	N	117	20	46	E	
Tijuana	Mexico	Rodriguez (I)	TIJ	MMTJ	499	32	32	27	N	116	58	12	W	
Tirana	Albania	Tirana Rinas	TIA	LATI	125	41	24	53	N	19	43	14	E	
Tivat	Yugoslavia	Tivat	TIV	LYTV	20	42	24	16	N	18	43	23	E	
Tokyo	Japan	New Tokyo (I)	NRT	RJAA	141	35	45	53	N	140	23	11	E	
Toluca	Mexico	Mateos (I)	TLC	MMTO	8458	19	20	13	N	99	33	57	W	
Tongatapu	Tonga	Fua Amotu (I)	TBU	NFTF	126	21	14	27	S	175	9	0	W	
Toronto	Canada	Lester B Pearson (I)	YYZ	CYYZ	569	43	40	38	N	79	37	50	W	
Tripoli	Libya	Tripoli (I)	TIP	HLLT	263	32	39	48	N	13	9	32	E	
Trondheim	Norway	Trondheim Vaernes	TRD	ENVA	56	63	27	27	N	10	56	23	E	

(continued)

City, Town	Country	Runway Airport Name	IATA	ICAO	Alt. (ft)	Latitude D	M	S		Longitude D	M	S		RW (ft)
Tucson	USA	Davis Monthan AFB	DMA	KDMA	2704	32	9	59	N	110	52	59	W	
Tucson	USA	Tucson (I)	TUS	KTUS	2643	32	6	58	N	110	56	29	W	
Tunis	Tunisia	Carthage	TUN	DTTA	22	36	51	3	N	10	13	37	E	
Tustin	USA	Tustin MCAF	NTK	KNTK	54	33	42	22	N	117	49	38	W	
29 Palms	USA	29 EAF	NXP	KNXP	2042	34	17	46	N	116	9	43	W	
Urumqi	China	Diwopu	URC	ZWWW	2126	43	54	25	N	87	28	27	E	
Valencia	Spain	Valencia	VLC	LEVC	225	39	29	21	N	0	28	53	W	
Valencia	Venezuela	Arturo Michelena (I)	VLN	SVVA	1411	10	9	29	N	67	28	36	W	
Valparaiso	USA	Eglin AFB	VPS	KVPS	87	30	28	59	N	86	55	31	W	
Vancouver	Canada	Vancouver (I)	YVR	CYVR	14	49	11	42	N	123	10	55	W	
Venice	Italy	Venezia Tessera	VCE	LIPZ	7	45	30	18	N	12	21	6	E	
Vienna	Austria	Schwechat	VIE	LOWW	600	48	6	37	N	16	34	11	E	
Vientiane	Laos	Wattay (I)	VTE	VLVT	564	17	59	17	N	102	33	47	E	
Vitoria	Brazil	Goiabeiras	VIX	SBVT	11	20	15	20	S	40	17	20	W	
Vladivostok	Russia	Knevichi	VVO	UHWW	43	43	23	56	N	132	9	5	E	
Warsaw	Poland	Okecie	WAW	EPWA	362	52	9	56	N	20	58	1	E	
Washington, DC	USA	Reagan Wash–Nat'l	DCA	KDCA	15	38	51	7	N	77	2	15	W	

City, Town	Country	Runway Airport Name	IATA	ICAO	Alt. (ft)	Latitude D	M	S	N/S	Longitude D	M	S	E/W	RW (ft)
Washington, DC	USA	Washington Dulles (I)	IAD	KIAD	313	38	56	40	N	77	27	20	W	
Wellington	New Zealand	Wellington (I)	WLG	NZWN	41	41	19	38	S	174	48	19	E	
Whidbey Isl.	USA	Whidbey Island NAS	NUW	KNUW	47	48	21	6	N	122	39	21	W	
White Sands	USA	Condron AAF	WSD	KWSD	3934	32	20	29	N	106	24	9	W	
Wichita	USA	Mc Connell AFB	IAB	KIAB	1371	37	37	22	N	97	16	2	W	
Willow Grve	USA	Willow Grove NAS JRB	NXX	KNXX	362	40	11	59	N	75	8	53	W	
Winnipeg	Canada	Winnipeg (I)	YWG	CYWG	783	49	54	36	N	97	14	4	W	
Wright	USA	Wright AAF	LHW	KLHW	48	31	53	20	N	81	33	44	W	
Wrightstown	USA	Mc Guire AFB	WRI	KWRI	133	40	0	56	N	74	35	37	W	
Wuhan	China	Tianhe	WUH	ZHHH	115	30	47	1	N	114	12	29	E	
Yangon	Myanmar	Yangon (I)	RGN	VYYY	109	16	54	26	N	96	7	59	E	
Yaounde	Cameroon	Yaounde	YAO	FKKY	2464	3	50	7	N	11	31	25	E	
Yap	Micronesia	Yap (I)	YAP	PTYA	91	9	29	55	N	138	4	56	E	
Yuma	USA	Yuma MCAS Yuma (I)	YUM	KYUM	213	32	39	23	N	114	36	21	W	
Zanzibar	Tanzania	Zanzibar	ZNZ	HTZA	54	6	13	19	S	39	13	29	E	
Zhengzhou	China	Xinzheng	CGO	ZHCC	495	34	31	10	N	113	50	27	E	
Zurich	Switzerland	Zurich	ZRH	LSZH	1416	47	27	53	N	8	32	57	E	

INDEX

SUPPORTING MATERIALS

Many of the topics introduced in this book are discussed in more detail in other AIAA publications. For a complete listing of titles in the AIAA Education Series, as well as other AIAA publications, please visit http://www.aiaa.org.

"High Flight"

Oh, I have slipped the surly bonds of earth.
And danced the skies on laughter-silvered wings,
Sunward I've climbed and joined the tumbling mirth
Of sun-split clouds - and done a hundred things
You have not dreamed of - wheeled and soared and swung
High in the sunlit silence. Hov'ring there,
I've chased the shouting wind along and flung
My eager craft through footless halls of air.
Up, up the long delirious, burning blue
I've topped the wind swept heights with easy grace,
Where never lark, or even eagle, flew;
And, while with silent, lifting mind I've trod
The high untrespassed sanctity of space,
Put out my hand, and touched the face of God.

John Gillespie Magee Jr.

Pilot's Prayer

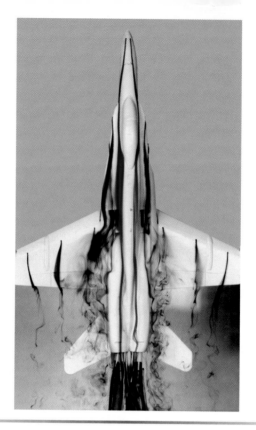

F-18 Vortices (NASA, Dryden Water Tunnel Facility)

Fig. 10.1 B-767 with Trailing Vortices (courtesy of Ray Nicolai)

Fig. 2.14   Vortices Shedding from F-18 LEX and F-22 Wing Leading Edge

Ch. 2   F-14 with Shock Condensation @ M = 0.9

Fig. 1.8   DC-3—Timeless Elegance

Appen. G.   Cessna 172 Skyhawk—Classic Design

Ch. 5 YF-16 and YF-17—LWF Competition Finalists

Fig. 1.9 YF-22 and YF-23—ATF Competitors

Fig. 1.11 X-32 and X-35—JSF Competitors

Appen. H   F-16 Fighting Falcon—Configured for Air-to-air Mission

F-18—Powered Approach to Aircraft Carrier (courtesy of John Stratton)

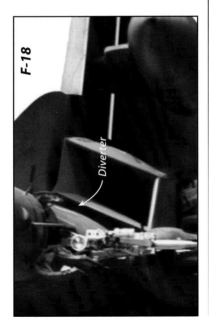

Fig. 15.13b   A Variety of Inlet Designs for Supersonic Missions

Ch. 3   SR-71—Kelly Johnson's Crowning Achievement

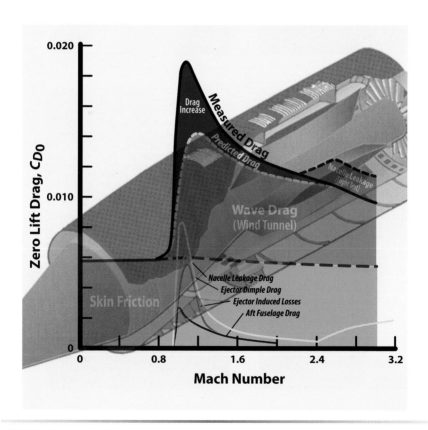

Ch. 16   Transonic C_{D_0} of the SR-71—A Surprise when Compared to Predictions

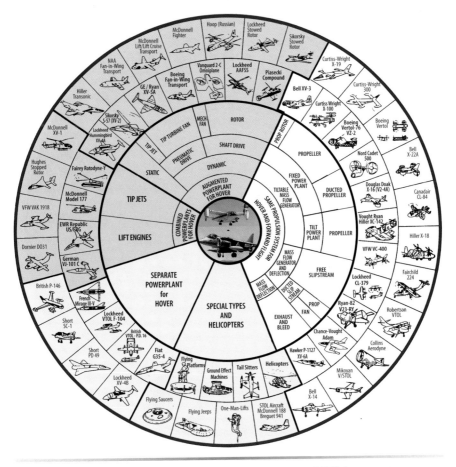

Fig. 9.31 V/STOL Aircraft Summary (1970s)

AV-8B Harrier Performs a Hover Maneuver

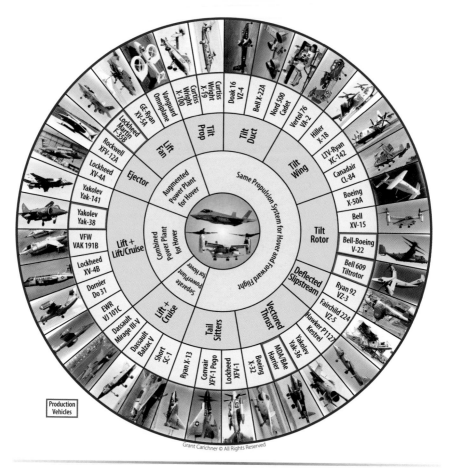

Grant Carichner © All Rights Reserved

Fig. 9.32 V/STOL Aircraft Summary (2008)

F-35B Transitions from Level Flight to Vertical Mode

Ch. 23   Canadian Geese Make STOL Landing

Appen. C   F-35B Lightning II Prepares for Vertical Landing

Ch. 12   F-117A and the "Father of Stealth" Ben Rich

Ch. 13   F-22 and F-117—Two Generations of Stealth

Fig. F.6   Polecat Unmanned Aircraft

Ch. 19   F-117—Built in Secrecy in Burbank, California

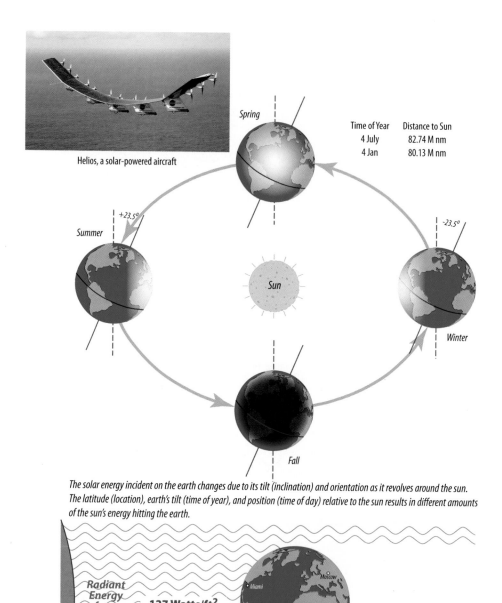

Helios, a solar-powered aircraft

Time of Year	Distance to Sun
4 July	82.74 M nm
4 Jan	80.13 M nm

The solar energy incident on the earth changes due to its tilt (inclination) and orientation as it revolves around the sun. The latitude (location), earth's tilt (time of year), and position (time of day) relative to the sun results in different amounts of the sun's energy hitting the earth.

Radiant Energy from Sun

127 Watts/ft^2

Fig. 18.5   Solar Energy Radiated to Earth

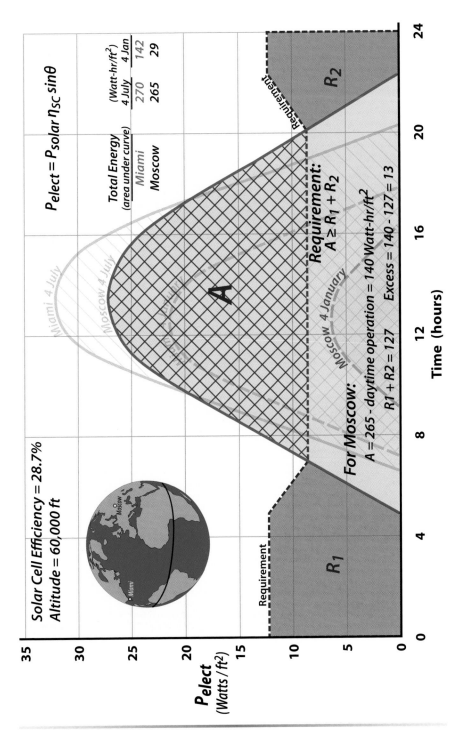

Fig. 18.8   Diurnal Energy Balance—Stationkeeping Over Moscow and Miami

Fig. 10.14   Mig 31, F-15, B-1B, and L1011—Takeoffs

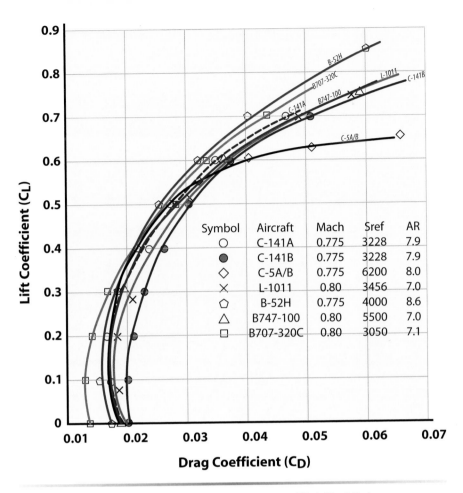

Symbol	Aircraft	Mach	Sref	AR
○	C-141A	0.775	3228	7.9
●	C-141B	0.775	3228	7.9
◇	C-5A/B	0.775	6200	8.0
✕	L-1011	0.80	3456	7.0
⬠	B-52H	0.775	4000	8.6
△	B747-100	0.80	5500	7.0
☐	B707-320C	0.80	3050	7.1

Fig. G.1   High-Subsonic Drag Polars—Flight Test Data

Boeing 747-400—Boeing's Gamble Pays Off (early 1970s)

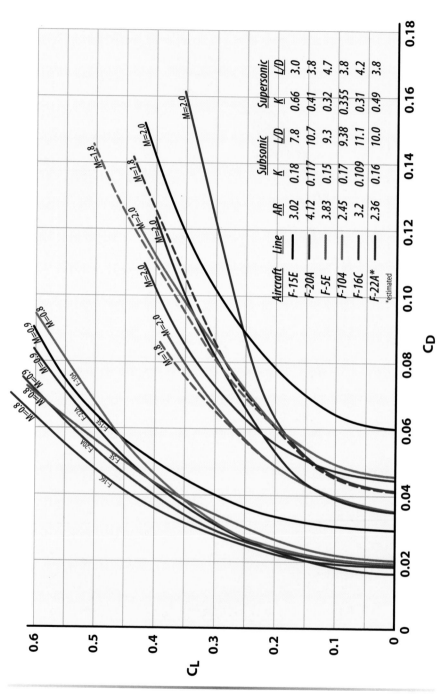

Fig. G.4 Drag Polars for Fighter Aircraft—Flight Test Data

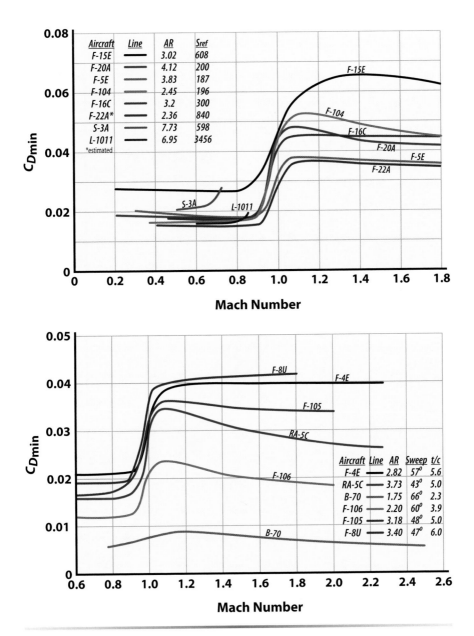

Fig. G.5 $C_{D_{min}}$ vs Mach Number—Flight Test Data

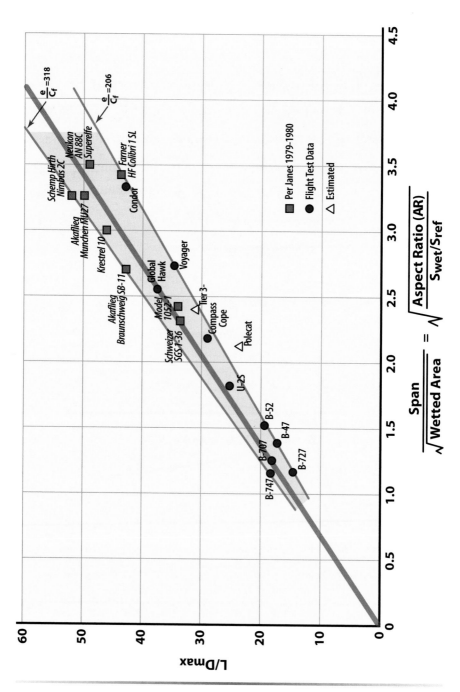

Fig. G.8   Max. L/D Correlation Curve—Flight Test Data

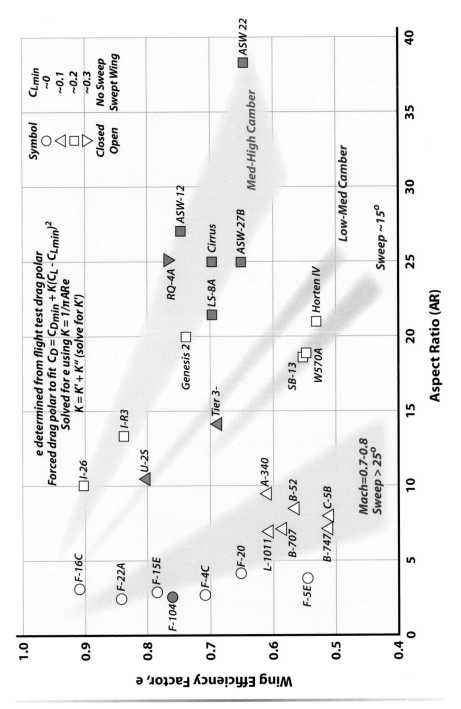

Fig. G.9   Subsonic Wing Efficiency vs Aspect Ratio

Ch. 21   Sopwith Camel—A Highly Maneuverable WWI Fighter

Ch. 18   Nemesis—A Formula One A-Class Racer

Fig. 18.10   Daedalus—Record Holder for Human-powered Distance

Appen. I   Eta Glider—With Today's Highest Maximum *L/D* of 73

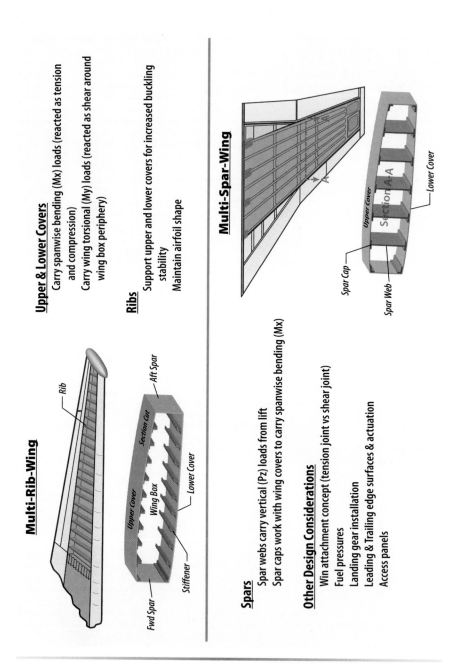

Multi-Rib-Wing

Upper & Lower Covers

Carry spanwise bending (Mx) loads (reacted as tension and compression)

Carry wing torsional (My) loads (reacted as shear around wing box periphery)

Ribs

Support upper and lower covers for increased buckling stability

Maintain airfoil shape

Multi-Spar-Wing

Spars

Spar webs carry vertical (Pz) loads from lift

Spar caps work with wing covers to carry spanwise bending (Mx)

Other Design Considerations

Win attachment concept (tension joint vs shear joint)

Fuel pressures

Landing gear installation

Leading & Trailing edge surfaces & actuation

Access panels

Fig. 19.16   Wing Structural Configurations